QUANTUM THEORY and SYMMETRIES

Proceedings of the 3rd International Symposium

QUANTUM THEORY and SYMMETRIES

Cincinnati, USA 10 – 14 September 2003

edited by

P C Argyres, T J Hodges, F Mansouri
J J Scanio, P Suranyi & L C R Wijewardhana
University of Cincinnati, USA

World Scientific

NEW JERSEY • LONDON • SINGAPORE • BEIJING • SHANGHAI • HONG KONG • TAIPEI • CHENNAI

Published by

World Scientific Publishing Co. Pte. Ltd.

5 Toh Tuck Link, Singapore 596224

USA office: 27 Warren Street, Suite 401-402, Hackensack, NJ 07601

UK office: 57 Shelton Street, Covent Garden, London WC2H 9HE

British Library Cataloguing-in-Publication Data
A catalogue record for this book is available from the British Library.

QUANTUM THEORY AND SYMMETRIES
Proceedings of the 3rd International Symposium

ISBN 981-256-068-8

Printed in Singapore.

FREYDOON MANSOURI
(1936-2003)

This volume is dedicated to the memory of Professor Freydoon Mansouri, who passed away on April 27, 2003. Professor Mansouri played the leading role in the organization of the third workshop on Quantum Theory and Symmetry in Cincinnati, and the success of the workshop is a gratifying measure of his tireless efforts. Professor Mansouri would have been very pleased to witness the breadth and diversity of the speakers and attendees at the workshop.

Professor Mansouri was born in Teheran, Iran on June 5, 1936. He remained there through his early years, until he obtained his secondary school diploma. Professor Mansouri then came to the United States for his college education, where he received his BA from the University of Pennsylvania in 1962. He received his MA from Temple University in 1964, and his Ph.D. from the Johns Hopkins University in 1969, where he worked with Dr. Gabor Domokos. This led to a lifelong professional relationship and personal friendship with Dr. Domokos. Professor Mansouri then accepted a position as postdoctoral associate at the Enrico Fermi Institute of the University of Chicago, where he worked in the research group of Dr. Yoichiro Nambu. There he met and began an extended collaboration with another postdoctoral associate, Dr. Lay Nam Chang. Together they wrote a seminal paper on string theory, in which the conformal gauge was discovered. The conformal gauge was used to establish rigorously that the spectrum and the gauge conditions of the then popular dual resonance models followed from the Nambu-Goto action for a relativistic string. This gave a significant boost to the development of string theory. Most of the concrete work in string theories of the last three decades has been carried out in the conformal gauge.

From 1974 until 1981 Professor Mansouri was at Yale University, where he made important contributions to the field of supergravity. In an important joint paper with Dr. Samuel MacDowell on the geometrical formulation of supergravity, the concept that is currently known as "MacDowell-Mansouri" gravity was introduced. Professor Mansouri joined the faculty of the Physics Department at the University of Cincinnati in 1981. He was promoted to Professor in 1985. During his tenure at the University of Cincinnati Professor Mansouri made a number of important contributions in the area of three-dimensional quantum gravity. Professor Mansouri mentored a number of graduate students and he was the author of about hundred journal articles.

People who knew Professor Mansouri enjoyed his company, and appreciated his wisdom and quiet and friendly nature. He was a soft spoken, kind, and caring individual. His colleagues at the University of Cincinnati and in the international physics community will sorely miss him. Professor Mansouri will live in our memory as a person to emulate both in his accomplishments and his personal conduct.

PREFACE

This book contains the Proceedings of the Third International Symposium on Quantum Theory and Symmetries, held in September, 2003 in Cincinnati, Ohio. The goal of this intensive one-week meeting was to gather physicists and mathematicians for an exchange of views and news. It was hosted by both the physics and mathematics departments at the University of Cincinnati.

The symposium comprised 135 talks, organized into morning plenary lectures, and afternoon parallel and special topics sessions, including a session of papers given in memory of Freydoon Mansouri, to whom the symposium was dedicated. Crucial to the success of the symposium were the efforts of three special session organizers: Michael Dine (string theory), Alan Kostelecký (Lorentz violation), and Lee Smolin (loop quantum gravity). The contributions in this volume are organized according to their special session, or have been placed into one of five (somewhat arbitrary) categories: algebras and representations; quantization and quantum gravity; $D < 4$ field theories and brane worldvolumes; $D > 3$ field theories and gravity; and applications.

QTS3 would not have been possible without the generous support of the University of Cincinnati, professional societies, and the funding agencies. In particular, we would like to thank the provost of the University of Cincinnati, the dean of its McMicken College of Arts and Sciences, and especially Frank Pinski, the head of the UC Physics Department, for valuable starting funds; the National Science Foundation and the Department of Energy for matching grants; and the Institute for Mathematics and its Applications, the Charles Phelps Taft Memorial Fund, and the Balinkin Fund of the UC Physics Department, for helpful support for travel and logistical expenses. Finally, we would also like to thank the secretarial, computing, and facilities staffs—Dexter Clark, Donna Deutenberg, John Markus, Elle Mengon, John Whitaker, Melody Whitlock, David Will, and Kirsten Woodard—for their hard work and cheerful efficiency.

Philip C. Argyres
Timothy J. Hodges
Freydoon Mansouri
Joseph J. Scanio
Peter Suranyi
L. C. Rohana Wijewardhana

PREFACE

[illegible] held in September 2003 in Cincinnati, Ohio. [illegible] of Cincinnati.

[illegible]

OPENING REMARKS

Dear Ladies and Gentlemen,

Dear Friends and Colleagues,

I welcome you to QTS3—the Third International Symposium on Quantum Theory and Symmetries—in Cincinnati as chairman of the Conference Board. This series of biannual events started with a symposium in 1999 in Goslar, Germany; it went in 2001 to Cracow, Poland; and jumped in 2003 to Cincinnati. The aim of the symposia is to arrange a meeting place for physicists, mathematical physicists, and mathematicians, for the presentation and discussion of recent developments in quantum theory which are related to those mathematical methods applicable to model symmetries in the widest sense, and to understand their geometrical background structures. This includes all parts of quantum theory together with existing and possible future applications, *e.g.*, foundations, extensions, quantum probability and information, nonrelativistic, finite dimensional, discrete and chaotic quantum systems, string theory, quantum gravity, quantum field theory, particle physics, discrete symmetries. The necessary mathematical fields cover: topology, algebra and its deformations, noncommutative geometry, differential geometry, group theory, number theory, functional analytic and approximation methods. Furthermore, numerical evaluation techniques for the necessary connection to experimental results are also included. Quantum theory is and was rapidly developing; the physical problems and the mathematical framework for model building and for their physical interpretations vary, and hence so do the topics of the symposia.

After the Symposium in 1999 in Goslar, participants proposed to continue this activity and to found a biannual symposium series. A kind of umbrella committee—the Conference Board of the series—with 12 members from 11 countries was elected. Freydoon Mansouri was one of the members. A Constitution was drafted and accepted, wherein the board's task is to ask our scientific community for proposals, and to decide on the proposals through a majority vote. Four years ago Fred and myself discussed possibilities to have one of the next symposia in the United States. Fred enthusiastically proposed the University of Cincinnati and its Department of Physics together with the Department of Mathematics. He and his colleagues presented an ambitious plan for QTS3 in 2003. The Board voted during the II. Symposium in Cracow 2001 in favour of Cincinnati. Fred and the local organizing Committee worked on an extremely interesting program with hot topics and an efficient and smooth organization. Boardmembers discussed part of the program and aspects of QTS3 with Fred. We were convinced that the preparation of the meeting was excellent.

The boardmembers including myself were deeply moved after we were informed that Fred suddenly passed away. This was almost impossible to believe, but as human beings we have to accept it. We mourn for Fred. Our deepest sympathy goes to his family. We remember and discuss his scientific work in the Freydoon Mansouri Memorial Session.

The local organizing committee continued the preparation forcefully. In the name of the Conference Board I express my appreciation and my thanks to the Local Organizing Committee. I understand that it was not an easy task. The final program is outstanding; it covers some of the most interesting developments in our field. The organization is very effective; we have all the necessary conditions for a scientifically successful and smooth meeting.

The III. Symposium is dedicated to the memory of Freydoon Mansouri. The results, the lectures and the discussion are a tribute to Fred—friend, colleague and physicist.

H.-D. Doebner

CONTENTS

III. Quantization and Quantum Gravity

IV. $D < 4$ Field Theories; String and Brane Worldvolumes

V. $D > 3$ Field Theories and Gravity

VI. String Theory

VII. Loop Quantum Gravity

VIII. Lorentz Violation

IX. Applications

I. FREYDOON MANSOURI MEMORIAL LECTURES

SOME CONSEQUENCES OF THE HYPOTHESIS OF MINIMAL LENGTHS

LAY NAM CHANG

Physics Department, Mail Code 0435, Virginia Tech
Blacksburg, Virginia 24061
email: laynam@vt.edu

Dedicated to the memory of Freydoon Mansouri

Freydoon Mansouri and I first met during the fall of 1970. I was in the second year of my postdoc at the Enrico Fermi Institute at the University of Chicago, and Freydoon was beginning his first postdoc there just after graduating from Johns Hopkins. As it turned out, I had been studying his work with Gabor Domokos, and when I was given the opportunity to share my office with him, I promptly accepted the offer. Thus began over thirty years of friendship, support, intellectual give-and-take, and warm memories. And a satisfying collaboration that lasted for almost ten years. But even though we went on to work with different collaborators, we both knew that we had established a lasting relationship, and that we could count on each other in times of problems both big and small. I shall miss him.

It is my honor to dedicate this essay to Freydoon Mansouri.

The subject of this paper is in many ways an outgrowth of what Freydoon and I did as our first joint research. We were examining the dynamical symmetries underlying duality, and in particular, to explore the extent to which strings could be the proper variables to use in embodying these symmetries. The basic entities in strings are extended structures, not points, and the position variables describing these structures do not necessarily commute. So the geometry underlying physical systems is non-commutative. I will be describing to you results that have been obtained in work carried out jointly with Djordje Minic, Tatsu Takeuchi, both faculty members at Virginia Tech, Naotoshi Okamura, a postdoc, and with Sandor Benczik and Saifuddin Rayyan, both graduate students in our group.

The precise nature of the geometry is uncertain, but there are clues from a perturbative treatment of strings. For distances large compared to the length scale implied by the string tension, excitations of the system could be described by postions and momenta, at least for energies for which pair creations are expected to be negligible. Perturbative treatments of strings suggest that, concomitant with the non-commutativity in position, the uncertainties in position and momenta are modified to [1]:

$$\Delta x \geq \frac{\hbar}{2}\left(\frac{1}{\Delta p} + \beta \Delta p\right) \tag{1}$$

which implies

$$\Delta x \geq \hbar\sqrt{\beta} \tag{2}$$

so a minimal length is embedded naturally in the system. Note the UV/IR mixing in the uncertainty relation, giving rise to possibilities of effects of short distance phenomena to be present at lower energies.

While the perturbative treatments cannot fix the precise communtation relations among canonical variables, the uncertainty relations suggest some simple forms for these relations. In what follows, we posit relations for the variables describing the positions and momenta of point particles as an abstraction of the underlying string dynamics which are consistent with these uncertainty relations [2]:

$$\begin{aligned}
[\hat{x}_i, \hat{p}_j] &= i\hbar(\delta_{ij} + \beta\hat{p}^2\delta_{ij} + \beta'\hat{p}_i\hat{p}_j \\
[\hat{p}_i, \hat{p}_j] &= 0 \\
[\hat{x}_i, \hat{x}_j] &= i\hbar\,\frac{(2\beta - \beta') + (2\beta + \beta')\beta\hat{p}^2}{(1 + \beta\hat{p}^2)}\,(\hat{p}_i\hat{x}_j - \hat{p}_j\hat{x}_i) \\
L_{ij} &= \frac{x_i p_j - x_j p_i}{(1 + \beta p^2)}
\end{aligned} \tag{3}$$

The parameters β and β' are relics of the string tension, and should be thought of as the expansion coefficients of expressions that are derivable from string theory. We have chosen to modify the canonical relationships first, as an expansion, and with the supposition that momenta commute, the noncommutativity of the position operators can then be deduced by appeal to the Jacobi identity. The relations are supposed to hold in D-dimensions. It can readily be checked that the operators L_{ij} satisfy the algebra of angular momentum.

An underlying assumption we shall be making in what follows is that the correspondence principle applies, and that the dynamical systems continue to be controlled by Hamiltonians that retain their forms unchanged from the limits $\beta = 0$ and $\beta' = 0$. With these caveats, let us first examine how the classical canonical structure gets modified. For definiteness, we will define the classical limit by replacing commutators with the corresponding Poisson brackets:

$$\{F, G\} = \left(\frac{\partial F}{\partial x_i} \frac{\partial G}{\partial p_j} - \frac{\partial F}{\partial p_i} \frac{\partial G}{\partial x_j} \right) \{x_i, p_j\} + \frac{\partial F}{\partial x_i} \frac{\partial G}{\partial x_j} \{x_i, x_j\} \tag{4}$$

The modified canonical equations of motion now take the form

$$\begin{aligned} \dot{x}_i &= \{x_i, H\} = \{x_i, p_j\} \frac{\partial H}{\partial p_j} + \{x_i, x_j\} \frac{\partial H}{\partial x_j} \\ \dot{p}_i &= \{p_i, H\} = - \{x_i, p_j\} \frac{\partial H}{\partial x_j} \end{aligned} \tag{5}$$

and Newton's force equations become

$$\begin{aligned} m\ddot{x}_i &= (1 + \beta p^2)(1 + 3\beta p^2) F_i \\ F_i &= -\frac{\partial V}{\partial x_i} \end{aligned} \tag{6}$$

For illustrative purposes, we set $\beta' = 0$ in the above. For the case of the $1/r^2$ gravitational force, the equation points to a breakdown of the equivalence principle.

To get a feel for what is happening, let us specialize to the case of central potentials, with Hamiltonians given by

$$H = \frac{p^2}{2m} + V(r) \tag{7}$$

where the potential has zero Poisson brackets with L_{ij}. Specifically,

$$\begin{aligned} \dot{x}_i &= \left[1 + (\beta + \beta')p^2\right] \frac{p_i}{m} + \left[(2\beta - \beta') + (2\beta + \beta')\beta p^2\right] \frac{F}{r} L_{ij} x_j \\ \dot{p}_i &= \left[(1 + \beta p^2) x_i + \beta' (p \cdot x) p_i\right] \frac{F}{r} \end{aligned} \tag{8}$$

with

$$\begin{aligned} \{x_k, L_{ij}\} &= x_i \delta_{jk} - x_j \delta_{ki} \\ \{p_k, L_{ij}\} &= p_i \delta_{jk} - p_j \delta_{ki} \end{aligned} \tag{9}$$

As in the normal case, motion occurs in a plane. But, for neither the simple harmonic potential, nor for the Coulomb potential are the orbits closed [3]. For simple harmonic motion, for the simpler case of $\beta' = 0$, the resultant precession is an advancement, contrary to the normal case, while for the Coulomb force, it is a retardation. We may attempt to place a limit on β through its effect on the orbit of Mercury coming solely from this retardation:

$$\hbar\sqrt{\beta} < 2.3 \times 10^{-68}\mathrm{m} \tag{10}$$

This limit is intended for illustrative purposes only, since modifications on other influences on this orbit have not been included.

The equations of motion do yield a modified Liouville theorem. It can be checked that the following is left invariant during evolution [4]:

$$\frac{d^D\mathbf{x}d^D\mathbf{p}}{\left[1+\beta p^2\right]^{D-1}\left[1+\left(\beta+\beta'\right)p^2\right]^{1-\beta'/2(\beta+\beta')}} \tag{11}$$

Using this volume, we can estimate the corrections to the zero point energy coming from the existence of a minimal length [4]. For $\beta' = 0$, and $D = 3$,

$$\begin{aligned}
\Lambda\left(m\right) &= \int \frac{d^3p}{(1+\beta p^2)^3}\left(\frac{1}{2}\sqrt{p^2+m^2}\right) \\
&= 2\pi \int_0^\infty \frac{p^2dp}{(1+\beta p^2)^3}\sqrt{p^2+m^2} \\
&= \frac{\pi}{2\beta^2} f\left(\beta m^2\right)
\end{aligned}$$

$$f(x) = \begin{Bmatrix} 1 + \frac{x}{2(1-x)} + \frac{x^2}{4(1-x)^{3/2}}\ln\left(\frac{1-\sqrt{1-x}}{1+\sqrt{1-x}}\right) \\ 1 - \frac{x}{2(x-1)} + \frac{x^2}{2(x-1)^{3/2}}\tan^{-1}\sqrt{x-1} \end{Bmatrix} \tag{12}$$

Asymptotically,

$$\begin{aligned}
f(x) &= \sqrt{x} \text{ for large } x, \\
&= (1+x)^{0.42} \quad 0 < x < 1, \\
\Lambda(0) &= \frac{\pi}{2\beta^2}.
\end{aligned} \tag{13}$$

As expected, the putative cosmological constant is finite, with higher frequencies being quenched by the minimal length. The corresponding effects on blackbody spectra are however quite slight.

We now examine some quantum consequences of the commutation relations posited above. We will first examine what happens when $D = 1$. For the sake of concreteness, we will examine the simple harmonic oscillator in 1-dimension. In many respects, this example is trivial, since the deformed relations can always be recast into the standard form, as we will demonstrate below. Since we are supposing that momentum variables remain commutative, we will work in momentum space throughout. In which case, the following representation reproduces the modified canonical relationship between position and momentum:

$$\hat{x} = i\hbar\left\{\left(1+\beta p^2\right)\frac{\partial}{\partial p} + \gamma p\right\},$$
$$\hat{p} = p \tag{14}$$

Hermiticity requires

$$\langle f|g\rangle = \int \frac{dp}{\left(1+\beta p^2\right)^{1-\alpha}} f^*(p)g(p)$$
$$\alpha = \frac{\gamma}{\beta} \tag{15}$$

The Hamiltonian is taken to be

$$\hat{H} = \frac{1}{2}\mu\omega^2\hat{x}^2 + \frac{1}{2\mu}\hat{p}^2 \tag{16}$$

The resulting Schrödinger equation cannot be easily compared with standard forms. However, it is straightforward to perform a change of variables so that it can be so compared. We define

$$\rho = \frac{1}{\sqrt{\beta}}\tan^{-1}\left(\sqrt{\beta}p\right), \qquad -\frac{\pi}{2\sqrt{\beta}} < \rho < \frac{\pi}{2\sqrt{\beta}}. \tag{17}$$

The resulting equation now takes the form:

$$\left(\frac{d^2}{d\xi^2} + 2\,\lambda\,\delta\,\frac{s}{c}\frac{d}{d\xi} - \left\{\frac{1}{\lambda^2} - \lambda^2\delta(1+\delta)\right\}\frac{s^2}{c^2} + (\varepsilon + \lambda^2\delta)\right)\Psi(\xi) = 0\,, \tag{18}$$

where

$$\xi \equiv \frac{\rho}{\sqrt{\mu\hbar\omega}}\,,\ \lambda \equiv \sqrt{\mu\hbar\omega\beta}\,,\ \delta \equiv \frac{\gamma}{\beta}\,,\ \varepsilon \equiv \frac{2E}{\hbar\omega}\,,\ c \equiv \cos\lambda\xi\,,\ s \equiv \sin\lambda\xi. \tag{19}$$

The equation resembles that for a particle trapped in a $\tan^2\theta$ potential, which is much steeper than the harmonic oscillator potential. Using standard techniques, the energy eigenvalues are given by

$$E_n = \hbar\omega\left\{\left(n+\frac{1}{2}\right)\sqrt{1+\frac{\beta^2 m^2\hbar^2\omega^2}{4}} + \left(n^2+n+\frac{1}{2}\right)\frac{\beta m\hbar\omega}{2}\right\}$$
$$n = 0,\ 1,\ \ldots \tag{20}$$

The dependence on n^2 is reminiscent of the eigenvalues for a square well potential, and is reflective of the steeper rise of the tranformed potential. Note that there is no dependence on the central parameter γ in the eigenvalues. The eigenfunctions are Gegenbauer polynomials [5]. The results are in agreement with those obtained by Kempf [2].

All of these can be generalized to D-dimensions. However, the non-commutativity of the position variables now plays a more important role, and it is not possible to handle general potentials in a straightforward manner. We begin by noting the momentum space representation for the position operator:

$$\hat{x}_i = i\hbar \left\{ \left(1+\beta p^2\right) \frac{\partial}{\partial p_i} + \beta' p_i p_j \frac{\partial}{\partial p_j} + \gamma p_i \right\}$$
$$\hat{p}_i = p_i \tag{21}$$

Hermiticity requires

$$\langle f|g\rangle = \int \frac{d^D p}{\left[1+(\beta+\beta')p^2\right]^{1-\alpha}} f^* g$$
$$\alpha = \frac{\gamma - \beta' \left(\frac{D-1}{2}\right)}{(\beta+\beta')} \tag{22}$$

For general potentials, the resultant Schrödinger equation is quite complicated. For central potentials, the existence of the angular momentum operator, and the resultant conservation enables us to separate variables, just as in the conventional case.

$$\sum_1^N \frac{\partial^2}{\partial p_i^2} = \frac{\partial^2}{\partial p^2} + \frac{D-1}{p}\frac{\partial}{\partial p} - \frac{L^2}{p^2}$$
$$\sum_1^N p_i \frac{\partial}{\partial p_i} = p\frac{\partial}{\partial p}$$
$$L^2 = \ell(\ell+D-2), \qquad \ell = 0,1,2,\ldots \tag{23}$$

The radial equation can now be analyzed by adapting the methods described above. In this fashion, we obtain that in D-dimensions, the simple harmonic oscillator eigenvalues are given by

$$E_{n\ell} = \hbar\omega \left\{ \left(n+\frac{D}{2}\right) \sqrt{1+\left(\beta^2 L^2 + \frac{(D\beta+\beta')^2}{4}\right)\epsilon^2} + (\beta+\beta')\left(n+\frac{D}{2}\right)^2 \right.$$
$$\left. + \left[(\beta+\beta')\left(L^2+\frac{D^2}{4}\right) + \beta'\frac{D}{2}\right]\frac{\epsilon}{2} \right\}$$
$$n = 0,\ 1,\ \ldots, \qquad \text{and } \epsilon = m\hbar\omega. \tag{24}$$

Once again, there is no dependence on γ. The corresponding eigenfunctions are given by Jacobi polynomials. Note that the accidental degeneracies are now lifted, and there is only the degeneracy associated with rotational symmetry. We should not be too surprised by this result of course, given that the classical orbits are not closed, and there are no Runge-Lenz vectors that we can define. In [5], we examined the effects of such deviations on oscillations of electrons in Penning traps. The effects are unfortunately quite small [5].

We might expect that a better bound on the minimal lengths and their associated effects might show up clearer in the many precision measurements associated with the Coulomb system. Knowledge of the Coulomb system, in addition, could provide ways for how spin is incorporated in the presence of minimal lengths.

We have looked for the general solutions for the 3-D hydrogen atom, but have not been able yet to obtain closed expressions for the energy eigenvalues and eigenfunctions. The problem has to do with the proper definition of the operator $\hat{r}^{-1}$. In the process, we have solved for the eigenvalues for the $\hat{r}^2$ operator and its associated eigenfunctions. By expanding the hydrogenic eigenfunctions in this basis, we have managed to solve for the eigenvalues and eigenfunctions numerically, with results that are consistent with the perturbative solutions obtained by Brau earlier [6].

I conclude by quoting the results for the eigenvalues for the $\hat{r}^2$ operator [7]:

$$\begin{aligned}
{\nu_{nl}}^2 &= \frac{{r_{nl}}^2}{\hbar^2(\beta+\beta')} \\
&= \left\{2n + \left(\frac{D}{2} + \ell + \sqrt{\frac{\{(D-1)\eta+1\}^2}{4} + \eta^2 L^2}\right)\right\}^2 \\
&\qquad - (1-\eta)^2\left\{L^2 + \frac{(D-1)^2}{4}\right\}, \\
\eta &\equiv \frac{\beta}{\beta+\beta'}, \qquad L^2 \equiv \ell(\ell+D-2).
\end{aligned} \tag{25}$$

The associated eigenfunctions are Jacobi polynomials. We will discuss a possible anomaly for the case of $\ell = 0$ in a later communication [8].

Acknowledgments

This work has been supported in part by a grant from the US Department of Energy, DE-FG05-92ER40709.

References

1. D. J. Gross and P. F. Mende, *Nucl. Phys.* **B303**, 407 (1988); D. J. Gross and P. F. Mende, *Phys. Lett.* **B197**, 129 (1987); D. Amati, M. Ciafaloni and G. Veneziano, *Phys. Lett.* **B216**, 41 (1989); D. Amati, M. Ciafaloni and G. Veneziano, *Int. J. Mod. Phys.* **A3**, 1615 (1988); D. Amati, M. Ciafaloni and G. Veneziano, *Phys. Lett.* **B197**, 81 (1987); E. Witten, *Phys. Today* **49**(4), 24 (1997).
2. A. Kempf, G. Mangano and R. B. Mann, *Phys. Rev.* **D52**, 1108 (1995), [arXiv:hep-th/9412167]; A. Kempf, *J. Phys.* **A30**, 2093 (1997) [arXiv:hep-th/9604045].
3. S. Benczik, L. N. Chang, D. Minic, N. Okamura, S. Rayyan and T. Takeuchi, *Phys. Rev.* **D66**, 026003 (2002) [arXiv:hep-th/0204049].
4. L. N. Chang, D. Minic, N. Okamura and T. Takeuchi, *Phys. Rev.* **D65**, 125028 (2002) [arXiv:hep-th/0201017].
5. L. N. Chang, D. Minic, N. Okamura and T. Takeuchi, *Phys. Rev.* **D65**, 125027 (2002) [arXiv:hep-th/0111181].
6. F. Brau, *J. Phys.* **A32**, 7691 (1999) [arXiv:quant-ph/9905033]; for special case of $\ell = 0$, see also R. Akhoury and Y. P. Yao, *Phys. Lett.* **B572**, 37 (2003) [arXiv:hep-ph/0302108].
7. T. Takeuchi and J. Slawny, *VPI-IPPAP-04-02.*
8. S. Benczik, L. N. Chang, D. Minic, T. Takeuchi, *in preparation.*

STRINGS AND BLACK HOLES: AN EMERGENCE OF PATTERNS?*

G. DOMOKOS
Department of Physics and Astronomy
The Johns Hopkins University
Baltimore, MD.

S. KOVESI-DOMOKOS
Department of Physics and Astronomy
The Johns Hopkins University
Baltimore, MD.

Two solutions have been proposed in order to resolve the conundrum of trans-GZK cosmic rays which have not been disproved yet by data: excitation of string-like structures with a characteristic energy scale of the order of a few TeV and the productionn of "mini black holes" with a comparable characteristic scale. We suggest that these two approaches are, in fact, equivalent: there is no physical distinction between a highly excited string state and a "mini black hole".

We dedicate this contribution to our friendship with Freydoon Mansouri, who has always been interested in discovering patterns others have not thought about.

1. Introduction

There is increasing evidence that some of the cosmic ray events observed by a variety of detectors appear to defy the famous Greisen-Zatsepin-Kuzmin (GZK) [1] cutoff in the cosmic ray spectrum. A careful analysis of those events has been given by Günter Sigl[2]. It is to be noted that the so-called trans-GZK events constitute so far a rather small sample of the order of

*Talk presented by G.D.

20 events or so and the teams operating the detectors are still discussing a number of technical questions regarding the determination of primary energies, *etc.* It is rather clear to us that a definitive answer to the puzzle of trans-GZK cosmic rays will not come until new detectors (the Pierre Auger Observatory, the Orbiting Wide Angle Light Collector (OWL) begin operation.) Nevertheless, even with the limited amount of data available, it is interesting to speculate about the origin of the trans-GZK cosmic rays[a]

Here we discuss only two approaches to the problem of the trans-GZK cosmic rays: the one proposed by us[5] and the one involving the production of "mini black holes"[6].

Both proposals have some common features. In particular, both approaches claim that the primaries of trans-GZK cosmic rays are neutrinos and both approaches invoke low scale (of the order of a few, or a few tens of a TeV) unification of interactions. Our approach in ref.[5] describes the process in terms of string excitations, whereas the approach in ref.[6] considers the formation of of black holes within the same general framework. We now want to show that both approaches are, in fact, equivalent, apart, perhaps, from some details.

A cautionary remark is in order. What we are going to show is that the string and mini black hole approaches are *asymptotically* equivalent, in the limit of high string excitations. It is well known, however that asymptotic formulae are, quite often, remarkably accurate.

Due to limitations of space, our list of references is obviously incomplete. We cited some of the original references and apologize to all those whose contribution to the literature has been omitted. A more complete list of references will be included in a version of this paper to be published.

2. Equivalence of the String and Black Hole Pictures

We illustrate our point on a simple example: the excitation of a high level string state. The inclusive single particle distribution considered by Amati and Russo[7] can be dealt with similarly. The levels of a string are labeled by a positive integer N such that the mass of the state isgiven by the formula

$$M^2 \sim N M_*^2 \qquad (N \gg 1) \tag{1}$$

[a]In a recent article[3] Bahcall and Waxman claim that there are no trans-GZK events. Unfortunately, that article contains a questionable manipulation of the data themselves, obsolete approximations, the neglect of interstellar and intergalactic magnetic fields, *etc.* Their conclusion is not borne out by a more detailed and careful analysis, see, for instance, ref.[4].

In eq. (1) M_* stands for the characteristic string scale. In general, for $N \gg 1$ the state is a a highly degenerate one. Let now $\mathcal{O}$ be an operator which creates a string state of index N from an initial state $|\, i\rangle$. (We have in mind, for instance, string excitation in ν-quark interactions, as conjectured in ref. [5]. In that case, $|\, i\rangle$ can stand for a quark state and $\mathcal{O}$ for the fermionic current corresponding to the neutrino.)

Omitting trivial factors, the transition probability is given by:

$$P \propto \sum_{\alpha} \langle i\,|\,\mathcal{O}^{\dagger}|\, N, \alpha\rangle\langle N, \alpha\,|\,\mathcal{O}|\, i\rangle\delta\,(E - NM_*) \tag{2}$$

In eq. (2) α stands for the collection of labels necessary for the full specification of states at level N: one has to sum over (most of) those, since the quantum numbers carried by $|\, i\rangle$ and $\mathcal{O}$ do not fully specify the substate at level N to which the transition takes place.

We recognize that the quantity in eq. (2),

$$\rho = \sum_{\alpha} |\, N, \alpha\rangle\langle N, \alpha\,|\,\delta\,(E - NM_*) \tag{3}$$

is just the (microcanonical) density matrix of the final state. (In statistical mechanics, a smoothing of the delta function is necessary in order to define a level density; in the present context, such a smoothing is automatic if the finite width of resonances is taken into account.) The summation over α leads to the information loss discussed by Amati [8].

It is well known that a *large* microcanonical ensemble can be well approximated by a canonical one. The effective temperature is determined by maximizing the entropy with the subsidiary condition that the value of the energy is fixed. The entropy is given by the logarithm of the trace of the density matrix:

$$S = \ln \mathrm{Tr}\rho \tag{4}$$

The temperature is then:

$$\frac{1}{T} = \frac{\partial S}{\partial E} \tag{5}$$

In the present context, the energy is the invariant center of mass energy of the interaction described by eq. (2), $E = \sqrt{s}$. An operational definition of a "large" microcanonical ensemble is that terms of $O(1/\sqrt{N})$ in the expression of the temperature are negligible[b].

---

[b] Amati and Russo *loc. cit.* use an adaptation of the Darwin-Fowler method and discuss some terms of $O(1/N)$ in the framework of specific string models. For $N \gg 1$, the asymptotic approach used here and the Darwin-Fowler method yield identical results.

Here we consider level densities of the form:

$$d(E) = \mathrm{Tr}\rho \sim CN^{-\gamma}\exp\left(\alpha N^{\delta}\right); \qquad N = \left(E/M_*\right)^2, (N \gg 1). \quad (6)$$

The quantities $C, \gamma, \alpha, \delta$ depend on the specific model considered. In known string models, $\delta = 1/2$. Correspondingly, the inverse temperature is:

$$\frac{1}{T} = \frac{\partial S}{\partial E} = 2\alpha\delta \frac{1}{M_*}\left(\frac{E}{M_*}\right)^{2\delta - 1} + O(1/E) \quad (7)$$

Notice that for $\delta = 1/2$, the temperature is asymptotically constant and, hence, it may be identified with a Hagedorn temperature, T_H. In what follows, we use this value of δ and neglect all corrections of $O(1/E)$. The expression of the Hagedorn temperature in terms of M_* is somewhat model dependent: in the calculations cited above, one has, $T_H = M_*/(a\pi)$, with a ranging, approximately, between 2 and 4. For the purposes of numerical estimates in this work, we adopt $a = 3$: this is equivalent, of course, to fixing the parameter α in eq. (7). Equation (7) establishes the connection between the string and (microscopic) black hole pictures: in fact, it means that – in a sense – a large and highly degenerate isolated system serves as its own thermal reservoir. Consequently, the terms "string" and "black hole" are used interchangeably in what follows.

We note in passing that the asymptotic estimate is insensitive to Kaluza-Klein excitations of compactified dimensions or to brane excitations in schemes where such excitations occur: all those modes have level densities growing as some power of E. Consequently, they contribute terms of $O(1/E)$ to the temperature.

The asymptotic estimate given here has far reaching consequences. In particular, the inclusive single particle distribution can be calculated using elementary statistical mechanics of a canonical ensemble as it was carried out in detail for specific string models by Amati and Russo, ref. [7]. We have in the rest frame of a decaying resonance:

$$dn = \frac{d\sigma}{\sigma} = \frac{1}{\exp(k_0/T) \pm 1}\frac{d^3k}{k_0}L^2, \quad (8)$$

where k_μ stands for the four momentum of the particle observed in the decay and, asymptotically, we may put $T = T_H$. The sign $\pm$ is valid for Fermi/Bose statistics of the observed particle, respectively. The quantity L is a characteristic length of the system. Tentatively we put it equal to a compactification radius in a symmetric toroidal compactification scheme. In what follows, we also put $k^2 = 0$. This is a permissible approximation except perhaps in the far infrared of the observed particle which, however, is not very relevant for the development of observable showers.

We now note that the expression of dn is Lorentz invariant. Consequently, one can replace the rest frame expression of k_0 by a Lorentz invariant quantity reducing to k_0 in that frame. This is elementary. Use the invariant:

$$K = \frac{(P \cdot k)}{\sqrt{P^2}} \tag{9}$$

in place of k_0 in eq. (8). Here, P stands for the total four momentum of the decaying resonance. In a scattering process, such as

$$\nu + q \longrightarrow \text{resonance} \longrightarrow (k) + X,$$

one can approximate $\sqrt{P^2} \approx \sqrt{s}$. (Here, "$(k)$" stands for the observed particle of four momentum k.) The inclusive distribution, eq. (8) with the replacement $k_0 \longrightarrow K$ shows the expected properties if evaluated in the laboratory frame. In the high energy limit, fermions and bosons have to be treated somewhat differently, due to their different statistics.

We evaluate the invariant K in the laboratory frame, where the target quark of an effective mass m is approximately at rest and the incident neutrino has energy E_ν. In the limit $E_\nu \gg m$ we have:

$$K \sim k_{0,L} \sqrt{\frac{E_\nu}{2m}} \left(\frac{m}{E_\nu} + 2 \sin^2 (\theta/2) \right). \tag{10}$$

Here, $k_{0,L}$ stands for the energy of the observed particle evaluated in the laboratory frame; θ is the angle between the incident neutrino and the observed particle.

There is a forward peak ($\theta = 0$) both for a boson and a fermion observed. For a boson, we have

$$\frac{1}{\sigma} \frac{k_0 d\sigma}{d^3k} \bigg|_{\theta=0} \sim \frac{T_H}{k_{0,L}} \sqrt{\frac{2E_\nu}{m}} L^2,$$

whereas for a fermion the term $\propto m/E_\nu$ can be neglected. One thus finds that the forward peak remains finite for a fermion:

$$\frac{1}{\sigma} \frac{k_0 d\sigma}{d^3k} \bigg|_{\theta=0} \sim \frac{L^2}{2}.$$

The different behavior of observed bosons and fermions is due, of course, to their different statistics.

We evaluate the invariant K in the laboratory frame, where the target quark of an effective mass m is approximately at rest and the incident neutrino has energy E_ν. In the limit $E_\nu \gg m$ we have:

$$K \sim k_{0,L}\sqrt{\frac{E_\nu}{2m}}\left(\frac{m}{E_\nu} + 2\sin^2(\theta/2)\right). \tag{11}$$

Here, $k_{0,L}$ stands for the energy of the observed particle evaluated in the laboratory frame; θ is the angle between the incident neutrino and the observed particle.

There is a forward peak ($\theta = 0$) both for a boson and a fermion observed. For a boson, we have

$$\left.\frac{1}{\sigma}\frac{k_0 d\sigma}{d^3k}\right|_{\theta=0} \sim \frac{T_H}{k_{0,L}}\sqrt{\frac{2E_\nu}{m}}L^2,$$

whereas for a fermion the term $\propto m/E_\nu$ can be neglected in eq. (10) and one finds that the forward peak remains finite:

$$\left.\frac{1}{\sigma}\frac{k_0 d\sigma}{d^3k}\right|_{\theta=0} \sim \frac{L^2}{2}.$$

The different behavior of observed bosons and fermions is due, of course, to their different statistics. Obviously, dn is exponentially small for a finite value of the emission angle θ, as expected: this is just the Boltzmann limit of the thermal distribution.

One can evaluate the total multiplicities by means of a straight forward integration. As stated before, neglecting the rest masses of the observed particles is a fair approximation. With that, one has:

$$I \equiv L^2 \int \frac{d^3k}{k_0}\frac{1}{\exp(k_0/T_H) - 1} = \frac{L^2}{12}T_H^2, \tag{12}$$

The corresponding integral for fermions is smaller by a factor of 1/2.

It follows that the total multiplicity of a particle of any kind is proportional to its statistical weight, *viz.*

$$n_B = \frac{1}{12}L^2T_H^2 g_B, \tag{13}$$

and

$$n_F = \frac{1}{24}L^2T_H^2 g_F, \tag{14}$$

respectively. *In the asymptotic limit the multiplicities are independent of the incident energy: an observable property of this scheme.*

As far as *unconfined particles* (γ, W, Z, leptons) are concerned, eqs. (13, 14) contain the final answer. However, for quarks and gluons, one has to consider their fragmentation into observable hadrons. Instead of carrying out a detailed analysis of the fragmentation, we adopt here an approximate procedure, which, however, should give us an idea about the order of magnitude of hadrons of any flavor produced.

We notice that popular analytical fits to parton fragmentation functions, on the average, give about $\langle n_H \rangle \approx 3$ hadrons produced from each parton, Consequently, in order to obtain the multiplicities of hadrons produced, we just multiply the quark and gluon multiplicities by 3.

Below, we list the statistical weights for each kind of particle considered, taking three families into account for quarks and leptons.

$$\begin{aligned} g_q + g_{\bar{q}} &= 72 \\ g_g &= 16 \\ g_l + g_{\bar{l}} &= 24 \\ g_W + g_{\bar{W}} + g_Z + g_\gamma &= 8 \end{aligned} \tag{15}$$

In compiling the statistical weights, we assumed that in the string regime symmetries are unbroken; in particular, W and Z are transverse. For the time being, we did not consider the various superpartners which may be produced in such reactions.

Finally, we have to determine the characteristic length, L, occurring in eq. (8). We use the ADD formula [9], assuming, for the sake of simplicity a symmetric toroidal compactification:

$$\frac{1}{L} = 2\pi M_* \left(\frac{M_*}{M_P}\right)^{2/n} \tag{16}$$

In eq. (16), M_P stands for the four dimensional Planck mass and n is the number of compactified dimensions. We also use $M_* = 80\text{TeV}$. (We shall comment on this choice of the characteristic scale shortly.) Using these ingredients, we estimate the total number of hadrons generated in the initial interaction. One has the estimate:

$$N_{had} \approx 3 \times (g_g + 1/2\,(g_q + g_{\bar{q}}))\, I \tag{17}$$

It is instructive to compare the multiplicity obtained from eq. (17) at a reasonable number of extra dimensions ($n = 6, 7, 8$) with what an incident proton or nucleus would produce[c]. In Table 1 we summarize the hadron multiplicities estimated in the black hole picture. The number of prompt

n	6	7	8
N_{had}	5×10^6	2×10^5	2×10^4

leptons can be calculated analogously.

Let us recall that according to available data, the trans-GZK air showers are "hadron-like": the showers develop similarly to hadron induced ones at lower energies, see Sigl, *loc. cit.*. Let us consider a primary energy, $E_L = 3 \times 10^{11}$GeV, about the highest energy observed, [2]. Assuming no new physics between present accelerator

energies and that energy, one can extrapolate the multiplicity of secondary hadrons in various reactions ($pp, p\bar{p}, e\bar{e}$). The energy dependence of the multiplicity is similar in all those reactions, *cf.* the tables in the Particle Data Group, [10]. Lumping all the data together, one obtains a simple fit:

$$N_{ch} = 2.60 - 0.18f + 1.23f^2, \tag{18}$$

where $f = s/1\text{GeV}^2$. Extrapolated to $E_L = 10^{11}$GeV in a pp interaction, one gets $N_{ch} \approx 170$. The total multiplicity, relevant for shower development, is estimated by multiplying this number by 3/2, since most of the secondaries are pions.

Clearly, the hadronic multiplicities produced by the microscopic black holes is much larger; thus, one may conclude that the event in question was initiated by a heavy nucleus. Adopting a simple superposition model of nuclear interactions[d], one can estimate the "effective atomic number" of the reaction by taking the ratio of the multiplicity produced by a black hole to that of the extrapolated hadronic multiplicity. The quantity directly measurable is, however, X_{max}, the position of the shower maximum in the absorber. Within the framework of the superposition model, the latter is given by

$$X_{\text{max}}(A) = X_{\text{max}}(\text{proton}) - X_0 \ln A,$$

[c] $n = 6, 7$ are theoretically motivated: superstrings and $n = 11$ SUGRA, respectively. We added $n = 8$ in order to keep possible future models in mind.

[d] assuming that the target nucleus is a Fermi gas of almost free nucleons.

where $X_0 \approx 34.5\text{g/cm}^2$. All such formulae are based on a number of assumptions and cannot be used for obtaining quantitative estimates. However, they are are adequate for the present purpose.

The difference in X_{max} between a proton induced and a black hole induced shower, expressed in units of g/cm^2 is given in Table 2. In a recent

n	6	7	8
$X_{\text{max}}(p) - X_{\text{max}}(bh)$	342	231	151

paper, Anchordoqui *et al.* conjectured that the "*Fly's Eye event*" [11] was initiated by a heavy nucleus; their MC simulation prefers a gold nucleus as the primary [12]. Using a superposition model, those authors conclude that X_{max} for the Fly's Eye event is about 100 g/cm^2 lower than it would be for a proton induced shower – hence their conclusion. This is in qualitative agreement with what we found for a black hole induced shower, see Table 2.

3. Discussion

The asymptotic equivalence of the string and black hole pictures of particle production at ultrahigh energies has some consequences which appear to be in qualitative agreement with the observation of trans-GZK cosmic rays. Work is in progress to examine this picture in a more quantitative way. It is to be pointed out, however that the two pictures as proposed in refs. [5] and [6] are not in complete agreement. In particular, in the papers on mini blackhole production a semiclassical formula is used for the production cross section, first proposed by Dimopoulos and Landsberg[14]. However, we argued that the characteristic string scale is not a TeV or so, but closer to 100TeV, *cf.* [13]. (Incidentally, this explains our choice of M_* in the previous Section.) If so, even at the highest energy event one is rather far from the semiclassical regime and one should rather think in terms of the resonance picture of the cross section as described in ref. [5].

A final comment is in order. It should be clear from what was said before that our considerations depend very weakly on specifics of any particular string model. This is fortunate: at present, no string model is known to be phenomenologically successful. Nevertheless, one believes that a successful forthcoming theory will incorporate some basic features of current day string models. Therefore, one is hoping that the scheme described here will translate into a real calculation within the framework of the theory of the future.

We thank the organizers of QTS3 for an invitation to what turned out to be a highly successful and enjoyable meeting.

References

1. K. Greisen, Phys. Rev. Lett. **16**, 748 (1966); G.T. Zatsepin and V.A. Kuzmin, Pisma Zh. Eksp. Teor. Fiz. **61**, 1028, (1966).
2. G. Sigl, Ann. Phys. (New York) **303**, 117 (2003. Sigl's review contains a comprehensive list of references to the observations as well as the proposed explanations.
3. J.N. Bahcall and E. Waxman, Phys. Lett. **B556**, 1 (2003).
4. T. Stanev, R. Engel, A. Mücke, R.J. Protheroe and J. Rachen, Phys. Rev. **D62**, 093005 (2000).
5. G. Domokos and S. Kovesi-Domokos, Phys. Rev. Lett. **82**, 1366 (1999).
6. L.A. Anchordoqui and Haim Goldberg, Phys. Rev. **D65**, 047502 (2002); L.A. Anchordoqui, J.L. Feng, H. Goldberg and A. Shapere, Phys.Rev. **D65**, 124027 (2002).
7. D. Amati and J.G. Russo, Phys. Letters **454**, 207 (1999).
8. D. Amati, in Proceedings of The Abdus Salam Memorial Meeting, edited J. Ellis, F. Hussain, T. Kibble, G. Thompson and M. Virasoro. World Scientific Publishing Co. Singapore (1997).
9. N. Arkani-Hamed, S. Dimopoulos and G. Dvali, Phys. Lett. **B429**, 263 (1998).
10. http://pdg.lbl.gov (K. Hagiwara *et al.*, Phys. Rev. **D66**, 010001 (2002).
11. D.J. Bird *et al.*, Astrophys. J. **441**, 144 (1995)
12. L.A. Anchordoqui, M.T. Dova, T.P. McCauley, T. Paul, S. Reucroft and J.D. Swain, Nuc. Phys **B** (Proc.Suppl.) **97**, 203 (2001).
13. W.S. Burgett, G. Domokos and S. Kovesi-Domokos, hep-ph/0209162. (This is an expanded version of a paper presented at ICHEP 2002, Amsterdam.)
14. S. Dimopoulos and G. Landsberg, Phys. Rev. Lett. **87**, 161602 (2001).

PHYSICS AND GEOMETRY

PETER G.O. FREUND

Enrico Fermi Institute and Department of Physics
University of Chicago, Chicago, IL 60637
Email: freund@theory.uchicago.edu

Our understanding of the four basic concepts of Physics — space, time, matter and force — has undergone radical change in the course of work on unification, starting with Maxwell's unification of electricity with magnetism all the way to current string theory. What started as four independent concepts, with space and time postulated and the possible forms of matter and force arbitrarily chosen, now appear as different aspects of a rich and novel dynamically determined structure.

Fred Mansouri and I became friends while he was with our group in Chicago in the early Seventies. For those of you who did not get to meet Fred. let me say, by way of description, that if the word "understated" had not made it already into the English dictionary, it would have had to be invented for Fred. There is much to be admired in his work. In particular the approach to gravity and supergravity proposed by Fred Mansouri and Sam MacDowell is a true gem. I feel both honored and saddened by my participation in this session.

Galileo and Newton, the founding fathers of modern Physics, have handed down to us four concepts: space, time, matter and force, in terms of which all of Physics is formulated. In terms of these concepts they set down what could now rightly be called the Galileo-Newton standard model of the seventeenth century. It involved

— a one-dimensional time continuum,

— a three dimensional commutative Euclidean space,

— arbitrarily chosen matter, and

— arbitrarily chosen forces

all constrained through Newton's Galilei-invariant equations of motion.

This model was very close to and in agreement with experiment.

Over the past three decades, a lot has been achieved in unifying these four concepts, to the point that now they appear as different aspects of one unifying fundamental concept. Here I wish to review these recent developments and to do so I will first briefly present some prerequisite older ideas.

Let me start with the concept of space. In its oldest form it was introduced axiomatically by the ancient Greeks as a two- or three-dimensional Euclidean space. This form involved the famous axiom of parallels, and was thought to be the only possible space for over two millenia. It underwent a first major revision at the hands of Bolyai and Lobachevsky in the nineteenth century. They discovered that certain symmetric spaces support a geometry which satisfies all of Euclid's axioms with the exception of the axiom of parallels. This revolutionary discovery ultimately led to Felix Klein's Erlangen Program, in which geometry is intrinsically related to group theory. According to Klein, if a group G acts on a space S, then the *geometry of* G on S is the study of G-invariant properties of the "figures" of S. For Klein G was to be a Lie group. For instance 3-dimensional Euclidean geometry is recovered by choosing $G = E_3$, the three-dimensional Euclidean group of translations and rotations. But this idea generalizes to any group, even to finite groups. We can speak of the geometry of a square or of a triangle, whose groups are obviously finite. Equally well, we can associate a geometry to less trivial finite groups, such as the 26 sporadic groups, even to the largest of these, the monster. This is the outcome of the work of Bueckenhout [1].

By contrast, Riemann proposed another way of going beyond Euclidean geometry, in which not the presence of a symmetry group, but that of a metric was to be the guiding principle. The spaces discovered by Riemann, are not symmetric spaces in general. At first sight they seem to defy the Erlangen program. The Riemannian and Kleinian ideas of what a geometry should be were finally reconciled through the introduction of the concept of a *connection* and the parallel transport it gives rise to. The symmetry is then discovered to reside in the tangent spaces of the Riemannian space.

Moving on to Physics, one first adds time as a fourth dimension, to obtain a 4-dimensional Minkowski space in the absence of gravity, or a full-fledged Riemann space upon the inclusion of gravity. The details of this Riemannian geometry are then determined — *not* postulated — through Einstein's equations from a knowledge of the otherwise arbitrary distribution of matter. This arbitrariness is the same as the one encountered already in the Galileo-Newton standard model, but here its appearance is much more jarring. The point — realized by Einstein already — is that something geometric, the Einstein tensor $G_{\mu\nu} = R_{\mu\nu} - \frac{1}{2}g_{\mu\nu}$ has to be proportional to something non-geometric, the energy-momentum tensor $\Theta_{\mu\nu}$.

This difficulty can be cured by turning the energy-momentum tensor itself into a geometric object. But, along with bosonic scalar and vector fields, this tensor also involves spinor fields, which obey Fermi-Dirac-Jordan statistics and this leads to the introduction of transformations which mix Fermi and Bose fields. This way we automatically land at the doorstep of *supersymmetry.*

Remarkably, by the time we require of this supersymmetry

i) that it contain the 4-dimensional Poincaré algebra

ii) that it respect the spin-statistics connection, and

iii) that it have nontrivial representations containing no fields with spin larger than two,

there are eight possible choices for it. They are labeled by an integer $\mathcal{N}$, in terms of which the number of Fermi generators of the supersymmetry algebra is $4\mathcal{N}$. In the maximal case $\mathcal{N} = 8$, there is a unique representation without spins larger than two, and all basic fields must fit into it. This is a remarkable property, for this model fully determines all possible forces and forms of matter, thus getting rid of the total arbitrariness concerning their choices encountered in the Galileo-Newton standard model. This is a major conceptual advance over the standard model of Particle Physics, where the choice of the gauge group and of the representations to which the matter fields must belong is also fraught with a great deal of arbitrariness, as the only constraints on this choice come from anomaly cancellation conditions.

The trouble with this $\mathcal{N} = 8$ supergravity in 4 dimensions is twofold. On phenomenological level it does not agree with experiment. On the conceptual level, its lagrangian, though well-determined, is very complicated and frustratingly unilluminating.

While the the phenomenological problem remains a serious obstacle, the conceptual problem can be considerably alleviated. Maybe the theory looks so complicated because we are looking at it in 4 dimensions, and this is in some way unnatural for it. Maybe its natural habitat is in a higher dimensional space. This brings us to a Nordström-Kaluza-Klein (NKK)type approach to supergravity [2], and requires investigating the possible supergravities in higher dimensions. In the supersymmetric context NKK theory is much more constrained than in the non-supersymmetric case, in which the higher dimension can be arbitrary. Supergravities only exist in dimensions $d \leq 11$. This is the counterpart in NKK theory of the $\mathcal{N} \leq 8$ of $d = 4$ supergravities. It is essentially due to the requirement that in a supersymmetric theory the number n_F of Fermi degrees of freedom must be equal to the number n_B of Bose degrees of freedom. But n_B increases polynomially with the dimension d of space-time, while n_F increases exponentially with d, so that as d increases, n_B cannot keep pace with n_F and beyond a maximal dimension $d_{max} = 11$, there are no supergravities. So we find a maximal 11-dimensional supergravity, and it is reasonable to investigate it.

The main surprise is that, while ordinarily we view the four-dimensionality of space-time as a given, a choice from infinitely many possibilities, in the supersymmetric case there are only eleven possible choices, and thereby we are presented with the opportunity of predicting, or equivalently of understanding the criteria for choosing, the dimension of space-time.

This maximal supergravity $SUGRA_{11}$ is simple and compelling. It contains a graviton, a gravitino and a a rank-three antisymmetric tensor field (3-form), all massless. Under gauge transformations. the 3-form transforms as a potential. Its curl, a 4-form, is a gauge invariant "field-strength." In the NKK spirit we now have to see how this theory compactifies to lower dimensions. But unlike NKK, we do not wish to postulate such a compactification, but rather derive it dynamically. To this end, we have to find classical solutions in which such compactification takes place. The simplest such solution sets the gravitino field to zero and identifies the field-strength 4-form with the volume form of a — then necessarily 4-dimensional — submanifold M_4 of the 11-dimensional space-time manifold M_{11}. The Einstein equations then require the structure of M_{11} to be $M_{11} = M_4 \times M_7$. with M_4 and M_7, both Einstein spaces, whose cosmological constants have opposite signs. Depending on which of the two has positive cosmological constant, the 7- or the 4-dimensional submanifold is the compact one. In the maximally symmetric case we thus obtain a compactification to $AdS_4 \times S^7$, or $AdS_7 \times S^4$. It is interesting that these compactifications prefer certain dimensions for the non-compact space, and that one of these two preferred dimensions is four, which is "experimentally" viable.

There are other problems though. The small size of the compact manifold dictates a very large (absolute value of the) cosmological constant of the non-compact manifold. Moreover, the 4-dimensional particle spectrum is non-chiral, which again runs against the experimental evidence. However recently it has been shown [3] that this is no longer the case if the compact 7-manifold has singularities, e.g. conical ones. In any case, these solutions will play an important role in what follows.

At a deeper level, this field theory is non-renormalizable, and as in the case of Einstein gravity, this calls for drastic modifications. The problem is that the point-like interaction vertices, allow for a precisely determined interaction event, which in turn leads to bad ultraviolet behavior and results in non-renormalizability. By moving on from a theory with interaction of point-like objects, to a theory in which the interacting objects are extended, the interaction gets smeared out and the ultraviolet behavior improves to the point that the theory becomes finite. We are thus led to strings [4]. Unlike theories of interacting point particles, interacting strings automatically dial certain "critical" dimensions, in which they can avoid the conformal anomaly. For bosonic strings this critical dimension is 26, whereas for superstrings it is 10. The bosonic strings exhibit a tachyon instability. The tachyon is eliminated in the superstring case. So, we end up in 10 dimensions, one dimension short of that of maximal supergravity. One could take a cynical attitude to this fact, after all we are talking only of a ten percent "correction" to the dimensionality of space-time. Yet in further developments the theory will find its way back into 11 dimensions, as we shall see.

More than one superstring theory exists. There are the

— open and closed type I superstrings,
— closed type IIA superstrings,
— closed type IIB superstrings,
— closed heterotic $SO(32)$) superstrings, and
— closed heterotic $E_8 \times E_8$ superstrings.

At first sight this seems disappointing, for if string theory is *the* ultimate physical theory, it should rightfully be unique. But it was soon realized that these five, on the face of it, different string theories are really but different aspects of one overall theory, and as such are connected by what are known as dualities.

There are various types of dualities, but I will concentrate here on the particular case of T-duality, and for simplicity I will consider the *closed* Bose string. Then the critical dimension, as was already said, is $d_c = 26$. Of the 25 space dimensions let us compactify one, say the 25-th, on a circle of radius R. The closed string can then *wind* on the compactification circle. The spectrum is

$$m^2 = \left(\frac{n}{R}\right)^2 + \left(\frac{wR}{\alpha'}\right)^2 + \frac{2}{\alpha'}(N + \bar{N} - 2).$$

The three terms here correspond to the NKK modes, the winding modes and the usual closed string oscillator modes respectively. This m^2 is invariant under the the replacements

$$R \longleftrightarrow \frac{\alpha'}{R} \qquad\qquad n \longleftrightarrow w$$

and so are the interactions. A string theory is then *dual* to another string theory with a different compactification radius, as required by this relation, and with the NKK and winding modes interchanged. Notice that for the special radius $R_{SD} = \sqrt{\alpha'}$, the theory is self-dual. This self-dual radius plays the role of a minimal length in the theory, for as the compactification radius keeps decreasing below the value R_{SD}, the compactification radius of the dual (and therefore equivalent) theory keeps increasing above the value R_{SD}. By the time the original compactification radius goes to zero, the dual one goes to infinity. So we can shrink the size of the 25-th dimension up to R_{SD}, but not beyond it, and we can not descend from 26 to 25 dimensions. The closed string stays in 26 dimensions.

Things change for the open string, for which there is obviously no winding number and therefore no exchange between NKK and winding modes can be envisioned. Therefore, there is no minimal length and letting the compactification radius shrink to zero, the 26-dimensional theory *does* reduce to a 25-dimensional one. But open and closed strings are made of the same stuff. This is clear, for just as one end of one string can attach itself to one end of another string to form a single open string at an open string vertex, so one end of one string can attach itself to the other end of the *same* string, to form a closed string. But as the compactification radius of one space dimension goes to zero, closed strings are trapped in 26-dimensional space-time, and therefore so is the stuff they are made of. It then stands to reason, that open strings experience the reduction of the space dimension only through their ends, which therefore must be confined to a 25-dimensional subspace of 26-dimensional space-time. This 25-dimensional subspace is called a D24-brane on account of the Dirichlet boundary conditions in x^{25}.

The D-branes are solitons and exist for superstrings as well. For type IIA superstrings there exist D0-branes, point-solitons on which strings can end. Their mass is

$$m_{D0} = \frac{1}{g\sqrt{\alpha'}},$$

where g is the string coupling. What is more remarkable is the existence of n D0-brane bound state configurations of mass $n\frac{1}{g\sqrt{\alpha'}}$. These states can in turn be viewed as spanning an NKK tower corresponding to a compactification of radius $g\sqrt{\alpha'}$. As the coupling g increases indefinitely, these states end up spanning a continuum. as if the string had caused space to "grow" a new dimension. Superstrings started in a 10-dimensional space-time, so that when they grow this extra dimension, we land in an 11-dimensional *M-theory.* Its low energy limit is the maximal 11-d SUGRA which we discussed earlier.

The D0-branes act as partons of this M-theory. Being zero-dimensional, they obey their own quantum mechanics, a dimensionally reduced — all the way to one dimension — form of 10-dimensional supersymmetric Yang-Mills theory. The coordinates of this quantum mechanics are then matrices, which in general do not commute. At small distances geometry then becomes *non-commutative* à la Connes.

I shall not discuss here the phenomenologically interesting string theory compactifications on 6-dimensional Calabi-Yau manifolds. Rather I will consider strings on $AdS_n \times K_{10-n}$, or M-theory on $AdS_n \times K_{11-n}$ background geometries, where K_m are compact m-dimensional manifolds. As discovered by Maldacena, this leads to a remarkable holographic connection with a conformal field theory (CFT) on $(n-1)$-dimensional Minkowski space M_{n-1}, the boundary of n-dimensional anti-de-Sitter space AdS_n.

An idea of this AdS/CFT correspondence is most readily obtained, by returning to the hadronic strings from which string theory got its start. These hadronic strings were abandoned in the wake of the great success of QCD. But here they re-emerge from QCD, which plays the role of CFT in the AdS/CFT correspondence.

To simplify matters as much as possible, let me consider maximally supersymmetric $\mathcal{N} = 4$ QCD, on 4-dimensional Minkowski space M_4 with gauge group $G = U(N)$. This theory is conformally symmetric: its couplings do not run, by default as it were. The 4-dimensional conformal symmetry of this theory is $SO(4,2)$, which is locally isomorphic to $SU(2,2)$. This theory also has a global $SO(6)$ symmetry, which is in turn locally isomorphic to $SU(4)$. The full superconformal symmetry is $SU(2,2|4)$.

For the string description, $SO(4,2)$ suggests an AdS_5 component to the background geometry. It is, after all, the isometry group of AdS_5. Similarly, $SO(6)$ is the isometry group of the 5-sphere, thus suggesting an overall $AdS_5 \times S^5$ background geometry. This geometry is the vacuum of the 10-dimensional IIB SUGRA, which is obtained in the same way as the $AdS_4 \times S^7$ vacuum of 11-dimensional SUGRA dicussed above. That it is now a 5- and not a 4-dimensional AdS space, is due to the simple fact that in IIB SUGRA, supersymmetry dictates the replacement of the 4-form field strength encountered in 11-dimensional SUGRA by a 5-form. So, the symmetries and supersymmetries of a string theory on an $AdS_5 \times S^5$ background and of a maximally supersymmetric gauge theory on 4-dimensional Minkowski space coincide. There is more to this, and in fact the two theories are equivalent: the Green functions of one can be obtained from those of the other. If both N — which appears in the gauge group $SU(N)$ — and its product with the square of the Yang-Mills coupling constant become very large, the correspondence becomes one between the gauge theory and the IIB SUGRA.

In either case, this result is most surprising, for it states that the Physics of a 4-dimensional quantum gauge field theory without gravity is the same as the Physics of a 10-dimensional theory *with* gravity, be that theory a string theory or, in the appropriate limit, a SUGRA. This is the *AdS/CFT correspondence.* The very presence of gravity and the dimension of space-time have become "relative": they depend on the description we choose.

As stated at the beginning of this paper, our picture of the four basic concepts of Physics has undergone a major revision.

The old seventeenth century Galileo-Newton standard model postulated a universal time, a 3-dimensional Euclidean commutative space, and arbitrary forms of matter moving in it under the influence of arbitrary forces. To its credit, this model was very close to experiment.

By contrast, at the geometric level the twenty-first century string-theoretic unified theory presents us with a $(1 + 3 + \delta)$-dimensional non-Euclidean, and in general non-commutative space-time, in which the number of extra space dimensions is "predicted" to obey $\delta \leq 7$. The precise dimension is "relative," it can change with the chosen description among holgraphically dual pairs. All forces are now unified, as are all forms of matter. Force and matter are themselves just different aspects of one and the same agency, the string, and they also determine the geometry, which unlike Galileo-Newton, is no longer postulated. In fact force, matter space and time, all the four basic concepts, now determine each other and we face a unified whole. To *its* credit, this theory of everything is very close to Mathematics.

References

1. F. Bueckenhout (editor) *Handbook of Incidence Geometry: Buildings and Foundations*, North-Holland, Amsterdam,1995.

2. The papers of Nordström, Kaluza and Klein, along with the English translations, where the original is in German, can be found in T. Appelquist, A.Chodos and P.G.O. Freund *Modern Kaluza-Klein Theories* Addison-Wesley, Menlo Park, 1987. This book also contains material and references for the topics on supersymmetry and supergravity discussed in this paper.

   Nordström's long neglected pre-general-relativity 1914 paper was the first to introduce the idea of higher-dimensional unification. The NKK label I use throughout the current paper is a belated, though alas still badly needed, attempt to do his contribution justice.
3. B.S. Acharya, F. Denef, C. Hofman and N Lambert, hep-th/0308046.
4. Much more on the string theory topics addressed in this paper can be found in the two monographs

   M. Green, J. Schwarz and E. Witten *Superstring Theory* volumes 1 and 2, Cambridge University Press, Cambridge, 1987, and

   J. Polchinski *String Theory* volumes I and II, Cambridge University Press, Cambridge 1998.

   For the newer topics I refer to the excellent review article of O. Aharony, S.S. Gubser, J.M. Maldacena, H. Ooguri and Y. Oz, Phys Rept. **323**, 183 (2000); hep-th/9905111.

NEUTRINOS: WINDOWS TO PLANCK PHYSICS*

P. RAMOND

Institute for Fundamental Theory,
Physics Department, University of Florida
P.O. Box 118440, Gainesville, FL 32611, USA
E-mail: ramond@phys.ufl.edu

After recalling some elegant contributions of the late Freydoon Mansouri, we turn to neutrino physics and use a *modicum* of grand unification to relate quark and lepton mixing matrices. We advocate an expansion for the MNS matrix, *à la* Wolfenstein, and argue that in a wide class of models, θ_{13} is a Cabibbo mixing effect. Also the large neutrino mixing angles reflect the mass patterns of the right-handed neutrinos near the Planck scale, and provide evidence for family structure at that scale.

1. A few words about Freyd

In my mind, one word captures Freyd Mansouri: elegance in physics and in life. I will mention two very beautiful papers and a recent one which make the point. The first is

• "*Dynamics Underlying Duality and Gauge Invariance in the Dual Resonance Model*" with L. N. Chang, Dec 1971.

In that paper, the authors use the conformal invariance of the Nambu-Goto action to linearize it, and obtain the well-known constraints. This was written at a time where most were unaware of the conformal symmetry of two-dimensional systems. These constraints were later explicitly solved in the classic GGRT paper. The second paper is

• "*Unified Geometric Gravity and Supergravity*" with S. MacDowell, Feb 1977,

where a beautiful geometric framework is introduced to describe both grav-

*This work was supported in part by the US Department of Energy under grant DE-FG02-97ER41029.

ity and supergravity. These papers did not get the publicity they deserved as Freyd always found it distasteful to promote his own work, a virtue not shared by most of his contemporaries. Finally I mention a more recent paper

• "*The Cosmological Constant Problem, The Spontaneously Broken Symmetry, And The Generalized Rest Energy*", Dec 2001,

where he derives the generalization of Einstein's famous formula in DeSitter space:

$$m^2c^4 \;=\; E_0(E_0 - 3\,E_\Lambda) + \; E_\Lambda^2(2 + s - s^2)$$

Then he goes on to say:

"*For particles of small mass, such as neutrinos, assuming that they are massive, such a deviation from the standard rest energy will significantly affect their kinematics at very high energies. It may be possible to test this proposal in not too distant a future*".

This brings me to the subject of this talk, neutrino physics.

2. Neutrino data

The results[1] of five years of experiments[2,3,4,5,6,7] on neutrinos can be summarized by the MNS neutrino mixing matrix

$$\mathcal{U}_{MNS} \;=\; \begin{pmatrix} \cos\phi & \sin\phi & \epsilon \\ -\cos\theta\;\sin\phi & \cos\theta\;\cos\phi & \sin\theta \\ \sin\theta\;\sin\phi & -\sin\theta\;\cos\phi & \cos\theta \end{pmatrix} ,$$

in which

$$\sin^2 2\theta \;>\; 0.85\;; \qquad 0.30 \;<\; \tan^2\phi \;<\; 0.65\;; \qquad |\,\epsilon\,|^2 \;<\; 0.005\;.$$

The most striking feature of this mixing matrix is the appearance of two large mixing angles, one of which may well be maximal, and only one small mixing angle. That angle has not yet been measured, its limiting value being set by the CHOOZ[5] reactor experiment. It is a strange situation where a matrix is diagonalized by two large and one small angle: it may hint (unlike the quark sector) at a non-Abelian structure in the mixing pattern. The following presents a framework in which to study such possibilities.

Experiments have also set limits on the mass values of the neutrinos. WMAP[8] is the only experiment to set a limit on the absolute value of the neutrino masses

$$\sum_i m_{\nu_i} < 0.71 \text{ eV}$$

The others, which detect neutrino oscillations, are only sensitive to mass differences with the results

$$\Delta m^2_{\odot} = \mid m^2_{\nu_1} - m^2_{\nu_2} \mid \sim 7. \times 10^{-5} \text{ eV}^2$$

$$\Delta m^2_{\oplus} = \mid m^2_{\nu_2} - m^2_{\nu_3} \mid \sim 3. \times 10^{-3} \text{ eV}^2$$

Since we have no absolute mass measurements, the present data are compatible with three possible mass patterns[9]:

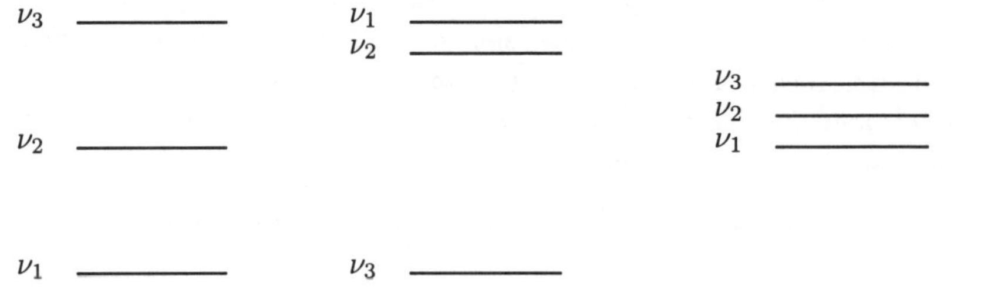

Hierarchy — $|m_{\nu_1}| \leq |m_{\nu_2}| \ll |m_{\nu_3}|$

Inverted — $|m_{\nu_1}| \simeq |m_{\nu_2}| \gg |m_{\nu_3}|$

Hyperfine — $|m_{\nu_1}| \simeq |m_{\nu_2}| \simeq |m_{\nu_3}|$

To determine which of these patterns Nature chooses will require the observation of neutrinoless double β decay.

3. Mixings in the Standard Model

In the Standard Model, masses and mixings of the quarks and charged leptons are the consequence of the spontaneous breakdown of electroweak symmetry, $\Delta I_w = 1/2$. This yields the Quark Yukawa Matrices

$$\mathcal{M}^{(2/3)} = \mathcal{U}_{2/3} \begin{pmatrix} m_u & 0 & 0 \\ 0 & m_c & 0 \\ 0 & 0 & m_t \end{pmatrix} \mathcal{V}^{\dagger}_{2/3} ,$$

$$\mathcal{M}^{(-1/3)} = \mathcal{U}_{-1/3} \begin{pmatrix} m_d & 0 & 0 \\ 0 & m_s & 0 \\ 0 & 0 & m_b \end{pmatrix} \mathcal{V}^{\dagger}_{-1/3} ,$$

from which the observable CKM quark mixing matrix is obtained

$$\mathcal{U}_{CKM} = \mathcal{U}^\dagger_{2/3}\mathcal{U}_{-1/3} .$$

Experimentally, it is nearly equal to the unit matrix, up to small powers of λ, the Cabibbo angle. Family mixing treats quarks of charge 2/3 and −1/3 almost alike. This is exploited in the Wolfenstein[10] expansion which uses the unit matrix as a starting point.

The same electroweak breaking produces the $\Delta I_{\rm w} = 1/2$ charged Leptons Yukawa Matrix

$$\mathcal{M}^{(-1)} = \mathcal{U}_{-1} \begin{pmatrix} m_e & 0 & 0 \\ 0 & m_\mu & 0 \\ 0 & 0 & m_\tau \end{pmatrix} \mathcal{V}^\dagger_{-1} .$$

In the original formulation of the Standard Model, these mixing matrices are unobservable and the neutrinos are massless. However by adding one right-handed neutrino per family, as suggested by Pati-Salam unification[11], we generate the $\Delta I_{\rm w} = 1/2$ neutral leptons Yukawa matrix

$$\mathcal{M}^{(0)}_{Dirac} = \mathcal{U}_0\,\mathcal{D}_0\,\mathcal{V}^\dagger_0 = \mathcal{U}_0 \begin{pmatrix} m_1 & 0 & 0 \\ 0 & m_2 & 0 \\ 0 & 0 & m_3 \end{pmatrix} \mathcal{V}^\dagger_0 .$$

In the absence of any other masses, this produces the Maki-Nakagawa-Sakata (MNS) matrix, the equivalent of the CKM matrix in the lepton sector

$$\mathcal{U}_{MNS} = \mathcal{U}^\dagger_{-1}\mathcal{U}_0 ,$$

of the same form as in the quark sector. However, unlike in the quark sectors, the right-handed neutrinos can have a $\Delta I_{\rm w} = 0$ Majorana mass, unprotected by the gauged electroweak symmetry, although it breaks the global lepton number symmetry

$$\mathcal{M}^{(0)}_{Majorana} \sim \Delta I_{\rm w} = 0 .$$

If the Standard Model has a natural cut-off Λ, we expect those masses to be of that order. If Λ is much larger than the scale of electroweak breaking, the seesaw mechanism[12] naturally suppresses the neutrino masses over that of their charged partners by the ratio of these scales:

$$\begin{aligned}\mathcal{M}^{(0)}_{Seesaw} &= \mathcal{U}_0\,\mathcal{D}_0\,\mathcal{V}^\dagger_0 \frac{1}{\mathcal{M}^{(0)}_{Majorana}} \mathcal{V}^*_0\,\mathcal{D}_0\,\mathcal{U}^T_0 \\ &\equiv \mathcal{U}_0\;\mathcal{C}\;\mathcal{U}^T_0 ,\end{aligned}$$

which defines the *Central Matrix*[13] $\mathcal{C}$, whose properties contain the added subtelties of the seesaw. It is diagonalized by a unitary matrix $\mathcal{F}$

$$\mathcal{C} = \mathcal{F}\mathcal{D}_\nu\, \mathcal{F}^T ,$$

its eigenvalues being the physical neutrino masses

$$\mathcal{D}_\nu = \begin{pmatrix} m_{\nu_1} & 0 & 0 \\ 0 & m_{\nu_2} & 0 \\ 0 & 0 & m_{\nu_3} \end{pmatrix} .$$

The MNS matrix now becomes

$$\mathcal{U}_{MNS} = \mathcal{U}^\dagger_{-1}\mathcal{U}_0\, \mathcal{F} ,$$

where $\mathcal{F}$ is the new $\Delta I_{\rm w} = 0$ ingredient that comes from the seesaw mechanism.

If the matrices from the $\Delta I_{\rm w} = 1/2$ sector contain only small angles, as the CKM matrix and mass hierarchies might suggest, it is natural to ask if the large angles come from $\mathcal{F}$. Furthermore, as the data indicates one small and two large angles, we may also ask if this arrangement is natural in a 3×3 matrix. Indeed, if mixing matrices are generated by textures of zeros[14], then such an arrangement is difficult to imagine as it would produce either three, two or no small angles. Any other arrangement would point to a symmetry beyond texture zeros[15]. It is therefore compelling to investigate the origin of the one small angle.

From the texture point of view, one large mixing angle is generic for a (3×3) matrix, and in a class of models, it is natural for $\mathcal{F}$ to contain only one large angle, with the second large angle hiding in the diagonalization of the *right-handed* quarks and charged leptons.

4. Hints from Grand Unification

To link the MNS and CKM matrices, we need to use some notions of Grand Unification. In its simplest form, it is well-known that the Standard Model families form $SU(5)$[16] and $SO(10)$[17] multiplets, the latter containing one right-handed neutrino per family (going up the exceptional group series, one arrives at E_6[18] with three right-handed neutrinos per family). The simplest Higgs structure of these models relates quark and lepton $\Delta I_{\rm w} = 1/2$ Yukawa matrices:

$$SU(5) \; : \qquad \mathcal{M}^{(-1/3)} \sim \mathcal{M}^{(-1)\,T} .$$

$$SO(10) : \qquad \mathcal{M}^{(2/3)} \sim \mathcal{M}^{(0)}_{Dirac} .$$

Applying these relations to the Yukawa matrices yields

$$\mathcal{U}_{-1/3} \sim \mathcal{V}^*_{-1} ; \qquad \mathcal{U}_{2/3} \sim \mathcal{U}_0 .$$

It follows that

$$\begin{aligned}\mathcal{U}_{MNS} &= \mathcal{U}^\dagger_{-1}\mathcal{U}_0\mathcal{F} = \mathcal{U}^\dagger_{-1}\mathcal{U}_{-1/3}\mathcal{U}^\dagger_{CKM}\mathcal{F} \\ &= \mathcal{V}^T_{-1/3}\mathcal{U}_{-1/3}\mathcal{U}^\dagger_{CKM}\mathcal{F} .\end{aligned}$$

This is the desired relation between the quark and lepton mixing matrices. They are seen to differ in two ways, first by the seesaw which requires $\mathcal{F}$ matrix, and also by the right-handed matrix that diagonalizes the charge $-1/3$ quark sector.

Accordingly we group models of Yukawa couplings in two categories, those for which the charge $-1/3$ Yukawa couplings are symmetric, and those for which they aren't.

- In models where $\mathcal{M}_{-1/3}$ is symmetric[15], we have

  $$\mathcal{U}_{-1/3} = \mathcal{V}^*_{-1/3} ,$$

  and the two matrices are simply related

  $$\mathcal{U}_{MNS} = \mathcal{U}^\dagger_{CKM}\mathcal{F} .$$

  In this case, $\mathcal{F}$ must contain two large mixing angles, and therefore be very special, implying perhaps a non-Abelian structure.
- In the other case, one or two large angles can hide in $\mathcal{V}_{-1/3}$. This occurs in models where asymmetric Yukawas are natural. We mention two possibilities. The first is that of *family cloning*[19] where there is one standard model gauge group for each family. It is broken down to one gauge group by a tri-chiral order parameter[20] which naturally yields one diagonal color group $SU(3)_1 \times SU(3)_2 \times SU(3)_3 \rightarrow SU(3)_{1+2+3}$ but can only break the weak $SU(2)$ to two groups $SU(2)_1 \times SU(2)_2 \times SU(2)_3 \rightarrow SU(2)_{1+2} \times SU(2)_3$, where the subscript denotes the family number. This results, at some intermediate scale, into the following gauge symmetry:

  $$SU(2)_{1+2} \times SU(2)_3 \times SU(3)_{1+2+3} ,$$

  which naturally produces asymmetric Yukawa matrices and one large angle in U_{-1}.

  Another, more generic class of model in which this appears, are of the Froggatt-Nielsen[21] type where Cabibbo suppression is related to the dimensions of the Yukawa operators. By expressing the charge $-1/3$ quark mass ratios as powers of the Cabibbo angle,

$$\frac{m_s}{m_b} \sim \lambda^2 , \qquad \frac{m_d}{m_b} \sim \lambda^4 ,$$

and assuming similar Cabibbo suppression in both $U_{2/3}$ and $U_{-1/3}$, the suggestive order of magnitude pattern appears

$$\mathcal{M}^{(-1/3)} = \begin{pmatrix} \lambda^4 & \lambda^3 & \lambda^3 \\ \lambda^3 & \lambda^2 & \lambda^2 \\ \lambda & 1 & 1 \end{pmatrix} .$$

The 32 matrix element is not suppressed compared to the 33, so that diagonalization requires a large angle *in the right-handed sector*, naturally predicting[22] unsuppressed mixing between mu- and tau-type neutrinos, although not explaining its near maximality. In this model, before Cabibbo corrections,

$$\mathcal{U}_{MNS} = \begin{pmatrix} 1 & 0 & 0 \\ 0 & \cos\theta & \sin\theta \\ 0 & -\sin\theta & \cos\theta \end{pmatrix} \mathcal{F} ,$$

and $\mathcal{F}$ need contain only one large angle, that responsible for the solar neutrino deficit. While not maximal, it is close to 30°. One amusing possibility is to posit a Central matrix of the form

$$\mathcal{C} = \begin{pmatrix} 0 & 1 & 0 \\ 1 & 1 & 0 \\ 0 & 0 & 1 \end{pmatrix} ,$$

which is diagonalized by the golden ratio angle[13]

$$\tan\eta = \frac{1-\sqrt{5}}{1+\sqrt{5}} ,$$

whose pleasing nature has been known to architects for millenia. Could it also be pleasing to *the* Architect?

In either case, we may insert the two large angles in the form

$$\mathcal{U}_{MNS} = \begin{pmatrix} 1 & 0 & 0 \\ 0 & \frac{1}{\sqrt{2}} & -\frac{1}{\sqrt{2}} \\ 0 & \frac{1}{\sqrt{2}} & \frac{1}{\sqrt{2}} \end{pmatrix} \begin{pmatrix} \cos\eta & \sin\eta & 0 \\ -\sin\eta & \cos\eta & 0 \\ 0 & 0 & 0 \end{pmatrix} .$$

In models with symmetric charge $-1/3$ Yukawas, the Cabibbo effects from the CKM matrix come by multiplication from the left, yielding the order of magnitude of the CHOOZ angle

$$\theta_{13} \sim \frac{\lambda}{\sqrt{2}} ,$$

which is satisfying near the experimental limit.

In models where the atmospheric angle hides in the right-handed sector, the CKM matrix is sandwiched between these two matrices; it follows that the CHOOZ angle is determined by the Cabibbo structure of the right-handed matrix $\mathcal{V}_{-1/3}$. The CKM matrix gives only a very small contribution of order λ^3 to θ_{13}, but

$$\theta_{13} \sim \frac{\lambda^p}{\sqrt{2}} ,$$

where λ^p is the order of magnitude of the 21 matrix element of $\mathcal{V}_{-1/3}$.

In these models, the value of that angle depends on the hithero unobservable matrix which diagonalizes the right-handed down quarks. Models with grand unified relations between the quark and lepton mixing matrices suggest a Wolfenstein parametrization of the MNS matrix, not using as a starting point a unit matrix as for the CKM matrix, but a matrix that contains the two large mixing angles. The CHOOZ angle will be determined as some power of the Cabibbo angle. Our favorite starting point is thus

$$\mathcal{U}_{MNS} = \begin{pmatrix} \cos\eta & \sin\eta & 0 \\ -\frac{\sin\eta}{\sqrt{2}} & \frac{\cos\eta}{\sqrt{2}} & -\frac{1}{\sqrt{2}} \\ -\frac{\sin\eta}{\sqrt{2}} & \frac{\cos\eta}{\sqrt{2}} & \frac{1}{\sqrt{2}} \end{pmatrix} + \cdots ,$$

where the dots contain Cabibbo-like effects (we could have chosen $\eta = \pi/6$ as well for this purpose, but we prefer the golden mean for æsthetic reasons).

5. The origin of large angles

As we have seen, $\mathcal{F}$ contains at least one large angle. Let us recall that it diagonalizes the Central matrix which is given by

$$\mathcal{C} \equiv \mathcal{D}_0 \mathcal{V}_0^\dagger \frac{1}{\mathcal{M}^{(0)}_{Majorana}} \mathcal{V}_0^* \mathcal{D}_0 ;$$

since we believe the $\Delta I_{\rm w} = 1/2$ neutral Dirac mass to be hierarchical, the large angle(s) in its diagonalization hints at some correlation between its

Majorana and Dirac components. Let us see how this works in a simplified (2×2) case. We start with hierarchical Dirac eigenvalues[13,23],

$$\mathcal{D}_0 = m\begin{pmatrix} a\,\lambda^\alpha & 0 \\ 0 & 1 \end{pmatrix},$$

from which it follows that

$$\mathcal{C} = \begin{pmatrix} (\frac{c^2}{M_1}+\frac{s^2}{M_2})\,a^2\,\lambda^{2\alpha} & (\frac{c\,s}{M_1}-\frac{c\,s}{M_2})\,a\,\lambda^\alpha \\ (\frac{c\,s}{M_1}-\frac{c\,s}{M_2})\,a\,\lambda^\alpha & (\frac{s^2}{M_1}+\frac{c^2}{M_2}) \end{pmatrix},$$

where $M_{1,2}$ are the eigenvalues of the Majorana matrix, and c, s stand for the sine and cosine of some angle (not the one that diagonalizes the matrix). It is diagonalized by a large mixing angle in two different cases:

- All its matrix elements are of the same order of magnitude, $\mathcal{C}_{11} \sim \mathcal{C}_{22} \sim \mathcal{C}_{12}$. This can happen as long as

$$s \sim b\,\lambda^\alpha\,, \qquad c \sim 1\,, \qquad \frac{M_1}{M_2} \sim \lambda^{2\beta}\,,$$

  implying a correlated hierarchy between the $\Delta I_{\rm w} = 1/2$ and $\Delta I_{\rm w} = 0$ sectors.
- Its diagonal matrix elements are much smaller than its off-diagonal ones (level-crossing case) $\mathcal{C}_{11}\,, \mathcal{C}_{22} \prec \mathcal{C}_{12}$. It is possible if

$$\lambda^\alpha \prec s \preceq \lambda^{\alpha-\beta}\,,$$

  yielding a Central matrix like

$$\frac{\lambda^\alpha\,m^2}{\sqrt{-M_1 M_2}}\begin{pmatrix} 0 & a \\ a & 0 \end{pmatrix},$$

  which displays maximal mixing. A way to arrange is is to take the two right-handed neutrinos to be Dirac partners.

In the first case, we are faced with a large hierarchy of the $\Delta I_{\rm w} = 0$ masses. If $\mathcal{F}$ contains two large angles, all three right-handed neutrinos show hierarchies. It follows that the hierarchy is present even at the Standard Model cut-off. As a welcome by-product, we note that leptogenesis requires right-handed neutrinos with mass of the order of 10^9 GeV, much less than Λ, implying a strong mass hierarchy.

Much remains to be done, but it is clear that searching for the origin of the large mixing angles brings us to consider GUT-scale physics as it relates to the masses of the right-handed neutrinos. The most exciting aspect of neutrino physics is that it brings to the Standard Model scales near the natural scale of gravity. To include gravity at the quantum level, we need superstrings which carry the message of extra symmetries associated with extra dimensions and supersymmetry between fermions and bosons. Since the right-handed neutrinos are a bridge between Planck and electroweak scale physics, model builders should forget about the Standard Model, and first incorporate right-handed neutrinos devoid of any electroweak quantum numbers into their favorite braneworld or extra dimension scenario. Having done so, electroweak physics will be added by completing the $SO(10)$ multiplets[24,25].

In these exciting times with a flow of new experimental information in particle physics, let us hope that neutrinos will prove the key to unravelling the long-standing riddle of flavor physics.

Acknowledgments

I wish to express my condoleances to Freyd's family, as well as my thanks to Professor Wijewardhana and his colleagues for organizing such an excellent conference under difficult circumstances. I also wish to thank A. Datta, L. Everett and F-S Ling for useful discussions.

References

1. G.L. Fogli *et al.*, *Phys. Rev. D* **66** (2002) 093008; G.L. Fogli *et al.*, *Phys. Rev. D* **67** (2003) 073002.
2. The Super-Kamiokande Collaboration, *Phys. Rev. Lett.* **85** (2000) 3999.
3. The Super-Kamiokande Collaboration, *Phys. Lett. B* **539** (2002) 179.
4. The SNO Collaboration, *Phys. Rev. Lett.* **89** (2002) 011301.
5. M. Apollonio *et al.*, *Phys. Lett. B* **338** (1998) 383; *Phys. Lett. B* **420** (1998) 397.
6. The KamLAND Collaboration, *Phys. Rev. Lett.* **90** (2003) 021802.
7. The K2K Collaboration, *Phys. Rev. Lett.* **90** (2003) 041801.
8. The WMAP Collaboration, astro-ph/0302207; astro-ph/0302209.
9. For a recent review, see A.Yu. Smirnov, *"Neutrino Physics: Open Theoretical Questions"* , Nov 2003. hep-ph/0311259.
10. L. Wolfenstein, *Phys. Rev. Lett.* **51** (1983) 1945.
11. J.C. Pati and A.Salam, *Phys. Rev. D* **10** (1974) 275.
12. M. Gell-Mann, P. Ramond, and R. Slansky in Sanibel Talk, CALT-68-709, Feb 1979 , hep-ph/9809459 (retroprint), and in *Supergravity* (North Holland, Amsterdam 1979). T. Yanagida, in *Proceedings of the Workshop on Unified Theory and Baryon Number of the Universe*, KEK, Japan, Feb 1979.
13. A. Datta, F-S Ling and P. Ramond, NPB **671** (2003) 383.
14. P. Ramond, R.G. Roberts and G.G. Ross, *Nucl. Phys. B* **406** (1993) 19.
15. S. King and G. G. Ross *Phys. Lett. B* **57** (2003) 239.

16. H. Georgi and S. Glashow, *Phys. Rev. Lett.* **32** (1974) 438.
17. H. Georgi, in *Particles and Fields-1974*, edited by C.E.Carlson, AIP Conference Proceedings No. 23 (American Institute of Physics, New York, 1975) p.575; H. Fritzsch and P. Minkowski, *Ann. Phys. NY* **93** (1975) 193.
18. F. Gürsey, P. Ramond, and P. Sikivie, *Phys. Lett. B* **60** (1975) 177.
19. Mark J. Bowick and P. Ramond, *Phys. Lett. B* **131** (1983) 367
20. F.-S. Ling and P. Ramond, *Phys. Rev. D* **67** (2003) 115010.
21. C. Froggatt and H.B. Nielsen, *Nucl. Phys. B* **147** (1979) 277.
22. N. Irges, S. Lavignac and P. Ramond, *Phys. Rev. D* **58** (1998) 035003.
23. A. Yu. Smirnov, *Phys. Rev. D* **48** (1993) 3264
24. C. H. Albright and S. M. Barr, *Phys. Lett. B* **461** (1999) 218
25. G. Altarelli, F. Feruglio and I. Masina, JHEP **0011** (2000) 040; W. Buchmüller, Acta Phys.Polon. B32 (2001) 3707-3718; T. Blazek, S. Raby, K. Tobe, *Phys. Rev. D* **62** (2000) 055001.

HAMILTONIANS OF ODD FERMION NUMBER AND PATH INTEGRALS

CHARLES M. SOMMERFIELD
Sloane Physics Laboratory, Yale University
P.O. Box 208120, New Haven, CT, 06520-8120, USA
E-mail: charles.sommerfield@yale.edu

I discuss the difficulties encountered when trying to write a transition amplitude in path-integral form for a theory whose hamiltonian is of odd fermion number. These difficulties are surmounted by introducing infinitesimal time steps that are fermionic in character.

1. Fred Mansouri

I believe that my colleague Samuel MacDowell was originally asked to give a talk including comments about Fred's sojourn at Yale. Unfortunately Sam is unable to be here and so I will try to pinch hit for him. I thank the organizers for giving me a chance to do so. I consider it a great honor.

In his eight or so years at Yale Fred was very much an interactive member of the high-energy theory group but, curiously, the only joint paper he wrote with another Yalie during this period was the one on the geometrical formulation of supergravity with Sam MacDowell. Sam has specifically asked me to say that he found working with Fred on this to be a really great pleasure.

In the 1970s the theorists at Yale formed a very cosmopolitan bunch, including not only Fred and Sam, but also Feza Gürsey, Itzhak Bars, Chia Tze, R. Shankar and visitors from the Middle East such as Meral Serdaroglu and Farhad Ardalan, not to mention the Americans. We all had offices on a common hallway and there were not only physics interactions but social ones involving our families as well. It was perhaps the most dynamic period of my association with Yale.

Fred invited me for a seminar here in Cincinnati not too long after he arrived and it was then that I had the pleasure of meeting Dewleen. Of course, I had known Marina as a child in New Haven and today I see her as a mature adult. I thank you both as well as Leila for suggesting that an appropriate way for us to remember Fred is through this memorial session.

To sum up my impression of Fred in just a few words I would say that he was a very smart guy and a very nice guy. He will be missed.

2. Introduction and motivation

Suppose we have a spin-$\frac{1}{2}$ system and we want to evaluate

$$\text{tr } e^{-i\lambda\sigma_3 T} = 2\cos\lambda T \tag{1}$$

by path-integral methods. One way would be to introduce the fermionic Fock-space operators a and $a^\dagger$ where

$$\{a, a^\dagger\} = 1, \quad a^2 = a^{\dagger 2} = 0 \tag{2}$$

which represent the three Pauli matrices as

$$\tfrac{1}{2}(\sigma_1 + i\sigma_2) = a^\dagger, \quad \tfrac{1}{2}(\sigma_1 - i\sigma_2) = a, \quad \sigma_3 = [a^\dagger, a]. \tag{3}$$

We then interpret the exponential as providing the time development of a system with the hamiltonian $\lambda[a^\dagger, a]$ which is of even fermion number. However, we notice that we would get the same answer for the trace in (1) if instead of σ_3 we used $\sigma_1 = \lambda(a + a^\dagger)$, which is of odd fermion number.

In this talk I will show what happens when you try to use path-integral methods on this latter representation of the hamiltonian. One might ask why anyone would want to do such a crazy thing and so I will start with a few words about how I got involved.

3. The vacuum state of supersymmetric matrix models

I have been trying to find the ground-state wave function for the $N = 16$ supersymmetric quantum-mechanical hamiltonian[1]

$$H = \tfrac{1}{2}(\pi_m^a)^2 + \tfrac{1}{4}g^2(f^{abc}\phi_m^b\phi_n^c)^2 - i\tfrac{1}{2}g\psi_\alpha^a f^{abc}(\Gamma_m)_{\alpha\beta}\phi_m^b\psi_\beta^c \tag{4}$$

which arises by dimensional reduction from an $N = 1$ supersymmetric field theory in 10 space-time dimensions. The canonical bosonic variables are ϕ_m^a and π_m^a and the fermionic variables are the hermitian ψ_β^c. The indices a, b, c... refer to an internal symmetry group with structure constants f^{abc}. The indices m, n... are relics of the spatial indices of the 10-dimensional theory and go from 1 to 9. The spinor indices α, β... go from 1 to 16. The 9 real symmetric 16×16 matrices Γ_m satisfy $\{\Gamma_m, \Gamma_n\} = 2\delta_{mn}$.

The supersymmetry is generated by the 16 operators

$$Q_\alpha = [(\Gamma_m)_{\alpha\beta}\pi_m^a + \tfrac{1}{2}gf^{abc}(\Gamma_m\Gamma_n)_{\alpha\beta}\phi_m^b\phi_n^c]\psi_\beta^a. \tag{5}$$

The anticommutator of these operators is

$$\{Q_\alpha, Q_\beta\} = 2H\delta_{\alpha\beta} + 2\Gamma^m_{\alpha\beta}\phi^a_m J^a \tag{6}$$

where

$$J^a = f^{abc}[\phi^b_m \pi^c_m - \tfrac{1}{2} i\psi^b_\alpha \psi^c_\alpha] \tag{7}$$

is the generator of the global internal symmetry group. Since as part of the gauge choice involved in the dimensional reduction, the first-class constraint $J^a|\Psi\rangle = 0$ must be imposed on any physical state $|\Psi\rangle$, a solution of the equation $Q_\alpha|\Omega\rangle = 0$ will be an eigenstate of H with eigenvalue 0.

The chief question is how do we write the vacuum wave function F in terms of the canonical variables. What are the fermionic canonical variables? The simplest procedure is to pair up the Majorana spinors ψ^a_α into Dirac spinors and their conjugates. Then F could be taken to be a function of ϕ^a_m and the Dirac spinors, but not the conjugates. However, there seems to be no simple, natural way to do the pairing.

A rather formidably complicated method would be to diagonalize the huge matrix of coefficients of the Majorana fermions in the hamiltonian. Another procedure, which would avoid the use of Dirac fermions, would be to write a projection operator onto the vacuum wave function.

Wisdom is often gained by studying simplifications of the theory of interest. To this end one can lower the number of supersymmetries or the rank of the symmetry group (which is supposed to be very large for the application to M theory). And in fact there has been nontrivial progress along these lines.[2]

In hunting for something new to try I thought that since the hamiltonian is quadratic in fermion operators, why not try to make it linear instead, as in the spin-$\frac{1}{2}$ problem. And that is how I got to the peculiar procedure I am about to describe.

4. Eigenstates of non-hermitian fermionic operators as basis states

We introduce coherent basis states[3] $\langle\psi|$ for the spin-$\frac{1}{2}$ problem which have the properties

$$\langle\psi|a = \psi\langle\psi|, \quad \langle\psi|a^\dagger = \frac{\partial_{\rm r}}{\partial\psi}\langle\psi|, \quad a|\psi\rangle = -|\psi\rangle\psi, \quad a^\dagger|\psi\rangle = -|\psi\rangle\frac{\partial_{\rm l}}{\partial\psi} \tag{8}$$

where r and l signify that the derivatives act to the right and left, respectively. The quantity ψ in (8) is a Grassmann number. Completeness is expressed by

$$\int |\psi\rangle\, d\psi\, \langle\psi| = 1, \quad \int d\psi = 0, \quad \int d\psi\, \psi = 1. \tag{9}$$

The role of a delta function is played by

$$\langle\psi|\psi'\rangle = \psi - \psi'. \tag{10}$$

There are corresponding equations for the conjugate basis states $\langle\bar{\psi}|$:

$$\langle\bar{\psi}|a^\dagger = -\bar{\psi}\langle\bar{\psi}|, \quad \langle\bar{\psi}|a = -\frac{\partial_\mathrm{r}}{\partial\bar{\psi}}\langle\bar{\psi}|, \quad a^\dagger|\bar{\psi}\rangle = |\bar{\psi}\rangle\bar{\psi}, \quad a|\bar{\psi}\rangle = |\bar{\psi}\rangle\frac{\partial_\mathrm{l}}{\partial\bar{\psi}} \tag{11}$$

$$\int |\bar{\psi}\rangle\, d\bar{\psi}\, \langle\bar{\psi}| = 1, \quad \int d\bar{\psi} = 0, \quad \int \bar{\psi}\, d\bar{\psi} = 1 \tag{12}$$

$$\langle\bar{\psi}|\bar{\psi}'\rangle = -(\bar{\psi} - \bar{\psi}'). \tag{13}$$

Integrals over ψ and $\bar{\psi}$ are equivalent to the derivatives $\frac{\partial_\mathrm{r}}{\partial\psi}$ and $\frac{\partial_\mathrm{l}}{\partial\bar{\psi}}$ respectively. Finally we have the transformation functions

$$\langle\psi|\bar{\psi}\rangle = 1 + \psi\bar{\psi}, \quad \langle\bar{\psi}|\psi\rangle = 1 + \bar{\psi}\psi. \tag{14}$$

There is one last formula that we will need, which is the trace over spin states of an operator Ω:

$$\mathrm{tr}\,\Omega = \int d\psi\, \langle\psi|\Omega|-\psi\rangle. \tag{15}$$

5. Fermionic path integral

To set up the fermionic path integral we write

$$X_i = e^{-i\lambda\sigma_i T} = \prod_{k=1}^{n} e^{-i\lambda\sigma_i T/n} \tag{16}$$

and proceed to find the matrix element $\langle\psi_\mathrm{f}|X_i|\psi_0\rangle$ by introducing completeness integrals over intermediate coherent states in the product. Writing the product of the resulting exponentials as the exponential of a sum and taking the limit as $n \to \infty$, we hope to obtain a fermionic path integral[4] of the exponential of an action, in complete analogy with what is done with bosonic variables.

It turns out that this is a straightforward procedure for $i = 3$, (that is, for the usual even-fermion-number hamiltonian) and we will not go through it here. But if we choose $i = 1$ we will find ourselves beset with difficulties, as illustrated in what follows.

We use the completeness relation to write

$$\langle\psi_{\rm f}|X_1|\psi_0\rangle = \langle\psi_{\rm f}|\prod_{k=1}^{n} e^{-i\lambda(a+a^\dagger)T/n}|\psi_0\rangle$$

$$= \langle\psi_{\rm f}|e^{-i\lambda_n(a+a^\dagger)}|\psi_{n-1}\rangle\left(\prod_{k=1}^{n-1}\int d\psi_k\,\langle\psi_k|e^{-i\lambda_k(a+a^\dagger)}|\psi_{k-1}\rangle\right)_k . \quad (17)$$

The notation $(\ldots)_k$ signifies a k-ordered product whereby the product is ordered so that k increases from right to left. The kth infinitesimal time interval has been denoted by $\lambda_k = \lambda T/n$. We need work only to first order in each λ_k, so that

$$\langle\psi_k|e^{-i\lambda_k(a+a^\dagger)}|\psi_{k-1}\rangle \approx \langle\psi_k|1 - i\lambda_k(a+a^\dagger)|\psi_{k-1}\rangle$$
$$= \psi_k - \psi_{k-1} - i\lambda_k(1 - \psi_k\psi_{k-1}). \quad (18)$$

It is a simple matter to carry out the products and integrals one at a time to find

$$\langle\psi_{\rm f}|X_1|\psi_0\rangle = \left[(\psi_{\rm f} - \psi_0)\,\overline{\cos\sum_{k=1}^{n}\lambda_k} - i(1 - \psi_{\rm f}\psi_0)\,\overline{\sin\sum_{k=1}^{n}\lambda_k}\right] . \quad (19)$$

The bar indicates a power series expansion in which only terms that are at most linear in each λ_k are kept. In the limit $n \to \infty$ the barred expressions become $\cos\lambda T$ and $\sin\lambda T$ and the subsequent calculation of the trace wherein ψ_0 is set equal to $-\psi_{\rm f}$ followed by integration over $\psi_{\rm f}$, gives the correct answer $2\cos\lambda T$.

This shows that our manipulations with coherent states are not inconsistent. However, we have not yet expressed $\langle\psi_{\rm f}|X_1|\psi_0\rangle$ as a path integral of the exponential of an action, that is of an integrated lagrangian.[5]

To attempt to remedy this situation we introduce the conjugate variables $\bar{\psi}$ and write

$$\langle\psi_k|1 - i\lambda_k(a+a^\dagger)|\psi_{k-1}\rangle \approx \langle\psi_k|(1 - i\lambda_k a)(1 - \lambda_k a^\dagger)|\psi_{k-1}\rangle$$
$$\approx \int \langle\psi_k|e^{-i\lambda_k a}|\bar{\psi}_k\rangle\, d\bar{\psi}_k\, \langle\bar{\psi}_k|e^{-i\lambda_k a^\dagger}|\psi_{k-1}\rangle$$
$$= \int e^{-i\lambda_k\psi_k} e^{\psi_k\bar{\psi}_k}\, d\bar{\psi}_k\, e^{i\lambda_k\bar{\psi}_k} e^{\bar{\psi}_k\psi_{k-1}}$$
$$= \int d\bar{\psi}_k\, e^{-\bar{\psi}_k(\psi_k - \psi_{k-1})} e^{i\lambda_k\psi_k} e^{i\lambda_k\bar{\psi}_k}$$
$$\approx \int d\bar{\psi}_k\, e^{-\bar{\psi}_k(\psi_k - \psi_{k-1})} e^{i\lambda_k(\psi_k + \bar{\psi}_k)}. \quad (20)$$

The product over all k yields

$$\langle\psi_{\rm f}|X_1|\psi_0\rangle = \int (\psi_{\rm f} - \psi_n)[d\psi\, d\bar{\psi}] e^{-\sum_{k=1}^{n} \bar{\psi}_k(\psi_k - \psi_{k-1})} \left(\prod_{k=1}^{n} e^{i\lambda_k(\psi_k+\bar{\psi}_k)}\right)_k . \quad (21)$$

where

$$[d\psi\, d\bar{\psi}] = \left(\prod_{k=1}^{n} d\psi_k\, d\bar{\psi}_k\right)_k . \quad (22)$$

We would like the right-hand part of the integrand to come out something like

$$\exp\left[-\sum_{k=1}^{n}[\bar{\psi}_k(\psi_k - \psi_{k-1}) - i\lambda_k(\psi_k + \bar{\psi}_k)]\right] \quad (23)$$

so that in a continuum limit it would be of the form

$$e^{i\int dt L(t)} = e^{i\int dt\left[i\bar{\psi}(t)\dot{\psi}(t)+\lambda[\psi(t)+\bar{\psi}(t)]\right]} . \quad (24)$$

But for two Grassmann numbers A and B the statement $e^A e^B = e^{A+B}$ is not true, even infinitesimally, unlike the case for commuting bosonic operators. Hence the step from (21) to (23) is not correct. And, in fact, doing the path integral over the integrand in the "desired" form of (24) will not give the correct answer (19), as can easily be demonstrated.

6. Infinitesimal time steps as Grassmann numbers

If we were able to treat the exponent $\lambda_k(\psi_k+\bar{\psi}_k)$ in (21) as even rather than odd in fermion number then we could write the product of the exponentials in (21) as the exponential of the sum in (23). This can be accomplished if we replace the λ_k by Grassmann numbers $\hat{\lambda}_k$. We then interpret $\lambda_k(a+a^\dagger)$ as the hermitian structure $\hat{\lambda}_k^\dagger a + a^\dagger\hat{\lambda}_k$ in terms of the new $\hat{\lambda}_k$. There clearly could be a sign ambiguity in this interpretation which will eventually be clarified. Below we will carry out the path integral incorporating these changes and then demonstrate that a certain transcription back to the c-number λ_k will yield the correct answer.

But first we have to establish a few technical points concerning the behavior of the new Grassmann numbers in matrix elements. The following are a set of consistent properties of the $\hat{\lambda}_k$ with respect to the coherent basis states:

$$\langle\psi|\hat{\lambda}_k = \hat{\lambda}\langle\psi|,\ \ \langle\psi|\hat{\lambda}_k^\dagger = \hat{\lambda}_k^\dagger\langle\psi|,\ \ \langle\bar{\psi}|\hat{\lambda}_k = -\hat{\lambda}_k\langle\bar{\psi}|,\ \ \langle\bar{\psi}|\hat{\lambda}_k^\dagger = -\hat{\lambda}_k^\dagger\langle\bar{\psi}|,$$
$$\hat{\lambda}_k|\psi\rangle = -|\psi\rangle\hat{\lambda}_k,\ \ \hat{\lambda}_k^\dagger|\psi\rangle = -|\psi\rangle\hat{\lambda}_k^\dagger,\ \ \hat{\lambda}_k|\bar{\psi}\rangle = |\bar{\psi}\rangle\hat{\lambda}_k,\ \ \hat{\lambda}_k^\dagger|\bar{\psi}\rangle = |\bar{\psi}\rangle\hat{\lambda}_k^\dagger . \quad (25)$$

An additional requirement consistency is to take $\hat{\lambda}_k^\dagger = -\hat{\lambda}_k$. The reason for this choice will also emerge eventually.

We repeat the calculation leading to (20) with the replacement

$$\lambda_k(a + a^\dagger) \to a\hat{\lambda}_k^\dagger + \hat{\lambda}_k a^\dagger = \hat{\lambda}_k(a + a^\dagger) \tag{26}$$

and find

$$\begin{aligned}
\langle\psi_k|e^{-i(a\hat{\lambda}_k^\dagger + \hat{\lambda}_k a^\dagger)}|\psi_{k-1}\rangle &= \langle\psi_k|(1 - i\hat{\lambda}_k a)(1 - i\hat{\lambda}_k a^\dagger)|\psi_{k-1}\rangle \\
&= \int \langle\psi_k|e^{-i\hat{\lambda}_k a}|\bar{\psi}_k\rangle\, d\bar{\psi}_k\, \langle\bar{\psi}_k|e^{-i\hat{\lambda}_k a^\dagger}|\psi_{k-1}\rangle \\
&= \int e^{-i\hat{\lambda}_k\psi_k} e^{\psi_k\bar{\psi}_k}\, d\bar{\psi}_k\, e^{-i\hat{\lambda}_k\bar{\psi}_k} e^{\bar{\psi}_k\psi_{k-1}} \\
&= \int d\bar{\psi}_k\, e^{-\bar{\psi}_k(\psi_k - \psi_{k-1})} e^{-i\hat{\lambda}_k\psi_k} e^{-i\hat{\lambda}_k\bar{\psi}_k} \\
&= \int d\bar{\psi}_k\, e^{-\bar{\psi}_k(\psi_k - \psi_{k-1})} e^{-i\hat{\lambda}_k(\psi_k + \bar{\psi}_k)}.
\end{aligned} \tag{27}$$

If before taking the product over k we do the integral over $\bar{\psi}_k$ we obtain

$$\langle\psi_k|e^{-i(a\hat{\lambda}_k^\dagger + \hat{\lambda}_k a^\dagger)}|\psi_{k-1}\rangle = \psi_k - \psi_{k-1} - i\hat{\lambda}_k(1 - \psi_k\psi_{k-1}) \tag{28}$$

which, if we now replace $\hat{\lambda}_k$ by the c-number λ_k, is just the same as what was obtained in (18). This justifies the sign choices we made, as referred to above.

Then using (27) without doing the $\bar{\psi}_k$ integral , we obtain in place of (21)

$$\langle\psi_{\rm f}|X_1|\psi_0\rangle = \int (\psi_{\rm f} - \psi_n)[d\psi\, d\bar{\psi}]\, e^{\sum_k[-\bar{\psi}_k(\psi_k - \psi_{k-1}) - i\hat{\lambda}_k(\psi_k + \bar{\psi}_k)]} \tag{29}$$

where the notation from now on is such that all sums over the index symbolized by k go from 1 to n. With the use of the fermionic integral representation

$$(\psi_{\rm f} - \psi_n) = \int d\bar{\beta}\, e^{-\bar{\beta}(\psi_{\rm f} - \psi_n)} \tag{30}$$

this can be written as a standard "gaussian" fermionic path integral of the form

$$\langle\psi_{\rm f}|X_1|\psi_0\rangle = \int d\bar{\beta}\, e^{-\bar{\beta}\psi_{\rm f}}\, [d\psi\, d\bar{\psi}]\, e^{-\sum_{k,l}\bar{\psi}_k C_{kl}\psi_l - \sum_k(\bar{\psi}_k B_k + A_k\psi_k)} \tag{31}$$

where

$$A_k = i\hat{\lambda}_k - \bar{\beta}\delta_{k,n}, \quad B_k = -i\hat{\lambda}_k - \psi_0\delta_{k,1}, \quad C_{kl} = \begin{cases} \delta_{k,l} - \delta_{k-1,l} & \text{if } k > 1 \\ \delta_{k,l} & \text{if } k = 1 \end{cases}. \tag{32}$$

The result of the integral is

$$\langle\psi_{\rm f}|X_1|\psi_0\rangle = \det(C)\int d\bar\beta\; e^{-\bar\beta\psi_{\rm f}} e^{\sum_{k,l} A_k (C^{-1})_{kl} B_l}. \tag{33}$$

In matrix form (with k decreasing to the right and downward) we have

$$C = \begin{pmatrix} 1 & -1 & 0 \dots & 0 & 0 \\ 0 & 1 & -1 \dots & 0 & 0 \\ 0 & 0 & 1 \dots & 0 & 0 \\ \dots & \dots & \dots & \dots & \dots \\ 0 & 0 & 0 \dots & 1 & -1 \\ 0 & 0 & 0 \dots & 0 & 1 \end{pmatrix}, \quad C^{-1} = \begin{pmatrix} 1 & 1 & 1 \dots & 1 & 1 \\ 0 & 1 & 1 \dots & 1 & 1 \\ 0 & 0 & 1 \dots & 1 & 1 \\ \dots & \dots & \dots & \dots & \dots \\ 0 & 0 & 0 \dots & 1 & 1 \\ 0 & 0 & 0 \dots & 0 & 1 \end{pmatrix} \tag{34}$$

and analytically

$$(C^{-1})_{kl} = \begin{cases} 1 \text{ if } k \geq l \\ 0 \text{ otherwise} \end{cases}, \qquad \det(C) = 1. \tag{35}$$

This leads to

$$\begin{aligned} \langle\psi_{\rm f}|X_1|\psi_0\rangle &= \int d\bar\beta\; e^{-\bar\beta\psi_{\rm f}} e^{\sum_{k>l}\hat\lambda_k\hat\lambda_l} e^{i\bar\beta\sum_k \hat\lambda_k} e^{-i\sum_k \hat\lambda_k\psi_0} e^{\bar\beta\psi_0} \\ &= \left(\psi_{\rm f} - \psi_0 - i\sum_k \hat\lambda_k\right)\left(1 - i\sum_k \hat\lambda_k\psi_0\right) e^{\sum_{k>l}\hat\lambda_k\hat\lambda_l} \\ &= \left(\psi_{\rm f} - \psi_0 - i(1-\psi_{\rm f}\psi_0)\sum_k \hat\lambda_k\right) e^{\sum_{k>l}\hat\lambda_k\hat\lambda_l}. \end{aligned} \tag{36}$$

And now for a bit of sleight of hand and mumbo-jumbo. We expand this result in powers of $\hat\lambda_k$ and then rearrange the resulting multilinear expression, keeping track of signs, so that it *anti-k* ordered. Then we replace each $\hat\lambda_k$ in place by λ_k.

For example, if $n = 4$ we have

$$\begin{aligned} & e^{\sum_{k>l}\hat\lambda_k\hat\lambda_l} = e^{\hat\lambda_4\hat\lambda_3+\hat\lambda_4\hat\lambda_2+\hat\lambda_4\hat\lambda_1+\hat\lambda_3\hat\lambda_2+\hat\lambda_3\hat\lambda_1+\hat\lambda_2\hat\lambda_1} \\ & = 1 + \hat\lambda_4\hat\lambda_3 + \hat\lambda_4\hat\lambda_2 + \hat\lambda_4\hat\lambda_1 + \hat\lambda_3\hat\lambda_2 + \hat\lambda_3\hat\lambda_1 + \hat\lambda_2\hat\lambda_1 \\ & + \tfrac{1}{2}(\hat\lambda_4\hat\lambda_3\hat\lambda_2\hat\lambda_1 + \hat\lambda_2\hat\lambda_1\hat\lambda_4\hat\lambda_3 + \hat\lambda_4\hat\lambda_2\hat\lambda_3\hat\lambda_1 + \hat\lambda_3\hat\lambda_1\hat\lambda_4\hat\lambda_2 + \hat\lambda_4\hat\lambda_1\hat\lambda_3\hat\lambda_2 + \hat\lambda_3\hat\lambda_2\hat\lambda_4\hat\lambda_1) \\ & = 1 - \hat\lambda_3\hat\lambda_4 - \hat\lambda_2\hat\lambda_4 - \hat\lambda_1\hat\lambda_4 - \hat\lambda_2\hat\lambda_3 - \hat\lambda_1\hat\lambda_3 - \hat\lambda_1\hat\lambda_2 + \hat\lambda_1\hat\lambda_2\hat\lambda_3\hat\lambda_4. \end{aligned} \tag{37}$$

When we make the replacement $\hat\lambda_k \to \lambda_k$, (37) becomes

$$1 - \lambda_3\lambda_4 - \lambda_2\lambda_4 - \lambda_1\lambda_4 - \lambda_2\lambda_3 - \lambda_1\lambda_3 - \lambda_1\lambda_2 + \lambda_1\lambda_2\lambda_3\lambda_4 = \overline{\cos\sum_{k=1}^{4}\lambda_k}. \tag{38}$$

Furthermore, we have

$$\left(\sum_k \hat{\lambda}_k\right) e^{\sum_{k>l}\hat{\lambda}_k\hat{\lambda}_l} = \hat{\lambda}_1 + \hat{\lambda}_2 + \hat{\lambda}_3 + \hat{\lambda}_4$$
$$-\hat{\lambda}_1\hat{\lambda}_3\hat{\lambda}_4 - \hat{\lambda}_1\hat{\lambda}_2\hat{\lambda}_4 - \hat{\lambda}_1\hat{\lambda}_2\hat{\lambda}_3 - \hat{\lambda}_2\hat{\lambda}_3\hat{\lambda}_4. \tag{39}$$

When we make the replacement $\hat{\lambda}_k \to \lambda_k$, (39) becomes

$$\lambda_1 + \lambda_2 + \lambda_3 + \lambda_4 - \lambda_1\lambda_3\lambda_4 - \lambda_1\lambda_2\lambda_4 - \lambda_1\lambda_2\lambda_3 - \lambda_2\lambda_3\lambda_4 = \overline{\sin\sum_{k=1}^{4}\lambda_k}. \tag{40}$$

It is not too hard to see that these replacements give

$$e^{\sum_{k>l}\hat{\lambda}_k\hat{\lambda}_l} \to \overline{\cos\sum_k \lambda_k}$$
$$\left(\sum_k \hat{\lambda}_k\right) e^{\sum_{k>l}\hat{\lambda}_k\hat{\lambda}_l} \to \overline{\sin\sum_k \lambda_k} \tag{41}$$

for arbitrary n and thus (36) becomes identical to (19).

When we take the continuum limit of (29) the exponent becomes

$$i\int \left(\bar{\psi}(t)\hat{dt}\frac{d}{\hat{dt}}\psi(t) - i\lambda\,\hat{dt}[\psi(t) + \bar{\psi}(t)]\right) \tag{42}$$

and the infinitesimal quantities $\hat{dt}$ must be thought of as a continuum of Grassmann numbers.

7. Conclusions

We have shown that the transition amplitude for a fermionic hamiltonian that describes a spin-$\frac{1}{2}$ system may be written in path-integral form when discrete time steps are taken, provided that these time intervals are treated as Grassmann variables. They are converted back to ordinary c-numbers in a well-defined way after the integrations have been performed.

One mystifying feature of the trickery presented here is that we end up having to anti-time-order the product of infinitesimals, even though we started off with time-ordered products. In this connection it is instructive to see what happens when we apply the method directly to the transition operator $e^{-i\lambda(a+a^\dagger)T}$. The fact that $(a + a^\dagger)^2 = 1$ allows us to expand the exponential in a power series and obtain

$$e^{-i\lambda(a+a^\dagger)T} = \cos\lambda T - i(a + a^\dagger)\sin\lambda T. \tag{43}$$

Using standard time steps, we can also write

$$e^{-i\lambda(a+a^\dagger)T} \approx \prod_k [1 - i\lambda_k(a+a^\dagger)]. \tag{44}$$

The terms with the product of an odd number of the λ_k will be multiplied by $a + a^\dagger$ while those with an even number will not be so multiplied. The result is that

$$e^{-i\lambda(a+a^\dagger)T} \approx \overline{\cos \sum_k \lambda_k} - i(a+a^\dagger)\,\overline{\sin \sum_k \lambda_k} \tag{45}$$

which is correct as the number of steps goes to infinity.

With fermionic time steps we get (44) with $\hat{\lambda}_k$ in place of λ_k and the product must now be k-ordered. In terms with the product of an even number, $2p$, factors of $\hat{\lambda}_k$ there will be a sign $(-1)^p$ coming from moving factors of $a + a^\dagger$ next to one another and no factor of $a + a^\dagger$ will survive. In terms with an odd number, $2p+1$, factors of $\hat{\lambda}_k$ a single factor of $a + a^\dagger$ will be left on the right, along with the sign $(-1)^p$. In both cases the signs $(-1)^p$ can be canceled by inversion of the order of the $\hat{\lambda}_k$. Thus the origin of the peculiar anti-k ordering lies in the fermionic character of all of the factors.

As a last word, I think it is unlikely that any of this will be useful for the matrix-model problem.

References

1. M. B. Claudson and M. Halpern, *Nucl. Phys.* **B250**, 689 (1985).
   R. Flume, *Ann. Phys.* **164**, 189 (1985).
   M. Baake, P. Reinieke and V. Rittenberg, *J. Math. Phys.* **26**, 1070 (1985).
2. M. B. Halpern and C. Schwartz, *Int. J. Mod. Phys.* **A13**, 4367 (1998). hep-th/9712133.
   J. Froehlich, G. M. Graf, D. Hasler, J. Hoppe and S.-T. Yau, *Nucl. Phys.* **B567**, 231 (2000). hep-th/9904182.
3. R. Floreanni and R. Jackiw, *Phys. Rev.* **D37**, 2206 (1988).
4. F. A. Berezin, *The Method of Second Quantization* (Academic Press, New York, 1966).
5. F. Cooper, A. Khare and U. Sukhatme, *Phys. Rept.* **251**, 267 (1995), hep-th/9405029.

II. ALGEBRAS AND REPRESENTATIONS

DAHA AND VERLINDE ALGEBRAS

IVAN CHEREDNIK*

Department of Mathematics
UNC Chapel Hill, NC 27599, USA
E-mail: chered@math.unc.edu

Double Affine Hecke algebra is a new powerful tool in Representation Theory. It will be introduced and discussed in the rank one case including the applications to the Verlinde algebras, Gaussian sums, and the combinatorics of two sets of variables (the diagonal coinvariants).

Contents

1. Origins of DAHA

The representation theory was greatly stimulated by (a) physics, (b) the theory of special functions, and (c) combinatorics. Historically, in the opposite order. However, I think, the demand from Physics played the major role.

(a) Harish-Chandra was a Dirac's student for some time and always expressed unreserved admiration for him according to Helgason's interesting recent note. The Lorentz group led him to the theory of infinite dimensional

*Partially supported by NSF grant DMS-0200276.

representations of semisimple Lie groups. However later the mathematical goals like the Plancherel formula became preponderant in his research.

(b) I think the theory of special and spherical functions was the main motivation for Gelfand in his studies of infinite dimensional representations, although he always emphasized the role of physics (and physicists). It was reflected in his program (1950-s) aimed at "adding" the hypergeometric function to the Lie theory.

(c) Before the Lie theory, the symmetric group was the main "representation" tool in the theory of functions. It still remains of fundamental importance. However using the symmetric group only is not sufficient to introduce and understand properly the differential equations and operators needed in the theory.

Concerning (a), not all representation theories are of physical importance. Unitary theories play the major role. There are four main sources of such theories:

(A) compact and finite groups,

(B) non-compact Lie groups,

(C) Heisenberg and Weyl algebras,

(D) von-Neumann's factors.

The Clifford algebra, super Lie algebras, and fermions must be also mentioned.

All have merits and demerits. Irreducible representations of (A) are finite dimensional. The representation theory for (B) is infinite dimensional (which is needed in modern physics), but the Harish-Chandra transform is not self-dual and its analytic theory is far from being complete. Also (B) is not very fruitful from the viewpoint of the theory of special functions. For instance, the spherical functions in the so-called group case ($k = 1$) "algebraically" coincide with the characters of compact Lie groups.

The Heisenberg/Weyl algebra is heavily used in physics ("bosonization") but it has essentially a unique irreducible representation, the Fock representation. Not too much. The factors actually have the same demerit. Only the pair "factor - subfactor" appeared good enough for combinatorially rich theory.

An open problem is creating a unitary theory based on the Kac-Moody algebras, what can be called the Kac-Moody harmonic analysis. These algebras play a well known role in modern theoretical physics. The left column of the first block of the diagram below shows the theory of compact Lie groups extended towards the orthogonal polynomials (the latter are generally beyond the Lie theory). Considering it as a sample harmonic analysis program, not much is known in the Kac-Moody case (the right column).

The theory of Kac-Moody characters is reasonably complete, in spite of combinatorial difficulties with the so-called string functions. The next level is supposed to be the theory of spherical functions. There are some results (let me mention Dale Peterson, Lian and Zuckerman, and the book by Etingof, Frenkel, Kirillov) but we have no satisfactory general theory so far with a reservation about the group case, where the technique of conformal blocks and some other approaches can be used. Extending the theory from the group case ($k = 1$) to arbitrary root multiplicities is a problem.

However, I think, a more serious problem is that the Kac-Moody characters are not pairwise orthogonal functions in a way which one can expect taking the classical theory (the left column) as a sample. Analytically, they are given in terms of theta-functions and cannot be integrated over non-compact regions unless special approaches are used. Nevertheless, there are important facts in this direction.

First, we can identify the characters of integrable representations of the Kac-Moody algebra of a given level with the algebra of classical characters treated as functions at roots of unity (Verlinde, Kazhdan, Lusztig, Finkelberg, and others). The latter is the Verlinde algebra. The fusion product corresponds to the multiplication under this identification, and the characters become pairwise orthogonal with respect to the Verlinde inner product.

The double affine Hecke algebra at roots of unity makes the theory uniform and simple and extends the theory to more general k.

Second, there is a theory of the "elliptic radial parts" due to Olshanetsky and Perelomov (the GL_n-case), Ochiai, Oshima, Sekiguchi (the BC-type), and my results for arbitrary (reduced) root systems based on the degenerate Hecke algebra. The second block of the diagram below is devoted to this algebra. The main its application so far is in the classical harmonic analysis. It generalizes the Harish-Chandra theory of spherical functions and, what seems the most important, adds to the Lie - Harish-Chandra theory the "nonsymmetric characters", namely, the Opdam-Macdonald polynomials.

Presumably it leads to an elliptic Harish-Chandra theory for an arbitrary cental charge. We note that the elliptic radial parts exist and are pairwise commutative only at critical center charge. Their nonsymmetric counterparts, the "elliptic" Dunkl operators, always commute. At the moment, there are some analytic problems to be clarified, so the elliptic theory is not finished yet.

The last block of the diagram shows the difference theory and the general Double Affine Hecke Algebra (DAHA). In contrast to the previous "Harish-Chandra level", the q-Fourier transform is self-dual as holds for the classical Fourier transform (and in the group case). DAHA has many applications. In physics, the main ones so far are to the Verlinde algebras, in the theory of the KZ-equations and the KZB-equations, and, more generally, in the Kac-Moody direction. The application to the diagonal coinvariants is presumably connected with the recent developments in the theory of matrix models (Kazakov and others), however it will not be touched upon.

Adding the parameters to the theory is always valuable. An extension of the theory of Kac-Moody algebras to general k (motivated by the theory of symmetric spaces and the Macdonald polynomials) has been already discussed. Concerning q, which is a root of unity in the Verlinde theory expressed in terms of the (integral) center charge of the corresponding Kac-Moody algebra, a variant of the Verlinde algebra at arbitrary q associated with more general non-integrable representations can be expected. However what was achieved via DAHA appeared more surprising.

Under minor technical restrictions, DAHA gives a flat deformation of the Verlinde algebra to arbitrary unimodular q preserving all properties and the dimensions, with a reservation about the integrality and positivity of the structural constants. The latter two are the only features of the Verlinde algebras which are lost for general q.

It was done using special k, rational numbers with the Coxeter number as the denominator. It triggers the question about the classification of all representations of DAHA for such special fractional k. Indeed, if a "non-compact unitary" Kac-Moody representation theory exists it could be associated with such representations in the same way as Verlinde algebras describe the integrable representations of the Kac-Moody algebras.

The theory at special fractional k seems a good candidate for an infinite dimensional theory of unitary type associated with the Kac-Moody algebras. It naturally unifies the latter with the Heisenberg and Weyl algebras and has intriguing connections with the von-Neumann's factors.

2. From Lie Theory to DAHA

LIE GROUPS	KAC-MOODY
Compact Characters	*KM – Characters*
⇓	⇓
Spherical Functions	*Symmetric Spaces?*
$(k = 1, 2, 1/2)$	*ellip radial parts* ‼
⇓	⇓
Orthog Polynomials	*Measure&Integrtn?*
(k *is arbitrary*)	*Verlinde Algebras* ‼
Degenerate DAHA	**Dunkl operators**
Nonsym Harish-Chandra	Elliptic Nonsym Theory
H-Ch transform : ↓	↓ **: not finished yet**
Rational-Difference	Nonsymmetric Theory

General DAHA & Difference theory

q-Fourier Transform becomes self-dual !

From this viewpoint, the harmonic Kac-Moody analysis or, in other words, a "unitary" counterpart of the Kac-Moody theory, could be expected an extension of the Verlinde algebras for $|q| = 1$. It must contain in some sense the Verlinde algebra if we do not want to "lose" integrable KM-representations.

Using DAHA, its unitary representations at $|q| = 1$ for special fractional k are natural candidates. Such representations must have additional structures, namely, a multiplication and a "good" restriction to the affine Hecke subalgebra similar to the classical decomposition with respect to the maximal compact subgroup. The existence of the multiplication indicates that the constituents of the polynomial representation of DAHA or its functional variants must be examined first. The most interesting AHA-restriction theory happens at special fractional k.

It is not surprising that DAHA is expected to play such an important role. It generalizes the Heisenberg and Weyl algebras, and directly generalizes the Hecke algebras which have fundamental applications in Mathematics.

Concerning the harmonic analysis, DAHA adds to the theory first order (Dunkl) operators and respectively the "nonsymmetric characters". Their symmetrizations are (k-deformations of) the radial parts and spherical functions. What seems the most important, it makes the Gaussian q-Fourier invariant! It is not invariant under the Harish-Chandra transform.

In the new theory, the Harish-Chandra transform is an analytic homomorphism from the differential-trigonometric polynomial representation of the degenerate DAHA to its difference-rational polynomial representation. There is the third elliptic-polynomial representation of DAHA which is analytically isomorphic to the same difference-rational one as for the Harish-Chandra transform. The corresponding map can be called the nonsymmetric elliptic Harish-Chandra transform.

Let me also mention that the one-dimensional Dunkl operator is the square root of the radial part of the Laplace operator in rank one case, similar to the Dirac operator. The one-dimensional DAHA is closely connected to the super Lie algebra $osp(2 \mid 1)$ and to its "even part" $sl(2)$. It gives another reason to expect much from DAHA.

Let us discuss the case of A_1 in detail.

3. Double Hecke Algebra

The most natural definition goes through the **elliptic braid group:** $\mathcal{B}_q \stackrel{\text{def}}{=\!=} \langle T, X, Y, q^{1/4}\rangle /$ relations:

$$TXT = X^{-1},\ TY^{-1}T = Y,$$
$$Y^{-1}X^{-1}YXT^2 \ = \ q^{-1/2}.$$

Here

$$\mathcal{B}_1 = \pi_1^{\mathrm{orb}}(\{E \setminus 0\}/\mathbf{S}_2)$$

$\cong \pi_1(\{E \times E \setminus \mathrm{diag}\}/\mathbf{S}_2)$, where E is an elliptic curve. Using the orbifold fundamental group here makes it possible to "divide" by the symmetric group $\mathbf{S}_2$ without removing the ramification, all four points of second order. In the case of A_n, it is sufficient to consider the product of $n+1$ copies of E and remove the "diagonal" before dividing by $\mathbf{S}_{n+1}$ instead of using the orbifold group (although the definition of the latter is simple). The corresponding braid group was calculated by Birman and Scott.

Finally, we impose the quadratic T-relation:

$$\mathcal{H\!H} \stackrel{\text{def}}{=} \mathbb{C}[\mathcal{B}_q]/((T - t^{1/2})(T + t^{-1/2})).$$

If $t = 1$, $\mathcal{H\!H}$ becomes the Weyl algebra extended by the reflection.

The **Fourier transform,** which plays a key role in the theory, corresponds to the outer automorphism of DAHA:

$$\sigma : X \mapsto Y^{-1}, Y \mapsto XT^2,\ T \mapsto T.$$

Thus the DAHA Fourier transforms has a conceptual relation to the transposition of the periods of the elliptic curve. It is not surprising from the viewpoint of CFT, KZB, and the Verlinde algebras. However this interpretation still remains challenging when we deal with the applications of DAHA in harmonic analysis.

The representations where σ acts (becomes inner) are called Fourier-invariant or self-dual. The nonsymmetric Verlinde algebra and the Schwartz space are examples.

More generally, the group $PSL_2(\mathbf{Z})$ acts projectively in $\mathcal{H\!H}$:

$$\tau_+ : Y \mapsto q^{-1/4}XY,\ X \mapsto X,\ T \mapsto T$$
$$\tau_- : X \mapsto q^{1/4}YX,\ \ Y \mapsto Y,\ T \mapsto T$$
$$\sigma = \tau_+\tau_-^{-1}\tau_+ = \tau_-^{-1}\tau_+\tau_-^{-1},$$

$$\text{where } \begin{pmatrix} 11 \\ 01 \end{pmatrix} \mapsto \tau_+,\ \begin{pmatrix} 10 \\ 11 \end{pmatrix} \mapsto \tau_-.$$

4. Polynomial Representation

$\mathcal{H\!H}$ acts in $\mathcal{L}$ = **Laurent polynomials** in terms of $X = q^x$. Namely,

$$T \mapsto t^{1/2}s + \frac{t^{1/2} - t^{-1/2}}{q^{2x} - 1}(s - 1),$$
$$Y \mapsto spT,\ sf(x) = f(-x),$$
$$pf(x) = f(x + 1/2),\ t = q^k.$$

The operator Y is called the **difference Dunkl Operator.**

It is important that τ_- preserves $\mathcal{L}$. On the other hand, τ_+ does not act there because of a very simple reason. Formally, it is the multiplication by the Gaussian q^{x^2}, which does not belong to $\mathcal{L}$.

The radial part appears as follows. One checks that $Y + Y^{-1}$ preserves $\mathcal{L}_{sym} \stackrel{\text{def}}{=}$ symmmetric (even) Laurent polynomials (the Bernstein Lemma in the theory of affine Hecke algebras). The restriction $H = Y + Y^{-1} \mid_{sym}$ is the q-radial part and can be readily calculated:

$$H = \frac{t^{1/2}X - t^{-1/2}X^{-1}}{X - X^{-1}} p + \frac{t^{1/2}X^{-1} - t^{-1/2}X}{X^{-1} - X} p^{-1}.$$

5. Nonsymmetric Verlinde Algebras

Let $q = \exp(2\pi i/N)$, $k < N/2$. Then the map $q^x(z) = q^z$ can be extended to a $\mathcal{H}$-homomorphism from $\mathcal{L} \to V \stackrel{\text{def}}{=}$ **Nonsymmetric Verlinde algebra** which is the algebra of functions of the set

$$\left\{ -\frac{N+k+1}{2}, \ldots, -\frac{k+1}{2}, -\frac{k}{2}, \frac{k+1}{2}, \ldots, \frac{N-k}{2} \right\}.$$

Note that X, Y, T become unitary operators in V with respect to the positive hermitian form which will be not discuss here (it generalizes the inner product for conformal blocks). The positivity requires choosing the "minimal" primitive N-root of unity q above, which is similar to the Verlinde theory.

The whole $PSL_2(\mathbf{Z})$ acts in V projectively as well as in the image V_{sym} of $\mathcal{L}_{sym}$. The latter image can be of course defined without any reference to the polynomial representation. A general definition is as follows: $V_{sym} = \{f \mid Tf = t^{1/2} f\}$. Here it simply means that the function $f(z)$ must be s-invariant (even) as far as s is defined (the above set is not s-invariant). Therefore:

$$\dim_{\mathbb{C}} V = 2N - 4k, \ \dim_{\mathbb{C}} V_{sym} = N - 2k + 1.$$

The nonsymmetric **characters** in V and the symmetric ones in V_{sym} are respectively the eigenfunctions of Y, and $Y + Y^{-1}$.

When $k = 0$, we come to the well known definitions in the theory of Weyl algebra. As $k = 1$, $V_{sym}^{k=1}$ is the usual **Verlinde algebra**; τ_+ becomes the T-operator, σ becomes the S-operator.

6. Deformation

We will need the **Little Nonsymmetric Verlinde algebra** $\widetilde{V}^k$ which is a natural irreducible quotient of V^k upon the restriction to **Little DAHA**

$\widetilde{\mathcal{H}} \stackrel{\text{def}}{=} \langle T, X^2, Y^2 \rangle$, corresponding to Laurent polynomials in terms of X^2. Its dimension is $\dim \widetilde{V}^k = \dim V^k/2 = N - 2k$. Note that $\widetilde{V}_{sym}^{k=1}$ is the Verlinde algebra for radical weights.

Let $N = 2n+1$, $q^{1/2} = -\exp(\frac{\pi i}{N})$. Such a choice of q is necessary for the positivity of the corresponding inner product.

We set $m = n - k$: $\dim_{\mathbb{C}} \widetilde{V}^k = N - 2k = 2m+1$, The aim is to deform $\widetilde{V}$, $\widetilde{V}_{sym}$, the projective action of $PSL_2(\mathbb{Z})$ and all other structures to arbitrary unimodular q.

Theorem 6.1. *For any q, $\bar{k} = -\frac{1}{2} - m$, $m \in \mathbb{Z}_+$, $\mathcal{L}$ considered as an $\mathcal{H}(\bar{k})$-module has a unique irreducible quotient $\bar{V} \neq 0$:*

$$\bar{V} = Funct\left(\frac{\bar{k}+1}{2}, \ldots, -\frac{\bar{k}+1}{2}, -\frac{\bar{k}}{2}\right),$$

$$dim_{\mathbb{C}} \bar{V} = 2m+1, \; dim_{\mathbb{C}} \bar{V}_{sym} = m+1 \; ,$$

It is unitary as $|q| = 1$, $0 < arg(q) < \pi/m$.

The representation $\bar{V}$ becomes $\widetilde{V}^k$ as

$$q = e^{\frac{2\pi i}{N}}, \; arg = \frac{2\pi}{2n+1},$$

$\bar{k} = k - 1/2 - n$, i.e., $m = n - k$.

Concerning the positivity of the inner product, note that $\frac{2\pi}{2n+1} < \pi/m$, since $k \geq 0$. The representation $\bar{V}$ is a q-deformation of an irreducible representation of $osp(2 \mid 1)$. Respectively, $\bar{V}_{sym}$ deforms the irreducible representation of $sl(2)$ of spin$=m/2$. We need the rational degeneration of DAHA to clarify it.

7. Rational Degeneration

The **Rational DAHA** is the following quasi-classical limit:

$$Y = e^{-\sqrt{\kappa}y/2}, \; X = e^{\sqrt{\kappa}x}, \; q = e^{\kappa}, \; \kappa \to 0$$

of $\mathcal{H}$. An explicit description is simple:

$$\mathcal{H}'' \stackrel{\text{def}}{=} \langle x, y, s \rangle / \text{relations} :$$

$$[y, x] = 1 + 2ks, \; s^2 = 1, \; sxs = -x, \; sys = -y.$$

It acts in the polynomial representation which is

$$\mathbb{C}[x],\ s(x) = -x,\ ,\text{ where}$$

$$y \mapsto D = \frac{d}{dx} - \frac{2k}{x}(s-1).$$

The square of the **Dunkl operator** is the radial part of the Laplace operator:

$$D^2\mid_{sym} = \frac{d^2}{dx^2} + \frac{2k}{x}\frac{d}{dx}.$$

The DAHA Fourier automorpism σ becomes the **Hankel transform.**

Let $\bar{k} = -\frac{1}{2} - m$. Then

$$\lim_{\kappa\to 0} \bar{V} = \bar{V}'' = \mathbb{C}[x]/(x^{2m+1}).$$

The corresponding symmetric part $\bar{V}''_{sym}$ is nothing else but the irreducible representation of $sl(2)$ of dim$= m+1$:

$$h = (xy+yx)/2,\ e = x^2, f = -y^2.$$

8. Gaussian Sums

Continuing the deformation construction, τ_- acts in $\bar{V}$ and $\bar{V}_{sym}$ since it acts in the polynomial representation $\mathcal{L}$. However now τ_+ also acts in $\bar{V}$ and its symmetric part. It is the multiplication by

$$q^{x^2} \stackrel{\textbf{def}}{=\!=} q^{\frac{(\pm\bar{k}+j)^2}{4}},$$

where $\pm = +$ for $j > 0$ and minus otherwise. Similarly, σ is the matrix $\left(\text{char}_i(\frac{\pm\bar{k}+j}{2})\right)$ for the nonsymmetric characters, up to a multiplication by a simple diagonal matrix.

The **Gaussian sum** (any q, $\bar{k} = -1/2 - m$) is defined as follows:

$$\sum_{j=0}^{m} q^{j^2-\bar{k}j}\frac{1-q^{2j+\bar{k}}}{1-q^{\bar{k}}}\prod_{l=1}^{2j}\frac{1-q^{l+2\bar{k}-1}}{1-q^l}.$$

It equals

$$\prod_{j=1}^{m}\frac{1-q^{2\bar{k}+2j}}{1+q^{\bar{k}+2j}},$$

which follows from the above interpretation of the Gaussian via τ_+.

When

$$N = 2n+1,\ m = n-k,\ 0 \le k \le n,$$

$$q^{1/2} = -\exp(\pi i/N),\ \text{and } \bar{V} = \widetilde{V}^k,$$

the product above can be somewhat simplified:

$$\prod_{j=1}^{m} \frac{1-q^{2\bar{k}+2j}}{1+q^{\bar{k}+2j}} = q^{-\frac{m(m+1)}{4}} \prod_{j=0}^{m-1} (1-q^{n-j}).$$

Now let $m{=}n$, i.e. the maximal possible, $k=0$, and $l=\frac{n}{2} \bmod N$. Then $\bar{k} = -1/2 - n = 0 \bmod N$, and we obtained the identity:

$$\sum_{j=0}^{N-1} q^{j^2} = q^{l^2} \prod_{j=1}^{n} (1-q^j).$$

This product can be readily calculate using the Galois theory. It equals $\sqrt{N}$ for $n - 2l$ and $i\sqrt{N}$ otherwise. It gives a new proof of tho classical Gauss formulas.

9. Diagonal Coinvariants

The previous theory can naturally generalized to the case of arbitrary Weyl group W acting in $\mathbb{R}^n$. Respectively, $w^{-1}f(x,y){=}f(wx,wy)$ for a function f in terms of $x, y \in \mathbb{R}^n$.

We set

$$\mathbf{I}_o \stackrel{\text{def}}{=\!=} \operatorname{Ker}(\mathbb{C}[x,y]_{sym} \ni f \mapsto f(0,0)),\ \mathcal{V} \stackrel{\text{def}}{=\!=} \mathbb{C}[x,y]/(\mathbb{C}[x,y]\mathbf{I}_o).$$

Haiman conjectured that $\mathcal{V}$ has a natural quotient $\mathbf{V}$ of dim$=(h+1)^n$ for the Coxeter number h. He checked that $\mathcal{V}{=}\mathbf{V}$ and dim$=(n+2)^n$ in the case of A_n, $W{=}\mathbf{S}_{n+1}$. The coincidence is a very special feature of A_n. Haiman's general conjecture was recently justified by Gordon:

$$\mathbf{V} = gr\bar{V}'' \text{ for } \bar{k} = -1/h - 1,$$

where $\bar{V}''$ is a straightforward W-generalization of the representation $\bar{V}''$ considered above. In the rank one case,

$$n = 1,\ h = 2,\ m = 1,\ \dim_{\mathbb{C}} \bar{V}'' = 2m+1 = 3.$$

The general q-construction is as follows:

Theorem 9.1. *a) There exist Lusztig type (exp-log) isomorphisms* $\bar{V}'' \rightleftharpoons \bar{V} \rightleftharpoons V^{\bullet}$ *for* $\bar{k}{=}{-}1/h - 1$. *Here* $V^{\bullet}$ *is a unique* $\mathcal{H}^{\bullet}$*-quotient of* $\mathcal{L} = \mathbb{C}[X, X^{-1}]$ *for the specialization:*

$$\bullet = \left\{ q = \exp\left(\frac{2\pi i}{N}\right), \; N \stackrel{\mathbf{def}}{=} h+1, \; t = q^{\bar{k}} = 1 \right\}.$$

b) Concerning the coinvariants,

$$\bar{V} = \mathcal{L}/\mathcal{H} I_o, \; I_o = Ker(\mathcal{H}_{sym} \ni H \mapsto H(\Delta) \in \bar{V}),$$

where Δ *is the discriminant. When* $n = 1$ *it equals* $t^{1/2}X - t^{-1/2}X^{-1}$.

Since $t = 1$ for the $\bullet$-reduction, $\mathcal{H}^{\bullet}$=(Weyl algebra)$\rtimes \mathbf{S}_2$. So $V^{\bullet}$ is its unique irreducible representation imposing the relations $X^N{=}1{=}Y^N$. It is well known: $V^{\bullet} \simeq \mathbb{C}[X, X^{-1}]/(X^N = 1)$. Its dimension is dim$=N^{rank}$, exactly what Haiman conjectured. In Gordon's proof, the $(h+1)^n$ formula requires a construction of the resolution of $\bar{V}''$ and more. The theorem gives an immediate explanation!

CONTRACTIONS AND DEFORMATIONS OF LIE ALGEBRAS

MARC DE MONTIGNY

Faculté Saint-Jean, University of Alberta,
Edmonton, Alberta T6C 4G9, Canada
Email: montigny@phys.ualberta.ca

This is a brief survey of contractions and deformations of Lie algebras. Various contraction methods are presented. Whereas the contractions typically allows one to punch zeros within the commutation table, the deformations replace zeros with non-trivial elements, thereby providing a more 'simple' algebraic structure.

1. Introduction

Contractions of Lie algebras explain, in terms of Lie groups, how some physical theories arise as limits of more exact theories. They were introduced fifty years ago by E.P. Wigner and E. Inönü[1]. Two years earlier, Segal had introduced sequences of groups[2]. The physical motivation was the transition from the Poincaré algebra to the Galilei algebra. The most down-to-earth example of contraction is the passage from the Lie algebra of the rotation group in three dimensions to the euclidean algebra in the plane. Similarly, the de Sitter algebra can be contracted to the Poincaré algebra. From these examples, one sees that the existence of an 'approximate', or 'anthropocentric', theory may be related to a contraction. Thus, when Nature plays such tricks on us, we are actually observing the result of a contraction.

The concept of deformation is somehow opposite to contractions[3]. The latter typically leads to a 'more abelian' Lie algebra, whereas the first procedure makes it more 'simple'. Both may be applied to finite or infinite Lie algebras, or Lie superalgebras. The purpose of various contraction methods[4−7] is to bring a greater variety of physical systems to a common understanding. To my knowledge, contractions have been studied mostly by physicists, whereas mathematicians rather seem to be interested in deformations. However, from the discussion above, better descriptions of Nature often correspond to the result of a deformation rather than a contraction. This point of view has been advocated at this conference by Professor Finkelstein[8]. For mathematicians, the situation is the opposite. Indeed, the theory of semisimple Lie algebras is extremely elegant and uniform, but the situation is quite different for non-semisimple algebras. Since many such Lie algebras are the results of contractions, the procedure could provide a way to better understand some characteristics of non-semisimple Lie algebras, such as representations, etc.

In Section 2, I present the traditional definition of contractions of Lie algebras. An original method, called 'graded contractions' of Lie algebras, is the subject of Section 3. Deformations are briefly discussed in Section 4.

2. Contractions of Lie algebras

Although contractions have never been very popular, it is a profound concept for working with groups and their representations in the formulation of new theories in physics, and the consequences may be fundamental. For instance, Einstein's special relativity consists mainly in replacing the Galilei algebra (contraction) by the Poincaré algebra (deformation), and their respective representations[1]. A nice textbook treatment is in Chapter 10 of Ref. [4].

The method has been generalized in 1961 by E. Saletan[5]. Hereafter, I borrow his definition of what I will refer to as 'traditional' contractions:

Consider a Lie algebra L, with Lie product $[\ ,\ ]$. Define a continuous family of ε-dependent automorphisms:

$$U_\varepsilon : \quad L \to L, \qquad 0 \le \varepsilon \le 1, \ \textit{with } U_{\varepsilon=1} = 1, \tag{1}$$

which are singular at $\varepsilon = \varepsilon_0$. We define a family of Lie algebras L^ε, with Lie product given by

$$[a, b]_\varepsilon = U_\varepsilon^{-1} \left([U_\varepsilon(a), U_\varepsilon(b)] \right), \qquad \textit{for all } a, b \in L, \tag{2}$$

which are isomorphic to L as long as $\varepsilon \neq \varepsilon_0$. If, at the singularity $\varepsilon = \varepsilon_0$, the commutators

$$[a, b]' = \lim_{\varepsilon \to \varepsilon_0} [a, b]_\varepsilon \tag{3}$$

are well defined, then they determine a new Lie algebra $L' \equiv L^{\varepsilon=\varepsilon_0}$, in general, non-isomorphic to L. The new algebra L' is called a 'contracted algebra', or 'contraction', of L. Obviously, the contracted Lie algebra has same dimension as the initial algebra.

The Inönü-Wigner contractions[1] are defined by

$$U_\varepsilon = \mathbf{1}_{L_0} \oplus \varepsilon \mathbf{1}_{L_1}. \tag{4}$$

The vector subspace L_0 turns out to be a Lie subalgebra. Therefore, in terms of a convenient notation, the Inönü-Wigner contractions consist in taking

$$L = L_0 + L_1, \qquad \text{with} \quad [L_0, L_0] \subseteq L_0, \tag{5}$$

$$L_0 \to L_0, \qquad L_1 \to \varepsilon L_1, \tag{6}$$

which means that all the elements of the subalgebra L_0 are left unchanged whereas those of the subspace L_1 are multiplied by the contraction parameter ε. The contracted Lie algebra has commutation relations[1]:

$$[L_0, L_0]' = L_0, \tag{7}$$

$$[L_0, L_1]' = L_1, \tag{8}$$

$$[L_1, L_1]' = 0. \tag{9}$$

Many generalizations are based on Saletan's definitions[6]. An interesting method is due to Weimar-Woods[7]. It consists in splitting up L as

$$L = L_0 + L_1 + \cdots + L_N \tag{10}$$

and then introduce the contraction parameter as follows:

$$U_\varepsilon \equiv \varepsilon^{n_0} \mathbf{1}_{L_0} + \varepsilon^{n_1} \mathbf{1}_{L_1} + \cdots + \varepsilon^{n_N} \mathbf{1}_{L_N}. \tag{11}$$

This means that all the elements of the subspace L_k are multiplied by ε^{n_k}. The exponents satisfy $n_0 \leq n_1 \leq \cdots \leq n_N$. Then a contraction is defined if $[L_j, L_k] \subset \oplus_l L_l$, with $n_l \leq n_j + n_k$, and the new commutators are given by $[L_j, L_k]' \subset L_l$, with $n_l = n_j + n_k$. The original Inönü-Wigner corresponds to $N = 1$, with $n_0 = 0$ and $n_1 = 1$.

An interesting application of contractions is the classification of the kinematical algebras of space-time[9].

3. Graded contractions of Lie algebras

The method of graded contractions[10,11,12] is radically different from the traditional contractions. They do not involve a change of basis. Instead, the contraction parameters are introduced directly within the commutators. Moreover one seeks to preserve some *grading* (i.e. automorphisms of finite order, such as the time-reversal or parity operations) throughout the limit process. A brief account is given hereafter, with more details in Refs. [10, 12].

A *grading* of a Lie algebra L is a splitting into vector subspaces

$$L = \bigoplus_{\mu \in \Gamma} L_\mu, \tag{12}$$

where μ takes on values in some index set Γ, called 'grading group', typically of the form $\mathbf{Z}_M$, $\mathbf{Z}_M \otimes \mathbf{Z}_N$, etc. such that[11]

$$[L_\mu, L_\nu] \subseteq L_{\mu+\nu}, \tag{13}$$

which means that if $x \in L_\mu$ and $y \in L_\nu$, then $[x, y]$ belongs to the subspace $L_{\mu+\nu}$. All the elements of a grading subspace L_μ are treated the same way. It is the commutation relations between grading subspaces that are modified. Moreover, the solutions of Eqs. (17), which determine the outcome of graded contractions, can be applied simultaneously to any Lie algebra (finite or infinite, or Lie superalgebra) which admits the chosen grading. This means that one does not need to specify the Lie algebra, or even to fix its dimension, in order to solve Eqs. (17). Another difference with the traditional contraction methods is that, whereas the Jacobi identity was automatically satisfied after a contraction, here we force the new commutators to satisfy the Jacobi identity, obtaining thereby the determining conditions on the contraction parameters. In all the non-trivial cases, graded contractions are controlled by many contraction parameters, whose number increases with the order of the grading group. This procedure is sometimes called 'Γ-contraction', and the resulting Lie algebras, 'Γ-contractions'.

The graded contractions are defined by modifying the commutators in Eq. (13) as[10]

$$[L_\mu, L_\nu]_\varepsilon \equiv \varepsilon_{\mu,\nu}[L_\mu, L_\nu], \tag{14}$$

while preserving the splitting of Eq. (12). The Jacobi identity is satisfied by the initial Lie algebra L:

$$[[L_\mu, L_\nu], L_\sigma] + [[L_\nu, L_\sigma], L_\mu]] + [[L_\sigma, L_\mu], L_\nu] = 0. \tag{15}$$

In order to define a new Lie algebra, the Jacobi identity must also be satisfied by the modified commutator $[,]_\varepsilon$, so that one finds

$$\begin{aligned} &\varepsilon_{\mu,\nu}\varepsilon_{\mu+\nu,\sigma}[[L_\mu, L_\nu], L_\sigma] + \varepsilon_{\nu,\sigma}\varepsilon_{\nu+\sigma,\mu}[[L_\nu, L_\sigma], L_\mu] \\ &\quad + \varepsilon_{\sigma,\mu}\varepsilon_{\sigma+\mu,\nu}[[L_\sigma, L_\mu], L_\nu] = 0. \end{aligned} \tag{16}$$

Eqs. (15) and (16) hold simultaneously, in the generic case, if

$$\varepsilon_{\mu,\nu}\varepsilon_{\mu+\nu,\sigma} = \varepsilon_{\nu,\sigma}\varepsilon_{\nu+\sigma,\mu} = \varepsilon_{\sigma,\mu}\varepsilon_{\sigma+\mu,\nu}, \tag{17}$$

which are the contractions defining equations. Therefore, given a specifc grading group Γ, the method consists in solving the appropriate system, Eq. (17), and then substituting the ε-solutions back into Eq. (14). Examples are given in Ref. [10]. The kinematical algebras have been shown to correspond to $\mathbf{Z}_2 \times \mathbf{Z}_2$ contractions of the de Sitter algebras[13].

From the previous discussion, one can see that the formalism of graded contractions does not allow the introduction of central charges in the new commutation relations. Indeed, Eq. (14) states that either a commutator will stay the same (up to a constant), or it will vanish. The only central charges that can be added are for the non-vanishing commutators, obtained by a rescaling of the remaining generators. Therefore, all the extensions are trivial. Motivated by the importance of central extensions in physics and, in particular, the role played by the central charges in the kinematical algebras[9], the definition of graded contractions, Eq. (14), has been generalized in order to allow the existence of central charges, while keeping the preservation of a grading as the crucial feature of the contraction. More details, as well as physical applications, are given in Ref. [12].

4. Deformations of Lie algebras

An early discussion of deformations by a physicist is in Ref. [3]. Just like contractions, there are many ways to define deformations. A widely used definition consists in considering an ε-dependent orbit in the space of Lie algebras of a fixed dimension:

$$f_\varepsilon(a,b) = F_0(a,b) + \varepsilon F_1(a,b) + \varepsilon^2 F_2(a,b) + \ldots \tag{18}$$

From Jacobi identity, we find, at order ε: $F_1(F_0(a,b),c) + F_0(F_1(a,b),c) +$ cycl. perm. In other words, $dF_1 = 0$ so that $F_1 \subset Z^2(L,L)$. Similarly, one finds, at order ε^2: $dF_2 = -F_1(F_1(a,b),c) +$ cycl. perm. $\subset B^3(L,L)$. Therefore, deformations are closely related to cohomological theory.

Let me just display an example relevant to conformal field theories: the global deformation of the Witt algebra into a Lie algebra of Krichever-Novikov type[14]:

$$V_{2N} \equiv z(z-\alpha)^N(z+\alpha)^N\partial_z,\ V_{2N+1} \equiv (z-\alpha)^{N+1}(z+\alpha)^{N+1}\partial_z, \tag{19}$$

N integer. The resulting commutation relations read:

$$[V_M, V_N] = \begin{cases} (N-M)V_{M+N} & M, N \text{ odd} \\ (N-M)(V_{M+N} + \alpha^2 V_{M+N-2}) & M, N \text{ even} \\ (N-M)V_{M+N} + (N-M-1)\alpha^2 V_{M+N-2} & M \text{ odd, } N \text{ even} \end{cases} \tag{20}$$

Further work is needed in order to understand the contractions and deformations of representations. This would allow more systematical investigations of deformations associated to the symmetries of various physical systems.

Acknowledgments

The author is grateful to the Natural Sciences and Engineering Research Council of Canada for financial support.

References

1. E. Inönü and E. P. Wigner, *Proc. Nat. Acad. Sc. USA* **39**, 510 (1953); id., **40**, 119 (1954); E. Inönü, in *Group Theoretical Concepts and Methods in Elementary Particle Physics*, Ed. Feza Gürsey, Gordon and Breach, New York, 391 (1962).
2. I.E. Segal, *Duke Math. J.* **18**, 221 (1951).
3. M. Lévy-Nahas, *J. Math. Phys.* **8**, 1211 (1967).
4. R. Gilmore *Lie Groups, Lie Algebras, and Some of Their Applications*, (1974) Wiley, New York.
5. E. Saletan, *J. Math. Phys.* **2**, 1 (1961).
6. H.D. Doebner and O. Melsheimer, *Nuov. Cim.* **A49**, 306 (1967); G.C. Hegerfeldt, *Nuov. Cim.* **B51**, 439 (1967); D. Arnal and J.C. Cortet, *J. Math. Phys.* **20**, 556 (1979); E. Celeghini and M. Tarlini, *Nuov. Cim.* **B61**, 265 (1981); id., **B65**, 172 (1981); id., **B68**, 133 (1982).
7. E. Weimar-Woods, *J. Math. Phys.* **32**, 2028 (1991); id., **32**, 2660 (1991); id., **36**, 4519 (1995); ib. *Rev. Math. Phys.* **12**, 1505 (2000).
8. D. Finkelstein, these proceedings.
9. H. Bacry H and J.M. Lévy-Leblond, *J. Math. Phys.* **9**, 1605 (1968).
10. M. de Montigny and J. Patera, *J. Phys. A: Math. Gen.* **21**, 525 (1991); R.V. Moody and J. Patera, *J. Phys. A: Math. Gen.* **21**, 2227 (1991).
11. J. Patera and H. Zassenhaus, *Lin. Alg. Appl.* **112**, 87 (1989); V.G. Kac, *Funct. Anal. Appl.* **3**, 252 (1969); J. Patera and H. Zassenhaus, *J. Math. Phys.* **29**, 665 (1988); J. Patera, *J. Math. Phys.* **30**, 2756 (1989).
12. H. de Guise and M. de Montigny, *J. Phys. A: Math. Gen.* **33**, 4039 (2000).
13. M. de Montigny, J. Patera and J. Tolar, *J. Math. Phys.* **35**, 405 (1994).
14. A. Fialowshi and M. Schlichenmaier, Global deformations of the Witt algebra of the Krichever-Novikov type, to appear in *Comm. Contemporary Math.*, math.QA/0206114

ON THE UNITARITY OF D=9,10,11 CONFORMAL SUPERSYMMETRY*

V.K. DOBREV

School of Informatics, University of Northumbria,
Newcastle-upon-Tyne NE1 8ST, UK
Permanent address: Institute of Nuclear Research and Nuclear Energy,
Bulgarian Academy of Sciences, Sofia, Bulgaria.

R.B. ZHANG

School of Mathematics and Statistics,
University of Sydney,
Sydney, New South Wales 2006, Australia

B.S. ZLATEV

Institute of Nuclear Research and Nuclear Energy,
Bulgarian Academy of Sciences, Sofia, Bulgaria

We consider the unitarity of $D = 9, 10, 11$ conformal supersymmetry using the recently established classification of the UIRs of the superalgebras $osp(1|2n, I\!R)$.

1. Introduction

Recently, applications in string theory require the knowledge of the UIRs of the conformal superalgebras for $D > 6$. In these applications most prominent role play the superalgebras $osp(1|\,2n)$, cf., e.g., [1–10]. Initially, the superalgebra $osp(1|\,32)$ was put forward for $D = 10$ [1]. Later it was realized that $osp(1|\,2n)$ would fit any dimension, though they are minimal only for $D = 3, 9, 10, 11$ (for $n = 2, 16, 16, 32$, resp.) [7]. In all cases one needs to find first the UIRs of $osp(1|\,2n, I\!R)$. This was done for general n in [11]. In the present paper we consider the implications for conformal supersymmetry for $D = 9, 10, 11$.

*Talk presented by V.K.D.

2. UIRs of the superalgebras $osp(1|2n,I\!R)$

The conformal superalgebras in $D = 9,10,11$ are $\mathfrak{g} = osp(1|2n,I\!R)$, $n = 16,16,32$, resp., cf. [1,7]. The even subalgebra of $osp(1|2n,I\!R)$ is the algebra $sp(2n,I\!R)$ with maximal compact subalgebra $\mathfrak{k} = u(n) \cong su(n) \oplus u(1)$. We label the relevant representations of $\mathfrak{g}$ by the signature:

$$\chi = [d; a_1, ..., a_{n-1}] \qquad (1)$$

where d is the conformal weight, and $a_1, ..., a_{n-1}$ are non-negative integers which are Dynkin labels of the finite-dimensional UIRs of the subalgebra $su(n)$ (the simple part of $\mathfrak{k}$). The positive energy UIRs of $\mathfrak{g}$ for any $n > 1$ are given in the following list [11]:

$$\begin{aligned}
d &\geq d_1 = n-1+\tfrac{1}{2}(a_1+\cdots+a_{n-1}), \quad \text{no restrictions on } a_j, \qquad (2)\\
d &= d_{12} = n-2+\tfrac{1}{2}(a_2+\cdots+a_{n-1}+1), \quad a_1 = 0,\\
&\dots\\
d &= d_{j-1,j} = n-j+\tfrac{1}{2}(a_j+\cdots+a_{n-1}+1), \quad a_1 = \dots = a_{j-1} = 0\\
&\dots\\
d &= d_{n-1,n} = \tfrac{1}{2}, \quad a_1 = \dots = a_{n-1} = 0.
\end{aligned}$$

These UIRs are realized as the irreducible quotients of Verma modules V^Λ of lowest weight $\Lambda = \Lambda(\chi)$. The weight is fixed by its products with the simple roots α_i $(i = 1,\dots,n)$ of $\mathfrak{g}$ [11]:

$$(\Lambda, \alpha_k^\vee) = (\Lambda, 2\alpha_k/(\alpha_k,\alpha_k)) = (\Lambda,\alpha_k) = -a_k, \quad k < n, \qquad (3a)$$
$$(\Lambda, \alpha_n^\vee) = 2(\Lambda,\alpha_n) = 2d + a_1 + \cdots + a_{n-1}. \qquad (3b)$$

3. Unitarity for the conformal subalgebras

From the prescription of [7] follows that the even subalgebra $sp(2n,I\!R)$, $n = 16,16,32$, resp., contains the conformal algebra $\mathfrak{C} = so(D,2)$, $D = 9,10,11$. Then $\mathfrak{k}$ contains the maximal compact subalgebra $so(D) \oplus so(2)$ of $\mathfrak{C}$, $so(2)$ being idenitified with the $u(1)$ factor of $\mathfrak{k}$, and $su(n)$ contains the algebra $so(D)$. The easiest way to describe the embeddings is via the root systems. The superalgebra $\mathfrak{g}$ is the split real form of the basic classical superalgebra $osp(1|2n)$ and has the same root system. The root system of $su(n)$, actually of $sl(n)$, is comprised of the even simple roots of $\mathfrak{g}$: $\alpha_i, i = 1,\dots,n-1$, with standard non-zero products: $(\alpha_i,\alpha_i) = 2$, $(i = 1,\dots,n-1)$, $(\alpha_i,\alpha_{i+1}) = -1$, $(i = 1,\dots,n-2)$. The root system of $so(D)$ is comprised of simple roots $\gamma_j, j = 1,\dots,\ell \equiv [D/2]$.

For even D the non-zero scalar products are: $(\gamma_j\,,\gamma_j) \;=\; 2\kappa$, $(j = 1,\ldots,\ell)$, $(\gamma_j\,,\gamma_{j+1}) \;=\; -\,\kappa$, $(j = 1,\ldots,\ell-2)$, $(\gamma_{\ell-2}\,,\gamma_\ell) \;=\; -\,\kappa$, where κ is a non-zero common multiple (it is inessential since it cancels in the Cartan matrix elements). In the case of interest $D = 10,\;\; \ell = 5,\;\; n = 16$:

$$\begin{aligned}
\gamma_1 &= \alpha_4 + \alpha_7 + \alpha_9 + \alpha_{12}\ , \\
\gamma_2 &= \alpha_3 + \alpha_6 + \alpha_{10} + \alpha_{13}\ , \\
\gamma_3 &= \alpha_2 + \alpha_4 + \alpha_5 + \alpha_6 + \alpha_7 + 2\alpha_8 + \alpha_9 + \alpha_{10} + \alpha_{11} + \alpha_{12} + \alpha_{14}\ , \\
\gamma_4 &= \alpha_3 + \alpha_5 + \alpha_7 + \alpha_{15}\ , \\
\gamma_5 &= \alpha_1 + \alpha_9 + \alpha_{11} + \alpha_{13}\ .
\end{aligned} \tag{4}$$

(The roots γ_i satisfy the prescribed products with $\kappa = 4$.) Correspondingly, the $so(10)$ Dynkin labels $r_k \;\equiv\; -\,\kappa(\Lambda,\gamma_k^\vee) \;=\; -\,(\Lambda,\gamma_k)$ are:

$$\begin{aligned}
r_1 &= a_4 + a_7 + a_9 + a_{12}\ , \\
r_2 &= a_3 + a_6 + a_{10} + a_{13}\ , \\
r_3 &= a_2 + a_4 + a_5 + a_6 + a_7 + 2a_8 + a_9 + a_{10} + a_{11} + a_{12} + a_{14}\ , \\
r_4 &= a_3 + a_5 + a_7 + a_{15}\ , \\
r_5 &= a_1 + a_9 + a_{11} + a_{13}\ .
\end{aligned} \tag{5}$$

The dimensions of the $so(10)$ UIRs is:

$$\dim\, L(r_1,...,r_5) \;=\; \prod_{1\leq s<t\leq 5} \frac{n_t^2 - n_s^2}{(t-s)\;(t+s-2)}\ , \tag{6}$$

where we use the additional parametrization:

$$\begin{aligned}
n_1 &= \tfrac{1}{2}(r_5 - r_4)\ , \qquad n_2 \;=\; \tfrac{1}{2}(r_5 + r_4) + 1 \\
n_s &= \tfrac{1}{2}(r_5 + r_4) + r_3 + \cdots + r_{6-s} + s - 1\ , \qquad s = 3,...,5\ .
\end{aligned} \tag{7}$$

The parameters n_s are either all integer or all half-integer obeying: $n_5 \;>\; n_4 \;>\; n_3 \;>\; n_2 \;>\; |n_1| \geq 0$.

It is known that the unitarity restrictions for a conformal superalgebra $\mathfrak{a}$ are stronger than those for the even subalgebra of $\mathfrak{a}$ (for $D = 4, 6$ cf. [12]). Here, in addition the conformal algebra $so(D,2)$ is smaller than the even subalgebra $sp(2n)$. Thus, the unitarity conditions for $so(D,2)$ are given only in terms of r_i. Thus, $so(D,2)$ unitarity would not require that all parameters a_k are non-negative integers - that would be required only for their combinations r_i. The unrestricted parameters a_k are combined in so-called tensorial charges [1]. We shall leave this together with more detailed analysis for a follow-up paper.

Here, due to the lack of space we only consider briefly the embedding of the fundamental irreps of $so(D,2)$ in those of $sp(2n)$.

Of course, the one-dimensional irreps, when $a_i = 0 = r_s$ for all i, s, coincide. In the Dynkin labelling of the 15 fundamental irreps Λ_k of $su(16)$ are characterized for fixed k by $a_i = \delta_{ik}$. The five fundamental irreps Λ^o_t of $so(10)$ are characterized for fixed k by $r_s = \delta_{st}$. Each of the two fundamental 16-dimensional $so(10)$ spinors when $r_4 = 1$ or $r_5 = 1$ are contained in four different cases, once without reduction: The 16-dimensional fundamental $su(16)$ UIR with $a_1 = 1$ gives the case with $r_5 = 1$, while the 16-dimensional fundamental $su(16)$ UIR with $a_{15} = 1$ gives the case with $r_4 = 1$. The 120-dimensional fundamental $su(16)$ UIRs with $a_2 = 1$ or $a_{14} = 1$ both give the 120-dimensional fundamental $so(10)$ with $r_3 = 1$. The 560-dimensional fundamental $su(16)$ UIRs with $a_3 = 1$ or $a_{13} = 1$ both give the 45-dimensional fundamental $so(10)$ with $r_2 = 1$, and in addition, the fundamental spinor of $so(10)$ with $r_4 = 1$, or the fundamental spinor of $so(10)$ with $r_5 = 1$, resp. The 1820-dimensional fundamental $su(16)$ UIRs with $a_4 = 1$ or $a_{12} = 1$ both give the 10-dimensional defining fundamental $so(10)$ with $r_1 = 1$, and the 120-dimensional fundamental $so(10)$ irrep with $r_3 = 1$. Similar analysis is done in the oscillator approach in [4].

For odd D the non-zero scalar products are: $(\gamma_j, \gamma_j) = 2\kappa$, $(j = 1, \ldots, \ell - 1)$, $(\gamma_\ell, \gamma_\ell) = \kappa$, $(\gamma_j, \gamma_{j+1}) = -\kappa$, $(j = 1, \ldots, \ell - 1)$.

In the case $D = 9$, $\ell = 4$, $n = 16$, we have:

$$\begin{aligned}
\gamma_1 &= \alpha_5 + 2\alpha_6 + \alpha_7 + \alpha_9 + 2\alpha_{10} + \alpha_{11} \,, \\
\gamma_2 &= \alpha_3 + 2\alpha_4 + \alpha_5 + \alpha_{11} + 2\alpha_{12} + \alpha_{13} \,, \\
\gamma_3 &= \alpha_2 + \alpha_6 + \alpha_7 + 2\alpha_8 + \alpha_9 + \alpha_{10} + \alpha_{14} \,, \\
\gamma_4 &= \tfrac{1}{2}(\alpha_1 + \alpha_3 + \alpha_5 + \alpha_7 + \alpha_9 + \alpha_{11} + \alpha_{13} + \alpha_{15}) \,,
\end{aligned} \tag{8}$$

(with $\kappa = 4$). Then, the $so(9)$ Dynkin labels and dimensions are:

$$\begin{aligned}
r_1 &= a_5 + 2a_6 + a_7 + a_9 + 2a_{10} + a_{11} \,, \\
r_2 &= a_3 + 2a_4 + a_5 + a_{11} + 2a_{12} + a_{13} \,, \\
r_3 &= a_2 + a_6 + a_7 + 2a_8 + a_9 + a_{10} + a_{14} \,, \\
r_4 &= a_1 + a_3 + a_5 + a_7 + a_9 + a_{11} + a_{13} + a_{15} \,,
\end{aligned} \tag{9}$$

$$\dim L(r_1, ..., r_4) = \frac{2^4 n_1 n_2 n_3 n_4}{7.5.3} \prod_{1 \le s < t \le 4} \frac{n_t^2 - n_s^2}{(t-s)\,(t+s-1)} \,, \tag{10}$$

$$\begin{aligned}
&n_1 = \tfrac{1}{2}(r_4 + 1) \,, \qquad n_s = \tfrac{1}{2}(r_4 - 1) + r_3 + \cdots + r_{5-s} + s \,, \quad s = 2,3,4, \\
&n_4 > n_3 > n_2 > n_1 > 0 \,.
\end{aligned} \tag{11}$$

The fundamental 16-dimensional $so(9)$ spinor - obtained for $r_4 = 1$ - is contained without reduction in both 16-dimensional fundamental $su(16)$ UIRs with $a_1 = 1$ and $a_{15} = 1$. It is contained also in six other fundamental $su(16)$ UIRs. The defining 9-dimensional irrep - obtained for $r_1 = 1$ - is contained in four fundamental $su(16)$ UIRs (with $a_5 = 1$, $a_7 = 1$, $a_9 = 1$, $a_{11} = 1$). The 36-,84-dimensional fundamental irrep, resp. - obtained for $r_2 = 1, r_3 = 1$, resp. - is contained in four, six, resp., fundamental $su(16)$ UIRs (with $a_3 = 1$, $a_5 = 1$, $a_{11} = 1$, $a_{13} = 1$, resp., $a_2 = 1$, $a_5 = 1$, $a_{11} = 1$, $a_{13} = 1$).

In the case $D = 11$, $\ell = 5$, $n = 32$, we have:

$$\begin{aligned}
\gamma_1 &= \alpha_7 + 2\alpha_8 + 2\alpha_9 + \alpha_{10} + \alpha_{13} + 2\alpha_{14} + 3\alpha_{15} + 2\alpha_{16} + \\
&\quad + \alpha_{17} + 2\alpha_{18} + \alpha_{19} + \alpha_{22} + 2\alpha_{23} + 2\alpha_{24} + \alpha_{25} \ , \\
\gamma_2 &= \alpha_5 + 2\alpha_6 + \alpha_7 + \alpha_9 + \alpha_{10} + \alpha_{11} + 2\alpha_{12} + \alpha_{13} + \\
&\quad + \alpha_{19} + 2\alpha_{20} + \alpha_{21} + \alpha_{22} + \alpha_{23} + \alpha_{25} + 2\alpha_{26} + \alpha_{27} \ , \\
\gamma_3 &= \alpha_3 + 2\alpha_4 + \alpha_5 + \alpha_{13} + 2\alpha_{14} + \alpha_{15} + 2\alpha_{16} + \\
&\quad + 3a_{17} + 2\alpha_{18} + \alpha_{19} + \alpha_{27} + 2\alpha_{28} + \alpha_{29} \ , \\
\gamma_4 &= \alpha_2 + \alpha_6 + \alpha_7 + 2\alpha_8 + \alpha_9 + 2\alpha_{10} + 2\alpha_{11} + 2a_{12} + \\
&\quad + \alpha_{13} + \alpha_{14} + \alpha_{15} + \alpha_{16} + \alpha_{18} + \alpha_{19} + 2a_{20} + 2\alpha_{21} + \\
&\quad + 2\alpha_{22} + \alpha_{23} + 2\alpha_{24} + a_{25} + \alpha_{26} + \alpha_{30} \ , \\
\gamma_5 &= \tfrac{1}{2}(\alpha_1 + \alpha_3 + \alpha_5 + \alpha_7 + \alpha_9 + \alpha_{11} + \alpha_{13} + \alpha_{15} + \\
&\quad + \alpha_{17} + \alpha_{19} + \alpha_{21} + \alpha_{23} + \alpha_{25} + \alpha_{27} + \alpha_{29} + \alpha_{31}) \ ,
\end{aligned} \tag{12}$$

(with $\kappa = 8$). Then, the $so(11)$ Dynkin labels and dimensions are:

$$\begin{aligned}
r_1 &= a_7 + 2a_8 + 2a_9 + a_{10} + a_{13} + 2a_{14} + 3a_{15} + 2a_{16} + \\
&\quad + a_{17} + 2a_{18} + a_{19} + a_{22} + 2a_{23} + 2a_{24} + a_{25} \ , \\
r_2 &= a_5 + 2a_6 + a_7 + a_9 + a_{10} + a_{11} + 2a_{12} + a_{13} + \\
&\quad + a_{19} + 2a_{20} + a_{21} + a_{22} + a_{23} + a_{25} + 2a_{26} + a_{27} \ , \\
r_3 &= a_3 + 2a_4 + a_5 + a_{13} + 2a_{14} + a_{15} + 2a_{16} + \\
&\quad + 3a_{17} + 2a_{18} + a_{19} + a_{27} + 2a_{28} + a_{29} \ , \\
r_4 &= a_2 + a_6 + a_7 + 2a_8 + a_9 + 2a_{10} + 2a_{11} + 2a_{12} + \\
&\quad + a_{13} + a_{14} + a_{15} + a_{16} + a_{18} + a_{19} + 2a_{20} + 2a_{21} + \\
&\quad + 2a_{22} + a_{23} + 2a_{24} + a_{25} + a_{26} + a_{30} \ , \\
r_5 &= a_1 + a_3 + a_5 + a_7 + a_9 + a_{11} + a_{13} + a_{15} + \\
&\quad + a_{17} + a_{19} + a_{21} + a_{23} + a_{25} + a_{27} + a_{29} + a_{31} \ ,
\end{aligned} \tag{13}$$

$$\dim L(r_1, ..., r_5) \ = \ \frac{2^5 \, n_1 n_2 n_3 n_4 n_5}{9.7.5.3} \prod_{1 \le s < t \le 5} \frac{n_t^2 - n_s^2}{(t-s)\,(t+s-1)} \ , \tag{14}$$

$$n_1 = \frac{1}{2}(r_5+1)\ , \quad n_s = \frac{1}{2}(r_5-1) + r_4 + \cdots + r_{6-s} + s\ , \quad s = 2, \ldots (15)$$
$$n_5 > n_4 > n_3 > n_2 > n_1 > 0\ . \qquad (16)$$

The fundamental 32-dimensional $so(11)$ spinor - obtained for $r_5 = 1$ - is contained without reduction in both 32-dimensional fundamental $su(32)$ UIRs with $a_1 = 1$ and $a_{31} = 1$. It is contained also in 14 other fundamental $su(32)$ UIRs. The defining 11-dimensional irrep - obtained for $r_1 = 1$ - is contained in seven fundamental $su(32)$ UIRs. The 55-,165-,330-dimensional fundamental irrep, resp. - obtained for $r_2 = 1, r_3 = 1, r_4 = 1$, resp. - is contained in 12,7,14, resp., fundamental $su(32)$ UIRs.

References

1. P.K. Townsend, P-brane democracy, PASCOS/Hopkins 0271-286 (1995) hep-th/9507048; Four lectures on M theory, in: *Trieste 1996, High energy physics and cosmology* 385-438, hep-th/9612121; Nucl. Phys. Proc. Suppl. **68**, 11-16 (1998) hep-th/9708034; J.P. Gauntlett, G.W. Gibbons, C.M. Hull and P.K. Townsend, Commun. Math. Phys. **216**, 431-459 (2001) hep-th/0001024.
2. P. Horava and E. Witten, Nucl. Phys. **B460**, 506-524 (1996) hep-th/9510209.
3. C.R. Preitschopf, T. Hurth, P. van Nieuwenhuizen and A. Waldron, Nucl. Phys. Proc. Suppl. **56B**, 310-317 (1997) hep-th/9702123.
4. M. Gunaydin, Nucl. Phys. **B528**, 432-450 (1998) hep-th/9803138.
5. I. Bandos and J. Lukierski, Mod. Phys. Lett. **A14**, 1257-1272 (1999) hep-th/9811022; I. Bandos, J. Lukierski, C. Preitschopf and D.P. Sorokin, Phys. Rev. **D61**, 065009 (2000) hep-th/9907113.
6. E. Bergshoeff and A. Van Proeyen, Class. Quant. Grav. **17**, 3277-3304 (2000) hep-th/0003261; The unifying superalgebra osp(1—32), Proc. Int. Conf. 'Quantization, Gauge Theory and Strings' dedicated to the memory of Efim Fradkin, Moscow, Russia, 5-10 Jun 2000, hep-th/0010194.
7. R. D'Auria, S. Ferrara, M.A. Lledo and V.S. Varadarajan, J. Geom. Phys. **40** 101-128, (2001) hep-th/0010124; R. D'Auria, S. Ferrara and M.A. Lledo, Lett. Math. Phys. **57**, 123-133 (2001) hep-th/0102060; M.A. Lledo and V.S. Varadarajan, in: Proc. 2nd International Symposium on Quantum Theory and Symmetries, Cracow, Poland, 18-21 Jul 2001, hep-th/0111105; S. Ferrara and M.A. Lledo, Rev. Math. Phys. **14**, 519-530 (2002) hep-th/0112177.
8. M. Banados, Phys. Rev. Lett. **88**, 031301 (2002) hep-th/0107214.
9. H. Saleur and B. Wehefritz-Kaufmann, Nucl. Phys. **B628**, 407-441 (2002) hep-th/0112095.
10. M. Bagnoud, L. Carlevaro and A. Bilal, Nucl. Phys. **B641**, 61-92 (2002) hep-th/0201183.
11. V.K. Dobrev and R.B. Zhang, Positive Energy Unitary Irreducible Representations of the Superalgebras $osp(1|2n, I\!R)$, Plenary talk (by VKD) at X International Conference on Symmetry Methods in Physics (Yerevan, 2003); to appear in: Physics of Atomic Nuclei (2004).
12. V.K. Dobrev and V.B. Petkova, Phys. Lett. **162B**, 127 (1985); V.K. Dobrev, J. Phys. **A35**, 7079 (2002) hep-th/0201076.

REALIZATIONS OF EXCEPTIONAL U-DUALITY GROUPS AS CONFORMAL AND QUASI-CONFORMAL GROUPS AND THEIR MINIMAL UNITARY REPRESENTATIONS

MURAT GUNAYDIN

Physics Department
Penn State University
University Park, PA 16802
e-mail: murat@phys.psu.edu

We review the novel quasiconformal realizations of exceptional U-duality groups whose "quantization" leads directly to their minimal unitary irreducible representations. The group $E_{8(8)}$ can be realized as a quasiconformal group in the 57 dimensional charge-entropy space of BPS black hole solutions of maximal $N = 8$ supergravity in four dimensions and leaves invariant "lightlike separations" in this space. Similarly $E_{7(7)}$ act as a conformal group in the 27 dimensional charge space of BPS black hole solutions in five dimensional maximal supergravity which leaves invariant the lightlike separations. For the exceptional $N = 2$ Maxwell-Einstein supergravity theory the corresponding quasiconformal and conformal groups are $E_{8(-24)}$ and $E_{7(-25)}$, respectively. These conformal and quasiconformal groups act as spectrum generating symmetry groups in five and four dimensions and are isomorphic to the U-duality groups in four and three dimensions, respectively. Hence the spectra of these theories are expected to form unitary representations of these groups, of which the minimal one plays a special role.

1. U-Duality Groups in Supergravity Theories

1.1. *Noncompact exceptional groups as symmetries of maximally extended supergravity theories*

Eleven dimensional supergravity [1] is the effective low energy theory of strongly coupled phase of M-theory [2]. Toroidal compactification of eleven dimensional supergravity theory down to d dimensions yields the maximally extended supergravity with a global non-compact symmetry group $E_{(11-d)(11-d)}$ [3]. It is believed that only the discrete subgroups $E_{(11-d)(11-d)}(\mathbb{Z})$ of these groups are the symmetries of the non-perturbative spectra of toroidally compactified M-theory [4]. We shall use the term U-duality for these discrete subgroups as well as for the global noncompact symmetry groups of supergravity theories.

In five dimensions $E_{6(6)}$ is a symmetry of the lagrangian of the maximal ($N = 8$) supergravity, under which the 27 vector fields of the theory transform irreducibly while the 42 scalar fields transform nonlinearly and parameterize the coset space $E_{6(6)}/USp(8)$. On the other hand the $E_{7(7)}$ symmetry of the maximally extended supergravity in $d = 4$ is an on-shell symmetry group. The field strengths of 28 vector fields of this theory together with their "magnetic" duals transform irreducibly in the 56 of $E_{7(7)}$ and 70 scalar fields parameterize the coset space $E_{7(7)}/SU(8)$. In three dimensions all the propagating bosonic degrees of the maximal $N = 16$ supergravity can be dualized to scalar fields, which transform nonlinearly under the symmetry group $E_{8(8)}$ and parameterize the coset space $E_{8(8)}/SO(16)$.

1.2. *Exceptional U-duality groups in matter coupled supergravity theories*

Non-compact global U-duality groups arise in matter coupled supergravity theories as well. In this talk I will focus on U-duality groups in $N = 2$ Maxwell-Einstein supergravity theories (MESGT) in $d = 5$ and the corresponding theories in four and three dimensions. These theories describe the coupling of an arbitrary number n of (Abelian) vector fields to $N = 2$ supergravity and five dimensional MESGT's were constructed in [5]. The bosonic part of the Lagrangian is given by [5]

$$e^{-1}\mathcal{L}_{\text{bosonic}} = -\frac{1}{2}R - \frac{1}{4}\overset{\circ}{a}_{IJ}F^{I}_{\mu\nu}F^{J\mu\nu} - \frac{1}{2}g_{xy}(\partial_{\mu}\varphi^{x})(\partial^{\mu}\varphi^{y})$$

$$+ \frac{e^{-1}}{6\sqrt{6}}C_{IJK}\varepsilon^{\mu\nu\rho\sigma\lambda}F^{I}_{\mu\nu}F^{J}_{\rho\sigma}A^{K}_{\lambda}, \qquad (1)$$

where e and R denote the fünfbein determinant and the scalar curvature, respectively, and $F^{I}_{\mu\nu}$ are the field strengths of the Abelian vector fields A^{I}_{μ}, ($I = 0, 1, 2\cdots, n$). The metric, g_{xy}, of the scalar manifold $\mathcal{M}$ and the "metric" $\overset{\circ}{a}_{IJ}$ of the kinetic energy term of the vector fields both depend on the scalar fields φ^{x}. The Abelian gauge invariance requires the completely symmetric tensor C_{IJK} to be constant. Remarkably, the entire $N = 2$, $d = 5$ MESGT is uniquely determined by the constant tensor C_{IJK} [5].

Denoting the symmetry group of the tensor C_{IJK} as G one finds that the full symmetry group of $N = 2$ MESGT in $d = 5$ is of the form $G \times SU(2)_R$, where $SU(2)_R$ denotes the local R-symmetry group of the $N = 2$ supersymmetry algebra. A MESGT is said to be *unified* if all the vector fields, including the graviphoton, transform in an irreducible representation of a *simple* symmetry group G. Of all the $N = 2$ MESGT's whose scalar manifolds are symmetric spaces only four are unified [5,6a]. If one defines a cubic form $\mathcal{N}(h) := C_{IJK}h^I h^J h^K$ using the constant tensor C_{IJK}, it was shown in [5] that the cubic forms associated with the four unified MESGT's can be identified with the norm forms of simple (Euclidean) Jordan algebras of degree three. There exist only four simple (Euclidean) Jordan algebras of degree three and they can be realized in terms of 3×3 hermitian matrices over the four division algebras with the product being one-half the anticommutator. They are denoted as $J_3^{\mathbb{A}}$, where $\mathbb{A}$ stands for the underlying division algebra, which can be real numbers $\mathbb{R}$, complex numbers $\mathbb{C}$, quaternions $\mathbb{H}$ and octonions $\mathbb{O}$. The supergravity theories defined by them were referred to as magical supergravity theories [5] since their symmetry groups in five, four and three dimensions correspond precisely to the symmetry groups of the famous Magic Square. The octonionic Jordan algebra $J_3^{\mathbb{O}}$ is called the exceptional Jordan algebra and consequently the $N = 2$ MESGT defined by it is called the exceptional supergravity theory [5]. In the table below we list the scalar manifolds G/H of the magical supergravity theories in five, four and three dimensions, where G is the global noncompact symmetry and H is its maximal compact subgroup.

$J\backslash d$	5	4	3
$J_3^{\mathbb{R}}$	$SL(3,\mathbb{R})/SO(3)$	$Sp(6,\mathbb{R})/U(3)$	$F_{4(4)}/USp(6) \times SU(2)$
$J_3^{\mathbb{C}}$	$SL(3,\mathbb{C})/U(3)$	$SU(3,3)/SU(3)^2 \times U(1)$	$E_{6(2)}/SU(6) \times SU(2)$
$J_3^{\mathbb{H}}$	$SU^*(6)/USp(6)$	$SO^*(12)/U(6)$	$E_{7(-5)}/SO(12) \times SU(2)$
$J_3^{\mathbb{O}}$	$E_{6(-26)}/F_4$	$E_{7(-25)}/E_6 \times U(1)$	$E_{8(-24)}/E_7 \times SU(2)$

Note that the exceptional $N = 2$ supergravity has $E_{6(-26)}, E_{7(-25)}$ and $E_{8(-24)}$ as its global symmetry group in five, four and three dimensions, respectively, whereas the maximally extended supergravity theory has the maximally split real forms $E_{6(6)}, E_{7(7)}$ and $E_{8(8)}$ as its symmetry in the respective dimensions.

2. U-duality Groups and Black Hole Entropy in Supergravity Theories

The entropy of black hole solutions in maximally extended supergravity as well as in matter coupled supergravity theories are invariant under the corresponding U-duality groups. For example in $d = 5$, $N = 8$ supergravity the entropy S of a black hole solution can be written in the form [8]

$$S = \alpha\sqrt{I_3} = \alpha\sqrt{C_{IJK}q^I q^J q^K} \tag{2}$$

[a] It has recently been shown that if one relaxes the condition that the scalar manifolds be homogeneous spaces then there exist three novel infinite families (plus an additional sporadic one) of unified MESGT's in $d = 5$ [7].

where α is some fixed constant and I_3 is the cubic invariant of $E_{6(6)}$. The $q^I, I = 1, 2, ..., 27$ are the charges coupling to the 27 vector fields of the theory. The BPS black hole solutions with $I_3 \neq 0$ preserve 1/8 susy [9]. The solutions with $I_3 = 0$, but with $\frac{1}{3}\partial_I I_3 = C_{IJK} q^J q^K \neq 0$ preserve 1/4 susy , while those solution with both $I_3 = 0$ and $C_{IJK} q^J q^K = 0$ preserve 1/2 susy [9]. The orbits of the black hole solutions of $N = 8$ supergravity in $d = 5$ under the action of $E_{6(6)}$ were classified in [10].

The entropy S of black hole solutions of $N = 8$ supergravity in $d = 4$ is given by the quartic invariant of the U-duality group $E_{7(7)}$ [8]

$$S = \beta\sqrt{I_4} = \beta\sqrt{d_{IJKL} q^I q^J q^K q^L} \tag{3}$$

where β is a fixed constant and $q^I, I = 1, 2, ..., 56$ represent the 28 electric and 28 magnetic charges. The orbits of the BPS black hole solutions of $N = 8$ supergravity preserving 1/8, 1/4 and 1/2 susy under the action of $E_{7(7)}$ were classified in [10]. The number of supersymmetries preserved by the extremal black hole solutions depend on whether or not I_4 , $\partial_J I_4$ and $\partial_J \partial_K I_4$ vanish [9].

The orbits of the BPS black hole solutions of $N = 2$ MESGT's in $d = 5$ and $d = 4$ with symmetric target spaces , including those of the exceptional supergravity, were also classified in [10].

The classification of the orbits of BPS black hole solutions of the $N = 8$ and the exceptional $N = 2$ theory in $d = 5$ as given in [10] associates with a given BPS black hole solution with charges q^I an element $J = \sum_{I=1}^{27} e_I q^I$ of the corresponding exceptional Jordan algebra with basis elements e_I. (Split exceptional Jordan algebra $J_3^{\mathbb{O}_s}$ for the $N = 8$ theory and the real exceptional Jordan algebra $J_3^{\mathbb{O}}$ for the $N = 2$ theory.) The cubic invariant $I_3(q^I)$ is then simply given by the norm form of the Jordan algebra

$$\mathcal{N}_3(J) = I_3(q^I) \tag{4}$$

Invariance group of the norm form (known as the reduced structure group in mathematics literature) is isomorphic to the U-duality group in these five dimensional theories. In the corresponding four dimensional supergravity theory the black hole solution is associated with an element of the Freudenthal triple system defined by the exceptional Jordan algebra and the U-duality group is isomorphic to the invariance group of its associated quartic form [10].

3. Conformal and Quasi-Conformal Realizations of Exceptional Groups

The linear fractional group of the exceptional Jordan algebra $J_3^{\mathbb{O}}(J_3^{\mathbb{O}_s})$ as defined by Koecher is the exceptional group $E_{7(-25)}(E_{7(7)})$ which can be interpreted as a generalized conformal group [14]. Acting on an element $J=\sum_{I=1}^{27} e_I q^I$ the conformal action of E_7 changes its norm and hence the entropy of the corresponding black hole solution. Thus one can regard $E_{7(-25)}$ ($E_{7(7)}$) as a spectrum generating symmetry in the charge space of black hole solutions of the exceptional $N=2$ ($N=8$) supergravity in five dimensions. If one defines a distance function between any two solutions with charges q^I and q'^I as

$$d(q,q') \equiv \mathcal{N}_3(J-J') \tag{5}$$

one finds that the light like separations are preserved under the conformal action of E_7 [11,12]. The explicit action of $E_{7(7)}$ and $E_{7(-25)}$ on the corresponding 27 dimensional spaces are given in [11] and [12] , respectively. Let us review briefly the conformal action of $E_{7(7)}$ given in [11]. The Lie algebra of $E_{7(7)}$ has a 3-graded decomposition

$$\mathbf{133} = \mathbf{27} \oplus (\mathbf{78} \oplus \mathbf{1}) \oplus \overline{\mathbf{27}} \tag{6}$$

under its $E_{6(6)} \times \mathcal{D}$ subgroup, where $\mathcal{D}$ represents the dilatation group $SO(1,1)$. Under its maximal compact subalgebra USp(8) the Lie algebra $E_{6(6)}$ decomposes as a symmetric tensor $\tilde{G}^{ij}$ in the adjoint **36** of USp(8) and a fully antisymmetric symplectic traceless tensor $\tilde{G}^{ijkl}$ transforming as the **42** of USp(8) (indices $1 \le i,j,\ldots \le 8$ are USp(8) indices). $\tilde{G}^{ijkl}$ is traceless with respect to the real symplectic metric $\Omega_{ij}=-\Omega_{ji}=-\Omega^{ij}$ (thus $\Omega_{ik}\Omega^{kj}=\delta_i^j$). The symplectic metric is used to raise and lower indices, with the convention that this is always to be done from the left. The other generators of conformal $E_{7(7)}$ consist of a dilatation generator $\tilde{H}$, translation generators $\tilde{E}^{ij}$ and the nonlinearly realized "special conformal" generators $\tilde{F}^{ij}$, transforming as **27** and $\overline{\mathbf{27}}$, respectively.

The fundamental **27** of $E_{6(6)}$ on which $E_{7(7)}$) acts nonlinearly can be represented as the symplectic traceless antisymmetric tensor $\tilde{Z}^{ij}$ transforming as [b]

$$\begin{aligned} \tilde{G}^i{}_j(\tilde{Z}^{kl}) &= 2\,\delta_j^{\overline{k}}\tilde{Z}^{\overline{i}l}\,, \\ \tilde{G}^{ijkl}(\tilde{Z}^{mn}) &= \tfrac{1}{24}\epsilon^{ijklmnpq}\,\tilde{Z}_{pq}\,, \end{aligned} \tag{7}$$

where $\tilde{Z}_{ij} := \Omega_{ik}\Omega_{jl}\tilde{Z}^{kl} = (\tilde{Z}^{ij})^*$ and $\Omega_{ij}\tilde{Z}^{ij} = 0$. The cubic invariant of $E_{6(6)}$ in the **27** is given by

$$\mathcal{N}_3(\tilde{Z}) := \tilde{Z}^{ij}\tilde{Z}_{jk}\tilde{Z}^{kl}\Omega_{il}\,. \tag{8}$$

The generators $\tilde{E}^{ij}$ act as translations on the space $\tilde{Z}_{ij}$:

$$\tilde{E}^{ij}(\tilde{Z}^{kl}) = -\Omega^{i[k}\Omega^{l]j} - \tfrac{1}{8}\Omega^{ij}\Omega^{kl} \tag{9}$$

[b]Throughout we use the convention that indices connected by a bracket are antisymmetrized with weight one.

and $\tilde{H}$ by dilatations

$$\tilde{H}(\tilde{Z}^{ij}) = \tilde{Z}^{ij} \,. \tag{10}$$

The "special conformal generators" $\tilde{F}^{ij}$ in the $\overline{\mathbf{27}}$ are realized nonlinearly:

$$\begin{aligned}\tilde{F}^{ij}(\tilde{Z}^{kl}) &:= -2\,\tilde{Z}^{ij}(\tilde{Z}^{kl}) + \Omega^{i[k}\Omega^{l]j}(\tilde{Z}^{mn}\tilde{Z}_{mn}) + \tfrac{1}{8}\,\Omega^{ij}\Omega^{kl}(\tilde{Z}^{mn}\tilde{Z}_{mn})\\ &\quad + 8\,\overbrace{\tilde{Z}^{km}\tilde{Z}_{mn}\Omega^{n[i}\Omega^{j]l} - \Omega^{kl}(\tilde{Z}^{im}\Omega_{mn}\tilde{Z}^{nj})}\end{aligned} \tag{11}$$

The norm form needed to define the $E_{7(7)}$ invariant "light cones" is constructed from the cubic invariant of $E_{6(6)}$. If we define the "distance" between $\tilde{X}$ and $\tilde{Y}$ as $\mathcal{N}_3(\tilde{X}-\tilde{Y})$ then it is manifestly invariant under $E_{6(6)}$ and under the translations $\tilde{E}^{ij}$. Under $\tilde{H}$ it transforms by a constant factor, whereas under the action of $\tilde{F}^{ij}$ we have

$$\tilde{F}^{ij}\Big(\mathcal{N}_3(\tilde{X}-\tilde{Y})\Big) = (\tilde{X}^{ij}+\tilde{Y}^{ij})\mathcal{N}(\tilde{X}-\tilde{Y})\,. \tag{12}$$

which proves that the light cones in $\mathbb{R}^{27}$ with base point $\tilde{Y}$

$$\mathcal{N}_3(\tilde{X}-\tilde{Y}) = 0 \tag{13}$$

are indeed invariant under $E_{7(7)}$.

The above formulas carry over in a straightforward manner to the conformal realization of $E_{7(-25)}$ on a 27 dimensional space. In this case the cubic form is invariant under $E_{6(-26)}$ which has $USp(6,2)$ as a maximal subgroup. The $USp(8)$ covariant formulas above for $E_{7(7)}$ are then replaced by $USp(6,2)$ covariant formulas [12].

The conformal groups $E_{7(7)}$ and $E_{7(-25)}$ acting on the 27 dimensional charge spaces of the $N=8$ and the exceptional $N=2$ supergravity in five dimensions are isomorphic to the U-duality groups of the corresponding four dimensional theories obtained by dimensional reduction. One may wander whether there exist conformal groups acting on the charge space of four dimensional supergravity theories that are isomorphic to the U-duality groups of the corresponding three dimensional theories obtained by dimensional reduction. This question was investigated in [11] and it was found that in the case of maximal supergravity, even though there is no conformal action of $E_{8(8)}$, it has a quasi-conformal group action on a 57 dimensional space which is an extension of the 56 dimensional charge space by an extra coordinate. For BPS black hole solutions this extra coordinate can be taken to be the entropy [11].

The realization of quasi-conformal action of $E_{8(8)}$ uses the 5-graded decomposition of its Lie algebra with respect to the Lie algebra of its $E_{7(7)} \times \mathcal{D}$ subgroup

$$\begin{array}{ccccc} \mathfrak{g}^{-2} \oplus \mathfrak{g}^{-1} \oplus & \mathfrak{g}^{0} & \oplus \mathfrak{g}^{+1} \oplus \mathfrak{g}^{+2} \\ \mathbf{1} \oplus \mathbf{56} \oplus & (\mathbf{133} \oplus \mathbf{1}) & \oplus \mathbf{56} \oplus \mathbf{1} \end{array} \tag{14}$$

with $\mathcal{D}$ representing dilatations, whose generator together with grade ± 2 elements generate an $SL(2,\mathbb{R})$ subgroup. It turns out to be very convenient to work in a basis covariant with respect to the SL$(8,\mathbb{R})$ subgroup of $E_{7(7)}$ [11]. Let us denote the SL$(8,\mathbb{R})$ covariant generators belonging to the grade $-2,-1,0,1$ and 2 subspaces in the above decomposition as follows:

$$E \oplus \{E^{ij}, E_{ij}\} \oplus \{G^{ijkl}, G^i{}_j\,;\, H\} \oplus \{F^{ij}, F_{ij}\} \oplus F \tag{15}$$

where $i,j,... = 1,2...,8$ are now SL$(8,\mathbb{R})$ indices.

Consider now a 57-dimensional real vector space with coordinates

$$\mathcal{X} := (X^{ij}, X_{ij}, x) \tag{16}$$

where X^{ij} and X_{ij} transform in the 28 and $\tilde{28}$ of SL$(8,\mathbb{R})$ and x is a singlet. The generators of $E_{7(7)}$ subalgebra act linearly on this space

$$\begin{aligned} G^i{}_j(X^{kl}) &= 2\,\delta^{k}_{j} X^{il} - \tfrac{1}{4}\delta^i_j X^{kl}\,, & G^{ijkl}(X^{mn}) &= \tfrac{1}{24}\epsilon^{ijklmnpq} X_{pq}\,, \\ G^i{}_j(X_{kl}) &= -2\,\delta^i_k X_{jl} + \tfrac{1}{4}\delta^i_j X_{kl}\,, & G^{ijkl}(X_{mn}) &= \delta^{[ij}_{mn} X^{kl]}\,, \\ G^i{}_j(x) &= 0\,, & G^{ijkl}(x) &= 0\,, \end{aligned} \tag{17}$$

The generator H of dilatations acts as

$$H(X^{ij}) = X^{ij}\,, \qquad H(X_{ij}) = X_{ij}\,, \qquad H(x) = 2\,x\,, \tag{18}$$

and the generator E acts as translations on x:

$$E(X^{ij}) = 0\,, \qquad E(X_{ij}) = 0\,, \qquad E(x) = 1\,. \tag{19}$$

The grade ± 1 generators act as

$$\begin{aligned} E^{ij}(X^{kl}) &= 0\,, & E^{ij}(X_{kl}) &= \delta^{ij}_{kl}\,, & E^{ij}(x) &= -X^{ij}\,, \\ E_{ij}(X^{kl}) &= \delta^{kl}_{ij}\,, & E_{ij}(X_{kl}) &= 0\,, & E_{ij}(x) &= X_{ij}\,. \end{aligned} \tag{20}$$

The positive grade generators are realized nonlinearly. The generator F acts as

$$\begin{aligned} F(X^{ij}) &= 4X^{ik}X_{kl}X^{lj} + X^{ij}X^{kl}X_{kl} \\ &\quad -\tfrac{1}{12}\epsilon^{ijklmnpq}X_{kl}X_{mn}X_{pq} + X^{ij}\,x\,, \\ F(X_{ij}) &= -4X_{ik}X^{kl}X_{lj} - X_{ij}X^{kl}X_{kl} \\ &\quad +\tfrac{1}{12}\epsilon_{ijklmnpq}X^{kl}X^{mn}X^{pq} + X_{ij}\,x\,, \\ F(x) &= 4\mathcal{I}_4(X^{ij}, X_{ij}) + x^2 \end{aligned} \tag{21}$$

where $\mathcal{I}_4$ is the quartic invariant of $E_{7(7)}$

$$\begin{aligned}\mathcal{I}_4 \equiv X^{ij}X_{jk}X^{kl}X_{li} - \tfrac{1}{4}X^{ij}X_{ij}X^{kl}X_{kl} + \tfrac{1}{96}\,\epsilon^{ijklmnpq}X_{ij}X_{kl}X_{mn}X_{pq} \\ + \tfrac{1}{96}\,\epsilon_{ijklmnpq}X^{ij}X^{kl}X^{mn}X^{pq}\end{aligned} \tag{22}$$

The action of the remaining generators of $E_{8(8)}$ are as follows:

$$\begin{aligned}
F^{ij}(X^{kl}) &= -4\,X^{i[k}X^{l]j} + \tfrac{1}{4}\,\epsilon^{ijklmnpq}X_{mn}X_{pq}\,,\\
F^{ij}(X_{kl}) &= +8\,\delta^{[i}_{k}X^{j]m}X_{ml} + \delta^{ij}_{kl}\,X^{mn}X_{mn} + 2\,X^{ij}X_{kl} - \delta^{ij}_{kl}\,x\,,\\
F_{ij}(X^{kl}) &= -8\,\delta^{k}_{[i}X_{j]m}X^{ml} + \delta^{kl}_{ij}\,X^{mn}X_{mn} - 2\,X_{ij}X^{kl} - \delta^{kl}_{ij}\,x\,,\\
F_{ij}(X_{kl}) &= 4\,X_{ki}X_{jl} - \tfrac{1}{4}\,\epsilon_{ijklmnpq}X^{mn}X^{pq}\,,\\
F^{ij}(x) &= 4\,X^{ik}X_{kl}X^{lj} + X^{ij}X^{kl}X_{kl}\\
&\quad -\tfrac{1}{12}\,\epsilon^{ijklmnpq}X_{kl}X_{mn}X_{pq} + X^{ij}\,x\,,\\
F_{ij}(x) &= 4\,X_{ik}X^{kl}X_{lj} + X_{ij}X^{kl}X_{kl}\\
&\quad -\tfrac{1}{12}\,\epsilon_{ijklmnpq}X^{kl}X^{mn}X^{pq} - X_{ij}\,x\,.
\end{aligned} \tag{23}$$

The above action of $E_{8(8)}$ was called quasiconformal in [11] since it leaves a certain norm invariant up to an overall factor. Since the standard difference $(\mathcal{X} - \mathcal{Y})$ of two vectors in the 57 dimensional space is not invariant under translations generated by (E^{ij}, E_{ij}), one defines a nonlinear difference that is invariant under these noncommuting translation generators as [11]

$$\delta(\mathcal{X},\mathcal{Y}) := (X^{ij} - Y^{ij}, X_{ij} - Y_{ij}\;;\; x - y + \langle X, Y\rangle) = -\delta(\mathcal{Y},\mathcal{X}) \tag{24}$$

where $\langle X, Y\rangle := X^{ij}Y_{ij} - X_{ij}Y^{ij}$. Defining the norm of a vector $\mathcal{X}$ as

$$\mathcal{N}_4(\mathcal{X}) \equiv \mathcal{N}_4(X^{ij}, X_{ij}; x) := \mathcal{I}_4(X) - x^2\,, \tag{25}$$

and the "distance" between any two vectors $\mathcal{X}$ and $\mathcal{Y}$ as $\mathcal{N}_4(\delta(\mathcal{X},\mathcal{Y}))$ one finds that

$$\begin{aligned}
F\Big(\mathcal{N}_4(\delta(\mathcal{X},\mathcal{Y}))\Big) &= 2\,(x+y)\,\mathcal{N}_4(\delta(\mathcal{X},\mathcal{Y}))\\
F^{ij}\Big(\mathcal{N}_4(\delta(\mathcal{X},\mathcal{Y}))\Big) &= 2\,(X^{ij}+Y^{ij})\,\mathcal{N}_4(\delta(\mathcal{X},\mathcal{Y}))\\
H\Big(\mathcal{N}_4(\delta(\mathcal{X},\mathcal{Y}))\Big) &= 4\,\mathcal{N}_4(\delta(\mathcal{X},\mathcal{Y}))
\end{aligned}$$

Therefore, for every $\mathcal{Y} \in \mathbb{R}^{57}$ the "light cone" with base point $\mathcal{Y}$, defined by the set of $\mathcal{X} \in \mathbb{R}^{57}$ obeying

$$\mathcal{N}_4(\delta(\mathcal{X},\mathcal{Y})) = 0\,, \tag{26}$$

is preserved by the full $E_{8(8)}$ group.

The quasiconformal realization of the other real noncompact form $E_{8(-24)}$ with the maximal compact subgroup $E_7 \times SU(2)$ is given in [12]. In going to $E_{8(-24)}$ the the subgroup $SL(8,\mathbb{R})$ of $E_{7(7)}$ is replaced by the subgroup $SU^*(8)$ of $E_{7(-25)}$. The quasiconformal groups $E_{8(8)}$ and $E_{8(-24)}$ are isomorphic to the U-duality groups of the maximal $N = 16$ supergravity and the $N = 4$ exceptional supergravity in three dimensions. Since their action changes the "norm" in the charge-entropy space of the corresponding four dimensional theories they can be interpreted as spectrum generating symmetry groups.

The quasiconformal realizations of $E_{8(8)}$ and $E_{8(-24)}$ can be consistently truncated to quasiconformal realizations of other exceptional groups that are subgroups. For a complete list of the real forms of these exceptional subgroups we refer the reader to [11,12].

4. The Minimal Unitary Representations of Exceptional Groups

To obtain unitary realizations of exceptional groups ,based on their quasiconformal realizations, over certain Hilbert spaces of square integrable functions one has to find the corresponding phase space realizations of their generators and quantize them. For $E_{8(8)}$ this was done in [13] and for $E_{8(-24)}$ in [12]. Here we shall summarize the results for $E_{8(8)}$.

Since the positive graded generators form an Heisenberg algebra one introduces 28 coordinates X^{ij} and 28 momenta $P_{ij} \equiv X_{ij}$, and one extra real coordinate y to represent the central term. By quantizing

$$[X^{ij}, P_{kl}] \;=\; \mathrm{i} \tag{27}$$

we can realize the positive grade generators of $E_{8(8)}$ as

$$E^{ij} \;:=\; y\,X^{ij}\,, \quad E_{ij} \;:=\; y\,P_{ij}\,, \quad E \;:=\; \tfrac{1}{2}\,y^2. \tag{28}$$

To realize the other generators of $E_{8(8)}$ one introduces a momentum conjugate to the coordinate y representing the central charge

$$[y,p] \;=\; \mathrm{i} \tag{29}$$

Then the remaining generators are given by

$$
\begin{aligned}
H &:= \tfrac{1}{2}(y\,p + p\,y)\,, \\
F^{ij} &:= -p\,X^{ij} + 2\mathrm{i}y^{-1}\,[X^{ij}\,,\,I_4(X,P)] \\
&= -4\,y^{-1}X^{ik}P_{kl}X^{lj} - \tfrac{1}{2}\,y^{-1}(X^{ij}P_{kl}X^{kl} + X^{kl}P_{kl}X^{ij}) \\
&\quad + \tfrac{1}{12}y^{-1}\epsilon^{ijklmnpq}P_{kl}P_{mn}P_{pq} - p\,X^{ij}\,, \\
F_{ij} &:= -p\,P_{ij} + 2\mathrm{i}y^{-1}\,[P_{ij}\,,\,I_4(X,P)] \\
&= 4\,y^{-1}P_{ik}X^{kl}P_{lj} + \tfrac{1}{2}\,y^{-1}(P_{ij}X^{kl}P_{kl} + P_{kl}X^{kl}P_{ij}) \\
&\quad - \tfrac{1}{12}y^{-1}\epsilon_{ijklmnpq}X^{kl}X^{mn}X^{pq} - p\,P_{ij}\,, \\
F &:= \tfrac{1}{2}p^2 + 2y^{-2}I_4(X,P) \\
G^i{}_j &:= 2\,X^{ik}P_{kj} + \tfrac{1}{4}X^{kl}P_{kl}\,\delta^i_j\,, \\
G^{ijkl} &:= -\tfrac{1}{2}X^{[ij}X^{kl]} + \tfrac{1}{48}\epsilon^{ijklmnpq}P_{mn}P_{pq}\,.
\end{aligned} \tag{30}
$$

The hermiticity of all generators is manifest. Here $I_4(X,P)$ is the fourth order differential operator

$$
\begin{aligned}
I_4(X,P) &:= -\tfrac{1}{2}(X^{ij}P_{jk}X^{kl}P_{li} + P_{ij}X^{jk}P_{kl}X^{li}) \\
&\quad + \tfrac{1}{8}(X^{ij}P_{ij}X^{kl}P_{kl} + P_{ij}X^{ij}P_{kl}X^{kl}) \\
&\quad - \tfrac{1}{96}\,\epsilon^{ijklmnpq}P_{ij}P_{kl}P_{mn}P_{pq} \\
&\quad - \tfrac{1}{96}\,\epsilon_{ijklmnpq}X^{ij}X^{kl}X^{mn}X^{pq} + \tfrac{547}{16}\,.
\end{aligned} \tag{31}
$$

and represents the quartic invariant of $E_{7(7)}$ because

$$
[G^i{}_j\,,\,I_4(X,P)] = [G^{ijkl}\,,\,I_4(X,P)] = 0\,. \tag{32}
$$

The quadratic Casimir operator of $E_{8(8)}$ reduces to a number for the above realization and one can show that all the higher Casimir operators must also reduce to numbers as required for an irreducible unitary representation. Thus by exponentiating the above generators we obtain the minimal unitary irreducible representation of $E_{8(8)}$ over the Hilbert space of square integrable functions in 29 variables.

In the minimal unitary realization of the other noncompact real form $E_{8(-24)}$ given explicitly in [12] the coordinates and momenta transform in the **28** and $\tilde{\mathbf{28}}$ of $SU^*(8)$ subgroup of $E_{7(-25)}$ and the above formulas for $E_{8(8)}$ carry over in a straightforward manner to those of $E_{8(-24)}$.

Acknowledgments

I thank the organizers of QTS3 for the opportunity to speak at the session dedicated to the memory of Freydoon Mansouri and would like to extend my sincerest condolences to his Family. I got to know Freydoon through my frequent visits to Yale University in the late seventies. I will always remember him as a polite, kind and caring person and for his support and encouragement during a difficult period in my career in mid eighties. This work was supported in part by the National Science Foundation under grant number PHY-0245337.

References

1. E. Cremmer, B. Julia and J. Scherk, "Supergravity Theory in 11 Dimensions", *Phys. Lett.* , **B76**, 409 (1978).
2. E. Witten, "String Theory Dynamics in various Dimensions", *Nucl. Phys.* **B443** , 85 (1995).
3. B Julia, "Group Disintegrations", in: *Superspace and Supergravity*, eds. S.W.Hawking and M. Rocek, Cambridge University Press, 1981.
4. C. Hull and P.K. Townsend, "Unity of Superstring Dualities", *Nucl.Phys.* **B438** , 109 (1995).
5. M. Günaydin, G. Sierra and P.K. Townsend, Nucl. Phys. **B242** (1984), 244 ; Phys. Lett. **133B** (1983) 72.
6. M. Günaydin, G. Sierra and P.K. Townsend, Class. Quantum Grav. **3** (1986) 763.
7. M. Günaydin and M. Zagermann, "Unified Maxwell-Einstein and Yang-Mills-Einstein supergravity theories in five dimensions," *JHEP* **0307**, 023 (2003)
8. See [9,10] for references on the subject.
9. See S. Ferrara and J. Maldacena, "Branes, central charges and U-duality invariant BPS conditions", *Class.Quant.Grav.* **15** 749 (1998) and the references therein.
10. S. Ferrara and M. Gunaydin, "Orbits of exceptional groups, duality and BPS states in string theory," *Int. J. Mod. Phys.* **A13**, 2075 (1998)
11. M. Gunaydin, K. Koepsell and H. Nicolai, "Conformal and quasiconformal realizations of exceptional Lie groups," *Commun. Math. Phys.* **221**, 57 (2001)
12. M. Gunaydin and O. Pavlyk, to appear.
13. M. Gunaydin, K. Koepsell and H. Nicolai, "The Minimal Unitary Representation of $E_8(8)$,"*Adv. Theor. Math. Phys.* **5**, 923 (2002)
14. See [11,13] for references on the subject.

FUSION PRODUCTS OF $\mathfrak{sl}_N$ SYMMETRIC POWER REPRESENTATIONS AND KOSTKA POLYNOMIALS

RINAT KEDEM

Department of Mathematics, University of Illinois, 1409 W. Green Street, Urbana, IL 61801, rinat@uiuc.edu

We explain the relation between the Feigin-Loktev fusion product and the graded multiplicities of Specht modules in the integer cohomology ring of the GL_N generalized flag manifold. We use only very basic notions, most notably the Schur-Weyl duality and the description of the cohomology ring as a quotient of the polynomial ring in N variables.

1. Introduction

Let $\mathfrak{g} = \mathfrak{sl}_n$, $\mu = (\mu_1, ..., \mu_n)$ with $\mu_1 \geq \mu_2 \geq \cdots \geq \mu_n \geq 0$, and V_μ the finite-dimensional irreducible $\mathfrak{sl}_n$-module with highest weight $\sum_{i=1}^{n-1}(\mu_i - mu_{i+1})\omega_i$, ω_i fundamental weights. Given a collection of m such representations $\{V_{\mu^{(i)}}\}_{i=1}^m$, their tensor product is completely reducible as a $\mathfrak{g}$-module, and can be decomposed as

$$V_{\boldsymbol{\mu}} := V_{\mu^{(1)}} \otimes \cdots \otimes V_{\mu^{(m)}} \simeq \bigoplus_\lambda V_\lambda \otimes M_{\lambda,\boldsymbol{\mu}}, \tag{1}$$

where $M_{\lambda,\boldsymbol{\mu}}$ is a multiplicity space on which $\mathfrak{g}$ acts trivially, of dimension $K_{\lambda,\boldsymbol{\mu}}$ which can be computed from the Clebsch-Gordan coefficients or the Littlewood-Richardson rule.

The Schur-Weyl duality concerns the case where $\mu^{(i)} = (1)$ for all i. In that case, there is an action of the symmetric group S_N on the tensor product by permutation of factors, which centralizes the action of $\mathfrak{sl}_n$. In this case, $M_{\lambda,\{1,...,1\}} \simeq W_\lambda$ is the irreducible Specht module of the symmetric group. In the case when $\mu^{(i)} = (\mu_i)$ are single-part partitions, $M!_{\lambda,\mu}$ is not an S_N-module but its dimension is known to be the Kostka number[5] $K_{\lambda,\mu}$. Here, $\mu = (\mu_1, ..., \mu_m)$, where without loss of generality we order $\mu_i \geq \mu_{i+1}$.

The Feigin-Loktev fusion product[1], inspired by the definition of the fusion action of the affine algebra $\widehat{\mathfrak{g}}$ on conformal blocks of WZW models in conformal field theory, is a $\mathfrak{g}$-equivariant grading on the tensor product

(1),

$$\mathcal{F}_\mu := V_{\mu^{(1)}} \star \cdots \star V_{\mu^{(m)}} \simeq \bigoplus_\lambda V_\lambda \otimes \mathcal{M}_{\lambda,\boldsymbol{\mu}}, \tag{2}$$

where $\mathcal{M}_{\lambda,\boldsymbol{\mu}}$ is a graded multiplicity space with graded components $\mathcal{M}_{\lambda,\boldsymbol{\mu}}[n]$, of total dimension $K_{\lambda\boldsymbol{\mu}}$. The problem is to compute the graded dimension of $\mathcal{M}_{\lambda,\boldsymbol{\mu}}$,

$$\overline{K}_{\lambda\boldsymbol{\mu}}(q) := \sum_{n\geq 0} q^n \dim \mathcal{M}_{\lambda,\boldsymbol{\mu}}[n].$$

In defining the fusion product[1], the authors conjectured that these q-multiplicities are the generalized Kostka polynomials[2]. Kostka polynomials[5] are the transition functions between Hall-Littlewood and Schur polynomials, and are defined in the case where $\mu^{(i)}$ are single part partitions. Generalized Kostka polynomials correspond to the case where $\mu^{(i)} = (m^j)$ corresponding to rectangular partitions, although the fusion product is defined even more generally. Here, we give a simple proof for the case of single-part partitions.

The exact statement is as follows. Let $\mu^{(i)} = (\mu_i)$ where $\mu_i \in \mathbb{Z}_{\geq 0}$. Using the commutativity of the fusion product (which follows from the description below), order the factors in the product (2) so that $\mu_i \geq \mu_{i+1}$. Let $\mu = (\mu_1, ..., \mu_m)$ denote the associated partition of $N = \sum_i \mu_i$. In this case, $\overline{K}_{\lambda,\boldsymbol{\mu}}(q) = \widetilde{K}_{\lambda,\mu}(q)$, where

$$\widetilde{K}_{\lambda,\mu}(q) = q^{n(\mu)} K_{\lambda,\mu}(1/q),$$

with $K_{\lambda,\mu}(q)$ the Kostka polynomial[5] and $n(\mu) = \sum_i (i-1)\mu_i$.

The Kostka polynomial also appears in the context of the integer cohomology ring[3] of the generalized flag manifold $Fl_{\mu'}$ of GL_N. This is a graded space and the Weyl group of GL_N, which is the symmetric group S_N, acts on the graded components. It is known that the cohomology ring is isomorphic as a graded space and as an S_N-module to the quotient ring $R_\mu = \mathbb{C}[z_1, ..., z_N]/I_\mu$ where I_μ is a graded, S_N-invariant ideal[3] (see below). The symmetric group S_N acts on $\mathbb{C}[z_1, ..., z_N]$, and on the quotient, by permuting the variables, thus preserving homogenous degree in z_i. Therefore there is a decomposition

$$R_\mu \simeq \bigoplus_{\lambda\vdash N} W_\lambda \otimes \overline{\mathcal{M}}_{\lambda,\mu}, \tag{3}$$

where W_λ is the irreducible S_N Specht module, and $\overline{\mathcal{M}}_{\lambda,\mu}$ is a graded multiplicity space on which S_N acts trivially. The q-dimension of $\overline{\mathcal{M}}_{\lambda,\mu}$ is $\widetilde{K}_{\lambda\mu}(q)^3$. Our goal is to identify this space with $\mathcal{M}_{\lambda,\mu}$. Details of this construction can be found in[4].

2. The fusion product

Let us recall the definition of the fusion product[1]. Given a simple Lie algebra $\mathfrak{g}$, let $\mathfrak{g}[t] \simeq \mathfrak{g} \otimes \mathbb{C}[t]$ Let z be a complex number or a formal variable, and define the action of $\mathfrak{g}[t]$ on the $\mathfrak{g}$-module V_μ by evaluation at the point z:

$$x \otimes f(t)v = f(z)xv, \qquad x \in \mathfrak{g},\ f(t) \in \mathbb{C}[t],\ v \in V_\mu.$$

The evaluation module of $\mathfrak{g}[t]$ is denoted by $V_\mu[z]$. Let $\{z_1, ..., z_m\}$ be distinct complex numbers or formal variables. The algebra $\mathfrak{g}[t]$ acts on the tensor product of evaluation modules $V_{\boldsymbol{\mu}}[\mathcal{Z}] := V_{\mu^{(1)}}[z_1] \otimes \cdots \otimes V_{\mu^{(m)}}[z_m]$. by the usual co-product.

Let v_i be the highest weight vector with respect to $\mathfrak{g}$ of the $\mathfrak{g}$-module $V_{\mu^{(i)}}[z_i]$, and define the following subspace of $V_{\boldsymbol{\mu}}[\mathcal{Z}]$:

$$W_{\boldsymbol{\mu}}[\mathcal{Z}] = U(\mathfrak{g}[t])v_1 \otimes \cdots \otimes v_m.$$

The algebra $\mathfrak{g}[t]$, and hence $U(\mathfrak{g}[t])$, is graded by degree in t. Let $U^{(n)}(\mathfrak{g}[t])$ denote the graded component, and $U^{(\leq n)}(\mathfrak{g}[t]) = \oplus_{i\leq n} U^{(i)}(\mathfrak{g}[t])$. Thus there is a filtration of $W_{\boldsymbol{\mu}}[\mathcal{Z}]$ with filtered components $W_{\boldsymbol{\mu}}[n] = U^{\leq n}(v_1 \otimes \cdots \otimes v_m)$. If we take z_i to be complex numbers, the associated graded space of $W_{\boldsymbol{\mu}}[\mathcal{Z}]$ is the fusion product:

$$\mathcal{F}_{\boldsymbol{\mu}} := \mathrm{gr}\ W_{\boldsymbol{\mu}}[\mathcal{Z}]. \tag{4}$$

Equivalently, let z_i be formal variables, and define the degree of z_i to be 1 for all i. Then $W_{\boldsymbol{\mu}}[\mathcal{Z}]$ is graded by homogenous degree in z_i, with graded components $W_{\boldsymbol{\mu}}^{(n)}$ and filtered components $W_{\boldsymbol{\mu}}[n] = \oplus_{i\leq n} W_{\boldsymbol{\mu}}^{(i)}$. Then $W'_{\boldsymbol{\mu}}[n] := (\mathbb{C}[z_1, ..., z_N] \otimes_{\mathbb{C}} W_{\boldsymbol{\mu}}[n-1]) \cap W_{\boldsymbol{\mu}}^{(n)} \subset W_{\boldsymbol{\mu}}^{(n)}$, and

$$\mathcal{F}_{\boldsymbol{\mu}} = \oplus_n W_{\boldsymbol{\mu}}^{(n)} / W'_{\boldsymbol{\mu}}[n].$$

The two definitions are equivalent.

The fusion product is a graded space, and since $\mathfrak{g}$ acts on homogeneous components of $W_{\boldsymbol{\mu}}[\mathcal{Z}]$, the grading is $\mathfrak{g}$ equivariant, and $\mathcal{F}_{\boldsymbol{\mu}}$ is isomorphic, as a $\mathfrak{g}$-module, to $V_{\boldsymbol{\mu}}$ in (1). Thus we have a decomposition of the form (2), with $\mathcal{M}_{\lambda,\boldsymbol{\mu}}$ graded multiplicity spaces on which $\mathfrak{g}$ acts trivially.

3. The case of fundamental representations

Consider the case where $\mu^{(i)} = (1)$ and let $\pi = V_{\omega_1}$ denote the basic representation of $\mathfrak{g} = \mathfrak{sl}_n$, with highest weight vector v_0. We have

$$V_N[\mathcal{Z}] := V_{(1)^N}[\mathcal{Z}] = \pi[z_1] \otimes \cdots \otimes \pi[z_N] \simeq \pi^{\otimes N} \otimes_{\mathbb{C}} \mathbb{C}[z_1, ..., z_N],$$

where we take z_i to be formal variables. We note that in this case we have the Schur-Weyl duality:

$$\pi^{\otimes N} \simeq \bigoplus_{\lambda:|\lambda|=N,\ell(\lambda)\leq n} V_\lambda \otimes W_\lambda. \tag{5}$$

The symmetric group S_N acts on $\pi^{\otimes N}$ by permuting factors, and it acts on $\mathbb{C}[z_1,...,z_N]$ by permuting variables. We consider the diagonal action of S_N on $V_N[\mathcal{Z}]$:

$$\Delta\sigma(v\otimes f) = \sigma(v)\otimes\sigma(f), \qquad v\in\pi^{\otimes N},\ f\in\mathbb{C}[z_1,...,z_N],\ \sigma\in S_N.$$

The subspace $W_N[\mathcal{Z}] := W_{(1^N)}[\mathcal{Z}] = U(\mathfrak{g}[t])v_0\otimes\cdots v_0$ is characterized in terms of the diagonal action of S_N, as follows. Since the cyclic vector $w_0 = v_0^{\otimes N}$ of $W_N[\mathcal{Z}]$ is invariant with respect to S_N, and the action of $\mathfrak{g}[t]$ on w_0 by the co-product commutes with the diagonal action of S_N, it is clear that

$$W_N[\mathcal{Z}] \subset V_N[\mathcal{Z}]^{S_N},$$

where by $V_N[\mathcal{Z}]^{S_N}$ we mean the subspace invariant with respect to the diagonal action of the symmetric group. Note that this means that the graded components $W_N^{(n)}$ are also S_N invariant.

The inclusion is definitely proper: for example, vectors of the form

$$v_0\otimes\cdots\otimes v_0\otimes f(z_1,...,z_N)$$

where v_0 is the highest weight vector of π and $f(z_1,...,z_N)$ is a symmetric polynomial of positive degree, are not in $W_N[\mathcal{Z}]$. However, the following is true[4]:

Theorem 3.1.

$$V_N[\mathcal{Z}]^{S_N} \simeq W_N[\mathcal{Z}]\otimes_{\mathbb{C}}\mathbb{C}[z_1,...,z_N]^{S_N}.$$

Now consider the fusion product. To compute the space $W'_N[n]$, note that since $W_N^{(n)}$ is S_N-invariant for each n,

$$W'_N[n] = (\mathbb{C}[z_1,...,z_N]\otimes W_N[n-1])\cap W_n^{(n)} = (\Lambda_N\otimes W_N[n-1])\cap W_N^{(n)},$$

where Λ_N is the ring of symmetric polynomials of positive degree. Using Theorem 3.1, we see that

$$(\Lambda_N\otimes_{\mathbb{C}} W_N[n-1])^{(n)} \subset (\Lambda_N\otimes_{\mathbb{C}} V_N[\mathcal{Z}]^{S_N})^{(n)} \subset (\Lambda_N\otimes_{\mathbb{C}} W_N[n-1])^{(n)},$$

where by $F^{(n)}$ we mean the homogenous graded component of the space F. Therefore we have the equality of ideals, and the fusion product is equal to $V_N[\mathcal{Z}]^{S_N}$ modulo the ideal generated by symmetric functions of positive degree, Λ_N:

Theorem 3.2.

$$\mathcal{F}_{(1^N)} := \oplus_n W_N[n]/W'_N[n] \simeq V_N[\mathcal{Z}]^{S_N}/\langle\Lambda_N\rangle \simeq (\pi^{\otimes N} \otimes R_{(1^N)})^{S_N},$$

where $R_{(1^N)} = \mathbb{C}[z_1, ..., z_N]/\langle\Lambda_N\rangle$.

The quotient $R_{(1^N)}$ is well-known to be isomorphic to the integer cohomology ring of the complete flag variety of GL_N, and has the decomposition (3) into irreducible representations of S_N, with $\mu = (1^N)$. Using the Schur-Weyl duality (5), recalling that the trivial representation of S_N occurs in the tensor product $W_\lambda \otimes W_\mu$ once iff $\lambda = \mu$, we have

$$\mathcal{F}_{(1^N)} \simeq \left(\left(\bigoplus_{\lambda:|\lambda|=N,\ell(\lambda)\leq n} V_\lambda \otimes W_\lambda\right) \otimes \left(\bigoplus_{|\nu|=N} W_\nu \otimes \overline{\mathcal{M}}_{\nu,(1^N)}\right)\right)^{S_N}$$

$$\simeq \bigoplus_{\lambda:|\lambda|=N,\ell(\lambda)\leq n} V_\lambda \otimes \overline{\mathcal{M}}_{\lambda,(1)^N}.$$

We conclude that $\overline{\mathcal{M}}_{\lambda,(1^N)} = \mathcal{M}_{\lambda,(1^N)}$.

4. The case of generic symmetric power representations

Recall that $V_{(m)} \simeq \sum_{\sigma\in S_m} \sigma(\pi^{\otimes N})$. Therefore,

$$V_\mu[\mathcal{A}] := V_{(\mu_1)}[a_1] \otimes \cdots \otimes V_{(\mu_m)}[a_m] \simeq Y_\mu \phi_\mu(\mathcal{A}) V_N[\mathcal{Z}].$$

where $Y_\mu = \sum_{\sigma\in S_\mu} \sigma$ is the symmetrization over the Young subgroup $S_\mu = S_{\mu_1} \times \cdots \times S_{\mu_m}$, and $\phi_\mu(\mathcal{A})$ is the evaluation map, sending the first μ_1 variables in $\mathbb{C}[z_1, ..., z_N]$ to a_1, the next μ_2 variables to a_2, etc, where a_i are distinct complex numbers. We denote the projection from $V_N[\mathcal{Z}]$ to $V_\mu[\mathcal{A}]$ by Π_μ.

Two facts simplify the case of symmetric power representations. First, the highest weight vector of $V_{(m)}$ is the image, under the projection from $\pi^{\otimes m}$, of the product of highest weight vectors of π. In addition, both $V_{(m)}$ and π are generated by the action of the abelian subalgebra $\mathfrak{a}$ on the highest weight vector, where $\mathfrak{a} = \langle f_{\alpha_{1j}},\ 1 \leq j < n\rangle$, with $\alpha_{1j} = \sum_{k=1}^{j} \alpha_k$ (we use the notation f_α to denote an element of the nilpotent subalgebra $\mathfrak{n}_-$). It follows that $W_\mu[\mathcal{A}] = \Pi_\mu W_N[\mathcal{Z}]$, and $\mathcal{F}_\mu \subset \Pi_\mu \mathcal{F}_{(1^N)}$ (coset representatives of $\mathcal{F}_\mu$ are a subset of the projection of coset representatives of $\mathcal{F}_{!(1^N)}$).

We need one simple result[4]:

Lemma 4.1.

$$Y_\mu \phi_\mu(\mathcal{A}) V_N^{S_N} = \phi_\mu(\mathcal{A}) V_N^{S_N}.$$

This is easy to see by considering the action on any properly symmetrized vector in $V_N^{S_N}$. We then have the series of equalities

$$\begin{aligned}
\mathcal{F}_\mu &:= \text{gr } W_\mu[\mathcal{A}] \\
&\simeq \text{gr } Y_\mu \phi_\mu(\mathcal{A}) W_N[\mathcal{Z}] \\
&= \text{gr } \phi_\mu(\mathcal{A}) V_N[\mathcal{Z}]^{S_N} \\
&\simeq \text{gr } \left(V_N[\mathcal{Z}]^{S_N} / \ker \phi_\mu(\mathcal{A})\right) \\
&= \text{gr } \left(V_N[\mathcal{Z}]^{S_N} / (\pi^{\otimes N} \otimes J_\mu)^{S_N}\right) \\
&= \text{gr } \left(\pi^{\otimes N} \otimes A_\mu\right)^{S_N} \\
&= (\pi^{\otimes N} \otimes R_\mu)^{S_N}.
\end{aligned}$$

The ideal $J_\mu \subset \mathbb{C}[z_1, ..., z_N]$ consists of functions which have the property $\phi_\mu(\mathcal{A})(\sigma(f(z_1, ..., z_N))) = 0$ for all $\sigma \in S_N$. by definition, $A_\mu = \mathbb{C}[z_1, ..., z_N]/J_\mu$. It is a theorem that[3] $R_\mu = \text{gr } A_\mu$. The isomorphism between the second and third lines uses Theorem 3.1 and the fact that the image of symmetric polynomials does not contribute to the associated graded space.

Again using the decomposition (3) of R_μ and the Schur-Weyl duality (5), we have

$$\mathcal{F}_\mu \simeq \bigoplus_{\lambda : |\lambda| = |\mu|, \ell(\lambda) \le n} V_\lambda \otimes \overline{\mathcal{M}}_{\lambda,\mu},$$

and we have the identification of $\mathcal{M}_{\lambda,\mu}$ with the graded multiplicity space of W_λ in R_μ.

Acknowledgments

I thank the organizers of the QTS3 Symposium for their kind invitation to present these results.

References

1. B. Feigin and S. Loktev, *Differential topology, infinite-dimensional Lie algebras, and applications*, *Amer. Math. Soc. Transl. Ser. 2* **194**, 61 (1999).
2. A. Schilling and O. Warnaar, *Comm. Math. Phys.***202**, 359 (1999).
3. A. M. Garcia and C. Procesi, *Adv. Math.* **94** 82 (1992).
4. R. Kedem, preprint (to appear in IMRN, 2003).
5. I. G. Macdonald, *Symmetric functions and Hall polynomials*, Oxford University Press, New York (1995).

STRUCTURE OF BI-IDEALS VIA DIFFERENTIAL CALCULI

V.K. KHARCHENKO

1-ro de mayo s/n, FES-C UNAM, Campo 1, Cuautitlán Izcalli, Edo. de México, MEXICO, vlad@servidor.unam.mx

In contrary with the classical case of universal enveloping algebras a bi-ideal of a primitively generated Hopf algebra is not necessary generated by its skew-primitive elements. Nevertheless under a weak homogeneity restriction every bi-ideal is generated by constants with respect to a suitable coordinate differential calculus.

According to the Heyneman-Radford theorem (see [3], or [9], or the very skew primitive version [5]) every nonzero bi-ideal of a Hopf algebra generated by skew primitive elements always have nonzero skew primitive elements. Nevertheless, in contrary with the classical case (of universal enveloping algebras), all skew primitive elements of an arbitrary bi-ideal may not generate this bi-ideal. For this reason the following notion of a combinatorial rank naturally arises.

Let H be Hopf algebra generated by skew-primitive semi-invariants a_i, $1 \leq i \leq n$ over the group G of grouplikes, that is

$$\Delta(a_i) = a_i \otimes 1 + g_i \otimes a_i, \quad g^{-1}ag = \chi^x(g)a, \ \Delta(g) = g \otimes g, \quad g \in G. \quad (1)$$

Recall that a variable x is called a *quantum variable* if an element g_x of G and a character $\chi^x \in G^*$ are associated with it. The parameters g_x and χ^x associated with a quantum variable say that an element a in a Hopf algebra H may be considered as a value of this quantum variable only if a is a skew primitive semi-invariant with the same parameters, that is

$$\Delta(a) = a \otimes 1 + g_x \otimes a, \quad g^{-1}ag = \chi^x(g)a, \quad g \in G. \quad (2)$$

Let us associate a quantum variable x_i with the parameters (χ^{a_i}, g_i) to a_i. For each word u in $x_1, \ldots, x_n$ we denote by g_u an element of G that appears from u by replacing of all x_i with g_{x_i}. In the same way we denote by χ^u a character that appears from u by replacing of all x_i with χ^{x_i}. Thus on the free algebra $\mathbf{k}\langle x_1, \ldots, x_n\rangle$ a grading by the group $G \times G^*$ is defined. For each pair of homogeneous elements u, v we fix the denotations $p_{uv} = \chi^u(g_v) = p(u,v)$.

The group G acts on $\mathbf{k}\langle x_1, \dots x_n\rangle$ by $g^{-1}ug = \chi^u(g)u$, where u is an arbitrary monomial in $x_1, \dots x_n$. The skew group algebra $G\langle X\rangle = \mathbf{k}\langle x_1, \dots, x_n\rangle * G$ with a natural Hopf algebra structure

$$\Delta(x_i) = x_i \otimes 1 + g_{x_i} \otimes x_i, \quad 1 \leq i \leq 1, \ \ \Delta(g) = g \otimes g, \ g \in G$$

is called a *free enveloping algebra* of X.

The map $x_i \to a_i$ has an extension to a homomorphism of Hopf algebras $\varphi : G\langle X\rangle \to H$. Denote by I the kernel of this homomorphism. If $I \neq 0$ then by the Heyneman–Radford theorem I has a non-zero skew primitive element. Let $C_1(I)$ be an ideal generated by all skew primitive elements of I. Clearly $C_1(I)$ is a Hopf ideal as well. Now consider the Hopf ideal $I/C_1(I)$ of the quotient Hopf algebra $G\langle X\rangle/C_1(I)$. This ideal also has non-zero skew primitive elements (provided $C_1(I) \neq I$). Denote by $C_2(I)/C_1(I)$ the ideal generated by all skew primitive elements of $I/C_1(I)$, with $C_2(I)$ being its preimage with respect to the projection $G\langle X\rangle \to G\langle X\rangle/C_1(I)$. Continuing the process we will find a strictly increasing, finite or infinite, chain of Hopf ideals of $G\langle X\rangle$:

$$0 = C_0(I) \subset C_1(I) \subset C_2(I) \subset \dots \subset C_m(I) \subset \dots \ , \ \bigcup_\alpha C_\alpha(I) = I. \quad (3)$$

Definition. The length $\kappa = \kappa(H)$ of (3) is called a *combinatorial rank* of H.

Some examples of character Hopf algebras of the combinatorial rank 2 have been constructed in [5] . The next result shows in particular that there exists character Hopf algebras of an arbitrary finite rank.

THEOREM 1. (V.K. Kharchenko, J. Cervantes Ojeda, A. Andrade Alvarez). *The combinatorial rank of the Frobenius-Lusztig kernel of type* A_n *equals* $\lceil 1 + \log_2 n \rceil$.

Recall that the *Frobenius-Lusztig kernel* is the finite dimensional Hopf algebra defined by G. Lusztig [7] , [8], see also [10] .

This theorem shows that bi-ideals in (character) Hopf algebras are far from being generated by theirs skew-primitives. For this reason it is interesting to find a more general type of elements sufficient for generation of all bi-ideals. This is possible to do in line with noncommutative differential calculi defined by Woronowicz [12]. Every character Hopf algebra with a weak homogeneity restriction has a noncommutative differential calculus of a special type (a first order differential calculus with right partial derivatives defined by a diagonal braiding, see below). Hence every bi-ideal is a differential ideal with respect to this calculus.

THEOREM 2. ([4]). *Every left differential ideal of an arbitrary finitely generated algebra with right coordinate differential calculus set up by a Yang-Baxter operator (not necessary invertible) is generated by constants (that is by elements with zero partial derivatives.)*

Recall that the general concept of the noncommutative differential calculus appears in the famous paper by S.L. Woronowicz [12]. A *differential calculus* on an associative unitary algebra S is defined by a linear map d from S into a bimodule M such that the Leibniz formula $d(uv) = du{\cdot}v + u{\cdot}dv$ is valid, see [6] . If S has a distinguished set of algebra generators y_i, $i \in I$ such that M is freely generated by the differentials $dy_i, i \in I$ as a right module then the calculus is called *(right) coordinate*, originally a *first order differential calculus with right partial derivatives*, while the generators may be considered as coordinate noncommutative functions. This type of calculi has been effectively used by R.H. Fox [2] for the investigation of various problems in topology and combinatorial group theory.

On S we may define right *partial derivatives* according to the formula

$$du = \sum_i dy_i \cdot \frac{\partial u}{\partial y_i}. \tag{4}$$

The bimodule structure on M defines commutation rules

$$v \cdot dy_k = \sum_s dy_s \cdot A_k^s(v). \tag{5}$$

The Leibniz formula is equivalent to the following relations for the partial derivatives

$$\frac{\partial(uv)}{\partial y_k} = \frac{\partial u}{\partial y_k} v + \sum_j A_j^k(u) \frac{\partial v}{\partial y_j}. \tag{6}$$

Conversely, if commutation rules (5) are given, then there exists at most one (right) coordinate differential calculus with these commutation rules (see details in [6], Section 12.3 or in these very notations in [1], Section 1 and Theorem 2.1). In Theorem 2 we suppose that $A_k^s(y_i) = \sum_j \alpha_{ik}^{sj} y_j$ and the operator $\alpha : y_i \otimes y_k \to \sum_{j,s} y_j \otimes y_s$ satisfies the braid relation. This type of calculus appears on a skew-primitively generated Hopf algebra H in the following way.

Denote by G the group of all group like elements of a Hopf algebra H. For $g \in G$ we put $L_g = \{h \in H | \Delta(h) = h \otimes 1 + g \otimes h\}, L = \sum_{g \in G} \oplus L_g$.

Suppose that H is generated over $\mathbf{k}[G]$ as an algebra by a Yetter-Drinfeld submodule $D \subseteq L$, that is $D = \sum_{g\in G} \oplus D_g$, $D_g \subseteq L_g$, $D^g \subseteq D$, $g \in G$. In this case we have a standard filtration

$$H = \bigcup_{k=0}^{\infty} H_k, \quad H_k = \mathbf{k}[G] + \mathbf{k}[G]D + \mathbf{k}[G]D^2 + \cdots + \mathbf{k}[G]D^k. \tag{7}$$

Assume that this filtration is homogeneous in the first two members (the weak homogeneity restriction), that is

$$H = \mathbf{k}[G] \oplus \mathbf{k}[G]D \oplus H^{(2)}, \quad H^{(2)} = \sum_{k=2}^{\infty} \mathbf{k}[G]D^k. \tag{8}$$

Denote by S an unitary subalgebra of H generated by D. The group G acts on S by conjugations. The condition (8) implies, in particular, that there exists a Hopf algebra projection $p : H \to \mathbf{k}[G]$, $p(D) = 0$. Therefore one may apply Radford theorem [11]. In the context this theorem says that S itself is a Yetter-Drinfeld module, $S = \sum_{g\in G} \oplus S_g$, $S_g^h = S_{h^{-1}gh}$, and S has a $\mathbf{k}[G]$ braided Hopf algebra structure with

$$\Delta^b(u) = \sum_{(u)} u_b^{(1)} \underline{\otimes} u_b^{(2)}, \text{ where } u_b^{(1)} = u^{(1)}\text{gr}(u^{(2)})^{-1}, \; u_b^{(2)} = u^{(2)}, \tag{9}$$

here we use the Sweedler notation $\Delta(u) = \sum_{(u)} u^{(1)} \otimes u^{(2)}$, and by definition $\text{gr}(u)$ or g_u is an element from G, such that $u \in S_{\text{gr}(u)}$. We denote by $\underline{\otimes}$ the tensor product of algebras in the braided category with the braiding $c(u \otimes v) = v^h \otimes u$, where $h = \text{gr}(u)^{-1}$, that is

$$(1\underline{\otimes}u)(v\underline{\otimes}1) = v^h\underline{\otimes}u, \quad h = \text{gr}(u)^{-1}; \;\; \Delta^b : S \to S\underline{\otimes}S. \tag{10}$$

The homogeneity restriction (8) for S reduces to

$$S = \mathbf{k} \oplus D \oplus S^{(2)}, \qquad S^{(2)} = \sum_{k=2}^{\infty} D^k. \tag{11}$$

Let us fix a set of linearly independent generators $y_1, y_2, \dots y_k, \dots$; $y_k \in D_{g_k}$, $g_k \in G$, $k \in I$.

Since D is invariant under conjugations by G, we have $g_k^{-1} y_s g_k = \sum_i \gamma_{sk}^i y_i$. The following commutation rules define the differential calculus to which we may apply theorem 2:

$$y_s \cdot dy_k = dy_k \cdot \sum_i \gamma_{sk}^i y_i. \tag{12}$$

Acknowledgments

The work supported by SNI México, exp. 18740, and CONACyT grant 44512-F.

References

1. A. Borowiec, V.K. Kharchenko, *Siberian Adv. Math.* **5(2)**, 10(1995).
2. R.H. Fox, *Annals of Mathematics* **57**,517 (1953); II, ibid. **59**, 196 (1954); III, ibid. **64**, 404 (1956); IV, ibid. **68**, 81 (1958); V, ibid. **71**, 408 (1960).
3. R.G. Heyneman, D.E. Radford, *Journal of Algebra* **29**, 215 (1974).
4. V.K. Kharchenko, *Journal of Algebra* **267**, 96 (2003).
5. V.K. Kharchenko, *Journal of Algebra* **238**, 534 (2001).
6. A. Klimyk, K. Schmüdgen, Springer-Verlag Berlin Heidelberg, 1997.
7. G. Lusztig, *Geom. Dedicata* **35**, 89 (1990).
8. G. Lusztig, *J. of Amer. math. Soc.* **3**, 257 (1990).
9. S. Montgomery, CBMS, 82, AMS, Providence, 1993.
10. E. Müller, *Journal of Algebra* **206**, 624 (1998).
11. D.E. Radford, *Journal of Algebra* **92**, 322 (1985).
12. S.L. Woronowicz, *Comm. Math. Phys.* **122**, 125 (1989).

VIRASORO-TYPE ALGEBRAS ASSOCIATED WITH HIGHER-RANK APERIODIC POINT SETS*

VOLODYMYR MAZORCHUK[1] and REIDUN TWAROCK[2]

[1] *Department of Mathematics, Uppsala University, SE-75106, Uppsala, Sweden*
[2] *Centre for Mathematical Sciences, City University, Northampton Square, London EC1V 0HB, England, r.twarock@city.ac.uk*

A method for the construction of infinite-dimensional Lie algebras of Virasoro-type is discussed, which uses aperiodic point sets as basic building blocks. The corresponding algebras have generators in a one-to-one correspondence with aperiodic point sets that are obtained via a projection formalism from higher dimensional lattices. They share structural similarities with the Virasoro algebra by construction, but exhibit different properties.

1. Introduction

The Virasoro algebra plays a crucial role in many areas of mathematical physics, and algebraic techniques for a deformation of this algebra are of interest, because they lead to perturbations of these theories. We present here a construction method for Virasoro-type algebras which is based on aperiodic point sets that can be obtained via a projection from a higher dimensional periodic lattice[1]. These aperiodic structures play an important role in the study of quasicrystals, that is alloys with noncrystallographic symmetries and long-range order[2], but the algebras presented here are not constructed for applications in this area but rather for potential applications in the theory of integrable systems as we discuss in the concluding remarks.

The main idea of the construction is to substitute the index set of the Virasoro algebra by a suitable aperiodic point set and to make modifications in the structure constants to ensure that the algebras remain Lie algebras. The construction leads to a family of infinite dimensional Lie algebras with properties that depend crucially on geometric properties of the associated aperiodic structures. The case of one-dimensional aperiodic point sets has

*Talk given by R.T.

been studied in a series of papers [3,4,5,6], and most recently, generalisations to two-dimensional aperiodic point sets have been achieved[7]. In these references, also the structure of highest weight modules and the irreducibility of Verma modules have been studied. Due to the fact that one-dimensional point sets are obtained from two-dimensional lattices by projection, each point is specified by two coordinates. Therefore the corresponding algebras also have structural resemblances with higher-rank Virasoro algebras[8,9,10].

It is the purpose of this contribution to review the construction principle with focus on recent developments and to direct the reader to the corresponding literature for further details. In particular, section 2 introduces the projection method for aperiodic point sets based on two explicit examples, and section 3 reviews the construction of the corresponding Virasoro-type algebras with emphasis on the most recent developments. In the concluding remarks, possible applications in the theory of integrable systems are pointed out.

2. Aperiodic point sets via the projection method

The construction principle is applicable to all aperiodic point sets that are model sets[1]. Here focus will be placed on two special cases, which correspond to a family of one- and a family of two-dimensional aperiodic point sets, respectively, because they will be used for the discussion of the Virasoro-type algebras in section 3.

2.1. *A one-dimensional example*

Let $\tau = \frac{1}{2}(1+\sqrt{5})$, and let $\mathbb{Z}[\tau] = \{a + \tau b | a, b \in \mathbb{Z}\}$ denote the ring of integers in the algebraic extension $\mathbb{Q}[\sqrt{5}]$ of the rational numbers by $\sqrt{5}$. Consider the Galois automorphism $' : \mathbb{Q}[\sqrt{5}] \to \mathbb{Q}[\sqrt{5}]$ defined via $(a+\sqrt{5}b)' = a - \sqrt{5}b$. Restricted to $\mathbb{Z}[\tau]$, this automorphism maps $a+\tau b$ to $a + \tau' b$ and links the two solutions of the equation $x^2 = x + 1$, namely the first solution τ and the second solution $\tau' = \frac{1}{2}(1-\sqrt{5})$. Examples of one-dimensional aperiodic point sets are then defined as follows.

Let Ω be a bounded interval in $\mathbb{R}$. Then

$$\Sigma(\Omega) := \{x \in \mathbb{Z}[\tau] | x' \in \Omega\} \tag{1}$$

is a one-dimensional aperiodic point set.

The interval Ω is called acceptance window because it controls how many points are admitted in the point set.

2.2. *A two-dimensional example*

In order to generalise this formalism to two dimensions, we consider a projection from the root lattice of A_4 onto two hyperplanes, each containing a copy of the root system of type H_2[11].

Let $\xi := \exp(\frac{2\pi i}{5})$. Then the root system $\Delta_2 \subset \mathbb{C}$ of type H_2 contains ten roots, which are given in terms of the simple roots $\alpha_1 := \xi^0$ and $\alpha_2 := \xi^2$ as follows:

$$\Delta_2 = \{\pm\alpha_1, \pm\alpha_2, \pm(\alpha_1 + \tau\alpha_2), \pm(\tau\alpha_1 + \alpha_2), \pm(\tau\alpha_1 + \tau\alpha_2)\}\,.$$

Denote by $\mathbb{Z}[\tau]\Delta_2$ the $\mathbb{Z}[\tau]$–lattice with basis $\{\alpha_1, \alpha_2\}$. The assignment $\xi^* = \xi^2$ uniquely extends to a $\mathbb{Z}[\tau]$–semilinear bijection * on $\mathbb{Z}[\tau]\Delta_2$. This map is usually called the *star map*[11] . Moreover, according to this reference, this map uniquely extends to a $\mathbb{Z}[\tau]$–semilinear bijection on $\mathbb{Q}[\tau]\Delta_2$, which we will also denote by *.

It is immediate that the star-map defined above acts like the original Galois automorphism $' : \mathbb{Q}[\sqrt{5}] \to \mathbb{Q}[\sqrt{5}]$ when restricted to $\mathbb{Z}[\tau]$. Indeed, $\tau^* = (-\xi^2 - \xi^3)^* = (-\xi^4 - \xi^6) = \tau'$. Obviously, the star map is a permutation on $\mathbb{Z}[\tau]\Delta_2$ (or $\mathbb{Q}[\tau]\Delta_2$) of order 4.

Let $\Omega \subset \mathbb{C}$ be a bounded set. Then

$$\Sigma(\Omega) = \{x \in \mathbb{Z}[\tau]\Delta_2 | x^* \in \Omega\} \tag{2}$$

is a planar aperiodic point set.

For applications to the algebraic setting, we consider in particular the case $\Omega \equiv \Omega_P^T := T\xi^0 + \Omega_P$, where Ω_P denotes the regular unital pentagon and Ω_P^T is a translation of Ω_P by $T\xi^0$, $T > 0$.

3. Virasoro-type algebras

In this section we introduce Lie algebras with generators in a one-to-one correspondence with the aperiodic point sets in (1) and (2).

3.1. *The one-dimensional case*

Witt-type algebras related to the point sets in (1) are given by the following Lemma[3]:

Lemma 3.1. *Let $\Sigma(\Omega)$ be a one-dimensional quasicrystal with acceptance window $[a, b]$, $(a, b]$, $[a, b)$, or (a, b), where $a \neq b \in \mathbb{R}$ and $0 \leq ab < \infty$. Then the algebra $Q(\Omega)$, defined as a linear span of*

$$B(Q(\Omega)) = \{L_n \mid n \in \Sigma(\Omega)\} \tag{3}$$

with commutation relations

$$[L_n, L_m] = \begin{array}{cl} (m-n)L_{n+m} & \textit{if}\ \ n+m \in \Sigma(\Omega) \\ 0 & \textit{if}\ \ n+m \notin \Sigma(\Omega) \end{array}, \tag{4}$$

is a Lie algebra.

Note that the Jacobi identity holds only provided that $ab \geq 0$ which implies a restriction on the aperiodic point sets that may be used as building blocks in the construction. Observe furthermore that in contrast to the Witt algebra (which is simple), many commutators in $Q(\Omega)$ vanish because $m, n \in \Sigma(\Omega)$ does not imply $m+n \in \Sigma(\Omega)$ in general. As a consequence, the algebras are locally finite, that is the closure of any finite set of generators from $B(Q(\Omega))$ under the Lie operation gives a finite dimensional subalgebra of $Q(\Omega)$.

Structure and existence of a central extension for these algebras depend on the choice of Ω and on the choice of the structure constants. It has been shown[5] that there exists a unique central extension for the algebras in Lemma 3.1 for $\Omega = [0, 1]$ if the structure constants are given in terms of the second component of points $m = m_1 + \tau m_2 \in \Sigma(\Omega)$, that is if one chooses the structure constants $(m_2 - n_2)$ instead of $(m - n)$. The corresponding algebra has been called the *Aperiodic Virasoro algebra* $AV(\Omega)$. Highest weight representations for this algebra have been discussed in this reference and a conjecture for an analog to the Kac-determinant formula[12] has been suggested. The latter has later been proved[6].

3.2. *The two-dimensional case*

In this section we discuss a generalisation of the results in section 3.1 to the aperiodic point sets in (2)[7]. In order to introduce the algebras, we need the following terminology.

Definition 3.1. An acceptance window, Ω, is called *admissible* if for each triple n, m, $k \in \Sigma(\Omega)$ one has the condition

$$n^* + m^* + k^* \in \Omega \text{ and } n^* + m^* \in \Omega \Rightarrow m^* + k^* \in \Omega \text{ and } n^* + k^* \in \Omega\,. \tag{5}$$

The concept of admissibility is necessary to ensure that the aperiodic point sets are such that the corresponding algebras fulfill the Jacobi identity.

In particular, one obtains the following planar generalisation of Lemma 3.1[7].

Lemma 3.2. *Let* $\mathbb{F}$ *be any number field such that* $\mathbb{F} \supset \mathbb{Q}[\tau]$ *and let* T *be such that* Ω_P^T *is admissible. Let furthermore* $\varphi : \Sigma(\Omega_P^T) \to \mathbb{F}$ *be a* $\mathbb{Z}$*-linear* *map. Then for each such T the $\mathbb{F}$–span of $\{L_n | n \in \Sigma(\Omega_P^T)\}$ with the Lie bracket defined by*

$$\begin{aligned}[L_n, L_m] &= \begin{cases} \varphi(m-n)L_{n+m} & \text{if } n+m \in \Sigma(\Omega_P^T) \\ 0 & \text{if } n+m \notin \Sigma(\Omega_P^T) \end{cases} \\ &= \chi_{\Omega_P^T}(n^* + m^*)\varphi(m-n)L_{n+m} \,. \end{aligned} \tag{6}$$

is a Lie algebra.

As in the one-dimensional case, the structural properties of these algebras depend on the geometry of the acceptance window Ω_P^T[7]. Moreover, as has been shown in this reference, these Lie algebras allow for a central extension. For a natural triangular decomposition the corresponding Verma modules have been studied.

In particular, if we denote by $\mathcal{V}(\Omega_P^T)^+$ the subalgebras generated by elements L_n, where $\Sigma(\Omega_P^T) \ni n = n_1\xi^0 + n_2\xi^2$ and $n_2' < 0$, and by $\mathcal{V}(\Omega_P^T)^-$ the subalgebras generated by elements L_n, where $\Sigma(\Omega_P^T) \ni n = n_1\xi^0 + n_2\xi^2$ and $n_2' > 0$, and call $\mathfrak{h}_T$ the subalgebra, generated by L_n, where $\Sigma(\Omega_P^T) \ni n = n_1\xi^0$, and by all central terms $\hat{c}_{v_\lambda}$ with $v_\lambda \in \mathbb{Z}[\tau] \cap [2T - \tau, T+1]$, then there is a triangular decomposition given by[7]

$$\mathcal{V}(\Omega_P^T) = \mathcal{V}(\Omega_P^T)^- \oplus \mathfrak{h}_T \oplus \mathcal{V}(\Omega_P^T)^+ \,.$$

Let furthermore $\lambda \in \mathfrak{h}_T^*$ and consider $\mathbb{F}_\lambda = \mathbb{F}$ as an $\mathfrak{h}_T$-module with

$$\begin{aligned} hz &= \lambda(h)z \quad \forall h \in \mathfrak{h}_T \,, z \in \mathbb{F} \,, \\ \mathcal{V}(\Omega_P^T)^+\mathbb{F}_\lambda &= 0 \end{aligned} \tag{7}$$

Then $\mathbb{F}_\lambda$ can be viewed as a $\mathfrak{b} = \mathfrak{h}_T \oplus \mathcal{V}(\Omega_P^T)^+$-module. Consider the induced module $M(\lambda) = U(\mathcal{V}(\Omega_P^T)) \otimes_{U(\mathfrak{b})} \mathbb{F}_\lambda$ and denote the unique simple quotient of $M(\lambda)$ by $L(\lambda)$. Then one has[7]:

Theorem 3.1.

(1) $M(\lambda)$ is always reducible.
(2) The module $L(\lambda)$ is one dimensional if and only if λ is zero on $\mathfrak{h}_T^{ess} := \langle \hat{c}_{v_\alpha}, L_{v_\alpha'} | v_\alpha \in \mathbb{Z}[\tau] \cap [2T - \tau, T+1] \rangle$. Otherwise, $L(\lambda)$ is infinite-dimensional.

4. Concluding remarks

The Virasoro algebra plays an important role in the framework of Calogero-Sutherland models[13,14,15]. In particular, the structure of its highest weight representations and especially its singular vectors allows one to construct explicit solutions[16,17]. It has been shown that a similar approach, based on the Aperiodic Virasoro algebra[18], that is for the Virasoro-type algebras related to one-dimensional aperiodic point sets, is possible. We expect that, along the same lines, the algebras related to two-dimensional aperiodic point sets that have been pointed out here should have applications in the framework of Calogero-Sutherland models.

Acknowledgments

We would like to thank the London Mathematical Society and The Royal Swedish Academy of Science for the financial support of the visit of the first author to City University, where this manuscript has been written. Furthermore, financial support of the second author in the framework of an EPSRC Advanced Research Fellowship is gratefully acknowledged.

References

1. R. V. Moody, Model sets: A survey, in: *From Quasicrystals to More Complex Systems*, eds F Axel, Dénoyer F, Gazeau J.-P. (Springer) 145 (2000).
2. D. Shechtman, I. Blech, D. Gratias and J. Cahn, *Phys. Rev. Lett.* **53**, 1951 (1984).
3. J. Patera, E. Pelantova and R. Twarock, *Phys. Lett. A* **246**, 209 (1998).
4. J. Patera and R. Twarock, *Phys. Rep.* **315**, 241 (1999).
5. R. Twarock, *J. Math. Phys.* **41**, 5088 (1999).
6. V. Mazorchuk V, *J. Math. Phys.* **43**, 2791 (2002).
7. V. Mazorchuk and R. Twarock, *J. Phys. A* **36**, 4363 (2003).
8. V. Mazorchuk, *Compositio Math.* **115**, 21 (1999).
9. J. Patera and H. Zassenhaus, *Comm. Math. Phys.* **136**, 1 (1991).
10. Y. Su, *Lett. Math. Phys.* **53**, 263 (2000).
11. L. Chen, R. V. Moody and J. Patera, Non-crystallographic root systems, in: *Quasicrystals and discrete geometry* (Toronto, ON, 1995), Fields Inst. Monogr., 10, Amer. Math. Soc., Providence, RI, 135 (1998).
12. V. Kac and D. Kazhdan, *Adv. in Math.* **34**, 97 (1979).
13. F. Calogero *J. Math. Phys.* **10**, 2197 (1969).
14. B. Sutherland *J. Math. Phys.* **12**, 246 (1971).
15. B. Sutherland *J. Math. Phys.* **12**, 251 (1971).
16. H. Awata, Y. Matsuo, S. Odake and J. Shiraishi *Phys. Lett.* **B 347**, 49 (1995).
17. H. Awata, Y. Matsuo, S. Odake and J. Shiraishi *Nucl. Phys.* **B 449**, 347 (1995).
18. R. Twarock, *Int. J. Mod. Phys. A* **15**, 4179 (2000).

REPRESENTATIONS OF THE POINCARÉ GROUP FROM POSITIVE ENERGY REPRESENTATIONS OF SO(2,3)

P. MOYLAN

Physics Department, The Pennsylvania State University,
Abington College, Abington, PA 19001, USA
E-mail: pjm11@psu.edu

We describe representations of the simply connected covering group of the Poincaré group, $\widetilde{\mathfrak{P}}$, which are associated with certain positive energy representations of $SO_0(2,3)^\sim$, the simply connected cover of $SO_0(2,3)$. The translation generators of these representations of $\widetilde{\mathfrak{P}}$ can be viewed as solutions of certain algebraic equations with coefficients from a commutative algebraic extension of the skew field of $\widetilde{\mathfrak{P}}$. These representations of $\widetilde{\mathfrak{P}}$ depend upon a parameter λ that is essentially the reciprocal of the radius of anti-deSitter space, and they go over into the Segal-Inönü-Wigner contractions of the corresponding representations of $SO_0(2,3)^\sim$ as $\lambda \to 0$. Explicit results are given for the Di and Rac representations and the representations of $SO_0(2,3)^\sim$ which extend to massless, unitary irreducible representations of $SU(2,2)$, the four-fold cover of the conformal group of Minkowski space-time.

1. Introduction

There are three well-known space-times of general relativity which carry local causal structures identical to that of Minkowski space $\mathfrak{M}_0$, and into which $\mathfrak{M}_0$ may be conformally embedded, namely, the Einstein universe and the two spaces of constant curvature $\mathfrak{M}_\lambda = \mathfrak{G}_\lambda/\mathfrak{L}$ where $\mathfrak{L}$ is $SO_0(1,3)$ and $\mathfrak{G}_\lambda$ is isomorphic to $SO_0(2,3)$ for $\lambda \in \mathbb{R}$ and $\mathfrak{G}_\lambda$ is isomorphic to $SO_0(1,4)$ for $\lambda \in i\,\mathbb{R}$. Here we concentrate on the second one. We consider positive energy representations $SO_0(2,3)^\sim$ and describe a relationship between these representations and certain representations of $\widetilde{\mathfrak{P}}$. The representations of $\widetilde{\mathfrak{P}}$ are constructed out of skew symmetric solutions $\mathbf{P}_\mu$ to eqns. (6), solutions which are explicitly described in Theorem 3.2. We have also obtained results identical to those in Theorems 3.2 and 3.3 for q deformations $U_q(so(p,q))$ with small p and q values, and applying our results to the representations of $U_q(so(p,q))$ (for q both generic and a root of unity), we get representations of classical Lie algebras which depend upon the deformation parameter q.[1] It seems that our results should be related to those of Curtright and Zachos[3] in the simplest case of type A_1 i.e. that associated with $U_q(so(2,1))$. We have also obtained some analogous results in the classical case for higher dimensional $SO_0(p,q)$ and associated Poincaré groups.[2]

2. $SO_0(2,3)$, the Poincaré group and their Lie algebras

Let $\beta_0 = \mathrm{diag}(1, -1, -1, -1, 1)$, where the right hand side of this equation denotes a diagonal matrix with diagonal entries as shown inside the parentheses. $\mathfrak{G} = SO_o(2,3)$ is the component connected to identity of the group

$$SO(2,3) = \{g \in SL(5,\mathbb{R}) \mid g\, \beta_0\, g^\dagger = \beta_0\} . \tag{1}$$

($\dagger$ denotes transpose of a matrix.) Denote by $\mathcal{G}$ the Lie algebra of $\mathfrak{G}$. A realization of $\mathcal{G}$ is provided by the set of all matrices (a_{ij}) $(0 \le i,j \le 4)$ such that $a_{ii} = 0$ $(0 \le i \le 4)$, $a_{ij} = -a_{ji}$ $(1 \le i \le j \le 3)$, $a_{0,j} = a_{j,0}$ $(1 \le j \le 3)$, $a_{4,j} = a_{j,4}$ $(1 \le j \le 3)$ and $a_{o4} = -a_{40}$. Let E_{ij} be the matrix such that the (i,j) component is equal to 1 and the other components are all equal to 0. Let $\mathbf{L}_{ij} = E_{ij} - E_{ji}$ $(0 \le i \le j \le 4)$, $\mathbf{L}_{0i} = E_{i,o} + E_{o,i}$ $(1 \le i \le 3)$, $\mathbf{L}_{04} = E_{i,4} + E_{4,i}$ $(1 \le i \le 3)$ and $\mathbf{L}_{04} = E_{04} - E_{40}$. The $\mathbf{L}_{ab}$ $(a,b = 0, 1, 2, 3, 4)$, viewed abstractly, are a basis for $\mathcal{G}$.

A Cartan-Weyl basis for $\mathcal{G}_{\mathbb{C}}$ is given by

$$-i\mathbf{H}_1 = 2\,\mathbf{L}_{21}\, , \; \mathbf{H}_2 = (\mathbf{L}_{40} - \mathbf{L}_{21})\, , \; -i\mathbf{X}_1^{\pm} = \mathbf{L}_{31} \mp i\mathbf{L}_{23}\, , \tag{2}$$

$$-i\,\mathbf{X}_2^{\pm} = \pm\,\frac{1}{2}\,(\mathbf{L}_{41} + \mathbf{L}_{20}) - \frac{i}{2}(\mathbf{L}_{10} - \mathbf{L}_{42})\, . \tag{3}$$

$\mathbf{H}_1$ and $\mathbf{H}_2$ are a basis for a Cartan subalgebra $\mathcal{H}_{\mathbb{C}}$ of $\mathcal{G}_{\mathbb{C}}$. The commutation relations for the basis vectors of the simple roots are

$$[\mathbf{H}_j\, , \mathbf{H}_k] = 0\, , \; [\mathbf{H}_j\, , \mathbf{X}_k^{\pm}] = \pm\, a_{jk}\mathbf{X}_k^{\pm}\, , \; [\mathbf{X}_j^{+}\, , \mathbf{X}_k^{-}] = \delta_{jk}\,\mathbf{H}_k \tag{4}$$

(No summation is intended in these equations.) $(a_{jk}) = 2\,\frac{(\alpha_j\, , \,\alpha_k)}{(\alpha_j\, , \,\alpha_j)}$ is the Cartan matrix of $so(5,\mathbb{C})$, and α_1, α_2 are the simple roots with products: $(\alpha_1, \alpha_1) = 2$, $(\alpha_2, \alpha_2) = 4$, $(\alpha_1, \alpha_2) = -2$. The Cartan-Weyl basis vectors corresponding to the non-simple roots are given by:

$$\mathbf{X}_3^{\pm} = \pm\,[\mathbf{X}_1^{\pm}\, , \mathbf{X}_2^{\pm}]\, , \; \mathbf{X}_4^{\pm} = \pm\,\frac{1}{2}\,[\mathbf{X}_1^{\pm}\, , \mathbf{X}_3^{\pm}]\, . \tag{5}$$

Let $\mathfrak{P}$ be the Poincaré group, and denote its Lie algebra by $\mathcal{P}$. A basis for the Poincaré Lie algebra is given by the $\mathbf{L}_{0i}$ and $\mathbf{L}_{ij}$ $(i,\ j\ =\ 1,\ 2,\ 3)$ of $so(1,3)$ together with a commuting Lorentz vector operator $\mathbf{P}_\mu$, which generates the translations.

Denote by $\mathcal{E}(\mathcal{G})$ and $\mathcal{E}(\mathcal{P})$ the universal enveloping algebras of $\mathcal{G}$ and $\mathcal{P}$, respectively. We introduce the following elements of $\mathcal{E}(\mathcal{G})$: $\mathbf{Q}_2 = \mathbf{L}_{01}^2 + \mathbf{L}_{02}^2 + \mathbf{L}_{03}^2 - \mathbf{L}_{12}^2 - \mathbf{L}_{23}^2 - \mathbf{L}_{31}^2$, $\mathbf{Q}_4 = (\mathbf{L}_{12}\mathbf{L}_{30} + \mathbf{L}_{23}\mathbf{L}_{10} + \mathbf{L}_{31}\mathbf{L}_{20})^2$, $\mathbf{C}_2 = -\mathbf{L}_{04}^2 + \mathbf{L}_{01}^2 + \mathbf{L}_{02}^2 + \mathbf{L}_{03}^2 - \mathbf{L}_{12}^2 - \mathbf{L}_{23}^2 - \mathbf{L}_{31}^2 + \mathbf{L}_{14}^2 + \mathbf{L}_{24}^2 + \mathbf{L}_{34}^2$ and $\mathbf{C}_4 = -\left[\frac{1}{4}(\lambda - \rho)\right]^2 - (\mathbf{L}_{12}\,\mathbf{L}_{30} + \mathbf{L}_{23}\,\mathbf{L}_{10} + \mathbf{L}_{31}\,\mathbf{L}_{20})^2 + \sum_{ijk\ell m\,=\,1}^{3} (\epsilon_{ijk}\{\frac{1}{2}\mathbf{L}_{04}\mathbf{L}_{jk} - \mathbf{L}_{0j}\mathbf{L}_{4k}\})(\epsilon_{i\ell m}\{\frac{1}{2}\mathbf{L}_{04}\mathbf{L}_{\ell m} - \mathbf{L}_{0\ell}\mathbf{L}_{4m}\})$ with $\frac{1}{4}(\lambda - \rho) = \mathbf{L}_{12}\,\mathbf{L}_{34} + \mathbf{L}_{23}\,\mathbf{L}_{14} + \mathbf{L}_{31}\mathbf{L}_{24}$. The center $\mathbb{Z}(\mathcal{G})$ of $\mathcal{E}(\mathcal{G})$ is generated by $\mathbf{C}_2$ and $\mathbf{C}_4$. The center $\mathbb{Z}(\mathcal{P})$ of $\mathcal{E}(\mathcal{P})$ is generated by the following set of elements: $\mathbf{P}^2 = \sum_{k=0}^{3} \mathbf{P}_k\,\mathbf{P}^k = \mathbf{P}_0^2 - \mathbf{P}_1^2 - \mathbf{P}_2^2 - \mathbf{P}_3^2$, and $\mathbf{W} = \sum_{\mu=0}^{3}\sum_{\nu=0}^{3}\sum_{\rho=0}^{3}(\mathbf{P}_\mu\,\mathbf{P}^\nu\,\mathbf{L}_{\nu\rho}\,\mathbf{L}^{\rho\mu} - \frac{1}{2}\mathbf{P}_\rho\,\mathbf{P}^\rho\,\mathbf{L}_{\mu\nu}\,\mathbf{L}^{\nu\mu})$.

3. Algebraic results

Let $\mathcal{K}(\mathcal{P})$ be the skew field of $\mathcal{P}$ and $\mathcal{K}(\mathcal{G})$ be the skew field of $\mathcal{G}$. We define the following commutative algebraic extensions of $\mathcal{K}(\mathcal{P})$ and $\mathcal{K}(\mathcal{G})$:[4] $\tilde{\mathcal{K}}(\mathcal{P}) = \left\{a + bY \,\middle|\, a,\, b \in \mathcal{K}(\mathcal{P}),\ [Y, a] = 0\,\forall\, a \in \mathcal{K}(\mathcal{P}),\ Y^2 = \mathbf{P}^2\right\}$, and $\tilde{\mathcal{K}}(\mathcal{G}) = \left\{a + b\,\tilde{Y} + c\,\tilde{Y}^2 + d\,\tilde{Y}^4 \,\middle|\, a,\, b,\, c,\, d \in \mathcal{K}(\mathcal{G})\right\}$ where $\tilde{Y}$ commutes with all elements of $\tilde{\mathcal{K}}(\mathcal{G})$ and satisfies the equation $\tilde{Y}^4 + \mathbf{C}_2'\,\tilde{Y}^2 + \mathbf{C}_4' = 0$ with $\mathbf{C}_2' = \left(\mathbf{C}_2 + \frac{5}{2}\,\mathbf{I}\right)$ and $\mathbf{C}_4' = \left(\mathbf{C}_4 + \frac{1}{4}\,\mathbf{C}_2 + \frac{9}{16}\,\mathbf{I}\right)$. ($\mathbf{I}$ is the identity in $\tilde{\mathcal{K}}(\mathcal{G})$.) Now define a mapping τ_λ from $\mathcal{G}$ to $\tilde{\mathcal{K}}(\mathcal{P})$ by

$$\tau_\lambda(\tilde{\mathbf{L}}_{\mu\nu}) = \mathbf{L}_{\mu\nu}\ ,\quad \tau_\lambda(\tilde{\mathbf{L}}_{4\mu}) = \frac{i\,\lambda}{2\,Y}\Big[\mathbf{Q}_2\,,\,\mathbf{P}_\mu\Big] + \mathbf{P}_\mu\ . \qquad (6)$$

The $\lambda^{-1}\tau_\lambda(\tilde{\mathbf{L}}_{4\mu})$ and $\tau_\lambda(\tilde{\mathbf{L}}_{\mu\nu})$ satisfy the commutation relations of the generators of $\mathfrak{G}$. The $\tau_\lambda(\tilde{\mathbf{L}}_{4\mu})$ and $\tau_\lambda(\tilde{\mathbf{L}}_{\mu\nu})$ are a basis for an isomorphic copy $\mathcal{G}_\lambda$ of $\mathcal{G}$, which differs from $\mathcal{G}$ by a scaling factor λ in the $\mathbf{L}_{4\mu}$ directions, and hence generate $\mathfrak{G}_\lambda$. We henceforth consider for simplicity the case $\lambda = 1$ and let $\tau_{\lambda=1}(\tilde{Y}) = Y$, then $\tau = \tau_{\lambda=1}$ can be extended to a homomorphism of $\tilde{\mathcal{K}}(\mathcal{G})$ into $\tilde{\mathcal{K}}(\mathcal{P})$ in an obvious way, which, because of Theorem 3.2, is actually surjective. Denote this extension also by τ. Elements of $\tilde{\mathcal{K}}(\mathcal{G})$ are denoted with a tilde to keep them distinct from elements of $\tilde{\mathcal{K}}(\mathcal{P})$.

Theorem 3.1. *Let $\mathcal{G}$ be the deformation of $\mathcal{P}$ having basis elements $\mathbf{L}_{ij} \in \mathcal{G}$ and $\mathbf{L}_{4\mu} \in \tilde{\mathcal{K}}(\mathcal{P})$ defined by eqns. (6). Then (for $\lambda = 1$)*

the following holds:

$$\mathbf{C}_2 \;=\; -Y^2 - \left[\frac{\mathbf{W}}{Y^2} + \frac{9}{4}\,\mathbf{I}\right] \;,\quad \mathbf{C}_4 \;=\; \left[Y^2 + \frac{1}{4}\right]\frac{\mathbf{W}}{Y^2} \;. \tag{7}$$

Now we view the second set of eqns. in (6) as algebraic equations in $\tilde{\mathcal{K}}(\mathcal{P})$ and solve them for the $\mathbf{P}_\mu$.

Theorem 3.2. *Solutions $\mathbf{P}_\mu$ to eqns. (6) ($\lambda = 1$) are given by:*

$$\mathbf{P}_\mu \;=\; \mathbf{D}^{-1}\,\mathbf{A}_\mu{}^\nu \mathbf{L}_{\nu 4} \tag{8}$$

with $\mathbf{A}_\mu{}^\nu = -\mathbf{C}_4'\,\delta_\mu^\nu + \frac{i}{2}\left[\left(\mathbf{Q}_2 + \frac{1}{4}\right)\delta_\mu^\nu - \frac{3}{2}\mathbf{L}_\mu^\nu - \mathbf{L}_{\mu\rho}\mathbf{L}^{\rho\nu} - \mathbf{Q}_4\epsilon_{\mu\rho\tau}^\nu\mathbf{L}^{\rho\tau}\right]Y - \left[\left(\mathbf{Q}_2 + \frac{1}{4} - \mathbf{C}_2'\right)\delta_\mu^\nu - \mathbf{L}_\mu^\nu - \mathbf{L}_{\mu\rho}\mathbf{L}^{\rho\nu}\right]Y^2 + i\left(\frac{1}{2}\delta_\mu^\nu - \mathbf{L}_\mu^\nu\right)Y^3$, *and* $\mathbf{D} = \mathbf{Q}_4 + \frac{1}{4}\mathbf{Q}_2 - \mathbf{C}_4' + \frac{3}{16}\mathbf{I} + i\left(\mathbf{Q}_2 + \frac{1}{2}\right)Y - \left(\mathbf{Q}_2 - \mathbf{C}_2' - \frac{1}{2}\right)Y^2 + 2iY^3$. *Furthermore Y^2 satisfies the equation*

$$Y^4 + \mathbf{C}_2' Y^2 + \mathbf{C}_4' \;=\; 0 \;. \tag{9}$$

An easy way to prove Theorem 3.2 is to show that $\mathbf{P}_0 \;=\; \mathbf{D}^{-1}\,\mathbf{A}_0{}^\nu \mathbf{L}_{\nu 4}$ satisfies eqns. (6), and then use $\mathbf{P}_i \;=\; \left[\mathbf{L}_{0i}\;,\;\mathbf{P}_0\right]$.

Lemma 3.1. $ker(\tau)\Big|_{\mathcal{E}(\mathcal{G})} \;=\; 0$.

Proof: Let $I = \left\{\tilde{u} \in \mathcal{E}(\mathcal{G}\middle|\tau(\tilde{u}) = 0\right\}$. I is a two sided ideal in $\mathcal{E}(\mathcal{G})$. Suppose $I \neq 0$. Then $\mathcal{G}$ semisimple $\Rightarrow I \cap \mathbb{Z}(\mathcal{G}) \neq \{0\}$. Thus $\exists \tilde{z} \in I \cap \mathbb{Z}(\mathcal{G}), \tilde{z} \neq 0, \tilde{z} = \sum \alpha_{ij}(\tilde{\mathbf{C}}_2')^i\,(\tilde{\mathbf{C}}_4')^j \;\ni\; \tau(z) = \sum\,\alpha_{ij}\,(-Y^2 - \frac{\mathbf{W}}{Y^2} + \frac{1}{4})^i\,(Y^2\left(\frac{\mathbf{W}}{Y^2} + \frac{1}{2}\right))^j = 0$. Now evaluate this last equation in an arbitrary representation π of $\mathcal{P}$. Since $d\pi$ is arbitrary it must be that the α_{ij} vanish identically, i.e. $\alpha_{ij} \;=\; 0$, and thus $ker(\tau)|_{\mathcal{E}(\mathcal{G})} \;=\; 0$.

Now we introduce $*$ structures on $\tilde{\mathcal{K}}(\mathcal{P})$ and $\tilde{\mathcal{K}}(\mathcal{G})$.

Lemma 3.2. *If* $\tilde{\mathbf{L}}_{\mu\nu}^\dagger = -\tilde{\mathbf{L}}_{\mu\nu}$, $\tilde{\mathbf{L}}_{4\mu}^\dagger = -\tilde{\mathbf{L}}_{4\mu}$ *and if* $\tilde{Y}^\dagger = \tilde{Y}$, *then* $\tilde{\mathbf{P}}_\mu = \left(\tilde{\mathbf{D}}^{-1}\tilde{\mathbf{A}}_\mu^\rho\tilde{\mathbf{L}}_{4\rho}\right)$ *and also* $\tilde{\mathbf{P}}_\mu^\dagger \;=\; \left(\tilde{\mathbf{D}}^{-1}\tilde{\mathbf{A}}_\mu^\rho\tilde{\mathbf{L}}_{4\rho}\right)^\dagger \;=\; \tilde{\mathbf{L}}_{4\rho}^\dagger\tilde{\mathbf{A}}_{\mu\rho}^\dagger(\tilde{\mathbf{D}}^\dagger)^{-1} = -\tilde{\mathbf{P}}_\mu$. *Furthermore* $\left[\hat{\tilde{\mathbf{P}}}_\mu\;,\;\hat{\tilde{\mathbf{P}}}_\nu\right] \;=\; 0$.

Proof: Only commutativity of the $\tilde{\mathbf{P}}_\mu$ is difficult: $\tilde{\mathbf{P}}_\mu\tilde{\mathbf{P}}_\nu - \tilde{\mathbf{P}}_\nu\tilde{\mathbf{P}}_\mu = -\left(\tilde{\mathbf{P}}_\mu\tilde{\mathbf{P}}_\nu^\dagger - \tilde{\mathbf{P}}_\nu\tilde{\mathbf{P}}_\mu^\dagger\right) = \tilde{\mathbf{D}}^{-1}\left(\tilde{\mathbf{F}}_{\mu\nu} - \tilde{\mathbf{F}}_{\mu\nu}\right)(\tilde{\mathbf{D}}^{-1})^\dagger$ where $\tilde{\mathbf{F}}_{\mu\nu} \;-\; \tilde{\tilde{\mathbf{F}}}_{\mu\nu} \in \mathcal{E}(\mathcal{G})$. Now since $\mathcal{E}(\mathcal{G})$ has no zero divisors, and neither $\tilde{\mathbf{D}}$ nor $\tilde{\mathbf{D}}^\dagger$ is the zero element of $\mathcal{E}(\mathcal{G})$ we have $\tau\left(\tilde{\mathbf{P}}_\mu\,\tilde{\mathbf{P}}_\nu - \tilde{\mathbf{P}}_\nu\,\tilde{\mathbf{P}}_\mu\right) \;=\; 0 \;\Rightarrow\; \tau\left(\tilde{\mathbf{F}}_{\mu\nu} - \tilde{\tilde{\mathbf{F}}}_{\mu\nu}\right) \;=\; 0 \Rightarrow \tilde{\mathbf{F}}_{\mu\nu} - \tilde{\tilde{\mathbf{F}}}_{\mu\nu} \;=\; 0$ by Lemma 3.1. Thus $\left[\tilde{\mathbf{P}}_\mu\;,\;\tilde{\mathbf{P}}_\nu\right] \;=\; 0$.

Although a representation of $\mathcal{G}$ always gives a representation of the enveloping algebra $\mathcal{E}(\mathcal{G})$, it does not necessarily give a representation of the skew field. However, the following follows immediately from Lemma 3.2.

Theorem 3.3. *Let $(d\pi, \mathcal{H})$ be an infinitesmally unitarizable representation of $\mathcal{G}$ on an Hilbert space $\mathcal{H}$, and let $\tilde{Y}$ be a self adjoint operator on on $\mathcal{H}$ which satisfies $\tilde{Y}^4 + d\pi(\tilde{\mathbf{C}}_2')\tilde{Y}^2 + d\pi(\tilde{\mathbf{C}}_4') = 0$. Then, if both $d\pi(\tilde{\mathbf{D}})^{-1}$ and $d\pi(\tilde{\mathbf{D}})^{\dagger^{-1}}$ exist on a suitable, dense domain in $\mathcal{H}$, there exists a skew symmetric representation $d\tilde{\pi}$ of $\mathcal{P}$ on $\mathcal{H}$ defined by: $d\tilde{\pi}(\mathbf{L}_{ij}) = d\pi(\tilde{\mathbf{L}}_{ij})$, $d\tilde{\pi}(\mathbf{P}_0) = d\pi(\tilde{\mathbf{D}})^{-1}\ d\pi(\sum_{i=0}^{3} \tilde{\mathbf{A}}_0^i\ \tilde{\mathbf{L}}_{n,i})$, and $d\tilde{\pi}(\mathbf{P}_i) = [d\tilde{\pi}(\mathbf{L}_{i0}), d\tilde{\pi}(\mathbf{P}_0)](i = 1, 2, 3)$,*

Notice it is crucial that both operators $d\pi(\tilde{\mathbf{D}})^{-1}$ and $d\pi(\tilde{\mathbf{D}})^{\dagger^{-1}}$ exist on the suitable dense domain in $\mathcal{H}$ in order to be assured of the existence of mutually commuting translation operators $d\tilde{\pi}(\mathbf{P}_\mu)$.

4. Representations

Definition 4.1. (For brevity, we let here and below $\tilde{X} = d\pi(\tilde{\mathbf{X}})$ for $\tilde{\mathbf{X}} \in \tilde{\mathcal{K}}(\mathcal{G})$) or $\mathcal{E}(\mathcal{G}_{\mathbb{C}})$.) Let V be an $\mathcal{E}(\mathcal{G}_{\mathbb{C}})$ module with action $d\pi$ of $\mathcal{E}(\mathcal{G}_{\mathbb{C}})$ on V. Let: i) $V = \sum_{\lambda}^{\oplus} V_\lambda$ with $V_\lambda = \left\{v \in V \mid H_i v = \lambda(\mathbf{H}_i)v\right\}$; 2) $\exists\ v_0 \ni$ a) $X_i^+(X_i^-)\ v_0 = 0$, b) $H_i\ v_0 = \Lambda(\mathbf{H}_i)\ v_0$, c) $\mathcal{E}(\mathcal{G}_{\mathbb{C}})v_0 = V$. Then, if $d\pi$ comes from a unitary representation of $\mathfrak{G}$, V is an infinitesmally unitarizable lowest (highest) weight representation with lowest (highest) weight Λ.

Definition 4.2. An infinitesmally unitarizable, irreducible representation $d\pi$ of $\mathcal{G}$ on an Hilbert space $\mathcal{H}$ is a representation with positive energy if $\left\{X \in \mathcal{G} \mid i\ d\pi(X) \text{ is a non}-\text{negative self adjoint operator}\right\}$ is a non-zero and proper cone in $\mathcal{G}$.

Theorem 4.1. *(Paneitz, Olshanski)*[5] *Let $d\pi$ be an irreducible, infinitesmally unitarizable representation of $\mathcal{G}$ on an Hilbert space $\mathcal{H}$. Then $d\pi$ is positive energy $\Leftrightarrow$ $d\pi$ is a lowest (highest) weight representation.*

A classification of the infinitesmally unitarizable, irreducible lowest weight representations of $\mathcal{G}$ using a modern representation theoretic approach involving Verma modules and singular vectors is given in Ref. 6. Similar classifications for other noncompact Lie groups and supergroups are given in Ref. 7. The results in Ref. 6 are: let $D(E_0, s_0)$ be a given such lowest weight representation with lowest weight $\Lambda \ni \Lambda(H_1) = s_0$

and $\Lambda(H_2) = E_0$. Then we have: i) $D(E_0, s_0) = D(\frac{1}{2}, 0)$ (Rac); ii) $D(E_0, s_0) = D(1, \frac{1}{2})$ (Di); iii) $D(E_0 > \frac{1}{2}, s_0 = 0)$; iv) $D(E_0 > 1, s_0 = \frac{1}{2})$; v) $D(E_0 \geq s_0 + 1, s_0 \geq 1)$. The massless representations are: $D(E_0 = s_0 + 1, s_0 \geq 1)$.

Proposition 4.1. *The massless representations $D(E_0 = s_0 + 1, s_0 \geq 1)$ give representations of $\mathcal{P}$, none of which are skew symmetric.*

Proof: We have $\tilde{C}_2 = 2(s_0^2 + 1)\cdot I$ and $\tilde{C}_4 = \left[s_0^2(s_0 + 1)(s_0 - 1)\right]\cdot I$. ($I$ is the identity in $\mathcal{H}$.) Thus $\tilde{Y}^4 + \tilde{C}_2' \tilde{Y}^2 + \tilde{C}_4' = 0 \Rightarrow \left[\tilde{Y}^2 + (s^2 + \frac{1}{4})\cdot I\right]^2 = s^2\cdot I \Rightarrow \tilde{Y}^2 + (s^2 + \frac{1}{4})\cdot I = \pm s\cdot I \Rightarrow \tilde{Y}^2 = -(s \pm \frac{1}{2})^2 \cdot I \Rightarrow$ according to Theorem 3.3 that $d\tilde{\pi}(\mathbf{P}_\mu)$ is not skew symmetric. However, $d\pi(\tilde{\mathbf{D}})^{-1}$ and $\left(d\pi(\tilde{\mathbf{D}})^\dagger\right)^{-1}$ exist as continuous operators on the representation spaces for all massless representations, so that we still get from Theorem 3.3 non-unitary representations of the Poincaré group.

Similar calculations for the other representations give essentially the same results, except in the case of the Di and Rac. What goes wrong in these cases is more serious. For both the Di and Rac is $\tilde{Y}^2 = -1, -\frac{1}{4}$ which implies not only that $d\pi(\tilde{\mathbf{P}}_\mu)$ are not skew-symmetric operators, but, e.g. for the Rac: $d\pi(\tilde{\mathbf{D}}) = 0$ (with $\tilde{Y} = -i$). Thus the $d\tilde{\pi}(\mathbf{P}_\mu)$ do not even exist as skew symmetric operators on the Hilbert space of the representation. Hence, the Di and the Rac do not have contractions to representations of the Poincaré group. This analysis sharpens the meaning of *kinematical confinement* by Flato, Fronsdal, Sternheimer and their coworkers[8], a notion which one must admit is one of the most fascinating and powerful ideas of theoretical physics in modern times.

References

1. P. Moylan in GROUP 24 editors: Eds. J-P Gazeau et al., *Institute of Physics Conference Series 173*, (2003), p. 230-235.
2. P. Moylan, in Proceedings of QTS 2, eds. E. Kapuścik, A. Horzela, World Scientific, (2001), p. 504-512.
3. T.L. Curtright, C.K. Zachos, *Phys. Lett B*, **243**, *3*, (1990) 237-244.
4. P. M. Cohn, *Skew Fields: The General Theory of Division Rings*, Encylopedia of Mathematics and Its Applications, **57**, Camb. Univ. Press, (1995).
5. S.M. Paneitz, *Jour. Funct. Anal.*, **54** (1983), 18-112.
6. V.K. Dobrev, E. Sezgin, *Lecture Notes in Physics*, **379**, (Springer, 1990), 227.
7. Murat Günaydin, *Jour. Math. Phys.*, **29**, (6), (1988), 1275.
8. M. Flato, C. Fronsdal, D. Sternheimer, *Singleton Physics*, arXiv:hep-th/9901043 v1 12 Jan 1999.

III. QUANTIZATION AND QUANTUM GRAVITY

THEORETICAL FRAMEWORK FOR RELATIVISTIC RESONANCES*

A. R. BOHM and Y. SATO
Department of Physics,
The University of Texas at Austin,
Austin, TX78712, USA
E-mail: satoyosh@physics.utexas.edu

To obtain a consistent mathematical theory that unifies resonance scattering and decay phenomena one has to modify the axiom of quantum mechanics that specifies the boundary conditions of the dynamical equations and choose Hardy spaces of the lower complex energy plane for the Schrodinger equation (of states) and of the uppor complex energy plane for the Heisenberg equation (of observables). This theory can be extended into the relativistic domain where it provides a unique definition of the lineshape and the mass and width for relativistic resonances.

The Particle Data Table[1] gives two values for the well-measured relativistic resonances; *e.g.* for the Z^0 boson these values are:

$M_Z = 91.1876 \pm 0.0021 GeV$	$\bar{M}_Z = 91.1541 \pm 0.0031 GeV$
$\Gamma_Z = 2.4952 \pm 0.0023 GeV$	$\bar{\Gamma}_Z = 2.4936 \pm 0.0024 GeV$

and they differ by 10 times the quoted experimental error $M_Z - \bar{M}_Z \simeq 34 MeV \simeq 10\Delta_{\text{exp}}$. A similar situation exists for the hadron resonances Δ_{33} and ρ[1,2,3].

Δ_{33}	$M_\Delta = 1233(1) MeV$	$\bar{M}_\Delta = 1210(1) MeV$
	$\Gamma_\Delta = 120(1) MeV$	$\bar{\Gamma}_\Delta = 100(1) MeV$
ρ	$M_\rho = 769(1) MeV$	$\bar{M}_\rho = 757(1) MeV$
	$\Gamma_\rho = 151(3) MeV$	$\bar{\Gamma}_\rho = 142(2) MeV$

These different values are extracted from the *same* sets of experimental data using *different* theoretical assumptions and *different definitions* for the resonance mass M and the resonance width Γ. This shows that one

*Talk presented by A.B.

needs a theoretical frame which uniquely defines the meaning of mass and width of relativistic resonances. For stable relativistic particles ($\Gamma = 0$) the mass M is defined by the irreducible unitary group representations of (restricted) Poincare transformations. How does one define M and Γ for relativistic resonances?

For non-relativistic resonances one had the Weisskopf-Wigner methods by which one derived for the decay probability $\mathcal{P}(t)$ of a prepared resonance state with (Breit-Wigner) width Γ that: $\mathcal{P}(t) \sim e^{-\Gamma t/\hbar} + something$.[4] From this many people conclude that (in the "approximation" $something \to 0$) the lifetime τ of a resonance (M, Γ) is given by $\tau = \hbar/\Gamma$ (or equivalently that the width of the Breit-Wigner energy distribution is equal to the total initial decay rate $\sum_\eta \frac{d\mathcal{P}_\eta(t)}{dt}|_{t=0} = \Gamma/\hbar$). However, to experts it was well known that " ... there does not exist a rigorous theory to which these (Weisskopf-Wigner) methods can be considered as *approximation*".[5]

If τ and inverse of Γ are the same in quantum theory there needs to be a quantum mechanical state which has both properties. One knows that in Hilbert space of conventional quantum mechanics such a state cannot exist (*e.g.* theorem on the deviation from the exponential laws). To obtain an exponentially decaying "Gamow" state in order to define the lifetime τ, and moreover that has a Breit-Wigner energy distribution to define Γ, one has to change the foundation of quantum theory. But a minor modification of only one of the axioms will suffice:[6] One has to change the Hilbert space axiom or asymptotic completeness

$$\{\text{in-states}\,\phi^+\} = \{\text{out-observables}\,\psi^-\} = \Phi \subset \mathcal{H} \qquad (A_{old})$$

to the new Hardy space axiom

$$\begin{aligned} \{\text{in-states}\,\phi^+\} &= \Phi_- \subset \mathcal{H} \subset \Phi_-^\times \qquad (A_{new}) \\ \{\text{out-observables}\,\psi^-\} &= \Phi_+ \subset \mathcal{H} \subset \Phi_+^\times \end{aligned}$$

Whereas according to (A_{old}) the set of prepared in-states $\{\phi^+\}$ and the set of detected out-particles $\{\psi^-\}$ is the same dense subspace Φ (Schwartz space) of the Hilbert space, the new axiom distinguishes meticulously between the in-states $\{\phi^+\}$ defined by the preparation apparatus, and the detected out-particles $\{\psi_\eta^-\}$ defined by the detector, and assigns them to different dense Hardy subspaces[7] of the same Hilbert space $\mathcal{H}$. The dual spaces $\Phi_\pm^\times$ of $\Phi_\pm$ are the spaces which contain the Lippmann-Schwinger(LS) kets $|s^\pm\rangle \in \Phi_\pm^\times$, which can be analytically continued into the upper and lower half-plane respectively. The dual of the Schwartz space contain the ordinary Dirac kets $|s\rangle \in \Phi^\times$, but not the LS kets $|s \pm i\epsilon^\pm\rangle$.

Our starting point for the definition of a resonance state is the pole on the second Riemann sheet of the S-matrix $S_j^{\eta^{'}\eta}(s)$ below the cut along the real axis $s_0 \leq s < +\infty$ where s is the (invariant) center of mass energy squared $s = (p_1 + p_2)_\mu (p_1 + p_2)^\mu$ and p_i are the momenta of the two in- or out- particles.

In the non-relativistic case it has been shown[8] that the resonance at the pole position $z_R = E_R - i\Gamma_R/2$ is given by a Breit-Wigner scattering amplitude and that to this Breit-Wigner lineshape corresponds an eigenket $|z_R, j, \eta^-\rangle$ with complex eigenvalue z_R of the (selfadjoint) Hamiltonian H:

$$a_j^{BW}(E) = \frac{r^\eta}{E - (E_R - i\Gamma_R/2)} \Longleftrightarrow \psi^G = |z_R, j, \eta^-\rangle \tag{1}$$

$$\begin{aligned} z_R &= E_R - i\Gamma_R/2, \\ j &= \text{angular momentum}, \\ \eta &= \text{particle species quantum number}. \end{aligned}$$

This ket, called Gamow vector, is defined as a functional $\psi^G(\psi^-) \equiv \langle\psi_\eta^-|\psi_{\eta_0}^G\rangle$ on the Hardy space $\{\psi^-\} = \Phi_+$ of all possible decay products ψ_η^-. This means $\psi^G \in \Phi_+^\times$, *i.e.*, it is a "more generalized" vector than the Dirac ket.

The time evolution of this vector $\psi^G(t) = e^{-iH^\times t}\psi^G$ can be proved to be given by

$$\begin{aligned} \langle\psi_\eta^-(t)|\psi^G\rangle &\equiv \langle e^{iHt}\psi_\eta^-|E_R - i\Gamma_R/2^-\rangle \equiv \langle\psi_\eta^-|e^{-iH^\times t}|E_R - i\Gamma_R/2^-\rangle \\ &= e^{-iE_R t} e^{-\frac{\Gamma_R}{2}t}\langle\psi_\eta^-|E_R - i\Gamma_R/2\rangle \\ &\forall\ \psi_\eta^- \in \Phi_+ \quad \text{for } t \geq 0, \end{aligned} \tag{2}$$

but only for $t \geq 0$. This shows that ψ^G is the exponentially decaying "state" with the lifetime, $\tau = 1/\Gamma_R$, where Γ_R is the width of its Breit-Wigner resonance line shape (1). In addition it obeys causality expressed by the time asymmetry $t \geq 0$.[9]

In the non-relativistic case the functional form for the scattering amplitude of the resonance *per se* is not in doubt and given by the Breit-Wigner (1). However, in the relativistic case various functions of s and $\sqrt{s}$ are in use. One writes the relativistic scattering amplitude $a_j(s)$ as the sum of a resonance amplitude $a_j^R(s)$ and a non-resonant background $B_j(s)$:

$$a_j(s) = a_j^R(s) + B_j(s). \tag{3}$$

Of the different function used for the resonance amplitude $a_j^R(s)$, we mention here two functions and three ways, (4) and (5) with (6a, 6b), to express $a_j^R(s)$ in terms of two parameters (M, Γ):[2,3,10]

(1) In the on-shell renormalization scheme[10] $a_j^R(s)$ is defined by the propagator (e.g., Z-boson propagator) and given by:

$$a_j^R(s) = a_j^{om}(s) = \frac{-\sqrt{s}\sqrt{\Gamma_e(s)\Gamma_f(s)}}{s - M_Z^2 + i\sqrt{s}\Gamma_Z(s)} \approx \frac{-M_Z B_e B_f \Gamma_Z}{s - M_Z^2 + i\frac{s}{M_Z}\Gamma_Z}$$
$$= \frac{R_Z}{s - M_Z^2 + i\frac{s}{M_Z}\Gamma_Z} \tag{4}$$

where $\Gamma_Z(s)$ is an "energy dependent width" and $\Gamma_Z \equiv \Gamma_Z(s = M_Z^2)$.

(2) Using the S-matrix definition of the resonance by a pole at $s = s_R$ leads to a unique function for the resonance amplitude $a_j^R(s)$, which we call the relativistic Breit-Wigner $a_j^{BW}(s)$ (Cauchy kernel in mathematics)

$$a_j^R(s) = a_j^{BW}(s) \equiv \frac{r_Z}{s - s_R}. \tag{5}$$

However this Breit-Wigner still allows for many different definitions of the real parameters, mass and width, of which $(\bar{M}_Z, \bar{\Gamma}_Z)$, (M_R, Γ_R) given by

$$s_R = \bar{M}_Z^2 - i\bar{M}_Z\bar{\Gamma}_Z \quad \text{or} \tag{6a}$$
$$s_R = (M_R - i\frac{\Gamma_R}{2})^2, \tag{6b}$$

are two possibilities among many more.

Using the lineshape $|a_j(s)|^2$ only one cannot phenomenologically distinguish between (4) and (5) since one can always shift the difference between them into the unknown $B(s)$. But even if one decides in favor of (5) (for theoretical reasons because for it (M, Γ) are gauge invariant[10]) one still is left with the choice between the (many) possible parameterizations (6a, 6b) of s_R in terms of two real parameters (M, Γ).

In order to fix this ambiguity one defines, in analogy to the non-relativistic case[8], a state vector for the resonance, the relativistic Gamow vector[11] with energy distribution (5):

$$\psi^G_{[s_R,j]} = |[s_R, j]\hat{\mathbf{p}}j_3^-\rangle = \frac{1}{2\pi}\int_{-\infty_{II}}^{+\infty} ds|[s,j]\hat{\mathbf{p}}j_3^-\rangle \frac{1}{s - s_R}. \tag{7}$$

It is derived like in the non-relativistic case[8] from the resonance pole of the (relativistic) S-matrix $S_j^{\eta'\eta}(s)$ using the Hardy space axiom (A_{new}). Here the $|[s,j]\hat{\mathbf{p}}j_3\eta^-\rangle$ are the Lippmann-Schwinger kets transforming very much like the Wigner kets under Poincare transformations.[12] As a result the Gamow kets $|[s_R, j]\hat{\mathbf{p}} j_3^-\rangle$ have a definite transformation property under Poincare *semigroup* transformations into the *forward light cone* only.[13] A special case of this is the time evolution at rest

$$\begin{aligned}\psi^G_{s_R}(t) &= e^{-iH^\times t}|[s_R, j]\hat{\mathbf{p}} = 0\, j_3^-\rangle \\ &= e^{-iM_R t} e^{-\Gamma_R t/2} |[s_R, j]\hat{\mathbf{p}} = 0\, j_3^-\rangle, \quad t \geq 0. \end{aligned} \tag{8}$$

From this we obtain that the decay rate is proportional to:

$$|\langle \psi_\eta^- | \psi^G_{s_R}(t)\rangle|^2 = e^{-\Gamma_R t} |\langle \psi_\eta^- | \psi^G_{s_R}(0)\rangle|^2, \quad t \geq 0. \tag{9}$$

Comparing this with the exponential decay rate $\sim e^{-t/\tau}$ by which the lifetime τ is defined and measured, we see that the Gamow state described by (7) with the relativistic Breit-Wigner lineshape, $\frac{1}{s-s_R}$, has the lifetime

$$\tau = \frac{\hbar}{\Gamma_R}, \quad \Gamma_R = -2\mathcal{I}m\sqrt{s_R}\,. \tag{10}$$

In summary, of the infinitely many possible mass and width parameters (M, Γ) and of the many possible lineshape formulas, only (M_R, Γ_R) of (6b) and (5) fulfill the exact lifetime-width relation (10). Only $|a_j^{BW}(s)|^2$ of (5) gives the lineshape of the relativistic resonance *per se* (defined by the pole in the s-plane); other lineshape formulas include background. To arrive at conclusion (10) one had to extend Wigner's definition of relativistic particles[12] to quasistable objects of complex mass s_R, and one had to generalize Breit-Wigner's non-relativistic lineshape (1) into the relativistic domain.

References

1. K. Hagiwara *et al.*, *Phys. Rev.* **D66**, 010001 (2002).
2. G. Höhler, *Euro. Phys. J.* **3**, 624 (1998)
3. A. Bernicha, G. Lopez Castro, J.Pestieau, *Phys. Rev.* **D50**, 4454 (1994); *Nucl. Phys.* **A597**, 623 (1996).
4. See, for example, M.L. Goldberger, K.M. Watson *Collision Theory*, Wiley, New York, 1964, chap. 8.
5. M. Levy, *Nuovo Cimento* **13**, 115 (1959).
6. A. Bohm, M. Loewe, B. Van de Ven, *Fortschr. Phys.* **51**, No.6 (2003). A. Bohm, M. Gadella, M. J. Mithaiwala, *Proceedings of the XXII Solvay Conference on Physics*, I. Antoniou *et al.* edts., World Scientific Publishing, page 117–137 (2003).
7. M. Gadella, *J. Math. Phys.* **24**, 1462 (1983).

8. A. Bohm, *J. Math. Phys.* **22**, 2813 (1981); *Lett. Math. Phys.* **3**, 455 (1978). A. R. Bohm, S. Maxson, M. Loewe, and M. Gadella, *Physica A* **236**, 485 (1997).
9. A. Bohm, *Phys. Rev.* **A60**, 861 (1999).
10. S. Willenbrock, G. Valencia, *Phys. Lett.* **B259**, 373 (1991). R.G. Stuart, *Phys. Lett.* **B262**, 113 (1991); **272**, 353 (1991). A. Sirlin, *Phys. Rev. Lett.* **67**, 2127 (1991); *Phys. Rev. Lett.* **B267**, 240 (1991)
11. A. Bohm, H. Kaldass, S. Wickramasekara, P. Kielanowski, *Phys. Lett.* **A264**, 425 (2000).
12. E. P. Wigner, *Ann. Math* **(2) 40**, 149 (1939); V. Bargmann, E. P. Wigner, *Proc. Nat. Acad. Sci. USA* **34**, 211 (1984). S. Weinberg, *The Quantum Theory of Fields*, Cambridge University Press, Vol.1. L.S. Schulman *Ann. Phys.* **59**, 201 (1970).
13. A. Bohm, H. Kaldass, S. Wickramasekara, *Fortschr. Phys.* **51**, 569–603 and 604–634 (2003) where the detailed derivations are given.

NONLINEAR SCHRÖDINGER EQUATIONS FOR PARTICLES AND ANTIPARTICLES*

H.-D. DOEBNER

Department of Physics and Material Science,
Technical University of Clausthal, Germany
E-mail: asi@pt.tu-claustal.de

GERALD A. GOLDIN

Departments of Mathematics and Physics, Rutgers University,
Piscataway, New Jersey 08854
E-mail: gagoldin@dimacs.rutgers.edu

We discuss certain nonlinear extensions of the Schrödinger quantum-mechanical time evolution. Our equations are motivated by a quantization method for particles localized in $\mathbb{R}^3$, or equivalently, by a group of nonlinear transformations among physically equivalent systems. We construct some solutions of equations in our class, applicable to the $1/r$ potential. We present an Ansatz for an equation describing antiparticles, and describe a possibly surprising property.

1. INTRODUCTION

Quantum mechanics is intimately based on linear structures: *pure states* are elements of a (linear) Hilbert space; *mixed states* are (linear) trace class operators; *observables* are described by (linear) self adjoint operators; and the *time-evolution equations* for states are (linear) Schrödinger or von Neumann equations. This remarkable reliance on linearity raises interesting questions. Is such linearity best seen as a property of nature, or an artifact of the standard formalism? If the latter, is there a natural extension of quantum mechanics based on Hilbert space, with nonlinear operators representing observables (e.g., with nonlinear Hamiltonians), as anticipated by Wigner?[1] Such a scenario could permit new physical effects, measurable via precision experiments,[2,3] allowing an experimental upper bound

*Talk presented by H.-D.D.

on the magnitude of the nonlinearity and suggesting consequences within the wider framework of our "usual" understanding of quantum mechanics.

However, in the standard framework there are important obstructions to such nonlinear extensions,[4,5,6,7] of which the following are but a partial list: [1] The evolution of an n-particle system requires some additional prescription, if it is to be obtained in a canonical way from n noninteracting particles evolving nonlinearly. [2] Such evolution is generically nonlocal—thus for $n = 2$, in correlated states, an external potential acting only on particle 1 influences the distribution of outcomes of positional measurements for particle 2. [3] Nonlinear operators are not self-adjoint; the probability interpretation based on spectral decomposition must be reformulated. [4] The time-evolution of mixed states is not uniquely defined by a nonlinear time-evolution of pure states.

Thus, to accommodate nonlinear operators, we need an extended quantum-mechanical formalism that includes new notions and interpretations. For example, one possible method for avoiding nonlocal effects in correlation experiments is to take an "open system" approach.[8] In commencing the construction of such an extended (nonrelativistic) theory, a certain class of "natural" nonlinear time-evolutions provides a kind of cornerstone. This permits calculation of the influence of "small" nonlinear terms in the Hamiltonian on the "spectrum" and the "eigenfunctions."

2. A FAMILY OF NONLINEAR SCHRÖDINGER EQUATIONS DESCRIBING PARTICLES

We derived, from fundamental principles described below, a special parameterized family of nonlinear Schrödinger equations (NLSEs).[9,10] Let $\Psi(\mathbf{x}, t)$ be a wave function for a single particle with mass m, moving under the influence of an external scalar potential $V(\mathbf{x})$, with positional probability density $\rho(\mathbf{x}, t) = \Psi(\mathbf{x}, t)^* \Psi(\mathbf{x}, t)$ and probability current density $\mathbf{J}(\mathbf{x}, t) = (\hbar/m)\widehat{\mathbf{J}}(\mathbf{x}, t) = (\hbar/2mi)[\Psi^* \nabla \Psi - (\nabla \Psi^*)\Psi]$. Define the homogeneous nonlinear functionals $R_1 = \nabla \cdot \widehat{\mathbf{J}}/\rho$, $R_2 = \Delta\rho/\rho$, $R_3 = \widehat{\mathbf{J}}^2/\rho^2$, $R_4 = \widehat{\mathbf{J}} \cdot \nabla\rho/\rho^2$, and $R_5 = (\nabla\rho)^2/\rho^2$. Our general equation then reads

$$i\hbar\partial_t \Psi = \{-\frac{\hbar^2}{2m}\Delta + V(x) + \frac{i}{2}\hbar D\, R_2\,[\Psi] + \hbar D' \sum_{j=1}^{5} c_j R_j[\Psi]\,\}\Psi\,, \quad (1)$$

where D and $D'c_j$ $(j = 1, \ldots, 5)$ have the dimensions of diffusion coefficients; the c_j are dimensionless real parameters. In an equivalent form,[11] with real parameters ν_1, ν_2, and $\mu_1, \ldots, \mu_5$, we have

$$i\partial_t \Psi = \left\{ i \sum_{j=1}^{2} \nu_j R_j[\Psi] + \sum_{j=1}^{5} \mu_j R_j[\Psi] + U \right\} \Psi \, ; \tag{2}$$

where $\nu_1 = -\hbar/2m$, $\nu_2 = D/2$, $\mu_1 = D'c_1$, $\mu_2 = -\hbar/4m + D'c_2$, $\mu_3 = \hbar/2m + D'c_3$, $\mu_4 = D'c_4$, $\mu_5 = \hbar/8m + D'c_5$, and $U(\mathbf{x}) = V(\mathbf{x})/\hbar$. This system of nonlinear equations (with singularities) has been studied by several authors. Stationary solutions (see Sec. 3 below) and time-dependent solutions for certain parameter values have been constructed.[12,13,14] The Lie symmetries have been found,[15,16,17] and results have been obtained for the Cauchy problem.[18] Possible applications are envisioned to include irreversible and dissipative quantum systems.[19] Applications have been made to atomic spectra, to quantum optics ("nonlinear photons"),[20] and to the quantum mechanics of D-branes.[21]

We can derive Eqs. (1)-(2) from two different sets of fundamental principles. First, consider a localized system described by a configuration space M. Choose as observables the generalized (spatially-averaged) positional densities Q and generalized momentum densities P. These span an infinite-dimensional kinematical Lie algebra $S(M)$ that is an inhomogeneous current algebra.[22] For a quantization, we specify the kinematics via an irreducible, essentially self adjoint representation of $S(M)$ in a Hilbert space $\mathcal{H}$ (or, correspondingly, a continuous unitary representation of the associated semidirect product group of scalar functions on M with diffeomorphisms of M).[23,24,25,26] Under some technical conditions, such representations are characterized up to unitary equivalence by a real constant D, a quantum number in the sense of Wigner. The case $D = 0$ is associated with standard quantum mechanics. To interpret the new representations, we assume conservation of probability expressed through the equation of continuity.[9] This constrains the time-dependence of the system, and leads to a Fokker-Planck type of equation relating the probability and current densities:

$$\frac{\partial \rho(\mathbf{x},t)}{\partial t} = -\nabla \cdot \mathbf{J}(\mathbf{x},t) + D\Delta\rho(\mathbf{x},t) \, ; \tag{3}$$

note the additional diffusion term proportional to D. Considered as a condition on the time-evolution of Ψ, Eq. (3) *requires* the presence of the imaginary nonlinear term $(i\hbar D/2)R_2[\Psi]\Psi$ on the right-hand side of Eq. (1). The possibilities for the real nonlinear functionals R_j, $j = 1, \ldots, 5$ are then chosen to be structurally analogous to the functional R_2.

A second derivation is to begin with the linear Schrödinger equation (LSE), and to introduce local, possibly nonlinear transformations N in $\mathcal{H}$

that preserve the positional probability density $\rho(\mathbf{x},t)$ for all $\mathbf{x}$ and t.[11,27] A plausible case has been made for the idea that all quantum measurements reduce to sequences of positional measurements (in general, at different times);[28] so that N preserves the outcome of all measurements and transforms among physically equivalent systems. We further take N to respect product states in the n-particle Hilbert space. The result is a 2-parameter group $\mathcal{N}$ of *nonlinear gauge transformations* (extending the usual, Abelian gauge group of quantum mechanics). Elements of $\mathcal{N}$ take the form

$$\Psi' = N_{(\gamma,\Lambda)}[\Psi] = |\Psi| \exp i\left(\gamma \ln|\Psi| + \Lambda \arg \Psi\right)\,, \tag{4}$$

where γ and Λ are real numbers, $\Lambda \neq 0$. Note that there is some non-uniqueness in Ψ' due to non-integer values of Λ. Now if Ψ obeys the linear Schrödinger equation, Ψ' obeys a uniquely defined, physically equivalent NLSE belonging to the class of Eqs. (1)-(2), but with coefficients which are constrained (i.e., related to each other). Thus we see that even the usual quantum mechanics can be written as a nonlinear theory. Following the procedure of *gauge generalization* we break the constraints on the coefficients, again apply N, and continue until the system of equations closes. In this way, we recover Eqs. (1)-(2) with arbitrary coefficients.

Elements of $\mathcal{N}$, labeled by the 2 parameters γ and Λ of Eq. (4), act on the space of Eqs. (2) labeled by the 7 coefficients $\nu_1, \nu_2, \mu_1, ..., \mu_5$. Thus one expects and finds 5 independent gauge-invariant parameters $\tau_1, ..., \tau_5$:

$$\tau_1 = \nu_2 - \mu_1/2\,, \quad \tau_2 = \nu_1\mu_2 - \nu_2\mu_1\,, \quad \tau_3 = \mu_3/\nu_1 \quad \tau_4 = \mu_4 - \mu_1\mu_3/\nu_1\,,$$

$$\tau_5 = \nu_1\mu_5 - \nu_2\mu_4 + \nu_2^2\mu_3/\nu_1. \tag{6}$$

Also $\hat{U} = -\nu_1 U = V/2m$ is invariant. The gauge-invariant τ_2 is identified with the physically observed value of $\hbar^2/8m^2$. The choice $\nu_1 = -\hbar/2m = -\sqrt{2\tau_2}$, and $\mu_1 = 0$ so that $\nu_2 = \tau_1 = D/2$, is called the Schrödinger gauge.

3. STATIONARY SOLUTIONS

As in the case of the linear Schrödinger equation, there are solutions of Eqs. (1)-(2) that separate $\mathbf{x}$ and t, and are *stationary* in that $\partial_t\rho(\mathbf{x},t) = 0$. Writing $\Psi(\mathbf{x},t) = \Psi_E(\mathbf{x})\,T_E(t)$ we have $\dot{T}_E(t) = -i\omega_E T(t)$, together with the nonlinear "eigenvalue" equation $\{\, i\sum_{j=1}^{2} \nu_j R_j[\Psi_E] + \sum_{j=1}^{5} \mu_j R_j[\Psi_E] + U\,\}\Psi_E = \omega_E \Psi_E$ for Ψ_E. With the Ansatz $\Psi_E(\mathbf{x}) = [\Phi_E(\mathbf{x})]^\alpha \exp[-i\alpha(\nu_2/\nu_1)\ln|\Phi_E(x)|^2]$, where $\Phi_E(\mathbf{x}) \geq 0$ is real (which actually characterizes states of zero orbital angular momentum), one has

$i\sum_{j=1}^{2}\nu_j R_j[\Psi_E] = 0$; then Φ_E satisfies the *affiliated* gauge-invariant LSE: $[(\tau_2)^2/(\tau_2+\tau_5)]\triangle\Phi_E - \hat{U}\Phi_E = \nu_1\omega_E\Phi_E$. The combination $\nu_1\omega_E = -E/2m$ is gauge invariant. Here Ψ_E is square integrable if Φ_E^α is square integrable.

Thus stationary solutions to our NLSE with potential $V(\mathbf{x})$ may be found, and it is a reasonable interpretation that E describes the energy levels. Writing the equation for Φ_E as $-(\hbar/2m^*)\triangle\Phi_E + V\Phi_E = E\Phi_E$, where m^* is an effective mass, we have $m^* = \beta m$ where $\beta = \nu_1^2[(\tau_2+\tau_5)/\tau_2^2]$. In the Schrödinger gauge, we have $\beta = 2(\tau_1 + \tau_5)/\tau_2$.

4. UPPER BOUND TO NONLINEARITY

The stationary states for a given potential $V(\mathbf{x})$—for example, the $1/r$ potential—may now be compared in the linear and the nonlinear cases.[20] With the above interpretation of the "eigenvalues" we obtain the predicted relative frequency shift for transitions between the ground state and the first excited state for the hydrogen atom, $\triangle_{10} = (\nu_{10}^L - \nu_{10}^{NL})/\nu_{10}^L$, where ν_{10}^L and ν_{10}^{NL} are the relative frequencies for the linear and the nonlinear case respectively. The above results yield $\triangle_{10} = 3\,(\beta - 1)\,/4$. Precision experiments for the hydrogen atom do not presently show significant deviation from the linear theory. The accuracy of measured relative frequencies is approximately one part in 10^{-14}. Supposing that $c_1 = ... = c_5 = 0$ while $D \neq 0$, this leads to an experimental bound where $|D| \leq 10^{-7}\hbar/m$.

But in applying Eq. (1) to a realistic system like the hydrogen atom, we must recall the difficulties with nonlinear evolutions. The two-particle system is not uniquely determined from a nonlinear evolution equation for each particle. Introduction of center of mass and relative coordinates, and quantization of the center of mass motion with a LSE and the relative motion with a NLSE, leads to a coupling between them. One expects this coupling to be extremely small; in numerical examples, it is indeed small compared to the energy shift.[7]

5. PARTICLE VS. ANTIPARTICLE SYSTEMS

Next consider nonrelativistic antihydrogen (or, more generally, an antiparticle system). Because the antiparticle is a relativistic notion, an obvious possibility is to start with a nonlinear Dirac equation and obtain via a nonlinear Pauli equation the family of Eqs. (1) as a nonrelativistic limit. We should then be able to derive the correct NLSE for an antiparticle system. However, the nonlinear Dirac analogue of Eq. (1) or (2) is not at hand. A simpler method adapted to a nonrelativistic context is to note that if CPT

invariance holds, charge conjugation should be equivalent to PT. The terms $R_1, \ldots, R_5$ are invariant under parity P; thus we may reasonably view the time-reversed wave function Ψ^T as the antiparticle wave function. The corresponding NLSE for Ψ^T again belongs to the family of Eq. (2), with coefficients: $\nu_1^T = -\nu_1$, $\nu_2^T = -\nu_2$, and $\mu_j^T = -\mu_j$ for $j = 1, \ldots, 5$. For the corresponding invariants, we find[11] $\tau_j^T = -\tau_j$ for $j = 1, 4$, and $\tau_j^T = +\tau_j$ for $j = 2, 3, 5$. Note that if we complex conjugate as well as time-reverse Eq. (2), we obtain the transformation $\nu_1' = -\nu_1$, $\nu_2' = -\nu_2$, $\mu_j' = -\mu_j$ for $j = 1, 4$, and $\mu_j' = \mu_j$ for $j = 2, 3, 5$. But the corresponding invariants τ_j' are the same as the invariants τ_j^T; thus these are gauge-equivalent descriptions of the antiparticle wave function.

But we saw in Sec. 3 that the coefficient of the affiliated LSE for the stationary state problem with potential V depends only on τ_2 and τ_5. These invariants *do not change* in the stationary solutions we constructed, when particles are replaced by antiparticles—the effective mass shift and energy spectrum shift due to nonlinearity is the same for particles and antiparticles. Hence our possibly surprising result is that nonlinearities of the type discussed here result in *no* change in the energy spectrum for antihydrogen (or, respectively, a general antiparticle system) as compared with hydrogen (or, respectively, a general particle system). To detect the effects of nonlinearity by comparing particles with antiparticles, it would be necessary to compare the results of experiments requiring solutions to the time-dependent problem.

Acknowledgments

We are indebted for discussions with J. Ellis (CERN) and N. Mavromatos (King's College, London) about the spectrum of the antihydrogen atom. GG thanks the Alexander von Humboldt Foundation for research support.

References

1. Wigner, E. P., *Ann. Math.* **40**, 149 (1939).
2. Weinberg, S., *Phys. Rev. Lett.* **62**, 485 (1989); *Ann. Phys.* **194**, 336 (1989).
3. Bollinger, J. J., Heinzin, D. J., Itano, W. M., Gilbert, S. L., Wineland, D. J., *Phys. Rev. Lett.* **63**, 1031 (1989).
4. Gisin, N., *Phys. Lett. A* **143**, 1 (1990).
5. Gisin, N., in H.-D. Doebner, V. K. Dobrev, P. Nattermann (eds.), *Nonlinear, Deformed and Irreversible Quantum Systems,* p. 109. Singapore: World Scientific (1995).
6. Goldin, G. A., Svetlichny, G. S., *J. Math. Phys.* **35**, 3322 (1994).

7. Lücke, W., "Gisin nonlocality of the Doebner-Goldin 2-particle equation," arxiv: quant-ph/9710033 (1997).
8. Czachor, M., Doebner, H.-D., *Phys. Lett. A* **301**, 139 (2002).
9. Doebner, H.-D., Goldin, G. A., *Phys. Lett. A* **162**, 397 (1992).
10. Doebner H.-D., Goldin G. A., *J. Phys. A: Math. Gen.* **27**, 1771 (1994).
11. Doebner, H.-D., Goldin, G. A., *Phys. Rev. A* **54**, 3764 (1996).
12. Goldin, G. A., *Int. J. Mod. Phys. B* (1992).
13. Ushveridze, A. G., *Phys. Lett A* **185**, 123 and 128 (1994).
14. Dodonov, V. V., Mizrahi, S. S., *Ann. Phys.* **37**, 226 (1995); *Physica A* **214**, 619 (1995).
15. Nattermann, P., *Rep. Math. Phys.* **36**, 387 (1995).
16. Nattermann, P., Zhdanov, R. *J. Phys. A: Math. Gen.* **29**, 2969 (1996).
17. Volkmann, J., Südland, N., Schmid, R., Engelmann, J., Baumann, G., *Procs. Institute of Mathematics of NAS of Ukraine* **43**, Part 1, 240 (2002).
18. Teismann, H., in H.-D. Doebner, P. Nattermann, and W. Scherer (eds.), *Group 21: Physical Applications and Mathematical Aspects of Geometry, Groups, and Algebras, Vol. 1,* p. 433. Singapore: World Scientific (1997).
19. G. A. Goldin, "From anyons to nonlinear quantum mechanics: Physical predictions from diffeomorphism group representations," in H.-D. Doebner, W. Scherer, and F. Schroeck, Jr. (eds.), *Classical and Quantum Systems: Foundations and Symmetries,* p. 38. Singapore: World Scientific (1993).
20. Doebner, H.-D., Manko, V. I., Scherer, W., *Phys. Lett. A* **268**, 17 (2000).
21. Mavromatos, N. E., Szabo, R. J., *Int. Mod. Phys. A* **16**, 209 (2001).
22. Dashen, R., Sharp, D. H., *Phys. Rev.* (1968).
23. Goldin, G. A., Sharp, D. H. "Lie algebras of local currents and their representations," in V. Bargmann (ed.), *Group Representations in Mathematics and Physics: Battelle-Seattle 1969 Rencontres,* p. 300. *Lecture Notes in Physics* **6**, Berlin: Springer (1970).
24. Goldin, G. A., *J. Math. Phys.* **12**, 462 (1971).
25. Angermann, B., Doebner, H.-D., Tolar, J., in S. I. Andersson and H.-D. Doebner (eds.), *Nonlinear Partial Differential Operators and Quantization Procedures,* p. 171. *Lecture Notes in Mathematics* **1037**, Berlin: Springer (1983).
26. Doebner, H.-D., Stovicek, P., Tolar, J., *Rev. Math. Phys.* **13**, 2246 (2001).
27. Doebner, H.-D., Goldin, G. A., Nattermann, P., *J. Math. Phys.* **40**, 49 (1999).
28. Mielnik, B., *Commun. Math. Phys.* **37**, 221 (1974).

EXPANDED QUANTUM LINEAR HARMONIC OSCILLATOR*

DAVID RITZ FINKELSTEIN
School of Physics, Georgia Institute of Technology
david.finkelstein@physics.gatech.edu

MOHSEN SHIRI-GARAKANI
School of Physics, Georgia Institute of Technology
mohsen.shiri@physics.gatech.edu

Canonical quantization is unstable. The most economical stabilization expands the canonical group to an orthogonal group, with two new quantum constants. The expanded theory of the linear harmonic oscillator is isomorphic to a quantum dipole rotator of large angular momentum quantum number l polarized so that the magnetic quantum number m has nearly its maximum magnitude $m \approx l$. The oscillator quantum number n is $n = l - |m|$. Expanded quantum variables all have discrete finite spectra.

Keywords: Expanded quantum, group contraction, harmonic oscillator, rotation group.

1. Introduction

Segal proposed that physical theories should be stable against small changes in their commutation relations and showed that this requires the group of a physical theory to be simple or semisimple [10]. Groups that are neither, and their Lie algebras, we term compound. They are unstable in the sense of being non-generic, non-robust, singular. They indicate an impending group expansion (inverse group contraction) akin to relativization and quantization, which also make singular theories more robust.

The Heisenberg group $\mathbf{H}(3)$ is compound but is made simple by a group expansion that supplements Planck's quantum constant $\hbar$ with two further

*Talk presented by D.R.F.

quantum constants $\hbar', \hbar''$ associated with an $SO(3)$ triplet [10, 3, 11, 12, 13, 7, 2]. This results in a quantum spacetime and a dynamical i, which we designate by $\breve{i}$. We call this expanded form of quantization Xq for short; X (Chi) is also a fundamental time quantum formed from the constants $c, \hbar, \hbar', \hbar''$.

2. Time-independent oscillator

Here we form the Xq theory of the time-independent linear harmonic oscillator. We preserve the relations among processes, not among variables, in expanded quantization. Before expansion we transform from hermitian variables q, p to skew-hermitian generators

$$\widehat{q} = iq, \quad \widehat{p} = -ip. \tag{1}$$

The quantum canonical commutation relations are then

$$[\widehat{q}, \widehat{p}] = \hbar i \tag{2}$$
$$[i, \widehat{q}] = 0, \tag{3}$$
$$[\widehat{p}, i] = 0, \tag{4}$$
$$i^2 = -1. \tag{5}$$

Then we expand these quantum relations [10, 1]. The three operators $\widehat{q}, \widehat{p}, i$ become three symmetrically related infinitesimal orthogonal-transformation generators $\breve{q}, \breve{p}, \breve{i}$ obeying

$$\begin{aligned} [\breve{q}, \breve{p}] &= \hbar \breve{i}, \\ [\breve{i}, \breve{q}] &= \hbar' \breve{p}, \\ [\breve{p}, \breve{i}] &= \hbar'' \breve{q}, \\ \breve{i}^2 &\approx -1. \end{aligned} \tag{6}$$

We suppose that $\hbar, \hbar' > 0$. If $\hbar'' > 0$ the orthogonal group is SO(3); if $\hbar'' < 0$, SO(2,1). We treat both cases in parallel where convenient.

In this way the Xq process introduces a new dynamically variable generator $\breve{i}$, somewhat as general-relativization introduced the new dynamical variable $g_{\mu\nu}$ the gravitational metric tensor field. The most primitive theory with a dynamical variable like $\breve{i}$ is quaternion quantum field theory [4]. There $\breve{i}$ generates rotations about the electric (or electromagnetic) axis in isospin space, defining a natural Higgs field. We suppose that the present generator $\breve{i}$ is also a Higgs field.

Except for scale factors the Xq relations are satisfied by the three components of an SO(3) or SO(2,1) quantum skew-angular-momentum operator-valued vector $\breve{\mathbf{L}} = \breve{\mathbf{L}} \times \breve{\mathbf{L}}$ for a dipole rotator in three dimensions, highly polarized along one axis, which we take be the third:

$$\begin{aligned}
[\breve{L}_1, \breve{L}_2] &= \breve{L}_3, \\
[\breve{L}_2, \breve{L}_3] &= \breve{L}_1, \\
[\breve{L}_3, \breve{L}_1] &= \breve{L}_2, \\
(\breve{L}_1)^2 + (\breve{L}_2)^2 + (\breve{L}_3)^2 &= -l(l+1), \\
(\breve{L}_3)^2 &\approx -l^2.
\end{aligned} \tag{7}$$

The canonical operators $\breve{x}, \breve{p}, \breve{\imath}$ are just components of the one operator-valued vector $\breve{\mathbf{L}}$, rescaled. This means that in the Xq theory all canonical variables have discrete spectra and transform under an orthogonal group, like rotator angular momenta. We must fix the eigenvalues of the casimirs of the orthogonal group to make the algebra simple. Then the spectra become bounded as well as discrete. The theory is genuinely finite [1].

To fix the scale factors and determine the spectra we set

$$\begin{aligned}
\breve{q} &= Q\breve{L}_1, \\
\breve{p} &= P\breve{L}_2, \\
\breve{\imath} &= J\breve{L}_3,
\end{aligned} \tag{8}$$

By (6),

$$\begin{aligned}
J &= \sqrt{\hbar'\hbar''} \;=\; 1/l, \\
Q &= \sqrt{\hbar\hbar'}, \\
P &= \sqrt{\hbar\hbar''}.
\end{aligned} \tag{9}$$

The commutation relations $\breve{\mathbf{L}} \times \breve{\mathbf{L}} = \breve{\mathbf{L}}$ and the angular momentum quantum number l, defined by $\breve{\mathbf{L}}^2 = -l(l+1)$, determine a simple algebra $\mathrm{Alg}(\mathbf{L}, l)$; where l can have any non-negative integer eigenvalue. The spectral spacing of the $\breve{L}_k$ is 1, so the Xq constants P, Q, J serve as "quanta" of $\breve{p}, \breve{q}, \breve{\imath}$. Since q, p are supposed to have continuous spectra in quantum theory, the constants Q, P are very small on the ordinary quantum scale of $h \sim 1$. It follows that $J = QP/\hbar$ is also very small on that scale.

For $\sqrt{l} \gg 1$, variations $\delta(\breve{\imath}^2) \leq O(l^{-1/2}) \ll 1$ about $(\breve{\imath})^2 = -1$ can be negligible at the same time as the spectral intervals $\delta p \leq P\sqrt{l}$ and $\delta q \leq Q\sqrt{l}$ for quasicontinuous $p, q \approx 0$.

Evidently for given Xq constants P, Q the linear harmonic oscillator with Hamiltonian of the form

$$H = \frac{E_0}{2}\left(\frac{p^2}{P^2} + \frac{q^2}{Q^2}\right) \tag{10}$$

with constant E_0 has a particularly simple Xq theory, with Hamiltonian

$$\breve{H} = \frac{E_0}{2}\left(l(l+1) + (\breve{L}_3)^2\right) \tag{11}$$

The usual oscillator quantum number n is simply

$$n = l - |m|. \tag{12}$$

The Xq energy spectrum is

$$E_n = \frac{E_0}{2}\left(l(l+1) - (l-n)^2\right) \approx lE_0\left(n + \frac{1}{2} - \frac{n^2}{2l}\right) \tag{13}$$

For $n \ll \sqrt{l}$ this reproduces the usual uniformly-spaced energy spectrum as closely as desired.

It is natural to suppose that this H is just part of a more accurate SO(3)-symmetric Xq Hamiltonian

$$H_U = -\frac{E_0}{2}\breve{\mathbf{L}}^2. \tag{14}$$

Changing from H to H_U does not affect the spectrum spacing for $n \ll \sqrt{l}$, but it affects the zero-point energy greatly. What quantum theory describes as a linear harmonic oscillator near the bottom of its n spectrum is actually a dipole rotator near the top of its m spectrum according to this Xq theory. Furthermore, the Newtonian and quantum concepts of energy are invariant under a shift in the zero point of energy, but the angular momentum relations $\mathbf{L} \times \mathbf{L} = \mathbf{L}$ are not invariant under a shift $\mathbf{L}' = \mathbf{L} + \Delta\mathbf{L}$ by any constant vector $\Delta\mathbf{L} \neq 0$. Therefore the Xq relations are not invariant under a shift in the zero-point energy. Therefore this zero-point energy likely contributes to the gravitational field. It will be interesting to estimate its contribution to the missing matter in our cosmos. For a consistent estimate of this effect, we should transquantize the space-time variables t, ∂_t as well as the field variables q, p, since the quantum theory has the same instability in both algebras. Here we take up the time-independent Xq theory only.

3. Clifford algebra representation

Let us first introduce our notation. If C, C' are free Clifford algebras over the respective vector spaces V, V' we write

$$\begin{aligned} C &= \mathbf{2}^V = \Sigma \otimes \Sigma^\dagger, \\ C' &= \mathbf{2}^{V'} = \Sigma' \otimes \Sigma'^\dagger, \\ V &= \triangle C, \quad \Sigma = \sqrt{C}, \end{aligned}$$

$$V' = \triangle C', \quad \Sigma' = \sqrt{C'},$$
$$CC' = \mathbf{2}^{V \oplus V'} = (\Sigma \otimes \Sigma') \otimes (\Sigma^{\dagger} \otimes \Sigma' adj) \tag{15}$$

$\triangle C$ represents the least difference or atom or bit of C.

We use $\triangle$ for bit much as one uses d for differential. $\triangle$ defines a factor in a product, the differential a term in a sum. If s is a variable in C then $\triangle s$ is an independent variable in $\triangle C$.

We call $\triangle S$ the bit (space) of the algebra $S = \mathbf{2}^{\triangle S}$. We write the corresponding systems as $\triangle$S and S. The bit system $\triangle$S is the atom, the system S is the quantum aggregate of atoms.

$C = \mathbf{2}^{\triangle C}$ expresses the system with mode space C as a set of atoms. $C = \Sigma \otimes \Sigma^{\dagger}$ expresses the generic transformation of the same system as a pair of a creation and an annihilation.

If C is a Clifford algebra we write

$$C' = \mathbf{2}^{C} \tag{16}$$

for the free Clifford algebra over the quadratic space consisting of the elements of C with their usual sums and products by scalars, and with the norm

$$\|x\| = \Re x^2, \tag{17}$$

where the operator $\Re$ takes the grade-0 part of its operand.

For typographical convenience we may write exponentials functionally:

$$A^B = A(B), \quad \mathbf{2}^X = \mathbf{2}(X), \quad ,\ldots \tag{18}$$

We are ready now to discuss the representation of the oscillator operators in Clifford algebra. The dynamical operators of the oscillator form the dynamical algebra A of the oscillator. We represent the Xq algebra $\breve{A}$ using a product of many replicas of the Clifford algebra $C(3)$ (or $C(2,1)$) appropriate to the orthogonal group of the oscillator.

We describe joint input and output (IO) operations on the oscillator [5] by elements of the co-algebra A^* dual to A, composed of the linear maps $A \to \mathbb{R}$. Elements of A^* are called co-operators.

This means that we represent I and O processes separately by spinors. We might instead represent I processes by Clifford elements. Then we would have to represent $\breve{q}$ and $\breve{p}$ by linear operators $C \to C$ rather than by elements of C.

Our prototype Clifford algebra $C(3)$ is the linear associative real algebra generated by the vectors v of the Euclidean vector-space R^3 with the positive-definite metric $\|v\| = (v^1)^2 + (v^2)^2 + (v^3)^2$, subject to the Clifford law in the form

$$v^2 = \|v\|. \tag{19}$$

We choose an orthonormal basis $\gamma_1, \gamma_2, \gamma_3$ for E^3, with many orthogonal (anticommuting) replicas $\gamma_k(n)$ $(k = 1, 2, 3; n = 1, 2, \ldots, N)$. We associate the axes 1 and 2 with the real and imaginary axes of the complex plane, and represent $\breve{\imath}$ using N replicas

$$\gamma_{12}(n) := \gamma_1(n)\gamma_2(n) \tag{20}$$

of $\gamma_{12} := \gamma_1\gamma_2$:

$$\breve{\imath} := \frac{1}{N}\sum_{n=1}^{N} \gamma_{12}(n). \tag{21}$$

We also associate the 2 axis with q and the 1 axis with p, so that

$$\begin{aligned} L_1 &= \sum_{n=1}^{N} \gamma_{13}(n), \\ L_2 &= \sum_{n=1}^{N} \gamma_{23}(n). \end{aligned} \tag{22}$$

This illustrates how Clifford algebra serves as the universal algebraic language of Xq theory. The Clifford algebraic theory of the time-dependent oscillator is under study.

Acknowledgments

Helpful discussions with James Bough, Eric Calren and Heinrich Saller are gratefully acknowledged.

References

1. Atakishiyev, N. M., G. S. Pogosyan, and K. B. Wolf. Contraction of the finite one-dimensional oscillator. *International Journal of Modern Physics* **A18**, 317-327 (2003).
2. J. Baugh, D. R. Finkelstein, A. Galiautdinov, and M. Shiri-Garakani. Transquantum dynamics .*Foundations of Physics* **33** 1267 (2003).
3. Carlen, E. and R. Vilela Mendes. Non-commutative space- time and the uncertainty principle. *arXiv: quant-ph/0106069* **1-12**, (Jun 2001).
4. Finkelstein, D., , J.M. Jauch, S. Schiminovich and D. Speiser. Principle of general Q-covariance, *Journal of Mathematical Physics* 4, 788-796 (1963)

5. D. Finkelstein. *Quantum Relativity.* Springer-Verlag, Heidelberg (1996).
6. Finkelstein, D., A. Galiautdinov. Cliffordons. *Journal of Mathematical Physics* **42**, 3299-3314 (2001).
7. Galiautdinov A.A., and D. R. Finkelstein. Chronon corrections to the Dirac equation. hep-th/0106273 [LANL]. *Journal of Mathematical Physics* **43**, 4741 (2002).
8. Kuzmich, A., N. P. Bigelow, and L. Mandel, *Europhysics Letters* **A 42**, 481 (1998). Kuzmich, A., L. Mandel, J. Janis, Y. E. Young, R. Ejnisman, and N. P. Bigelow, *Physical Review A* **60**, 2346 (1999). Kuzmich, A., L. Mandel, and N. P. Bigelow, *Physical Review Letters* **85**, 1594 (2000).
9. G. 'tHooft, *International Jornal of Theoretical Physics*, **42-2**, 355-361 (Feb 2003).
10. Segal, I. E. . A class of operator algebras which are determined by groups. *Duke Mathematics Journal* **18** 221 (1951).
11. Vilela Mendes, R. *Journal Of Physics A: Math, Gen*, **27**, 8091 (1994).
12. Vilela Mendes, R. *Physics Letters* **A210**, 232 (1996).
13. Vilela Mendes, R. *J. Math. Phys.* **41**, 156 (2000).

THERMODYNAMIC LIMITS, NON-COMMUTATIVE PROBABILITY, AND QUANTUM ENTANGLEMENT

J. F. JOHNSON

Mathematics Department,
Villanova University
Villanova, PA 19085, USA
E-mail: joseph.f.johnson@villanova.edu

We construct a rigourous model of quantum measurement. A two-state model of a negative temperature amplifier, such as a laser, is taken to a classical thermodynamic limit. In the limit, it becomes a classical measurement apparatus obeying the stochastic axioms of quantum mechanics. Thus we derive the probabilities from a deterministic Schroedinger's equation by procedures analogous to those of classical statistical mechanics. This requires making precise the notion of 'macroscopic.'

1. Introduction

Although Quantum Mechanics, in the form discovered by Heisenberg, Schroedinger, and Dirac in 1925, is the most successful physics theory in history, explaining almost every phenomenon within laboratory dimensions, especially including all of chemistry and biology, there has always been some dissatisfaction with it. Nobel prize-winners such as Einstein, de Broglie, Planck, and Schroedinger himself have all ended up dissenting from the usual interpretation of Quantum Mechanics. At issue is not the formal, mathematical methods of calculating experimentally verifiable quantities, but rather what philosophical interpretation or mental picture of the wave-particle duality can be considered true, or even sensible? Wigner[1] analysed this as the problem of Quantum Measurement. As long as we do not measure anything, all Nature is governed by a deterministic wave equation, Schroedinger's equation, and what the wave will do in the future is completely determined by what it did in the past. All calculations of experimentally verifiable quantities involve calculations involving this wave and using the wave equation. But whenever we measure a system, its wave makes some kind of jump which is a violation of the wave equation, is random and unpredictable, and our measurement always seems to be the detection of a particle (the quantum of the wave's field). This is no logical contradiction. Different views on the subject make no practical difference to actual experimental design. So, for all practical purposes so far, this puzzle is irrelevant. On the other hand, advances in mesoscopic engineering and quantum computation may upset this quiescent state of affairs. Nevertheless, as the famous particle physicist, John S. Bell remarked more than once, "as long as the [jump] is an essential component, and as long as we do not know exactly when and how it takes over from the Schroedinger equation, we do not have an exact and unambiguous formulation of our most fundamental theory."[2]

2. The precise problem

The main difference between Quantum Mechanics and Classical Mechanics, ever since the time of de Broglie, has been that a particle is described by a wave and a wave equation. So is a two-particle system. In fact, any system, no matter how large, is, if it is a closed system, described by a wave function and obeys a wave equation, Schroedinger's equation. The next most important difference, and it only was realised a few years later, by Dirac, Heisenberg, and Born, is that if a closed system is going to be measured, then it is not closed anymore! It is being impacted by the measuring apparatus in an unavoidable, essential way. Therefore, a jump may occur, and the laws of Quantum Mechanics for measurement processes no longer is a wave equation, but a way of calculating the probabilities of the different jumps and the different results of the measurement, as registered by different pointer positions on the macroscopic measuring apparatus for example. Although these new rules use the wave function, they violate Schroedinger's equation, and are needed in order to agree with experiment.

This is new, novel and counter-intuitive, but it is not illogical or inconsistent. Most practicing physicists especially in the thirties could tell when they were measuring a microscopic system and when they were not, so they "just knew" when to use Schroedinger's equation and when to use the jump rules, which were, in those days, referred to as 'the reduction of the wave packet.' The convenient consensus grew up, that actually, Schroedinger's equation only applied to microscopic systems that were so small that they could not be measured except by amplifying devices. And that measuring devices such as amplifying devices—this includes photographic plates, photo-amplfiiers, lasers, cloud chambers, bubble chambers, Geiger counters, just about everything—*had* to be macroscopic and so large that Quantum Mechanics was not really applicable, one could use classical Newtonian Mechanics for them. Of course this was just a dodge to avoid dealing with an intractable problem. Recent advances in technology have eroded this consensus: macroscopic quantum tunnelling has been exhibited in the laboratory, mesoscopic quantum chaos has been detected, and it is no longer believed that one can draw a line between the microscopic realm and the realm of measurement apparati. They must be governed by the same laws of Nature.

Wigner's fundamental article[1] draws out the logic of this situation quite clearly. According to Quantum Mechanics, there are two seemingly incompatible ways to analyse the following physical process: Suppose we have a particle being detected by a Geiger counter or any measurement apparatus that detects particles. Suppose further that the counter is tuned so that it will detect the particle if it is vertically polarised, but not detect it if it is horizontally polarised. Now the laws of Quantum Mechanics tell us that the wave function of the incident particle can be either purely vertically polarised, or purely horizontally polarised, or some superposition (it i.e., linear combination) of the two. On the one hand, this physical process is a measurement. The combined system of incident particle and measurement apparatus is a measurement, and therefore the laws of Quantum Mechanics about measurements say that the particle will have some probability of being detected and jumping into a state of pure vertical polarisation, and will have the complementary probability of being absorbed and *not* detected. The probability depends on how much of the pure vertical state is present in the superposition.

On the other hand, this combined system of particle plus apparatus is a physical, closed system, governed by the laws of Nature, which are Quantum Mechanics, which means Schroedinger's equation. There must be *one* wave function which describes the *joint* behaviour of the particle *cum* apparatus. This total wave function evolves according to the appropriate wave equation of the total combined system, and does not jump, and there are no probabilities: it evolves completely deterministically.

Wigner's own solution to this philosophical puzzle has not been widely accepted. It is that Schroedinger's equation is true as long as consciousness does not *look* at the result of the measurement, but when consciousness is involved, some non-linear equation modifies the behaviour of the total wave, with unpredictable results. He could not propose an actual equation, so this solution need not detain us.

Now, there is no way to compare a wave function of a large object, such as an apparatus, with the sense perceptions which we have of that large object, for example, our conscious visual observation of its pointer position. The wave equation allows for the object to have one possible state, say the pointer position being "detected!" and another, quite different, state, say, the pointer points to "not detected" but it also allows for the apparatus to be in a superposition of states. No one knows what such a superposition of states would look like, and current opinion is divided on whether they can exist, or whether they do exist but decay rapidly via ad hoc super-selection rules into sharp states when observed or jiggled or something. In superconductivity phenomena, such states have been detected experimentally (but invisibly, so we still don't know what they 'look' like).

Since there are no rules about what such superpositions look like, or mean macroscopically, there is no logical contradiction between the two different ways to analyse the same physical measurement process. Heretofore, there has been no way to compare the view which cuts off the apparatus from the incident particle, describing one by a wave function but the other classically, with definite pointer positions at all times and never quantum superpositions—with the view which does not cut Nature anywhere, but studies the joint wave function in its continuous deterministic evolution. Since there is no way to confront the two ways with each other, there is no contradiction. The puzzle has been the status of their co-existence.

This paper introduces the first precise notion of comparison of macroscopic with state vectors. It seems that there ought to be a way to compare the macroscopic description of some body, some object, such as a measurement apparatus or our sense-perceptions of it and of its pointer positions, with the microscopic description of the same object, which consists simply and solely of its wave function. Heretofore, there has not been a precise definition of macroscopic, and the rough-and-ready thirties style definition is unconvincing. Nor has there been a precise definition, in terms of the wave equation, of measurement process. In a recent paper,[3] I have proposed a solution to these two questions which solves the problem of quantum measurement, or at least that part of it which work in the last decade has concentrated on (the jumps have usually been regarded as[4] a merely subjective selection from the mixture, few now believe that they really happen).

Macroscopic entanglement is a fact of life (*pace* W. Zurek[5]). Therefore, we take Schroedinger's equation to be exactly valid over laboratory dimensions, including the amplifying apparatus.

Then the problem becomes how to explain mathematically how the wave of the measured system, when interacting with the wave of the amplifying apparatus, produces intermittent particle events such as the click of a Geiger counter. The production of a discontinuous transition from continuous inputs is often the result of an interchange of limits phenomenon, and that is what we have here.

3. Coherent superposition and Quantum entanglement

It is the possibility of coherent superposition which makes quantum interference effects possible, possibly the most sensational effects of the so-called wave-particle duality. In Quantum Mechanics, a particle, such as an electron, is modelled by a wave function. Many physicists have thought that the wave is real, and the wave function represents something real. There have been many distinguished dissenters, however. For the sake of discussion, let us assume that the wave is at least equally real as the particle, since this makes the problem more difficult to solve, not easier.

Abstractly, the two-slit experiment boils down to the following, if we put a photodetector directly in one of the two slits. The state space $\mathcal{H}_o$ of the particle is spanned by ψ_1, meaning it definitely passes through the slit where the detector is, and ψ_o, meaning it definitely passes through the other slit and fails to be detected. But any coherent superposition $\psi = c_o\psi_o + c_1\psi_1$ is possible. The c_i are called the quantum amplitudes. If the detector M_n has n degrees of freedom call its Hilbert space $\mathcal{H}_n$. If we assume, as Wigner, von Neumann, and practically everybody else do, that the strong topology is physically relevant to the 'pointer variable' of M_n, then we may assume that one position of the pointer variable, 'detection,' the loud click or bright flash, correponds to $|0\rangle \in \mathcal{H}_n$. Failure to go off, remaining in the charged state of population inversion, is $|1\rangle$. Then no matter what the joint Hamiltonian H^{com} on $\mathcal{H}_o \otimes \mathcal{H}_n$ is, it must take $\psi_o \otimes |1\rangle$ to $\psi_o \otimes |1\rangle$ and $\psi_1 \otimes |1\rangle$ to $\psi_1 \otimes |0\rangle$. By linearity, it is forced to take $(c_o\psi_o + c_1\psi_1) \otimes |1\rangle$ to $c_o\psi_o \otimes |1\rangle + c_1\psi_1 \otimes |0\rangle$ which is not a decomposable state. The apparatus and the particle are "entangled." It is no longer acceptable to introduce *ad hoc* extra postulates to eliminate this entanglement.

The superposition $c_o\psi_o \otimes |1\rangle + c_1\psi_1 \otimes |0\rangle$ is called coherent. Bell has described its evolution into the statistical mixture (called incoherent)

$$|c_o|^2\chi_{\{\psi_o\otimes|1\rangle\}} + |c_1|^2\chi_{\{\psi_1\otimes|0\rangle\}}$$

as "the philosopher's stone" of quantum measurement. Here χ means the characteristic function of a set, and our mixture is a atomic probability density on a two point space. Hence, the $|c_i|^2$ are classical probabilities.

Thus the problem of Quantum Measurement is as follows: how do the quantum amplitudes ever get turned into classical probabilities? Schroedinger's equation is a linear wave equation, a superposition of two states ψ_1 and ψ_2 with quantum amplitudes c_1 and c_2, evolves forever for all future time as a coherent superposition with the same quantum amplitudes of two continuously, smoothly evolving quantum states. Hence, if the incident particle is in such a superposition, so is the combined system of particle *cum* apparatus, and it remains such, forever. It *never* changes the quantum amplitudes, not even into different quantum amplitudes, much less into probabilities or jumps. So how can we explain why, when we analyse the same physical process differently, we obtain that $|c_1|^2$ and $|c_2|^2$ are the probabilities that the measurement apparatus can be observed to be in either one classical pointer position or in the other?

It would go against the grain of the experimental evidence (that entanglement persists) to try to derive this as a time evolution. We, instead, derive this as a thermodynamic limit.

Quantum amplitudes are essentially quantum mechanical, wave phenomena. They are preserved by Schroedinger's equation. Classical probabilities can be calculated from the amplitudes, and are observed when we measure, but only some sort of 'cut' between the use of the wave equation and the use of the probability rules can tell us when to use one and when to use the other.

4. Where do we put the cut?

Wigner said, put the cut to separate matter from consciousness. Bohr and von Neumann said, put the cut in between the system being measured and the classically describable, macroscopic measuring apparatus. Von Neumann went further and said that if you had another measuring apparatus to measure the first measuring apparatus, you could go ahead and shift the cut so that the joint wave function of incident particle *cum* first apparatus was being treated as a wave, smoothly, but any measurement of it by the second apparatus introduced the jumps. The praxis of the average physicist in the thirties just said, put the cut somewhere in the mesoscopic realm so it divides small molecules from our amplifiers.

The solution which I have proposed is as follows. We shift the cut out to infinity. It only occurs in the sort of idealised limit which is used in Classical Thermodynamics and probability theory. No real object is macroscopic, nevertheless, if the number of particles or degrees of freedom in it is quite large, the idealised limit of macroscopy is a physically valid approximation. No real object behaves exactly according to the axioms governing measuring processes. But the approximation is, as usual in Classical Thermodynamics, exceedingly good. An idealised limit object would behave exactly according to the axioms, and we can derive its properties mathematically from Schroedinger's equation simply by passing to the thermodynamic limit in the way usual in logically careful accounts of classical Statistical Mechanics such as those by Darwin–Fowler,[6] Khintchine,[7] and Ford–Kac–Mazur.[8,9] Thus we can derive the probability rules of Quantum Mechanics from the deterministic rules of Quantum Mechanics by passing to the thermodynamic limit. The program, as has existed for a long time now, is to write down the wave equation for the joint system of incident particle plus amplifying detector.[3] As one lets n, the number of particles in the detector, go to infinity, the pointer positions of the detector behave approximately classically, and the approximation improves the more n increases. In the limit, the detector is a classical object, there are no superpositions of pointer positions, and the usual measurement rules of Quantum Mechanics are exactly true. But at every point short of the limit, Schroedinger's equation is instead exactly true, and the continuous, deterministic evolution of the total wave is also true. But this is the same as in Ford–Kac–Mazur's paper (except that they were not thinking about Quantum Measurement, so they were not thinking about amplifying devices, so they only studied positive temperature and not negative temperature), it is a standard feature of classical Statistical Mechanics. Which is what led Einstein to ask, once, whether the probabilities of Quantum Mechanics did not arise out of the deterministic dynamics in an analogous way to the way they did in classical Statistical Mechanics (including Thermodynamics as a special case). He seems to have thought that Quantum Mechanics was incomplete[10], and would have to be completed in some radical fashion for this to be true, but Schroedinger did not think so. And it turns out that it is not *necessary* to alter the usual wave equation in order to answer Einstein's question in the affirmative. If future experimental phenomena uncover new physical laws which require the modification of Schroedinger's equation, that would be a different matter. We need not postulate unmotivated modifications to Schroedinger's equation just to solve the problem of Quantum Measurement. Any new equations discovered to explain new physical forces will probably be susceptible to the same thermodynamic limit treatment, allowing the same sort of solution to the problem of measurement.

We cannot, naively, pass directly to the limit object by taking the tensor product of an infinite number of copies of $\mathcal{H}_o$. Although this non-separable space does have separable Hilbert subspaces, they are not preserved by the dynamics of the measurement process. Nor is this limit 'classical'. Techniques[11] useful for infinite volume quantum limits are not applicable to this physical situation. Farhi, Goldstone, and Gutmann[12] have attempted to derive the probabilistic axioms from the deterministic axioms, without using statistical mechanical techniques, also leaving aside the reduction of the wave packet. They do not model the measurement process and their methods are purely formal, without physical content. The formalities are invalid, resting on both a simple statistics fallacy and crucial use of the same symbol with two incompatible meanings.

5. The physical content of the solution

We take Quantum Mechanics as a complete description of Nature, and Schroedinger's linear wave-equation as exact, yielding a unitary deterministic time-evolution for any system not subject to outside influences. Since nothing physical is outside Nature, anytime we wish to improve our approximation of an actual physical system, all we have to do is include more of the forces acting on it. This includes the analysis of measurement processes. All systems are taken as closed systems, and only time-independent Hamiltonians are considered. All states are taken as pure.

Measuring devices have to be large enough so that they may be approximately treated as macroscopic, and have to involve amplification and a coupling to the system they are measuring. Amplification of a microscopic effect means that the apparatus has to be in an unstable state of population inversion (this insight is due to Schroedinger and Green[13]) like a laser or cloud chamber, ready to undergo a sort of chain reaction at the smallest touch, provided it is the right sort of touch to touch off the chain reaction! (This insight is due to Daneri, Loinger, and Prosperi.[14]) (Schwinger[15] has also remarked that the negative temperature amplification of a laser amplifies quantum motion to the classical level.) Sometimes such a reaction is called discontinuous and irreversible, but this is not true at any finite point short of the thermodynamic limit, which is an idealisation never to be attained. (In Classical Thermodynamics, too, the discontinuity of phase transitions are only artifacts of passing to the infinite limit. I would like to thank Goldstone for helping me see the relevance of this. Even according to Classical Thermodynamics, there isn't really any such thing as water or steam. If you looked at the individual molecule, you could not tell for sure which state it was in. Neither temperature nor phases really exist, they are artifacts of the approximation procedure involved in passing to the thermodynamic limit. They are still physically valid conceptual constructs.)

This approximation does not take place in the strong topology, and neither does ours. Just which quantum-mechanical property or observable of the incident particle is measured depends explicitly on the physics of the coupling. We thus *derive* non-commutative probability from ordinary, commutative, Kolmogoroffian probability. That is, for a fixed experimental set up, the abelian probabilities of ordinary classical statistical mechanics are implemented as an algebra generated by one two-by-two matrix—thus abelian. But choice of a second, alternative, incompatible experimental set-up yields a different abelian algebra which would not have commuted with the first alternative. The two non-commuting quantum probability 'variables' can never be both realised at the same time. Hence quantum probability is of strictly *derivative* status from commutative probability and the physics of the coupling.

Only some features of the measuring apparatus are what we define as macroscopically observable, quantum states which differ by negligible numbers of occupation numbers (by negligible, we mean in some vanishing proportion in the thermodynamic limit) are not distinguishable, at least in the limit, by macroscopic observables. (This is the exact opposite of Hepp's notion[16] of macroscopic observable ...) Some macroscopic observables are not very relevant to the analysis of measurement, those that are we call macroscopic pointer variables. Very few such exist, and for a given coupling, only inessential variations on a canonical representative seem to exist. In the thermodynamic limit, these pointer variables pass into classical probability measures on a classical object, the limit of the quantum system.

There are only waves. There are no particles. The fact that measurements seem to detect particles is an artifact of the process of amplification. The seeming 'particle-events' that occur when we measure a field and detect a quantum are the product of the interaction of the wave functions of the measuring apparatus and the microscopic system being detected.

Since thermodynamic limits do not actually exist in Nature (no amplifying device with an actual infinite number of degrees of freedom *relevantly coupled* to the incident particle have yet been observed in Nature), none of the jumps physically occur, and the probabilities we calculate are merely convenient ways to talk about the results of experiments, approximately valid because of the usefulness of thermodynamic limits in calculating good approximations to expectation values or time averages.

6. Mathematical content

The only new feature is the rigourous definition of macroscopic observable. In keeping with our program, these are classical dynamical variables, i.e., abelian functions on the phase space. Let $\mathcal{H}_n = \mathcal{H}_{n-1} \otimes \mathcal{H}_0$ be a sequence of state spaces with an increasing number of degrees of freedom. Let f_n be a real-valued measurable function on $\mathcal{H}_n$. Let $\langle f_n \rangle_t$ be the time average of f_n over a trajectory and we assume that this is almost independent of the initial condition chosen in the space of accessible states. The sequence $\{f_n\}$ is a macroscopic observable if it is, so to speak, a tail event. More precisely, we require that if $\{v_n\}$ is a sequence of elements of $\mathcal{H}_o$ (of norm one) such that $\lim_{n\to\infty} \langle f_n(v_o \otimes v_1 \otimes \cdots \otimes v_n)\rangle_t$ exists, then this limit is independent of finite substitutions in the v_i. This implements the intuition that naked eye observation cannot detect microscopic variations in the state and can only detect properties which pass to the thermodynamic limit in which Planck's constant is negligible. Details will appear elsewhere.

Pointer positions are examples of a macroscopic observable.

Because of the coupling, all other macroscopic observables are either trivial or are functions of the two pointer positions. Therefore, in the limit, the values of all macroscopic observables can be obtained as the expectations of probability distributions on a two point space, $\{P_o, P_1\}$, representing the two pointer positions.

References

1. E. Wigner, *Am. J. Phys.* **31**, 6 (1963).
2. J. S. Bell, *Helv. Phys. Acta* **48**, 447 (1975); *Phys. World* **3**, 33 (1980).
3. J. F. Johnson, in *Proc. VIIIth Inter. Wigner Symposium, N. Y. 2003,* edited by S. Catto and B. Nicolescu (to be published).
4. K. Hannabuss, *Helv. Phys. Acta* **57**, 610 (1984); *Ann. Phys.* **239**, 296 (1995).
5. W. Zurek, *Phys. Rev.* **D24**, 1516 (1981); *Physica Scripta* **76**, 186 (1998).
6. C. G. Darwin and R. H. Fowler, *Phil. Mag.* **44**, 450, 823 (1922).
7. A. Khinchin, Mathematical Foundations of Statistical Mechanics, N. Y. 1949.
8. G. W. Ford, M. Kac and P. Mazur, *J. Math. Phys.* **6**, 504 (1965).
9. J. von Plato, in *Causation, Chance, and Credence,*, edited by B. Harper and S. Skyrms, Dordrecht, 1988.
10. P. A. M. Dirac, *Directions in Physics*, N. Y., 1978.
11. F. Benatti, *Deterministic Chaos in Infinite Quantum Systems*, Trieste 1993.
12. E. Farhi, J. Goldstone and S. Gutmann, *Ann. Phys.* **192**, 368 (1989).
13. H. S. Green, *Nuovo Cimento* **9**, 880 (1958).

14. A. Daneri, D. Loinger and G. Prosperi, *Nucl. Phys.* **33**, 297 (1962).
15. J. Schwinger, *J. Math. Phys.* **2**, 407 (1961).
16. K. Hepp, *Helv. Phys. Acta* **45**, 237 (1972).

CANONICAL REDUCTION OF GRAVITY: FROM GENERAL COVARIANCE TO DIRAC OBSERVABLES AND POST-MINKOWSKIAN BACKGROUND-INDEPENDENT GRAVITATIONAL WAVES

LUCA LUSANNA

Sezione INFN di Firenze, Polo Scientifico
v.Sansone 1, 50019 Sesto Fiorentino, Italy
lusanna@fi.infn.it

In general relativity the metric tensor has a double role: it is the potential of the gravitational field and simultaneously describes the chrono-geometrical structure of the space-time in a dynamical way by means of the line element. It teaches to all the other fields relativistic causality: for instance it selects the paths to be followed by the rays of light in the geometrical optic approximation. This aspect of the metric tensor is completely lost every time it is split in a background metric plus a perturbation. This implies that, like in special relativity, we introduce a background non-dynamical (absolute) chrono-geometrical structure and the perturbation (like every other massless field) propagates along the fixed background light-cone. As a consequence, since we do not know realistic solutions of Einstein's equations for macroscopic bodies, the only existing theory of measurement in general relativity is the axiomatic one of Ehlers, Pirani and Schild [1], which utilizes *test* massive particles and rays of light in place of their *dynamical* counterparts. Moreover, we have that in the approaches to quantum gravity based on the introduction of a background metric (effective quantum field theory, string theory) gravity is simulated by gravitons with the same space-time behaviour as photons and gluons except for the spin and no method is know to rebuild the dynamical chrno-geometrical structure.

On the other hand the only sufficiently developed background-independent approach to quantum gravity, namely loop quantum gravity, gives rise to a suggestive quantum geometry for the 3-space but has problems with time evolution and does not correspond to a Fock space. As a consequence, it is not known how to introduce electro-magnetism (not to speak of the standard model of elementary particles) so to make a comparison with ordinary perturbative quantum electro-dynamics and understand its modifications induced by gravity.

This state of affairs motivated an attempt [2,3,4] to revisit classic metric gravity [2] and its ADM Hamiltonian formulation to see whether it is possible to define a model of general relativity able to incorporate fields and particles and oriented to a background-independent quantization. First of all to include fermions it is natural to resolve the metric tensor in terms of cotetrad fields [3,4] [$g_{\mu\nu}(x) = E^{(\alpha)}_{\mu}(x)\, \eta_{(\alpha)(\beta)}\, E^{(\beta)}_{\nu}(x)$; $\eta_{(\alpha)(\beta)}$ is the flat Minkowski metric in Cartesian coordinates] and to reinterpret the gravitational field as a *theory of time-like observers endowed with tetrads*, whose dynamics is controlled by the ADM action thought as a function of the cotetrad fields. Since the standard model of elementary particles and its extensions are a chapter of the theory of representations of the Poincare' group on the non-compact Minkowski space-time and we look for a Hamiltonian decription, the mathematical pseudo-Riemannian 4-manifold M^4 introduced to describe space-time is assumed to be non-compact and globally hyperbolic. This means that it admits 3+1 splittings with foliations whose leaves are space-like Cauchy 3-surfaces assumed diffeomorphic to R^3 (so that any two points on them are joined by a unique 3-geodesic). As discussed in Ref.[5] these 3-surfaces are also *simultaneity surfaces*, namely a convention for the synchronization of clocks (in general different from Einstein's one, valid in inertial systems in special relativity). If τ is the mathematical time labeling these 3-surfaces, Σ_τ, and $\vec{\sigma}$ are 3-coordinates (with respect to an arbitrary observer, a centroid $x^\mu(\tau)$, chosen as origin) on them, then $\sigma^A = (\tau, \vec{\sigma})$ can be interpreted as Lorentz-scalar radar 4-coordinates and the surfaces Σ_τ are described by embedding functions $x^\mu = z^\mu(\tau, \vec{\sigma})$. In these coordinates the metric is $g_{AB}(\tau, \vec{\sigma}) = z^\mu_A(\tau, \vec{\sigma})\, g_{\mu\nu}(z(\tau, \vec{\sigma}))\, z^\nu_B(\tau, \vec{\sigma})$ [$z^\mu_A = \partial z^\mu / \partial \sigma^A$]. Since the 3-surfaces Σ_τ are *equal time* 3-spaces with all the clocks synchronized, the spatial distance between two equal-time events will be $dl_{12} = \int_1^2 dl \sqrt{{}^3g_{rs}(\tau, \vec{\sigma}(l))\, \frac{d\sigma^r(l)}{dl}\, \frac{d\sigma^s(l)}{dl}}$ [$\vec{\sigma}(l)$ is a parametrization of the 3-geodesic γ_{12} joining the two events on Σ_τ]. Moreover, by using test rays of light we can define the *one-way* velocity of light between events on Σ_τ. Therefore, the Hamiltonian description has naturally built in the tools (essentially the 3+1 splitting) to make contact with experiments in a relativistic framework, where simultaneity is a frame-dependent property. The manifestly covariant description using Einstein's equations is the natural one for the search of exact solutions, but is inadequate to describe experiments.

Other requirements [3,4] on the Cauchy and simultaneity 3-surfaces Σ_τ induced by particle physics are:

i) Each Σ_τ must be a Lichnerowitz 3-manifold [6], namely it must admit an involution so that a generalized Fourier transform can be defined and the notion of positive and negative frequencies can be introduced (otherwise the notion of particle is missing like it happens in quantum field theory in arbitrary curved space-times).

ii) Both the cotetrad fields (and the metric tensor) and the fields of the standard model of elementary particles must belong to the same family of suitable weighted Sobolev spaces so that simultaneously there are no Killing vector fields on the space-time (this avoids the cone-over-cone structure of singularities in the space of metrics) and no Gribov ambiguity (either gauge symmetries or gauge copies [7]) in the particle sectors; in both cases no well defined Hamiltonian description is available.

iii) The space-time must be *asymptotically flat at spatial infinity* and with boundary conditions there attained in a way independent from the direction (like it is needed to define the non-Abelian charges in Yang-Mills theory [7]). This eliminates the supertranslations (the obstruction to define angular momentum in general relativity) and reduces the *spi group* of asymptotic symmetries to the ADM Poincare' group. The constant ADM Poincare' generators should become the standard conserved Poincare' generators of the standard model of elementary particles when gravity is turned off and the space-time becomes the Minkowski one. As a consequence, the *admissible foliations* of the space-time must have the simultaneity surfaces Σ_τ tending in a direction-independent way to Minkowski space-like hyper-planes at spatial infinity, where they must be orthogonal to the ADM 4-momentum. But these are the conditions satisfied by the Christodoulou-Klainermann space-times [8] , which are near Minkowski space-time in a norm sense. Therefore the surfaces Σ_τ define the *rest frame* of the τ-slice of the universe and there are asymptotic inertial observers to be identified with the *fixed stars* (the standard origin of rotations to study the precession of gyroscopes in space). In this class of space-times there is an *asymptotic Minkowski metric* (asymptotic background), which allows to define weak gravitational field configurations *without splitting the metric* in a background one plus a perturbation and without being a bimetric theory of gravity.

The absence of supertranslations also implies that the lapse and shift functions have the form $N(\tau,\vec{\sigma}) = \epsilon + n(\tau,\vec{\sigma})$, $N_r(\tau,\vec{\sigma}) = n_r(\tau,\vec{\sigma})$ [$\epsilon = +1$

with signature (+—) and $\epsilon = -1$ with signature (-+++)], with $n(\tau, \vec{\sigma})$, $n_r(\tau, \vec{\sigma})$ tending to zero at spatial infinity.

In ADM tetrad gravity there are 16 configuration variables: the cotetrad fields can be parametrized in terms of $n(\tau, \vec{\sigma})$, $n_r(\tau, \vec{\sigma})$, 3 boost parameters $\varphi_{(a)}(\tau, \vec{\sigma})$, 3 angles $\alpha_{(a)}(\tau, \vec{\sigma})$ and cotriads $e_{(a)r}(\tau, \vec{\sigma})$ on Σ_τ $[a = 1, 2, 3]$. There are 14 first class constraints $\pi_n(\tau, \vec{\sigma}) \approx 0$, $\pi_{n_r}(\tau, \vec{\sigma}) \approx 0$, $\pi_{\varphi_{(a)}}(\tau, \vec{\sigma}) \approx 0$ (the generators of local Lorentz boosts), $M_{(a)}(\tau, \vec{\sigma}) \approx 0$ (the generators of local rotations), $\Theta^r(\tau, \vec{\sigma}) \approx 0$ (the generators of the changes of 3-coordinates on Σ_τ) and $\mathcal{H}(\tau, \vec{\sigma}) \approx 0$ (the superhamiltonian constraint).

It can be shown [2,4] that the addition of the DeWitt surface term to the Dirac Hamiltonian (needed to make the Hamiltonian theory well defined in the non-compact case) implies that the Hamiltonian does not vanish on the constraint surface (no frozen reduced phase space picture in this model of general relativity) but is proportional to the *weak ADM energy* (i.e. its form as a volume integral), which governs the τ-evolution [9]. Moreover the Hamiltonian gauge transformations generated by the superhamiltonian constraint do not have the Wheeler-DeWitt interpretation (evolution in local time), but transform an admissible 3+1 splitting into another admissible one (so that all the admissible notions of simultaneity are gauge equivalent). See Refs.[2,4] for the scheme of addition of gauge fixings and, in particular, for the bibliography showing that a completely fixed Hamiltonian gauge becomes a unique 4-coordinate system on space-time only on the solutions of Einstein's equations (on-shell), so that it describes an extended physical laboratory. Finally it can be shown that Lichnerowicz's identification of the conformal factor of the 3-metric on Σ_τ ($\phi = [det\, {}^3g]^{1/12}$) as the unknown in the superhamiltonian constraint is the correct one: as a consequence the gauge variable describing the normal deformations of the simultaneity surfaces Σ_τ is the momentum $\pi_\phi(\tau, \vec{\sigma})$ canonically conjugate to $\phi(\tau, \vec{\sigma})$ (and not the trace of the extrinsic curvature of Σ_τ, the so called intrinsic York time).

In Ref.[10] there is a review of the various implications of Einstein's Hole Argument,namely of the property of general covariance. The invariance of Einstein's equations under active diffeomorphisms (the widest local symmetry group of general relativity, whose passive reinterpretation by Bergmann and Komar contains the ordinary coordinate transformations and the Legendre pull-back of the Hamiltonian gauge transformations) imply: i) the absence of determinism (only two of Einstein's equations contain dynamical information: four are restrictions on initial data and four are void due to Bianchi identities), i.e. the presence of arbitrary gauge variables;

ii) absence of a physical individuation of the mathematical points as physical point-events of space-time. On one side such an individuation can be achieved by formulating four of the gauge fixing constraints as the requirement that a set of ordinary 4-coordinates coincides with four suitable scalar functions of the four eigenvalues of the Weyl tensor (Bergmann-Komar intrinsic pseudo-coordinates). This implies that in a sense the *space-time is the gravitational field itself* and that each 4-coordinate system has a *non-commutative structure* associated to it already at the classical level.

On the other side, following Dirac, we have to identify a canonical basis $r_{\bar{a}}(\tau,\vec{\sigma})$, $\pi_{\bar{a}}(\tau,\vec{\sigma})$, $\bar{a}=1,2$, of *Dirac observables* (DO) as the physical degrees of freedom of the gravitational field, i.e. a canonical basis of predictable gauge-invariant quantities satisfying deterministic Hamilton equations governed by the weak ADM energy. This can be achieved by means of a Shanmugadhasan canonical transformation adapted to 13 of the 14 first class constraints (not to the superhamiltonian one), which turns out to be a *point* canonical transformation as a consequence of the form of the finite gauge transformations. In the new canonical basis 13 new momenta vanish due to the 13 constraints and their 13 conjugate configuration variables are *abelianized gauge variables* (as already said the 14th one is π_ϕ). While the 14 gauge variables describe *generalized inertial effects*, the DO describe *generalized tidal effects.* Since the transformation is a point one, the old momenta are linear functionals of the new ones, with the kernels determined by a set of elliptic partial differential equations. In this way for the first time we can identify a canonical basis of non-local and non-tensorial DO, which remains canonical in the class of gauges where $\pi_\phi(\tau,\vec{\sigma})\approx 0$, even if no one knows how to solve the superhamiltonian constraint, i.e. the Lichnerowicz equation for the conformal factor.

A special completely fixed Hamiltonian gauge in this class can be defined by adding other 13 suitable gauge fixing constraints. In this gauge, which turns out to be non-harmonic in the weak field regime, the 3-metric on Σ_τ is *diagonal* and it corresponds to a unique 3-orthogonal 4-coordinate system on space-time (with an associated admissible 3+1 splitting with well defined simultaneity leaves) on the solutions of Hamilton equations. In this gauge it is possible to give a *background-independent* definition of a weak gravitational field: the DO $r_{\bar{a}}(\tau,\vec{\sigma})$, $\pi_{\bar{a}}(\tau,\vec{\sigma})$ should be slowly varying on a wavelength of the resulting post-Minkowskian gravitational wave, with the configurational DO $r_{\bar{a}}$ replacing the two polarizations of the harmonic gauges. A *Hamiltonian linearization* is defined in the following way [11]:

i) Assuming $ln\,\phi(\tau,\vec{\sigma})=O(r_{\bar{a}})$, the Lichnerowitz equation can be linearized and for the first time a non-trivial solution for ϕ can be found. Using this solution all the other constraints and the elliptic canonicity conditions can be linearized and solved. By putting these solutions in the integrand of the weak ADM energy, we get a well defined form for the energy density in this gauge in terms of the DO, i.e. the physical degrees of freedom of the gravitational field.

ii) The resulting ADM energy is approximated with the terms *quadratic* in the DO and the resulting linearized Hamilton equations are studied and solved. It is explicitly checked that the linearized Einstein's equations are satisfied by this solution. Even if the gauge is not harmonic, the wave equation $\Box r_{\bar{a}}(\tau, \vec{\sigma}) = 0$ is implied by the Hamilton equations and solutions satisfying the universe rest-frame condition are found (they cannot be transverse waves in the rest frame). These are the *post-Minkowskian background-independent gravitational waves.* The deformation patterns of a sphere of test particles induced by $r_{\bar{1}}$ and $r_{\bar{2}}$ are determined by studying the geodesic deviation equation.

We are now studying tetrad gravity coupled to a perfect fluid described by a suitable singular Lagrangian [12]. The Hamiltonian linearization in the special 3-orthogonal gauge, together with an adaptation to our formalism of the theory of Dixon's multipoles [13], will allow to find the post-Minkowskian (without any post-Newtonian approximation!) generalization of the quadrupole emission formula and the explicit form in this gauge of the action-at-a-distance Newton and gravito-magnetic potentials inside the fluid together with its tidal interactions. The resulting formalism should help to find a description of binary systems in a post-Minkowskian regime, where the post-Newtonian approximations fail. Moreover, the two-body problem of general relativity in the post-Minkowskian weak field regime will be studied by using a new semi-classical regularization of the self-energies, implying the $i \neq j$ rule like it appens in the electro-magnetic case [14]. Also tetrad gravity coupled to Klein-Gordon, electro-magnetic and Dirac fields is under investigation.

It will be explored the possibility of defining a scheme of Hamiltonian numerical relativity, based on expansions in the Newton constant G (the so called post-Minkowskian approximations), to study the strong field regime of tetrad gravity.

Finally we will try to find the Hamiltonian re-formulation of the Newman-Penrose formalism. This would allow to look for a Shanmugadhasan canonical basis in which the physical degrees of freedom of the gravitational field are described by *Bergmann observables* (BO), namely a set of coordinate-independent scalar DO. Such a basis would allow to start a new program of quantization of gravity, based on the idea of quantizing only the tidal effects (the BO) and not the inertial ones (the gauge variables), since the latter describe only the *appearances* of the phenomena. A prototype of this quantization is under study in special relativity to arrive at a formulation of atomic physics in non-inertial systems: while for relativistic particles (and their non-relativistic limit) there are already preliminary results [15], for the inclusion of the electro-magnetic field we have to find a way out from the Torre-Varadarajan no-go theorem, the obstruction to the Tomonaga-Schwinger formalism. Moreover, if the weak ADM energy in a completely fixed Hamiltonian gauge can be expressed in terms of BO, this would help to clarify the problem of the coordinate-dependence of the energy density in general relativity, which we think is a preliminary step for a correct understanding of the cosmological constant and dark energy problems.

References

1. J.Ehlers, F.A.E.Pirani and A.Schild, *The Geometry of Free-Fall and Light Propagation* in *General Relativity, Papers in Honor of J.L.Synge*, ed. L.O'Raifeartaigh (Oxford Univ.Press, London, 1972).
2. L.Lusanna, *The Rest-Frame Instant Form of Metric Gravity*, Gen.Rel.Grav. **33**, 1579 (2001) (gr-qc/0101048).
3. L.Lusanna and S.Russo, *A New Parametrization for Tetrad Gravity*, Gen.Rel.Grav. **34**, 189 (2002)(gr-qc/0102074).
4. R.De Pietri, L.Lusanna, L.Martucci and S.Russo, *Dirac's Observables for the Rest-Frame Instant Form of Tetrad Gravity in a Completely Fixed 3-Orthogonal Gauge*, Gen.Rel.Grav. **34**, 877 (2002) (gr-qc/0105084).
5. D.Alba and L.Lusanna, *Simultaneity, Radar 4-Coordinates and the 3+1 Point of View about Accelerated Observers in Special Relativity* (gr-qc/0311058).
6. A.Lichnerowicz, *Propagateurs, Commutateurs et Anticommutateurs en Relativite Generale*, in Les Houches 1963, *Relativity, Groups and Topology*, eds. C.DeWitt and B.DeWitt (Gordon and Breach, New York, 1964).
   C.Moreno, *On the Spaces of Positive and Negative Frequency Solutions of the Klein-Gordon Equation in Curved Space-Times*, Rep.Math.Phys. **17**, 333 (1980).
7. L.Lusanna, Int.J.Mod.Phys. **A10**, 3531 and 3675 (1995). L.Lusanna, *Towards a Unified Description of the Four Interactions in Terms of Dirac-Bergmann Observables*, invited contribution to the book *Quantum Field Theory: a 20th Century Profile* of the Indian National Science Academy, ed. A.N.Mitra (Hindustan Book Agency, New Delhi, 2000) (hep-th/9907081).
8. D.Christodoulou and S.Klainerman, *The Global Nonlinear Stability of the Minkowski Space* (Princeton, Princeton, 1993).

9. Y.Choquet-Bruhat, A.Fischer and J.E.Marsden, *Maximal Hypersurfaces and Positivity of Mass*, LXVII E.Fermi Summer School of Physics *Isolated Gravitating Systems in General Relativity*, ed. J.Ehlers (North-Holland, Amsterdam, 1979).
10. L.Lusanna and M.Pauri, *General Covariance and the Objectivity of Space-Time Point-Events: The Physical Role of Gravitational and Gauge Degrees of Freedom in General Relativity* (gr-qc/0301040).
11. J.Agresti, R.DePietri, L.Lusanna and L.Martucci, *Hamiltonian Linearization of the Rest-Frame Instant Form of Tetrad Gravity in a Completely Fixed 3-Orthogonal Gauge: a Radiation Gauge for Background-Independent Gravitational Waves in a Post-Minkowskian Einstein Space-Time* (gr-qc/0302084), to appear in Gen.Rel.Grav.; *Background Independent Gravitational Waves* (gr-qc/0302085)(short review of the results).
12. J. D. Brown, Class. Quantum Grav. **10**, 1579 (1993).
    L. Lusanna and D. Nowak-Szczepaniak, Int. J. Mod. Phys. **A15**, 4943 (2000).
    D.Alba and L.Lusanna, *Generalized Eulerian Coordinates for Relativistic Fluids: Hamiltonian Rest-Frame Instant Form, Relative Variables, Rotational Kinematics* (hep-th/0209032), to appear in Int.J.Mod.Phys. A.
13. W.G.Dixon, J.Math.Phys. **8**, 1591 (1967).
    D.Alba, L.Lusanna and M.Pauri, *Multipolar Expansions for Closed and Open Systems of Relativistic Particles*, (hep-th/0103092).
    D. Alba, L. Lusanna and M. Pauri, *Centers of Mass and Rotational Kinematics for the Relativistic N-Body Problem in the Rest-Frame Instant Form*, J. Math. Phys. **43**, 1677 (2002) (hep-th/0102087).
14. H.Crater and L.Lusanna, Ann.Phys. (NY) **289**, 87 (2001) (hep-th/0001046).
    D. Alba, H. Crater and L. Lusanna, Int. J. Mod.Phys. **A16**, 3365 (2001) (hep-th/0103109).
15. D.Alba and L.Lusanna, *From Parametrized Minkowski Theories to Quantum Mechanics in Non-Inertial Frames in Absence of Gravity: I) Relativistic and Non-Relativistic Particles*, in preparation.

WHERE ARE THE STATES OF A BLACK HOLE?

SAMIR D. MATHUR
Department of Physics, Ohio State University
Columbus, OH 43210, USA
mathur@mps.ohio-state.edu

We argue that bound states of branes have a size that is of the same order as the horizon radius of the corresponding black hole. Thus the interior of a black hole is not 'empty space with a central singularity', and Hawking radiation can pick up information from the degrees of freedom of the hole.

1. INTRODUCTION

Let us recall the puzzles arising in the quantum physics of black holes. Gedanken experiments show that we must associate an entropy[1]

$$S = \frac{A}{4G} \tag{1}$$

with a black hole, where A is the area of the horizon. Statistical mechanics then suggests that there should be e^S states of the hole. But the geometry of the hole appears to be completely determined by its mass, charge and angular momentum, implying an entropy $S = \ln 1 = 0$, which contradicts (1). Closely related is the 'information paradox': Hawking radiation from the hole is created by the progressive dilation of vacuum modes near the horizon, and if the geometry here is determined only by the above conserved quantities then the radiation will carry no details of the matter which went in to make the hole. We thus get a violation of the unitarity of quantum mechanics[2].

Clearly, the key to resolving these long standing problems is to understand where the states of the hole are located; in other words, to understand the 'hair' that differentiate different states of the hole. In this talk we propose that the interior of the horizon is not described by the conventional picture of Fig.1(a) where we have 'empty space with a central singularity'. Rather the differences between the e^S states are manifested throughout the interior of the 'horizon' (Fig.1(b)), there is in general no special point to play the role of the singularity, and the horizon is just the boundary of the region where the typical states differ from each other.

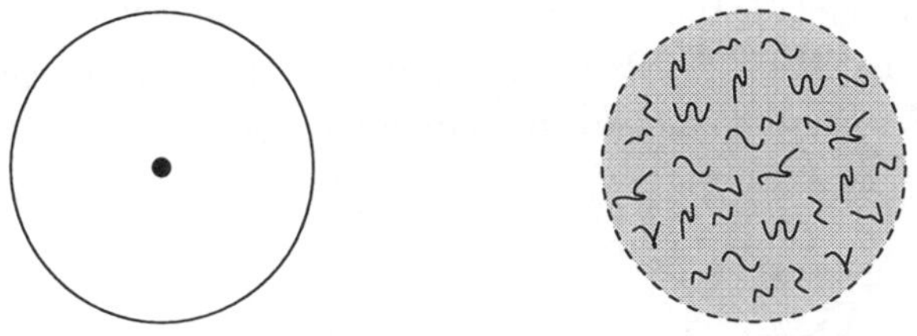

1(a) 1(b)

Figure 1. (a) The conventional picture of a black hole (b) the proposed picture – state information is distributed throughout the 'fuzzball'.

We will first give an abstract argument for the latter picture. Next, we show that the simplest string theoretic system with entropy – the 2-charge extremal D1-D5 system – behaves like Fig.1(b) rather than Fig.1(a); here we can construct all the e^S states explicitly in the gravity description. We then construct one state of the 3-charge extremal system – a state with large D1, D5 charges and just one unit of momentum P. This state also has the structure of Fig.1(b). Lastly, we describe the physics which might underlie this large departure from the conventional picture Fig.1(a). String states exhibit entropy when different types of branes bind and the degrees of freedom are carried by 'fractional brane excitations'. But such fractional branes have very low tension and can stretch upto distances of the order of the horizon radius, destroying the naive picture of the black hole interior.

2. AN ABSTRACT ARGUMENT ABOUT 'HAIR'

A large set of Gedanken experiments have argued for a general link between entropy and horizons – whenever we find a horizon with area A then we must associate an entropy $S = \frac{A}{4G}$ with the region behind that horizon. But now consider the e^S microstates that give the entropy of a black hole. Some attempts to locate the 'hair' have looked for small perturbations near the horizon of Fig.1(a). Other approaches would suggest that the microstates differ only within a planck distance of the central singularity. In either case each microstate looks pretty much like Fig.1(a), in that it has a horizon with an area $\approx A$.

But if each of the e^S microstates has such a horizon, then it must itself represent $\sim e^S$ different states. This makes no sense, since we wanted the microstates to explain the entropy, not have further entropy themselves. We conclude that *the e^S microstates should have no horizons individually – the notion of a horizon should arise only after 'coarse-graining' over these microstates.*

3. THE D1-D5 SYSTEM

This looks like a tall order since to avoid having a horizon the microstates must differ from the naive black hole geometry not just within planck distance of the singularity but all the way upto the horizon. Remarkably, however, for the extremal D1-D5 system we found exactly this structure. We take type IIB string theory compactified on $T^4 \times S^1$. The volume of T^4 is $(2\pi)^4 V$ and the length of S^1 is $2\pi R$. We wrap n_5 D5 branes on $T^4 \times S^1$ and n_1 D1 branes on S^1. The metric usually written for this system (which we will call the 'naive metric') is

$$ds^2 = \frac{[-dt^2 + dy^2]}{\sqrt{(1+\frac{Q_1}{r^2})(1+\frac{Q_5}{r^2})}} + \sqrt{(1+\frac{Q_1}{r^2})(1+\frac{Q_5}{r^2})}dx_i dx_i + \sqrt{\frac{1+\frac{Q_1}{r^2}}{1+\frac{Q_5}{r^2}}}dz_a dz_a \tag{2}$$

and is sketched in Fig.2(a). One knows that this geometry cannot be the complete story, since the branes wrapped as above give a 1+1 dimensional CFT with fermions in the Ramond sector, and the ground state here has degeneracy $e^{2\sqrt{2}\pi\sqrt{n_1 n_5}}$. We must therefore see a corresponding degeneracy in the dual gravity description. By a sequence of S,T dualities we map the D1-D5 system to the FP system – the fundamental string (F) wrapped n_5 times on S^1, carrying n_1 units of momentum (P) along S^1. Since we are counting the *bound* states of the FP system the P charge is carried on the F string as traveling waves. But the F string has no longitudinal vibrations, so it must bend away from its central axis in a *transverse* direction to carry the momentum. The n_5 strands of the F string separate under this deformation and spread over a nonzero transverse region, thus cutting off the naive FP metric before $r = 0$. Dualizing back we get (classically) a family of geomeries for the D1-D5 system characterized by $\vec{F}(v)$, the vibration profile of the F string[3]

$$ds^2 = \frac{-(dt - A_i dx^i)^2 + (dy + B_i dx^i)^2}{\sqrt{[(1+K)/H]}} + \sqrt{\frac{1+K}{H}} dx_i dx_i + \sqrt{H(1+K)} dz_a dz_a \tag{3}$$

$$e^{2\Phi} = H(1+K), \quad C^{(2)}_{ti} = \frac{B_i}{1+K}, \quad C^{(2)}_{ty} = -\frac{K}{1+K}$$
$$C^{(2)}_{iy} = -\frac{A_i}{1+K}, \quad C^{(2)}_{ij} = C_{ij} + \frac{A_i B_j - A_j B_i}{1+K} \tag{4}$$

where B_i, C_{ij} are given by

$$dB = - *_4 dA, \quad dC = - *_4 dH^{-1} \tag{5}$$

and $*_4$ is the duality operation in the 4-d transverse space $x_1 \dots x_4$ using the flat metric $dx_i dx_i$. The functions H, K, A_i are generated by $\vec{F}$

$$\frac{1}{H} = 1 + \frac{Q}{L} \int_0^L \frac{dv}{|\vec{x} - \vec{F}(v)|^2}, \quad K = \frac{Q}{L} \int_0^L \frac{dv(\dot{F}(v))^2}{|\vec{x} - \vec{F}(v)|^2}, \quad A_i = -\frac{Q}{L} \int_0^L \frac{dv \dot{F}_i(v)}{|\vec{x} - \vec{F}(v)|^2} \tag{6}$$

The geometries, pictured in Fig.2(b), have no horizons and no singularity (at points $\vec{x} = \vec{F}(v)$ we have just the coordinate singularity arising at the center of Kaluza-Klein monopoles[4]). If we 'coarse-grain' by drawing a boundary to enclose the region where these geometries differ significantly from each other (the dashed line in Fig.2(b)) then from the area A of this boundary we find[5]

$$S \equiv \frac{A}{4G} \sim \sqrt{n_1 n_5} \sim S_{micro} \tag{7}$$

Thus the Bekenstein 'area entropy' arises directly as a 'coarse-graining over hair'. It is important that the geometries (3) can be distinguished from each other by dynamical experiments. The CFT state dual to the geometry described by $\vec{F}(v)$ is given by the map

$$(a^{i_1}_{n_1})^\dagger \dots (a^{i_k}_{n_k})^\dagger |0\rangle \;\rightarrow\; \sigma^{i_1}_{n_1} \dots \sigma^{i_k}_{n_k} |0\rangle_{NS} \tag{8}$$

where the vibration profile $\vec{F}(v)$ of the F string is described at the quantum level by transverse oscillators $(a^i_n)^\dagger$ and σ^i_n are twist operators of the CFT which is a sigma model with target space the orbifold $(T^4)^N/S^N$. A quantum thrown into the 'throat' region of one of these geometries bounces off the end and returns in a time Δt_{sugra} which depends on the geometry. The CFT dual of this process has a pair of vibrations traveling around the 'multiwound' circle produced by the twist operators, and this takes a time

Δt_{CFT}. We find $\Delta t_{sugra} = \Delta t_{CFT}$, and observe that in this process the backreaction of the test quantum of the geometry remains small[3]. Thus we can reliably probe the structure of the 'hair'.

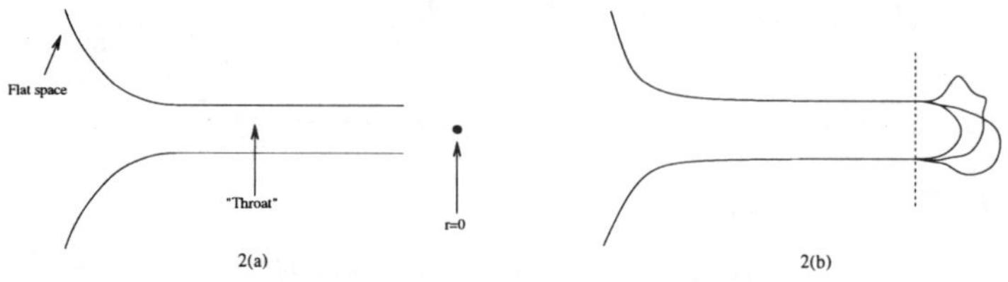

Figure 2. (a) The naive geometry of extremal D1-D5 (b) the actual geometries; the dashed line gives a 'horizon' whose area gives the entropy.

4. THE THREE CHARGE D1-D5-P SYSTEM

The 'naive' geometry of the 3-charge system also has an infinite length throat extending to a horizon at $r = 0$, and this geometry can be extended to a region 'inside' the horizon. But the 2-charge results suggest that the throat gets capped off before $r = 0$, with different caps giving different microstates as in Fig.2(b). While we were unable to construct the generic 3-charge state we note that if our picture were true then we should be able to add a small perturbation to any state of the 2-charge extremal system such that we get a 3-charge extremal system with one unit of P. It is nontrivial that such a perturbation exist because we demand smoothness everywhere and decay at infinity. Such a perturbation was indeed constructed[6] (by matching interior and exterior solutions to several orders) giving the geometry dual to the CFT state

$$|\psi'\rangle = J^-_{-1}|\psi\rangle_R, \quad (L_0 - \bar{L}_0)|\psi'\rangle = 1 \tag{9}$$

where $|\psi\rangle_R$ is a particular Ramond ground state of the 2-charge system. We thus get the analogue of Fig.2(b) rather than Fig.2(a) for at least this 3-charge state. If the throat is similarly 'capped' for generic 3-charge states then we cannot just fall through the horizon at $r = 0$ into the black hole interior in the manner implied by the naive geometry analogous to Fig.1(a).

5. FRACTIONATION

Consider a string wrapped on a circle of length $2\pi R$. The minimum excitation (with no net P charge) comes from one left and one right mover, with $\Delta E = \frac{1}{R} + \frac{1}{R} = \frac{2}{R}$. But if we have a *bound* state of n strings, then we get effectively one string of length $2\pi nR$. The momentum modes then come in fractional units $\frac{1}{nR}$ and the threshold drops to[7] $\Delta E = \frac{2}{nR}$. Similarly, the threshold for vibrating a D1-D5 bound state is[8] $\frac{2}{n_1 n_5 R}$. We see that bound states made out of many branes have very light 'fractional' excitations.

To see the physical consequence of such excitations, consider an extremal 3-charge D1-D5-P bound state and imagine a test quantum brought to a distance r from this state. This needs a minimum localization energy $\sim 1/r$, which is small if r is macroscopic. But the fractional excitations of the bound state are light too, and we ask if using the energy $1/r$ a pair of fractional brane excitations can be created that extend from the origin to the location r of the test quantum. We want this process to be probable and not just possible, so we demand that the entropy gained by creating the fractional pair be at least $\Delta S - 1$ (thus the phase space increases by a factor $e \approx 2.7$ upon creation of the pair). A standard entropy computation then yields that $\Delta S > 1$ for $r < r_{crit}$ where[9]

$$r_{crit} \sim \left[\frac{g^2 \alpha'^4 \sqrt{n_1 n_5 n_p}}{VR}\right]^{\frac{1}{3}} \sim r_H \tag{10}$$

where r_H is the horizon radius of the D1-D5-P bound state! This estimate supports the 'fuzzball' picture Fig.1(b), and is in line with the above construction of 'hair' for D1-D5 and D1-D5+(1 unit of P).

As with all string theory constructions of black holes, we expect that if extremal holes behave like Fig.1(b) then near extremal and generic black holes will be qualitatively similar. The entropy for general black holes in 5-D is reproduced *exactly* if we partition the mass optimally among brane-antibrane pairs[10]

$$S = 2\pi(\sqrt{n_1} + \sqrt{\bar{n}_1})(\sqrt{n_5} + \sqrt{\bar{n}_5})(\sqrt{n_p} + \sqrt{\bar{n}_p}) \tag{11}$$

Assuming such a microstructure the size estimate along the lines of (10) again gives $r_{crit} \sim r_H$.

6. CONCLUSIONS

Hawking's calculation of radiation showing information loss is so robust because it uses no details of the physics at the planck scale. Resolving the information paradox thus needs an explicit nonlocality over macroscopic distances. String theory has succeeded in matching entropy and radiation rates of brane bound states with corresponding black holes[11]. Using these same systems we have argued that bound states swell upto a size of order the horizon radius; thus the interior of the horizon is not just 'empty space with a central singularity'. This makes it possible for radiation from the hole to pick up information from the 'hair' and avoid the information paradox.

It will be interesting to address dynamical questions with the above picture. It is possible that infalling matter falls straight through the 'fuzzball' towards $r = 0$ (as if it were falling through a conventional horizon), but over the Hawking evaporation time information is transferred to the 'light fractional modes' and into the radiation.

ACKNOWLEDGMENTS

Much of the work described here is in collaboration with O. Lunin, A. Saxena and Y. Srivastava. I thank for discussions I. Bena, S. Das, A. Dabholkar, V. Hubeny, C. Johnson, P. Kraus, D. Kutasov, E. Martinec, S. Shenker and L. Susskind. I thank A. Saxena for making the figures. This work was supported in part by DOE grant DE-FG02-91ER-40690.

REFERENCES

1. J. D. Bekenstein, Phys. Rev. D **7**, 2333 (1973).
2. S. W. Hawking, Commun. Math. Phys. **43**, 199 (1975).
3. O. Lunin and S. D. Mathur, Nucl. Phys. B **623**, 342 (2002), hep-th/0109154.
4. O. Lunin, J. Maldacena and L. Maoz, hep-th/0212210.
5. O. Lunin and S. D. Mathur, Phys. Rev. Lett. **88**, 211303 (2002), hep-th/0202072.
6. S. D. Mathur, A. Saxena and Y. K. Srivastava, arXiv:hep-th/0311092.
7. S. R. Das and S. D. Mathur, Phys. Lett. B **375**, 103 (1996) [arXiv:hep-th/9601152].
8. J. M. Maldacena and L. Susskind, Nucl. Phys. B **475**, 679 (1996) [arXiv:hep-th/9604042].
9. S. D. Mathur, Nucl. Phys. B **529**, 295 (1998) [arXiv:hep-th/9706151].
10. G. T. Horowitz, J. M. Maldacena and A. Strominger, Phys. Lett. B **383**, 151 (1996) [arXiv:hep-th/9603109].
11. A. Strominger and C. Vafa, Phys. Lett. B **379**, 99 (1996), hep-th/9601029, C. G. . Callan and J. M. Maldacena, Nucl. Phys. B **472**, 591 (1996) [arXiv:hep-th/9602043], S. R. Das and S. D. Mathur, Nucl. Phys. B **478**, 561 (1996) [arXiv:hep-th/9606185], J. M. Maldacena and A. Strominger, Phys. Rev. D **55**, 861 (1997) [arXiv:hep-th/9609026].

WHAT IS QUANTUM THEORY OF GRAVITY?*

DJORDJE MINIC and CHIA-HSIUNG TZE
Institute for Particle Physics and Astrophysics
Department of Physics, Virginia Tech
Blacksburg, VA 24061, U.S.A.
E-mail: dminic@vt.edu, kahong@vt.edu

We present a line by line derivation of canonical quantum mechanics stemming from the compatibility of the statistical geometry of distinguishable observations with the canonical Poisson structure of Hamiltonian dynamics. This viewpoint can be naturally extended to provide a conceptually novel, non-perturbative formulation of quantum gravity. Possible observational implications of this new approach are briefly mentioned.

1. What is Quantum Mechanics?

In this talk we would like to 1) give a line by line derivation of canonical quantum mechanics (QM) founded on the compatibility of the statistical geometry of distinguishable observations with the canonical Poisson structure of Hamiltonian dynamics and 2) describe a natural extension of this viewpoint so as to provide a conceptually novel approach to the problem of a non-perturbative formulation of quantum gravity, one which should reduce in the correspondence limit to general relativity (GR) coupled to matter degrees of freedom. The presentation is based on our recent work [1].

To understand the fundamental structure of QM we reason as follows: Assume that individual quantum events are statistical and statistically distinguishable (Postulate I). (This premise is of course a huge conceptual leap in comparison to classical physics, but it is absolutely crucial for the structure of QM.) On the space of probability distributions there is a natural metric, called Fisher metric, which provides a geometric measure of

*Talk given by D.M.

statistical distinguishability [2]

$$ds^2 = \sum_i \frac{dp_i^2}{p_i}, \quad \sum_i p_i = 1, \quad p_i \geq 0. \tag{1}$$

(This distance naturally arises as follows: to estimate probabilities p_i from frequencies f_i, given N samples, when N is large, the central limit theorem states that the probability for the frequencies is given by the Gaussian distribution $\exp(-\frac{N}{2}\frac{(p_i-f_i)^2}{p_i})$. Thus a probability distribution $p_i^{(1)}$ can be distinguished from a given distribution p_i provided the Gaussian $\exp(-\frac{N}{2}\frac{(p_i^{(1)}-p_i)^2}{p_i})$ is small. Hence the quadratic form $\frac{(p_i^{(1)}-p_i)^2}{p_i}$, or its infinitesimal form $\sum_i \frac{dp_i^2}{p_i}$, the Fisher distance, is the natural measure of distinguishability.) Now, upon setting $p_i = x_i^2$, making p_i manifestly non-negative, the Fisher distance reads $ds^2 = \sum_i dx_i^2$ with $\sum x_i^2 = 1$ i.e. the Euclidean metric on a sphere. Therefore the latter distance is nothing but the shortest distance along this unit sphere [2]

$$ds_{12} = cos^{-1}\left(\sum_i \sqrt{p_{1i}}\sqrt{p_{2i}}\right). \tag{2}$$

Next, we demand that on this metric space of probabilities one can define a canonical Hamiltonian flow (Postulate II). So the dimensionality of such a symplectic space of x_i must be *even* (hence, $\sum x_i^2 = 1$ defines an odd-dimensional sphere). Then the Hamiltonian flow is given (locally) as

$$\frac{df(x_i)}{dt} = \omega_{ij}\frac{\partial h(x_i)}{\partial x_i}\frac{\partial f(x_i)}{\partial x_j} \equiv \{h, f\}, \tag{3}$$

where ω is a closed non-degenerate 2-form. The compatibility of the symplectic form ω and the metric g allows for the introduction of an almost complex structure $J \equiv \omega g^{-1}$ (in the matrix notation), $J^2 = -1$ (since the compatibility demands $\omega_{ij} g^{jk} \omega_{kl} = g_{il}$). Given this constant complex structure we may introduce complex coordinates on this space ψ_a (and their conjugates ψ_a^*), so that $\sum_i x_i^2 \equiv \sum_a \psi_a^* \psi_a = 1$, and thus $p_a = \psi_a^* \psi_a$. This statistical distance is invariant under $\psi \to e^{J\alpha}\psi$, J being the above integrable almost complex structure. Thus ψ can be identified with $e^{J\alpha}\psi$. Indeed an odd dimensional sphere can be viewed as a $U(1)$ Hopf -fibration of a complex projective space $CP(n)$, a coset space $\frac{U(n+1)}{U(n)\times U(1)}$. $CP(n)$ is a homogeneous, isotropic and simply connected Kahler manifold with a constant holomorphic sectional curvature. The unique metric on $CP(n)$ is the Fubini-Study (FS) metric (which is but the above statistical Fisher metric up to a multiplicative constant, the Planck constant $\hbar$). In the Dirac notation (using (2) and the derived Born rule, $p_a = \psi_a^* \psi_a$), this FS metric reads:

$$ds_{12}^2 = 4(cos^{-1}|\langle\psi_1|\psi_2\rangle|)^2 = 4(1-|\langle\psi_1|\psi_2\rangle|^2) \equiv 4(\langle d\psi|d\psi\rangle - \langle d\psi|\psi\rangle\langle\psi|d\psi\rangle), \quad (4)$$

Thus $CP(n)$ is the underlying manifold of statistical events with a well defined Hamiltonian flow and as such provides a kinematical background on which a Hamiltonian dynamics is defined. The only Hamiltonian flow compatible with the isometries of $CP(n)$ (which are the unitaries $U(n+1)$) is given by a quadratic function of x_i or, alternatively, a quadratic form in the pair $q_a \equiv Re(\psi_a)$ and $p_a \equiv Im(\psi_a)$), $h = \frac{1}{2}\sum_a[(p^a)^2 + (q_a)^2]\omega_a$, or in the usual notation, $h = \langle\hat{H}\rangle$, ω_a being the eigenvalues of $\hat{H}$. The Hamiltonian equations for the ψ_a and its conjugate then give the linear evolution equation (Schrödinger equation), $\frac{dp_a}{dt} = \{h, p_a\}$, $\frac{dq^a}{dt} = \{h, q^a\}$, that is $J\frac{d|\psi\rangle}{dt} = H|\psi\rangle$. Any observable, consistent with the isometries of the underlying space of events, is given as a quadratic function in the q_a, p_a. These are just the usual expectation values of linear operators.

Everything we know about quantum mechanics is in fact contained in the geometry of $CP(n)$[3,4]: entanglements come from the embeddings of the products of two complex projective spaces in a higher dimensional one; geometric phases stem from the symplectic structure of $CP(n)$, quantum logic, algebraic approaches to QM etc, are all contained in the geometric and symplectic structure of complex projective spaces [3,4]. (While we only consider here the finite dimensional case, the same geometric approach is extendable to generic infinite dimensional quantum mechanical systems, including field theory.) Finally, the following three lemmas are important for the material of the next section: (A) The Fisher-Fubini-Study quantum metric (4) in the $\hbar \to 0$ limit becomes a spatial metric provided the configuration space for the quantum system under consideration *is* space. For example, consider a particle moving in 3-dimensional Euclidean space. Then the quantum metric for the Gaussian coherent state $\psi_l(x) \sim \exp(-\frac{(\vec{x}-\vec{l})^2}{\delta l^2})$ yields the natural metric in the configuration space, in the $\hbar \to 0$ limit, $ds^2 = \frac{d\vec{l}^2}{\delta l^2}$. (B) Similarly, the time parameter of the evolution equation can be related to the quantum metric via

$$\hbar ds = \Delta E dt, \quad \Delta E^2 \equiv \langle\psi H^2 \psi\rangle - (\langle\psi H \psi\rangle)^2 \quad (5)$$

(C) Finally, the Schrodinger equation can be viewed as a geodesic equation on a $CP(n) = \frac{U(n+1)}{U(n)\times U(1)}$

$$\frac{du^a}{ds} + \Gamma^a_{bc} u^b u^c = \frac{1}{2\Delta E} Tr(HF^a_b) u^b. \qquad (6)$$

Here $u^a = \frac{dz^a}{ds}$ where z^a denote the complex coordinates on $CP(n)$, Γ^a_{bc} is the connection obtained from the FS metric, and F_{ab} is the canonical curvature 2-form valued in the holonomy gauge group $U(n) \times U(1)$.

The above geometric structure describing canonical QM, beautifully tested in numerous experiments, is also very robust from the purely geometric point of view [1]. A consistent generalization of QM would doubtlessly be interesting from both the experimental and theoretical viewpoints. Unlike various generalizations proposed in the past (which in many instances have lead to difficult conceptual problems) the one put forward in [1] extends the kinematical structure so that it is compatible with the generalized dynamical structure! *The quantum symplectic and metric structure, and therefore the almost complex structure become fully dynamical.* The underlying physical reason for such a more general dynamical framework of the above geometric formalism is found in the need for a quantum version of the equivalence principle, a fundamental physical underpinning of a non-perturbative formulation of quantum theory of gravity.

2. And What is Quantum Theory of Gravity?

The main intuition behind a quantum version of the equivalence principle is to demand the validity of the canonical QM, as laid out in the previous section, in every local neighborhood, *in the space of quantum events.* Here we envision a larger geometric structure whose tangent spaces, viewed as vector spaces by definition, are just canonical Hilbert spaces. (In this picture [1] the tangent *spatial* transverse metric emerges from the *quantum* metric, as in the lemma (A), by assuming that the underlying configuration space is space and time appears as a measure of the geodesic distance in this general space of statistical events (the events do not have to be necessarily distinguishable!), as in lemma (B). Finally, the longitudinal spatial coordinate corresponds to the dimensionality of the tangent Hilbert spaces.) The crucial point is to allow for any metric and symplectic form in the geodesic version of the evolution equation (lemma (C)). These are in turn determined dynamically [1].

Our postulates (I) and (II) as stated above can indeed be naturally extended by allowing both the metric and symplectic form on the space of quantum events to be no longer rigid but fully dynamical entities. In this process, just as in the case of spacetime in GR, the space of quantum events becomes dynamical and only individual quantum events make sense observationally. Specifically, we do so by relaxing the first postulate to allow for *any* statistical (information) metric while insisting on the compatibility of this metric with the symplectic structure underlying the second postulate. Physics is therefore required to be diffeomorphism invariant in the sense of information geometry such that the information geometric and symplectic structures remain compatible, requiring only a *strictly* (i.e non-integrable) almost complex structure J. This extended framework readily implies that the wave functions labeling the event space, while still unobservable, are no longer relevant. They are in fact as meaningless as coordinates in GR. There are no longer issues related to reductions of wave packets and associated measurement problems. At the basic level of our scheme, there are only dynamical correlations of quantum events. The observables are furnished by diffeomorphism invariant quantities in the space of quantum events.

To find the kinematical arena for this generalized framework we seek as coset of $Diff(C^{n+1})$ such that locally the latter looks like $CP(n)$ and allows for a compatibility of its metric and symplectic structures, expressed in the existence of a (generally non-integrable) almost complex structure. The following nonlinear Grassmannian

$$Gr(C^{n+1}) = Diff(C^{n+1})/Diff(C^{n+1}, C^n \times \{0\}), \tag{7}$$

with $n = \infty$ fulfills these requirements [5]. $Gr(C^{n+1})$ is a nonlinear analog of a complex Grassmannian since it is the space of (real) co-dimension 2 submanifolds, namely a hyperplane $C^n \times [0]$ passing through the origin in C^{n+1}. Its holonomy group $Diff(C^{n+1}, C^n \times \{0\})$ is the group of diffeomorphisms preserving the hyperplane $C^n \times \{0\}$ in C^{n+1}. Just as $CP(n)$ is a co-adjoint orbit of $U(n+1)$, $Gr(C^{n+1})$ is a coadjoint orbit of the group of volume preserving diffeomorphisms of C^{n+1}. As such it is a symplectic manifold with a canonical Kirillov-Kostant-Souriau symplectic two-form Ω which is closed ($d\Omega = 0$) but not exact. Indeed the latter 2-form integrated over the submanifold is nonzero; its de Rham cohomology class is integral. This means that there is a principal 1-sphere, a $U(1)$ or line bundle over $Gr(C^{n+1})$ with curvature Ω. This is the counterpart of the $U(1)$-bundle of S^{2n+1} over $CP(n)$ of quantum mechanics. It is also known that there is an almost complex structure given by a 90 degree rotation in the two dimensional normal bundle to the submanifold. While $CP(n)$ has an integrable almost complex structure and is therefore a complex manifold, in fact a Kahler manifold, this is *not* the case with $Gr(C^{n+1})$. Its almost complex structure J is strictly *not* integrable in spite of its formally vanishing Nijenhius tensor. While the vanishing of the latter implies integrability in the finite dimensional case, such a conclusion no longer holds in the infinite dimensional setting. However what we do have in $Gr(C^{n+1})$ is a strictly (i.e. non-Kahler) almost Kahler manifold since there is by way of the almost structure J a compatibility between the closed symplectic 2-form Ω and the Riemannian metric g which *locally* is given by $g^{-1}\Omega = J$. Clearly, it would be very interesting to understand how unique is the structure of $Gr(C^{n+1})$.

Just as in standard geometric QM, the probabilistic interpretation should be found in the definition of geodesic length on the new space of quantum states/events. Notably since $Gr(C^{n+1})$ is only a strictly almost complex, its J is only locally complex. This fact translates into the existence of only local time and local metric on the space of quantum events. The local temporal evolution equation is a geodesic equation on the space of quantum events $\frac{du^a}{d\tau} + \Gamma^a_{bc}u^b u^c = \frac{1}{2E_p}Tr(HF^a_b)u^b$ where now τ is given through the metric $\hbar d\tau = 2E_p dt$, where E_p is the Planck energy. Γ^a_{bc} is the affine connection associated with this general metric g_{ab} and F_{ab} is a general curvature 2-form in the holonomy gauge group $Diff(C^{n+1}, C^n \times \{0\})$. This geodesic equation follows from the conservation of the energy-momentum tensor $\nabla_a T^{ab} = 0$ with $T_{ab} = Tr(F^{ac}g_{cd}F^{cb} - \frac{1}{4}g_{ab}F_{cd}F^{cd} + \frac{1}{2E_p}Hu_a u_b)$. Since both the metrical and symplectic data are also contained in H and are $\hbar \to 0$ limits of their quantum counterparts [4,1], we have here a non-linear "bootstrap" between the space of quantum events and the generator of dynamics. The diffeomorphism invariance of the new quantum phase space is explicitly taken into account in the following dynamical scheme [1]:

$$R_{ab} - \frac{1}{2}g_{ab}R - \lambda g_{ab} = T_{ab} \tag{8}$$

($\lambda = \frac{n+1}{\hbar}$ for $CP(n)$; in that case $E_p \to \infty$). Moreover we demand for compatibility

$$\nabla_a F^{ab} = \frac{1}{E_p}Hu^b. \tag{9}$$

The last two equations imply via the Bianchi identity a conserved energy-momentum tensor, $\nabla_a T^{ab} = 0$. The latter, taken together with the conserved "current" $j^b \equiv \frac{1}{2E_p}Hu^b$, i.e. $\nabla_a j^a = 0$, results in the generalized geodesic Schrödinger equation. As in GR it will be crucial to understand both the local and global features of various solutions to the above dynamical equations. The kinematical structure of ordinary QM is compatible with our general dynamical formulation and thus we naturally expect that our general formalism is compatible with all known cases in which quantum theories of gravity have been non-perturbatively defined, albeit in *fixed* asymptotic backgrounds (such as string theory in asymptotically AdS spaces).

What determines the form of the Hamiltonian H? The only requirement is that H should define a canonical quantum mechanical system whose configuration space *is* space and whose dynamics defines a consistent quantum gravity in an asymptotically flat background. We are aware of only one example satisfying this criterion: Matrix theory [6]! *Thus our proposal defines a background independent, non-perturbative, holographic formulation of Matrix theory* [1]. This choice of the Hamiltonian might at first appear artificial given the generality of our proposal. Yet we note that, from the suggestive geodesic form of the Schrödinger equation, H can be viewed as a "charge", and thus may be determined in a quantum theory of gravity by being encoded in the non-trivial symplectic topology of the space of quantum events. Such a realization may well be possible here since our non linear Grassmannian is non-simply connected [5].

Finally, among the possible observational implications of our proposal, those we are currently attempting to understand are: 1) possible deviations from the Planck law involving primordial gravitational waves, or Hawking radiation; 2) deviations from the classic QM formula for the vacuum energy (which underlies the cosmological constant problem); 3) relativity of the double slit experiment: once we relax postulate I, so that any information metric is allowed, the relativity (observer dependence) of canonical QM experiments (such as the double-slit experiment) becomes possible; 4) highly constrained deviations from linearity (superposition principle), and canonical QM entanglement; 5) the fact that in our proposal the generalized geometric phase is in $Diff(C^{n+1}, C^n \times \{0\})$, is also amenable to experimental tests.

Acknowledgments

We thank Tom Banks, Raphael Bousso, Ram Brunstein, Lay Nam Chang, Marty Einhorn, Willy Fischler, Jim Hartle, Bob McNees, Joan Simon and Tatsu Takeuchi for recent discussions of our work. D.M. would like to thank the organizers of the QTS3 conference for providing him with an opportunity to present this work. DM was supported in part by the U.S. Department of Energy under contract DE-FG05-92ER40677 and by NSF grant PHY-9907949 at KITP, Santa Barbara, the stimulating atmosphere of which is gratefully acknowledged.

References

1. D. Minic and C. H. Tze, Phys. Rev. D **68**, 061501 (2003) [arXiv:hep-th/0305193]. D. Minic and C. H. Tze, arXiv:hep-th/0309239, to appear in Phys. Lett. B. This work is a continuation of D. Minic and C. H. Tze, Phys. Lett. B **536**, 305 (2002) [arXiv:hep-th/0202173] motivated by the problem of a covariant formulation of Matrix theory as explained in H. Awata, M. Li, D. Minic and T. Yoneya, JHEP **0102**, 013 (2001) [arXiv:hep-th/9906248]; D. Minic, arXiv:hep-th/9909022; D. Minic, [arXiv:hep-th/0009131].
2. W. K. Wootters, Phys. Rev. D **23**, 357 (1981). Also, S. L. Braunstein and C. M. Caves, Phys. Rev. Lett. **72** 3439 (1994).
3. N. P. Landsman, *Mathematical topics between classical and quantum mechanics*, Springer, 1998. Also, quant-phys/9604008; Int. Jour. Mod. Phys. **B 10**, 1545 (1996); Rev. Math. Phys. **9**, 29, (1997).
4. B. Mielnik, Comm. Math. Phys. (1974). T. W. Kibble, Commun. Math. Phys. **65**, 189 (1979); J. P. Provost and G. Vallee, Comm. Math. Phys. **76** (1980); A. Heslot, Phys. Rev. D **31**, 1341 (1985); A. M. Bloch, Phys. Lett. A **116**, 353 (1986); J. Anandan and Y. Aharonov, Phys. Rev. Lett. **65**, 1697 (1990); J. Anandan, Found. Phys. **21**, 1265 (1991); R. Cirelli, A. Mania and I. Pizzochero, J. Math. Phys. 31, , 2891-2903 (1990); S. Weinberg, Annals Phys. **194**, 336 (1989); J. Polchinski, Phys. Rev. Lett. **66**, 397 (1991); G. W. Gibbons, J. Geom. Phys. **8**, 147 (1992); G. W. Gibbons and H. J. Pohle, Nuc. Phys. **B410**, 117, (1993); talk by Gibbons at a celebration of the 60th birthday of James Hartle, http://www.newton.cam.ac.uk/webseminars/hartle60/; A. Ashtekar and T. A. Schilling, [gr-qc/9706069] ; D. C. Brody and L. P. Hughston, J. Geom. Phys. **38** (2001) 19 [quant-ph/9906086]; and references therein.
5. S. Haller and C. Vizman, math.DG/0305089.
6. T. Banks, W. Fischler, S. H. Shenker and L. Susskind, Phys. Rev. D **55**, 5112 (1997) [arXiv:hep-th/9610043].

QUANTIZING GRAVITATIONAL COLLAPSE

CENALO VAZ
University of Cincinnati, Cincinnati, OH 45221-0011
and
Universidade do Algarve, Faro, Portugal.

1. Introduction

It is generally believed that the following is true: the classical gravitational field of a sufficiently massive star will overcome its neutron degeneracy pressure and, barring quantum gravitational effects in the final stages of collapse, the star will form a singularity of space-time. The initial conditions determine the precise nature of the singularity of the collapse: either the singularity will be covered, in which case a black hole is formed, or it will be naked, *i.e.*, not covered by a horizon. Naked singularities lead to violations of causality, which encouraged Penrose to propose the Cosmic Censorship Hypothesis (CCH)[1]. The CCH simply states that physically reasonable initial data cannot produce naked singularities. Nevertheless, to date there is little evidence that it is true in general relativity, even though most of our current emphasis on black hole physics is based on the validity of this hypothesis.

The LeMaître-Tolman-Bondi (LTB) collapse of spherical, inhomogeneous dust[2] is (arguably) the simplest model that allows for the formation both of black holes and of naked singularities. It is defined by specifying two functions, *viz.*, a mass function, $F(\rho)$, and an energy function, $f(\rho)$, where ρ is a shell label coordinate. The former represents the weighted mass contained within the matter shell labeled by ρ, and the latter is related to the velocity profile within the collapsing cloud at the initial time. The configuration space for all LTB models consists of the dust proper time, τ, and the area radius, R. The classical collapse may end in a black hole or in a naked singularity depending on the behavior of $F(\rho)$ and $f(\rho)$ near the center[3]. When it ends in a black hole, a significant fraction of the star is expected to evaporate via Hawking radiation[4] during the semi-classical phase. However, the back-reaction of the space-time prevents us from making definitive predictions at the final stages. A different result is obtained when the collapse is toward a naked singularity[5]. It is quantum mechanically unstable (the radiated power behaves as $P \sim (U_0 - U)^{-\alpha}$ for some – model dependent – value of $\alpha > 0$) but, during the validity of the semi-classical approximation (curvatures should be less than Planck scale), the collapsing cloud emits only about one Planck unit of energy[6]. Because the back-reaction does not become important so long as gravity can be treated classically, the future evolution of the star is governed *exclusively* by quantum gravity and it is impossible to say, from the semi-classical approximation, whether the star radiates away its energy on a short time scale or settles down into a black hole state. Quantum gravitational effects are therefore expected to modify the very nature of the classical singularities that result from the collapse of a relativistic star, serving even as the Cosmic Censor. However, the true end state of the collapse can only be fully understood by an application of a quantum theory of gravity.

2. Quantization

The canonical dynamics of the collapsing cloud is described by embedding the spherically symmetric ADM[7] 4-metric in the LTB space time, and by casting the action for the Einstein-dust system in canonical form. Performing a version of the canonical transformation developed by Kuchař[8], one may use the chart consisting of $(\tau, R, F, P_\tau, P_R, P_F)$ whence the time evolution of the wave-functional, $\Psi[\tau, R, F]$, is determined by the functional Schroedinger equation,

$$i\frac{\delta\Psi}{\delta\tau} = \hat{h}\Psi = \sqrt{\pm c^2 \frac{\delta^2}{\delta R_*^2} \pm m_p^2 c^4 \frac{F'^2}{4|\mathcal{F}|}}\, \Psi[\tau, R, F], \tag{1}$$

the upper sign within the square-root referring to the region outside the horizon and the lower sign to the region inside. Invariance under spatial diffeomorphisms is implemented by the momentum constraint,

$$\left[\tau' \frac{\delta}{\delta\tau} + R'_* \frac{\delta}{\delta R_*} + F' \frac{\delta}{\delta F}\right] \Psi[\tau, R, F] = 0 \tag{2}$$

and an inner product on the Hilbert space of wave-functionals may be defined in a natural way, by exploiting the fact that the DeWitt supermetric[9] is manifestly flat in the configuration space (τ, R_*), as the functional integral[10]

$$\langle\Psi_1|\Psi_2\rangle = \int_{R_*(0)}^{\infty} \mathcal{D}R_* \Psi_1^\dagger \Psi_2 \ . \tag{3}$$

This inner product is defined on $\tau=$ constant hypersurfaces and the set-up in equations (1)–(3) implies a specific choice, albeit a natural one, of operator ordering. For a solution to the quantum constraints, we note that momentum constraint is obeyed by *any* functional that is a spatial scalar. In particular the functional

$$\Psi = A\exp\left[-\frac{ic}{2l_p^2}\int_0^\infty drF'\left[\tau+\mathcal{U}(R,F)\right]\right], \tag{4}$$

where $A=A(F)$, is a spatial scalar if $\mathcal{U}(R,F)$ has no explicit dependence on r and describes a stationary state of proper energy $F'/2$. Throughout we will use this as a solution ansatz[11].

3. The "eternal" black hole: Mass spectrum and entropy

A mass function describing an eternal black hole of mass M may be taken to be that of a spherical shell $F(r)=2M\theta(r)$, where $\theta(x)$ is the Heaviside function. Equation (4) then tells us that the problem of a single shell is essentially quantum *mechanical*,

$$\tilde{\Psi}[\tau,R,F]=\exp\left[M\mathcal{W}_0(\tau,R,M)\right], \tag{5}$$

where $\tau=\tau(0), R=R(0)$ and $F(0)=2M$ represent, respectively, the proper time, the radial coordinate and the total mass of the single shell. In the WKB approximation, the stationary states of the black hole are easily described: in the interior they are a superposition of ingoing and outgoing waves,[11]

$$\tilde{\Psi}[\tau,R_*]=A_\pm e^{-iM(\tau\pm R_*)}, \qquad -\pi M<R_*<\pi M \tag{6}$$

and in the exterior, they are exponentially decaying

$$\begin{aligned}\tilde{\Psi}[\tau,R_*]&=Be^{-iM(\tau-iR_*)}, && R_*>\pi M\\ &=Ce^{-iM(\tau+iR_*)}, && R_*<-\pi M,\end{aligned} \tag{7}$$

which is in keeping with the fact that there are no wave solutions in vacuum spherical gravity. Matching the wave function and its derivative at the horizon, one finds that the energy (mass) squared of the black hole is quantized in half integer units,

$$M^2=\left(n+\frac{1}{2}\right)M_p^2, \ \ \forall\ \ n\in\mathbf{N}\cup\{0\} \tag{8}$$

where M_p is the Planck mass. This is just the mass spectrum originally proposed by Bekenstein[12].

The above considerations may be extended to the case of many shells. A simple generalization of the mass function above, which describes many shells, is $F(r) = 2\sum_{j=1}^{N} \mu_j \theta(r - r_j)$ and a straightforward application of the boundary conditions appropriate to each shell gives the following quantization condition for the states of shell k,[13]

$$\mu_k M_k = \left(n_k + \frac{1}{2}\right) M_p^2. \tag{9}$$

where M_k represents the total mass contained within shell k. These conditions, if applied recursively, show that the mass of shell k is determined by k quantum numbers. Thus the total mass depends on N quantum numbers for a black hole formed out of N quantum shells. The appearance of N quantum numbers means a quantum black hole is not simply described by its total mass: such a description ignores the manner in which the mass is distributed among the shells. The entropy counts the number of distributions for a given total mass. For an *eternal* black hole, M_k in (9) should be replaced by M, the mass of the hole. The total mass (squared) of the hole continues quantized as before and the problem of counting the number of distributions is precisely the problem of asking for the number of ways in which N integers may be added to give another integer[13]. This result depends on the number of shells that have collapsed to form the black hole, which we do not know but which can be independently determined by maximizing the entropy with respect to N. When both N and M/M_p are large, one readily finds, to leading order,

$$S \approx 0.962 \times \frac{\mathcal{A}}{4} \tag{10}$$

in units of Planck area, which agrees well with the Bekenstein-Hawking value.

4. Hawking radiation

When applied to a matter distribution that is appropriate to a massive black hole surrounded by dust whose total energy is small compared with the mass of the black hole, the WKB approximation describes Hawking radiation. For this purpose, let us assume that the mass function $F(r)$ is of the form[14] $F(r) = 2M\theta(r) + f(r)$, where $\theta(r)$ is the Heaviside step-function, and $f(r)$ is differentiable, representing a dust distribution with $f(r)/2M \ll 1$. It can be interpreted as the presence of a Schwarzschild black hole of mass M at the origin, and $f(r)$ is a dust matter perturbation on the black hole, which can be related to Hawking radiation. In a sense, it plays the role of the quantum field used in standard derivations of Hawking radiation, except that now it is quantized along with the gravitational field.

Inserting this mass function in the wave-functional (4) we have $\Psi = \Psi_{\rm bh} \times \Psi_f$ where $\Psi_{\rm bh}$ represents the black hole and

$$\Psi_f \sim \exp\left[\frac{1}{2}\int_0^\infty dr f' \mathcal{W}^f(\tau(r), R(r), M)\right].$$

The solution for $\mathcal{W}^f$ outside the apparent horizon and written in terms of Schwarzschild time is given by

$$\mathcal{W}^f_{\rm out} = -i\left[T + 4\sqrt{2M}\left(\sqrt{R} - \frac{\sqrt{2M}}{2}\ln\left[\frac{\sqrt{R}+\sqrt{2M}}{\sqrt{R}-\sqrt{2M}}\right]\right)\right]. \qquad (12)$$

The projection of our solution (12) on the negative frequency modes of an outgoing basis on $\mathcal{I}^+$ represents the negative frequency modes present in the solution. So we compute $|\langle \Psi^\dagger_\omega | \Psi^{\rm hor}_f \rangle|^2$ because these are the analogs of the Bogoliubov coefficients near the apparent horizon, $|\beta(f,\omega)|^2$. We find[14]

$$|\beta(f,\omega)|^2 = \prod_r \frac{2\pi M}{\Delta f}\left[\frac{1}{e^{8\pi M \Delta f} - 1}\right]. \qquad (13)$$

This is interpreted as the eternal black hole being in equilibrium with a thermal bath at the Hawking temperature $(8\pi M)^{-1}$. Our derivation provides a functional Schrödinger picture for dust Hawking radiation, consistent with the WKB wave-functional which solves the Wheeler-DeWitt equation.

5. Beyond the WKB approximation

Because of Hawking radiation, we do not expect to have a sharp boundary between the collapsing matter and an external vacuum (as is generally considered in the classical studies) owing to the evaporation process. Thus $F(r)$ will increase with increasing $r \in [0, \infty)$ and we could equivalently label shells by $F(r)$ instead of r. The Hamiltonian constraint is ill-defined and requires regularization, which we perform on a lattice (for details on the lattice construction, see[15]). If j represent lattice points, the wavefunctional can be expressed as the product

$$\Psi[\tau, R] = \prod_j \Psi_j(\tau_j, R_j) = \prod_j \exp\left[-\omega_j\left[\tau_j + \mathcal{U}_j(R_j, F_j)\right]\right]. \qquad (14)$$

where $\varepsilon_j = \hbar\omega_j$ is the energy of shell j and the Wheeler-DeWitt equation is to be solved independently at each label. Each lattice point physically represents a shell of matter. Thus we get an infinite set of formally identical equations, one for each shell. After carrying out the time derivative , defining the dimensionless variables $z_j = R_j/F_j$ and $\gamma_j = F_j\omega_j/c$, denoting $y_j \equiv \tilde{\Psi}_j$ (the time independent shell wave function, and suppressing the subscript j throughout, the equation for each shell takes the form

$$z(z-1)^2\frac{d^2y}{dz^2} + \frac{(z-1)}{2}\frac{dy}{dz} + \gamma^2 z^2 y = 0 \tag{15}$$

both in the exterior ($z > 1$) and in the interior ($z < 1$). Every shell's evolution is therefore completely determined by the total mass contained within it.

Solutions to Ref. 15 can be obtained as a power series about the non-essential singularities at the origin ($z = 0$) and the horizon ($z = 1$). It is possible to show[15] that the Hawking spectrum itself arises from retaining only the lowest order terms and dropping all terms of $\mathcal{O}(\hbar)$. When we include the first order correction (in $\mathcal{O}(\hbar)$) we find,

$$|\beta(\omega,\omega')|^2 \approx 2\pi^2 F^2 \sqrt{\frac{2c}{F\omega'}}\frac{kT_H}{\varepsilon}\frac{1}{e^{\frac{\varepsilon}{kT_H}}-1}\left[1-\frac{1}{2}\ln\left(\frac{\pi k T_H}{\varepsilon}\right)\right], \tag{16}$$

which completes the first correction to the WKB approximation. The correction term cannot be accounted for by modifying the Hawking temperature and one must conclude that it renders the Hawking radiation non-thermal.

6. Conclusions

There are two points worth noting: (a) equation (16) does not represent a correction to the Hawking temperature (equivalently, the Bekenstein-Hawking entropy). It shows instead that the spectrum is *non-thermal*, and (b) this non-thermal spectrum *is associated with the horizon* because we have considered only the near horizon wave-functional in our calculation. It is not, therefore, related to the non-thermal spectrum observed at large distances from the horizon, e.g., at spatial infinity, due to grey body factors arising from the scattering of the radiation against the geometry.

Within the context and limitations of the canonical theory, our result indicates that unitarity *is* preserved in quantum gravitational evolution[16]. Many open questions, such as what meaning is to be assigned to the (logarithm of) the number of states, or the statistical entropy, of the black hole in the absence of a thermal spectrum and what is the final fate of black holes and naked singularities, remain to be examined.

Acknowledgments

Much of the work reported here was done in collaboration with Louis Witten and T.P. Singh. I am very grateful for our close collaboration and many discussions. I also acknowledge the financial support from the *Fundação para a Ciência e a Tecnologia* (FCT) under contract number POCTI/32694/FIS/2000 (*III Quadro Comunitário de Apoio*).

References

1. Roger Penrose, Riv. Nuovo Cimento **1** (1969) 252; *ibid.* in *General Relativity, An Einstein Centenary Survey*, ed. S. W. Hawking and W. Israel, Cambridge Univ. Press, Cambridge, England, (1979) 581.
2. G. LeMaître, Ann. Soc. Sci. Bruxelles I, **A53** (1933) 51; R.C. Tolman (1934) Proc. Nat. Acad. Sci. USA **20** 169; H. Bondi (1947) Mon. Not. Astron. Soc. **107** 410.
3. see, for example, P. S. Joshi, *Global Aspects in Gravitation and Cosmology*, Clarendon Press, Oxford, (1993).
4. S.W. Hawking, Comm. Math. Phys. **43** (1975) 199; S.W. Hawking, Phys. Rev. **D14** (1976) 2460; W.G. Unruh, Phys. Rev. **D14** (1976) 870.
5. Cenalo Vaz and Louis Witten, Nucl. Phys. **B487** (1997) 409; Cenalo Vaz and Louis Witten, Phys. Lett. **B442** (1998) 90; Cenalo Vaz and T.P. Singh, Phys. Rev. **D61** (2000) 124005.
6. T. Harada, H. Iguchi, K. Nakao, T.P. Singh, T. Tanaka and Cenalo Vaz, Phys. Rev. **D64** (2001) 041501.
7. R. Arnowitt, S. Deser and C.W. Misner in *Gravitation: An Introduction to Current Research*, ed. Louis Witten, Wiley, New York (1962).
8. K.V. Kuchař, Phys. Rev. **D50** (1994) 3961.
9. B.S. DeWitt, Phys. Rev. **160** (1967) 1113.
10. Cenalo Vaz, L. Witten and T.P. Singh, Phys. Rev. **D63** (2001) 104020.
11. Cenalo Vaz and Louis Witten, Phys. Rev. **D60** (1999) 024009.
12. J.D. Bekenstein, Lett. Nuovo Cimento **11** (1974) 467; *ibid*, Phys. Rev. Lett. **70** (1993) 3680; *ibid*, Phys. Lett. **B360** (1995) 7.
13. Cenalo Vaz and Louis Witten, Phys. Rev. **D63** (2001) 024008.
14. Cenalo Vaz, Claus Kiefer, Louis Witten and T.P. Singh, Phys. Rev. **D67** (2003) 024014.
15. Cenalo Vaz, L. Witten and T.P. Singh, "Exact Quantum State of Collapse and Black Hole Evaporation", [arXiv:gr-qc/0306045]
16. P. Hájíček, C. Kiefer, Nucl.Phys. B603 (2001) 531-554; P. Hájíček, Nucl. Phys. **B603** (2001) 555-577.

IV. $D < 4$ FIELD THEORIES; STRING AND BRANE WORLDVOLUMES

PHASE STRUCTURE OF NON-COMPACT QED3 AND THE ABELIAN HIGGS MODEL *

THOMAS APPELQUIST

Department of Physics, Yale University, New Haven, CT 06511

L.C.R. WIJEWARDHANA

Department of Physics, University of Cincinnati, Cincinnati, OH 45221

We review the phase structure of a three-dimensional, non-compact Abelian gauge theory (QED3) as a function of the number N of 4-component massless fermions. There is a critical N_c up to which there is dynamical fermion mass generation and an associated global symmetry breaking. We discuss various approaches to the determination of N_c, which lead to estimates ranging from $N_c = 1$ to $N_c = 4$. This theory with $N = 2$ has been employed as an effective continuum theory for the 2D quantum antiferromagnet where the observed Neel ordering corresponds to dynamical fermion mass generation. Thus the value of N_c is of some physical interest. We also consider the phase structure of the model with a finite gauge boson mass (the Abelian Higgs model).

1. Introduction

We review the dynamical generation of fermion mass in a three-dimensional, non-compact Abelian gauge theory (QED3) and its generalization to the case of a finite gauge boson mass. Attention was first drawn to this problem by Pisarski [1]. Some years ago, a study concluded that with the gauge symmetry unbroken, mass generation will occur, corresponding to the breaking of a certain global "chiral" symmetry, if and only if the number N of four-component fermions is no larger than a critical value N_c, estimated to be 4 [2, 3]. Since then, most studies have agreed that there is indeed such an N_c, but there has been little agreement as to its magnitude. Estimates have ranged from that of Refs. 2 and 3 to as low as $N_c = 1$ [4]. This uncertainty is typical of our rudimentary understanding of most strongly coupled quantum field theories.

*Talk presented by T.A.

The value of N_c is a potentially important real-world question. It has been suggested that QED3 with $N = 2$ massless fermions can provide a continuum description of a 2D quantum antiferromagnet [5, 6]. Marston [7], Laughlin and Zou [8], and others have noted that the observed Neel ordering corresponds to dynamical fermion mass generation and its associated chiral symmetry breaking in QED3 with $N = 2$. More recent discussions of these ideas are provided by Kim and Lee [9], and Kleinert, Nogueira, and Sudbo [10]. For another application of non-compact $QED3$ to condensed matter systems see Ref. [11]. The suggestion that (non-compact) QED3 provides a continuum description of the 2D quantum antiferromagnet has, however, been challenged in a recent series of papers by Herbut et al [12].

Several approaches have been brought to bear on the determination of N_c in QED3. The original study [2, 3] was based on a questionable use of the $1/N$ expansion along with a continuum gap equation. Continuing to non-integral values of N, it led to the estimate $N_c \approx 128/3\pi^2 \approx 4.3$. This approach also led to some speculation on the nature of the phase transition as a function of N [13], namely that because of the long-range force it is of infinite order. A lower value of N_c is suggested by a conjectured inequality based on the counting of thermodynamic degrees of freedom [4]. It indicates that N_c can be no greater than 3/2. It is likely that the determination of N_c will be settled only by numerical lattice studies. These studies have so far suggested that N_c could be as low as 2 [14, 15], but a conclusive determination will need more refined simulations.

We first review the features of QED3 and discuss its symmetry properties and behavior for large N where the $1/N$ expansion may be reliably employed. We discuss fermion mass generation, which takes place at low values of N. The conjectured constraint of Ref. 4 is then described and applied to QED3. We review recent lattice computations, discuss their sensitivity to finite size effects, and to provide some perspective on these simulations include a rough numerical estimate using the continuum gap equation. We also discuss dynamical fermion mass generation in a related theory, the Abelian Higgs model in three-dimensions with the same fermion content as QED3, but where the gauge field is massive. We then summarize, revisit the relevance of these models to condensed matter systems, and describe some open questions.

2. QED3 and its Symmetry Breaking

The Lagrangian of the model is

$$L = \sum_{j=1}^{N} \bar{\psi}_j (i \not{D}) \psi_j - \frac{1}{4} F_{\mu\nu} F^{\mu\nu}, \tag{1}$$

where $D_\mu = i\partial_\mu + eA_\mu$ is the covariant derivative, e is the gauge coupling, and ψ_j is a set of N 4-component fermion fields. To explore the phase structure as a function of N, it is convenient to keep fixed the dimensionful quantity $\alpha \equiv \frac{e^2 N}{8}$. Being super-renormalizable, this theory is UV-complete, rapidly damped at momentum scales beyond α. As an effective theory at lower momenta, one could nevertheless take the view that higher momentum physics is unknown, being integrated out into a tower of higher-dimension operators associated with the scale α. We do not include such terms, which could modify the critical value of N for chiral symmetry breaking. We return to this question in the summary when we discuss possible condensed matter applications of QED3.

For large N, the theory remains weakly coupled at all momentum scales. This can be seen by computing the gauge boson propagator in the large-N limit and extracting from it an effective, dimensionless running coupling $\bar{\alpha}(k)$. The large-N form of the Euclidean propagator is $(k^2 + \alpha k)^{-1}$, where k is the magnitude of the Euclidean three-momentum and the second term arises from the N fermion loops. Multiplying the propagator by the coupling e^2 and by one power of k to make a dimensionless quantity, we have

$$\bar{\alpha}(k) \equiv \frac{e^2 k}{k^2 + (e^2 N/8)k} = \frac{8}{N} \frac{\alpha}{\alpha + k}. \tag{2}$$

This expression exhibits asymptotic freedom at large momentum, and conformal symmetry with an IR fixed point of strength $8/N$ as $k \to 0$. For large N, the coupling is always weak, and no dynamical fermion mass generation is expected.

For finite N, on the other hand, the infrared coupling becomes strong, governed by a strong IR fixed point, and fermion mass generation becomes possible. In the absence of fermion mass, the global symmetry associated with the N four-component fermions is $U(2N)$. The dynamical generation of an equal mass for all the fermion flavors would break the global symmetry to $U(N) \otimes U(N)$ giving rise to $2N^2$ Goldstone bosons [1]. More generally, mass terms for the Dirac spinors can violate parity (P) and time reversal symmetry (T), and the gauge field admits a corresponding P- and T-violating Chern-Simons mass term. These are not generated spontaneously [16]. We concentrate here on the parity conserving case.

Several studies over the years have led to the conclusion that a parity conserving mass is indeed generated dynamically for this theory providing that N is no larger than a critical value N_c [2]. An upper limit is expected, since, as we have noted, the theory is weakly coupled at all momentum scales in the large-N limit. Since the Lagrangian of Eq. 1 is UV-complete, damped rapidly at momentum scales above α, the value of N_c is completely determined by the conformal, IR behavior.

3. A Conjectured Constraint

A conjectured constraint on the infrared structure of asymptotically free gauge theories [4] can be utilized to analyze the phase structure of QED3. This constraint, which takes the form of an inequality, states that for a wide class of such theories, the number of IR "degrees of freedom", defined using the thermal free energy per unit volume $F(T)$ of the theory, is less than or equal to the number of corresponding UV degrees of freedom.

These degree-of-freedom counts are defined by the quantities f_{IR} and f_{UV}, given in terms of $F(T)$ by taking the the zero- and infinite-T limits respectively of the quantity $-\frac{F(T)}{T^d} f(d)$. Here $f(d)$ is a function of the number of space time dimensions d, defined such that the contribution from a free bosonic field is 1. The inequality is then

$$f_{IR} \leq f_{UV}. \tag{3}$$

In general, f_{IR} and f_{UV} will exist providing that the theory is governed by IR and UV fixed points. The conjectured inequality was restricted, however, to asymptotically free theories, in which f_{UV} can be computed from free field theory. It has been applied to a variety of four-dimensional theories, and found to be satisfied whenever f_{IR} can be reliably computed [4]. This includes both IR-free theories and theories with weak IR fixed points.

For QED3, the asymptotic freedom leads to $f_{UV} = \frac{3}{4}(4N) + 1$ [4]. The second term counts the single bosonic degree of freedom associated with a massless gauge boson in three space-time dimensions. In the first term, the $4N$ counts the number of fermionic degrees of freedom associated with the N four-component fermions, and the $3/4$ represents the Boltzman weighting of fermions in three space-time dimensions.

The value of f_{IR} depends on whether the global symmetry is spontaneously broken. For large N, breaking is not expected, and only the massless fermions along with the one gauge degree of freedom remain in the infrared spectrum. From Eq. 2, one sees that the infrared theory is described by a weak (O(1/N)) fixed point. The free-field result, $f_{IR} = \frac{3}{4}(4N) + 1 (= f_{UV})$, can then be corrected perturbatively in $1/N$, the next-to-leading term being $O(1)$. The computation, similar in some ways to the corresponding perturbative computation in a 4D gauge theory, has not yet been done. The 4D result is negative, and if the same is true of the 3D $1/N$ computation, the inequality $f_{IR} \leq f_{UV}$ will be satisfied.

Now consider the more interesting possibility that for some finite N the global symmetry is broken. The fermions become massive and $2N^2$ Goldstone bosons are formed. The infrared (massless) theory consists of only the Goldstone bosons and the gauge boson. Thus $f_{IR} = 2N^2 + 1$ and the inequality will be satisfied only if $N_c \leq 3/2$. If this upper limit on N_c is correct, then QED3 cannot be in the broken phase if N is larger then $\frac{3}{2}$, meaning that the continuum gap equation [2] overestimates N_c.

A natural question about this application of the inequality $f_{IR} \leq f_{UV}$ has to do with the Mermin-Wagner-Coleman theorem [17]. We have determined f_{IR} in the broken phase by computing the free energy at a temperature $0 < T << \alpha$, counting the Nambu-Goldstone bosons plus the gauge field as the relevant, non-interacting degrees of freedom, multiplying by $1/T^3$ and an appropriate constant, and taking the limit $T \to 0$. Now at finite T, the IR behavior of this theory is that of the corresponding 2D theory. But the Mermin-Wagner-Coleman theorem states that there can be no spontaneous symmetry breaking, with its Nambu-Goldstone bosons, in 2D. This question was addressed by Rosenstein, Warr, and Park [18], who concluded that the symmetry is indeed unbroken at any non-zero T, and that for small T the zero-temperature Nambu-Goldstone bosons develop small masses. In our notation, one finds $m_{NG}^2 \sim T^2 exp(-\alpha/T)$, arising from the derivative interactions among the (pseudo) Nambu-Goldstone bosons. Since this mass vanishes more rapidly than T as $T \to 0$, it has no effect on the computation of f_{IR}.

4. Lattice Studies

Lattice studies may be the most reliable method to analyze dynamical mass generation and directly determine N_c in QED3. Numerical simulations began over a decade ago, with a study in the quenched approximation giving preliminary evidence that the global chiral symmetry is spontaneously broken [19]. Recent advances in computing power have led to improved studies of mass generation in QED3. Simulations of 2-flavor QED3 by Hands,

Kogut and Strouthos [14] on lattices of up to 50^3 sites report that the condensate is two orders of magnitude smaller than the quenched condensate value. They find that the value of the dimensionless condensate, defined by

$$\sigma = \frac{<\bar{\psi}\psi>}{e^4}, \tag{4}$$

where e is the dimensionful gauge coupling constant, is bounded above by 5×10^{-5}. They also analyze finite-size effects and find that this bound is stable for lattice sizes ranging from 10^3 to 50^3. They conclude that there is no decisive signal for chiral symmetry breaking for $N \geq 2$.

This conclusion is consistent with the conjectured inequality constraint described in the previous section, which led to $N_c \leq 3/2$ for QED3. On the other hand, this interpretation of the lattice results is likely premature. It has been noted by Gusynin and Reenders [20] that even a lattice of size 50^3 provides a sufficient IR cutoff to affect substantially the value of N_c. They use the continuum gap equation first employed to note the existence of an N_c, modeling the effect of a finite lattice size by imposing an IR cutoff μ on the integral in the gap equation. Because of the scale-invariant form of the gap equation following from the dominance by the IR fixed point, the equation is logarithmic in character. A measure of the importance of the cutoff μ relative to the effective UV cutoff α is therefore $ln\ \alpha/\mu$. As this quantity decreases from infinity, a growing portion of momentum space is truncated, and therefore a stronger gauge coupling is required to break the symmetry. Thus, N_c should drop. Gusynin and Reenders find, for example, that with $ln\ \alpha/\mu$ still as large as 6 ($\alpha/\mu \approx 400$), N_c drops by more than 30%. A lattice of size 50^3 corresponds to a much smaller α/μ, indicating that chiral symmetry breaking should not take place for $N = 2$.

If this analysis is qualitatively correct, that is, if the continuum IR cutoff correctly models the finite size of a lattice, then only a larger lattice would be able to determine accurately the value of N_c for a theory such as QED3, governed by a (conformal) IR fixed point. This same remark would apply to studies of the conformal phase transition as a function of the number of fermion species in four-dimensional gauge theories.

5. Gap Equation Estimates

Suppose that it becomes possible to carry out simulations with such large lattices that finite size effects are no longer important, and suppose that the results of the simulations for σ (Eq. 4) with $N = 2$ continue to be bounded as in Ref. [14]. What would one conclude? Here we present a rough argument using the continuum gap equation indicating that even then it could be unclear whether $N = 2$ is in the symmetric or broken phase.

The spirit of the gap equation approach is to use the large-N form of the kernel of this equation (the large-N gauge-boson propagator), and then to continue to finite N. The reliability of this approach, which first suggested the existence of a finite N_c ($\simeq 128/3\pi^2$), is not clear because the higher terms in the kernel are not parametrically small. There is, however, some evidence that the corrections are small numerically [3]. We use this approach to estimate the condensate for $N = 2$, expecting only that the estimate is order-of-magnitude.

The theory is UV-complete, rapidly damped at momentum scales beyond α. Thus, to a good approximation, the gap equation can be written down employing the $\alpha \to \infty$ form of the large-N kernel, with α then used as a UV cutoff. The resulting equation is

$$\Sigma(p) = \frac{16}{3N\pi^2 p} \int^{\alpha} \frac{kdk\Sigma(k)}{k^2 + \Sigma^2(k)} \left[p + k - |p - k|\right], \tag{5}$$

where we make use of the non-local Kondo-Nakatani gauge for which the leading large-N form of the wave function renormalization is unity. The critical N determined from this equation is $N_c = 128/3\pi^2$.

Numerical and analytical solutions of this equation [21] show that for a range of N such that $\frac{N_c}{N} > 1$, the overall scale of the solution $\Sigma(p)$ of Eq.(5), set by $\Sigma(0)$, is much less than α, and for $N = 2$,

$$\frac{\Sigma(0)}{\alpha} \approx 1.7 \times 10^{-2}. \tag{6}$$

In the range of momentum $\Sigma(p) < p < \alpha$ the solution has the following approximate form

$$\Sigma(p) = \frac{\Sigma(0)^{\frac{3}{2}}}{p^{\frac{1}{2}}} \, sin \left[\frac{1}{2}\sqrt{\left(\frac{N_c}{N} - 1\right)} \left(ln\left(\frac{p}{\Sigma(0)}\right) + \delta\right)\right], \tag{7}$$

where $\Sigma(0)$ is used to scale the log and δ is a phase.

We use these formulae to estimate the dimensionless fermion condensate of Eq. 4. The fermion condensate $<\bar{\psi}\psi>$ is given by the integral

$$<\bar{\psi}\psi> = \int^{\alpha} \frac{d^3k}{(2\pi)^3} \frac{4\Sigma(k)}{k^2} = \frac{2}{\pi^2} \int^{\alpha} dk\Sigma(k). \tag{8}$$

Noting that the sine function is bounded above by 1 in the region of integration up to α , we see that the dimensionless condensate σ of Eq. 4

$(=< N^2\bar{\psi}\psi/64\alpha^2 >)$ is bounded above by

$$\frac{N^2}{16\pi^2}(\frac{\Sigma(0)}{\alpha})^{\frac{3}{2}}.$$

With $N = 2$, $N_c = 128/3\pi^2$ and $\frac{\Sigma(0)}{\alpha} \approx 1.7\times10^{-2}$, this expression becomes 6×10^{-5} which is slightly above the upper bound of 5×10^{-5} found in lattice simulations [14]. We have numerically estimated the actual value of the condensate to be 4.4×10^{-5} which falls below the lattice bound [22].

This suggests that a dimensionless condensate (4) of order 10^{-5} may naturally arise in QED3 for N of order 2 even when N_c is as large as $128/3\pi^2 \approx 4$. It indicates that even if the numerical simulations are performed on a very large lattice, so that finite size effects are unimportant, a precise computation will be required to decide whether $N = 2$ is in the broken or symmetric phase.

6. Abelian Higgs Model

We conclude by discussing fermion mass generation when the Abelian gauge field has a non-zero mass. This theory has been used as a continuum description of the competition between long range antiferromagnetic order and superconducting order in planar cuprate systems [23], [24]. It can also be used to interpolate between QED3 and the 2+1 dimensional Thirring model, the latter having been studied in recent lattice simulations [14]. (Related studies, using continuum gap equation methods and invoking the idea of hidden local gauge symmetry may be found in references [25] [26]). Finally, it can be used to study the effect of an IR cutoff on QED3 [24], similarly to the use of a simple IR cutoff as described in Section 3 [20].

The Lagrangian of the model is

$$\mathcal{L} = \mathcal{L}_{QED3} + \frac{1}{2}D_\mu\Phi^* D^\mu\Phi - \lambda(\Phi^*\Phi - Nv)^2 \tag{9}$$

where $\mathcal{L}_{QED3}$ is given in equation (1), and Φ is a complex scalar field with vacuum expectation value v. The addition of the scalar field does not change the global symmetry structure of the theory. We remove the Higgs boson from the spectrum by taking the limit $\lambda \to \infty$ with v fixed. This leaves the theory UV-complete. The gauge boson mass is $M = e\sqrt{Nv}$. To explore the phase structure as a function of N, we keep fixed the quantities $\alpha \equiv \frac{e^2N}{8}$ and v^2. The gauge boson mass M is then also fixed.

As in QED3, we define a dimensionless running gauge coupling $\bar{\alpha}(k)$. The coefficient of $g_{\mu\nu}$ in the Euclidean gauge-boson propagator takes the form $(k^2+M^2+\alpha k)^{-1}$ to leading order in $1/N$, where k is the magnitude of the Euclidean three-momentum and the last term arises from the N fermion loops. The $k_\mu k_\nu$ term of the propagator is dropped since it couples to a conserved current. Multiplying the propagator by the coupling e^2 and by k to make it dimensionless , we have

$$\bar{\alpha}(k) \equiv \frac{e^2 k}{k^2 + M^2 + (e^2N/8)k} = \frac{8}{N}\frac{k}{(k^2/\alpha) + G^{-1} + k}, \tag{10}$$

where $G \equiv \alpha/M^2$. This effective coupling vanishes in both the UV and IR limits. It reaches a maximum no larger than $8/N$, attainable only when $M << \alpha$. As $\frac{M}{\alpha}$ increases, the maximum value decreases monotonically. Thus, for any M/α, the model is weakly coupled in the large-N limit at all momentum scales.

It is reasonable to expect that, as in QED3, a parity conserving mass is generated for N below some critical value N_c. Assuming an N_c to exist, what can it depend on? Since the forces are strongly damped at scales beyond α, no additional cutoff is necessary to define the theory. Thus N_c can be a function of only the dimensionless ratio M/α. This dependence describes the boundary of a phase diagram for the theory.

Consider first the limit $M/\alpha \to 0$. On the one hand, this may be thought of as taking $M \to 0$ with α fixed, giving QED3. As discussed earlier, estimates of N_c in this limit have ranged from 3/2 to approximately 4. The limit $M/\alpha \to 0$ may also be taken by letting $\alpha \to \infty$ with $G \equiv \alpha/M^2$ fixed. This leads to the 2+1 dimensional Thirring model, meaning that N_c for this model is the same as that for QED3. This assumes that the Thirring model is treated with any UV cutoff taken very large relative to G^{-1}. For finite N, this can be implemented using lattice techniques. (For large N, these two ways of taking the limit $M/\alpha \to 0$ may be considered by examining $\bar{\alpha}(k)$ (Eq. 11). In the case $M \to 0$ with α fixed (QED3), there appears an IR ($k << \alpha$) fixed point of strength $8/N$. In the case $\alpha \to \infty$ with G^{-1} fixed , there is a UV ($k >> G^{-1}$) fixed point of the same strength).

As M/α is increased from zero, low momentum components are damped out. If the force driving the symmetry breaking is attractive at all scales (as in the large N approximation), a stronger infrared coupling is required to trigger dynamical symmetry breaking, and therefore N_c will decrease. At some value of M/α, N_c drops below unity meaning that symmetry breaking will not take place at all [23]. This critical curve should be similar to that determined by Gusynin and Reenders [20], with the gauge boson mass replacing the explicit IR cutoff. The character of the phase transition along this critical curve is also of interest. For $M/\alpha = 0$, as discussed earlier, because the force is of infinite range, the continuum gap equation suggests that the transition is of infinite order [13]. With $M > 0$, the force becomes of finite range, and the transition is of second order.

Does the conjectured inequality $f_{IR} \leq f_{UV}$ provide any information about N_c as a function of M/α? With $\alpha \equiv \frac{e^2 N}{8}$ finite, the theory is asymptotically free. A free-field computation then gives $f_{UV} = \frac{3}{4}(4N) + 2 = 3N + 2$ [4]. The second term counts the 2 bosonic degrees of freedom associated with a massive gauge boson in three space-time dimensions. As in QED3, the value of f_{IR} depends on whether the global symmetry is spontaneously broken. For large N, breaking is not expected, and, if M is nonzero, the infrared theory consists of only the (non-interacting) massless fermions. Then $f_{IR} = \frac{3}{4}(4N)$, and the inequality is satisfied. Now suppose that for some finite N the global symmetry is broken. The fermions become massive, and for finite gauge boson mass M the IR theory consists of only the $2N^2$ non-interacting Goldstone degrees of freedom. Thus $f_{IR} = 2N^2$, and the inequality demands that N_c be no greater than 2. But the inequality already demands that N_c be no greater than 3/2 when $M = 0$, and we expect that N_c will decrease as M/α increases. So it would seem that the conjectured inequality has nothing useful to say about the shape of this critical curve.

Lattice simulations of the Abelian Higgs model for an appropriate range of values of M/α could determine the shape of this critical curve. Simulations have been done only for QED3 and the Thirring model, which correspond to the limit $M/\alpha \to 0$. We discussed in Section 3 the sensitivity of the QED3 simulation to finite size effects. A numerical study of the Thrring model was reported by Hands and Lucini [27]. We noted above that providing this model is treated with a UV cutoff large compared to the inverse four-fermion coupling G^{-1}, the value of the critical coupling N_c should approach that of QED3. The lattice simulations of Ref. [27] do take the ultraviolet cutoff larger than G^{-1}, and find chiral symmetry breaking for $N \approx 3$. Their results remain sensitive to the UV cutoff, but we anticipate that N_c will only increase as the cutoff is made larger (as the continuum limit is approached). This suggests that $N_c > 3$ for the continuum Thirring model, and therefore also for QED3.

We end this section by digressing to use the large-N Thirring model to explain why the inequality [4] was restricted to asymptotically free theories. The model becomes non-interacting in the infrared ($k << G^{-1}$) so $f_{IR} = 3N$. Having taken $\alpha \to \infty$, the resultant weak UV fixed point means that

f_{UV} exists and may be computed in the $1/N$ expansion. To leading order, $f_{UV} = 3N$. The computation of the next-to-leading corrections should be the same as that of f_{IR} in QED3, since the latter is governed by an IR fixed point identical in strength to the UV fixed point of the Thirring model. If the result is negative as anticipated, then $f_{IR} > f_{UV}$ for this model. Another 2+1 dimensional model with a UV fixed point, leading to $f_{IR} > f_{UV}$, was studied by Sachdev [28]. In light of these examples, the conjectured inequalty was restricted to asymptotically free theories.

7. Summary and Discussion

A key question for a non-compact Abelian gauge theory in three space-time dimensions is the critical number N_c of four-component fermions below which there is fermion mass generation with an associated chiral symmetry breaking, and above which the fermions remain massless. Over the years, estimates of N_c with the gauge symmetry unbroken (QED3) have ranged from infinity to one. Recent studies agree that it is finite and about 4, but some estimates continue to be as low as 1. Since this theory is UV-complete, damping rapidly at momentum scales above $\alpha \equiv e^2N/8$, the value of N_c depends only of the IR behavior of the theory, where conformal symmetry sets in.

This question has a potential real-world interest, because of the possibility that QED3 with $N = 2$ can describe the physics of the planar anti-ferromagnet [5] [6]. The observed antiferromagnetic (Neel) ordering corresponds to chiral symmetry breaking in QED3 [7] [8]. Numerical simulations [29] of the planar antiferromagnet also indicate Neel ordering. Herbut and collaborators [12] have challenged the idea that non-compact QED3 is relevant to the planar antiferromagnet, noting that the underlying theory is compact QED3 with its topological (instanton) structure. Some argue, however, that for large enough N, instantons and anti-instantons are bound at distance scales of order $1/\alpha$ by a logarithmic term in the action [10]. Then for $N = 2$ the non-compact theory could be employed at larger distances. But Herbut et al [12] claim that the finite density of instantons and anti-instantons screen the logarithmic term in the action and that instantons are therefore unsuppressed at all N. If this is the case, the non-compact theory cannot be used for this condensed matter system [30].

In general, the question of whether the value of N_c as determined by (UV-complete) QED3 is relevant to condensed matter systems depends on the role of new, short-distance physics. With the physical lattice spacing of order $1/\alpha = N/8e^2$, physics at this scale can be represented in the effective theory at momentum scales below α by the addition of a set of higher-dimension operators. The importance of these terms depends on their strength. We have analyzed this question in the framework of the continuum gap equation in 4-dimensions [31], which has much the same structure as QED3 treated in the $1/N$ expansion. With a four-fermion additive term, there is a critical four-fermion coupling (attractive) of order unity above which this term by itself can trigger spontaneous chiral symmetry breaking. As long as the magnitude of this coupling, either attractive or repulsive, is less than one quarter of its critical value [32, 33], the additive term is unimportant, not affecting the value of N_c. This limit may not be quantitatively accurate, but it seems reasonable that a limit of this order exists. Whether this limit is satisfied for condensed matter systems such as the planar antiferromagnet is not clear. Thus even if the non-compact theory does provide a continuum descriptiion of the antiferromagnet, the value of N_c for this system may or may not be determined accurately by QED3 itself.

Returning to our summary of QED3, we have briefly reviewed the various approaches to the determination of N_c. A recently conjectured inequality constraining asymptotically free quantum field theories [4], when applied to QED3, suggests that $N_c \leq 3/2$. This bound disagrees with the estimate $N_c \approx 4$ emerging from the continuum gap equation approach which originally suggested the existence of a finite N_c.

Recent lattice simulations of $N = 2$ QED3 [14] indicate that the fermion condensate is very small, possibly signaling that $QED3$ is in the symmetric phase, and therefore that $N_c < 2$. This conclusion depends, however, on the importance of finite-size effects. A recent analysis of Gusynin and Reenders [20] using the continuum gap equation indicates that a lattice size of 50^3, used in the simulations, provides a strong enough suppression of IR physics to reduce N_c below 2, thus accounting for the obseravation of a small upper bound for the condensate. Using the solution of the gap equation with no IR cutoff, we have made a rough estimate of the fermion condensate, assuming that N_c is larger – approximately 4 as suggested by the continuum gap equation studies [3]. Interestingly, even though $N = 2$ is then well into the broken phase, the condensate is estimated to be very small, below the upper bound of Ref. [14]. This indicates that even if simulations are conducted on a large enough lattice to make finite size effects unimportant, a precise determination of the condensate is needed to decide whether QED3 with $N = 2$ is in the symmetric or broken phase.

The value of N_c in the Abelian Higgs model is also of interest. This model leads to either QED3 or the 2+1-dimensional Thirring model in the limit $M/\alpha \to 0$ depending on how the limit is taken. But since N_c depends on only the ratio M/α the value of N_c must be the same in both models. Lattice simulations of the Thirring model on a 12^3 lattice lead to $N_c \approx 3$ [27]. This value remains sensitive to the UV cutoff (the lattice spacing), and should increase toward the N_c of QED3 as the continuum limit is approached. On the other hand the inequality (3) applied to the Abelian Higgs model yields $N_c \leq 2$.

To summarize, after more than fifteen years there is still no definitive answer to the question of the critical number of four-component fermions in QED3 (and its generalization to the Abelian Higgs model) marking the boundary between broken and unbroken chiral symmetry. This question is of interest because it could be relevant to the behavior of certain condensed matter systems and because the answer requires an understanding of a strongly coupled, UV-complete quantum field theory.

Acknowledgments

The work described here was supported in part by funds provided by the U.S. Department of Energy under Contracts DE-AC02-76ER0 3075 and DE-FG02-84ER40153. We thank Simon Hands and Subir Sachdev for valuable conversations, Frank Pinski for his assistance with computing and V. P. Gusynin for helpful correspondence.

References

1. R.Pisarski, Phys. Rev. D29. 2423 (1984).
2. Thomas Appelquist, Daniel Nash and L.C.R. Wijewardhana , Phys. Rev. Lett. 60, 2575 (1988).
3. D. Nash, Phys. Lett. 62, 3024 (1989).
4. T. Appelquist, A. Cohen, and M. Schmaltz, Phys. Rev. D60, 045003 (1999).
5. I. Affleck and J. B. Marston, Phys. Rev B39, 11538 (1989), L.B.Ioffe and A. I. Larkin, Phys. Rev. B39, 8988 (1989).
6. M.C.Diamantini, P.Sodano and G.Semenoff, Nuc.Phys.B406,595(1993) .
7. J. B. Marston, Phys. Rev. Lett. 64, 1166 (1990).
8. R.Laughlin and Z.Zou, Phys. Rev. B41, 664 (1990).
9. D. H. Kim and Patrick A. Lee, cond-matt/9810130 published in Ann. Phys. (N. Y.) 272, 130 (1999).
10. H. Kleinert, F. Nogueira, and A. Sudbo, Phys. Rev. Lett. 88 (2002) 232001, hep-th/0201168.
11. Z. Tesanovic, O. Vafek and M. Franz, Phys. Rev. Lett. 87, 257003 (2001) ,

Phys. Rev. B 65, 180511 (2002), M. Franz, Z. Tesanovic and O. Vafek, Phys. Rev. B 66, 054535 (2002).
12. I. Herbut and B. Seradjeh, Phys. Rev. Lett. 91, 171601, 2003 , cond-mat/0305296, M. Case, B. Seradjeh, and I. Herbut, Nucl. Phys. B676: 572-586, 2004, e-print archive cond-mat/0308260.
13. T. Appelquist, J. Terning, and L. C. R. Wijewardhana, Phys. Rev. Lett. 75, 2081 (1995).
14. S. Hands, J. Kogut and K. Strouthos, Nucl. Phys. B645: 321-336, 2002; e-Print Archive: hep-lat/0208030.
15. D. Lee, P. Maris, Phys. Rev. D67: 076002, 2003, e-print archive hep-lat/0212033.
16. C. Vafa and E. Witten, Commun. Math. Phys. 95, 257 (1984); Phys. Rev. Lett. 53, 535 (1984) and T. Appelquist, M. Bowick, D. Karabali and L. C. R. Wijewardhana, Phys. Rev. D33, 3704, 1986.
17. By N.D. Mermin, H. Wagner, Phys. Rev. Lett.17: 1133-1136,1966, S.Coleman, Commun. Math. Phys.31 :259-264, 1973.
18. B. Rosenstein, B. Warr, and S. Park, Nuclear Physics B336, 435 (1990).
19. S. Hands and J. Kogut, Nuc. Phys. B335 (1990) 455.
20. V.P.Gusynin and M.Reenders, Phys. Rev. D68, 025017 (2003), hep-ph/0304302.
21. Thomas Appelquist, Mark J. Bowick, Eugene Cohler, L.C.R. Wijewardhana Phys. Rev. Lett. 55:1715, 1985, Thomas W. Appelquist, Mark J. Bowick, Dimitra Karabali, L.C.R. Wijewardhana Phys. Rev. D33, 3704,1986.
22. V.Gusynin has also obtained a similar estimate for the two flavor condensate, using analytical solutions to the gap equation described in Ref. [20].
23. Guo-Zhu Liu and Geng Cheng, Phys. Rev. D67, 065010, 2003 e-Print Archive: hep-th/0211231.
24. T. Pereg-Barnea and M. Franz, Phys. Rev. B67 060503 (2003), cond-mat/0209301.
25. Taichi Itoh, Yoonbai Kim, Masaki Sugiura, Koichi Yamawaki, Prog. Theor. Phys. 93: 417-440, 1995, hep-th/9411201.
26. M.Sugiura, Prog. Theor. Phys. 97, 311(1997), hep-th/9611198.
27. Simon Hands, Biagio Lucini, Phys. Lett. B461 :263-269,1999, e-Print Archive: hep-lat/9906008.
28. Subir Sachdev, Phys. Lett. B309, 285-288, 1993, e-Print Archive: hep-th/9305131.
29. U.-J. Wiese and H.-P. Ying, cond-mat/9212006 and B. B. Beard and U.-J. Wiese, Phys. Rev. Lett. 77 , 5130 (1996).
30. Some discussion of these issues may be found in T. Senthil, L. Balents, S. Sachdev, A. Vishwanath, and M. P. A. Fisher, cond-mat/0312617.
31. T. Appelquist, J. Terning, and L. C. R. Wijewardhana, Phys. Rev. D44, 871 (1991).
32. T.Appelauist, M.Soldate, T.Takeuchi and L.C.R.Wijewardhana, Proceedings of the Johns Hopkins workshop on current problems in physics, edited by G.Domokos and S.Kovesi-Domokos, World Scientific, Singapore 1988, Keiichi Kondo , Hidetoshi Mino, Koichi Yamawaki, Phys. Rev. D39, 2430,1989.

33. M. Carena, T.E.Clark and C.E.M. Wagner Nucl. Phys. B356,117-153,1991, U. Mahanta Phys. Rev. D44, 3356-3359, 1991.

PLANE WAVE AND VORTEX-LIKE SOLUTIONS OF NONCOMMUTATIVE MAXWELL-CHERN-SIMONS THEORY*

DANIEL ARNAUDON

Laboratoire d'Annecy-le-Vieux de Physique Théorique LAPTH
CNRS, UMR 5108, Université de Savoie
B.P. 110, F-74941 Annecy-le-Vieux Cedex, France

GARNIK ALEXANIAN and M. B. PARANJAPE

Groupe de physique des particules, Université de Montréal
C.P. 6128, succ. centre-ville, Montréal, Québec, Canada H3C 3J7

We investigate the spectrum of noncommutative Yang-Mills-Chern-Simons theory. We find noncommutative massive "plane wave" solutions with polarization dependent on the magnitude of the wave-vector and with a radially symmetric ansatz a vortex-like solution.

1. Introduction

Since their reappearance noncommutative (NC) field theories have been a subject of much attention [2]. There was an interesting proposal by Susskind [6] on the possible relation between NC Chern-Simons theory and the Quantum Hall (QH) effect. It would be interesting if the NC description would lead to some specific experimentally observable differentiating phenomena. We will study possible classical solutions of the corresponding action.

2. Noncommutative Yand-Mills-Chern-Simons theory

The CS action that we are going to work with reads

$$S = \frac{2\lambda}{\theta^{\frac{3}{2}}}\left(2\pi\theta^{\frac{3}{2}}\right)\int d\tilde{t}\mathrm{tr}\left(-[D,\bar{D}]+1\right)D_0\,. \qquad (1)$$

*Talk given by M.B.P.

The gauge theory action is

$$S_{\rm YM} = (2\pi\theta^{\frac{3}{2}}) \int d\tilde{t}{\rm tr}\frac{-2}{4g^2\,\theta^2}\left(2[D_0, D][D_0, \bar{D}] + [D, \bar{D}][D, \bar{D}]\right). \qquad (2)$$

This yields the total action expression

$$\frac{(2\pi\theta^{3/2})}{2g^2\,\theta^2}\int_{\tilde{t}} {\rm tr}\left\{(-2[D_0, D][D_0, \bar{D}] - [D, \bar{D}][D, \bar{D}]) + 2\tilde{\lambda}\left(-[D, \bar{D}] + 1\right) D_0\right\}, \qquad (3)$$

where

$$\tilde{\lambda} = 2\,g^2\,\theta^{1/2}\,\lambda\,. \qquad (4)$$

Varying (3) with respect to D_0 one gets

$$-[D, [D_0, \bar{D}]] - [\bar{D}, [D_0, D]] = \tilde{\lambda}([D, \bar{D}] - 1)\,. \qquad (5)$$

In turn, for the $\delta\bar{D}$ variation the equation is

$$[D_0, [D_0, D]] + [D, [D, \bar{D}]] = \tilde{\lambda}[D_0, D]\,. \qquad (6)$$

3. Noncommutative plane waves

Let us now describe the plane wave solutions of the theory. Generic form of the equations of motion is given by

$$-\ddot{D} + [D[D, D^\dagger]] = -i\,\lambda\,\dot{D} \qquad (7)$$

$$i\,[D, \dot{D}^\dagger] + i\,[D^\dagger, \dot{D}] = \lambda[D, D^\dagger] - \lambda \qquad (8)$$

(to simplify notation we use λ instead of $\tilde{\lambda}$ in what follows). The "vacuum" solution is given by $D = a$, $D^\dagger = a^\dagger$, where $[a, a^\dagger] = 1$ — standard creation and annihilation operators for the harmonic oscillator. Let us look for the solution of the equations (7) and (8) using the following ansatz:

$$D = a + e^{i\,\omega\,t} R_1\left(k\,a + k^*\,a^\dagger\right) + e^{-i\,\omega\,t} R_2\left(k\,a + k^*\,a^\dagger\right), \qquad (9)$$

where k, k^* are complex numbers and $R_1(x)$ and $R_2(x)$ are some (possibly complex) functions of the argument that can be expanded in series. The solution satisfies the following dispersion relation

$$\omega^2(\omega^2 - \lambda^2 - 2|k|^2) = 0\,. \qquad (10)$$

For the $\omega \neq 0$ solution one then has the "massive" modes, given by

$$\omega = \pm\sqrt{\lambda^2 + 2|k|^2}\,. \qquad (11)$$

Finding the eigenvectors corresponding to these eigenvalues gives

$$R_2^* = \left(\frac{\omega - \lambda}{\omega + \lambda}\right) e^{2i\phi} R_1$$
$$k = |k|\, e^{i\phi}\,, \qquad \omega^2 = \lambda^2 + 2|k|^2\,. \tag{12}$$

However, this solution only satisfies equation (7) so far. One has to check that it is going to be compatible with equation (8) as well. Quite remarkably, upon substituting ansatz (9), this equation reduces to the derivative of condition (12).

$$R_2^{*'} = \left(\frac{\omega - \lambda}{\omega + \lambda}\right) e^{2i\phi} R_1'\,. \tag{13}$$

Thus (12) is a solution. The dispersion relation is massive but it shows some interesting properties — unlike the usual (i.e. commutative) plane waves there is a dependence of the polarization on the magnitude of the wave vector, k (see [33] for the noncommutative QED case).

4. Static gauge solutions

In this section we will look for vortex type solutions taking the "static" gauge ansatz, defined by the following condition:

$$D_t = \phi(N)\,,$$

where N is the number operator, and ϕ is a well defined function of N. The corresponding ansatz for the spatial part is given by

$$D = f(N)a \qquad D^\dagger = a^\dagger f^\dagger(N)\,.$$

Now let us proceed to the equations (7) and (8) in terms of these variables. The Ampère law becomes

$$[D_t, [D_t, D]] + \big[D, \big[D, D^\dagger\big]\big] = \lambda[D_t, D] \tag{14}$$

while the Gauss law becomes

$$-\big[D^\dagger, [D_t, D]\big] - \big[D, \big[D_t, D^\dagger\big]\big] = \lambda(\big[D, D^\dagger\big] + 1)\,. \tag{15}$$

Using definition

$$Q(N) = \phi(N) - \phi(N+1)$$

we get the following operator equations for (14) and (15) the Ampère law and the Gauss law, respectively,

$$\lambda f(N)Q(N)a = f(N)(Q(N))^2 a + f(N)(B(N+1) - B(N))a \tag{16}$$

$$\lambda(B(N) - 1) = 2\Big((N+1)f^\dagger(N)f(N)Q(N) - Nf^\dagger(N-1)f(N-1)Q(N-1))\Big) . \tag{17}$$

The Ampere law, equation (16) yields

$$\frac{1}{4}(2Q(N) - \lambda)^2 + \nabla^2\left((N+1)f^\dagger(N)f(N)\right) = \frac{\lambda^2}{4} , \tag{18}$$

where ∇^2 is the appropriate discrete laplacian operator. The Gauss law, equation (17) actually be solved as an operator equation, to give

$$(2Q(N) - \lambda)f^\dagger(N)f(N) = -\lambda .. \tag{19}$$

Replacing for $(2Q(N) - \lambda)$ in equation (18) yields

$$\frac{1}{4}\left(\frac{\lambda}{f^\dagger(N)f(N)}\right)^2 + \nabla^2\Big((N+1)f^\dagger(N)f(N)\Big) = \frac{\lambda^2}{4} . \tag{20}$$

Defining $u(N) = (N+1)f^\dagger(N)f(N)$ and $g^2 = \lambda^2/4$ we get the nonlinear recurrence relation (in the spectral representation)

$$u_{n+1} - 2u_n + u_{n-1} + g^2\left(\frac{n+1}{u_n}\right)^2 = g^2 . \tag{21}$$

u_{-1} is taken to be zero as can be seen from consistency with its definition. Hence specifying the value of u_0 is sufficient to determine all subsequent values of u_n for $n \geq 1$.

Let us study the simplest solution taking $Q(N) = 0$ and $f^\dagger(N)f(N) = 1$ since the (discrete) laplacian of $N+1$ evidently vanishes. This corresponds to the "vacuum" solution of the quantum Hall system of Susskind and its gauge transformations.

Numerical analysis of the recurrence relation (21) yields the following intriguing behavior, some of which is shown on figure 1. If u_0 is taken to be one, we get the above mentioned vacuum solution (data with $u_0 = 1.0$). However if u_0 is perturbed greater than one, values for subsequent u_n become large (data with $u_0 = 1.005$). This solution is asymptotic to the solution found by [38] for the Yang-Mills-Chern-Simons theory with the pure D^3 expression for the Chern-Simons term. If u_0 is perturbed slightly less than one, the behavior is very interesting but always pathological. It seems the vacuum solution $u_0 = 1$ is unstable.

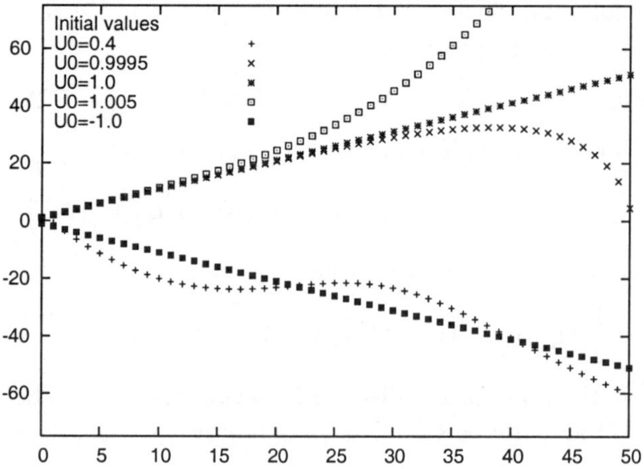

Figure 1. Behavior of the solutions of the equation (21) for the different values of u_0.

Acknowledgments

We would like to thank C. Burgess, B. Dolan, A. Khare, R.B. MacKenzie and V.P. Nair for useful discussions and NSERC of Canada for financial support.

References

1. H. S. Snyder, Phys. Rev. **71**, 38 (1947).
2. M. R. Douglas and N. A. Nekrasov, Rev. Mod. Phys. **73**, 977 (2001) [arXiv:hep-th/0106048].
3. A. Connes, arXiv:math.qa/0011193.
4. A. P. Balachandran, Pramana **59**, 359 (2002) [arXiv:hep-th/0203259].
5. J. Madore, Lect. Notes Phys. **543**, 231 (2000).
6. L. Susskind, arXiv:hep-th/0101029.
7. C. P. Burgess and B. P. Dolan, arXiv:hep-th/0010246.
8. C. P. Burgess and B. P. Dolan, Phys. Rev. B **65**, 155323 (2002) [arXiv:cond-mat/0105621].
9. A. P. Polychronakos, Phys. Lett. B **495**, 407 (2000) [arXiv:hep-th/0007043]; JHEP **0011**, 008 (2000) [arXiv:hep-th/0010264].
10. V. P. Nair and A. P. Polychronakos, Phys. Rev. Lett. **87**, 030403 (2001) [arXiv:hep-th/0102181].
11. D. Bak, K. M. Lee and J. H. Park, Phys. Rev. Lett. **87**, 030402 (2001) [arXiv:hep-th/0102188].
12. A. P. Polychronakos, JHEP **0104**, 011 (2001) [arXiv:hep-th/0103013].
13. S. Hellerman and M. Van Raamsdonk, JHEP **0110**, 039 (2001) [arXiv:hep-th/0103179].

14. D. Karabali and B. Sakita, Phys. Rev. B **65**, 075304 (2002) [arXiv:hep-th/0107168].
15. D. Karabali and B. Sakita, Phys. Rev. B **64**, 245316 (2001) [arXiv:hep-th/0106016].
16. A. Pinzul and A. Stern, JHEP **0111**, 023 (2001) [arXiv:hep-th/0107179].
17. G. S. Lozano, E. F. Moreno and F. A. Schaposnik, JHEP **0102**, 036 (2001) [arXiv:hep-th/0012266].
18. E. Fradkin, V. Jejjala and R. G. Leigh, Nucl. Phys. B **642**, 483 (2002) [arXiv:cond-mat/0205653].
19. A. P. Balachandran, K. S. Gupta and S. Kurkcuoglu, JHEP **0309**, 007 (2003) [arXiv:hep-th/0306255].
20. A. Pinzul and A. Stern, Mod. Phys. Lett. A **18**, 2509 (2003) [arXiv:hep-th/0307234].
21. J. L. F. Barbon and A. Paredes, Int. J. Mod. Phys. A **17**, 3589 (2002) [arXiv:hep-th/0112185].
22. A. Kokado, T. Okamura and T. Saito, arXiv:hep-th/0307120.
23. O. F. Dayi and A. Jellal, J. Math. Phys. **43**, 4592 (2002) [arXiv:hep-th/0111267].
24. B. Morariu and A. P. Polychronakos, JHEP **0107**, 006 (2001) [arXiv:hep-th/0106072].
25. A. P. Polychronakos, JHEP **0106**, 070 (2001) [arXiv:hep-th/0106011].
26. H. J. Groenewold, Physica **12**, 405 (1946).
27. J. E. Moyal, Proc. Cambridge Phil. Soc. **45**, 99 (1949).
28. C. K. Zachos, J. Math. Phys. **41**, 5129 (2000) [arXiv:hep-th/9912238].
29. H. Grosse and P. Presnajder, Lett. Math. Phys. **28**, 239 (1993).
30. M. Kontsevich, arXiv:q-alg/9709040.
31. A. P. Balachandran, B. P. Dolan, J. H. Lee, X. Martin and D. O'Connor, J. Geom. Phys. **43**, 184 (2002) [arXiv:hep-th/0107099].
32. G. Alexanian, A. Pinzul and A. Stern, Nucl. Phys. B **600**, 531 (2001) [arXiv:hep-th/0010187].
33. Z. Guralnik, R. Jackiw, S. Y. Pi and A. P. Polychronakos, Phys. Lett. B **517**, 450 (2001) [arXiv:hep-th/0106044].
34. R. G. Cai, Phys. Lett. B **517**, 457 (2001) [arXiv:hep-th/0106047].
35. Y. Abe, R. Banerjee and I. Tsutsui, Phys. Lett. B **573**, 248 (2003) [arXiv:hep-th/0306272].
36. I. Y. Aref'eva, D. M. Belov and A. S. Koshelev, Phys. Lett. B **476**, 431 (2000) [arXiv:hep-th/9912075].
37. S. Deser, R. Jackiw and S. Templeton, Annals Phys. **140**, 372 (1982) [Erratum-ibid. **185**, 406.1988 APNYA,281,409 (1988 APNYA,281,409-449.2000)].
38. A. Khare and M. B. Paranjape, JHEP **0104**, 002 (2001)[arXiv:hep-th/0102016].

QUANTUM HALL PHYSICS IN STRING THEORY

OREN BERGMAN
Department of Physics
Technion, Israel Institute of Technology
Haifa 32000, Israel

In certain backgrounds string theory exhibits quantum Hall-like behavior. These backgrounds provide an explicit realization of the effective non-commutative gauge theory description of the fractional quantum Hall effect (FQHE), and of the corresponding large N matrix model. I review results on the string theory realization of the two-dimensional fractional quantum Hall fluid (FQHF), and describe new results on the stringy description of higher-dimensional analogs.

1. Introduction

Two of the most exciting developments in physics in the last 20 years have been the (experimental) discovery of the fractional quantum Hall effect (FQHE) [1], and the (theoretical) discovery of superstring theory [2]. The two disciplines, and even more so their practitioners, could not have less to do with each other however. Or so it seemed. In this lecture I will describe configurations in superstring theory which exhibit the fractional quantum Hall effect, thereby establishing a connection between these two fields of physics. This connection suggests a new direction for string theory, namely as an *effective* theory of the FQHE. In no way do I mean to imply that string theory should replace the Coulomb Hamiltonian as the microscopic physics underlying the FQHE. I simply propose that string theory, in the particular configurations described below, may present an improvement over other theories as an effective long-wavelength description of the FQHE. String theory in fact implies the effective gauge theory of the FQHE. In particular, I will show that the string theory picture provides a derivation of Susskind's recent proposal that the ground state of the FQHF is described by a non-commutative Chern-Simons (NCCS) gauge theory [3]. This, together with the direct derivation of the quantization of the inverse filling fraction, and the fractionally charged quasiparticles, serves as evidence for my proposal.

Most of this lecture follows work with John Brodie and Yuji Okawa [4], and a continuing collaboration with John Brodie [5].

2. Branes and the 2d FQHE

2.1. *The D0-brane quantum Hall fluid*

The configuration for the two-dimensional FQHF consists of a D2-brane in the background of an even number $2k$ of D8-branes, and a B-field b along the D2-brane (Figure 1a). The D8-brane background implies that the low-energy effective spacetime theory is massive IIA supergravity [6], where the cosmological constant is given by the RR 0-form field strength $G_0 = k$ [7]. The B-field has two effects: it induces D6-brane charge in the D8-branes, and D0-brane charge in the D2-brane. This is a consequence of the Chern-Simons term in the Dp-brane world-volume theory,

$$S_{CS} = \int C \wedge e^{F+B} \,, \tag{1}$$

where C represents a formal sum of all the odd-degree RR potentials. The induced D0-brane and D6-brane charge densities are therefore given by $\rho_0 = b$ and $\rho_6 = 2kb$, respectively. The D6-branes are a source of a uniform magnetic RR field perpendicular to the D2-brane, $G_2 = dC_1 = kb$. We therefore end up with a system of charged particles, the D0-branes, moving in two dimensions, *i.e.* in the D2-brane, in a uniform background magnetic field. The Landau level filling fraction is given by

$$\nu = \frac{\rho_0}{G_2} = \frac{1}{k} \,. \tag{2}$$

Furthermore, it can be shown that the two-dimensional statistics of the D0-branes are fermionic for odd k and bosonic for even k. This configuration therefore describes the primary Laughlin state of the FQHF at filling fraction $1/k$ [8], with the D0-branes playing the role of the electrons.

2.2. *The D2-brane effective gauge theory*

The above system can also be studied from the point of view of the D2-brane world-volume. The low energy dynamics of the D2-brane is described by the three-dimensional world-volume gauge theory,

$$S_{D2} = S_{DBI} + S_{CS} + \frac{k}{4\pi}\int A \wedge F \,, \tag{3}$$

where S_{DBI} is the Dirac-Born-Infeld action, and the additional Chern-Simons term is a result of integrating out massive fermions coming from

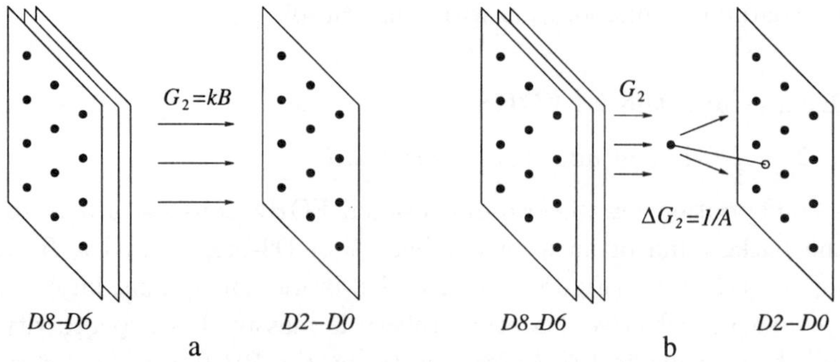

Figure 1. (a) The D0-brane quantum Hall fluid: RR charged D0-branes in two dimensions with a uniform magnetic RR field G_2. (b) A quasiparticle excitation.

the D2-D8 strings. In order to decouple the world-volume theory from the bulk we need to scale the various parameters and fields in a particular way as we take the low-energy limit $\alpha' \to 0$. There is actually some freedom here, and we choose the following scaling behavior:

$$b \sim 1 \ , \ C_n \sim (\alpha')^{n/2} \ , \ g_s \sim (\alpha')^{3/2+\epsilon} \ , \ G_{ij} \sim (\alpha')^{2+\delta} \ \ (i,j=1,2) \ , \quad (4)$$

where $\delta > \epsilon > 0$. Apart from the contributions of the background D2-brane and D0-brane charges, the action reduces to

$$S_{D2} = \frac{k}{4\pi} \int A \wedge F + \frac{1}{2\pi} \int C_1 \wedge F \, . \quad (5)$$

What we have obtained in (5) is known as the *effective hydrodynamic gauge theory* of the quantum Hall fluid [9, 10]. The world-volume gauge field A plays the role of the *hydrodynamic gauge field*, which describes the fluctuations of the fluid, and the RR field C_1 plays the role of the electromagnetic gauge field. The second term corresponds simply to the coupling of the charge to the external electromagnetic field. The special properties of the fluid are encoded in the first term, which is an ordinary Abelian CS action.

The scaling behavior described above is a generalization of the Seiberg-Witten limit [11], which corresponds to $\delta = \epsilon = 0$. In this case the world-volume gauge theory is most naturally expressed as a *non-commutative gauge theory*, in terms of a new gauge field $\widehat{A}$, and a new kind of product called the Moyal star product. In our case the CS action is re-expressed as a non-commutative CS action,

$$\hat{S}_{NCCS} = \frac{k}{4\pi} \int \left(\hat{A} \star d\hat{A} + \frac{2i}{3} \hat{A} \star \hat{A} \star \hat{A} \right) . \quad (6)$$

This reproduces Susskind's improved proposal that, in order to capture the graininess of the quantum Hall fluid, one needs to elevate the effective hydrodynamic gauge theory to a non-commutative gauge theory [3].

2.3. *The quasihole/quasiparticle excitations*

The lowest lying excitations of the fractional quantum Hall fluid are fractionally charged states which carry a unit of magnetic flux. These are known as quasiparticles or quasiholes, depending on the sign of the charge and flux. We create such an excitation in the string theory picture by taking an additional D6-brane across the D2-brane (Figure 1b). The D6-brane is completely orthogonal to the D2-brane, in other words the two branes are "linked" in the sense of [12]. This move has two physical consequences: it changes the G_2 flux by one unit, and it creates a string between the D6-brane and the D2-brane. The end of the string actually carries a fractional D0-brane charge $1/k$ in the D2-brane. This can be shown by extending Strominger's charge conservation argument [13] to D-branes in massive IIA supergravity [5]. This means that the D0-brane density is now

$$\rho_0(x) = \left(B + \frac{1}{kA}\right) - \frac{1}{k}\delta^{(2)}(x) \ , \tag{7}$$

where A is the area of the D2-brane. The shift in the uniform component of the density is due to the change in G_2, and this ensures that the filling fraction of the fluid remains $1/k$. The localized component is the fractionally charged quasiparticle (or rather quasihole in this case).

2.4. *The matrix model*

The role of the electrons in our quantum Hall fluid is played by D0-branes. We can therefore describe the fluid alternatively in terms of a large N D0-brane matrix model, which in the scaling limit (4) reduces to

$$S_{D0} = \int dt \, \mathrm{Tr}\left[G_2 \, X^1 D_0 X^2 - kA_0 + i\psi_a^\dagger D_0 \psi_a\right] \ . \tag{8}$$

The first term is the Lorentz interaction of the D0-branes with the RR 2-form field strength, the second term is a one-dimensional CS term which comes from the coupling of the D0-branes to the RR 0-form field strength, and the last term corresponds to the D0-D8 fermions. This is precisely the matrix model proposed by Polychronakos as an alternative formulation of the NCCS theory of the FQHF [14]. The allowed D0-brane configurations are solutions of the constraint

$$G_2[X^1, X^2] + i\psi^\dagger\psi = ik\mathbf{1} . \tag{9}$$

A D2-brane is a large N solution for which $\mathrm{Tr}[X^1, X^2] = iA$. In particular, the ground state of the FQHF is given by

$$[X^1, X^2] = \frac{iA}{N}\mathbf{1} , \ \psi^\dagger\psi = 0 , \tag{10}$$

and a FQHF with a quasihole excitation is given by

$$[X^1, X^2] = \frac{iA}{N + \frac{1}{k}}\left(\mathbf{1} + \frac{1}{k}|0\rangle\langle 0|\right) , \ \psi^\dagger\psi = -|0\rangle\langle 0| . \tag{11}$$

3. Branes and higher-dimensional FQHFs

Higher dimensional analogs of the FQHE have recently been proposed. These include non-Abelian particles moving on S^4 in a background $SU(2)$ instanton gauge field [15], as well as ordinary charged particles on CP^3 [16], and more generally CP^n [17], in a background $U(1)$ magnetic field. In the large volume limit the latter system becomes a planar $2n$-dimensional quantum Hall fluid with a uniform magnetic field in each of the n independent planes. The main properties of these generalized QHFs are:

(1) Quantized inverse filling fraction $\nu = 1/k^n$, k odd.
(2) Fractionally charged quasiholes/quasiparticles, charge $\pm\nu$.
(3) Brane excitations with fractional statistics.

String theory provides a natural generalization of the planar QHF to four and six dimensions, by replacing the D2-brane with a D4 or D6-brane, respectively. Let me focus on the four-dimensional case, the generalization to six dimensions should be straightforward. The configuration consists of a D4-brane in the background of $2k$ D8-branes and *two* non-trivial components of the B-field, b, b', along the two planes of the D4-brane. This will induce both D2-branes and D0-branes in the D4-brane, and D6-branes and D4-branes in the D8-branes (Figure 2a). The 4-dimensional analog of the Landau level filling fraction is given by the ratio of the D0-brane density and the product of the two RR magnetic fields due to the D6-branes,

$$\nu = \frac{\rho_0}{G_2 G_2'} = \frac{bb'}{(kb)(kb')} = \frac{1}{k^2} , \tag{12}$$

which agrees with property 1 above.

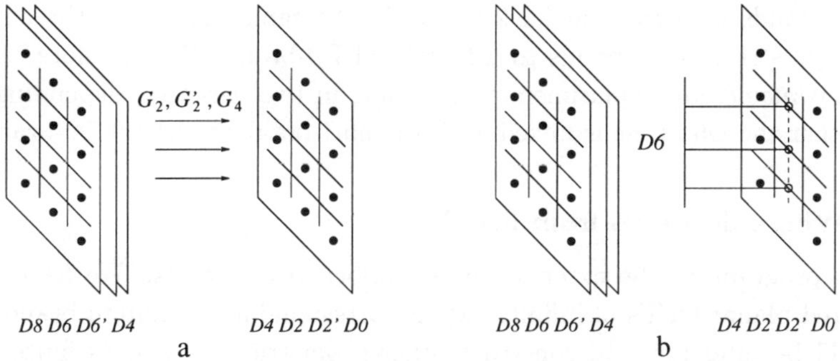

Figure 2. (a) The 4-dimensional D0-brane quantum Hall fluid. The vertical and horizontal lines represent the two kinds of D2-branes (D6-branes) in the D4-brane (D8-branes), and the dots represent the D0-branes (D4-branes) in the the D4-brane (D8-branes). (b) A quasimembrane excitation of the 4-dimensional FQHF.

This system can also be described in terms of the low-energy effective gauge theory on the D4-brane world-volume. Adapting the scaling limit (4) to the D4-brane gives

$$S_{D4} = \frac{k}{24\pi^2}\int A\wedge F^2 + \frac{1}{(2\pi)^2}\int C_1\wedge F^2 + \frac{1}{2\pi}\int C_3\wedge F\,. \tag{13}$$

This is our proposal for the effective gauge theory of the four-dimensional quantum Hall fluid at filling fraction $1/k^2$. As in the two-dimensional case, A corresponds to the hydrodynamic gauge field, and C_1 to the electromagnetic gauge field. The new ingredient here is the 3-form gauge field C_3, whose field strength G_4 was also part of the background. This field did not appear in the original proposals for the higher-dimensional QHFs, but it is a very natural extension.

3.1. *Quasimembranes*

The 3-form couples naturally to the membrane excitations of the 4-dimensional QHF [16]. Let me briefly describe how these membranes appear in the stringy construction. Consider a process similar to the one discussed in the context of the quasiparticles in the two-dimensional case, namely a D6-brane moving across the D4-brane (Figure 2b). The D6-brane is "linked" with the horizontal D2-branes in the D4-brane, so strings are created between the D6-brane and these D2-branes. One ends up with a 2-dimensional array of strings, each of which carries a $1/k$ D0-brane charge in the D4-brane. This array forms a membrane-like object in the D4-brane. It turns out that this object also carries $1/k$ charge under C_3. It therefore corresponds to a fractional D2-brane, in the same sense that the end of a single string corresponds to a fractional D0-brane. We therefore call it a *quasimembrane*. Quasimembranes come in two varieties, depending on whether the D6-brane is linked with the horizontal or vertical D2-branes.

4. Where do we go from here?

This program can be extended in a number of directions. We have considered planar QHFs of infinite extent corresponding to infinite branes. It would be interesting to construct other geometries, as well as finite systems. For example, a QHF on a strip should correspond to a D2-brane suspended between two NS5-branes. Such a system should exhibit massless edge excitations, in addition to the massive quasiparticles. We would also like to understand how the hierarchy of filling fractions arises in the stringy picture. There are a couple of different phenomenological models of the hierarchy [18, 19], and the question is which, if any, is realized by the stringy QHF.

Brane realizations of gauge theories in string theory have led to many new insights into the dynamics of gauge theories, especially through the large N gauge/gravity duality. The FQHE can also described by an (effective) gauge theory - one which we were able to realize using branes in string theory. Is there a FQHE/gravity duality? To answer this question one must first find the supergravity solution corresponding to the brane configuration I described. This has not been done yet. The real test of the usefulness of the string theory/QHE connection is whether string theory teaches us something *new* about the fractional quantum Hall effect. This is yet to be determined.

Acknowledgments

This work was supported in part by the Israel Science Foundation under grant no. 101/01-1.

References

1. D. C. Tsui, H. L. Stormer and A. C. Gossard, Phys. Rev. Lett. **48**, 1559 (1982).
2. M. B. Green and J. H. Schwarz, Phys. Lett. B **149**, 117 (1984).
3. L. Susskind, arXiv:hep-th/0101029.
4. O. Bergman, Y. Okawa and J. H. Brodie, JHEP **0111**, 019 (2001) [arXiv:hep-th/0107178].

5. O. Bergman and J. H. Brodie, to appear.
6. L. J. Romans, Phys. Lett. B **169**, 374 (1986).
7. J. Polchinski, Phys. Rev. Lett. **75**, 4724 (1995) [arXiv:hep-th/9510017].
8. R. B. Laughlin, Phys. Rev. Lett. **50**, 1395 (1983).
9. S. Bahcall and L. Susskind, Int. J. Mod. Phys. B **5**, 2735 (1991).
10. E. Fradkin, V. Jejjala and R. G. Leigh, Nucl. Phys. B **642**, 483 (2002) [arXiv:cond-mat/0205653].
11. N. Seiberg and E. Witten, JHEP **9909**, 032 (1999) [arXiv:hep-th/9908142].
12. A. Hanany and E. Witten, Nucl. Phys. B **492**, 152 (1997) [arXiv:hep-th/9611230].
13. A. Strominger, Phys. Lett. B **383**, 44 (1996) [arXiv:hep-th/9512059].
14. A. P. Polychronakos, JHEP **0104**, 011 (2001) [arXiv:hep-th/0103013].
15. S. C. Zhang and J. p. Hu, Science **294**, 823 (2001) [arXiv:cond-mat/0110572].
16. B. A. Bernevig, C. H. Chern, J. P. Hu, N. Toumbas and S. C. Zhang, Annals Phys. **300**, 185 (2002) [arXiv:cond-mat/0206164].
17. D. Karabali and V. P. Nair, Nucl. Phys. B **641**, 533 (2002) [arXiv:hep-th/0203264].
18. F. D. M. Haldane, Phys. Rev. Lett. **51**, 605 (1983).
19. J. K. Jain, Phys. Rev. Lett. **63** (1989) 199.

BRANES, STRINGS, AND ODD QUANTUM NAMBU BRACKETS*

T L CURTRIGHT

Department of Physics, University of Miami,
Coral Gables, FL 33124-8046, USA
E-mail: curtright@physics.miami.edu

C K ZACHOS

High Energy Physics Division,
Argonne National Laboratory,
Argonne, IL 60439-4815, USA
E-mail: zachos@anl.gov

The dynamics of topological open branes is controlled by Nambu Brackets. Thus, they might be quantized through the consistent quantization of the underlying Nambu brackets, including odd ones: these are reachable systematically from even brackets, whose more tractable properties have been detailed before.

1. Classical Nambu Dynamics

The classical motion of topological open membranes is controlled by Nambu Brackets, the multilinear generalization of Poisson Brackets.[1] We have addressed even branes before, in Refs 2, 3, 4, but not odd ones, treated here, essentialy by reduction to even ones, in a parallel solution of the problem.

Consider the "vortex string" 2-form action,[5,6,7,8,4]

$$S = \int \left(z^1 \; dz^2 \wedge dz^3 - H \; dG \wedge dt \right) , \tag{1}$$

originating in an exact 3-form,

$$d\omega_2 = dz^1 \wedge dz^2 \wedge dz^3 - dH \wedge dG \wedge dt \; , \tag{2}$$

(in analogy to the Hamilton-Poincaré symplectic 2-form, $d\omega_1 = dx \wedge dp_x + dH \wedge dt$). The 3-integral of this form on an open 3-surface yields the above

*Talks presented by T.L.C and K.Z.

2-form action evaluated on the 2-boundary of that surface, *i.e.*, a Cartan integral invariant, analogous to the (1 + 1)-dimensional σ-model WZW topological interaction terms or Chern-Simons actions.

More explicitly, in world-sheet coordinates t, σ, the action reads

$$S = \int dt d\sigma \left(\frac{\epsilon^{ijk}}{3} z^i \partial_t z^j \partial_\sigma z^k + H \partial_\sigma G \right) . \tag{3}$$

For such systems, Dirac quantization procedures[a] have been outlined in, *e.g.*, Refs 6, 10, 11; but they are cumbersome to complete, insofar as they rely on Kontsevich's generalization[12] of Moyal Brackets to nonflat Poisson manifolds, if they are to lead to explicit consistent results, and will not be considered here. Instead, we utilize the linkage of the above action (and actions of this broad type) to Nambu Brackets (NB).

The classical variational equations of motion of the action resulting from δz^i are

$$0 = \left(\epsilon^{ijk} \partial_t z^j + \partial_i H \partial_k G - \partial_k H \partial_i G \right) \partial_\sigma z^k . \tag{4}$$

Motion along the string (along $\partial_\sigma z^i$) may be gauge-fixed by virtue of σ reparameterization invariance,[7] so that this reduces to

$$\dot{z}^i = \epsilon^{ijk} \partial_j H \partial_k G \ , \tag{5}$$

and these amount to Nambu's dynamical equations,[1]

$$\dot{z}^i = \frac{\partial(z^i, H, G)}{\partial(z^1, z^2, z^3)} \equiv \{z^i, H, G\}. \tag{6}$$

In this context, this Jacobian determinant (volume element), defines the Nambu 3-bracket, which generalizes and supplants the Poisson Bracket, and is likewise linear and antisymmetric in its arguments —here, with two "Hamiltonians", instead of one. Thus, for an arbitrary function $f(z^i)$,

$$\frac{df}{dt} = \{f, H, G\} \ . \tag{7}$$

---

[a]The Lagrangean in this action is singular,[6] since it is linear in the velocities, yielding three primary constraints $\phi^i = \pi^i + \frac{\epsilon^{ijk}}{3} z^j \partial_\sigma z^k = 0$, involving the canonical momentum densities $\pi^i(\sigma)$ conjugate to the z^is. Taking Poisson brackets (PB) of these constraints with the Hamiltonian (which is minus the second term in the Lagrangean) yields further, secondary, constraints,[9] *etc.*, in an arduous iterative procedure. The mutual PBs for some of these constraints vanish, but not for some others: the ensuing second class constraints could be projected out through Dirac brackets (DB), which specify novel (non-flat) Poisson structures, which may then lead to quantization.[9,6]

As manifest from this, H and G are time-invariant, and the above velocity is divergenceless, $\partial_i \dot{z}^i = 0$, (solenoidal flow, *cf.* the Appendix). In form language, the "Cauchy characteristics"[13] are directly read off the 3-form (2), whose first variation yields the above equations, since

$$d\omega_2 = (dz^1 - \{z^1, H, G\}dt) \wedge (dz^2 - \{z^2, H, G\}dt) \wedge (dz^3 - \{z^3, H, G\}dt). \quad (8)$$

The generalization of the above 3-form illustration to an arbitrary exact p-form, $d\omega_{p-1} = dz^1 \wedge ... \wedge dz^p - dt \wedge dI_1 \wedge ... \wedge dI_{p-1}$, hence a $(p-2)$-brane, is straightforward.[5,8,14,15] The I_is are invariants ("Hamiltonians") entering into p-NBs.[4] (Formally, it may describe $(p-2)$-branes moving in p-dimensional spacetimes; $p = 2$ reduces to Hamiltonian particle mechanics in phase space and Poisson Brackets.) For such topological systems,[4] "open membranes" is a bit of a misnomer, only adhered to for historical reasons. They are akin to D-branes, as they represent the dynamics of sets of points z^i which do not really influence the motion of each other: the membrane coordinates (string coordinate σ above) are only implicit in the z^is; and their number may only be inferred from the above action whose formulation they expedited—but they do not enter explicitly in Nambu's equation of motion,

$$\frac{df}{dt} = \{f, I_1, ..., I_{p-1}\} . \quad (9)$$

Now, consider introducing another variable, z^4, in the 3-NB (7), to convert it to a 4-NB,

$$\frac{df}{dt} = \{f, H, G\} = \frac{\partial(f, H, G, z^4)}{\partial(z^1, z^2, z^3, z^4)} \equiv \{f, H, G, z^4\}. \quad (10)$$

In general, one may thus promote any p-NB to a $(p+1)$-NB,[10] but we will focus on the case of *odd* p upgrading to an *even* $p+1$.

The reason is that even-p-NBs are "nicer".[3,18] One may think of the $p = 2n$ variables z^i as paired phase-space variables q^i, p^i of a system of n degrees of freedom. The $2n$-NBs always resolve to a fully antisymmetrized sum of strings of PBs in that phase space, of all pairs of the NB arguments.[3,2,4] For example, for a 4-NB,

$$\{I_1, I_2, I_3, I_4\} = \{I_1, I_2\}\{I_3, I_4\} - \{I_1, I_3\}\{I_2, I_4\} - \{I_1, I_4\}\{I_3, I_2\}, \quad (11)$$

$(= \epsilon^{ijkl}\{I_i, I_j\}\{I_k, I_l\}/8$. K Bering has observed that, in general, the PB resolution of the $p = 2n$-NB amounts to the Pfaffian of the antisymmetric matrix with elements $\{I_i, I_j\}$.)

Perhaps equally importantly, $2n$-NBs automatically describe classical motions in phase space for all Hamiltonian maximally superintegrable systems with n degrees of freedom, *i.e.*, systems with $2n-1$ algebraically independent integrals of motion.[16,2,3] Here, motion is confined in phase space on the constant surfaces specified by these integrals, and thus their collective intersection: so that the phase-space velocity $\mathbf{v} = (\dot{\mathbf{q}}, \dot{\mathbf{p}})$ is always perpendicular to the $2n$-dimensional phase-space gradients $\nabla = (\partial_{\mathbf{q}}, \partial_{\mathbf{p}})$ of these integrals of the motion. Consequently, the phase-space velocity must be proportional to the generalized cross-product of all those gradients, and for any phase-space function $f(\mathbf{q}, \mathbf{p})$, motion is fully specified by the NB of eqn (9), with a prefactor V

$$\frac{df}{dt} = \nabla f \cdot \mathbf{v} = V \; \partial_{i_1} f \; \epsilon^{i_1 i_2 \ldots i_{2n}} \; \partial_{i_2} I_1 \ldots \partial_{i_{2n}} I_{2n-1} \; . \tag{12}$$

The proportionality constant V is known to be time-invariant[2,3] and is a function of the invariants I_i, so that $\nabla \cdot \mathbf{v} = 0$ (solenoidal flow).

Thus, there is an abundance of simple classical symmetric systems controlled by NBs, such as multioscillator systems, chiral models, or free motion on spheres, even if they are also describable by Hamiltonian dynamics.[2,3,16,17] Through the embedding conversion of the type (10), it is then evident that this extends to odd-NBs as well.

As an illustration of (10), consider a simple system in this phase-space language, $(z^1 = x,\; z^2 = p_x,\; z^3 = y,\; z^4 = p_y)$, *cf.* Fig 1,

$$H = \frac{p_x^2 + x^2}{2} \; , \qquad G = i \ln(p_x + ix) + y \; . \tag{13}$$

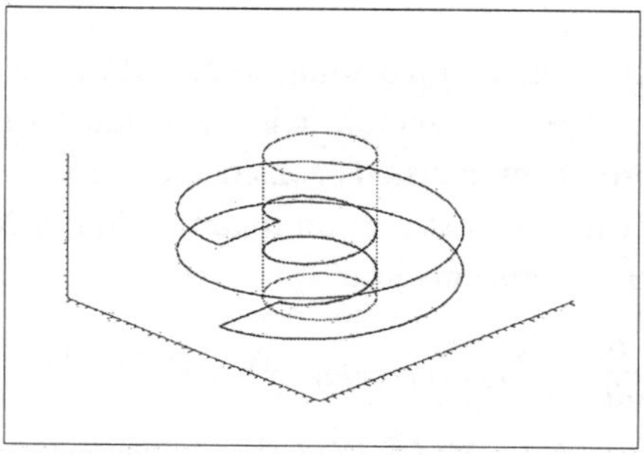

Figure 1. A cylinder of constant H intersects a helical ramp of constant Re G to define a trajectory.

Since a determinant is unaffected under combination of its columns, the 4-NB in eqn (10) is the same when the extraneous coordinate p_y is introduced into a modification, $\tilde{H} \equiv H + p_y$, (so that $\{\tilde{H}, G\} = 0 = \{\tilde{H}, p_y\}$.) Then, by eqn (11),

$$\frac{df}{dt} = \{f, \tilde{H}\}, \tag{14}$$

and so yields a superintegrable Hamiltonian system.

By contrast, suppose one took instead,

$$\tilde{H} = \frac{p_x^2 + x^2}{2} + p_y \ , \quad \tilde{G} = (p_x + ix)\exp(-iy) \ , \tag{15}$$

hence

$$\frac{df}{dt} = -i\tilde{G}\{f, \tilde{H}\} = -i\{\tilde{G}f, \tilde{H}\}, \tag{16}$$

which is *not* a Hamiltonian system. For a Hamiltonian system, the four Hamilton's equations have 6 cross-consistency conditions among them. For this system, though, even though $\nabla \cdot \mathbf{v} = 0$, nevertheless

$$\partial_x \dot{x} + \partial_{p_x} \dot{p}_x - \partial_y \dot{y} - \partial_{p_y} \dot{p}_y = 2\tilde{G} \neq 0, \tag{17}$$

in violation of Hamiltonian consistency. The trajectories in phase space are the same as above, but the speed along each trajectory depends on the value of $\tilde{G}$ on that trajectory.

The evolution (16) in this auxiliary phase space specifies a non-flat Poisson structure, $\overleftarrow{\partial}_i \alpha^{ij} \overrightarrow{\partial}_j$ (which satisfies the Jacobi identity,[16] hence $d(\alpha^{-1}) = 0$). Such Poisson manifolds are defined *in general for all*[16] *NB evolution laws* (12). Like the curved Poisson manifold evolutions of constrained Hamiltonian systems in $2p$-dim phase space discussed earlier, these flows could also be quantized in phase space, in principle, by Kontsevich's[12] elaborate deformation algorithm leading to Moyal brackets (which also satisfy the Jacobi identity). In general, it is not evident that such results do not differ from those of the methods considered here.

We will consider in the next section quantization of NB evolutions simplifying through PB resolutions into

$$\frac{df}{dt} = V\{f, I_1, ..., I_{2n-1}\} = W\{f, H\}, \tag{18}$$

for some functions $H(I_i), V(I_i), W(I_i)$. W=constant (*e.g.* eqn (14)) amounts to maximally superintegrable Hamiltonian systems, quantized and illustrated extensively in Refs 2, 3, 4.

Classically, by linearity in all derivatives involved, all three members of this equation are derivations, *i.e.* satisfy Leibniz's rule, $\delta(fg) = \delta(f)\, g + f\, \delta(g)$.

By contrast, eqn (16) has $V = 1, W = -i\tilde{G}$. To be sure, $W \neq$const systems might well be convertible to hamiltonian ones, by virtue of non-canonical transformations—necessarily, if the hamiltonian structure is *not* to be preserved. *E.g.*, for the system (15), the four equations of motion resulting from (16) can be rewritten through the change of variables,

$$p_w \equiv \tilde{G}, \qquad w \equiv iy\ \frac{p_x^2 + x^2}{2\tilde{G}} , \tag{19}$$

to

$$\dot{x} = p_x p_w \ , \qquad \dot{p}_x = -x p_w \ , \qquad \dot{p}_w = 0 \ , \qquad \dot{w} = \frac{p_x^2 + x^2}{2} \ , \tag{20}$$

which are derivable from a Hamiltonian, $\mathcal{H} = p_w(p_x^2 + x^2)/2$.

2. Quantum Nambu Dynamics

To quantize eqn (18), we seek a map to operators in Hilbert space corresponding conventionally to the classical quantities, which preserves as much of the structure and the symmetries of it as possible,[3] or, equivalently, a deformation in phase space relying on a $\star$-product which acts on the same phase-space functions.[2] (When W=constant, the answer must coincide with results provided by Hamiltonian quantization.) We have found that several consistency complications are not debilitating.

Let us adopt Nambu's proposal[1] of a fully antisymmetric multilinear generalization of Heisenberg commutators, now called quantum NBs (QNB):

$$[A, B] \equiv AB - BA, \tag{21}$$

$$[A, B, C] \equiv ABC - ACB + BCA - BAC + CAB - CBA, \tag{22}$$

$$\begin{aligned}[A, B, C, D] &\equiv A[B, C, D] - B\,[C, D, A] + C\,[D, A, B] - D\,[A, B, C] \\ &= [A, B][C, D] + [A, C][D, B] + [A, D][B, C] \\ &\quad + [C, D][A, B] + [D, B][A, C] + [B, C][A, D] \ ,\end{aligned} \tag{23}$$

etc. As for classical NBs, all even-QNBs resolve into strings of commutators,[3] with all suitable inequivalent orderings required for full antisymmetry, ($\propto \epsilon^{ijkl\ldots}[I_i, I_j]\ [I_k, I_l]\ldots$): a "quantum Pfaffian". Consequently, they manifestly have the proper classical limit[3] to the classical NBs of the previous section,

$$[I_1, ..., I_{2n}] \quad \rightarrow \quad n! \ (i\hbar)^n \ \{I_1, ..., I_{2n}\}. \tag{24}$$

By contrast, odd-QNBs have a bad classical limit[b], underlying our conversion to even-NBs at the classical level, (10). Thus, *e.g.*, taking $f = p_y$ and eqn (13), it is easy to see that a naive quantum counterpart of (10) fails,

$$[f, H, G, p_y] \neq \text{const} \ [f, H, G]. \tag{25}$$

It is the 4-QNB on the left-hand side which we choose for quantization below, and its even generalizations throughout. Even though the extraneous variable p_y does not enter in the description of the classical trajectories, its operator counterpart is unavoidable in Hilbert space, as it is the commutator conjugate to the operator y.

A further complication of odd-QNBs[1] is also remedied by the odd-to-even embedding adopted. In quantizing (18), if the dynamical variable (whose time evolution involves its entering in a QNB) is taken to be the identity operator (e.g. $f \propto 1\!\!1$), its time derivative vanishes. Whereas, *e.g.* for eqn (22),

$$[1\!\!1, B, C] = [B, C] \neq 0 \ , \tag{26}$$

which restricts the specification of df/dt through odd QNBs, in general[c].

In contrast, this *does* result in the vanishing of all even-QNBs, identically, such as (23), by virtue of their commutator resolution,

$$[1\!\!1, I_1, ..., I_{2n-1}] = 0 \ . \tag{27}$$

For other aspects of this odd-even dimorphism see Refs 19, 18, 3.

The associative strings of operators making up the $2n$-QNBs, satisfy the proper fully antisymmetric Generalized Jacobi Identity.[19,18,3] *E.g.*, for 4-QNBs, (23),

$$\epsilon^{ijklmnr}[[I_i, I_j, I_k, I_l], I_m, I_n, I_r] = 0. \tag{28}$$

However, the above $2n$-QNBs forfeit Leibniz's derivation property (beyond the quantum commutator, the 2-QNB), in general. As a result, they

---

[b]This is seen most directly in phase-space quantization, as $\hbar \rightarrow 0$, since the $\star$-products intercalated among all functions in, *e.g.*, (22), yield an even number of derivatives—not an odd one, as required by the classical NBs such as eqn (6).

[c]For privileged cases, *e.g.* $[B, C] = 0$, there is no such problem, *cf.* the Appendix.

do not satisfy the classical identities arising from the derivation property (sometimes referred to as "fundamental identities";[20,8] for the above 4-QNBs, those would have 5, instead of the above $7!/4!3! = 35$ terms.) This may only be a subjective shortcoming, dependent on the specific application context: if a choice is forced, associativity trumps the derivation property.[18]

For example, in several Hamiltonian maximally superintegrable physical systems,[2,3,4] with W=constant, and arbitrary V in eqn (18), quantization was demonstrated explicitly to be consistent. The QNBs failure to be a derivation is in accord with its being equal to operator expressions corresponding to the classical quantities

$$\frac{1}{V}\frac{df}{dt} . \tag{29}$$

Upon quantization, such expressions involve entwinements of derivatives of operators df/dt with other operators, such as

$$\frac{1}{V}\frac{df}{dt} + \frac{df}{dt}\frac{1}{V} , \tag{30}$$

but often with longer, more involved strings of operators.[3] Such strings also fail to be derivations, although they can be demonstrated to provide consistent results equivalent to Hamiltonian quantization. The formal Jordan-Kurosh spectral problem[21] involving such long associative operator strings may be involved technically, but yields consistent answers.[3]

Of course, in exceptional cases (such as those with W=const, V=const), $2n$-QNBs can be derivations, if only because they are equivalent to quantum commutators. For example, quantization of eqn (14) by (25) trivially yields, by the commutator resolution (23),

$$\frac{df}{dt} = -\frac{1}{2\hbar^2}[f, H, G, p_y] = \frac{1}{i\hbar}[f, \tilde{H}], \tag{31}$$

where the operator p_y corresponding to the extraneous variable is now featured in the effective hamiltonian. (The eigenstates of $\tilde{H}$ are thus $\psi(x, y, n, k) \propto H_n(x/\sqrt{\hbar})\ \exp(iky - x^2/\hbar)$.)[d]

Our conclusion, then, is that the embedding of odd to even brackets of rank one higher enables the corresponding membranes to rely on the consistent results on even-QNBs, as described, for their quantization. The

[d] To be sure, not all classical features of Nambu dynamics transfer over to the quantum domain without complication. A more technical treatment of departures from formal quantum operator solenoidal flow is provided in the Appendix.

toolbox provided in Ref 3 may guide quantization of more general systems beyond Hamiltonian ones, *i.e.* systems with nontrivial W. A further important question is whether the results (such as spectra) of the QNB quantization discussed here coincide with those obtained from simple systems quantized through Kontsevich's product—but in those cases explicit answers are normally hard to come by.

Acknowledgments

We thank the organizers of this conference for their hard work and hospitality, and also Y Nambu, D Fairlie, Y Nutku, J de Azcárraga, A Stern, A Polychronakos, K Bering, V S Nair, and V Gueoruiev for insights.

This work was supported by the US Department of Energy, Division of High Energy Physics, Contract W-31-109-ENG-38, and the NSF Award 0303550.

Appendix A. Solenoidal Flow

Preservation of phase-space volume through solenoidal flow, $\nabla \cdot \mathbf{v} = 0$, (Liouville's theorem) was a guiding principle for Nambu,[1] who confirmed it for the classical dynamics of his original *maximal* NB. We show that all flows generated by NBs, even submaximal ones, are solenoidal, but, in general, the ones formally generated by QNBs are not.

For even classical maximal brackets in $2n$-dimensional phase space, it was made evident for (12) that $\nabla \cdot \mathbf{v} = 0$, since

$$v^{i_1} = \left\{z^{i_1}, I_1, \cdots, I_{2n-1}\right\} = \epsilon^{i_1 \cdots i_{2n}}\ \partial_{i_2} I_1 \cdots \partial_{i_{2n}} I_{2n-1}. \tag{A.1}$$

Even sub-maximal NBs are obtained from maximal ones for $2k-1$ fixed Is with $k < n$, through symplectic tracing of maximal brackets,[3]

$$\begin{aligned} &\left\{z^j, I_1, \cdots, I_{2k-1}\right\} \\ &= \frac{1}{(n-k)!} \sum_{\text{all } i=1}^{n} \left\{z^j, I_1, \cdots, I_{2k-1}, x_{i_1}, p_{i_1}, \cdots, x_{i_{n-k}}, p_{i_{n-k}}\right\}. \end{aligned} \tag{A.2}$$

Hence, since each term in the sum produces a solenoidal flow, the sum of terms is also solenoidal, $\partial_j\ \left\{z^j, I_1, \cdots, I_{2k-1}\right\} = 0$.

Odd-NBs are obtainable through the embedding conversion described, and can likewise be made submaximal, hence they generate solenoidal flows as well,

$$0 = \partial_j \left\{ z^j, I_1, \cdots, I_{2k-2} \right\} \tag{A.3}$$
$$\equiv \frac{1}{(n-k)!} \sum_{\text{all } i=1}^{n} \partial_j \left\{ z^j, I_1, \cdots, I_{2k-2}, x_{i_1}, p_{i_1}, \cdots, x_{i_{n-k}}, p_{i_{n-k}}, p_{2n} \right\} .$$

The direct quantum operator analog of the above classical $\nabla \cdot \mathbf{v}$ (reducing to it in the classical limit) is

$$K(I_i, ...) \equiv J_{ij}[z^i, [z^j, I_1, I_2, \cdots]], \tag{A.4}$$

where J_{ij} is the standard symplectic metric for the canonical variables, $[z^i, z^j] = i\hbar J_{ij}$. In general, K does *not* vanish, even though in special circumstances, such as (15), it can vanish. That is, *e.g.* for 3-QNBs,

$$K = J_{ij}[z^i, [z^j, H, G]] = \frac{1}{2} J_{ij} \, [z^i, z^j, H, G] - J_{ij} z^i [z^j, [H, G]] \; . \tag{A.5}$$

For $\tilde{H}$ and $\tilde{G}$, both their commutator and the quantum symplectic trace $(J_{ij}[z^i, z^j, \tilde{H}, \tilde{G}]] = 0)$ vanish.

In general, however, even for privileged flows for which $[H, G] = 0$ (so that they leave the unit operator invariant, by eqn (26): $[\mathbb{1}, H, G] = [H, G] = 0$), the above symplectic trace is non-vanishing, as can be seen by taking brackets of exponentials linear in the canonical variables:

$$J_{lj} \left[z^l, z^j, \exp(i\alpha \cdot z), \exp(i\beta \cdot z) \right] \tag{A.6}$$
$$= i\hbar \left(4n - \frac{2\hbar \alpha \wedge \beta}{\tan\left(\frac{\hbar}{2} \alpha \wedge \beta\right)} \right) \left[\exp(i\alpha \cdot z), \exp(i\beta \cdot z) \right] ,$$

where $\alpha \wedge \beta \equiv \alpha^i J_{ij} \beta^j$. Operator Fourier analysis of general H and G evinces the quantum symplectic trace to not be necessarily proportional to $[H, G]$.

In general, K reduces to the antisymmetric *derivator*[3] for canonical variables. The derivator for any given string of As is defined for all B and C as

$$(B \, , C \mid A_1, \cdots, A_k) \tag{A.7}$$
$$\equiv [BC, A_1, \cdots, A_k] - B \, [C, A_1, \cdots, A_k] - [B, A_1, \cdots, A_k] \, C,$$

so that it vanishes for *all* B and C iff the $(k+1)$-QNB-generated action of the string of As is a derivation.

A sufficient condition for formal "operator solenoidal flow" follows from the identity,

$$\begin{aligned} & J_{ij}\ \left[z^i, \left[z^j, A_1, \cdots, A_k\right]\right] \\ & = i\hbar n\ [\mathbb{1}, A_1, \cdots, A_k] - J_{ij}\ \left(z^i, z^j \mid A_1, \cdots, A_k\right). \end{aligned} \tag{A.8}$$

For even $k+1 \equiv 2m$, by eqn (27), the first term on the r.h.s. drops out, and K amounts to minus the derivator.

For odd QNBs, even $k = 2m$, as for (26), $[\mathbb{1}, A_1, \cdots, A_{2m}] = [A_1, \cdots, A_{2m}]$ may not vanish in general, but one might consider the privileged "flows" for which it does. Thus, for such privileged odd-QNB-generated operator flows and for all even-QNB-generated flows,

$$K = J_{ij}\ \left[z^i, \left[z^j, A_1, \cdots, A_k\right]\right] = -J_{ij}\ \left(z^i, z^j \mid A_1, \cdots, A_k\right). \tag{A.9}$$

Quantum flows are thus solenoidal if the action of the $A_1, \cdots, A_k$ in an $(k+1)$-QNB is a derivation.

However, it is not necessary for the general derivator to vanish to have operator solenoidal flow—just this particular derivator. For the example considered above, $\tilde{H}$, $\tilde{G}$, a 3-QNB based on them is not a derivation, in general: $\left(B, C \mid \tilde{H}, \tilde{G}\right) \neq 0$ for all B and C. Nonetheless, as noted, K vanishes for this case, as well as for the 3-QNB system with $\tilde{H}, G$, eqs (13, 15), and the 4-QNB system with $\tilde{H}, \tilde{G}, p_y$, eqs (13, 31).

References

1. Y Nambu, *Phys Rev* **D7**, 2405-2412 (1973).
2. T Curtright and C Zachos, *New J Phys* **4**, 83.1-83.16 (2002) [hep-th/0205063].
3. T Curtright and C Zachos, *Phys Rev* **D68**, 085001 (2003) [hep-th/0212267].
4. T Curtright and C Zachos, *Short Distance Behavior of Fundamental Interactions: 31st Coral Gables Conference on High Energy Physics and Cosmology*, B Kursunoglu et al, eds, (AIP Conference Proceedings 672, 2003) pp 165-182 [hep-th/0303088]; C Zachos, *Phys Lett* **B570**, 82-88 (2003) [hep-th/0306222]; T Curtright, [hep-th/0307121].
5. F Estabrook, *Phys Rev* **8**, 2740-2743 (1973).
6. M Rasetti and T Regge, *Physica* **80A**, 217-233 (1975).
7. F Lund and T Regge, *Phys Rev* **D14**, 1524-1535 (1976).
8. L Takhtajan, *Commun Math Phys* **160**, 295-316 (1994).
9. P A M Dirac, *Can J Math* **3**, 1 (1951); *Lectures on Quantum Mechanics*, Yeshiva University 1964 (reprinted by Dover, 2001).
10. N Mukunda and E Sudarshan, *Phys Rev* **D13**, 2846-2850 (1976).
11. F Bayen and M Flato, *Phys Rev* **D11**, 3049-3053 (1975); A Kálnay and R Tascón, *Phys Rev* **D17**, 1552-1562 (1978).
12. M Kontsevich (1997) [q-alg/9709040].

13. R Bryant, S Chern, R Gardner, H Goldschmidt, and P Griffiths, *Exterior differential systems* (Springer-Verlag, New York, 1991).
14. Y Matsuo and Y Shibusa, *JHEP02*, 006 (2001) [hep-th/0010040];
A Speliotopoulos, *J Phys* **A35**, 8859-8866 (2002).
15. B Pioline, *Phys Rev* **D66**, 025010 (2002);
D Minic and C-H Tze, *Phys Lett* **B536**, 305-314 (2002) [hep-th/0202173];
C Hofman and J-S Park, [hep-th/0209148];
S Pandit and A Gangal, *J Phys* **A31**, 2899-2912 (1998);
D Fairlie and T Ueno, *J Phys* **A34**, 3037 (2001) [hep-th/0011076];
I Bengtsson, N Barros e Sá, and M Zabzine, *Phys Rev* **D62** (2000) 066005 [hep-th/0005092].
16. C Gonera and Y Nutku, *Phys Lett* **A285**, 301-306 (2001); Y Nutku, *J Phys* **A36**, 7559-7567 (2003) [quant-ph/0306059].
17. R Chatterjee, *Lett Math Phys* **36**, 117-126 (1996);
A Teğmen and A Verçin, [math-ph/0212070].
18. J de Azcárraga, A Perelomov, and J Pérez Bueno, *J Phys* **A29**, 7993-8010 (1996) [hep-th/9605067]; J de Azcárraga, and J Pérez Bueno, *Commun Math Phys* **184**, 669-681 (1997) [hep-th/9605213]; J de Azcárraga, J Izquierdo, and J Pérez Bueno, *J Phys* **A30**, L607-L616 (1997) [hep-th/9703019]; *Rev R Acad Cien Serie A Mat* **95**, 225-248 (2001) [physics/9803046].
19. P Hanlon and M Wachs, *Adv Math* **113**, 206-236 (1995).
20. V T Filippov, *Sib Math Jou*, **26**, 879-891 (1985);
D Sahoo and M Valsakumar, *Phys Rev* **A46**, 4410-4412 (1992); *Mod Phys Lett* **A9**, 2727-2732 (1994);
J Hoppe, *Helv Phys Acta* **70**, 302-317 (1997) [hep-th/9602020].
21. A G Kurosh, Russian Math Surveys **24**, 1-13 (1969).

D BRANES IN 2D STRING THEORY AND CLASSICAL LIMITS

SUMIT R. DAS

Department of Physics, University of Kentucky, Lexington, KY 40506. USA.

In the matrix model formulation of two dimensional noncritical string theory, a D0 brane is identified with a single eigenvalue excitation. In terms of open string quantities (i.e fermionic eigenvalues) the classical limit of a macroscopically large number of D0 branes has a smooth classical limit : they are described by a filled region of phase space whose size is $O(1)$ and disconnected from the Fermi sea. We show that while this has a proper description in terms of a *single* bosonic field at the quantum level, the classical limit is rather nontrivial. The quantum dispersions of bosonic quantities *survive in the classical limit* and appear as additional fields in a semiclassical description. This reinforces the fact that while the open string field theory description of these D-branes (i.e. in terms of fermions) has a smooth classical limit, a closed string field theory description (in terms of a single boson) does not.

1. Overview

D branes of string theory are usually described in terms of excitations of open strings. In recent years it has become clear that they also appear as classical solutions of open string field theory. Their description in a closed string theory is however rather obscure. In almost all cases of interest we do not have a tractable off shell description of closed strings. The most interesting exception is noncritical string theory in two space-time dimensions [1]. Here there is a complete nonperturbative description in terms of the double scaling limit of matrix quantum mechanics. The space of eigenvalues of the matrix in fact provides the space dimension in which the string moves [2]. This allows a theory in $0+1$ dimensions to be dual to a theory which contains gravity in $1+1$ dimensions, thus providing the earliest example of holographic correspondence. In this case, the string theory is rather trivial : for the bosonic string the only dynamical degree of freedom is a single massless scalar, which is related to the collective field [3] - the density of eigenvalues. Recall that the eigenvalues of this theory behave as fermions and the collective field may be regarded as a bosonization of these fermions [4], albeit the bosonization is rather complicated and subtle. Thus we have a well defined closed string field theory as well as a well defined holographic dual.

In a recent interesting development [7], matrix quantum mechanics was reinterpreted as a theory of N D0 branes whose worldsheet description was discovered in [6]. A single unstable D0 brane is interpreted as a single eigenvalue excitation over the filled Fermi sea, whose worldsheet description is a product of the D0 boundary state with Sen's rolling tachyon [8,9]. The model with only one side of the potential filled is the usual bosonic string theory, while the model with both sides filled is the 0B theory [10]. In fact it has been known that the latter theory is the only one which is nonperturbatively stable [5], though its physical meaning was never clear. There is a related matrix model for the Type 0A theory as well [11]. Single eigenvalue excitations have played an important role in the subject, e.g. in the discovery that nonperturbative effects in string theory behave as e^{-1/g_s} rather than e^{-1/g_s^2} [12]. Single particle solitonic excitations also appear as states in exact collective field theory as a separate branch [13]. Explicit single particle states, particularly corresponding to tunnelling, have been studied in [14] and [15].

In the modern interpretation, the matrix degrees of freedom, or equivalently noninteracting fermions, are the open string degrees of freedom and the collective field is the closed string degree of freedom. Open-closed duality is then bosonization [16]. In the asymptotic region the bosons are free massless particles. The late time decay product of a D0 brane can be seen to agree with a coherent state of these massless particles, as expected from the worldhsheet picture [8,17].

This note deals with the nature of classical limit of D-brane states from the point of view of closed string theory, largely based on work done in collaboration with S.D. Mathur and P. Mukhopadhyay.

In the limit $g_s \to 0$ each of the fermions may be thought of as particles moving classically in the inverted harmonic oscillator potential, subject to the Pauli principle [18]. In the phase space of fermions, the ground state is the filled Fermi sea. In the bosonized theory, this limit corresponds to the classical limit of the field theory. Thus small deformations of the Fermi sea correspond to classical waves of the bosonic theory with small amplitudes. The classical dynamics of such small ripples is given by classical solutions of collective field theory, which may be obtained directly from the dynamics of the phase space density of fermions. The time evolution of a single unstable D-brane is then represented by the trajectory of a single fermion which is *disconnected* from the Fermi sea. A state with many D0 branes is represented as a disconnected blob of fermi fluid. As we will soon see, and as has been noted earlier [15], the resulting density of fermions *do not* obey the classical equations of collective field theory. In fact the density of fermions $u(x, p, t)$ in phase space is equivalent to an infinite number of fields in configuration space, e.g. the various moments $\int dpp^n u(x, p, t)$ and for a generic configuration these fields are independent. For quadratic profiles of the Fermi surface, i.e. for profiles in which a constant x line intersects the Fermi surface precisely twice, these various moments are not independent of each other and the configuration can be determined by the collective field and its conjugate momentum [19]. This would seem to indicate that the "closed string" description of D0 branes necessarily involve an infinite number of bosonic fields at the classical level.

We will argue, however, that at the full *quantum* level these states are still described by a *single scalar field*, using a key result of [20] about the physics of formation of folds on the fermi surface. If we start out with some ripple on the Fermi surface - which is represented by a classical configuration of the collective field - time evolution generically results in folds which make the Fermi profile non-quadratic. In [20] it was argued that this is still described as a state in a quantum theory involving a *single* scalar field. What happens is that an initial coherent state representing a classical wave evolves into a state in which quantum dispersions become $O(1)$ rather than $O(\hbar)$ after a certain time. This is precisely the time at which a fold forms on the Fermi surface.

For D0 branes the fermi surface has disconnected pieces to begin with and the same phenomenon occurs. For a single D0 brane we find that quantum flutuations of closed string quantities diverge in the classical limit. This is entirely expected. Just as in critical string theory, we do not expect to describe a state of a small number of D-branes in terms of a classical configuration of closed string fields. On the other hand one would expect that the state of a *large* number of D0 branes (i.e.$O(1/g_s)$ in number) should be a classical state of the closed string theory, i.e. the bosonic field should have vanishing quantum dispersion in the $g_s \to 0$ limit. We will show, however, that this is not true. Quantum fluctuations of quantities in the bosonic field theory do not vanish in the $g_s \to 0$ limit and are in fact of the same order as the expectation values of the fields themselves. Consequently, these fluctuations enter the effective classical equations as independent fields. It would be interesting to find the relationship between the description of D0 brane states in terms of bosons as discussed above and the solitonic states found in an exact treatment of collective field theory in [13].

The upshot of this is that while the open string field theory description of D0 branes in this model has a smooth classical limit in which Ehrenfest's theorem holds (as argued in [16]), a description in terms of closed strings necessarily involves states which have large quantum fluctuations.

2. Preliminaries

In the double scaling limit, matrix quantum mechanics may be written in terms of a second quantized fermionic field $\psi(x,t)$ with a Hamiltonian

$$H = \int dx \left[\frac{1}{2\mu} \partial_x \psi^\dagger \partial_x \psi - \frac{1}{2} \psi^\dagger \psi \right] \tag{1}$$

This form of the hamiltonian clearly shows that the coupling constant $g_s = \frac{1}{\mu}$ acts like a $\hbar$ so that the classical limit is $g_s \to 0$. The singlet sector states are best described in terms of the density of fermions or the collective field $\rho(x,t)$. The effective theory for ρ can be obtained directly from the matrix hamiltonian [3,2] :

$$H = \frac{1}{2} \int dx \left[\frac{1}{3}(P_+^3 - P_-^3) - (x^2 - 2\mu)(P_+ - P_-) \right] \tag{2}$$

where $P_\pm$ are

$$P_\pm = \partial_x \Pi \pm \pi \rho \tag{3}$$

Here Π is the momentum conjugate to ρ. The classical solution is given by

$$\Pi = 0 \qquad \rho = \frac{1}{\pi} P_0 = \frac{1}{\pi} \sqrt{x^2 - 2\mu} \tag{4}$$

Expanding around the classical solution as

$$P_\pm = \pm P_0 + \frac{1}{\sqrt{2}\, P_0} [\Pi_\phi \pm \partial_q \phi] \tag{5}$$

where we have defined a new space variable q by $\partial_q = P_0 \partial_x$ The quadratic part of the hamiltonian is

$$H_2 = \frac{1}{2} \int dq [\Pi_\phi^2 + (\partial_q \phi)^2] \tag{6}$$

which identifies ϕ as a massless scalar field.

3. Classical bosonization

At the classical level the collective Hamiltonian can be alternatively obtained by considering the phase space dynamics of the fermions. Let $u(x,p,t)$ denote the density of fermions in phase space. In the classical limit $g_s \to 0$ Pauli principle dictates that $u = \frac{1}{2\pi g_s}$ in regions where the fermions are present and $u = 0$ elsewhere. The ground state is thus given by

$$u(p,x,t) = \frac{1}{2\pi g_s}\theta(x^2 - p^2 - 2\mu) \tag{7}$$

Consider now a small ripple on the Fermi sea which corresponds to a *quadratic* profile. This means that a $x=$ constant line intersects the Fermi surface $x^2 - p^2 = 2\mu$ precisely *twice*, at $P_\pm(x,t)$. The density of fermions in space at some given time $\rho(x,t)$ and the momentum density $\nu(x,t)$ may be obtained as [18]

$$\begin{aligned}\rho(x,t) &= \int dp\; u(x,p,t) = \frac{1}{2\pi g_s}[P_+(x,t) - P_-(x,t)] \\ \nu(x,t) &= \int dp\; p\; u(x,p,t) = \frac{1}{2\pi g_s}\frac{1}{2}[P_+^2 - P_-^2]\end{aligned} \tag{8}$$

Identifying $\nu = \partial_x \Pi\, \rho$ one gets the standard relationship between collective field (and its conjugate momentum) with $P_\pm$. The hamiltonian (2) then follows by noting that

$$H = \int dxdp \frac{1}{2}[p^2 - x^2 + 2\mu]u(x,p,t) \tag{9}$$

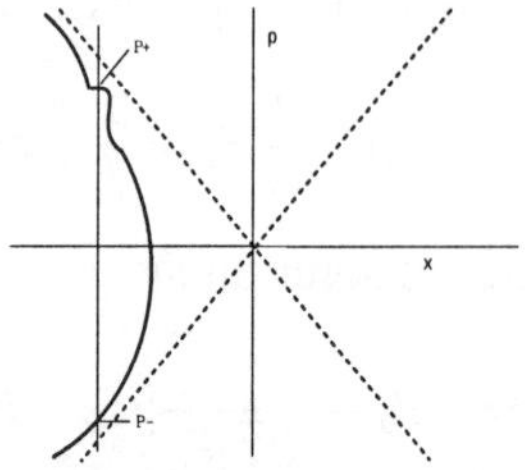

Figure 1. *A quadratic profile of the Fermi surface.*

As is well known, however, the quadratic profile approximation made above is valid only for very small and shallow ripples on the fermi sea. A generic intial quadratic deformation would inevitably form "folds" rendering the profile non-quadratic. Of particular interest are initial states which correspond to non-quadratic profiles. Single eigenvalue excitations or excitations of a bunch of eigenvalues are examples of these. The classical dynamics is still described by the equation of motion for the phase space density $u(x,p,t)$

$$(\partial_t + x\partial_p + p\partial_x)u(x,p,t) = 0 \tag{10}$$

However this equation cannot be any longer reduced to classical collective field theory equations [15]. This seems to indicate that at the classical level there are an infinite number of bosonic fields which may be defined as follows [21]

$$\begin{aligned}
\int dp\ u(x,p,t) &= \frac{1}{2\pi g_s}[\beta_+(x,t) - \beta_-(x,t)] \\
\int dp\ p\ u(x,p,t) &= \frac{1}{2\pi g_s}\left[\frac{1}{2}(\beta_+^2(x,t) - \beta_-^2(x,t)) + (w_{1+}(x,t) - w_{1-}(x,t))\right] \\
\int dp\ p^2\ u(x,p,t) &= \frac{1}{2\pi g_s}\left[\left(\frac{1}{3}\beta_+^3(x,t) + \beta_+ w_{1+}(x,t) + w_{2+}(x,t)\right)\right. \\
&\quad \left. - \left(\frac{1}{3}\beta_-^3(x,t) + \beta_- w_{1-}(x,t) + w_{2-}(x,t)\right)\right]
\end{aligned} \tag{11}$$

The $\beta_\pm, w_{n\pm}$ satsify w_∞ algebras under usual Poisson brackets.

In this parametrization, the presence of nonzero w_n signifies that the profile is not quadratic. For example if the fermi sea consists of two disconnected pieces (Fig.2) and a given $x =$ constant line intesects the fermi surface at four points $P_{i\pm}$ and $P_{2\pm}$ one has

$$\begin{aligned}
\beta_+ &= P_{1+} - P_{1-} + P_{2+} & \beta_- &= P_{2-} \\
w_{1+} &= (P_{1+} - P_{1-})(P_{1-} - P_{2+}) & w_{1-} &= 0 \quad \cdots
\end{aligned} \tag{12}$$

The equations of motion for these various fields $\beta_\pm(x,t), w_{\pm i}(x,t)$ are coupled. As a result, even if we start at $t = 0$ with all the $w_{\pm i} = 0$ time evolution may result in a nonzero $w_{\pm i}$ at a later time. This is the physics of fold formation from unfolded configurations.

The semiclassical description of a single $D0$ brane at some instant of time is a filled region in phase space of size $O(g_s)$ which is *disconnected* from the fermi sea, just like in Fig 2. Equation (12) then implies that w_{1+} is of $O(g_s)$ and hence subleading in the classical limit. Note that each of $P_{1\pm}$ are of $O(1)$ so that this state has an energy of order $O(1/g_s)$. As expected, a single $D0$ brane is a highly quantum state. However, when we have a macroscopically large number of $D0$ branes we expect a classical state in terms of bosons. Indeed in this case the disconnected region has a size of $O(1)$, the various w_i are also of $O(1)$ and contribute to the classical limit.

In terms of an open string (fermionic) description this collection of $D0$ branes has a smooth classical limit since the phase space density $u(x, p, t)$ has small quantum fluctuation. The closed string (bosonic) description also has a classical limit. However, it appears that such a state necessarily involves an infinite number of bosonic fields.

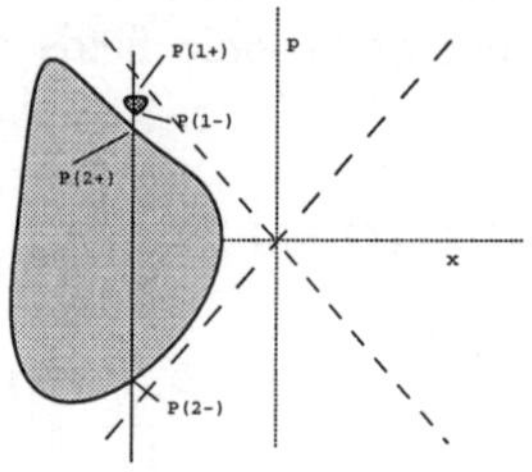

Figure 2. *A semiclassical description of D0 branes.*

4. Quantum bosonization

The conclusion of the preceding section is puzzling. The discussion of classical bosonization using phase space can be repeated for *relativistic* fermions as well. Consider for example the right moving part of a massless relativistic fermion. Now the fermi sea has only one edge - all states with negative momentum are filled. This in fact simplifies the discussion of classical bosonization. Instead of having two copies of each field labelled by $\pm$ we have a single copy. For a state with some disconnected filled region, we have Figure (3) instead of Figure (2).

The formulae are the same as in (11) except that we have to set $\beta_- = w_{i-} = 0$ and rename $\beta_+ \to \beta$ and $w_{i+} \to w_i$.

$$\int dp\, u(x,p,t) = \frac{1}{2\pi g_s}\, \beta(x,t), \quad \int dp\, p\; u(x,p,t) = \frac{1}{2\pi g_s}\left[\frac{1}{2}\beta^2 + w_1(x,t)\right] \tag{13}$$

and so on. The *dynamics* is of course different, governed by a hamiltonian

$$H_{rel} = \int dp dx\; p\; u(x,p,t) \tag{14}$$

This in fact ensures that the equations for β and w do not mix up so that if we have a state with $w_i = 0$ at some time, wW_i's remain zero at all future times, implying that folds cannot form out of unfolded configuration.

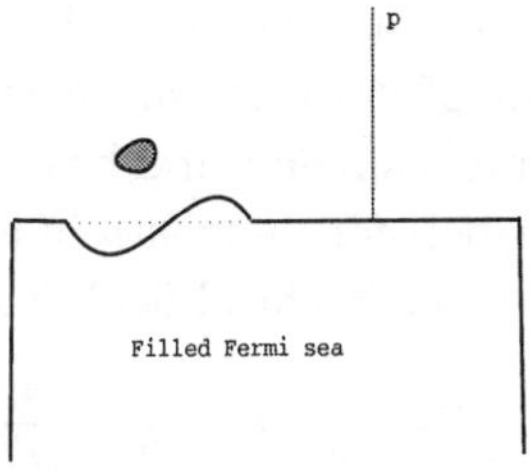

Figure 3. *An excitation of a set of right moving relativistic fermions.*

However, we can certainly have a state which corresponds to Figure (3) to begin with and we would conclude - as in the previous section - that we need an infinite number of bosonic fields to describe this classically. On the other hand we certainly have an exact bosonization at the full quantum level in terms of single (chiral) boson. All states of the theory can be written down exactly in terms of this *single* boson. The puzzle is: how is it that one needs an infinite number of fields when one passes to the classical limit?

4.1. *Bosonization at the algebraic level*

To see this let us review the bosonization formula at an algebraic level. Suppose there are an infinite set of (Schrodinger picture) fermionic operators ψ_n and $\psi_n^\dagger$, with n a positive or negative integer or zero, which satisfy the anticommutation relations

$$\{\psi_m^\dagger, \psi_n\} = \delta_{mn} \qquad \{\psi_m^\dagger, \psi_n\} = 0 \qquad \{\psi_m, \psi_n\} = 0 \tag{15}$$

Suppose we also have a vacuum state $|0>$ which obeys

$$\psi_n|0>=0 \;\; \text{for } n > 0 \qquad \psi_n^\dagger|0>=0 \;\; \text{for } n \leq 0 \tag{16}$$

which allows the definition of a normal ordering prescription

$$:\psi_m^\dagger \psi_n := \psi_m^\dagger \psi_n \quad (n > 0) \qquad :\psi_m^\dagger \psi_n := -\psi_n \psi_m^\dagger \quad (n \leq 0) \tag{17}$$

Then the following bosonic operators

$$\alpha_m = \sum_n :\psi^\dagger_{-m+n}\psi_n: \tag{18}$$

satisfy the commutation relations

$$[\alpha_m, \alpha_n] = m\delta_{m+n,0} \tag{19}$$

Note that the above facts do not depend on the dynamics of the theory.

The simplest application of the above algebraic relations is bosonization of a single chiral fermion into a single chiral boson. In this case the integers m, n may be regarded as labels for momenta in a box. Then the operators ψ_n may be combined into a single chiral field $\psi(x)$

$$\psi(x) = \frac{1}{\sqrt{L}} \sum \psi_m \, e^{\frac{2\pi i m x}{L}} \qquad \psi^\dagger(x) = \frac{1}{\sqrt{L}} \sum \psi_m^\dagger \, e^{-\frac{2\pi i m x}{L}} \tag{20}$$

while the oscillators α_n can be combined into a single chiral boson field $\phi(x)$

$$\partial_x \phi(x) = \frac{1}{\sqrt{L}} \sum \alpha_m \, e^{\frac{2\pi i m x}{L}} \tag{21}$$

The bosonization relationship then becomes

$$\alpha(x) \equiv \partial_x \phi(x) =: \psi^\dagger(x)\psi(x) : \tag{22}$$

The vacuum $|0>$ is the Fock vacuum of a free theory. We have chosen the fermion to be right moving so that the energy of a single free fermion is $E = p$, where p is the momentum. The vacuum chosen above then corresponds to the standard filled Dirac sea. In the $L \to \infty$ limit the momentum becomes continuous. The above relations have the standard limits. $\psi^\dagger(k)$ is a creation operator for $k > 0$ and an annihilation operator for $k \leq 0$.

The fermion field may be also expressed in terms of the bosonic field by the relationship

$$\psi(x) =: e^{i\alpha(x)} : \tag{23}$$

where normal ordering in the bosonic theory is standard, i.e. all the oscillators with negative indices have to be pushed to the right. Using (23) one can express other fermion bilinears in terms of α. One such relation which will be crucial in what follows is

$$-\frac{i}{2} : [\psi^\dagger \partial_x \psi - (\partial_x \psi^\dagger)\psi] := \pi : \alpha^2(x) : \tag{24}$$

4.2. *Nonrelativistic fermions in an inverted harmonic oscillator potential*

The key ingredient in the above is the presence of an *infinite* filled Fermi (or Dirac) sea. This is the reason why the above bosonization does not work for free non-relativistic fermions. The energy of a single fermion is $h = \frac{1}{2m}p^2$ and now the momentum space Fermi sea has a *upper* and a *lower* edge. For example the fermionic modes $\psi(k)$ are creation operators for $-k_F < k < k_F$ where k_F is the Fermi momentum. It is now clear that the momentum modes of the field $\alpha(x) =: \psi^\dagger\psi :$ do not satisfy the standard bosonic oscillator algebra. It is only for momenta *close to the Fermi sea* that we can use the standard bosonization formulae. But this is simply because for such momenta nonrelativistic fermions become effective relativistic with an effective velocity of light (k_F/m).

The situation is, however, quite different for nonrelativistic fermions in an invereted harmonic oscillator potential. In this case the fermion field may be expanded in terms of Parabolic cylinder functions $\chi^\pm_\nu(x)$ with a continuous index ν which runs from $-\infty$ to $+\infty$.

$$\psi(x) = \int d\nu[\psi^+_\nu \chi^+_\nu(x) + \psi^-_\nu \chi^-_\nu(x)] \tag{25}$$

where the modes ψ_ν satisfy the standard fermionic anticommutation relations. The functions $\chi^\pm_\nu(x)$ are eigenstates of the single particle Hamiltonian

$$h = \frac{1}{2}[p^2 - x^2] \tag{26}$$

The vacuum is defined by $\psi_\nu|0>= 0$ for $\nu > -\mu$ and $\psi^\dagger_\nu|0>= 0$ for $-\infty < \nu < -\mu$. Thus the Fermi sea is infinite, pretty much like a relativistic fermion. A bosonization based on the exact eigenstates have been recently used in [22] to investigate interactions involving D0 branes. At the classical level this is clear from the fact that there is a canonical transformation in phase space which maps the problem to that of a relativistic fermion. For negative energy orbits on the left side of the potential is this achieved by

$$x = -\sqrt{2\xi}\cosh\,\eta \qquad p = -\sqrt{2\xi}\sinh\,\eta \tag{27}$$

while for positive energy orbits we have a similar paramertization. Here $-\infty < \eta < \infty$ and $0 < \xi < \infty$. In this parametrization the single particle Hamiltonian becomes $h = -\xi$ and the Fermi level is $\xi = \mu$. Here η is a coordinate and ξ is a momentum. In the ground state the filled Fermi sea is $\mu < \xi < \infty$.

In any case, what this shows that there is an *exact* bosonization for fermions in an inverted harmonic oscillator potential in terms of a *single* bosonic field since the filled Fermi sea is infinite. The natural position space bosonic field is of course not x, but - at least in the semiclassical level - η. This is clear from the fact that if we regard x as the coordinate there is an upper as well as a lower edge of the Fermi sea. When $\eta \to \pm\infty$ motion in

η becomes the same as motion in x so that the bosonization yields a local field in x space.

Such a canonical transformation does not exist for *free* non-relativistic fermions.

5. Disconnected fermi fluids and quantum fluctuations

Let us return to the question posed in the beginning of the previous section: how is it that at the classical level one finds that there has to be an infinite number of bosonic fields ? Our discussion of the previous section shows that for the present purpose we can treat free relativistic fermions and non-relativistic fermions in a $-x^2$ potential in a unified way. Let us denote the coordinates in phase space by p and q. For relativistic fermions q coordinate is the standard position x , while for the fermions coming from matrix model q stands for the conjugate to the quantum number ν - which may be considered to the coordinate η of (27) at the semiclassical level. With this understanding the bosonization formulae are given by equations (22 - 24) with the variable x replaced by q. Our discussion below has nothing to do with the dynamics of the theory.

Consider now a state $|\Psi>$ of the system which is represented at the semiclassical limit by a single disconnected blob of fermi fluid on top of the infinite filled bulk of the fermi sea, as depicted in Figure 3. Note that the boson $\alpha(q)$ is simply the density of fermions in q space, while the left hand side of (24) is the *momentum density* of fermions. Therefore we can combine the operator statements in (22 - 24) with the phase space definitions of semiclassical fields in (13) to obtain

$$< \Psi| : \psi^\dagger(q)\psi(q) : |\Psi> = < \Psi|\alpha(q)|\Psi> = \int dp\ u(q,p,t) = \frac{1}{2\pi g_s}\tilde{\beta}(q,t) \tag{28}$$

$$-\frac{ig_s}{2} < \Psi| : [\psi^\dagger\partial_x\psi - (\partial_x\psi^\dagger)\psi] : |\Psi> = \pi g_s < \Psi| : \alpha^2(q) : |\Psi>$$
$$= \int dp\ p\ u(x,p,t) = \frac{1}{2\pi g_s}[\frac{1}{2}\tilde{\beta}^2 + w_1(q,t)] \tag{29}$$

where $\tilde{\beta}$ denotes the quantities where the ground state values have been subtracted.

This shows that w_1 is related to the quantum dispersion of the field $\alpha(q)$

$$< \Psi| : \alpha^2(q) : |\Psi> - < \Psi|\alpha(q)|\Psi>^2 = \frac{1}{2\pi^2 g_s^2}w_1(q) \tag{30}$$

A state with $w_1 \sim O(1)$ (as in Fig. 3) therefore has a rather unconventional property ; the quantum dispersion $\Delta\alpha$ in such a state is of the same order as the value of $< \alpha >$.

This state represents a large number of fermions which are all excited above the fermi level leaving a gap. From the point of view of fermions, such a state certainly has smooth semiclassical limit since the position and momentum of each fermion is determined upto $O(g_s)$. This means that the quantum spread of $u(p,q,t)$ is controlled as usual by g_s and vanishes in the $g_s \to 0$ limit. However quantities which are natural to the bosonized theory, like expectation values of α have quantum dispersions which do not vanish in the $g_s \to 0$ limit. Furthermore these quantum dispersions like $< \alpha^n > - < \alpha >^n$ are precisely the extra infinite set of fields w_n which appear in a semiclassical description of the state in terms of bosons.

In contrast, a state with no diconnected pieces or folds - a smooth shallow ripple - is a coherent state of the field α for which dispersions of operators constructed from α vanish in the $g_s \to 0$ limit. From the point of view of the bosonic theory these are "classical states". However a state with no folds generically time evloves into a state with folds if the dynamics is appropriate. In [20] it was shown (for free nonrelativistic fermions) tha this happens because the dispersions of α, while initially of $O(g_s)$ become $O(1)$ after a certain time. This time is precisely the time when a fold forms.

It is now clear why states with non-quadratic profiles will not evolve according to the classical equations of motion of the bosonic field theory even in the semiclassical limit. The reason is that while there could be an exact quantum bosonization, the states being considered are not those which have a smooth classical limit for natural bosonic variables. In other words Ehernfest's theorem does not hold for the equations of motion of the closed string field theory, which is the quantum bosonic theory in this case. Ehernfest's theorem of course holds for the equations of motion of the open string field theory, which is the theory of eigenvalues, simply because in the classical limit the fermions have trajectories in phase space with a fuzziness which is at most of order g_s.

6. Explicit calculations

In this section we will perform simple calculations which illustrate the above point. The system we consider is the infinite set of fermionic oscillators as in equations (15)-(19). As argued in the previous section, this system can refer to either a single chiral relativistic fermion or nonrelativistic fermions which arise in the $c = 1$ matrix model or in fact any theory of fermions where the gorund state consists of an infinite Fermi sea. The meaning of the integer index is of course different in each case. With this understanding we can then use the bosonization formulae (20 - 26) where x , now denoted by q, is the position coordinate conjugate to the quantum number ν which characterizes the eigenstates in the inverted harmonic oscillator potential, where we have put this conjugate position in a box of size L, thus rendering ν integral. In the semiclassical limit this is the coordinate η defined in (27).

A ripple on the Fermi sea is a coherent state of the bosonic field

$$|C(q)>= N \prod_{n=1}^{\infty} e^{\frac{C_n \alpha_{-n}}{2\pi n g_s}} |0> \tag{31}$$

where the C_n are the fourier components of the function $C(q)$

$$C(q) = \frac{1}{\sqrt{L}} \sum_n C_n \; e^{\frac{2\pi i n q}{L}} \tag{32}$$

and N is a normalization factor which ensures that $< C(q)|C(q) >= 1$. Since $|C(q) >$ is an eigenstate of α_n for $n > 0$ with eigenvalue $C_n/2\pi g_s$ it is clear that

$$< C(q)|\alpha(q)|C(q) >= \frac{1}{2\pi g_s} C(q) \qquad < C(q)| : \alpha^2(q) : |C(q) >= \frac{1}{4\pi^2 g_s^2} C^2(q) \tag{33}$$

Comparing with (28)-(29) it is clear that we have $\tilde{\beta}(q) = C(q)$ and $w_i(q) = 0$. This then corresponds to a quadratic profile. In fact the state $|C(q) >$ has been chosen so that the quantum dispersions of all bosonic operators which do not involve g_s explicitly vanish in the $g_s \to 0$ limit. These "ripple" states thus have smooth classical limits from the point of view of the bosonic theory.

Consider now a state consisting of M D0 branes, each of which are in a definite energy eigenstate, and hence delocalized in q. The D0 branes occupy successive energy levels between $n = n_A (> 0)$ and $n = n_B = n_A + M - 1$

$$|n_A, M >= \psi^\dagger_{n_A} \psi^\dagger_{n_A+1} \cdots \psi^\dagger_{n_B} \; |0> \tag{34}$$

Let γ_n denote the fourier modes of $: \alpha^2(q) :$

$$: \alpha^2(q) := \frac{1}{L^2} \sum_n \gamma_n \; e^{-\frac{2\pi i n q}{L}} \tag{35}$$

The following relationships then follow from the basic bosonization relation (18)

$$[\alpha_n, \psi_m^\dagger] = \psi^\dagger_{-n+m} \qquad [\alpha_n, \psi_m] = -\psi_{n+m} \tag{36}$$

$$[\gamma_n, \psi_m^\dagger] = (2m-n)\psi^\dagger_{-n+m} \qquad [\gamma_n, \psi_m] = -(2m+n)\psi_{n+m} \tag{37}$$

Note that the ground state obeys

$$\alpha_n|0>=0 \qquad \gamma_n|0>=0 \qquad \text{for } n \geq 0 \tag{38}$$

These relations lead to the following results

$$\begin{aligned} < n_A, M|\alpha(q)|n_A, M > &= \frac{M}{L} \\ < n_A, M| : \alpha^2(q) : |n_A, M > &= \frac{1}{L^2}(2n_A + M - 1) \end{aligned} \tag{39}$$

The result for $< \alpha >$ simply reflects the fact that we have M fermions which are completely delocalized in q space.

The physical momenta in q space corresponding to the lowest and the highest filled level above the Fermi sea are given by

$$p_A = \frac{2\pi g_s n_A}{L} \qquad p_B = \frac{2\pi g_s(n_A + M - 1)}{L} \tag{40}$$

In terms of these quantities

$$\begin{aligned} < n_A, M|\alpha(q)|n_A, M > &= \frac{1}{2\pi g_s}(p_B - p_A) + \frac{1}{L} \\ < n_A, M| : \alpha^2(q) : |n_A, M > &= \frac{1}{4\pi^2 g_s^2}(p_B^2 - p_A^2) + \frac{1}{2\pi g_s L}(p_A + p_B) \end{aligned} \tag{41}$$

In the limit $L \to \infty$ the physical momenta become continuous. Furthermore in the classical limit $g_s \to 0$ the momenta become $O(1)$ if $n_A \to \infty$ keeping p_A, p_B fixed. In this limit we can ignore the $O(1/L)$ terms and the above results reproduce the expectations from classical Fermi fluid picture. In particular, the quantities defined in (28-29) we have

$$\begin{aligned} \tilde{\beta} &= (p_B - p_A) \\ w_1 &= (p_B - p_A)p_A \end{aligned} \tag{42}$$

as expected.

When the number of D0 branes $M \sim O(1)$, we have $w_1 \sim \frac{g_s}{L}$ so that the quantum dispersion of α is large

$$\frac{\Delta\alpha}{\alpha} \sim \frac{\sqrt{w_1}}{\tilde{\beta}} \sim \sqrt{\frac{Lp_A}{g_s M}} \tag{43}$$

This clearly diverges in the classical limit $g_s \to \infty$. This is entirely expected. As in critical string theory a state of a finite number of D branes is a highly quantum state from the point of view of closed string theory - we have simply reproduced this result.

In critical string theory, however, closed string theory provides a classical description of a large number of D-branes in the limit $g_s \to 0$ and $M \to \infty$ with $g_s M$ held fixed. This is the limit in which the collection of D branes is described in terms of a classical gravitational background. In the present case this corresponds to the limit in which the quantity $(p_B - p_A) \sim O(1)$. In the $L \to \infty$ limit this is described by a continuous disconnected region in the Fermi fluid picture. Since we have considered D0 branes in energy eigenstates, this is a band rather than a blob. Contrary to what one might have expected, this corresponds to a limit in which the quantum dispersions of the bosonic field $\frac{\Delta\alpha}{\alpha} \sim O(1)$ rather than $O(g_s)$! A classical description of such configurations of a large number of D0 branes involves an infinite number of fields, while an exact quantum description involves a single bosonic field. The additional fields are nothing but quantum dispersions which survive the classical limit. In other words, while an open string description of D0 branes has a smooth classical limit in accordance with Ehernfest's theorem [16], a closed string description does not.

In order to describe disconnected blobs in phase space corresponding to a large number of D0 branes approximately localized in space at any given time we need to consider wavepackets of individual fermions. It is clear from the preceeding discussion that the conclusion remains unchanged for such configurations, though the calculations are more complicated. These calculations are useful in an understanding of gravitational effects of D-branes [23].

Acknowledgments

I would like to thank the organizers of QTS3 for their invitation to present this talk. I also thank Samir D. Mathur and Partha Mukhopadhyay for discussions and collaboration, and A. Jevicki, I. Klebanov and S.R. Wadia for discussions. This work was supported by National Science Foundation grant PHY-0244811 and the Department of Energy garnt No. DE-FG01-00ER45832.

References

1. For reviews and references to original literature, see I. Klebanov, hep-th9108019, S.R. Das, hep-th/9211085, A. Jevicki, hep-th/9309115, P. Ginsparg and G. Moore, hep-th/9304011, J. Polchinski, hep-th/9411028.
2. S.R.Das and A. Jevicki, *Mod. Phys. Lett.* **A5** (1990) 1639.
3. A. Jevicki and B. Sakita,*Nucl. Phys.* **B165** (1980) 511.
4. D. Gross and I. Klebanov, *Nucl.Phys.* **B352** (1991) 671 ; A.M. Sengupta and S.R. Wadia, *Int. J. Mod. Phys.* **A6** (1991) 1961.
5. G.Moore, *Nucl. Phys.* **B368** (1992) 557; G. Moore, R. Plesser and S. Ramgoolam, *Nucl. Phys.* **B 377** (1992) 143, hep-th/9111035; J. Polchinski, hep-th/9411028; A. Dhar, G. Mandal and S.R. Wadia, *Nucl. Phys.* **B454** (1995) hep-th/9507041
6. V. Fateev, A.B. Zamolodchikov and Al.B. Zamolodchikov, hep-th/0001012; J. Teschner, hep-th/0009138; A.B. Zamolodchikov and Al. B. Zamolodchikov, hep-th/0101152.
7. J. McGreevy and H. Verlinde, hep-th/0304224
8. I.R. Klebanov, J. Maldacena and N. Seiberg, *JHEP* **0307** (2003) 045, hep-th/0305159.
9. J. McGreevy, J. Teschner and H. Verlinde, hep-th/0305194.
10. T. Takayanagi and N. Toumbas, hep-th/0307083; M. Douglas, I. Klebanov, D. Kutasov, J. Maldacena, E. Martinec and N. Seiberg, hep-th/0307195
11. I.R. Klebanov, J. Maldacena and N. Seiberg, hep-th/0309168.
12. S. Shenker, in *Cargese 1990, Proceedings, Random surfaces and quantum gravity* 191-200.*
13. A. Jevicki, *Nucl. Phys.* **376** (1992) 75.
14. J. Lee and P. Mende, *Phys.Lett.* **B312** (1993) 433, hep-th/9211049.
15. A. Dhar, G. Mandal and S.R. Wadia, *Int.J.Mod.Phys.* **A8** (1993) 3811, hep-th/9212027.
16. For a detailed discussion of this aspect see A. Sen, hep-th/0308068.
17. M. Gutperle and P. Kraus, hep-th/0308047.
18. J. Polchinski, *Nucl.Phys.* **B362** (1991) 125.
19. A. Dhar, G. Mandal and S.R. Wadia, *Int.J.Mod.Phys.* **A8** (1993) 325, hep-th/9204028; *Mod.Phys.Lett.* **A7** (1992) 3129, hep-th/9207011.
20. S.R. Das and S.D. Mathur, *Phys. Lett.* **B 365** (1996) 79, hep-th/9507141
21. J. Avan and A. Jevicki, *Nucl. Phys.* **B397** (1993) 672, hep-th/9209036.; A. Jevicki, hep-th/9302106.
22. J. de Boer, A. Sinkovics, E. Verlinde and J.T. Yee, hep-th/0312135
23. S.R. Das, S.D. Mathur and P. Mukhopadhyay, *work in progress*

ASPECTS OF DIFFEOMORPHISM INVARIANT THEORY OF EXTENDED OBJECTS

V. G. GUEORGUIEV *†

Department of Physics and Astronomy,
Louisiana State University, Baton Rouge, LA 70803
E-mail: vesselin@phys.lsu.edu

The structure of a diffeomorphism invariant Lagrangians for an extended object W embedded in a bulk space M is discussed by following a close analogy with the relativistic particle in electromagnetic field as a system that is reparametrization-invariant. The current construction naturally contains, relativistic point particle, string theory, and Dirac–Nambu–Goto Lagrangians with Wess–Zumino terms. For Lorentzian metric field, the non-relativistic theory of an integrally submerged W-brane is well defined provided that the brane does not alter the background interaction fields. A natural time gauge is fixed by the integral submergence (sub-manifold structure) within a Lorentzian signature structure. A generally covariant relativistic theory for the discussed brane Lagrangians is also discussed. The mass-shell constraint and the Klein–Gordon equation are shown to be universal when gravity-like interaction is present. A construction of the Dirac equation for the W-brane that circumvents some of the problems associated with diffeomorphism invariance of such Lagrangians by promoting the velocity coordinates into a non-commuting gamma variables is present.

Introduction. The Hamiltonian and Lagrangian formulation[1,2] are two very useful approaches in physics. In general, these two approaches are related by the Legander transformation. For a reparametrization-invariant theory, however, there are problems in changing from the Lagrangian to the Hamiltonian approach.[2,3,4,5] In this paper the focus is on the properties of reparametrization-invariant matter systems such as, the relativistic particle and its extended object (brane) generalization within the Lagrangian

*On leave of absence from Institute of Nuclear Research and Nuclear Energy, Bulgarian Academy of Sciences, Sofia 1784, Bulgaria.

†Research in part supported by the U.S. National Science Foundation under Grant No. PHY 0140300, the *3rd International Symposium on Quantum Theory and Symmetries*, and the *Argonne Workshop on Branes and Generalized Dynamics*.

approach. We try to answer the question: "What is the Lagrangian for an extended 'matter' object?"

Matter Lagrangian for relativistic particle. The action for a massive relativistic particle has a nice geometrical meaning: it is the distance along the particle trajectory[6] provided that the units are such that $x^0 = ct$ and the particle moves with a constant 4-velocity ($g_{\mu\nu}v^\mu v^\nu = 1$):

$$S_1 = \int d\tau L_1(x, v) = \int d\tau \sqrt{g_{\mu\nu}v^\mu v^\nu} \rightarrow \int d\tau. \tag{1}$$

For a massless particle, such as a photon, the length of the 4-velocity is zero ($g_{\mu\nu}v^\mu v^\nu = 0$) and the appropriate 'good' action[6] is:

$$S_2 = \int L_2(x, v) d\tau = \int g_{\mu\nu}v^\mu v^\nu d\tau. \tag{2}$$

The Euler–Lagrange equations obtained from S_1 and S_2 are equivalent, even more, they are equivalent to the geodesic equation as well:

$$\frac{d}{d\tau}\vec{v} = D_{\vec{v}}\vec{v} = v^\beta \nabla_\beta \vec{v} = 0 \tag{3}$$

Since the Levi–Civita connection ∇ preserves the length of the vectors[6] ($\nabla g(\vec{v}, \vec{v}) = 0$) this equivalence is not surprising because the Lagrangians in (1) and (2) are functions of the preserved arc length $g(\vec{v}, \vec{v}) = \vec{v}^2$. However, the equivalence between S_1 and S_2 has a much deeper roots.

Homogeneous Lagrangians. Since L_2 is a homogeneous function of order 2 with respect to $\vec{v}$, the corresponding Hamiltonian function ($h = v\partial L/\partial v - L$) is exactly equal to L_2. Thus L_2 is conserved, and so is the length of $\vec{v}$. Any homogeneous Lagrangian in $\vec{v}$ of order $n \neq 1$ is conserved because $h = (n-1)L$. If $dL/d\tau = 0$, then the Euler–Lagrange equations for L and $\tilde{L} = f(L)$ are equivalent. This is an equivalence that applies to homogeneous Lagrangians in particular. It is different from the usual equivalence $L \rightarrow \tilde{L} = L + d\Lambda/d\tau$ or the more general equivalence discussed by Hojman and Harleston Ref. 7. Any solution of the Euler–Lagrange equation for $\tilde{L} = L^\alpha$ would conserve $L = L_1$ since $\tilde{h} = (\alpha - 1)L^\alpha$. All these solutions are solutions of the Euler–Lagrange equation for L as well, thus $L^\alpha \subset L$. In general, conservation of L_1 is not guaranteed since $L_1 \rightarrow L_1 + d\Lambda/d\tau$ is also a homogeneous Lagrangian of order one equivalent to L_1. This suggests that there may be a choice of Λ, a "gauge fixing", such that $L_1 + d\Lambda/d\tau$ is conserved even if L_1 is not. The above discussion applies to a more general homogeneous Lagrangians as well.[8]

In the example of the relativistic particle, the Lagrangian and the trajectory parameterization have a geometrical meaning. In general, however, parameterization of a trajectory is quite arbitrary for any observer. If there is no preferred trajectory parameterization in a smooth space-time, then we are free to choose the standard of distance (time, using natural units $c = 1$). Thus, *our theory should not depend on the choice of parameterization.* By inspection of the Euler–Lagrange equations, any homogeneous Lagrangian of order n ($L(x, \alpha\vec{v}) = \alpha^n L(x, \vec{v})$) provides a reparametrization invariant equations ($\tau \to \tau/\alpha, \vec{v} \to \alpha\vec{v}$). Next, note that the action S involves an integration that is a natural structure for orientable manifolds (M) with an n-form of the volume. Since a trajectory is a one-dimensional object, then what we are looking at is an embedding $\phi : \mathbb{R}^1 \to M$. This means that we push forward the tangential space $\phi_* : T(\mathbb{R}^1) = \mathbb{R}^1 \to T(M)$, and pull back the cotangent space $\phi^* : T(\mathbb{R}^1) = \mathbb{R}^1 \leftarrow T^*(M)$. Thus a 1-form ω on M that is in $T^*(M)$ ($\omega = A_\mu(x)\, dx^\mu$) will be pulled back on $\mathbb{R}^1$ ($\phi^*(\omega)$) and there it should be proportional to the volume form on $\mathbb{R}^1$ ($\phi^*(\omega) = A_\mu(x)\,(dx^\mu/d\tau)d\tau \sim d\tau$), allowing us to integrate $\int \phi^*(\omega)$:

$$\int \phi^*(\omega) = \int L d\tau = \int A_\mu(x)\, v^\mu d\tau.$$

Therefore, by selecting a 1-form $\omega = A_\mu(x)\, dx^\mu$ on M and using $L = A_\mu(x)\, v^\mu$ we are actually solving for the embedding $\phi : \mathbb{R}^1 \to M$ using a chart on M with coordinates $x : M \to \mathbb{R}^n$. The Lagrangian obtained this way is homogeneous of first order in v with a very simple dynamics. The corresponding Euler–Lagrange equation is $F_{\nu\mu}v^\mu = 0$ where F is a 2-form ($F = dA$) – the Faraday's tensor. If the assumption that L is a pulled back 1-form is relaxed and instead one assumes that it is just a homogeneous Lagrangian of order one, then the corresponding reparametrization-invariant theory may have an interesting dynamics.

First order homogeneous Lagrangians – canonical form. Now we define what we mean by the *canonical form of the first order homogeneous Lagrangian* and why do we prefer this mathematical expression. Let $S_{\alpha_1\alpha_2\ldots\alpha_n}$ be a symmetric tensor of rank n which defines a homogeneous function of order n ($S_n(\vec{v}, ..., \vec{v}) = S_{\alpha_1\alpha_2\ldots\alpha_n} v^{\alpha_1} ... v^{\alpha_n}$). The symmetric tensor of rank two is denoted by $g_{\alpha\beta}$. Using this notation, the canonical form of the first order homogeneous Lagrangian is defined as:

$$L(\vec{x}, \vec{v}) = \sum_{n=1}^{\infty} \sqrt[n]{S_n(\vec{v}, ..., \vec{v})} = A_\alpha v^\alpha + \sqrt{g_{\alpha\beta} v^\alpha v^\beta} + ... \sqrt[m]{S_m(\vec{v}, ..., \vec{v})}. \quad (4)$$

Any Lagrangian for the matter should involve interaction fields that couple with the velocity $\vec{v}$ to a scalar. When the matter action is combined with the action for the interaction fields ($\mathcal{S} = \int \mathcal{L}dV$), one obtains a full *background independent theory.* Then the corresponding Euler–Lagrange equations contain "dynamical derivatives" on the left hand side and sources on the right hand side:

$$\partial_\gamma \left(\frac{\delta \mathcal{L}}{\delta(\partial_\gamma \Psi^\alpha)} \right) = \frac{\delta \mathcal{L}}{\delta \Psi^\alpha} + \frac{\partial L_{matter}}{\partial \Psi^\alpha}.$$

The advantage of the canonical form of the first order homogeneous Lagrangian (4) is that each interaction field, which is associated with a symmetric tensor, has a unique matter source that is a monomial in the velocities:

$$\frac{\partial L}{\partial S_{\alpha_1 \alpha_2 \dots \alpha_n}} = \frac{1}{n} \left(S_n(\vec{v}, \dots, \vec{v}) \right)^{\frac{1-n}{n}} v^{\alpha_1} \dots v^{\alpha_n}. \tag{5}$$

There are many other ways one can write first-order homogeneous functions.[4] For example, one can consider the following expression $L(\vec{x}, \vec{v}) = \left(h_{\alpha\beta} v^\alpha v^\beta \right) \left(g_{\alpha\beta} v^\alpha v^\beta \right)^{-1/2}$. However, each of the fields h and g has the same source type ($\sim v^\alpha v^\beta$). At this stage, however, it is not clear why the same source type should produce different fields. Therefore, the canonical form (4) seems more appropriate for the current discussion.

Extended objects. In the previous sections, the classical mechanics of a point-like particle have been discussed as a problem concerned with the embedding $\phi : \mathbb{R}^1 \to M$. The map ϕ provides the trajectory (the word line) of the particle in the target space M. In this sense, we are dealing with a one dimensional object, the world-line of the particle (one dimensional W-brane). We think of an extended object as a manifold W with dimension denoted by D. In this sense, we have to solve for $\phi : W \to M$ such that some action integral is minimized. From this point of view, we are dealing with mechanics of a brane. In other words, how is this D-dimensional extended object submerged in M, and what are the relevant interaction fields? Following the relativistic point particle discussion, we consider the space of the D-forms over the manifold M, denoted by $\Lambda^D(M)$, that has dimension $\binom{m}{D} = \frac{m!}{D!(m-D)!}$. An element Ω in $\Lambda^D(M)$ has the form $\Omega = \Omega_{\alpha_1 \dots \alpha_m} dx^{\alpha_1} \wedge dx^{\alpha_2} \wedge \dots dx^{\alpha_m}$. We use the label Γ to index different D-forms over $M, \Gamma = 1, 2, \dots, \binom{m}{D}$; thus $\Omega \to \Omega^\Gamma = \Omega^\Gamma_{\alpha_1 \dots \alpha_m} dx^{\alpha_1} \wedge dx^{\alpha_2} \wedge \dots dx^{\alpha_m}$. Next we introduce "*generalized velocity vectors*" with components ω^Γ :

$$\omega^\Gamma = \frac{\Omega^\Gamma}{dz} = \Omega^\Gamma_{\alpha_1 \dots \alpha_D} \frac{\partial (x^{\alpha_1} x^{\alpha_2} \dots x^{\alpha_D})}{\partial (z^1 z^2 \dots z^D)}, \quad dz = dz^1 \wedge dz^2 \wedge \dots \wedge dz^D.$$

In the above expression, $\frac{\partial (x^{\alpha_1} x^{\alpha_2} \dots x^{\alpha_D})}{\partial (z^1 z^2 \dots z^D)}$ represents the Jacobian of the transformation from coordinates $\{x^\alpha\}$ over the manifold M to coordinates $\{z^a\}$

over the brane. The pull back of a D-form Ω^Γ must be proportional to the volume form over the brane:

$$\phi^*\left(\Omega^\Gamma\right) = \omega^\Gamma dz^1 \wedge ... \wedge dz^D = \Omega^\Gamma_{\alpha_1...\alpha_D} \frac{\partial\left(x^{\alpha_1}...x^{\alpha_D}\right)}{\partial(z^1...z^D)} dz^1 \wedge ... \wedge dz^D.$$

Thus, it is suitable for integration over the W-manifold and the action is:

$$S[\phi] = \int_W L\left(\vec{\phi},\vec{\omega}\right) dz = \int_W \phi^*(\Omega) = \int_W A_\Gamma(\vec{\phi})\omega^\Gamma dz.$$

This is a homogeneous function in ω and is reparametrization (diffeomorphism) invariant with respect to the diffeomorphisms of the W-manifold. If we relax the linearity $L(\vec{\phi},\vec{\omega}) = \phi^*(\Omega) = A_\Gamma(\vec{\phi})\omega^\Gamma$ in $\vec{\omega}$, then the canonical expression for the first order homogeneous Lagrangian is:

$$L\left(\vec{\phi},\vec{\omega}\right) = \sum_{n=1}^{\infty} \sqrt[n]{S_n\left(\vec{\omega},...,\vec{\omega}\right)} = A_\Gamma\omega^\Gamma + \sqrt{g_{\Gamma_1\Gamma_2}\omega^{\Gamma_1}\omega^{\Gamma_2}} + ... \tag{6}$$

At this point, there is a strong analogy between the relativistic point particle and the extended object. Some specific examples of W-brane theories correspond to the following familiar Lagrangians:

Lagrangian for a relativistic point particle in an electromagnetic field: $\dim W = 1$ (World-line) with $\omega^\Gamma \to v^\alpha = \frac{dx^\alpha}{d\tau}$

$$L\left(\vec{\phi},\vec{\omega}\right) = A_\Gamma\omega^\Gamma + \sqrt{g_{\Gamma_1\Gamma_2}\omega^{\Gamma_1}\omega^{\Gamma_2}} \to qA_\alpha v^\alpha + m\sqrt{g_{\alpha\beta}v^\alpha v^\beta}.$$

Lagrangian for strings: $\dim W = 2$ (World-sheet)
$L\left(x^\alpha, \partial_i x^\beta\right) = \sqrt{Y^{\alpha\beta}Y_{\alpha\beta}}$, with the following notation:

$$\omega^\Gamma \to Y^{\alpha\beta} = \frac{\partial(x^\alpha,x^\beta)}{\partial(\tau,\sigma)} = \det\begin{pmatrix}\partial_\tau x^\alpha & \partial_\sigma x^\alpha\\ \partial_\tau x^\beta & \partial_\sigma x^\beta\end{pmatrix} = \partial_\tau x^\alpha\partial_\sigma x^\beta - \partial_\sigma x^\alpha\partial_\tau x^\beta.$$

Dirac–Nambu–Goto Lagrangian (DNG)[9]: $L\left(x^\alpha, \partial_W x^\beta\right) = \sqrt{Y^\Gamma Y_\Gamma}$.

The corresponding electromagnetic interaction term for W-banes is know as Wess–Zumino term[10] in string theory.

From the expressions (4) and (6), one can see that the corresponding matter Lagrangians (L), in their canonical form, contain electromagnetic (A) and gravitational (g) interactions, as well as interactions that are not clearly identified yet ($S_n, n > 2$), if present at all in nature. At this stage, we have a theory with background fields since we don't know the equations for the interaction fields A, g, and S_n. To complete the theory, we need to introduce actions for these interaction fields. If one is going to study the new interaction fields $S_n, n > 2$, then some guiding principles for writing field Lagrangians are needed.

One approach is to apply the above discussion and view the S_n fields as related to an embedding of the M-manifold into the manifold of symmetric tensors over M. Another approach would be to use the external derivative d, external multiplication $\wedge$, and Hodge dual $*$ operations in the external algebra $\Lambda(T^*M)$ over M to construct objects proportional to the volume form over M. For example, for any n-form (A) the expressions $A \wedge *A$ and $dA \wedge *dA$ are forms proportional to the volume form. The next important principle comes from the symmetry in the matter equation. That is, if there is a transformation $A \to A'$ that leaves the matter equations unchanged, then there is no way to distinguish A and A'. Thus the action for the field A should obey the same gauge symmetry. For the electromagnetic field ($A \to A' = A + \mathrm{d}f$) this leads uniquely to the field Lagrangian $\mathcal{L} = dA \wedge *dA = F \wedge *F$, when for gravity[11] it leads to the Cartan–Einstein action[12] $S[R] = \int R_{\alpha\beta} \wedge *(\mathrm{d}x^\alpha \wedge dx^\beta)$.

Non-relativistic limit. For a W-brane we assume the existence a local coordinate frame where one component of the generalized velocity can be set to 1 ($\omega^0 = 1$). This generalized velocity component is associated with the brane "time coordinate." In fact, $\omega^0 = 1$ means that there is an integral embedding of the brane in the target space M, and the image of the brane is a sub-manifold of M. If the coordinates of M are labeled so that $x^i = z^i, i = 1, ..., D$, then x^i are internal coordinates that can collapse to only one coordinate – the "world line". This provides a gauge-fixing that allows one to do canonical quantization. This approach is mainly concerned with the choice of a coordinate time that is used as the trajectory parameter.[13,14,15,16] Such choice removes the reparametrization invariance of the theory.

In a local coordinate system where $\omega^0 = 1$ and the metric is a "one-time-metric" we have:

$$L = A_\Gamma \omega^\Gamma + \sqrt{g_{\Gamma_1\Gamma_2}\omega^{\Gamma_1}\omega^{\Gamma_2}} + ... + \sqrt[m]{S_m(\vec{\omega}, ..., \vec{\omega})} \to$$

$$\to A_0 + A_i\omega^i + \sqrt{1 - g_{ii}\omega^i\omega^i} + ... \approx A_0 + A_i\omega^i + 1 - \frac{1}{2}g_{ii}\omega^i\omega^i + ...$$

Thus the Hamiltonian function is not zero anymore, so we can do canonical quantization, and the Hilbert space consists of the functions $\Psi(x) \to \Psi(z, \tilde{x})$ where $\tilde{x} = x^i, i = D + 1, ..., m$. The brane coordinates z should be treated as t in quantum mechanics in the sense that the scalar product should be an integral over the space coordinates $\tilde{x}$. For W-branes the one-time coordinate reflects separation of the internal from the external coordinates when the W-brane is considered as a sub-manifold of the target space manifold M.

Even though canonical quantization can be applied after the above gauge fixing, one is not usually happy because the covariance of the theory is lost and time is a privileged coordinate. In general, there are well developed procedures for covariant quantization.[13,15,17,18,19] In this paper, however, we are not going to discuss these methods. Instead, we will employ a different quantization strategy[20], but before that we will discuss the mass-shell and Klein–Gordon equations.

The mass-shell and Klein–Gordon equation. Since the functional form of the canonical Lagrangian is the same for any W-brane, we use v, but it could be ω as well. We define the momentum p and generalized momentum π for our canonical Lagrangian as follow:

$$p_\Gamma = \frac{\delta L(\phi,\omega)}{\delta\omega^\Gamma} = eA_\Gamma + m\frac{g_{\Gamma\Sigma}\omega^\Sigma}{\sqrt{g(\vec{\omega},\vec{\omega})}} + ... + \frac{S_{\Gamma\Sigma_1...\Sigma_n}\omega^{\Sigma_1}...\omega^{\Sigma_n}}{(S(\omega,...,\omega))^{1-1/n}} + ...,$$

$$\pi_\alpha = p_\alpha - eA_\alpha - ...\frac{S_{\alpha\beta_1...\beta_n}v^{\beta_1}...v^{\beta_n}}{(S(v,...,v))^{n/(n+1)}}... = m\frac{g_{\alpha\beta}v^\beta}{\sqrt{g(\vec{v},\vec{v})}}.$$

In the second equation we have used v instead of ω for simplicity. Notice that this generalized momentum (π) is consistent with the usual quantum mechanical procedure $p \to p - eA$ that is used in Yang–Mills theories, as well as with the usual GR expression $p_\alpha = mg_{\alpha\beta}v^\beta$. Now it is easy to recognize the mass-shell constraint as a mathematical identity:

$$\frac{\vec{v}}{\sqrt{\vec{v}^2}}\cdot\frac{\vec{v}}{\sqrt{\vec{v}^2}} = 1 \Rightarrow \pi_\alpha\pi^\alpha = m^2 \Rightarrow \left(\vec{p} - e\vec{A} - \vec{S}_3(v) - \vec{S}_4(v) - ...\right)^2\Psi = m^2\Psi.$$

Notice that "gravity" as represented by the metric is gone, while the Klein–Gordon equation appears. The v dependence in the S terms reminds us about the problem related to the change of coordinates $(x,v) \to (x,p)$. So, at this stage we may proceed with the Klein–Gordon equation, if we wish.

Dirac equation from H=0. An interesting approach to the Dirac equation has been suggested by H. Rund.[4] The idea uses Hamiltonian linear in the momentum ($H = \gamma^\alpha p_\alpha$) and the base manifold principle group G. To have the Hamiltonian H invariant under G-transformations, the γ objects should transform appropriately and provide also a realization of the generators of G. Since we want γ and p to transform as vectors, it is clear that p should be a covariant derivative, but what is its structure? Consider a homogeneous Lagrangian that can be written as $L(\phi,\omega) = \omega^\Gamma p_\Gamma = \omega^\Gamma \partial L(\phi,\omega)/\partial\omega^\Gamma$ with a Hamiltonian function that is identically zero: $h = \omega^\Gamma \partial L(\phi,\omega)/\partial\omega^\Gamma - L(\phi,\omega) \equiv 0$. Notice that ω^Γ is the determinant of a matrix (the Jacobian of a transformation[21]); thus $\omega^\Gamma \to \gamma^\Gamma$ seems an interesting option for quantization. Even more, for the Dirac theory we know that γ^α are the 'velocities' ($dx/d\tau = \partial H/\partial p$).

If we quantize using ($h \to H$), then the space of physical states should satisfy: $H\Psi = 0$. By applying $\omega^\Gamma \to \gamma^\Gamma$, which means that the (generalized) velocity is considered as a vector with non-commutative components, we have $\left(\gamma^\Gamma p_\Gamma - L(\phi,\gamma)\right)\Psi = 0$. For a point particle, using the canonical form of the Lagrangian (4) and the algebra of the γ matrices following Run's approach[4] this gives:

$$
\begin{aligned}
H &= \gamma^\alpha p_\alpha - L(\phi,\gamma) = \gamma^\alpha p_\alpha - eA_\alpha\gamma^\alpha - m\sqrt{g_{\alpha\beta}\gamma^\alpha\gamma^\beta} - \ldots \sqrt[m]{S_m(\vec{\gamma},\ldots,\vec{\gamma})},\\
&\to \gamma^\alpha p_\alpha - eA_\alpha\gamma^\alpha - m - \ldots \sqrt[2m]{S_{2m}g^m} - \ldots \sqrt[2n+1]{S_{2n+1}g^n\gamma}\ldots
\end{aligned}
$$

Since $g_{\alpha\beta}$ is a symmetric tensor such that $\{\gamma^\alpha,\gamma^\beta\} \sim g^{\alpha\beta}$, then $g_{\alpha\beta}\gamma^\alpha\gamma^\beta \sim g_{\alpha\beta}\{\gamma^\alpha,\gamma^\beta\} \sim g_{\alpha\beta}g^{\alpha\beta} \sim 1$. Therefore, gravity seems to leave the picture again. The symmetric structure of the extra terms S_m can be used to reintroduce g and to reduce the powers of γ. Thus the higher even order terms contribute to the mass m, making it variable[22] with $\vec{x}$.

Summary. We have discussed the structure of the matter Lagrangian for extended objects. Imposing reparametrization invariance of the action S naturally leads to a first order homogeneous Lagrangian. In its canonical form, L contains electromagnetic and gravitational interactions, as well as interactions that are not yet identified. The non-relativistic limit for a brane has been defined as those coordinates where the brane is an integral sub-manifold of the target space. This gauge can be used to remove reparametrization invariance of the action S and make the Hamiltonian function suitable for canonical quantization. The existence of a mass-shell constraint is universal. It is essentially due to the gravitational (quadratic in velocities) type interaction in the Lagrangian and always leads to a Klein–Gordon like equation. Once the algebraic properties of the γ-matrices are defined, one can use $v \to \gamma$ quantization in the Hamiltonian function $h = pv - L(x,v)$ to obtain the Dirac equation.

References

1. C. W. Kilmister, *Lagrangian Dynamics: an Introduction for Students* (Plenum Press, New York, 1967).

2. H. Goldstein, *Classical Mechanics* (Addison-Wesley, Massachusetts, 1980).
3. X. Gràcia and J. M. Pons, *J. Phys.* **A34**, 3047 (2001).
4. H. Rund, *The Hamilton–Jacobi theory in the calculus of variations: its role in mathematics and physics* (Van Nostrand, Huntington N.Y., 1966).
5. C. Lanczos, *The variational principles of mechanics* (University of Toronto Press, Toronto, 1970).
6. W. Pauli, *Theory of relativity* (Pergamon Press, New York, 1958).
7. S. Hojman and H. Harleston, *J. Math. Phys.* **22**, 1414 (1981).
8. D. B. Fairlie, *Phys. Lett.* **B456**, 141 (1999) (hep-th/9902204); L. M. Baker and D. B. Fairlie (hep-th/9908157); L.M. Baker and D.B. Fairlie, *Nucl. Phys.* **B596**, 348 (2001) (hep-th/0003048).
9. M. Pavsic, *The landscape of theoretical physics* (Kluwer Academic Publishers, Boston, 2001).
10. P. Bozhilov, *Nucl. Phys.* **B656**, 199 (2003) (hep-th/0211181).
11. V. G. Gueorguiev, *The relativistic particle and its d-brane cousins* in Gravity, Astrophysics, and Strings02, eds. P. P. Fiziev and M. D. Todorov (St. Kliment Ohridski University Press, Sofia, 2003) (math-ph/0210021).
12. M. Adak, T. Dereli and L. H. Ryder, *Class. Quantum Grav.* **18**, 1503 (2001).
13. I. Nikitin, *Introduction to String Theory*, http://viswiz.gmd.de/~nikitin/.
14. P. A. M. Dirac, *Proc. Roy. Soc.* **A246**, 333 (1958).
15. M. Henneaux and C. Teitelboim, *Quantization of gauge systems* (Princeton University, Princeton N.J., 1992).
16. J. Schwinger, *Phys. Rev.* **130**, 1253 (1963).
17. P. A. M. Dirac, *Proc. Roy. Soc.* **A246**, 326 (1958).
18. C. Teitelboim, *Phys. Rev.* **D25**, 3159 (1982).
19. K. Sundermeyer, *Constrained dynamics* (Springer-Verlag, New York, 1982).
20. V. G. Gueorguiev, *Matter, Fields and Reparametrization-Invariant Systems* in 4th Conference on Geometry, Integrability and Quantization (Coral Press, Sofia, Bulgaria, 2003) (math-ph/0210022).
21. D. B. Fairlie and T. Ueno, *J. Phys.* **A34**, 3037 (2001).
22. J. D. Bekenstein, *Phys. Rev.* **D48**, 3641 (1993).

INTERACTION BETWEEN VORTICES IN MODELS WITH TWO ORDER PARAMETERS*

R. MACKENZIE, M.-A. VACHON, U. F. WICHOSKI

Laboratoire René-J.-A.-Lévesque, Université de Montréal
C.P. 6128, Succ. Centre-ville, Montréal, QC H3C 3J7

The interaction energy and force between widely separated strings is analyzed in a field theory having applications in many branches of physics. This field theory has two order parameters, one of which is broken in the vacuum (giving rise to strings), the other of which is unbroken in the vacuum but which could nonetheless be broken in the core of the string. If this does occur, there is an effect on the energetics of widely separated strings.

1. Introduction

In several different contexts, nonlinear effects in field theories with spontaneous symmetry breaking give rise to topological solitons wherein a second order parameter (unbroken in the vacuum) attains a nonzero expectation value in the core of the soliton. Examples include bosonic superconducting cosmic strings [1], optical vortex solitons in nonlinear optics (see [2] for references and review), and vortices in the SO(5) model of high-temperature superconductivity [3, 4].

In this paper, we will study a field theory which applies (with minor modifications) to any of these situations, and will study the energetics of two widely-separated vortices. In the cosmic string context this has an influence on the dynamics of a network of strings [5, 6, 7]; in optics this affects the stability of solitonic waveguides [2]; and in the superconducting case this is one way to study the type of superconductor described by a given model, which provides another test of the SO(5) model [8, 9].

In the next section, we will establish notation and briefly review previous work (expressed in the language of superconductivity, but easily translated into the other contexts), wherein the effect of the AF phase on the magnetic

*Talk presented by R.M.

behaviour was studied, and the conditions under which SO(5) superconductors are type I or II were determined, in two ways: by studying the energy of vortices as a function of their winding number [8] and by studying the critical fields [9].

A third way of differentiating between type I and type II superconductors is through the force between two widely-separated vortices: they attract or repel for superconductors of type I or II, respectively. This approach is very intuitive, and is also applicable to both the cosmic strings and nonlinear optics contexts. In this talk, we will discuss this third approach, adopting for the most part the language of SO(5) superconductivity, though much of the discussion can easily be exported to other contexts such as those mentioned above. These results are derived in detail in [10].

2. Review of previous work

The model we wish to consider consists of a complex scalar field ϕ and a real triplet $\mathbf{n}$, described by the following two-dimensional free energy (written in terms of dimensionless variables):

$$F = \frac{1}{2}\int d^2s \left\{ (\nabla \times \mathbf{A})^2 + \frac{1}{\kappa^2}\left(|(-i\nabla - \kappa\mathbf{A})\phi|^2 + (\nabla\boldsymbol{\eta})^2 \right) \right.$$
$$\left. -\phi^2 - \beta\eta^2 + \frac{1}{2}(\phi^2 + \eta^2)^2 + \frac{1}{2} \right\}, \qquad (1)$$

where $\phi = |\phi|$ and $\eta = |\boldsymbol{\eta}|$. We see that the behaviour of the model is completely determined by two dimensionless parameters: the usual Ginzburg-Landau parameter κ familiar from conventional superconductivity, and a new parameter β, which is the ratio of the quadratic coefficients of η and ϕ.

Since the complex field ϕ attains a nonzero VEV, vortex solutions exist, wherein the phase of ϕ changes by 2π around a large circle in space. One can also consider configurations of higher winding number; a rotationally symmetric ansatz of winding number m ("m-vortex") is essentially the conventional one for ϕ and $\mathbf{A}$ together with a rotationally invariant ansatz for $\boldsymbol{\eta}$, whose orientation is taken to be a fixed, arbitrary unit vector $\hat{\mathbf{e}}$:

$$\phi(\mathbf{s}) = \varphi(s)e^{im\theta}, \quad A_i(\mathbf{s}) = \epsilon_{ij}\frac{s_j}{s}A(s), \quad \boldsymbol{\eta}(\mathbf{s}) = \hat{\mathbf{e}}\eta(s). \qquad (2)$$

As in the conventional case, these solutions carry a magnetic flux proportional to m (specifically, the dimensionless flux is $\Phi = -2\pi m/\kappa$).

As mentioned in the Introduction, a non-vanishing η field may appear inside the core of the vortex due to the competition between the potential and gradient energy terms. By examining the potential, one can see qualitatively that for fixed κ, as β increases, the impetus for η to be nonzero in the core of a vortex grows, since the potential energy savings gained by having $\eta \neq 0$ increases with β, while the kinetic energy cost is independent of β. Thus, we can define a critical value of β, above which the core of a vortex is AF and below which it is normal. This critical value (written β_{AF}) depends on κ, as well as the winding number m of the vortex.

The the model reduces to the SO(2) model if $\eta = 0$, a fact which enables us to easily make contact with that model; we need only set β to zero, since then there is never any reason for η to be nonzero in all situations we will consider here.

Substitution of (2) into (1) results in the following equations of motion:

$$\frac{1}{\kappa^2}\left[\varphi'' + \frac{1}{s}\varphi' - \left(\frac{m}{s} + \kappa A\right)^2 \varphi\right] + \varphi(1 - \varphi^2 - \eta^2) = 0, \tag{3}$$

$$\frac{1}{\kappa^2}\left(\eta'' + \frac{1}{s}\eta'\right) + \eta(\beta - \varphi^2 - \eta^2) = 0, \tag{4}$$

$$A'' + \frac{1}{s}A' - \frac{1}{s^2}A - \left(\frac{m}{ks} + A\right)\varphi^2 = 0. \tag{5}$$

These equations cannot be solved analytically; however, asymptotic solutions for large s can be found. Defining

$$\varphi(s) = 1 - f(s) \quad \text{and} \quad A(s) = -\frac{m}{\kappa s} + a(s), \tag{6}$$

the fields f, a and η approach zero exponentially as $s \to \infty$. The linearized equations in these fields have solutions

$$f(s) = C_f K_0(\sqrt{2}\kappa s), \quad a(s) = C_a K_1(s), \quad \eta(s) = C_\eta K_0(\sqrt{1-\beta}\kappa s) \tag{7}$$

where K_0 and K_1 are modified Bessel functions of the second kind and the C's are constants not determined by the linear equations. These results will be useful for deriving the potential energy between a pair of widely-separated vortices in the next section.

By studying either the energetics of vortices as a function of their winding number [8] or by studying the critical magnetic fields [9], one can determine a curve in the β-κ plane indicating the boundary between type I and type II behaviour; this boundary is

$$\kappa_c(\beta) = (1/\sqrt{2})\sqrt{(1+\beta)/(1-\beta)}\,. \tag{8}$$

This line, as well as the lines separating the normal/AF core boundary for $m = 1$ and $m = 2$ vortices, lines C, A, and B respectively, are shown in Figure 1.

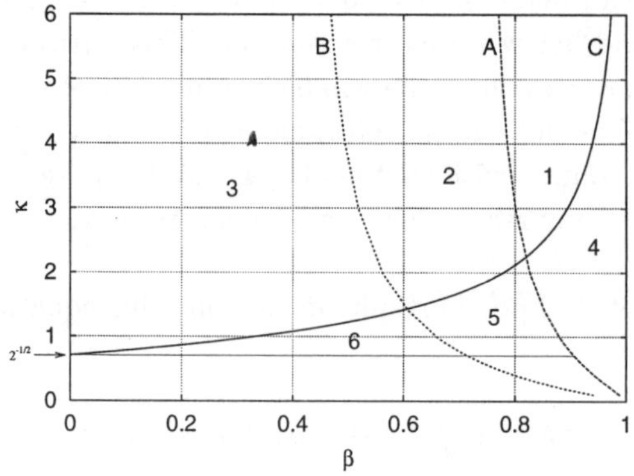

Figure 1. Stability of vortex lattice in the SO(5) model. Curves A and B separate the normal core region (left) from the AF core region (right) in the case of $m = 1$ and $m = 2$ vortices, respectively. Curve C depicts the border between type I and type II behaviour (see (8)). These curves divide the β-κ plane into several regions, the most interesting of which are regions 5 and 6, where the SC is type I yet vortices (of winding number 1) repel one another.

3. Intervortex force

A third way of determining the magnetic nature of a superconductor is through the interaction energy and force between vortices. We will study the interaction of widely-separated vortices in the SO(5) model, following a method due to Speight [11]. In conventional superconductivity, vortices of any separation are attractive or repulsive in the case of type I or type II superconductors, respectively, as was shown in great detail numerically by Jacobs and Rebbi [12]. We would thus expect the intervortex force in the SO(5) model to change from repulsive to attractive as we cross curve C in Figure 1 from left to right. This turns out to be not always the case. The reason is that vortex energetics does not always reflect the type of superconductor in the SO(5) model, as a comparison of the following two facts demonstrates. On the one hand, the type of an SO(5) superconductor depends critically on the AF sector (see [9]). On the other hand, when vortices have normal cores, their behaviour (their interaction energy, in particular) is blind to the AF sector of the model, as was noted in the previous section. Thus, vortices with normal cores cannot be expected to necessarily behave in the fashion dictated by the type of superconductor involved. In particular, it is possible that vortices in a type I (in the sense of bulk thermodynamic properties) superconductor repel one another.

Consider the interaction of ordinary vortices (that is, vortices of winding number 1). The core of these vortices is normal or AF to the left or right of curve A of Figure 1, respectively. Their interaction energy will be shown to obey the following behaviour. If the core is normal (to the left of curve A), the energetics is exactly as dictated by the SO(2) model: attractive for $\kappa < 1/\sqrt{2}$ and repulsive for $\kappa > 1/\sqrt{2}$. If the core is AF (to the right of curve A), then (except for a very thin transition zone wherein the AF core turns on) the energetics correctly reflects the type of superconductor involved (repulsive or attractive to the left or right of curve C).

If we imagine fixing κ and increasing β, then there are three cases. First, if $\kappa < 1/\sqrt{2}$ nothing dramatic happens: for all β the model describes type I superconductivity and vortices attract. Eventually curve A is crossed, so the vortices develop AF cores; they remain attractive, however.

Second, if κ lies above its value at the point of intersection of curves A and C ($\kappa > 2.25$), the situation is slightly more interesting: the core turns AF to the right of curve A, but the vortices remain repulsive until curve C is crossed, after which the superconductor is type I and the vortices attract.

The third, intermediate case, where curve C is crossed before curve A as β increases ($1/\sqrt{2} < \kappa < 2.25$), is the most interesting. When curve C is crossed, the superconductor becomes type I, so the vortices *ought to* attract. However, their cores are normal, their energetics is as in the SO(2) model, and they repel, since $\kappa > 1/\sqrt{2}$. This anomalous behaviour persists until curve A is crossed, when the core becomes AF and the vortices start to attract one another – the expected behaviour for a type I superconductor.

In summary, curve C delineates the boundary between type I (to the right) and type II (to the left) superconductivity; however, in regions 5 and 6 the superconductor has a sort of identity crisis: it is type I, but its vortices behave as type II vortices.[a]

---

[a]Perhaps some clarification on what is meant by type I vs type II superconductivity is warranted. In conventional superconductivity, type II superconductivity (for instance) is characterized by $\kappa > 1/\sqrt{2}$, negative surface energy at a normal/superconducting boundary, repulsive vortices, and $H_{c2} > H_c$. These are so inextricably connected that any of these features could be used as a definition of type II; the rest would follow. As we have seen, and as will be argued below (see also [9]), these features no longer imply

One can also study the interaction energy of vortices of higher winding number. The behaviour is qualitatively similar to that for 1-vortices, although the region of disagreement between vortex energetics and the type of superconductor is reduced. For 2-vortices, for example, the core turns AF at a lower value of β for any given κ (curve B rather than A), and consequently it is only in region 6 that one finds disagreement. This trend continues as the winding number is increased.

While the attraction or repulsion of vortices is clearly of interest in any physical situation described by the above work, the unexpected behaviour uncovered here is unlikely to be seen in HTSC, as it occurs in the wrong region of parameter space.

Acknowledgments

This work was supported by the Natural Science and Engineering Research Council of Canada and by the Fonds de Recherche sur la Nature et les technologies du Québec.

References

1. Edward Witten. Superconducting strings. *Nucl. Phys.*, B249:557–592, 1985.
2. Yu. S. Kivshar and B. Luther-Davies. Dark optical solitons: Physics and applications. *Phys. Rep.*, 298:81, 1998.
3. Shou-Cheng Zhang. SO(5) quantum nonlinear sigma model theory of the high Tc superconductivity. *Science*, 275:1089, 1997.
4. Daniel P. Arovas, A.J. Berlinsky, C. Kallin, and Shou-Cheng Zhang. Superconducting vortex with antiferromagnetic core. *Phys. Rev. Lett.*, 79:2871–2874, 1997.
5. L. M. A. Bettencourt and T. W. B. Kibble. Nonintercommuting configurations in the collisions of type I U(1) cosmic strings. *Phys. Lett. B*, 332:297–304, 1994.
6. L. M. A. Bettencourt and R. J. Rivers. Interactions between U(1) cosmic strings: An analytical study. *Phys. Rev. D*, 51:1842–1853, 1995.
7. Luis M. A. Bettencourt, Pablo Laguna, and Richard A. Matzner. Nonintercommuting cosmic strings. *Phys. Rev. Lett.*, 78:2066–2069, 1997.
8. M. Juneau, R. MacKenzie, M. A. Vachon, and J. M. Cline. Vortex lattice stability in the SO(5) model. *Phys. Rev. B*, 65:140512, 2002.
9. M. Juneau, R. MacKenzie, and M. A. Vachon. Magnetic properties of SO(5) superconductivity. *Ann. Phys.*, 298:421, 2002.

---

one another in SO(5) superconductivity. We refer to a superconductor as type I or II according to whether the surface energy is negative or positive, or equivalently, according to the relative value of H_c and H_{c2}.

10. R. MacKenzie, M. A. Vachon, and U. F. Wichoski. Interaction between vortices in models with two order parameters. *Phys. Rev.*, D67:105024, 2003.
11. J. M. Speight. Static intervortex forces. *Phys. Rev. D*, 55:3830–3835, 1997.
12. Laurence Jacobs and Claudio Rebbi. Interaction energy of superconducting vortices. *Phys. Rev. B*, 19:4486–4494, 1979.

SUPERSYMMETRIC BOUNDARY CONDITIONS FOR THE $\mathcal{N} = 2$ SIGMA MODEL*

I.V. MELNIKOV, M.R. PLESSER and S. RINKE

Center for Geometry and Theoretical Physics,
Duke University, Durham, NC 27708.
Email: lmel,plesser,rinke@cgtp.duke.edu

We clarify the discussion of $\mathcal{N} = 2$ supersymmetric boundary conditions for the classical $d = 2$, $\mathcal{N} = (2, 2)$ Non-Linear Sigma Model on an infinite strip. Our conclusions about the supersymmetric cycles match the results found in the literature. However, we find a constraint on the boundary action that is not satisfied by many boundary actions used in the literature.

1. Introduction

This work is part of the effort to extend the worldsheet approach to aspects of D-brane physics in the $g_s \to 0$ limit. We study the classical worldsheet theory for an open string whose endpoints are attached to D-branes in a Calabi-Yau manifold. We classify all boundary conditions and boundary actions which classically preserve an $\mathcal{N} = 2$ superconformal symmetry.

To extend the Non-Linear Sigma Model (NLSM) description to include open strings, we modify the $\mathcal{N} = (2, 2)$ NLSM by adding a local boundary action. We choose boundary conditions for the fields so that the bulk equations of motion continue smoothly to the boundary. To preserve an $\mathcal{N} = 2$ SUSY we must ensure that the boundary conditions and the action are invariant under the SUSY variations.

This analysis has been carried out by several groups [1–7]. Our results agree with the consensus, but we clarify the role of the "locality" [1] constraint and the necessity for the action *and* the boundary conditions to be invariant under the preserved supersymmetry. We find that the standard supersymmetric boundary due to Callan *et al* [8] does not satisfy the locality constraint! Space limitations do not permit us to include many details.

*Talk given by M.R.P.

These can be found in [10].

2. Symmetries and Boundaries in Classical Field Theory

To set notation, we begin with the Noether theorem. Consider a field theory with action

$$S = \int_{\mathbb{R}^2} d^2x \, \mathcal{L}\left(\varphi^i, \partial_\mu \varphi^i\right) . \tag{1}$$

Let $\delta_\eta \varphi^i = \eta f^i(\varphi, \partial_\mu \varphi, \ldots)$ be a symmetry of S. In other words, the variation of the Lagrangian is a boundary term: $\delta_\eta \mathcal{L} = -\eta \partial_\mu F^\mu$ for some $F^\mu(\varphi, \ldots)$. Then, the current

$$j^\mu = F^\mu + f^i \frac{\partial \mathcal{L}}{\partial(\partial_\mu \varphi^i)} \tag{2}$$

is conserved up to the equations of motion, and $Q = \int dx^1 j^0$ is the corresponding conserved charge. Now consider a theory with the same bulk Lagrangian but on a strip, Σ, with $-\infty < x^0 < \infty$, and $0 \leq x^1 \leq \pi$, and a boundary action

$$S_{\partial\Sigma} = \int_{\partial\Sigma} dx^0 \mathcal{M}\left(\varphi^i, \partial_0 \varphi^i\right) . \tag{3}$$

Note, $\mathcal{M}$ does not depend on $\partial_1 \phi$, as discussed in [10]. To have a well-posed initial value problem, we must specify a set of boundary conditions for the fields: $B_n(\varphi^i\big|_{\partial\Sigma}, \partial_\mu \varphi^i\big|_{\partial\Sigma}) = 0$. The choice of B_n will restrict both the values and the variations of φ^i on the boundary: the equations of motion follow by performing only *allowed* variations—those satisfying $\delta B_n = 0$. The B_n are required to satisfy a locality constraint. For any allowed variation the boundary term in δS has the form

$$\delta S = \int_{\partial\Sigma} dx^0 \left\{ \left[\frac{\partial \mathcal{M}}{\partial \varphi^i} - \partial_0 \left(\frac{\partial \mathcal{M}}{\partial(\partial_0 \varphi^i)} \right) + \frac{\partial \mathcal{L}}{\partial(\partial_1 \varphi^i)} \right] \delta \varphi^i \right\} . \tag{4}$$

If the bulk equations of motion are to extend smoothly to the boundary, the B_n must be chosen so that $\{\ldots\}$ vanishes for any allowed variation. Assuming that B_n are chosen accordingly, the symmetry variation of the action is given by $\delta_\eta S = -\eta \int_{\partial\Sigma} dx^0 \; j^1$. If the action is invariant, then $j^1\big|_{\partial\Sigma} = \partial_0 K$ for some $K(\varphi, \partial_0 \varphi)$ defined on the boundary, and $\widetilde{Q} = Q + K$ is conserved.

We now proceed to the $\mathcal{N} = (2,2)$ NLSM. We add boundaries, consider locality, and ensure that the boundary conditions and action are supersymmetric under an $\mathcal{N} = 2$ subalgebra.

3. The NLSM with Boundaries

We restrict attention to the compact part of the worldsheet theory, the NLSM of maps $\Phi : \Sigma \to X$ from a worldsheet Σ to the Kähler target space X of real dimension d. Locally we can specify these by a set of functions $\phi^I(x)$, $I = 1 \ldots d$. We also consider worldsheet Majorana-Weyl fermions, $\psi^I_\pm(x)$. We take our action to be $S = S_\Sigma + S_{\partial\Sigma}$. The bulk action, S_Σ, is the familiar closed string $\mathcal{N} = (2,2)$ NLSM. The most general form for the boundary action, $S_{\partial\Sigma}$, that is invariant (classically) under boundary reparametrizations and scale transformations is

$$S_{\partial\Sigma} = \int_{\partial\Sigma} dx^\mu V_\mu(\phi, \psi_-, \psi_+) + \int_{\partial\Sigma} \sqrt{|dx^\mu dx^\nu \eta_{\mu\nu}|}\, W(\phi, \psi_+, \psi_-), \quad (5)$$

where

$$\begin{aligned} V_- &= A_I(\phi)\partial_-\phi^I + iC_{IJ}(\phi)\psi^I_-\psi^J_-, \\ V_+ &= A_I(\phi)\partial_+\phi^I + i\widetilde{C}_{IJ}(\phi)\psi^I_+\psi^J_+, \\ W &= iD_{IJ}(\phi)\psi^I_+\psi^J_-. \end{aligned} \quad (6)$$

In what follows the worldsheet Σ will be a strip.

We assume that the end-point of the string moves along a submanifold M in X. This represents a D-brane wrapping M. By varying S, we find the most general allowed boundary conditions for the bosonic and fermionic fields, as well as constraints on the background fields. These are most easily written in a vielbein adapted to the splitting $TX|_U = TM \oplus NM$ for an open set $U \subset M$, with e^I_α a basis for $TX|_U$, and α split into the tangential $(a, b, c, \ldots)$, and normal $(z, y, x, \ldots)$ indices. The fermions must satisfy

$$\psi^\alpha_+ = \widetilde{R}^\alpha{}_\beta(\phi)\psi^\beta_- \quad (7)$$

with $\widetilde{R}_{\alpha\beta}\widetilde{R}_{\alpha\gamma} = \delta_{\beta\gamma}$, while the bosons satisfy

$$\partial_+\phi^\alpha = R^\alpha{}_\beta\partial_-\phi^\beta + T^\alpha_\beta V^\beta, \quad (8)$$

where R is an orthogonal matrix

$$R = \begin{pmatrix} \left(\frac{1+F}{1-F}\right)_{ab} & 0 \\ 0 & -\delta_{zy} \end{pmatrix}, \quad (9)$$

with $F_{ab} = A_{b,a} - A_{a,b} + B_{ab}$, $T_{\alpha\beta} = \frac{1}{2}\left(\delta_{\alpha\beta} + R_{\alpha\beta}\right)$, and

$$V_\alpha = i\widetilde{R}_{\mu\beta;\alpha}\widetilde{R}_{\mu\delta}\psi^\beta_-\psi^\delta_-. \quad (10)$$

The boundary couplings must satisfy

$$\widetilde{C} = \widetilde{R}D^T - D\widetilde{R}^T - \widetilde{R}^T C\widetilde{R}. \quad (11)$$

Note that V_α in Eq. (10) is independent of the boundary fermion couplings C,$\widetilde{C}$, and D: once the fermion boundary conditions are imposed, the fermion bilinear term in the boundary action is identically zero. The locality constraint, Eq.(11), is the only constraint on these couplings.

4. Conditions for $\mathcal{N} = 2$ Supersymmetry

In this section we find restrictions on M and the boundary couplings which lead to an unbroken $\mathcal{N} = 2$ SUSY.

The bulk action is supersymmetric up to a boundary term under $\mathcal{N} = (2,2)$ supersymmetry. Parametrizing these variations by $\epsilon_\pm^{1,2}$, there are essentially two choices for the $\mathcal{N} = 2$ subalgebra preserved by the boundary (We refer to [10] for explicit form of the variations.). These are labelled as **A** and **B**, and parametrized by ϵ and ϵ'. The $\mathcal{N} = (2,2)$ parameters are given by

$$\begin{aligned} \epsilon_+^2 &= -\epsilon_-^2 = \epsilon, \\ \epsilon_+^1 &= \eta\epsilon_-^1 = \epsilon', \end{aligned} \tag{12}$$

where $\eta = +1$ for **A** SUSY and $\eta = -1$ for **B** SUSY.

Computing the SUSY variation of the fermion boundary condition, we find the following non-trivial conditions for SUSY:

$$\widetilde{R} = R, \tag{13}$$

$$RJ + \eta JR = 0. \tag{14}$$

Other—at first sight non-trivial—conditions reduce to the Bianchi identity for the gauge-invariant field strength F.

A detailed calculation shows that the boson boundary condition is supersymmetric if the above conditions hold, and the bulk fermion equations of motion hold on the boundary. Classically, the use of the equations of motion is justified, but we will need to find a modification for the quantum problem.

Finally, the SUSY of the action leads to a single requirement that reduces to the Bianchi identity for F.

The local constraints on the geometry of M are contained in Eq. (14). Using the explicit form of R, Eq. (9), we find that for **A**-type supersymmetry, J and F must satisfy the following:

$$J_{zy} = 0; \qquad J_{ab} = F_{ac}J_{cd}F_{db}; \qquad F_{ab}J_{bx} = 0. \tag{15}$$

The first equation implies that the **A**-type supersymmetric cycle is locally a co-isotropic submanifold. The last two equations have solutions if and only if $\dim M = \frac{1}{2}\dim X + 2k$, where k is a non-negative integer. A cycle is Lagrangian if and only if $F_{ab} = 0$. Note that if $B \neq 0$, F is not a curvature associated to a connection on a line bundle!

For **B**-type supersymmetry, we find a different set of conditions:

$$J_{az} = 0; \qquad J_{ab}F_{bc} = F_{ab}J_{bc}. \tag{16}$$

Thus M is a holomorphic submanifold of X, and F is a $(1,1)$ form with respect to the complex structure M inherits from X.

5. Discussion

We have seen that the fermion boundary couplings have no effect on the supersymmetric cycles. If the boundary couplings are chosen to satisfy locality (Eqs. (11,8)), the fermion boundary coupling drops out of the action entirely. However, in order for locality to be satisfied, $C, D, \widetilde{C}$ must satisfy

$$\widetilde{C} = RD^T - DR^T - R^T CR, \tag{17}$$

where we have used the SUSY condition $\widetilde{R} = R$. A common fermion boundary coupling used in the literature is

$$F_{IJ}\left(\psi_+^I + \psi_-^I\right)\left(\psi_+^J + \psi_-^J\right). \tag{18}$$

It was motivated by Callan *et al* as the exponentiation of the open string photon vertex operator [8]. This natural term does not satisfy the locality requirement! A fermion boundary action that naturally emerges from superspace considerations and does satisfy locality is [4, 10]:

$$-\frac{i}{4}\left[F_{IJ}\left(\psi_+^I + \psi_-^I\right)\left(\psi_+^J + \psi_-^J\right) + 2G_{IJ}\psi_+^I\psi_-^J\right]. \tag{19}$$

The superspace approach is appropriate for dealing with the vexing problem of supersymmetry of the bosonic boundary conditions. The basic idea is to supplement the boundary conditions necessary for a well defined initial value problem with subsidiary conditions [9]. However, the form of these subsidiary conditions is not uniquely determined, and thus, the generalisation of the boundary conditions and action necessary for a proper quantum treatment is (at this time) not uniquely determined.

Acknowledgments

We would like to thank P.S. Aspinwall, R. Bryant, C. Curto, D. Fox, A. Kapustin, R. Karp, D.R. Morrison, K. Narayan, M. Roček and M. Stern for useful discussions. One of us (IVM) would like to thank the organizers of TASI 2003, where some of this work was completed. MRP and SR thank KITP and the organisers of the Geometry, Topology, and Strings workshop (MP03) where some of this work was completed. Their participation was supported in part by the National Science Foundation under Grant No. PHY99-07949. This work is partially supported by NSF grant DMS-0074072.

References

1. K. Hori, A. Iqbal, and C. Vafa, **hep-th/0005247** (2000).
2. H. Ooguri, Y. Oz, and Z. Yin, *Nucl. Phys.*, **B477**,407 (1996).
3. A. Kapustin and D. Orlov, **hep-th/0109098** (2001).
4. S. F. Hassan, **hep-th/0308201** (2003).
5. C. Albertsson, U. Lindström, and M. Zabzine, *Commun. Math. Phys.*, **233**, 403 (2003).
6. C. Albertsson, U. Lindström, and M. Zabzine. *hep-th/0202069* (2002)
7. U. Lindström and M. Zabzine. *JHEP*, **0302**, 6 (2003).
8. C.G. Callan, C. Lovelace, C.R. Nappi, and S.A. Yost. *Nucl. Phys.*, **B308**,221 (1988).
9. U. Lindström, M. Roček, and P. van Nieuwenhuizen, *Nucl. Phys.* **B662**, 147 (2003).
10. I.V. Melnikov, M.R. Plesser, and S. Rinke. **hep-th/0309223** (2003).

YANG-MILLS THEORY, FROM $2+1$ TO $3+1$ DIMENSIONS

V.P. NAIR

Physics Department, City College of the CUNY
New York, NY 10031, USA
vpn@sci.ccny.cuny.edu

The Hamiltonian analysis of (2+1)-dimensional Yang-Mills (YM_{2+1}) theory via the use of gauge-invariant matrix variables is reviewed. The vacuum wavefunction, string tension, the propagator mass for gluons are discussed. A possible method for extending this approach to $3+1$ dimensions is briefly considered. The possibility of using YM_{2+1} as a comparison theory for some results in YM_{3+1} is also discussed.

1. Introduction

In this talk I shall review a Hamiltonian approach to Yang-Mills theory on which I have been working for the last few years. In this analysis, for the Yang-Mills theory in $2+1$ dimensions, it is possible to carry out some nonperturbative calculations to the extent that results on mass gap and string tension can be compared with lattice simulations of the theory. I shall also discuss recent attempts to extend this to $3+1$ dimensions. The work I shall report on was done in collaboration with D. Karabali and Chanju Kim [1,2,3] and more recently with A. Yelnikov [4]. I have recently reviewed the $2+1$ dimensional case elsewhere [5] and so I shall consider this theory briefly and concentrate on YM_{3+1}.

Our analysis will involve the following steps. We will start with a parametrization of fields for YM_{2+1} and go to an exact calculation of the volume element of the gauge-invariant configuration space. This will define the inner product for the wave functions. A conformal field theory (CFT) argument will then indicate how the Hamiltonian can be simplified. Obtaining the Hamiltonian, we will solve the Schrödinger equation for the vacuum wave function and calculate the string tension. We will comment on the propagator mass and other results. We will then introduce a parametriztion of field for YM_{3+1} and obtain some results by comparison to YM_{2+1}.

2. The $2+1$ dimensional theory

Consider the Yang-Mills theory in 2+1 dimensions with group $G = SU(N)$. Since we do a Hamiltonian analysis it is convenient to take the $A_0 = 0$ gauge. The gauge potential can be written as $A_i = -it^a A_i^a$, $i = 1, 2$, where t^a are hermitian $N \times N$-matrices which form a basis of the Lie algebra of $SU(N)$ with $[t^a, t^b] = if^{abc}t^c$, $\mathrm{Tr}(t^a t^b) = \frac{1}{2}\delta^{ab}$.

The spatial coordinates x_1, x_2 will be combined into the complex combinations z, $\bar{z}$ with the corresponding components for the potential $A \equiv A_z = \frac{1}{2}(A_1 + iA_2)$, $\bar{A} \equiv A_{\bar{z}} = \frac{1}{2}(A_1 - iA_2) = -(A_z)^\dagger$. The parametrization of fields is given by $A_z = -\partial_z M M^{-1}$, $A_{\bar{z}} = M^{\dagger -1}\partial_{\bar{z}} M^\dagger$ Here M, $M^\dagger$ are complex matrices, not unitary in general. (If they are unitary, the potential is a pure gauge.) This parametrization is standard in many discussions of two-dimensional gauge fields. A particular advantage of this parametrization is the way gauge transformations are realized. A gauge transformation $A_i \to A_i^{(g)} = g^{-1}A_i g + g^{-1}\partial_i g$, $g(x) \in SU(N)$ is obtained by the transformation $M \to M^{(g)} = gM$. The gauge-invariant degrees of freedom are parametrized by the hermitian matrix $H = M^\dagger M$. Physical state wavefunctions are functions of H.

The measure of integration over the fields $A, \bar{A}$ is $d\mu(\mathcal{C}) = d\mu(\mathcal{A})/vol(\mathcal{G}_*)$ where $d\mu(\mathcal{A}) = \prod_{x,a} dA^a(x) d\bar{A}^a(x)$ is the Euclidean volume element on the space of gauge potentials $\mathcal{A}$ and $vol(\mathcal{G}_*)$ is the volume of gauge transformations, viz., volume of $SU(N)$-valued functions on space. Our parametrization of the fields gives $\delta A = -D(\delta M M^{-1})$, $\delta\bar{A} = \bar{D}(M^{\dagger -1}\delta M^\dagger)$ which gives $d\mu(\mathcal{A}) = (\det D\bar{D})\; d\mu(M, M^\dagger)$, where $d\mu(M, M^\dagger)$ is the volume for the complex matrices $M, M^\dagger$, associated with the metric $ds_M^2 = 8\int \mathrm{Tr}(\delta M M^{-1}\; M^{\dagger -1}\delta M^\dagger)$. Using $M = U\rho$, $H = \rho^2 = M^\dagger M$, this may be simplified as $d\mu(M, M^\dagger) = vol(\mathcal{G}_*) \prod_x d\mu(H)$ where $d\mu(H) = \epsilon_{a_1 \ldots a_n}(H^{-1}dH)_{a_1} \ldots (H^{-1}dH)_{a_n}$ and $vol(\mathcal{G}_*) = \prod_x d\mu(U)$. $d\mu(U)$ is the standard Haar measure for $SU(N)$.

The determinant of the two-dimensional operator $D\bar{D}$ is well known and is given by the Wess-Zumino-Witten (WZW) action as [6]

$$(\det D\bar{D}) = \left[\frac{\det' \partial\bar{\partial}}{\int d^2x}\right]^{\mathrm{dim}G} \exp\left[2c_A\; \mathcal{S}(H)\right] \tag{1}$$

where $c_A\delta^{ab} = f^{amn}f^{bmn}$; it is equal to N for $SU(N)$. $\mathcal{S}(H)$ is the WZW action for the hermitian matrix field H [7]. The volume element for the gauge-invariant configurations can now be written as $d\mu(\mathcal{C}) = d\mu(H)\exp(2c_A\mathcal{S}(H))$.

The inner product for physical (gauge-invariant) states $|1\rangle$ and $|2\rangle$, represented by the wavefunctions Ψ_1 and Ψ_2, can now be written as [8]

$$\langle 1|2\rangle = \int d\mu(H) e^{2c_A\ \mathcal{S}(H)}\ \ \Psi_1^* \Psi_2 \tag{2}$$

Notice that this effectively reduces any matrix element in YM_{2+1} to a correlator of a hermitian WZW model.

The wavefunctions, being gauge-invariant, are expected to be functionals of H, but actually they are more restricted; they can be taken as functionals of the current $J = (c_A/\pi)\partial_z H\ H^{-1}$ of the WZW model defined by $2c_A S(H)$. This follows from consideration of the correlators of the hermitian model [1,2]. This means that we can transform the Hamiltonian $T + V$ to express it in terms of J and functional derivatives with respect to J. The result is

$$T = m\left[\int_u J^a(u)\frac{\delta}{\delta J^a(u)} + \int \Omega_{ab}(u,v)\frac{\delta}{\delta J^a(u)}\frac{\delta}{\delta J^b(v)}\right] \tag{3}$$

$$V = \frac{\pi}{mc_A}\int \bar{\partial} J_a(\vec{x})\bar{\partial} J_a(\vec{x}) \tag{4}$$

$$\Omega_{ab}(u,v) = \frac{c_A}{\pi^2}\frac{\delta_{ab}}{(u-v)^2} \quad - \ i\frac{f_{abc}J^c(v)}{\pi(u-v)} \tag{5}$$

where $m = e^2 c_A/2\pi$. The first term in T shows that every power of J in the wavefunction gives a value m to the energy, suggesting the existence of a mass gap.

We have solved the Schrödinger equation for the vacuum wavefunction, $(T + V)\Psi_0 = 0$. to obtain $\Psi_0 = \exp(P)$, where

$$P = -\frac{1}{2e^2}\int B^a(x_1)\left[\frac{1}{(m+\sqrt{m^2-\nabla^2})}\right]_{x_1,x_2} B^a(x_2) + \cdots \tag{6}$$

where the ellipsis indicate terms involving higher powers of the current, see reference 3. P has the correct (perturbative) high momentum limit. The terms with higher number of J's can be shown to be small for the low momentum limit and for the high momentum limit.

The Wilson loop operator is given as $W(C) = \mathrm{Tr}P\ e^{-\oint_C (Adz+\bar{A}d\bar{z})}$. For the fundamental representation, its expectation value, with the vacuum wave function (6) can be calculated as $\langle W_F(C)\rangle = \text{constant}\quad \exp\left[-\sigma\mathcal{A}_C\right]$ where

$$\sqrt{\sigma} = e^2\sqrt{\frac{N^2-1}{8\pi}} \tag{7}$$

$\mathcal{A}_C$ is the area of the loop C; σ is the string tension. This is a prediction of our analysis starting from first principles. Notice that the dependence on e^2 and N is in agreement with large-N expectations, with σ depending only on the combination e^2N as $N \to \infty$. The values of $\sqrt{\sigma}/e^2$ given by (7) is within $\sim 3\%$ of the lattice estimates [9].

Using (3, 4), one can also check that $(T+V)\ J^a = \sqrt{m^2 - \nabla^2}\ J^a + \cdots$, showing that J^a is an approximate eigenstate with mass m and the correct relativistic dispersion relation. It may be considered as the gauge-invariant definition of the gluon. There are also other ways to see this, as explained in references 1 and 5. The propagator mass $m = e^2 c_A / 2\pi$ is related to the magnetic mass for YM_{3+1} at finite temperature.

For $SU(2)$ our result is $m/e^2 \approx 0.32$. which is roughly in agreement with resummation approaches (0.38 [10] and 0.28 [11]) and lattice estimates (0.31 to 0.40 [12] and 0.44 to 0.46 in different gauges [13]). An unambiguous lattice evaluation of this would require a gauge-invariant definition a gluon operator and its correlator. Philipsen has recently given such a definition; preliminary numerical estimates give 0.37 [14].

While excited states are much more difficult, some results can be found in reference 2. We have also extended this Hamiltonian analysis to the Yang-Mills-Chern-Simons theory [15].

3. The $3+1$ dimensional theory

The success of our analysis for YM_{2+1} depended largely on the parametrization of the potentials given in the last section. So, as a first step towards a generalization to YM_{3+1}, we will consider a similar parametrization for a gauge potential A_i^a, $i = 1, 2, 3$, on $\mathbf{R}^3$ [4].

For a $U(N)$ gauge field, the matrix M in terms of which the potentials can be parametrized must take values in $U(2N, \mathbf{C})$; this leads to more redundancy than in the $2+1$ dimensional case and will require further constraints. Specifically, let t^a be a basis of the Lie lagebra of $U(N)$ (considered as $N \times N$ matrices), and let σ_i denote the Pauli matrices. Then a basis for $U(2N, \mathbf{C})$ is given by $\{\mathbf{1} \otimes t^a, \mathbf{1} \otimes it^a, \sigma_i \otimes t^a, \sigma_i \otimes it^a\}$. One can then show that for every gauge potential A_i^a, one can construct a $U(2N, \mathbf{C})$-valued matrix M by solving the equation $\sigma \cdot A = -\sigma \cdot \partial M\ M^{-1}$. Conversely, if an arbitrary $M \in U(2N, \mathbf{C})$ is given, one can obtain a gauge potential A_i^a from the above equation, provided M obeys the further constraints $\mathrm{Tr}(t^a \sigma \cdot \partial M\ M^{-1} - \mathrm{h.c.}) = 0$ and $\partial_i(M^\dagger \sigma_i M) = 0$. These results hold for a neighbourhood of the flat potential in the space of potentials. If we expand

M as $M \approx 1 + it^a\varphi^a + i\sigma_k t^a \Theta_k^a + \cdots$, it is easy to see that these reproduce known Abelian gauge field parametrizations.

We may write the matrix M as $M = (U \otimes 1)\, N$ where U is an $N \times N$ unitary matrix and N can be obtained as an algebraic function of the gauge-invariant quantities $H = M^\dagger M$ and $W_i = M^\dagger \sigma_i M$. U can be removed by a gauge choice. The invariant measure of integration is then given by

$$\frac{d\mu(\mathcal{A})}{vol(\mathcal{G}_*)} = d\mu(H, W) e^{\Gamma + \bar{\Gamma}} \delta[\sigma \cdot \partial N N^{-1} + \text{h.c.}] \delta[\text{Tr} t^a \sigma \cdot \partial N N^{-1} - \text{h.c.}] \quad (8)$$

$d\mu(H, W)$ is the product over the spatial points of the volume on $U(2N, \mathbf{C})/U(N)$ and can be explicitly given [4]. Unlike the lower dimensional case, Γ can only be calculated in powers of F_{ij}; to the lowest order it is given by $\Gamma + \bar{\Gamma} = -(c_A/128) \int F_{ij}^a \left[(-(\partial + A)^2)^{-\frac{1}{2}}\right]^{ab} F_{ij}^b + \cdots$.

One could now try to proceed in a way similar to what we did for YM_{2+1}; however, it is interesting to look ahead and see what we can say about YM_{3+1} by comparison to YM_{2+1}. First of all, notice that the result (7) may be rewritten, based on the full Euclidean invariance of the Wick rotated theory, as $\langle W_F(C)\rangle = \int d\mu(\mathcal{C})\, \exp\left(-(1/4e^2) \int F^2\right)\, W_F(C)$ This version may in turn be interpreted as the equal time correlator in the $(3+1)$-dimensional theory with a vacuum wave function of the form $\sim \exp(-\int F^2/8e^2)$. Thus if the $(3+1)$-dimensional theory has a vacuum wave function

$$\Psi_0 \sim \exp\left(-\int \frac{F^2}{8\Lambda}\right) \quad (9)$$

then we get confinement and a string tension $\sqrt{\sigma} = \Lambda\sqrt{N^2 - 1/8\pi}$ Let us therefore assume that the wave function has the form (9), at least for the long wave length modes, and ask what it may imply. The mass of a 0^{++} glueball in the lower dimensional theory is given by

$$\langle B^2(x) B^2(0)\rangle = \int d\mu(\mathcal{C})\, \exp\left(-\int \frac{F^2}{4e^2}\right) \sim \exp(-M_{0^{++}}|x|) \quad (10)$$

for large separations $|x|$. The mass $M_{0^{++}}$ should be proportional to $e^2 N$, $M_{0^{++}} = \alpha e^2 N$, where α is, in principle, calculable in the Hamiltonian formulation of the $(2+1)$-dimensional theory. An explicit calculation has been difficult; lattice data show that $\alpha \approx 0.808$ [9] as $N \to \infty$. The result (10) may also be interpreted as an equal time correlator in YM_{3+1}, for the wave function (9) (with $e^2 \to \Lambda$), in which case the glueball mass is given by $M_{0^{++}} = \alpha \Lambda N$. This means that, if the wave function (9) is a good description in YM_{3+1}, the ratio $M_{0^{++}}/\sqrt{\sigma}$ is the same in YM_{2+1}

and YM_{3+1}; there is a certain dimension-independence. Collecting results $M_{0^{++}}/\sqrt{\sigma} = \alpha N\sqrt{8\pi/N^2-1} \approx 4.05\ N/\sqrt{N^2-1}$ where we have used the lattice value for the $(2+1)$-dimensional theory. This is then a prediction, based on the premise of (9), for the $(3+1)$-dimensional theory. The lattice estimate of this quantity for the $(3+1)$-dimensional theory is approximately 3.37 as $N \to \infty$ [16]; the discrepancy is about 20%. Thus equation (9) may be considered to be a reasonable ansatz for a first approximation to the wave function. Can we derive it by solving the Schrödinger equation? This is under study. Arguments with some similarity to ours for the dimension-independence have been presented before [17,14].

Acknowledgments

This work was supported in part by NSF grant number PHY-0244873 and by a PSC-CUNY grant.

References

1. D. Karabali and V.P. Nair, *Nucl. Phys.* **B464** (1996) 135; *Phys. Lett.* **B379** (1996) 141; *Int. J. Mod. Phys.* **A12** (1997) 1161.
2. D. Karabali, Chanju Kim and V.P. Nair, *Nucl. Phys.* **B524** (1998) 661.
3. D. Karabali, Chanju Kim and V.P. Nair, *Phys. Lett.* **B434** (1998) 103.
4. V.P. Nair and A. Yelnikov, hep-th/0302176.
5. V.P. Nair, hep-th/0309061.
6. A.M. Polyakov and P.B. Wiegmann, *Phys.Lett.* **B141** (1984) 223.
7. E. Witten, *Commun. Math. Phys.* **92** (1984) 455; S.P. Novikov, *Usp. Mat. Nauk.* **37** (1982) 3.
8. K. Gawedzki and A. Kupiainen, *Phys. Lett.* **B215** (1988) 119; *Nucl.Phys.* **B320** (1989) 649; M. Bos and V.P. Nair, *Int.J.Mod.Phys.* **A5** (1990) 959.
9. M. Teper, Phys. Rev. **D59** (1999) 014512 and references therein; B. Lucini and M. Teper, Phys. Rev. **D66** (2002) 097502.
10. G. Alexanian and V.P. Nair, *Phys.Lett.* **B352** (1995) 435;
11. W. Buchmüller and O. Philipsen, *Nucl.Phys.* **B443** (1995) 47; R. Jackiw and S.Y. Pi, *Phys.Lett.* **B368** (1996) 131; *ibid* **B403** (1997) 297.
12. W. Buchmüller and O. Philipsen, *Phys. Lett.* **B397** (1997) 112.
13. F. Karsch *et al*, *Nucl. Phys.* **B474** (1996) 217; F. Karsch, M. Oevers and P. Petreczky, *Phys. Lett.* **B442** (1998) 291.
14. O. Philipsen, *Phys. Lett.* **B521** (2001) 273; Nucl. Phys. Proc. Suppl. **106** (2002) 242; Nucl. Phys. **B 628** (2002) 167.
15. D. Karabali, Chanju Kim and V.P. Nair, *Nucl. Phys.* **B566** (2000) 331.
16. B. Lucini and M. Teper, JHEP **0106** (2001) 050.
17. S. Samuel, Phys. Rev. **D55** (1997) 4189.

THREE DIMENSIONAL SIMPLICIAL YANG-MILLS THEORY: AN APPROACH TO THE MASS GAP

S. G. RAJEEV

Department of Physics and Astronomy
University of Rochester,Rochester, NY14627, USA
E-mail: rajeev@pas.rochester.edu

We show how to formulate Yang-Mills Theory in $2+1$ dimensions as a hamitonian system within a simplicial regularization and construct its quantization, with special attention to the mass gap. An approximate conformal invariance of the hamiltonian of this theory is useful to construct a continuum limit.

There has been substantial progress in recent years in understanding non-abelian Yang-Mills theories. The case of $1+1$ dimensional Yang-Mills theory is particularly simple and has been solved in the lattice [1], hamiltonian [2] and path integral[3] formalisms. We will study here the case of two space and one time dimension, where Karabali et. al.[4] have made remarkable progress; for example, they have computed the string tension and given an argument that the 'magnetic mass' is non-zero. (The magnetic mass is a quantity that determines the decay of certain gluon correlation functions at spatial infinity).

It is our aim here to compute another, more direct, notion of a mass gap: the difference between the lowest two eigenvalues of the hamiltonian. We will do so rigorously within an approximation where the magnetic energy is ignored. It has been argued heuristically[5] that the mass gap computed within this approximation is in fact the exact answer for the complete theory. This reduced theory is invariant under conformal transformations of the spatial metric. Our regularization has a discrete version of this symmetry, which allows us to take the continuum limit.(A conformal symmetry also plays an important role in the work of Karabali et al.) Our calculation gives answers that seem to agree with numerical simulations[6].

The analogous problem in three space and one time dimension is of course more interesting, but substantially more involved, because the ultraviolet divergences are worse; indeed it remains as one of the great challenges of mathematical physics[7].

1. Hamiltonian Formulation of Yang-Mills Theory

Let G be a semi-simple Lie group and X a Riemann manifold of dimension d. We will consider gauge fields as connections on the topologically trivial principal bundle $X \times G$. These may be thought of as one-forms on X valued in $\underline{G}$. If $G = R$, the abelian group of real numbers, the whole theory should reduce to electromagnetism.

A gauge transformation $g : X \to G$ acts on such a one-form $A \mapsto gAg^{-1} + gdg^{-1}$. The curvature $B(A) = dA + A \wedge A$ transforms homogenously under this transformation, and is the generalization of the magnetic field. The Electric field is a $\underline{G}$-valued $(d-1)$-form, canonically conjugate to A (modulo some constraints we will list in a minute.)

The lagrangian of Yang–Mills theory may be written as

$$L = -\int \mathrm{tr}\; E * \left[\frac{\partial A}{\partial t} - d_A A_0\right] + \frac{1}{2}\int \mathrm{tr}\; [\alpha E * E + \frac{1}{\alpha} B(A) * B(A)] \quad (1)$$

Here, A_0 is a $\underline{G}$-valued scalar which generalizes the electrostatic potential. Also, tr denotes the trace in the defining representation of G in $U(N)$, which induces a negative-definite bilinear on $\underline{G}$. The covariant exterior derivative is defined as $d_A A_0 = dA_0 + AA_0 - A_0 A$.

Variation with respect to A_0 induces the constant on the pair (A, E), $d_A E = 0$. These are first class constraints that encode the gauge invariance of the theory. Following the usual philosophy of constrained dynamics, the true phase space of Yang-Mills theory thus consists of pairs (A, E) of Lie-algebra valued forms satisfying the above constraint, modulo the gauge equivalence relation $(A, E) \sim (gAg^{-1}gdg^{-1}, gEg^{-1})$.

The dynamics of the theory is given by the hamiltonian $H = -\frac{1}{2}\int \mathrm{tr}\; [\alpha E * E + \frac{1}{\alpha} B(A) * B(A)]$. The quantity α is the gauge coupling constant; it measures the departure of the theory from being a free theory. For, we could re-scale $A \to \sqrt{\alpha} A, E \to \frac{1}{\sqrt{\alpha}} E$ without losing the fact that A and E are canonically conjugate. In the abelian theory, this constant would then drop out. But in the non-abelian theory it would multiply the terms in the hamiltonian that are not quadratic: $\frac{1}{\alpha} B(A) * B(A) \to [dA + \sqrt{\alpha}[A, A]]^2$, thus becoming a measure of the strength of the non-linearity*.

*In 'natural' units in which $\hbar = c = 1$, α has the unit of mass for $d = 2$.

2. Conformal and Symplectic Invariance

There are some special symmetries for this theory for low values of the dimension of space-time, which should be useful in obtaining some exact results.

First of all, note that the constraint equations are invariant under all diffeomorphisms of space X. The metric tensor makes its appearance only in the formula for the hamiltonian. If $d = 2$, the electric energy $-\int \mathrm{tr}\ E * E$ is invariant under conformal transformations of the metric on X. To see this we just have to write it explicitly in tensor notation: $H_0 = -\frac{1}{2}\alpha \int \mathrm{tr}\ E * E = -\frac{1}{2}\alpha \int \mathrm{tr}\ E_i E_j g^{ij} \sqrt{g} d^2x$, the combination $\sqrt{g} g^{ij}$ being invariant under $g_{ij} \rightarrow e^{\phi} g_{ij}$. Thus this term in invariant under all conformal Killing vectors: an infinite dimensional Lie algebra for the case of two dimensional manifolds.

On the other hand, the magnetic energy (potential energy) $V = -\frac{1}{2}\frac{1}{\alpha} \int \mathrm{tr}\ B(A) * B(A)$ depends only on the area form induced by the metric: the Hodge dual of a two form is a scalar obtained by contracting with the inverse of the area form, $*B = \frac{1}{2}\frac{1}{\sqrt{g}} \epsilon^{kl} B_{kl}$. Thus the magnetic energy is invariant all area-preserving transformations, another infinite dimensional Lie algebra. The total hamiltonian is invariant only under the intersection of these two Lie algebras, which is finite dimensional.

3. Reduced Yang–Mills Theory

he hamiltonian divided by α is dimensionless; the gauge coupling constant measures the relative strengths of the electric and magnetic energies.

What if we ignore V and consider an approximation to Yang–Mills theory for which the hamiltonian is just H_0 ('Strong Coupling limit')? Since we preserve the non-abelian gauge invariance,we should expect many phenomena such as confinement to survive. More remarkably, it has been argued that [5] the mass gap of quantum Yang-Mills theory can be calculated exactly within this approximation, provided that it is non-zero: due to its scaling property, the potential energy will make a negligible contribution to this gap as the area of space tends to infinite.

To be a bit more concrete, suppose we start with an 'Infra-Red regularized' version of the quantum theory, in which X is compact but of large diameter R (more precisely, αR is large). If there are massless particles in the spectrum of H_0 ('no gap'), we should expect that the difference between the lowest two eigenvalues of H_0 will be of order R^{-2}. For example, this is the case for the abelian theory with $G = R$. If there is a gap in the spectrum (so that the lightest particle in the truncated theory is massive), this difference will remain non-zero even if we scale the metric so that $\alpha R \to \infty$: the energy of the first excited state is the mass of the lightest particle.

Under this scaling we expect the potential energy to behave like $(\alpha R)^{-2}$, so that it is negligible for large R *in the case where* H_0 *has a gap*. No such conclusion can be made for the case of no gap. Also, the other energy differences will behave like R^{-2} even when all particles are massive: for example, the energy of the second excited state minus the first excited state just corresponds to allowing the lightest particle to move with a small kinetic energy. Thus, the effect of V on them can not be ignored even for large R. Indeed V is the kinetic energy of these 'glueballs'.

We will now show the reduced theory with hamiltonian H_0 is an exactly solvable quantum theory. Moreover, the mass-gap of this reduced theory is non-zero for non-abelian gauge theories. We expect that the value of this quantity calculated within this approximation is exactly what it would be in Yang–Mills theory, although we do not provide a rigorous proof here. In this approximation, the classical theory describes geodesics in the space of connections modulo gauge transformations. The metric induced on this space of gauge equivalence classes is curved and it is not at all a simple matter to understand its dynamics fully [5]. In the corresponding reduced quantum theory, the hamiltonian is the Laplacian on this infinite dimensional curved manifold. There have been attempts before to show that this manifold has positive sectional curvature which could imply the existence of a mass -gap. However, these approaches are plagued by various ultra-violet divergences, as is typical in quantum field theory.

In order to make any precise statement about a quantum field theory we need an ultra-violet regularization, preferably one that does not break its important symmetries. It looks at first impossible to regularize the reduced Yang–Mills theory without breaking scale invariance; for example a lattice regularization will introduce a constant that measures the distance between nearest neighbor points. We will instead use a modification of the lattice regularization that does preserve a discrete version of conformal invariance; the cost will be that we must allow 'lattices' that no longer have equal spacing between nearest neighbors. We will construct a regularized version of Yang–Mills theory and determine its spectrum in the limit of ignoring the magnetic energy. Then we will show that the gap is unchanged in the continuum limit. A discrete version of the conformal invariance (invariance under subdivision of simplices) will make it possible to pass to the continuum limit.

4. Simplicial Sets and Complexes

Now let us build an analogue of Yang–Mills theory on a discrete approximation **S** to space X. A convenient language is provided by the notion of simplicial set [8] in algebraic topology. This is an abstraction of the more obvious notion of a triangulation of X. It will be convenient to formulate a discrete version of a gauge theory using only the abstract notion of a simplicial set.

A simplicial complex $\tilde{K}$ is a set of finite subsets of X (called simplices) of X such that every subset of an element of $\tilde{K}$ is also an element of $\tilde{K}$. We are to think of each set $\{x_0, x_1 \cdots x_r\}$ as an r-simplex (r-dimensional version of a triangular pyramid) with vertices at $x_0, x_1, \cdots x_r$. Thus, 0-simplices are just points; 1-simplices are to be thought of as lines (edges) connecting $\{x_0, x_1\}$, 2-simplices are triangles (faces) with vertices at $\{x_0, x_1, x_2\}$, 3-simplices are tetrahedrons (volumes) with vertices at $\{x_0, x_1, x_2, x_3\}$ and so on. A subset of a simplex in $\tilde{K}$ is itself a simplex in it; thus the edges of a triangle or the faces of a tetrahedron in $\tilde{K}$ are also in $\tilde{K}$. Because the vertices in a simplex are not ordered, we don't have any ordering of the vertices. To deal with non-abelian gauge theories, we will need a slightly more refined notion that has such an ordering as well: a *simplicial set*, which we define next.

Let K_r be the set of all sequences $(x_0, x_1, \cdots x_r)$ such that the set $\{x_0, x_1 \cdots x_r\}$ is an m-simplex in $\tilde{K}$ for some $m \leq r$. If some of the elements in the sequence are equal, the corresponding set will have fewer than r elements; then it will define a simplex of some lower dimension m.

There are maps $\partial_i : K_r \to K_{r-1}$ for $i = 0, 1, \cdots r$ which pick out the ith face of K_r by omitting the ith vertex.

$$\partial_i(x_0, \cdots x_r) = (x_0, x_1, \cdots, x_{i-1}, x_{i+1}, \cdots x_r).$$

The face maps satisfy the relation

$$\partial_i \partial_j = \partial_{j-1} \partial_i, \quad \text{if } i < j. \tag{2}$$

There are also certain maps

$$s_i : K_r \to K_{r+1} s_i(x_0, \cdots x_r) = (x_0, x_1, \cdots x_i, x_i, \cdots x_r)$$

which are inverses of the face maps. (We don't need them very much in what follows.) The K_r together form a *simplicial set*, which means that the maps ∂_i, s_i satisfy the axioms (easily verified although we don't use them)

$$\begin{aligned} s_i s_j &= s_{j+1} s_i, \quad \text{if } i \leq j, \quad \partial_i s_j = s_{j-1} \partial_i, \quad \text{if } i < j, \\ \partial_j s_j &= \text{identity} = \partial_{j+1} s_j, \quad \partial_i s_j = s_j \partial_{i-1}, \quad \text{if } i > j+1 \end{aligned} \tag{3}$$

in addition to (2).

For example, an edge $e = (x_0, x_1)$ has as its boundaries $\partial_0 e = x_1$ and $\partial_1 e = x_0$, its ending and starting points. (Note that ∂_0 removes the starting vertex hence yielding the ending vertex.) A face $f = (x_0, x_1, x_2)$ is to be thought of as a triangle with vortices at x_0, x_1, x_2, and has boundaries which are the three edges of this triangle:

$$\begin{aligned} \partial_0(x_0, x_1, x_2) &= (x_1, x_2), \\ \partial_1(x_0, x_1, x_2) &= (x_0, x_2), \\ \partial_2(x_0, x_1, x_2) &= (x_0, x_1). \end{aligned} \tag{4}$$

We can verify $\partial_0\partial_2 f = x_1 = \partial_1\partial_0 f$ etc.

5. Simplicial Gauge Theory

We can now give the discrete version of a gauge theory on such a simplicial set. A gauge transformation is a map $g : K_0 \to G$. A connection is a map $U : K_1 \to G$, satisfying the condition $U(x_0, x_1) = U(x_1, x_0)^{-1}$. The action of a gauge transformation on a connection is

$$U(x_0, x_1) \mapsto g(x_0)U(x_0, x_1)g^{-1}(x_1).$$

In other words, $U(e) \mapsto g(\partial_1 e)U(e)g(\partial_0 e)^{-1}$. The curvature of a connection is a map $\Phi : K_2 \to G$ given by $\Phi(x_0, x_1, x_2) = U(x_0, x_1)U(x_1, x_2)U(x_2, x_0)$ or equivalently, $\Phi(f) = U(\partial_2 f)U(\partial_0 f)U(\partial_1 f)^{-1}$. Under a gauge transformation, $\Phi(x_0, x_1, x_2) \mapsto g(x_0)\Phi(x_0, x_1, x_2)g^{-1}(x_0)$. This can also be verified using just the axioms of simplicial sets:

$$\begin{aligned} \Phi(f) \mapsto\ & g(\partial_1\partial_2 f)U(\partial_2 f)g(\partial_0\partial_2 f)^{-1}g(\partial_1\partial_0 f)U(\partial_0 f)g(\partial_0\partial_0 f)^{-1} \\ & \left[g(\partial_1\partial_1 f)U(\partial_1 f)g(\partial_0\partial_1 f)^{-1}\right]^{-1} \\ =\ & g(\partial_1\partial_2 f)\Phi(f)g(\partial_1\partial_2 f)^{-1} \end{aligned} \tag{5}$$

using the relation (2). Thus the curvature transforms in the adjoint representation at the starting point and hence tr $\Phi(f)$ is gauge invariant. Moreover the curvature satisfies the discrete analogue of the Bianchi identity: $\Phi(x_0, x_1, x_2)\Phi(x_0, x_2, x_3)\Phi(x_0, x_3, x_1) = U(x_0, x_1)\Phi(x_1, x_2, x_3)U(x_1, x_0)$, which may also be written in the language of simplicial sets:

$$\Phi(\partial_3 v)\Phi(\partial_1 v)\Phi(\partial_2 v) = U(\partial_2\partial_3 v)\Phi(\partial_0 v)U(\partial_2\partial_3 v)^{-1}, \quad \text{for } v \in K_3. \tag{6}$$

The electric field is a map † $E : K_1 \to \underline{G}$ such that $E(x_0, x_1) = -E(x_1, x_0)$. We assign the gauge transformation $E(x_0, x_1) \mapsto g(x_0)E(x_0, x_1)g(x_0)^{-1}$ [or alternately $E(e) \mapsto \text{ad}\, g(\partial_1 e)E(e)$], so that tr E^2 is invariant.

---

†It is more natural to let the Electric field take values in the dual of the Lie algebra; since we are using a non-degenerate invariant bilinear tr throughout, we identify the Lie algebra with its dual.

The lagrangian of simplicial gauge theory is then

$$L = -\sum_{e\in K_1} \text{tr } E(e)\left[\frac{\partial U(e)}{\partial t}U(e)^{-1} + A_0(\partial_1 e) - U(e)A_0(\partial_0 e)U(e)^{-1}\right]$$
$$+\frac{1}{2}\left[a\sum_{e\in K_1} \text{tr } E(e)^2 + b\sum_{f\in K_2} \text{tr } \Phi(f)\right]. \tag{7}$$

Here a and b are 'coupling constants' which will have to be tuned to zero at some specific rate as we let the number of vertices edges and faces grow to infinity, in order to get the continuum limit. It is part of our problem to determine these rates.

This Lagrangian is invariant under time dependent gauge transformations $g : K_0 \times R \to G$ acting on the variables U, E as above and on A_0 as follows: $A_0(x) \mapsto g(x)A_0(x)g(x)^{-1} + g(x)\frac{\partial}{\partial t}g(x)^{-1}$.

The constraint implied by varying A_0 is a 'conservation law' (the discrete analogue of the divergence being zero):

$$\sum_{\partial_1 e=x} E(e) = \sum_{\partial_0 e=x} U(e)E(e)U(e)^{-1}, \quad \text{for } x\in K_0. \tag{8}$$

It says that the sum of the electric fields over all edges starting at the vertex x is equal to the sum over all those ending at x (except that we must parallel transport the latter using the connection to make both sides transform the same way).

The equation of motion $aE(e) = \frac{\partial U(e)}{\partial t}U(e)^{-1}$ obtained by varying $E(e)$ shows that the electric field is the generator of left translations on the variables $U(e)$.

Thus our system is analogous to a set of coupled rigid bodies, one on each edge of the simplicial set, with $E(e)$ playing the role of angular momentum. The hamiltonian is $H = \frac{1}{2}\left[a\sum_{e\in K_1} \text{tr } E(e)^2 + b\sum_{f\in K_2} \text{tr } \Phi(f)\right]$. The first term is the kinetic energy and the second the potential energy if we think of this as a set of coupled rigid bodies. This analogy allows us to pass to the quantum theory easily.

6. Quantization of Simplicial Gauge Theory

In the quantum theory, the wave function is any function of the variables $U(e), e\in K_1$ satisfying the constraint of gauge invariance:

$$\psi\left(g(\partial_1 e)U(e)g(\partial_0 e)^{-1}\right) = \psi(U(e)). \tag{9}$$

This is just the quantum analogue of the constraint (8) on the 'angular momenta' $E(e)$. Also, $\hat{H} = \frac{1}{2}a\sum_{e\in K_1}\Delta(e) + \frac{1}{2}b\sum_{f\in K_2}\text{tr}\ \Phi(f)$ is the quantum hamiltonian, where $\Delta(e)$ is the Laplace operator on the group G, acting on the variable $U(e)$.

By the Peter-Weyl theorem, we can expand the function of a compact Lie group in terms of the matrices of the irreducible representations of the group. Applied to our case, we find that a solution to the constraints above are given by the following process:

A. Decorate each edge e with an irreducible representation $\rho(e)$ of G.

B. Attach an invariant map $\tilde{\psi}(x) : \bigotimes_{\partial_1 e=x}\rho(e)\otimes\bigotimes_{\partial_0 e=x}\rho(e)^* \to C$ to each vertex x. The product of these $\tilde{\psi}(x)$ defines a solution of the $U(e)$. The most general solution of the constraint is a linear combination of these 'invariant polynomials."

What we are doing is in fact a kind of non-abelian analogue of network theory. If the group G is the real line, the constraint (8) reduces to Kirchoff's law for conservation of current. Our condition (B) above is the non abelian analogue of this conservation law: the tensor product of the incoming representations and the conjugate of the outgoing representations must contain the trivial representation.

The simplest solution is the 'vacuum': the function is a constant and all the representations $\rho(e)$ are trivial. The next simplest is to choose the representations to be trivial everywhere except on an edge loop that doesn't intersect itself anywhere, on which it is the fundamental representation. The trace of product of traces of the connection along this loop is an invariant ("Wilson loop") and is a possible wave function.

Let us start to understand the spectral problem of the hamiltonian by ignoring the magnetic energy; we will recover this term in perturbation theory later:

$$H = H_0 + H_1, \quad H_0 = \frac{1}{2}a\sum_{e\in K_1}\Delta(e), \quad H_1 = \frac{1}{2}b\sum_{f\in K_2}\text{tr}\ \Phi(f) \qquad (10)$$

The ground state of the sum of H_0 is simply the constant function. Simple representation theory shows that for an edge loop γ that doesn't intersect itself anywhere, $W(\gamma) = \text{tr}\prod_{e\in\gamma}U(e)$ is an eigenfunction of the Laplace operator, $\sum_e\Delta(e) = C_2 l(\gamma)W(\gamma)$ where $l(\gamma)$ is the number of edges in the loop and C_2 the quadratic Casimir of the defining representation.

Thus the first excited state of the Laplacian is the Wilson loop of the shortest length, formed by the edges around a face: nothing but tr $\Phi(f)$: $H_0\text{tr}\ \Phi(f) = 3aC_2\text{tr}\ \Phi(f)$. There are as many such states as there are faces in our triangulation of X.

7. Subdivision of a Face

The discrete analogue of conformal invariance is the invariance under subdivision of the triangulation: we can subdivide a face without changing the rest of the triangulation, which corresponds to a local conformal transformation. Each triangulation induces a metric, the distance between two vertices being just the minimum number of edges in a path connecting them. When we subdivide a particular face f, we change this metric only locally, without affecting it for all the vertices not in f. What is remarkable about the hamiltonian H_0 is that its low lying eigenvalues are unchanged under such a subdivision, provided that the constant a is not changed: the ground state energy is still zero corresponding to the state where every link carries the trivial representation. The first excited state will still be a loop around any face, including possibly one of the new faces added by the subdivision; its energy will still be $3C_2a$ where C_2 is the quadratic Casimir of the fundamental representation. In the dimensionless units we use, this energy eigenvalue does not change. The degeneracy of the eigenvalue does change however, since now there are more faces. Indeed this is true of most of the the spectrum as a whole: the eigenvalues and eigenstates are unchanged under the subdivision, only the degeneracies change except when the size of the loop is comparable to the whole lattice. Because then, we might, by subdividing a face, create a loop of a length longer than was possible before. But these are the states that will tend to have an energy of the order of the cut-off, and so we should ignore them.

In this way we see that the free energy $W(\beta) = -\frac{1}{\#(\text{ faces })} \log \operatorname{tr} e^{-\beta H_0}$ tends to a finite limit as we subdivide faces repeatedly. Only loops of length of order one will contribute substantially in this sum, as we take the limit of a large number of faces. Their degeneracies are large (of the order of the number of faces). Under each subdivision of a face, the change in these degeneracy is of order one, so in the limit it does not change the free energy. This is how we can construct the continuum limit of reduced Yang–Mills theory. As a corollary of this construction we see that the gap in the spectrum is $3C_2a$ even in the continuum limit. The quantity a is just the gauge coupling constant of the continuum theory.

It may also be possible to construct this continuum limit as a path integral on the space of connections modulo gauge transformations more directly along the lines of the construction of the volume measure of this space in ^{3}.

The eigenstates of H_0 described above are nothing like the particles we expect in a quantum field theory: they describe loops ('glue rings') that are frozen in space. Of course we can take linear combinations of states with the same energy eigenvalue, to get other eigenstates, for example those that are momentum eigenstates. But the energy is independent of momentum; it is as if the particles are infinitely heavy. This is of course because we have ignored the magnetic energy. We can show that the magnetic energy of the Yang–Mills theory (which would be the potential energy in the language of gluons) serves as the kinetic energy term of these glue-rings. This should restore translation invariance and Lorentz invariance in the continuum limit, as well as justify the identification of the gap in the eigenvalues of H_0 as the mass of the lightest particle in Yang-Mills theory. We will give a detailed argument in a later publication, noting here only that second order degenerate perturbation theory bears out this interpretation.

Acknowledgments

I thank V. P. Nair for discussions on the work cited above and for reviving my interest in lower dimensional gauge theories. Also, I thank Rohana Wijewardhene for the opportunity to present this work at the Third Conference on Quantum Theory and Symmetries at the University of Cincinnati, Sep 10-14 2003.

References

1. A. A. Migdal, *Sov.Phys.JETP* **42** 413 (1975).
2. S. G. Rajeev, *Phys.Lett.* **B212** 203 (1988).
3. E. Witten *Commun.Math.Phys.* **141** 153 (1991); D.S. Fine *Commun.Math.Phys.* **140** 321 (1991).
4. D. Karabali, C-J Kim and V.P. Nair *Phys.Lett.* **B434** 103 (1998) [hep-th/9804132]; *Nucl.Phys.* **B524** 661 (1998)[hep-th/9705087].
5. R. P. Feynman, *Nucl.Phys.* **B188** 479 (1981); I. Singer, *Phys.Scripta***24** 817 (1981).
6. M. Teper hep-lat/0206027; hep-lat/9804008 .
7. A. Jaffe and E. Witten, http://www.claymath.org/Millennium_Prize_Problems/Yang-Mills_Theory/_objects/Official_Problem_Description.pdf
8. J. Peter May, *Simplicial Objects in Algebraic Topology*, U. of Chicago Press, Chicago, Il (1967)[MR **93m**:55025].

V. $D > 3$ FIELD THEORIES AND GRAVITY

CROSS-ORDER RELATIONS IN $\mathcal{N} = 4$ SUPERSYMMETRIC GAUGE THEORIES*

C. ANASTASIOU and L. J. DIXON

Stanford Linear Accelerator Center, Stanford University, Stanford, CA 94309

Z. BERN

Department of Physics and Astronomy, UCLA, Los Angeles, CA 90095-1547

D. A. KOSOWER

Service de Physique, CEA–Saclay, F–91191 Gif-sur-Yvette cedex, France

The anti-de Sitter/conformal field theory duality conjecture raises the question of how the perturbative expansion in the conformal field theory can resum to a simple function. We exhibit a relation between the one-loop and two-loop amplitudes whose generalization to higher-point and higher-loop amplitudes would answer this question. We also provide evidence for the first of these generalizations.

1. Introduction

Thirty years ago, shortly after the discovery of asymptotic freedom, 't Hooft[1] studied the large-order behavior of gauge theories. He used the double-line notation, studying the theory in the limit of a large number of colors N_c. In this limit, the insertion of a gluon in a diagram, corresponding to increasing the number of loops by one, costs a power of g^2 but gains a power of N_c. It is thus natural to study the theory in the regime where $g^2 N_c$ is held fixed. If this coupling is of order unity, higher-loop contributions are just as important as lower-order ones. We can think of this as filling in a mesh for any given process, so that we obtain a string worldsheet.

This connection was formulated into a duality conjecture by Maldacena[2]. In its most studied form, the anti-de Sitter/conformal field theory conjecture posits a duality between $\mathcal{N} = 4$ supersymmetric gauge theory (the conformal field theory) at strong coupling and large number of

*Talks presented by Z.B. and D.K.

colors, and the appropriate limit of string theory (perturbative supergravity) on an anti-de Sitter background. This duality has passed numerous tests[3] on quantities protected by supersymmetry, and more recently[5] on unprotected quantities allowing another expansion parameter (the large-'spin' limit of BMN[4] operators).

From the viewpoint of 't Hooft's original work, the AdS/CFT duality gives rise to another puzzle: how can the entirety of the complicated perturbative expansion of a quantum field theory be simple enough to be expressed at fixed order in another field theory? The computation we summarize here suggests that part of the explanation may lie in special, unexpected relations between different orders of perturbation theory in the $\mathcal{N} = 4$ gauge theory. Eden et al.[6] have alluded to such relationships.

The operator Green functions that have been studied in ref. 3 are off shell. In contrast to our intuition that manifestly Lorentz- and supersymmetry-invariant calculations should be much harder with increasing numbers of supersymmetries, the opposite is in fact true for on-shell quantities, in particular scattering amplitudes. Indeed, the $\mathcal{N} = 4$ calculations[7,8,9,10,11] have been pushed further earlier than the corresponding ones in QCD. Such calculations thus allow us to probe further into the perturbative expansion.

Most of these calculations were done using the unitarity-based method first described in ref. 9. To compute a one-loop amplitude, one computes tree amplitudes corresponding to the various cuts of the desired amplitude. In general, one must do these calculations with the cut-crossing legs taken in $D = 4 - 2\epsilon$ dimensions, in order to avoid subtraction ambiguities[12] in the reconstruction of the full loop amplitude from its cuts. At one loop, only two-particle cuts arise. For higher-loop amplitudes, we also compute their various cuts. The amplitudes on either side of the cut now include loop amplitudes, and for an l-loop amplitude, we must consider up to $(l + 1)$-particle cuts. In practice, it suffices to compute the cuts of these amplitudes as well, which are again tree amplitudes. For the four-point function at two loops, for example, we have to consider three-particle cuts, with five-point tree amplitudes on either side of the cut, and two-particle "double cuts", each involving a product of three tree-level four-point amplitudes. The construction of the cuts for the $\mathcal{N} = 4$ amplitude is surprisingly simple.[10]

Combining the cuts yields the integrand of a Feynman integral for the entire amplitude. In order to complete such a calculation, we also need to perform the integrals. At one loop, this is a straightforward task, and a general method for doing so in dimensional regularization was written down a decade ago.[13] At two loops, the inability to compute multi-scale integrals was a barrier to completing such calculations. Thanks to recent work on two-loop integrals[18,19] and general methods of tensor reductions at higher loops,[20] the barrier has been surmounted. Many two-loop amplitudes relevant to experiments have since been computed.[21]

The leading-color (planar) contributions to the two-loop amplitude in the $\mathcal{N}=4$ theory are given entirely[10] in terms of the planar double box,[18]

$$-stA_4^{\text{tree}}\Big\{ s$$
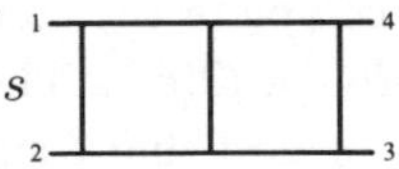

$$+\, t$$
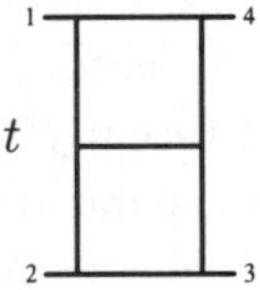

$$\Big\}. \qquad (1)$$

(The $\mathcal{N}=4$ supersymmetry relates all non-vanishing four-point amplitudes to each other order by order.)

Define the ratio $M_n^{(L)}(\epsilon) \equiv A_n^{(L)}/A_n^{\text{tree}}$ of the l-loop leading-color amplitude to the tree-level one. The explicit computation from eqn. (1) then reveals that

$$M_4^{(2)}(\epsilon) = \frac{1}{2}\Big(M_4^{(1)}(\epsilon)\Big)^2 + f(\epsilon)\, M_4^{(1)}(2\epsilon) - \frac{5}{4}\zeta_4 + \mathcal{O}(\epsilon)\,, \qquad (2)$$

where $f(\epsilon) \equiv (\psi(1-\epsilon) - \psi(1))/\epsilon = -(\zeta_2 + \zeta_3\epsilon + \zeta_4\epsilon^2 + \cdots)$. The relation between different loop orders expressed in this equation requires the use of polylogarithmic identities, and involves a non-trivial cancellation of terms between the two contributing planar double-box integrals. Note that terms through $\mathcal{O}(\epsilon^2)$ in the one-loop amplitude contribute to the $\mathcal{O}(\epsilon^0)$ terms on the right-hand side, since they can multiply the $1/\epsilon^2$ terms.

We may naturally expect a similar relation to hold between the n-point amplitudes $M_n^{(2)}$ and $M_n^{(1)}$. We can test such a generalization by studying the collinear limits of the amplitudes. The behavior of gauge-theory amplitudes in such limits may be described by splitting amplitudes whose ratio to the tree-level splitting amplitude we denote by $r_S^{(L)}$. The one- and two-loop amplitudes factorize as follows,[22]

$$M_n^{(1)}(\epsilon) \to M_{n-1}^{(1)}(\epsilon) + r_S^{(1)}(\epsilon)\,, \qquad (3)$$

$$M_n^{(2)}(\epsilon) \to M_{n-1}^{(2)}(\epsilon) + r_S^{(1)}(\epsilon) M_{n-1}^{(1)}(\epsilon) + r_S^{(2)}(\epsilon)\,. \qquad (4)$$

For the conjecture (2) to hold for n-point amplitudes as well, $r_S^{(2)}$ must be related to $r_S^{(1)}$. To see what relation is required, we can use the n-point analog of eqn. (2) to rewrite the two-loop quantities in eqn. (4) in terms of one-loop quantities and then use the one-loop collinear limit (3). Using the method of ref. 24, we find[23] that

$$r_S^{(2)}(\epsilon) = \frac{1}{2}\left(r_S^{(1)}\right)^2 + f(\epsilon) r_S^{(1)}(2\epsilon), \tag{5}$$

which is exactly what is required for the generalization of eqn. (2) to hold. This does not prove it, because of possible terms which are finite in the collinear limits; but it provides strong evidence that it is correct.

The form of the $\mathcal{N} = 4$ four-point integrand is known at three loops[10]; only two integrals are needed to complete the computation. One of these has been computed recently[25] through $\mathcal{O}(\epsilon^0)$. The relative simplicity and regular structure of the integrand suggest that a similar identity to eqn. (2) may hold at higher loops. Another indication comes from the structure of infrared-singular terms, as given by Catani[14] at two loops and Sterman and Tejeda-Yeomans[15] at three loops. The structure at two loops is an excellent guide to full relation (2) between one- and two-loop amplitudes. If the same holds true at higher loops, we expect a relation of the form,

$$M_4^{(L)}(\epsilon; s, t) = \frac{1}{L!}\left[M_4^{(1)}(\epsilon; s, t)\right]^L + \text{lower powers of } M_4^{(1)}(m\epsilon; s, t), \tag{6}$$

where the form and coefficient of the leading term are essentially determined by leading-log resummation.

We can also use the known structure of the infrared-singular terms, which must cancel against real emission contributions in any 'physical' quantity, to isolate the remaining finite terms; for these, the two-loop relation based on eqn. (2) is $F_4^{(2)} = \frac{1}{2}\left[F_4^{(1)}\right]^2 - \zeta_2 F_4^{(1)} - \frac{21}{8}\zeta_4$, and a higher-loop relation would then take the form,

$$F_4^{(L)}(s, t) = \text{Polynomial}(F_4^{(1)}(s, t)) = \frac{1}{L!}\left[F_4^{(1)}(s, t)\right]^L + \text{lower powers}. \tag{7}$$

It is worth noting that eqn. (2) does *not* hold beyond order ϵ^0. That is, the relation only holds as $D \to 4$, where the theory is conformal. The relation also appears to be special to the planar or leading-N_c contribution, as expected from the Maldacena conjecture.

Several open questions merit further study. It would be desirable, and is feasible, to compute the two-loop five-point amplitude explicitly in order to verify the generalization of eqn. (2). Likewise, knowledge of the three-loop four-point amplitude would clarify the structure of higher-loop generalizations. It would be interesting to identify a symmetry, presumably related to superconformal invariance, underlying the relation. Witten's new formulation[26] in terms of topological string theory may offer new insight.

Finally, it would be nice to connect the on-shell $\mathcal{N} = 4$ amplitudes to off-shell investigations of the duality conjecture. In this respect, it is worth noting that the tree-level collinear splitting amplitudes are closely related to the leading-order Altarelli–Parisi kernel, whose Mellin moments are the one-loop anomalous dimensions of leading-twist operators. Similarly, the next-to-leading order Altarelli–Parisi kernel can be computed[27] from the one-loop splitting amplitudes (and other ingredients), and we expect that the two-loop splitting amplitude can be used to compute next-to-next-to-leading order anomalous dimensions in the $\mathcal{N} = 4$ gauge theory.

There are also implications for the higher-loop structure of $\mathcal{N} = 8$ supergravity. Against prior expectations, this theory was argued[11] to be ultraviolet finite through four loops, based on the cut-by-cut absence of ultraviolet divergences in the four-point amplitude. At five loops, finiteness would require non-trivial cancellations between different contributions. The work outlined in this talk shows that such cancellations occur in maximally supersymmetric gauge theories. Given the close connection between the integrands of maximally supersymmetric gauge and gravity theories, as used in ref. 11, it is quite possible that non-trivial cancellations would take place in the gravitational theory.

Acknowledgments

We thank M. Staudacher for helpful discussions. This research was partially supported by the US Department of Energy under contracts DE-FG03-91ER40662 and DE-AC03-76SF00515.

References

1. G. 't Hooft, Nucl. Phys. B **72** (1974) 461.
2. J. M. Maldacena, Adv. Theor. Math. Phys. **2**, 231 (1998) [Int. J. Theor. Phys. **38**, 1113 (1999)] [hep-th/9711200]; S. S. Gubser, I. R. Klebanov and A. M. Polyakov, Phys. Lett. B **428**, 105 (1998) [hep-th/9802109]; O. Aharony, S. S. Gubser, J. M. Maldacena, H. Ooguri and Y. Oz, Phys. Rept. **323**, 183 (2000) [hep-th/9905111].
3. E. D'Hoker and D. Z. Freedman, hep-th/0201253.
4. D. Berenstein, J. M. Maldacena and H. Nastase, JHEP **0204**, 013 (2002).
5. J. A. Minahan and K. Zarembo, JHEP **0303**, 013 (2003) [hep-th/0212208]; N. Beisert, C. Kristjansen and M. Staudacher, Nucl. Phys. B **664**, 131 (2003) [hep-th/0303060]. N. Beisert, S. Frolov, M. Staudacher and A. A. Tseytlin, JHEP **0310**, 037 (2003) [hep-th/0308117], and references therein; N. Beisert, J. A. Minahan, M. Staudacher and K. Zarembo, JHEP **0309**, 010 (2003) [hep-th/0306139].

6. B. Eden, C. Schubert and E. Sokatchev, hep-th/0010005.
7. M.B. Green, J.H. Schwarz and L. Brink, Nucl. Phys. B **198**, 474 (1982).
8. Z. Bern, L.J. Dixon, D.C. Dunbar and D. A. Kosower, Nucl. Phys. B **425**, 217 (1994) [hep-ph/9403226].
9. Z. Bern, L.J. Dixon, D.C. Dunbar and D. A. Kosower, Nucl. Phys. B **435**, 59 (1995) [hep-ph/9409265].
10. Z. Bern, J.S. Rozowsky and B. Yan, Phys. Lett. B **401**, 273 (1997) [hep-ph/9702424];
11. Z. Bern, L.J. Dixon, D.C. Dunbar, M. Perelstein and J.S. Rozowsky, Nucl. Phys. B **530**, 401 (1998) [hep-th/9802162]; P. S. Howe and K. S. Stelle, Phys. Lett. B **554**, 190 (2003) [hep-th/0211279].
12. W. L. van Neerven, Nucl. Phys. B **268**, 453 (1986).
13. Z. Bern, L. J. Dixon and D. A. Kosower, Nucl. Phys. B **412**, 751 (1994) [hep-ph/9306240].
14. S. Catani, Phys. Lett. B **427**, 161 (1998) [hep-ph/9802439];
15. G. Sterman and M. E. Tejeda-Yeomans, Phys. Lett. B **552**, 48 (2003) [hep-ph/0210130].
16. T. Kinoshita, J. Math. Phys. **3**, 650 (1962); T.D. Lee and M. Nauenberg, Phys. Rev. **133**, B1549 (1964).
17. Z. Bern, G. Chalmers, L.J. Dixon and D. A. Kosower, Phys. Rev. Lett. **72**, 2134 (1994) [hep-ph/9312333].
18. V. A. Smirnov, Phys. Lett. B **460**, 397 (1999) [hep-ph/9905323].
19. J. B. Tausk, Phys. Lett. B469:225 (1999) [hep-ph/9909506]
20. V. A. Smirnov, Nucl. Phys. Proc. Suppl. **116**, 417 (2003) [hep-ph/0209295]; T. Gehrmann, Nucl. Phys. Proc. Suppl. **116**, 13 (2003) [hep-ph/0210157], and references therein.
21. Z. Bern, Nucl. Phys. Proc. Suppl. **117**, 260 (2003) [hep-ph/0212406]; E. W. N. Glover, Nucl. Phys. Proc. Suppl. **116**, 3 (2003) [hep-ph/0211412], and references therein.
22. Z. Bern, V. Del Duca and C. R. Schmidt, Phys. Lett. B **445**, 168 (1998) [hep-ph/9810409]; D. A. Kosower and P. Uwer, Nucl. Phys. B **563**, 477 (1999) [hep-ph/9903515]; Z. Bern, V. Del Duca, W.B. Kilgore and C.R. Schmidt, Phys. Rev. D **60**, 116001 (1999) [hep-ph/9903516].
23. C. Anastasiou, Z. Bern, L. J. Dixon and D. A. Kosower, Phys. Rev. Lett. **91**, 251602 (2003) [hep-th/0309040].
24. D. A. Kosower, Nucl. Phys. B **552**, 319 (1999) [hep-ph/9901201].
25. V. A. Smirnov, Phys. Lett. B **567**, 193 (2003) [hep-ph/0305142].
26. E. Witten, hep-th/0312171.
27. D. A. Kosower and P. Uwer, Nucl. Phys. B **674**, 365 (2003) [hep-ph/0307031].

BPS DOMAIN WALLS IN MASSIVE HYPER-KÄHLER SIGMA MODELS*

MASATO ARAI
Institute of Physics, AS CR, 182 21, Praha 8, Czech Republic

MUNETO NITTA†
Department of Physics, Purdue University, West Lafayette, IN 47907-1396, USA

NORISUKE SAKAI
Department of Physics, Tokyo Institute of Technology, Tokyo 152-8551, JAPAN

We discuss hypermultiplets admitting degenerate discrete vacua and BPS domain walls interpolating them. This talk is based on the original papers.[1,2]

1. Introduction

There exists a long history for searching topological defects such as domain walls. Lots of their properties in particular localization of massless or light fields on them have been investigated. Recent interest in the brane-world scenario[3] has brought us an idea to realize our world as the effective field theory of light localized fields on the brane. Supersymmetry (SUSY) is combined with this idea by considering the BPS solitons in SUSY field theories. They typically break the half of the original SUSY spontaneously. BPS domain walls are investigated in detail in $D = 4$, $\mathcal{N} = 1$ SUSY theories (with four supercharges), on which effective field theories have two unbroken supercharges. However in order to realize $D = 4$, $\mathcal{N} = 1$ SUSY theory on the world-volume, we need a higher-dimensional theory with eight supercharges. In these theories scalar multiplets are hypermultiplets. They must parameterize curved hyper-Kähler (HK) manifolds[4,5] with a scalar

*Talk given by M.N.
†Address after December 11: Tokyo Institute of Technology.

potential admitting at least two discrete degenerate vacua. These models are called the massive HK nonlinear sigma models (NLSM).

In this talk, the HK quotient method[6,7] to construct HK manifolds is shown to be generalized to the massive models and the BPS domain wall in the simplest case of the Eguchi-Hanson target space is given. Keeping essential properties of eight supercharges, we discuss a simpler and familiar case of $D = 4$, $\mathcal{N} = 2$ SUSY theories. Also we use the $\mathcal{N} = 1$ superspace formalism. A fully off-shell $\mathcal{N} = 2$ superspace (the Harmonic superspace) formalism is discussed in the original paper.[1]

2. Kähler Sigma Models and Walls

Here we recall BPS walls in $\mathcal{N} = 1$ SUSY theories, because we formulate $\mathcal{N} = 2$ SUSY models in terms of $\mathcal{N} = 1$ superfields in the following sections. Scalar fields belong to chiral superfields $\Phi^i(y,\theta) = \phi^i(y) + \sqrt{2}\theta\psi^i(y) + \theta\theta F^i(y)$ with $y^\mu = x^\mu + i\theta\sigma^\mu\bar{\theta}$. The Lagrangian for the most general $\mathcal{N} = 1$ SUSY Lagrangian for chiral superfields (the generalized Wess-Zumino model) is given by[8]

$$\mathcal{L} = \int d^4\theta K(\Phi, \Phi^\dagger) + \left[\int d^2\theta W(\Phi) + \text{c.c.}\right] \tag{1}$$

with K and W real and holomorphic functions, called the Kähler potential and the superpotential, respectively. After elimination of auxiliary fields by their equations of motion $F^i = -g^{ij^*}\partial_{j^*}W^* + \text{fermions}$, the bosonic part is calculated as

$$\mathcal{L}_{\text{boson}} = -g_{ij^*}\partial_\mu\phi^i\partial^\mu\phi^{*j} - g^{ij^*}\partial_i W\partial_{j^*}W^* \tag{2}$$

with $g_{ij^*} \equiv \partial_i\partial_{j^*}K$ the Kähler metric. (We denote $\partial_i = \frac{\partial}{\partial\phi^i}$.) The target manifold must be a Kähler manifold and therefore these models are called Kähler sigma models.

Assuming a domain wall configuration perpendicular to the third axis $x^3 = z$, its energy density per unit area in the x-y plane is given by

$$\begin{aligned} E &= \int dz(g_{ij^*}\partial_z\phi^i\partial_z\phi^{*j} + g^{ij^*}\partial_i W\partial_{j^*}W^*) \\ &= \int dz|\partial_z\phi^i - e^{i\alpha}g^{ik^*}\partial_{k^*}W^*|^2 + \int dz(e^{i\alpha}\partial_z\phi^i\partial_i W + \text{c.c.}) \\ &\geq \int dz(\partial_z\phi^i\partial_i W + \text{c.c.}) = 2\text{Re}(e^{i\alpha}\Delta W) \end{aligned} \tag{3}$$

with the norm defined by $|V^i|^2 \equiv g_{ij^*}V^iV^{*j}$, $\Delta W \equiv W|_{z=\infty} - W|_{z=-\infty}$ and α arbitrary real constant. Since we obtain the best bound at $e^{-i\alpha} =$

$\Delta W/|\Delta W|$, we derive the BPS bound $E \geq 2|\Delta W|$ saturated by solutions of the BPS equation

$$\partial_z \phi^i = e^{-i\alpha} g^{ij^*} \partial_{j^*} W^*, \qquad e^{-i\alpha} = \Delta W/|\Delta W| \,. \tag{4}$$

The SUSY transformation on the fermion ψ^i with a BPS wall background is calculated as $\delta_\epsilon \psi^i = \sqrt{2}(i\sigma^\mu \bar{\epsilon} \partial_\mu \phi^i + \epsilon F^i) = \sqrt{2}(i\sigma^z \bar{\epsilon} e^{-i\alpha} - \epsilon) g^{ij^*} \partial_{j^*} W^*$. Therefore two SUSYs satisfying $i e^{-i\alpha} \sigma^z \bar{\epsilon} = \epsilon$ out of four are preserved and so the solutions are called 1/2 BPS states.

The BPS domain walls (and their junction) in $D = 4$, $\mathcal{N} = 1$ SUSY NLSMs including runaway vacua and singularity of the metric were discussed.[9] It is easy to find $\mathcal{N} = 1$ SUSY models admitting wall solutions, because K and W are arbitrary and independent to each other, but they are not for $\mathcal{N} = 2$ SUSY as we will see in the following sections.

3. Massive Hyper-Kähler Sigma Models and Walls

We discuss hypermultiplets with potential terms. The on-shell component Lagrangian for massive HK model is well known[5], whose bosonic part is

$$\mathcal{L}_{\text{boson}} = -g_{ij^*} \partial_\mu \phi^i \partial^\mu \phi^{*j} - |\mu|^2 g_{ij^*} k^i k^{*j} \tag{5}$$

with g_{ij^*} the target HK metric, μ a complex mass parameter and $k^i(\phi, \phi^*)$ a tri-holomorphic Killing vector on the target HK manifold.[a] Therefore an isometry on the manifold is required for a nontrivial potential to exist and vacua are given by fixed points of its action.

We take μ real because phase can be absorbed into the definition of k^i. Energy density for a wall perpendicular to the z-axis is

$$\begin{aligned} E &= \int dz (g_{ij^*} \partial_z \phi^i \partial_z \phi^{*j} + \mu^2 g_{ij^*} k^i k^{*j}) \\ &= \int dz |\partial_z \phi^i - \mu k^i|^2 + \int dz (\mu g_{ij^*} k^i \partial_z \phi^{*j} + \text{c.c.}) \\ &\geq \int dz (\mu g_{ij^*} k^i \partial_z \phi^{*j} + \text{c.c.}) = \mu \Delta D \end{aligned} \tag{6}$$

where $k^i = g^{ij^*} \partial_{j^*} D$ with $D(\phi, \phi^*)$ a real function called the Killing potential (moment map)[b] and $\Delta D \equiv D|_{z=\infty} - D|_{z=-\infty}$. We thus obtain the BPS bound $E \geq \mu \Delta D$ saturated by the BPS equation $\partial_z \phi^i = \mu k^i$.[11]

---

[a]The potential term can be interpreted by the Sherk-Schwarz (SS) dimensional reduction[10] from six-dimensions, where HK sigma models are massless. Then the SS reduction to four (five) space-time dimensions is defined by $-i(\partial_5 + i\partial_6)\phi^i = \mu k^i$ with $\mu \in \mathbf{C}$ ($\partial_6 \phi^i = \mu k^i$ with $\mu \in \mathbf{R}$).

[b]It is in general difficult to find D for given manifold and Killing vector.

The Eguchi-Hanson space $T^*\mathbf{C}P^1$ admits $SU(2)$ tri-holomorphic isometry, one of whose three generators, say σ_3, can be used to obtain the potential. A rotation around the third axis on the base $S^2 \simeq \mathbf{C}P^1$ has two fixed points on the North and South poles, both of which are vacua. The BPS domain wall interpolating these vacua was firstly obtained by Abraham and Townsend[11] in the component formalism. Lots of interesting BPS solitons were constructed in toric HK manifolds.[12,13]

4. Massive Hyper-Kähler Models from $\mathcal{N} = 2$ SUSY QCD

Let (Φ, Ψ) be $\mathcal{N} = 2$ hypermultiplets with Φ and Ψ being $N \times M$ and $M \times N$ matrix chiral superfields. To obtain nontrivial vacua we need an $U(M)$ gauge symmetry introducing $\mathcal{N} = 2$ vector multiplets (V, Σ) with V an $M \times M$ matrix vector superfield and Σ an $M \times M$ matrix chiral superfield. We work out for the $U(M)$ gauge group in which $U(1)$ part is essential to obtain discrete vacua. We consider the Higgs branch of the theory taking the strong coupling limit $g \to \infty$ of gauge interactions, which eliminates the kinetic terms for V and Σ. The gauge invariant Lagrangian is given by

$$\mathcal{L} = \int d^4\theta \left[\mathrm{tr}\,(\Phi^\dagger \Phi e^V) + \mathrm{tr}\,(\Psi\Psi^\dagger e^{-V}) - c\,\mathrm{tr}\,V\right] + \left[\int d^2\theta \left\{\mathrm{tr}\,\{\Sigma(\Psi\Phi - b\mathbf{1}_M)\} + \sum_{a=1}^{N-1} m_a \mathrm{tr}\,(\Psi H_a \Phi)\right\} + \mathrm{c.c.}\right], \quad (7)$$

with $b \in \mathbf{C}$ and $c \in \mathbf{R}$ constituting a triplet of the Fayet-Iliopoulos parameters, m_a complex mass and H_a Cartan generators of $SU(N)$.[c] Eliminating superfields V and Σ using their algebraic equations of motion, we obtain the Lagrangian in terms of independent superfields, in which the Kähler potential is

$$K = c\,\mathrm{tr}\sqrt{\mathbf{1}_M + \frac{4}{c^2}\Phi^\dagger\Phi\Psi\Psi^\dagger} - c\,\mathrm{tr}\,\log\left(\mathbf{1}_M + \sqrt{\mathbf{1}_M + \frac{4}{c^2}\Phi^\dagger\Phi\Psi\Psi^\dagger}\right) + c\,\mathrm{tr}\,\log\Phi^\dagger\Phi\,, \quad (8)$$

with a gauge fixing[d]

$$\Phi = \begin{pmatrix}\mathbf{1}_M \\ \varphi\end{pmatrix} Q\,, \quad \Psi = Q(\mathbf{1}_M, \psi)\,, \quad Q = \sqrt{b}(\mathbf{1}_M + \psi\varphi)^{-\frac{1}{2}}\,, \quad (9)$$

[c] Flavor symmetry $\Phi \to \Phi' = g\Phi$, $\Psi \to \Psi' = \Psi g^{-1}$ with $g \in SU(N)$ in the massless limit $m_a = 0$ is explicitly broken by the mass to its Cartan $U(1)^{N-1}$ generated by H_a.

[d] We discuss the $b \neq 0$ case here. The $b = 0$ case must be discussed independently.[1]

with φ (ψ) an $(N-M)\times M$ $[M\times(N-M)]$ matrix chiral superfield, and the superpotential is

$$W = b\sum_a m_a \mathrm{tr}\left[H_a \begin{pmatrix}\mathbf{1}_M\\ \varphi\end{pmatrix}(\mathbf{1}_M+\psi\varphi)^{-1}(\mathbf{1}_M,\psi)\right]. \tag{10}$$

This is the massive extension of the HK NLSM on the cotangent bundle over the Grassmann manifold, $T^*G_{N,M}$, found by Lindström and Roček.[6] This model contains ${}_NC_M = N!/M!(N-M)!$ discrete degenerate vacua corresponding to independent gauge fixing conditions (9).[1]

5. Domain Wall Solutions

As seen in Section 2, the tension of the BPS domain wall is given by superpotential. This fact implies that we should take $b \neq 0$ in $\mathcal{N}=1$ superfields. The $M=1$ case of $U(1)$ gauge symmetry reduces to $T^*\mathbf{C}P^{N-1}$ with the superpotential, which admits N parallel domain walls.[13] Moreover if we take $N=2$ and $M=1$, the target space $T^*\mathbf{C}P^1$ is the Eguchi-Hanson space with the superpotential $W = b\frac{\mu}{1+\varphi\psi}$ ($\mu = m_1$). The BPS eq. (4) in $\mathcal{N}=1$ superfields can be solved to give[2]

$$\varphi = \psi^* = e^{|\mu|(z-z_0)}e^{i\delta}, \tag{11}$$

where z_0 and δ are integral constants. They correspond to zero modes arising from spontaneously broken translational invariance perpendicular to domain wall configuration and $U(1)$ isometry σ_3 in the *internal* space.

6. Conclusions

We have constructed massive HK NLSM on the cotangent bundle over $G_{N,M}$ in $\mathcal{N}=1$ superfields, which is the massive extension of Lindström and Roček. This model contains ${}_NC_M = N!/M!(N-M)!$ discrete degenerate vacua. A BPS Domain wall solution in the simplest $T^*\mathbf{C}P^1$ has been given.

Constructing domain walls in non-Abelian gauge group remains as an interesting future work. Turning on the gauge coupling does not change vacua. BPS walls should also be similar as shown in the $M=1$ case.[13,14] Coupling to supergravity is possible as in the $M=1$ case.[15]

Acknowledgments

We would like to thank Masashi Naganuma for discussions in the early stage of this work. M. N. is grateful to the organizers in QTS3 and the

Argonne brane workshop. This work is supported in part by Grant-in-Aid for Scientific Research from the Japan Ministry of Education, Science and Culture 13640269 (NS). The work of M. N. was supported by the U. S. Department of Energy under grant DE-FG02-91ER40681 (Task B).

References

1. M. Arai, M. Nitta and N. Sakai, [hep-th/0307274].
2. M. Arai, M. Naganuma, M. Nitta and N. Sakai, *Nucl. Phys.* **B652** (2003) 35, [hep-th/0211103]; in Garden of Quanta - In honor of Hiroshi Ezawa, Eds. by J. Arafune et al. (World Scientific Publishing Co. Pte. Ltd. Singapore, 2003) pp 299-325, [hep-th/0302028].
3. N. Arkani-Hamed, S. Dimopoulos and G. Dvali, *Phys. Lett.* **429B** (1998) 263, [hep-ph/9803315]. I. Antoniadis, N. Arkani-Hamed, S. Dimopoulos and G. Dvali, *Phys. Lett.* **436B** (1998) 257, [hep-ph/9804398]. L. Randall and R. Sundrum, *Phys. Rev. Lett.* **83** (1999) 3370, [hep-ph/9905221]; *Phys. Rev. Lett.* **83** (1999) 4690, [hep-th/9906064].
4. L. Alvarez-Gaumé and D. Z. Freedman, *Comm. Math. Phys.* **80** (1981) 443.
5. L. Alvarez-Gaumé and D. Z. Freedman, *Comm. Math. Phys.* **91** (1983) 87.
6. U. Lindström and M. Roček, *Nucl. Phys.* **B222** (1983) 285.
7. N. J. Hitchin, A. Karlhede, U. Lindström and M. Roček, *Comm. Math. Phys.* **108** (1987) 535.
8. J. Wess and J. Bagger, "Supersymmetry and Supergravity (2nd Ed.)", (1991), Princeton University Press.
9. M. Naganuma, M. Nitta and N. Sakai, *Phys. Rev.* **D65** (2001) 045016, [hep-th/0108179]; in the Proceedings of the third international Sakharov conference on Physics Vol.2, (World Scientific Pub., 2003) pp.535-549, [hep-th/0210205].
10. J. Scherk and J. H. Schwarz, *Phys. Lett.* **B82** (1979) 60.
11. E. Abraham and P. K. Townsend, *Phys. Lett.* **B 291** (1992) 85.
12. J. P. Gauntlett, R. Portugues, D. Tong and P.K. Townsend, *Phys. Rev.* **D63** (2001) 085002, [hep-th/0008221]. J. P. Gauntlett, D. Tong and P.K. Townsend, *Phys. Rev.* **D63** (2001) 085001, [hep-th/0007124]; *Phys. Rev.* **D64** (2001) 025010, [hep-th/0012178]. M. Naganuma, M. Nitta and N. Sakai, *Grav. Cosmol.* **8** (2002) 129, [hep-th/0108133]. R. Portugues and P. K. Townsend, *JHEP* **0204** (2002) 039, [hep-th/0203181]. M. Arai, E. Ivanov and J. Niederle, to be published in *Nucl. Phys.* **B**, [hep-th/0312037].
13. D. Tong, *Phys. Rev.* **D66** (2002) 025013, [hep-th/0202012]. K. S. M. Lee, *Phys. Rev.* **D67** (2003) 045009, [hep-th/0211058].
14. K. Kakimoto and N. Sakai, *Phys. Rev.* **D68** (2003) 065005, [hep-th/0306077]. Y. Isozumi, K. Ohashi and N. Sakai, *JHEP* **0311** (2003) 061, [hep-th/0310130]; *JHEP* **0311** (2003) 060, [hep-th/0310189].
15. M. Arai, S. Fujita, M. Naganuma and N. Sakai, *Phys. Lett.* **B556** (2003) 192, [hep-th/0212175]; [hep-th/0311210]. M. Eto, S. Fujita, M. Naganuma and N. Sakai, to be published in *Phys. Rev.* **D**, [hep-th/0306198].

HIGHER DERIVATIVE TERMS IN N=2 SUSY EFFECTIVE ACTIONS*

PHILIP C. ARGYRES, GREGORY A. BRAUN, F. PAUL ESPOSITO
Physics Department, University of Cincinnati, Cincinnati, OH 45221-0011
argyres, braun, esposito@physics.uc.edu

ADEL AWAD
Department of Physics, Faculty of Science, Ain Shams University
Cairo 11566, Egypt
awad.adel@usa.net

Harmonic superspace can be used to construct higher derivative terms in $\mathcal{N} = 2$ supersymmetric effective actions despite the infinite redundancy in their description due to the infinite number of auxiliary fields. We are able to write down all of the 3- and 4-derivative terms on the Higgs, Coulomb, and mixed branches, modulo the possible existence of superspace Chern-Simons-like terms, which we discuss. Many of the terms we find are holomorphic, and at least one is shown to not receive quantum corrections.

1. Introduction

Higher derivative terms in $\mathcal{N} = 2$ supersymmetric effective actions are of interest because some of them satisfy non-renormalization theorems[1,2]. Among these higher derivative terms are the Wess-Zumino terms, which have been indirectly shown to exist in $\mathcal{N} = 2$ theories[3,4]. We have performed a systematic exploration of higher derivative terms at the 3- and 4-derivative levels[5], which we summarize here.

2. Effective Actions

Low energy effective actions give the approximate behavior of a system at low energies/long wavelengths. We take the energy scale to be lower than any other scale in the theory, so that only massless particles appear. Among

*Talk presented by G.A.B.

the bosons we are then left with only $U(1)$ vectors, A^μ, and neutral scalars, ϕ. Charged scalars get masses via the Higgs mechanism, and so need not be generically included.

At these low energies, we expand in the number of derivatives, since terms with fewer derivatives will be more relevant at long wavelengths. To do this expansion, we make use of a derivative dimension counting. The leading term in the expansion will be of the form $g(\phi)\partial_\mu\partial^\mu\phi$. Since we want this to be of definite dimension, the function g, and therefore ϕ, must have dimension zero. Spacetime derivatives have dimension one (since it is they we wish to count), and supersymmetry then demands that vectors are dimensionless while spinors and Grassmann measures $d\theta$ (and equivalently Grassmann derivatives) have dimension $\frac{1}{2}$. Note that this derivative dimension is not the scaling dimension often used; we introduce it here just to count derivatives.

3. $\mathcal{N} = 2$ Supersymmetry

These theories have two multiplets. The hypermultiplet, with superfield q^+, and the vector multiplet, with field strength superfield W. In a generic vacuum, the expectation values of the two neutral massless scalar components of q^+'s give the Higgs branches, while those of the (neutral) complex scalars in the vector multiplet give the Coulomb branches. We want to find possible higher dimension terms on each of these branches, as well as on the mixed branches, which contain non-zero vevs for both q^+ and W.

(Note that the field strength W of the Coulomb branch satisfies a constraint. The unconstrained fields are the vector potential superfields $V^{\pm\pm}$, but are gauge variant and have negative dimension and so are difficult to deal with. This is because at each derivative level an inifinte number of negative derivative dimension terms may appear. So, we assume for now that all terms can be written in terms of the gauge invariant field strength W. The possible exception to this, what we call superspace Chern-Simons-like terms, will be discussed later.)

A difficulty is that in any $\mathcal{N} \geq 2$ superspace we need an infinite number of auxilliary fields. This leads to an infinite number of terms at each derivative level. Since all the auxiliary fields must integrate out when substituted for using their equations of motion, this infinite number of terms is a unwanted redundancy.

This problem has a solution in the derivative expansion we are using. When we do this expansion, we are expanding aroung the solution to the 2-

derivative, or kinetic term (ignoring Fayet-Illiopoulos terms). Thus, to look at 3- or 4-derivative (or higher) terms we can use the 2-derivative equations of motion to eliminate the auxiliary terms in favor of propagating fields. Any corrections to this will be of higher derivative order, and therefore unimportant. This demonstrates that auxiliary fields remain auxiliary at higher derivative order, and even though they may appear with derivatives, they do not become propagating.

4. Harmonic Superspace

We make use of harmonic superspace[6] to organize the infinite number of fields in terms of unconstrained superfields. Harmonic superspace extends the usual superspace by an auxiliary space $SU(2)/U(1) \sim S^2$. This is done by adding auxiliary bosonic $SU(2)$ coordinates $u^\pm$ to the usual superspace, which already has eight fermionic coordinates: $\theta^\pm, \overline{\theta}^\pm$. The +/- gives the $U(1)$ charge, so to mod out the $U(1)$ group we look at expressions of definite +/- charge.

In addition, there are derivatives involving only the $u^\pm$: D^{++} and D^{--}. These derivatives and the $u^\pm$ all have derivative dimension zero. These $u^\pm$ variables can be thought of as living on a 2-sphere. They are unphysical, and must be integrated out so as not to appear in any physical quantity.

The fields can be written in terms of two different "halves" of the superspace, each satisfying a constraint. A chiral superfield satisfies the constraints $\overline{D}^\pm_\alpha W = 0$, and is a function of only x_C, θ^+, and θ^-:

$$W = W(x_C, \theta^+, \theta^-). \tag{1}$$

An analytic superfield satisfies the similar constraints $D^+_\alpha q^+ = \overline{D}^+_{\dot\alpha} q^+ = 0$, and is therefore a function of only x_A, θ^+, and $\overline{\theta}^+$:

$$q^+ = q^+(x_A, \theta^+, \overline{\theta}^+). \tag{2}$$

Supersymmetric actions can be made by integrating over one of these halves of superspace:

$$\int d^2\theta^+\, d^2\theta^-\; L(\theta^+, \theta^-), \qquad \text{or} \qquad \int d^2\theta^+\, d^2\overline{\theta}^+\; L(\theta^+, \overline{\theta}^+), \tag{3}$$

or, over the union of the two, three-fourths of superpace:

$$\int d^2\theta^+\, d^2\theta^-\, d^2\overline{\theta}^+\; L(\theta^+, \theta^-, \overline{\theta}^+), \tag{4}$$

or finally, over all of superspace:

$$\int d^8\theta \, L(\theta^+, \theta^-, \overline{\theta}^+, \overline{\theta}^-). \tag{5}$$

Note that we must also, of course, integrate over the unphysical du, as well as d^4x to get an action.

By looking at all possible terms of these forms, we are able to get all of the 3- and 4-derivative terms, modulo the possible superspace Chern-Simons-like terms, mentioned earlier and discussed later.

5. Results

On the Higgs branches we find only 4-derivative terms, written as an integral of a Grassmann analytic function $B(q^+; u^\pm, D^{++})$:

$$S_4^H = \int du \, d^4x \, d^2\theta^+ \, d^2\overline{\theta}^+ \; \partial^\mu q^+ \partial_\mu q^+ B(q^+; u^\pm, D^{++}). \tag{6}$$

Each of the $d\theta$ in the measure has derivative dimension $\frac{1}{2}$, and each space-time derivative has dimension one, and so the term as a whole is dimension four. Note that this term is holomorphic, in that it depends only on q^+, and not its conjugate q^-.

On the Coulomb branch, we also find only 4-derivative terms, in the form of an integral over a chiral function $\mathcal{G}(W)$:

$$S_4^C = \int (du) \, d^4x \, d^2\theta^+ \, d^2\theta^- \; \partial^\mu W \partial_\mu W \, \mathcal{G}(W). \tag{7}$$

Once again, this terms is holomorphic, since it depends only on W, not $\overline{W}$.

On the mixed branches we find a number of terms. First we have a 4-derivative term that is a simple integral over all of superspace:

$$S_{4c}^M = \int du \, d^4x \;\; d^8\theta \; H(q^+, q^-, W, \overline{W}; u^\pm, D^{\pm\pm}). \tag{8}$$

There is no restriction on H; it need not be holomorphic in any of its variables. Note that there are terms on both the Higgs and Coulomb branches which are integrals over all of superspace, but we consider them to be special case of S_{4c}^M.

We also find 4-derivative terms that are written as an integral over three-quarters of harmonic superspace. Each of these has two spinor derivatives, as well as six $d\theta$'s in the Grassmann measure, for a total derivative dimension of four.

$$S^M_{4a} = \int du\, d^4x\, d^2\theta^+ d^2\overline{\theta}^+ d^2\theta^- D^+W_a \cdot D^+W_b F^{ab}(q^+, W; u^\pm, D^{++}), \quad (9)$$

$$S^M_{4b} = \int du\, d^4x\, d^2\theta^+ d^2\overline{\theta}^+ d^2\theta^- \overline{D}^-(D^{++})^n q^+_I \cdot \overline{D}^-(D^{++})^m q^+_J$$
$$G^{IJ}_{nm}(q^+, W; u^\pm, D^{++}). \quad (10)$$

Notice that these functions F and G are holomorphic.

Finally, we also find a 3-derivative term,

$$S^M_3 = \int du\, d^4x\, d^2\theta^+\, d^2\overline{\theta}^+\, d^2\theta^-\; G(q^+, W; u^\pm, D^{++}). \quad (11)$$

This term is important, for it is the leading contribution (after the kinetic term) to the low energy physics.

These terms may still seem plagued with the infinite redundancy due to the infinite number of possible D^{++} and D^{--} derivatives acting on the q^+'s. This is not, however, the case. Recall that the second order (2-derivative) equations of motion can be used to eliminate the auxiliary fields at higher derivative order. It can be shown[5] that these second order equations of motion can reduced the number of $D^{\pm\pm}$ derivatives to a finite number, thus making the expressions local in the $u^\pm$'s. This number of derivatives that it is necessary to include depends on the specific 2-derivative term in the theory, but is always finite. For example, if the 2-derivative term for the q^+ describes a free hypermultiplet, then only five (plus five more complex conjugate) combinations of $D^{\pm\pm}$ and q^+ appear after auxiliary fields are solved for.

As mentioned, all of these terms except S^M_{4c} are holomorphic functions of their fields. This limits the types of quantum corrections that may appear. In $\mathcal{N} = 2$ supersymmetric QCD the strong coupling scale Λ can be thought of as the lowest component of a field strength superfield W. Thus terms of the form S^H_4 can receive no Λ-dependent corrections, since W cannot appear; any such correction would break supersymmetry. The other holomorphic terms can only get corrections holomorphic in W: one loop and instanton corrections.

It can be shown[5] that the Wess-Zumino terms are in the S^M_{4c}-type mixed branch terms. The other terms cannot have four antisymmetric spacetime derivatives.

6. Superspace Chern-Simons-like Terms

We must also consider the possibility of terms on the Coulomb branch that cannot be written only in terms of the field strength superfield W, but require the use of the potential superfields $V^{\pm\pm}$. We refer to these as superspace Chern-Simons-like terms in analogy with regular Chern-Simons terms—gauge invariant terms which cannot be written using only the gauge invaraint field strength $F^{\mu\nu}$, but require also the gauge variant potential A^{μ}. (Though Chern-Simons terms only exist in an odd number of spacetime dimensions, there is no such restriction for their supersymmetric cousins; in particular, superspace Chern-Simons-like terms do *not* have regular Chern-Simons terms in their component expansions.)

Superspace Chern-Simons-like terms are known to exist in $\mathcal{N} = 3$ supersymmetric Yang-Mills in four dimensions[6]. Their existence is still an open question in $\mathcal{N} = 2$ and $\mathcal{N} = 1$ superspaces.

7. Conclusions

We have shown that harmonic superspace can be used to make a systematic derivative expansion in $\mathcal{N} = 2$ supersymmetric theories. This expansion uses equations of motion from the 2-derivative term to eliminate auxiliary fields at higher derivative order. We have found all terms at the 3- and 4-derivative level, and have located the Wess-Zumino term. This classification of terms is modulo the possible presence of superspace Chern-Simons-like terms, the existence of which we are currently investigating.

Acknowledgments

This work was supported in part by DOE grant DOE-FG02-84ER-40153. G.A.B. was supported in part by a University of Cincinnati Research Council's Summer Fellowship.

References

1. M. Dine and N. Seiberg, *Comments on higher derivative operators in some SUSY field theories, Nucl. Phys.* **B 409** [hep-th/9705057].
2. M. Dine and J. Gray, *Non-renormalization theorems for operators with arbitrary numbers of derivatives in N=4 Yang-Mills theory, J. High Energy Phys.* [hep-th/9909020].
3. A.A. Tseytlin and K. Zarembo, *Magnetic interactions of D-branes and Wess-Zumino terms in superYang-Mills effective actions, Phys. Lett.* **B 474** [hep-th/9911246].
4. K.A. Intriligator, *Anomaly matching and a Hopf-Wess-Zumino term in 6d N=(2,0) field theories, Nucl. Phys.* **B 581** [hep-th/0001205].
5. P.C. Argyres, A. Awad, G. Braun and F.P. Esposito, *Higher Derivative Terms in N=2 Supersymmetric Effective Actions*, [hep-th/0306118].

6. A.S. Galperin, E.A. Ivanov, V.I. Ogievetsky, and E.S. Sokatchev, *Harmonic Superspace*, Cambridge University Press, Cambridge 2001.

A SHORT INTRODUCTION TO THE FATE OF THE α-VACUUM

HAEL COLLINS *

Department of Physics
Carnegie Mellon University
Pittsburgh, PA 15213, USA
E-mail: hcollins@andrew.cmu.edu

A free scalar field propagating in de Sitter space has a one parameter family of invariant states called the α-vacua. In an interacting theory, all except a unique state, the Bunch-Davies vacuum, produce non-renormalizable divergences. This talk provides a brief introduction to the origin and the form of these divergences.

A striking difference between de Sitter and flat space-time is the richer vacuum structure of the former. For a free, massive scalar field in Minkowski space, there exists a unique Poincaré invariant vacuum state. Flat space also has a global time-like Killing vector, which allows us to define a Hamiltonian with respect to which the vacuum is the the lowest energy eigenstate. In contrast, de Sitter space lacks any such global time-like Killing vector and so no globally conserved energy can be defined. It is still possible to find states which are invariant under the connected part of the isometry group of de Sitter space, $SO(4,1)$, but there no longer is only a single such invariant state. Instead, for a free scalar field there exists an infinite family of invariant states[1] called the α-vacua, since they can be distinguished by a single complex number α. The α-vacua have been used to explore in a wide set of problems associated with de Sitter space, such as the transplanckian problem of inflation[2] and the dS/CFT correspondence[3].

This talk describes a pathology that arises for an interacting scalar field theory in an α-vacuum.[4,5] Its origin lies in the structure of the Green's function chosen to describe propagation in an α-vacuum. In a generic loop correction, the interference among different propagators produces terms in

*This work was done in collaboration with Rich Holman and Matthew Martin and with support by the Department of Energy (DE-FG03-91-ER40682).

which no oscillatory suppression damps the high momentum region of the loop integration. Loops which contain two or three propagators then lead to linear and logarithmic divergences respectively which cannot be removed by a counterterm of the form of those already present in the theory. Only a unique choice of these vacua—known alternately as the Euclidean or Bunch-Davies vacuum[6]—evades this class of divergences and yields completely renormalizable interacting theories in de Sitter space.

We begin with a free scalar field $\Phi(x)$ propagating in a classical de Sitter background coordinatized by

$$ds^2 = \frac{d\eta^2 - d\vec{x}^2}{\eta^2}, \qquad \eta \in [-\infty, 0]. \tag{1}$$

We can define a vacuum state $|E\rangle$ to be the state annihilated by the operators $a^E_{\vec{k}}$ obtained by expanding the field in terms of its eigenmodes,

$$\Phi(\eta, \vec{x}) = \int \frac{d^3\vec{k}}{(2\pi)^3} \left[U^E_k(\eta) e^{i\vec{k}\cdot\vec{x}}\, a^E_{\vec{k}} + U^{E*}_k(\eta) e^{-i\vec{k}\cdot\vec{x}}\, a^{E\dagger}_{\vec{k}} \right]. \tag{2}$$

The mode functions $U^E_k(\eta)$ are the solutions to the second order Klein-Gordon equation. One of the constants of integration is fixed by the requirement that $\Phi(x)$ and its conjugate momentum $\Pi(x) = \eta^{-2}\partial_\eta\Phi(x)$ satisfy the canonical commutation relation. In flat space, the second constant of integration can be fixed unambiguously since the space-time has a global time-like Killing vector with respect to which we define positive and negative frequencies. At short distances with respect to the natural curvature length of de Sitter space, the geometry also appears to be flat and we can thus use a prescription in which the mode functions are those which match onto the flat space modes with positive frequencies at short distances. This prescription defines the Euclidean. or Bunch-Davies,[6] vacuum,

$$U^E_k(\eta) = \frac{\sqrt{\pi}}{2}\eta^{3/2} H^{(2)}_\nu(k\eta), \qquad \nu = \left[\tfrac{9}{4} - m^2\right]^{1/2} \tag{3}$$

with $k \equiv |\vec{k}|$. m is the effective mass of $\Phi(x)$ in units where the Hubble constant is unity. The Wightman function for this vacuum, $G_E(x, x') = \langle E|\Phi(x)\Phi(x')|E\rangle$, only depends on the de Sitter invariant distance between x and x'.

In de Sitter space, the statement as to which are the positive and negative frequency modes cannot be made globally so the Euclidean vacuum is not the unique invariant state. Mottola and Allen, as well as earlier authors, demonstrated[1] that the state $|\alpha\rangle$ annihilated by the Bogolubov transform,

$$a_{\vec{k}}^{\alpha} = N_\alpha \left[a_{\vec{k}}^{E} - e^{\alpha^*} a_{-\vec{k}}^{E\dagger}\right], \qquad N_\alpha = \left[1 - e^{\alpha+\alpha^*}\right]^{-1/2}, \tag{4}$$

with $\mathrm{Re}\,\alpha < 0$, also yields an invariant Wightman function,

$$G_\alpha(x,x') = \langle\alpha|\Phi(x)\Phi(x')|\alpha\rangle = \int \frac{d^3\vec{k}}{(2\pi)^3}\, e^{i\vec{k}\cdot(\vec{x}-\vec{x}')}\, G_k^{\alpha}(\eta,\eta'). \tag{5}$$

Equation (4) induces a transformation of the mode functions, selecting a different linear combination of the solutions to the Klein-Gordon equation,

$$U_k^{\alpha}(\eta) = N_\alpha \frac{\sqrt{\pi}}{2}\eta^{3/2}\left[H_\nu^{(2)}(k\eta) + e^{\alpha} H_\nu^{(1)}(k\eta)\right]. \tag{6}$$

These are the mode functions associated with an expansion of $\Phi(x)$ with respect to the α-vacuum, $|\alpha\rangle$. Note that the Euclidean vacuum is itself an α-vacuum, corresponding to the limiting case where $\alpha \to -\infty$.

The construction of the propagator in an α-vacuum is more subtle. The most straightforward definition of the Feynman propagator,

$$G_F(x,x') \equiv i\langle\alpha|T(\Phi(x)\Phi(x'))|\alpha\rangle = \int \frac{d^3\vec{x}}{(2\pi)^3}\, e^{i\vec{k}\cdot(\vec{x}-\vec{x}')} G_k^F(\eta,\eta'), \tag{7}$$

is to define it to be the Green's function associated with a single point source,

$$\left[\nabla_x^2 + m^2\right] G_F(x,x') = \frac{\delta^4(x-x')}{\sqrt{-g(x)}}. \tag{8}$$

In terms of the Wightman functions, the propagator is given by

$$G_k^F(\eta,\eta') = \Theta(\eta-\eta')\, G_k^{>}(\eta,\eta') + \Theta(\eta'-\eta)\, G_k^{<}(\eta,\eta'), \tag{9}$$

with $G_k^{>}(\eta,\eta') = G_k^{<}(\eta',\eta) = i\, G_k^{\alpha}(\eta,\eta')$. For a general value of $\alpha \neq -\infty$, this propagator produces divergences in the perturbative corrections of an interacting theory which cannot be canceled by a simple set of countert-erms.[a]

From the perspective of the Euclidean vacuum, the transformation defining the α-vacuum in Eq. (4) introduces excitations to arbitrarily high momenta which, unlike a thermal distribution, are not damped as $k \to \infty$. This modification of the theory for high momenta leads to a new set of

[a]These divergences could indicate a fundamental pathology of the general α-vacuum, but they may also signal that we should apply a different construction for the propagator. In particular, adding a source at the antipode of x in Eq. (8) produces a theory without any non-renormalizable divergences.[7]

divergences for an interacting theory in an α-vacuum which appear in loop corrections. At arbitrarily short distances the positive and negative frequency modes interfere so as to cancel the rapidly oscillating phases among some of the terms within the loop integral. Without such a cancellation, these phases would have damped the contribution from the high momentum region through an appropriate $i\epsilon$ prescription. This interference of phases is a specific feature of the propagator in the α-vacuum and does not occur in the Euclidean case.

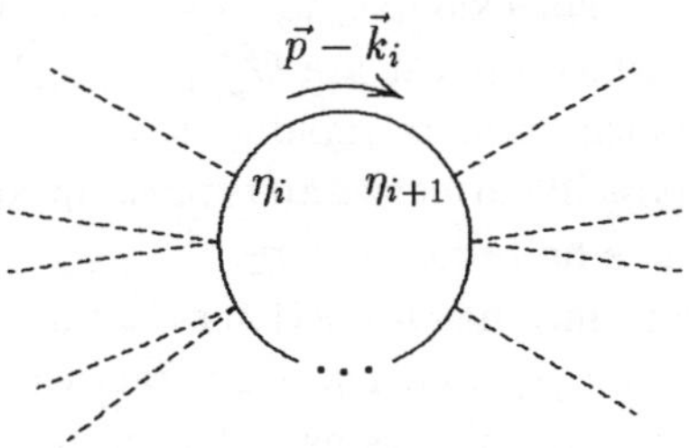

Figure 1. A general loop graph in an interacting scalar theory.

To understand the origin of these divergences, consider a general one loop diagram as shown in Figure 1. The loop contains n vertices connected by n propagators through which flow a common loop momentum $\vec{p}$. The external legs are shown as dashed lines and contain the appropriate number of lines for the types of interactions included in the theory. If the momentum leaving the loop at the ith vertex is $\vec{k}_i - \vec{k}_{i-1}$, then the momentum flowing through the propagator connecting the ith and the $(i+1)^{\text{th}}$ vertices is $\vec{p}-\vec{k}_i$ as shown. Expanding the Hankel functions in Eq. (6) in the UV limit, the leading p-dependence of the Wightman functions associated with this propagator is

$$\begin{aligned} G^{>}_{p-k_i}(\eta_i,\eta_{i+1}) & \qquad\qquad (10) \\ \to iN_\alpha^2 \frac{\eta_i\eta_{i+1}}{2|\vec{p}-\vec{k}_i|} \Big[& e^{-i|\vec{p}-\vec{k}_i|(\eta_i-\eta_{i+1})} + e^{\alpha+\alpha^*} e^{i|\vec{p}-\vec{k}_i|(\eta_i-\eta_{i+1})} \\ & -ie^{\alpha}e^{-i\pi\nu}e^{i|\vec{p}-\vec{k}_i|(\eta_i+\eta_{i+1})} + ie^{\alpha^*}e^{i\pi\nu}e^{-i|\vec{p}-\vec{k}_i|(\eta_i+\eta_{i+1})}\Big]. \end{aligned}$$

The prefactor determines the naïve UV behavior of the loop integral; since each propagator scales as $1/p$ for $p \gg k_i$, the integral of a loop containing

n propagators scales in the UV as

$$\int^{\Lambda} d^3\vec{p} \prod_{i=1}^{n} G^{\alpha}_{p-k_i}(\eta_i, \eta_{i+1}) \sim \int^{\Lambda} \frac{dp}{p^{n-2}}. \tag{11}$$

We only encounter a possible divergence if $n \leq 3$. Note that the $n = 1$ case can be removed by a counterterm since in this case the loop is independent of any of the external momenta.

In the Euclidean case, the Θ-functions in the propagators prevent the p-dependence of the phases from ever canceling in the UV in the product of all the propagators.[8] For example, for $n = 2$ the only products of loop Wightman functions which appear are $G^{>}_{p-k_1}(\eta_1, \eta_2)G^{<}_{p-k_2}(\eta_2, \eta_1)$ and its complex conjugate which oscillate rapidly as $p \to \infty$. For the Euclidean vacuum, it is always possible to choose an $i\epsilon$ prescription so that the integral over three-momentum is finite for $\eta_1 \neq \eta_2$.

For a generic α-vacuum, the Θ-function structure is identical; but since the phases depending on $p\eta_i$ and $p\eta_{i+1}$ appear with all possible relative signs in Eq. (10), there will always exist terms in the product of n loop propagators in which the p-dependent part of the phase completely cancels. In the $n = 2$ example, the product

$$\begin{aligned} &G^{>}_{p-k_1}(\eta_1, \eta_2)G^{<}_{p-k_2}(\eta_2, \eta_1) \\ &\quad \to e^{\alpha+\alpha^*} N_\alpha^4 \frac{(\eta_1\eta_2)^2}{4|\vec{p}-\vec{k}_1||\vec{p}-\vec{k}_2|} \\ &\qquad \times \left[e^{-i[|\vec{p}-\vec{k}_1|-|\vec{p}-\vec{k}_2|](\eta_1-\eta_2)} + e^{i[|\vec{p}-\vec{k}_1|-|\vec{p}-\vec{k}_2|](\eta_1+\eta_2)} + \text{c.c.} + \cdots\right] \end{aligned} \tag{12}$$

contains terms whose p-dependent phases cancel for high loop momenta, $p \gg k_1, k_2$. The loop integrand for these terms has no oscillatory suppression so that, when combined with the $p^2\,dp$ from the measure, the loop integral is linearly divergent. The explicit $e^{\alpha+\alpha^*}$ prefactor shows why these divergences did not arise in the Euclidean $(\alpha \to -\infty)$ vacuum. Similarly, in the $n = 3$ case we encounter a logarithmic divergence.

The dependence of the divergent part of the loop on the external momenta $\{k_1, \ldots k_n\}$ flowing into the loop is not the same as that of a counterterm that would generate a graph with the same external structure as Figure 1 but with the loop shrunk to a point.[4] It is in this sense that interacting theories in the generic α-vacuum are non-renormalizable.

This result is shown in full detail for the example of the number density of Euclidean particles in the α-vacuum for a theory with a Φ^3 interaction in [4]. There we show that the expectation value of this number density produces a non-renormalizable linear divergence in the one loop correction.

The source of the divergences for an interacting theory in an α-vacuum lies in the asymmetric treatment of the Wightman function and the Feynman propagator. The Bogolubov transformation that produces the α-vacuum can be regarded as a globally non-local transformation that defines the α mode functions to be a linear superposition of Euclidean mode functions evaluated at a point and at its antipode. This non-locality is inherited by the Wightman function. The time-ordering used to construct the Feynman propagator, indicated by the Θ-functions in Eq. (9), does not contain any of this non-locality and is identical in the Euclidean and the general α cases. This inconsistency suggests that the time-ordering should also be modified and that a renormalizable interacting field theory only results when the construction of the propagator consistently treats the global structure of de Sitter space.[7]

References

1. E. Mottola, Phys. Rev. D **31**, 754 (1985); B. Allen, Phys. Rev. D **32**, 3136 (1985); N. A. Chernikov and E. A. Tagirov, Annales Poincare Phys. Theor. A **9**, 109 (1968); E. A. Tagirov, Annals Phys. **76**, 561 (1973).
2. U. H. Danielsson, Phys. Rev. D **66**, 023511 (2002); A. Kempf, Phys. Rev. D **63**, 083514 (2001); A. Kempf and J. C. Niemeyer, Phys. Rev. D **64**, 103501 (2001); R. Easther, B. R. Greene, W. H. Kinney and G. Shiu, Phys. Rev. D **64**, 103502 (2001), R. Easther, B. R. Greene, W. H. Kinney and G. Shiu, Phys. Rev. D **67**, 063508 (2003), R. Easther, B. R. Greene, W. H. Kinney and G. Shiu, Phys. Rev. D **66**, 023518 (2002); N. Kaloper, M. Kleban, A. Lawrence, S. Shenker and L. Susskind, JHEP **0211**, 037 (2002); S. Shankaranarayanan, Class. Quant. Grav. **20**, 75 (2003); J. Martin and R. Brandenberger, Phys. Rev. D **68**, 063513 (2003); C. P. Burgess, J. M. Cline, F. Lemieux and R. Holman, JHEP **0302**, 048 (2003); C. P. Burgess, J. M. Cline and R. Holman, JCAP **0310**, 004 (2003); H. Collins and M. R. Martin, hep-ph/0309265.
3. R. Bousso, A. Maloney and A. Strominger, Phys. Rev. D **65**, 104039 (2002).
4. H. Collins, R. Holman and M. R. Martin, hep-th/0306028.
5. M. B. Einhorn and F. Larsen, Phys. Rev. D **67**, 024001 (2003); T. Banks and L. Mannelli, Phys. Rev. D **67**, 065009 (2003).
6. T. S. Bunch and P. C. Davies, Proc. Roy. Soc. Lond. A **360**, 117 (1978).
7. K. Goldstein and D. A. Lowe, hep-th/0308135; M. B. Einhorn and F. Larsen, Phys. Rev. D **68**, 064002 (2003); H. Collins and R. Holman, hep-th/0312143.
8. J. S. Schwinger, J. Math. Phys. **2**, 407 (1961); L. V. Keldysh, Zh. Eksp. Teor. Fiz. **47**, 1515 (1964) [Sov. Phys. JETP **20**, 1018 (1965)].

YANGIAN SYMMETRY IN D=4 SUPERCONFORMAL YANG-MILLS THEORY*

LOUISE DOLAN†
Department of Physics
University of North Carolina, Chapel Hill, NC 27599-3255

CHIARA R. NAPPI‡
Department of Physics, Jadwin Hall
Princeton University, Princeton, NJ 08540

EDWARD WITTEN§
School of Natural Sciences, Institute for Advanced Study
Olden Lane, Princeton, NJ 08540

We will discuss an integrable structure for weakly coupled superconformal Yang-Mills theories, describe certain equivalences for the Yangian algebra, and fill a technical gap in our previous study of this subject.

1. Symmetry Generators and Anomalous Dimensions

In [1] it was shown that the classical Green-Schwarz superstring action for $AdS^5 \times S^5$ possesses a hierarchy of non-local symmetries of the type that exist in integrable field theories [2, 3]. This is due to the fact that the Green-Schwarz superstring in $AdS^5 \times S^5$ can be interpreted as a coset theory where the fields take values in the coset superspace

$$\frac{PSU(2,2|4)}{SO(4,1) \times SO(5)} . \tag{1}$$

This coset theory (even though the target is not a symmetric space [4]) admits non-local currents which give rise to charges satisfying a Yangian

*Talks presented by L.D., C.R.N., and E.W.
†Work partially supported by DOE grant DE-FG02-03ER-41262.
‡This work is supported by NSFGrants PHY-0140311 and PHY-0243680.
§This work is supported by NSF Grant PHY-0070928.

algebra. A Yangian algebra $Y(G)$ is an associative Hopf algebra [5, 6, 2, 7] generated by the elements J^A and Q^B with

$$[J^A, J^B] = f^{AB}_C J^C\,, \qquad [J^A, Q^B] = f^{AB}_C Q^C\,, \tag{2}$$

and the Serre relations

$$\begin{aligned}
&[[Q^A,[Q^B,J^C]] + [Q^B,[Q^C,J^A]] + [Q^C,[Q^A,J^B]] \\
&= \frac{1}{24} f^{ADK} f^{BEL} f^{CFM} f_{KLM} \{J_D, J_E, J_F\}\,,
\end{aligned} \tag{3}$$

$$\begin{aligned}
&[Q^A,Q^B],[J^C,Q^D]] + [[Q^C,Q^D],[J^A,Q^B]]\,, \\
&= \frac{1}{24}(f^{AGL} f^{BEM} f^{KFN} f_{LMN} f^{CD}_K \\
&\qquad + f^{CGL} f^{DEM} f^{KFN} f_{LMN} f^{AB}_K)\{J_G, J_E, J_F\}\,,
\end{aligned} \tag{4}$$

for J^A taking values in the Lie algebra of an arbitrary semi-simple Lie group G. (Lie algebra indices A, B, C are raised and lowered with an invariant, nondegenerate metric tensor g_{AB} or g^{AB}.) The symbol $\{A, B, C\}$ denotes the symmetrized product of three operators A, B, and C. Under repeated commutators, the Q^A generate an infinite-dimensional symmetry algebra that has been called the Yangian. The Yangian has a basis $\mathcal{J}^A_n$ where $\mathcal{J}^A_0 = J^A$, $\mathcal{J}^A_1 = Q^A$, and $\mathcal{J}^A_n$ is an n-local operator that arises in the $(n-1)$-form commutator of Q's. Since we will work in this paper mainly with the generators J^A and Q^A, we have given them those special names. The Yangian relations as written above are redundant in the following sense. For $SU(2)$ the relation (3) is trivial. For other cases such as $SU(N)$ with $N \geq 3$, the relation (3) implies the following one (4).

In the superstring on $AdS^5 \times S^5$, the J^A will be generators of $G = PSU(2,2|4)$, the brackets will generalize to denote either commutators or anticommutators, and f^C_{AB} become the structure constants of $PSU(2,2|4)$. The Q^A are the new non-local charges whose existence was found in [1]. If the AdS/CFT correspondence is correct, then on the CFT side we must have the same Yangian symmetry. The question arises of what could be the Q^A charges in the super Yang-Mills side. We will answer this question in the extreme weak coupling limit, that is, the opposite limit from that which is considered in [1]. In order to justify our guess for Q^A, we will review how non-local symmetries arise in two-dimensional sigma models.

Let us consider a model with a group G of symmetries; the Lie algebra of G has generators T_A obeying $[T_A, T_B] = f^C_{AB} T_C$. The action of G is generated by a current $j^{\mu\,A}$ that is conserved, $\partial_\mu j^{\mu\,A} = 0$. Non-local charges arise if, in addition, the Lie algebra valued current $j_\mu = \sum_A j^A_\mu T_A$ can be interpreted as a flat connection,

$$\partial_\mu j_\nu - \partial_\nu j_\mu + [j_\mu, j_\nu] = 0. \tag{5}$$

(Indices of j_μ are raised and lowered using the Lorentz metric in two dimensions.) The conservation of j_μ leads in the usual fashion to the existence of conserved charges that generate the action of G:

$$J^A = \int_{-\infty}^{\infty} dx\, j^{0\,A}(x,t). \tag{6}$$

In addition, a short computation using (5) reveals that

$$Q^A = f^A_{BC} \int_{-\infty}^{\infty} dx \int_x^{\infty} dy\, j^{0\,B}(x,t)\, j^{0\,C}(y,t) - 2\int_{-\infty}^{\infty} dx\, j_1^A(x,t) \tag{7}$$

is also conserved. The charges J^A and Q^A generate a Yangian algebra [2, 3], even though the infinitesimal transformations generated by them generate half of a Kac-Moody algebra [8].

There are also discrete spin systems, that is systems in which the dynamical variables live on a one-dimensional lattice rather than on the real line, that similarly have Yangian symmetry. The lattice definition of J^A is clear. We assume that the spins at each site i have G symmetry, and transform in some representation $\mathcal{R}$. We let J_i^A be the symmetry operators at the ith site. The total charge generator for the whole system is then

$$J^A = \sum_i J_i^A. \tag{8}$$

What about Q^A? For general G and $\mathcal{R}$, there is no satisfactory definition of Q^A. However, for $G = SU(N)$, a Q^A can be defined for any $\mathcal{R}$. For certain representations, the requisite formula for Q^A is particularly simple. One just takes the obvious discretization of the bilocal part of (7):

$$Q^A = f^A_{BC} \sum_{i<j} J_i^B J_j^C. \tag{9}$$

This is the right formula in many of the most commonly studied lattice integrable systems.

In (9), one has made no attempt to discretize the second term in (7). In fact, for many choices of $\mathcal{R}$, discretizing that second term is impossible for elementary reasons. One would expect a discretization of $\int dx\, j_1^A$ to be of the form $\sum_i j_i^A$ where j_i^A acts on the ith site and transforms in the adjoint representation. If $\mathcal{R}$ is such that the adjoint representation of G appears only once in the decomposition of $\mathcal{R} \otimes \overline{\mathcal{R}}$, then j_i^A would have to be a multiple of J_i^A. This is so, for example, if $\mathcal{R}$ is the fundamental representation of $SU(N)$ or $SU(N|M)$ (and more generally if $\mathcal{R}$ is the representation of kth rank antisymmetric tensors, for any k). But taking j_i^A to be a multiple of J_i^A just adds to Q^A a multiple of J^A; this is an outer automorphism of the Yangian algebra, and so does not help (or hurt) in obeying the Serre relations. We will argue that (9) is the correct formula for the Q^A in the weak coupling limit of Yang-Mills theory. This will fill a technical gap in our previous paper [9], where we showed that Q^A as defined by this formula commutes with the one-loop anomalous dimension operator, leading to an infinity of conservation laws, but we did not show that it obeys the Serre relations, which is needed to ensure that the resulting integrable structure is the standard Yangian algebra.

Another justification for the proposal (9) is that it has an analog in gauge field theory at $g^2N = 0$. In order to make contact with conventional Noether current symmetry analysis, we give the expression for the non-local charge (9) in terms of the elementary fields of the super Yang-Mills Lagrangian

$$\mathcal{L} = \frac{1}{g_{YM}^2}\,\mathrm{Tr}\left(\frac{1}{2}F_{\mu\nu}F^{\mu\nu} + D_\mu\phi^I D^\mu\phi^I - \frac{1}{2}[\phi^I,\phi^J][\phi^I,\phi^J] + \text{fermions}\right). \tag{10}$$

For simplicity, we will only consider $A \in so(2,4)$. In the classical theory, the symmetry currents for the conformal group are given in terms of the improved energy-momentum tensor by

$$j^{A\mu}(x) = \kappa_\nu^A \theta^{\mu\nu}(x)\,, \tag{11}$$

where κ_μ^A are the conformal Killing vectors, and

$$\begin{aligned}\theta^{\mu\nu} &= 2\mathrm{Tr}F^{\mu\rho}F^{\nu}_{\ \rho} + 2\mathrm{Tr}D^\mu\phi^I D^\nu\phi^I - g^{\mu\nu}\mathcal{L} - \frac{1}{3}\mathrm{Tr}(D^\mu D^\nu - g^{\mu\nu}D_\rho D^\rho)\phi^I\phi^I\\ &\quad + \text{fermions}\,.\end{aligned} \tag{12}$$

The currents (11) are conserved at any g^2N using the classical interacting equations of motion. If we set $g^2N = 0$, we note that the *untraced* matrix

$$\begin{aligned}(\theta^{\mu\nu})_i^{\ j} &= F^{\mu\rho}F^{\nu}_{\ \rho} + F^{\nu\rho}F^{\mu}_{\ \rho} + \partial^\mu\phi^I\partial^\nu\phi^I\\ &\quad + \partial^\nu\phi^I\partial^\mu\phi^I - g^{\mu\nu}(\frac{1}{2}F_{\rho\sigma}F^{\rho\sigma} + \partial_\mu\phi^I\partial^\mu\phi^I)\\ &\quad - \frac{1}{3}(\partial^\mu\partial^\nu - g^{\mu\nu}\partial_\rho\partial^\rho)\phi^I\phi^I + \text{fermions}\end{aligned} \tag{13}$$

is also conserved, as is $\kappa_\nu^A(\theta^{\mu\nu})_i^{\ j}$. Here i,j are the matrix labels of the gauge group generators $(T^A)_i^{\ j}$. It follows that we can construct non-local conserved charges by

$$Q_0^{AB\dots} = \int_M \kappa_\nu^A (\theta^{0\nu})_i^j \int_M \kappa_\rho^B (\theta^{0\rho})_j^k \dots , \tag{14}$$

where M is an initial value surface in spacetime. In free field theory, this acts on a chain of partons rather as (9) does, but we have no idea how to extend the definition to $g^2N \neq 0$.

Inspired by the above examples, our basic assumption in [9] is that in $\mathcal{N} = 4$ super Yang-Mills theory at $g^2N = 0$, with J_i^A understood as the $PSU(2,2|4)$ generators of the ith parton, (9) is the correct formula for the Yangian generators Q^A. Our assumption, in other words, is that the bilocal symmetry deduced from [1] goes over to (9) for $g^2N \to 0$. Of course, in any case (8) is the appropriate free field formula for the J^A, so we do not need to state any hypothesis for these generators. And no further assumption is needed for the higher charges in the Yangian; they are generated by repeated commutators of the Q^A. So our hypothesis about Q^A completely determines the form of the Yangian generators in the free-field limit.

Though our goal in these notes is to fill the above-mentioned technical gap in the previous analysis, for completeness we here sketch some of the reasoning in our previous paper. (We return in the next section to the question of why the simple bilocal formula does give a representation of the Yangian.) Having made an ansatz for how the Yangian algebra is realized at $g^2N = 0$, we consider what happens when g^2N is not quite zero. Some generators of the Yangian do not receive quantum corrections. For example, the spatial translation symmetries and the Lorentz generators are uncorrected, because the theory can be regularized in a way that preserves them. But the dilatation operator D – the generator of scale transformations – certainly is corrected. The corrections to the eigenvalues of D are called anomalous dimensions.

We assume, in view of [1], that the $\mathcal{N} = 4$ Yang-Mills theory in the planar limit does have Yangian symmetry for all g^2N. If so, the corrections modify the form of the generators, but preserve the commutation relations. One of the commutation relations says that Q^A transforms in the adjoint representation of the global group $PSU(2,2|4)$ generated by J^A: $[J^A, Q^B] = f^{ABC}Q^C$. We will write J^A and Q^A for the charges at $g^2N = 0$, and δJ^A and δQ^A for the corrections to them of order g^2N. We write $\tilde{J}^A$ and $\tilde{Q}^A$ for the exact generators (which depend on g^2N), so $\tilde{J}^A = J^A + (g^2N)\delta J^A + \mathcal{O}((g^2N)^2)$, and likewise for $\tilde{Q}^A$. To preserve the commutation relations, we have

$$[\delta J^A, Q^B] + [J^A, \delta Q^B] = f^{ABC}\delta Q^C. \tag{15}$$

We are now going to make an argument for the Yangian that parallels one used in [10] for the $PSU(2,2|4)$ generators. We consider the special case of this relation in which A is chosen so that J^A is the dilatation operator D. We also pick a basis Q^B of the Q's to diagonalize the action of D, so the $PSU(2,2|4)$ algebra reads in part $[D, Q^B] = \lambda^B Q^B$, where λ^B is the bare conformal dimension of Q^B. Then (15) gives us

$$[\delta D, Q^B] + [D, \delta Q^B] = \lambda^B \delta Q^B. \tag{16}$$

However, in perturbation theory, operators only mix with other operators of the same classical dimension. So just as $[D, Q^B] = \lambda^B Q^B$, we have $[D, \delta Q^B] = \lambda^B \delta Q^B$. Combining this with (15), we have therefore

$$[\delta D, Q^B] = 0. \tag{17}$$

Precisely the same argument was used in [10] to show that $[\delta D, J^A] = 0$; this was a step in determining δD. Combining this with (17), we see that δD must commute with the $g^2N = 0$ limit of the whole Yangian.

The structure of perturbation theory implies in addition that the operator δD is a sum of operators local along the chain; this fact has been exploited in [11] and many subsequent papers. (In fact, δD, as described explicitly in [10], is a sum of operators that act on nearest neighbor pairs.) The operators of this type that commute with the Yangian – where here we mean the Yangian representation most commonly studied in lattice integrable models, which for us is the one generated at $g^2N = 0$ by J^A and Q^A – are called the Hamiltonians of the integrable spin chain. Thus, from our assumption about the free-field limit of the Yangian, we are able, starting with the nonlocal symmetries found in [1], to deduce the basic conclusion of [12], found earlier in a special case in [13], that δD is a Hamiltonian of an integrable spin chain.

In our paper [9], we verify this picture by proving directly, using formulas for the one loop operator computed in [10], [12], that it is true that δD commutes with the Yangian. Since its commutativity with J^A was already used in [10] to compute δD, the only novelty is to verify that $[\delta D, Q^A] = 0$.

From what we have said, it is clear that the appearance of a Hamiltonian that commutes with the Yangian depends on expanding to first order near $g^2N = 0$. In the exact theory, at a nonzero value of g^2N, one would simply say that the exact dilatation operator $\mathcal{D}$, which of course depends on g^2N, is one of the generators of the Yangian. It is not the case in the exact theory that one has a Yangian algebra and also a dilatation operator that commutes with it.

This result is highly non trivial, and is heavily based on non-trivial properties of the loop correction to the dilaton operator. Hence, it is a good step in the direction of proving that (9) is the correct Yangian charge. To strengthen our guess for the Q^A charges as expressed in (9), we will show that they satisfy the Serre relation, or equivalently the nesting relation given in (3). We first explain the key steps for $SU(N)$, and then outline how these steps change and generalize in the case of $PSU(2,2|4)$ under consideration.

2. Yangian Relations for $SU(N)$

We return to the question of showing that the bilinear ansatz (9) does give, under certain conditions, a solution of the Serre relations. We will show explicitly that (as indicated in [14]) the standard relations for the Yangian $Y(G)$, which are valid for any Lie group G, are equivalent when G is $SU(N)$ (or $U(N)$) to a matrix form of the commutation relations. We then use this to show that for certain types of representation $\mathcal{R}$, the formula (9) does give a solution of the Serre relations.

For $G = SU(N)$, the Lie algebra is the space of traceless $N \times N$ matrices. Instead of describing the Lie algebra in terms of an abstract basis J^A, as one could do for any Lie group G, it is useful for $SU(N)$ to describe it in terms of generators $J^a{}_b$, $a, b = 1, \ldots, N$, with $\sum_a J^a{}_a = 0$, and obeying

$$
\begin{aligned}
[J^a{}_b, J^c{}_d] &= \delta^c{}_b J^a{}_d - \delta^a{}_d J^c{}_b\,, \\
[J^a{}_b, Q^c{}_d] &= \delta^c{}_b\, Q^a{}_d - \delta^a{}_d\, Q^c{}_b\,.
\end{aligned} \tag{18}
$$

Similarly, the generators Q^A of the Yangian are rewritten as a traceless $N \times N$ matrix of operators Q^a_b. The Serre relation then becomes

$$
\begin{aligned}
&[J^a{}_b, [Q^c{}_d, Q^e{}_f]] - [Q^a{}_b, [J^c{}_d, Q^e{}_f]] \\
&= \frac{h^2}{4} \sum_{p,q} \left([J^a{}_b, [J^c{}_p J^p{}_d, J^e{}_q J^q{}_f]] - [J^a{}_p J^p{}_b, [J^c{}_d, J^e{}_q J^q{}_f]] \right).
\end{aligned} \tag{19}
$$

This matrix form of the Yangian relations is the most familiar one in integrable systems, and will be useful in the generalization to the superalgebra. We will prove below that (19) is equivalent to (3).

To do this, it is useful to work out in more detail how general Lie algebra notation simplifies for $SU(N)$. The generators T^A, $A = 1, \ldots, N^2 - 1$, of $SU(N)$ can be regarded as $N \times N$ matrices in the fundamental representation of $SU(N)$. We use the conventions $[T^A, T^B] = f_C^{AB} T^C$, $f_{ABC} f_{ABE} = 2N\delta_{CE}$,

$$\begin{aligned} \mathrm{Tr} T^A T^B &= -\delta^{AB} \\ \mathrm{Tr} T^A T^B T^C &= -\frac{1}{2}(f_{ABC} - i d_{ABC}) \,. \end{aligned} \tag{20}$$

We define the totally symmetric invariant tensor $id_{ABC} = Tr(\{T^A, T^B\} T^C)$, and use the fact that the generators in this representation, together with the identity matrix, span the space of all complex NxN matrices. We get:

$$\begin{aligned} T^A T^B &= \frac{1}{2}(f_{ABC} T^C + \{T^A, T^B\}) = -\frac{1}{N}\delta_{AB} + \frac{1}{2}(f_{ABC} - i d_{ABC}) T^C \\ [T^A, T^B] &= f_{ABC} T^C \\ \{T^A, T^B\} &= -\frac{2}{N}\delta_{AB} - i d_{ABC} T^C \,. \end{aligned} \tag{21}$$

Explicitly, we write the matrix elements of the matrix T^A, in the fundamental representation, as $T^{Aa}{}_b$, $a, b = 1 \ldots N$. We define

$$J^A = -T^{Ab}{}_a J^a_b \,, \qquad Q^A = -T^{Ab}{}_a Q^a_b \,, \tag{22}$$

where the two-index generators in (22) are traceless $\sum_a J^a_a = 0 = \sum Q^a_a$, and we can invert $J^a_b = J^A T^{Aa}_b$ and $Q^a_b = Q^A T^{Aa}_b$. Note that although T^{Aa}_b is in the N of $SU(N)$, J^A and consequently J^a_b can be in an arbitrary representation. The relation (19), which is sometimes called the nesting relation, reduces to (3) as follows. Multiplying (19) by $T^{Ab}_a T^{Bd}_c T^{Cf}_e$, we find

$$\begin{aligned} &[Q^A, [Q^B, J^C]] + [Q^B, [Q^C, J^A]] + [Q^C, [Q^A, J^B]] \\ &= -\frac{h^2}{4}\big((\mathrm{Tr} T^B T^D T^E) \,\mathrm{Tr}(T^C T^F T^G) \,[J^A, [J^D J^E, J^F J^G]]] \\ &\qquad -(\mathrm{Tr} T^A T^D T^E) \,\mathrm{Tr}(T^C T^F T^G) \,[J^D J^E, [J^B, J^F J^G]] \big)] \\ &= \frac{h^2}{16}\big(-f_{ACG} f_{GLM}(d_{MDE} d_{BLF} + d_{MFE} d_{BLD}) \\ &\qquad + f_{ABG} f_{GLM}(d_{MDE} d_{CLF} + d_{MFE} d_{CLD}) \\ &\qquad + f_{BCG} f_{GLM}(d_{MDE} d_{ALF} + d_{MFE} d_{ALD})\big) J^D J^E J^F \,. \end{aligned} \tag{23}$$

To evaluate the products of d symbols and structure constants solely in terms of the structure constants as in (3), we use the following identities. In addition to the Jacobi identity, there is a similar formula $[\{T^A, T^B\}, T^C] + [\{T^C, T^A\}, T^B] + [\{T^B, T^C\}, T^A] = 0$, which reduces to

$$d_{ABE} f_{ECD} + d_{CAE} f_{EBD} + d_{BCE} f_{EAD} = 0\,. \tag{24}$$

Another identity $[[T^A, T^B], T^C] + \{\{T^C, T^A\}, T^B\} - \{\{T^B, T^C\}, T^A\} = 0$ results in

$$f_{ABE} f_{CDE} = \frac{4}{N}(\delta_{AC}\delta_{BD} - \delta_{BC}\delta_{AD}) + d_{ACE} d_{BDE} - d_{BCE} d_{ADE}\,, \tag{25}$$

and

$$d_{ABC} d_{ABE} = 2(N - \frac{4}{N})\,\delta_{CE}\,. \tag{26}$$

From the Jacobi identity, we find the triple product

$$f_{DMA} f_{ABE} f_{ECD} = -N f_{MBC}\,, \tag{27}$$

from (24) we find

$$d_{ABE} f_{ECD} f_{DFA} = -N d_{BCF} \tag{28}$$

and from (25), and $\sum_A d_{AAB} = 0$, we have

$$\begin{aligned} d_{ACE} d_{EBD} f_{DMA} &= (N - \frac{4}{N}) f_{CBM} \\ d_{DMA} d_{ACE} d_{EBD} &= (N - \frac{12}{N}) d_{MCB}\,. \end{aligned} \tag{29}$$

Note that (25) expresses the difference of two products of d symbols in terms of the structure constants. We used (24) in deriving (23). It is convenient to symmetrize on the D, E, F indices in (23) and use (28,29). Then (23) becomes

$$\begin{aligned}
&[Q^A, [Q^B, J^C]] + [Q^B, [Q^C, J^A]] + [Q^C, [Q^A, J^B]] \\
&= \frac{h^2}{48}(f_{ABG} f_{GLM} d_{MDE} d_{CLF} \\
&\qquad - (A \leftrightarrow C) - (B \leftrightarrow C))\{J^D, J^E, J^F\}\,, \qquad * \\
&= \frac{h^2}{48}((f_{ALG} f_{GBM} + f_{BLG} f_{GAM}) d_{MDE} d_{CLF} \\
&\qquad - (A \leftrightarrow C) - (B \leftrightarrow C))\{J^D, J^E, J^F\} \\
&= \frac{h^2}{24}(((f_{FAL} f_{DBM} d_{CLG} d_{GME} - (A \leftrightarrow B)) - (A \leftrightarrow C) - (B \leftrightarrow C)) \\
&\qquad + 2(f_{CAL} f_{DBM} d_{LFG} d_{GME} - (A \leftrightarrow B) - (B \leftrightarrow C)))\{J^D, J^E, J^F\} \\
&= \frac{h^2}{24}(((f_{FAL} f_{DBM} d_{CLG} d_{GME} - (A \leftrightarrow B)) - (A \leftrightarrow C) - (B \leftrightarrow C)) \\
&\qquad - (f_{BAL} f_{GLM} d_{CFG} d_{MDE} - (A \leftrightarrow C) - (B \leftrightarrow C)))\{J^D, J^E, J^F\}\; *
\end{aligned} \tag{30}$$

where $\{J^D, J^E, J^F\}$ is the totally symmetrized product,

$$\begin{aligned}\{J^D, J^E, J^F\} &= J^D J^E J^F + J^E J^D J^F + J^F J^E J^D \\ &\quad + J^E J^F J^D + J^D J^F J^E + J^F J^D J^E \,. \end{aligned} \tag{31}$$

We observe that the two starred lines are proportional, so we have

$$\begin{aligned}&3\,\frac{h^2}{48}\,(f_{ABG}f_{GLM}d_{MDE}d_{CLF} - (A \leftrightarrow C) - (B \leftrightarrow C))\{J^D, J^E, J^F\} \\ &= \frac{h^2}{24}\,((f_{FAL}f_{DBM}d_{CLG}d_{GME} - (A \leftrightarrow B)) - (A \leftrightarrow C) - (B \leftrightarrow C)) \\ &\qquad \{J^D, J^E, J^F\} \\ &= \frac{h^2}{8}\, f^{ADK} f^{BEL} f^{CFM} f_{KLM} \{J^D, J^E, J^F\}\,. \end{aligned} \tag{32}$$

It follows that

$$\begin{aligned}&[Q^A, [Q^B, J^C]] + [Q^B, [Q^C, J^A]] + [Q^C, [Q^A, J^B]] \\ &= \frac{h^2}{24}\, f^{ADK} f^{BEL} f^{CFM} f_{KLM} \{J_D, J_E, J_F\}. \end{aligned} \tag{33}$$

This equation is totally antisymmetric in A, B, C, and for $h = 1$ is the equation (3).

A Useful Criterion

The point of this lengthy analysis is that although it is difficult to find a solution of the Yangian relation (3), it is much easier to find a solution of (19). The basic case is the case of just a single spin. We want to find a criterion under which, for $G = SU(N)$ and some irreducible representation $\mathcal{R}$ at a single site, we can obey the Yangian algebra with the simple choice $Q^A = Q^a{}_b = 0$. In (3), it is unclear when this works, but in (19), we can easily find a simple criterion.

Since the left hand side of (19) obviously vanishes for $Q^A = 0$, we need a criterion for vanishing of the right hand side. The object $J^c{}_p J^p{}_d$ that appears in (19) is a linear combination of pieces that transform as the singlet of $SU(N)$ and the adjoint. If $\mathcal{R}$ is irreducible, the singlet piece is a multiple of the identity and does not contribute in the commutator. If moreover $\mathcal{R}$ is such that the adjoint only appears once in the decomposition of $\mathcal{R} \otimes \overline{\mathcal{R}}$, then (modulo the irrelevant c-number) $J^c{}_p J^p{}_d$ is a multiple of $J^c{}_d$. Similarly, $J^e{}_q J^q{}_f$ can be replaced by the same multiple of $J^e{}_f$, and $J^a{}_p J^p{}_b$ by the same multiple of $J^a{}_b$. Once this is done, the right hand side of (19) vanishes.

So we have a criterion for finding irreducible representations $\mathcal{R}$ of $SU(N)$ such that for a single-spin system, the Yangian algebra is obeyed with $Q^A = Q^a{}_b = 0$. This criterion is not obeyed for all representations. For example, if $\mathcal{R}$ is the adjoint representation, then the criterion is not obeyed, since the adjoint then appears twice in the decomposition of $\mathcal{R} \otimes \overline{\mathcal{R}}$ (which in this example is the same as $\mathcal{R} \otimes \mathcal{R}$).

However, many important examples arise from representations $\mathcal{R}$ that *do* obey the criterion. Basic examples are the fundamental representation, and more generally the representation of antisymmetric k^{th} rank tensors, for any k.

A Chain Of Spins

Now let us explain how to go from this single-spin result to a representation of the Yangian algebra on a whole chain of spins.

The important property of the Yangian algebra is that it is a Hopf algebra, which means that there is a natural recipe for defining a tensor product of representations. If $\mathcal{A}$ is an algebra, a "coproduct" is a map $\Delta : \mathcal{A} \to \mathcal{A} \otimes \mathcal{A}$ that is a homomorphism of algebras (and obeys some additional axioms of which we explain the relevant one later). For our purposes, this means that the coproduct maps operators that represent $\mathcal{A}$ in a single-spin representation $\mathcal{R}$ to operators that represent $\mathcal{A}$ in the Hilbert space $\mathcal{R} \otimes \mathcal{R}$ of a two-spin system. Moreover, one can repeat the process, using the homomorphism $\Delta \otimes 1 : \mathcal{A} \otimes \mathcal{A} \to \mathcal{A} \otimes \mathcal{A} \otimes \mathcal{A}$, or alternatively the homomorphism $1 \otimes \Delta : \mathcal{A} \otimes \mathcal{A} \to \mathcal{A} \otimes \mathcal{A} \otimes \mathcal{A}$, to map a representation in the two-spin system to a representation in the three-spin system. (Here, for example, $1 \otimes \Delta$ is the operator that acts as the identity on the first spin and acts by Δ to map the Hilbert space of the second spin in a two-spin system to that of the last two spins in a three-spin system. Similarly $\Delta \otimes 1$ acts as the identity on the second or last spin while mapping the Hilbert space of the first to that of a two-spin system.) Repeating the process, one gets a representation of $\mathcal{A}$ in an n-spin system, for any n. Apart from being a homomorphism of algebras, the key axiom obeyed by Δ is "coassociativity," $\Delta \otimes 1(\Delta) = 1 \otimes \Delta(\Delta)$, which for our purposes says that starting with a given representation of $\mathcal{A}$ in the single-spin system, the representation that one arrives at in the n-spin system does not depend on the the precise route by which one applies these formulas.

For the Yangian, the explicit formula for the coproduct is

$$\begin{aligned} \Delta(J^A) &= J^A \otimes 1 + 1 \otimes J^A \\ \Delta(Q^A) &= Q^A \otimes 1 + 1 \otimes Q^A + f^A_{BC} J^B \otimes J^C \,. \end{aligned} \tag{34}$$

Using this coproduct, we can determine, given a single-spin representation of the Yangian with $Q^A = 0$, what the representation should be for a multi-spin system. Consider first the two-spin system. The two-spin representation of J^A is $\Delta(J^A) = J^A \otimes 1 + 1 \otimes J^A$. This is a fancy notation for writing the result that we would expect naively, since $J^A \otimes 1$ and $1 \otimes J^A$ are simply in a more typical physics notation the operators J_1^A and J_2^A that act by J^A on the first or second spin. So $\Delta(J^A) = J_1^A + J_2^A$, saying simply that the group generators of the two-spin system are the sum of the single-particle generators. If the single-spin representation of Q^A is zero, then the two-spin representation of Q^A reduces to $\Delta(Q^A) = f^A_{BC} J^B \otimes J^C$, which in the alternative notation is $f^A_{BC} J_1^B J_2^C$. This agrees with the two-spin case of (9). More generally, by repeated application of the coproduct, one learns that whenever for the one-spin system one can obey the Yangian algebra with $Q^A = 0$, the formula (9) supplies a representation of the Yangian algebra for a chain of spins.

If the adjoint representation appears more than once in the decomposition of $\mathcal{R} \otimes \overline{\mathcal{R}}$, then for the one-spin system, one cannot generally obey the Yangian algebra with $Q^A = 0$. However, for $SU(N)$, the form (19) of the relations implies that one can always, for any representation $\mathcal{R}$ of $SU(N)$, obey the Yangian algebra with $Q^a{}_b = J^a{}_p J^p{}_b / 2$. This contrasts with the situation for more general symmetry groups G, where for generic $\mathcal{R}$ there is no choice of Q^A that obeys the Serre relation.

3. Yangian Superalgebra

The analysis in the previous section would have worked in just the same way if we replace the simple group $SU(N)$ by the non-simple group $U(N)$. It similarly works for the supergroup $U(N|M)$, and for $SU(N|M)$ if $N \neq M$. The case $N = M$, however, requires a further study.

This fact is relevant for us because $PSU(2,2|4)$, which is a real form of $PSU(4|4)$, is the symmetry of $\mathcal{N} = 4$ super Yang-Mills theory, which thus involves the exceptional case $N = M = 4$. (For our purposes, the signature is not important, as we will be carrying out purely algebraic manipulations; we need not distinguish $PSU(4|4)$ from $PSU(2,2|4)$.)

First we give a few facts about the Lie superalgebras $U(n|n), SU(n|n)$ and $PSU(n|n)$ (see *eg.* [15]). The Lie superalgebra $U(n|m)$ has generators that can be represented by matrices of the form $x = \begin{pmatrix} a & b \\ c & d \end{pmatrix}$, where a and d are $n \times n$ and $m \times m$ bosonic hermitian matrices, and b and c are $n \times m$ and $m \times n$ fermionic matrices and are hermitian conjugates. The supertrace of

x is Str x = Tr a − Tr d. If we divide by multiples of the identity, we get a superalgebra $PU(n|m)$. If we restrict to x such that Str x= 0, we get a superalgebra of one less dimension that is called $SU(n|n)$. For $n \neq m$, at the Lie algebra level, requiring the trace of x to be zero removes the identity matrix and hence $PU(n|m)$ and $SU(n|m)$ are the same at the Lie algebra level (the global structure of the groups is different). For $n = m$, the identity matrix has zero supertrace, so requiring the trace to be zero does not remove the identity matrix. If we require x to be traceless and further identify any two $x's$ that differ by an additive scalar, we get a superalgebra that is called $PSU(n|n)$ or $A(n-1|n-1)$ and has two dimensions less than $U(n|n)$.

The generators of $PSU(n|n)$ can be represented by matrices $x = \begin{pmatrix} a & b \\ c & d \end{pmatrix}$, with Tr a = Tr d = 0. The bosonic part of $PSU(n|n)$ is $SU(n) \times SU(n)$, generated by a and d. We are here writing x in what we may call the $n|n$ representation. In any representation, the fermionic generators of $PSU(n|n)$, here represented by the matrices b and c, transform as $n \otimes \bar{n} \oplus \bar{n} \otimes n$ under $SU(n) \times SU(n)$.

The superconformal symmetry group of $\mathcal{N} = 4$ super Yang-Mills theory is a real form of $PSU(4|4)$, whose bosonic part is $SU(4) \times SU(4)$. It will be helpful to compare $PSU(4|4)$ to its cousins $SU(4|4)$ and $U(4|4)$. We have for the Lie algebras

$$\begin{aligned} SU(4|4) &= PSU(4|4) \oplus R \\ U(4|4) &= PSU(4|4) \oplus K \oplus R\,. \end{aligned} \tag{35}$$

(This is an additive decomposition of the Lie algebras; the commutation relations do not preserve this decomposition.) K is the Lie algebra generated by the identity matrix (which we also write as K). R is the Lie algebra of a $U(1)$ R-symmmetry group that is not contained in $PSU(4|4)$ and is not a symmetry of $\mathcal{N} = 4$ super Yang-Mills theory; we also call its generator R. In the $4|4$ representation, we take $Q_R = R$ where

$$R = \frac{1}{2}\begin{pmatrix} 1 & 0 \\ 0 & -1 \end{pmatrix}, \tag{36}$$

So commutation with R multiplies the blocks b and c of a generator x by 1 or −1 and annihilates a and d. The supertraces are

$$\mathrm{Str}\, K^2 = \mathrm{Str}\, R^2 = 0, \;\; \mathrm{Str}\, RK = 4. \tag{37}$$

We define the $U(4|4)$ structure constants by

$$[J_A, J_B\} = f_{AB}^{C} J_C\,, \tag{38}$$

where the brackets denote either commutators or anticommutators. Then $f^A_{KB} = 0$ for all A, B since K is central and commutes with everything, and $f^R_{AB} = 0$ for all B, C, since the $U(1)$ R-symmetry generator R never appears on the right hand side of the commutation relations. (It is precisely because R never appears on the right hand side of the commutation relations that there can exist a theory, such as $\mathcal{N} = 4$ super Yang-Mills theory, that has $PSU(4|4)$ symmetry but not the additional $U(1)$ R-symmetry.)

The formula (38) is the first formula in this paper in which it is important to carefully distinguish whether the "A" index of a Lie algebra generator such as J_A is "down" or "up." At the outset of these notes, we merely asserted that there is an invariant, nondegenerate metric g that is used to raise and lower indices, and in many formulas we have done so without comment. In the present example, we can take the metric for $U(4|4)$ to be $g_{AB} = \frac{1}{2}\mathrm{Str}\, J_A J_B$. So $g_{KK} = g_{RR} = 0$, $g_{KR} \neq 0$; and when $A \neq K, R$ then $g_{AB} = 2\delta_{AB}$ and $g_{KA} = g_{RA} = 0$. It follows that when we raise and lower indices, K and R are exchanged, so the assertions in the last paragraph become

$$f_A{}^{RB} = 0 = f_K^{AB}. \tag{39}$$

For $U(4|4)$, the analysis in the last section applies and shows that the simple bilinear formula (9) gives a representation of the Yangian algebra as long as the single-spin representation $\mathcal{R}$ has the property that the adjoint representation only appears once in the decomposition of $\mathcal{R}\otimes\overline{\mathcal{R}}$. For $\mathcal{N} = 4$ super Yang-Mills theory, we take $\mathcal{R}$ to be the representation consisting of the one-particle states of the free vector multiplet. The $U(1)$ R-symmetry generator R does act on this representation (though it is not a symmetry of the gauge theory), and we take K to act on the representation by $K = 0$. In this way we interpret the representation $\mathcal{R}$ as a representation of the extended group $U(4|4)$.

This representation does have the property that the adjoint representation only appears once in the decomposition of $\mathcal{R} \otimes \overline{\mathcal{R}}$. So we can use the familiar bilinear formula to get a multi-spin representation of the Yangian of $U(4|4)$:

$$\begin{aligned} Q_C &= g_{CC'}Q^{C'} = g_{CC'} f^{C'}_{AB} \sum_{i<j} J^A_i J^B_j \\ &= g_{CC'} f^{C'}_{AB} g^{AA'} g^{BB'} \sum_{i<j} (J_i)_{A'} (J_j)_{B'}. \end{aligned} \tag{40}$$

Here we have been careful in raising and lowering of indices to ensure that the second generator Q_C of the Yangian transforms like the $PSU(4|4)$ generators J_C.

Since we actually want to represent the Yangian of $PSU(4|4)$, not that of $U(4|4)$, we need a few more observations. From (40), we see that in the representation of the $PSU(4|4)$ Yangian, $Q_K = 0$ and $Q_R = 2f^K_{AB}g^{AA'}g^{BB'}\sum_{i<j}(J_i)_{A'}(J_j)_{B'}$. What about the generators Q_C where C corresponds to a generator of $PSU(4|4)$? They do not depend on K, since $K = 0$ in our chosen representation, and they do not depend on R, since $f^{C'}_{KB} = 0$. It follows that the Q_C's are given by the same formula as if we had evaluated the bilinear formula (9) directly for $PSU(4|4)$. We have thus established that this bilinear formula does give a representation of the Yangian algebra for $PSU(4|4)$.

References

1. I. Bena, J. Polchinski and R. Roiban, "Hidden Symmetries of the AdS(5) x S^5 Superstring," arXiv:hep-th/0305116.
2. D. Bernard, "Hidden Yangians in 2D Massive Current Algebras", Comm. Math. Phys. **137**, 191 (1991).
3. N. MacKay, "On the Classical Origins of Yangian Symmetry in Integrable Field Theory", Phys. Lett. **B 281** 90 (1992); Erratum-ibid. B **308**, 444 (1993).
4. G. Mandal, N. V. Suryanarayana, and S.R. Wadia, "Aspects of Semiclassical Strings in AdS(5)," Phys. Lett. **B543**, 81 (2002), arXiv:hep-th/0206103.
5. V. Drinfel'd, "Hopf Algebras and the Quantum Yang-Baxter Equation", Sov. Math. Dokl. **32** 254 (1985).
6. V. Drinfel'd, "A New Realization of Yangians and Quantized Affine Algebras", Sov. Math. Dokl. **36** 212 (1988).
7. F. Haldane, Z. Ha, J. Talstra, D. Bernard, and V. Pasquier, "Yangian Symmetry of Integrable Quantum Chains with Long-Range Interactions and a New Description of States in Conformal Field Theory", Phys. Rev. Lett. **69**, 2021 (1992).
8. L. Dolan, "Kac-Moody Algebra is Hidden Symmetry of Chiral Models," Phys. Rev. Lett. **47** 1371 (1981).
9. L. Dolan, C. Nappi, and E. Witten, "A Relation Between Approaches to Integrability in Superconformal Yang-Mills Theory", JHEP**0310**, 017 (2003) arXiv:hep-th/0308089.
10. N. Beisert, "The Complete One-loop Dilatation operator of N = 4 super Yang-Mills Theory," arXiv:hep-th/0307015.
11. D. Berenstein, J. Maldacena, and H. Nastase, "Strings in Flat Space and pp Waves from $N = 4$ Super Yang Mills," JHEP **204**, 013 (2002). arXiv:hep-th/0202021.
12. N. Beisert and M. Staudacher, "The N = 4 SYM Integrable Super Spin Chain," arXiv:hep-th/0307042.

13. J. A. Minahan and K. Zarembo, "The Bethe-ansatz for N = 4 super Yang-Mills," JHEP **0303**, 013 (2003) arXiv:hep-th/0212208.
14. D. Bernard, "An Introduction to Yangian Symmetries," Int. J. Mod. Phys. **B7**, 3517 (1993) arXiv:hep-th/9211133.
15. N. Berkovits, C. Vafa, and E. Witten, "'Conformal field theory of AdS background with Ramond-Ramond flux," JHEP **9903**, 018 (1999), arXiv:hep-th/9902098.
16. Nazarov, M., "Quantum Berezinian and the Classical Capelli Identity", Lett. Math. Phys. **21** 123 (1991).
17. F. Haldane, "Physics of the Ideal Semion Gas: Spinons and Quantum Symmetries of the Integrable Hadane-Shastry Spin Chain", cond-mat/9401001.
18. R. Zhang, "Representations of Super Yangians", J. Math. Phys. **36** 3854 (1995); "The $gl(M|N)$ Super Yangian and its Finite Dimensional Representations", Lett. Math. Phys. **37** 419 (1996).
19. T. Curtright and C. Zachos, "Supersymmetry and the Nonlocal Yangian Deformation Symmetry", Nucl. Phys. B **402**, 604 (1993), arXiv:hep-th/9210060.
20. E. Witten, "Perturbative gauge theory as a string theory in twistor space," arXiv:hep-th/0312171.

SELF-DUALITY, HELICITY AND HIGHER-LOOP EULER-HEISENBERG EFFECTIVE ACTIONS*

GERALD V. DUNNE
Department of Physics
University of Connecticut
Storrs, CT 06269-3046 USA

CHRISTIAN SCHUBERT
Department of Physics
University of Texas Pan American
Edinburg, TX 78539 USA

The Euler-Heisenberg effective action in a self-dual background is remarkably simple at two-loop. This simplicity is due to the inter-relationship between self-duality, helicity and supersymmetry. Applications include two-loop helicity amplitudes, beta-functions and nonperturbative effects. The two-loop Euler-Heisenberg effective Lagrangian for QED in a self-dual background field is naturally expressed in terms of one-loop quantities. This mirrors similar behavior recently found in two-loop amplitudes in N=4 SUSY Yang-Mills theory.

1. One-Loop Euler-Heisenberg Effective Action

In computing effective actions in gauge theories and in gravity, one can use the external field as a probe of the vacuum structure of the quantum theory[1]. A great deal has been learned from the seminal work of Heisenberg and Euler[2], who already in 1936 produced the paradigm for this entire approach by computing the nonperturbative, renormalized, one-loop effective action for QED in a background of constant field strength $F_{\mu\nu}$. This special soluble case of a constant field strength leads to several important insights and applications:

• Light-Light scattering. The Euler-Heisenberg effective action is nonlinear in the electromagnetic fields, and the quartic and higher terms represent new nonlinear interactions, the first of which is light-light scattering,

*Talk presented by G.D.

which does not occur in the tree level Maxwell action. Expanding the Euler-Heisenberg answer to quartic order we find

$$S^{(1)} = \frac{e^4}{360\pi^2 m^4}\int d^4x \left[(\vec{E}^2 - \vec{B}^2)^2 + 7(\vec{E}\cdot\vec{B})^2\right] + \dots \tag{1}$$

which gives the low energy limit (since the field strength was constant) of the amplitude for light-light scattering.

• Low-energy effective field theory. In more modern language, the expansion of the Euler-Heisenberg expression in powers of the electromagnetic fields is an example of a low-energy effective action, which describes the interactions of the light degrees of freedom which remain after integrating out the massive electron degrees of freedom. Indeed, the explicit expansion for a background magnetic field is

$$\mathcal{L}^{(1)} = -\hbar c \frac{2}{\pi^2}\left(\frac{mc}{\hbar}\right)^4 \sum_{n=0}^{\infty} \frac{2^{2n}\mathcal{B}_{2n+4}}{(2n+4)(2n+3)(2n+2)}\left(\frac{eB\hbar}{m^2c^3}\right)^{2n+4} \tag{2}$$

where $\mathcal{B}_{2n}$ are Bernoulli numbers. The dimensionless expansion parameter is the only gauge and Lorentz invariant combination of the background field, B^2, divided by compensating powers of the heavy mass scale m. Physically, this is the ratio of the cyclotron energy scale of the magnetic field to the electron rest energy scale, or the square of the ratio of the Compton wavelength to the magnetic length. The expansion also shows that the perturbative series is divergent, as the Bernoulli numbers grow factorially fast in magnitude.

• Pair-production in an electric field. The presence of a background electric field accelerates and splits the virtual vacuum dipole pairs, leading to e^+e^- particle production. This instability of the vacuum was realized already by Heisenberg and Euler in 1936, and later formalized in the language of QED by Schwinger[3]. The leading imaginary part of the one-loop effective Lagrangian in a weak electric field is

$$Im\,\mathcal{L}^{(1)} \sim \frac{e^2E^2}{8\pi^3}\exp\left[-\frac{m^2\pi}{eE}\right] \tag{3}$$

This imaginary part gives the rate of vacuum non-persistence due to pair production. The rate is more appreciable when the E field approaches a critical value $E_c \sim \frac{m^2c^3}{e\hbar} \sim 10^{16}\,\mathrm{Vcm}^{-1}$, where the work done accelerating a virtual pair apart by a Compton wavelength is of the order of the rest mass energy for the pair. Even though the condition of a constant electric field is rather unrealistic, the result (3) provides the starting point for more detailed analyses which incorporate time-dependent electric fields[5].

• Strong-field limit. The strong-field limit also contains important physical information, as will be discussed in detail below. For example, for a strong background magnetic field,

$$\mathcal{L}^{(1)} \sim \left(\frac{e^2}{12\pi^2}\right)\frac{B^2}{2}\ln\left(\frac{eB}{m^2}\right) \tag{4}$$

and the coefficient tells us the first coefficient of the QED β-function.

2. Two-Loop Euler-Heisenberg Effective Action

In this talk, I ask: what do we know about such effective actions beyond the one-loop level, and what can we learn from them?

In principle, the computation of the two-loop Euler-Heisenberg effective Lagrangian is completely straightforward, as we only need to compute a single vacuum diagram with an internal photon line and a single fermion (or scalar) loop, where these propagators are those in the presence of the background field. These propagators have been known in closed-form for a long time[4,3]. For example, the scalar propagator in coordinate space is (η is a gauge-dependent phase factor)

$$G_{\rm scalar}(x,x') = -i\frac{e^{-i\eta}}{(4\pi)^2}\int_0^\infty \frac{dT}{T^2}\exp\left[-im^2T - L(T) + \frac{i}{4}z\beta(T)z\right]$$

$$\begin{aligned} z_\mu &\equiv x_\mu - x'_\mu \\ \beta_{\mu\nu} &\equiv \left[eF\coth(eFT)\right]_{\mu\nu} \\ L &\equiv \frac{1}{2}\mathrm{tr}\ln\left(\frac{\sinh(eFT)}{eFT}\right) \end{aligned} \tag{5}$$

Here F is the constant field strength matrix. The new feature that makes two-loop nontrivial is the necessity of mass renormalization. In a *tour de force*, Ritus[6] found explicit expressions for the renormalized two-loop effective Lagrangian for both spinor and scalar QED. Unfortunately, these expressions are extremely complicated two-parameter integrals (roughly speaking, there is one parameter integral left for each spinor/scalar propagator), and it is difficult to extract detailed physical information beyond their leading behavior. This is the case even if one restricts to the less general case of a constant magnetic field or a constant electric field background [6,1,7].

The main new development[10,11,12] is that if the background field is chosen to be self-dual, so that

$$F_{\mu\nu} = \tilde{F}_{\mu\nu} \tag{6}$$

then the two-loop Euler-Heisenberg effective Lagrangians take the remarkably simple closed forms :

$$\mathcal{L}^{(2)}_{\rm spinor} = -\alpha^2 \frac{f^2}{2\pi^2}\left[3\,\xi^2(\kappa) - \xi'(\kappa)\right] \tag{7}$$

$$\mathcal{L}^{(2)}_{\rm scalar} = \alpha^2 \frac{f^2}{4\pi^2}\left[\frac{3}{2}\,\xi^2(\kappa) - \xi'(\kappa)\right] \tag{8}$$

where f is the field strength, $\frac{1}{4}F_{\mu\nu}F^{\mu\nu} = f^2$, and κ is the natural dimensionless parameter

$$\kappa \equiv \frac{m^2}{2ef} \tag{9}$$

The ubiquitous function $\xi(\kappa)$ is simply related to the Euler digamma function $\psi(\kappa)$:

$$\xi(\kappa) \equiv -\kappa\left(\psi(\kappa) - \ln(\kappa) + \frac{1}{2\kappa}\right) = -\frac{1}{2}\int_0^\infty ds\, e^{-2\kappa s}\left(\frac{1}{\sinh^2 s} - \frac{1}{s^2}\right) \tag{10}$$

The subtraction of the first two terms of the asymptotic expansion of $\psi(\kappa)$ correspond to renormalization subtractions.

The dramatic simplicity of the results (7,8), compared to Ritus's general expressions, raises two questions:

- why are these expressions so simple?
- why are these expressions so similar?

The answer lies in the three-way relationship between self-duality, helicity and supersymmetry.

Self-dual fields have definite helicity:

$$\sigma_{\mu\nu}F_{\mu\nu}\left(\frac{\mathbf{1}+\gamma_5}{2}\right) = 0 \tag{11}$$

It is well-known that amplitudes for external field lines with like helicities are particularly simple, in both gauge and gravitational theories, and these simplicities are thought to be deeply related to integrability conditions and the self-dual Yang-Mills equations[13,14]. Since the effective action for a self-dual field is the generating function for like-helicity amplitudes, it is consistent that the effective action in a self-dual background should be simple. However, most of the amplitude results are for massless particles, while here we see a generalization to massive particles. For example, a new prediction is that for massive QED the low energy limit of the two-loop amplitudes with all external helicities + behave as:

$$\Gamma^{(2)}[k_1,\epsilon_1^+;k_2,\epsilon_2^+;\ldots;k_N,\epsilon_N^+] = \alpha\pi\frac{(2e)^N}{(4\pi)2m^{2N-4}}\,c^{(2)}_{N/2}\,\chi_N \tag{12}$$

where the χ_N are standard helicity-basis factors[12], and the numerical coefficients $c_n^{(2)}$ come from the weak field expansion of the two-loop effective Lagrangians for a self-dual background:

$$c_n^{(2)} = \frac{1}{(2\pi)^2}\left\{\frac{2n-3}{2n-2}\mathcal{B}_{2n-2} + \frac{3}{2}\sum_{k=1}^{n-1}\frac{\mathcal{B}_{2k}}{2k}\frac{\mathcal{B}_{2n-2k}}{(2n-2k)}\right\}$$

These results have since been generalized[15] to all helicity combinations (up to 10 external legs) by making an explicit helicity expansion of Ritus's general result.

<u>Self-dual fields and SUSY:</u> self-duality implies that the corresponding Dirac operator has a quantum mechanical supersymmetry. Therefore, apart from zero modes, the Dirac operator has the same spectrum (but with a multiplicity of 4) as the corresponding scalar Klein-Gordon operator[16,17]. At one-loop this has the simple consequence

$$\mathcal{L}^{(1)}_{\text{spinor}} = -2\mathcal{L}^{(1)}_{\text{scalar}} + \frac{1}{2}\left(\frac{ef}{2\pi}\right)^2\ln\left(\frac{m^2}{\mu^2}\right) \tag{13}$$

where $N_0 = \left(\frac{ef}{2\pi}\right)^2$ is the zero mode number density. At two-loop the effective action is not simply a log determinant, but the quantum mechanical SUSY of the Dirac operator relates the spinor propagator to the scalar propagator via simple helicity projections, which means that the two-loop effective action can be written as the sum of two terms involving matrix elements of the scalar propagator, with different numerical coefficients in the case of spinor and scalar QED. This explains why (7) and (8) have similar forms[18].

A more prosaic reason for the simplicity of the two-loop expressions (7) and (8) is that for a self-dual field, $F_{\mu\nu}F_{\nu\rho} = -f^2\delta_{\mu\rho}$, which dramatically simplifies the propagators of spinors or scalars in the background field. For example, for a scalar particle, the Lorentz matrix structure in (5) disappears, leaving

$$G_{\text{scalar}}(x,x') = \left(\frac{ef}{4\pi}\right)^2\int_0^\infty \frac{dt}{\sinh^2(eft)}\,e^{-m^2t-\frac{ef}{4}(x-x')^2\coth(eft)} \tag{14}$$

$$G_{\text{scalar}}(p) = \int_0^\infty \frac{dt}{\cosh^2(eft)}\,e^{-m^2t-\frac{p^2}{ef}\tanh(eft)} \tag{15}$$

The simplification is even more dramatic for massless particles, where for scalars

$$G_{\rm scalar}(x,x') = \frac{e^{-\frac{ef}{4}(x-x')^2}}{4\pi^2(x-x')^2}\,; \qquad G_{\rm scalar}(p) = \frac{1-e^{-\frac{p^2}{ef}}}{p^2} \tag{16}$$

QED β-functions. Another application of background field propagators is to the computation of the β-function, which characterizes the scale dependence of the running coupling. This is a much simpler problem than finding the full renormalized effective Lagrangians (7, 8), as we only need the leading strong field behavior of $\mathcal{L}$. In this approach, one uses the external field as a probing scale, instead of the external momentum of a self-energy diagram (which is the usual text-book approach). The Gell-Mann Low renormalization group argument for $\Pi_{\mu\sigma}(q^2,\mu^2)$ can be adapted to the effective Lagrangian $\mathcal{L}(eF,\mu^2)$ to relate the strong-field limit of $\mathcal{L}$ to the β-function coefficients[6]. A similar idea was used at two loop by Shifman and Vainshtein to compute the QED β-function using the operator product expansion[8], and has recently been extended to the three-loop QCD β-function[0]. There is an immediate combinatorial advantage to this background field approach, as there are many fewer diagrams at a given order, since there are no external lines. Furthermore, each diagram has fewer vertices and propagators. The price, of course, is that the spinor (or scalar) propagators are not free ones, but are in the background field. However, as we have seen in (14,15,16), in a self-dual background the propagators are still rather simple, so this potentially tips the balance back in favor of this approach.

At n-loop order, the β-function coefficient β_n is related to the coefficient b_n from the strong field limit

$$\mathcal{L}^{(n)} \sim b_n\, F^2 \ln\left(\frac{e|F|}{m^2}\right) \tag{17}$$

For example, for a background magnetic field, from the Euler-Heisenberg and Ritus expressions we find the leading strong field behavior:

$$\mathcal{L}^{(1)}_{\rm scalar} \sim \frac{e^2}{48\pi^2}\frac{B^2}{2}\ln\left(\frac{eB}{m^2}\right), \qquad \mathcal{L}^{(1)}_{\rm spinor} \sim \frac{e^2}{12\pi^2}\frac{B^2}{2}\ln\left(\frac{eB}{m^2}\right)$$

$$\mathcal{L}^{(2)}_{\rm scalar} \sim \frac{e^4}{64\pi^4}\frac{B^2}{2}\ln\left(\frac{eB}{m^2}\right), \qquad \mathcal{L}^{(2)}_{\rm spinor} \sim \frac{e^4}{64\pi^4}\frac{B^2}{2}\ln\left(\frac{eB}{m^2}\right) \tag{18}$$

From the coefficients we can read off the QED β-functions :

$$\beta_{\rm scalar} = \frac{e^3}{48\pi^2} + \frac{e^5}{64\pi^4} + \ldots\,; \qquad \beta_{\rm spinor} = \frac{e^3}{12\pi^2} + \frac{e^5}{64\pi^4} + \ldots \tag{19}$$

Since in some sense the self-dual background is the "simplest", we now test this approach for a self-dual field. From the one-loop Euler-Heisenberg expression, and the closed-form two-loop expressions (7, 8) we find

$$\mathcal{L}^{(1)}_{\text{scalar}} \sim \frac{e^2}{48\pi^2} f^2 \ln\left(\frac{ef}{m^2}\right), \qquad \mathcal{L}^{(1)}_{\text{spinor}} \sim -\frac{e^2}{24\pi^2} f^2 \ln\left(\frac{ef}{m^2}\right)$$

$$\mathcal{L}^{(2)}_{\text{scalar}} \sim \frac{e^4}{64\pi^4} f^2 \ln\left(\frac{ef}{m^2}\right), \qquad \mathcal{L}^{(2)}_{\text{spinor}} \sim -\frac{e^4}{32\pi^4} f^2 \ln\left(\frac{ef}{m^2}\right)$$

So the argument works for scalar QED but fails for spinor QED, even at one-loop. The reason is that in a self-dual background spinor QED has zero modes, and these preclude the massless limit on which is based the argument[6] relating the β-function and the strong field limit of $\mathcal{L}$. This argument must be modified to account for the zero modes[18]. At one loop the mismatch is due to the infrared divergence of the unrenormalized $\mathcal{L}_{\text{spinor}}$, as is familiar from instanton physics[16,17]. But at two loop there is no IR divergence and the zero modes actually enter via the mass renormalization[18]. Thus, to continue this strategy to n^{th} order we find that for scalar QED one simply needs the coefficient of the leading single-log term as in (17), while for spinor QED one also needs the $(n-1)^{th}$ loop anomalous mass dimension coefficient.

Beyond two loops: To go beyond two loops one should take advantage of the great progress that has been made recently in understanding the structure of higher-loop quantum field theory (without background fields). The general strategy is to manipulate diagrams to reduce the number to a set of so-called "master diagrams" which need to be computed. This has led, for example, to many new two-loop results for QCD scattering amplitudes[19]. Kreimer[20] and collaborators have discovered an elegant Hopf algebra structure underlying the seemingly messy jumble of ordinary Feynman diagram perturbation theory. I conclude this talk with some brief comments and speculations about how these techniques might be extended to incorporate background fields.

It is a simple exercise in free QED (i.e. no background field) to show that in 4 dimensions

spinor QED :  $= -6\,e^2$ []^2 (20)

scalar QED :  $= \frac{3}{2}\,e^2$ []^2 (21)

(The loop on the RHS is a scalar loop in each case.) Thus, the two loop vacuum diagram can be expressed in terms of a simpler one loop diagram.

These results can either be derived by computing each side using dimensional regularization, or a quicker proof follows from an integration-by-parts argument. Now reconsider the two-loop effective Lagrangian results (7) and (8) in this diagrammatic language. For a self-dual background, the scalar propagator loops are evaluated using (15), and turn out to be simply related to the $\xi(\kappa)$ function:

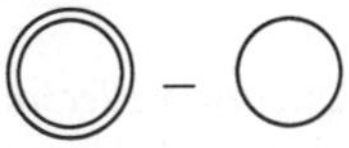

$$\equiv \int \frac{d^4p}{(2\pi)^4}\,[G(p) - G_0(p)] = -\frac{m^2}{(4\pi)^2}\,\frac{\xi(\kappa)}{\kappa} \qquad (22)$$

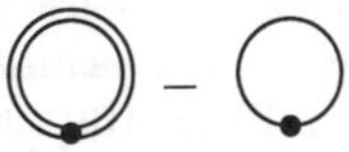

$$\equiv \int \frac{d^4p}{(2\pi)^4}\,[(G(p))^2 - (G_0(p))^2] = \frac{\xi'(\kappa)}{(4\pi)^2} \qquad (23)$$

Here the double lines refer to scalar propagators in the self-dual background and the single line is the free scalar propagator, while the dot on a propagator refers to the propagator being squared.

Thus, the results (7) and (8) for the renormalized two-loop effective Lagrangians can be written diagrammatically as

(24)

(25)

The comparison with the free vacuum diagram relations (20, 21) is striking. The entire effect of the background field is quite minimal, and the two loop diagrams can still be expressed simply in terms of one-loop diagrams. Note that this is a highly nontrivial result, as both mass and charge renormalization have been taken into account in (24, 25). If an algebraic, rather than a direct computational, understanding could be found for these relations, as has been done in the simpler theories without background fields, then the possibility of extending to higher loops becomes much greater. A first step in this direction has been taken in Ref.[21], which translates the results of [10,11,12,18] into dimensional regularization, the standard regularization technique for higher-loop analyses. Another tantalizing relation is the structural similarity between the results (24, 25) for the effective Lagrangians in a self-dual background, and recent results relating two-loop and one-loop amplitudes in maximally SUSY Yang-Mills theory[22]. This is a very different theory, and yet the two-loop amplitudes appear as squares of one-loop amplitudes, plus a one-loop correction, which is reminiscent of (24, 25).

Acknowledgment

GD thanks the Rockefeller Foundation for a Summer 2003 Residency at the Bellagio Center and the DOE for support through Grant DE-FG02-92ER40716.

References

1. W. Dittrich and M. Reuter, *Effective Lagrangians In Quantum Electrodynamics*, Springer Lect. Notes Phys. **220** (1985); W. Dittrich and H. Gies, *Probing the quantum vacuum. Perturbative effective action approach in quantum electrodynamics and its applications*, Springer Tracts Mod. Phys. **166** (2000).
2. W. Heisenberg and H. Euler, Z. Phys. **98** (1936) 714; V. Weisskopf, *Kong. Dans. Vid. Selsk. Math-fys. Medd.* **XIV** No. 6 (1936).
3. J. Schwinger, "On gauge invariance and vacuum polarization", Phys. Rev. **82** (1951) 664.
4. V. Fock, "Proper Time In Classical And Quantum Mechanics," Phys. Z. Sow. **12**, 404 (1937); Y. Nambu, "The Use Of The Proper Time In Quantum Electrodynamics," Prog. Theor. Phys. **5**, 82 (1950).
5. E. Brezin and C. Itzykson, "Pair Production In Vacuum By An Alternating Field," Phys. Rev. D **2**, 1191 (1970); V. S. Popov and M. S. Marinov, "E+ E- Pair Production In Variable Electric Field," Yad. Fiz. **16**, 809 (1972); "Electron - Positron Pair Creation From Vacuum Induced By Variable Electric Field," Fortsch. Phys. **25**, 373 (1977).
6. V. I. Ritus, "Lagrangian of an intense electromagnetic field and quantum electrodynamics at short distances", Sov. Phys. JETP **42** (1975) 774, "Connection between strong-field quantum electrodynamics with short-distance quantum electrodynamics", *ibid* **46** (1977) 423. "The Lagrangian Function of an Intense Electromagnetic Field", in *Proc. Lebedev Phys. Inst.* Vol. **168**, *Issues in Intense-field Quantum Electrodynamics*, V. I. Ginzburg, ed., (Nova Science Pub., NY 1987).
7. G. Dunne and C. Schubert, "Two-loop Euler-Heisenberg QED pair-production rate," Nucl. Phys. B **564**, 591 (2000) [arXiv:hep-th/9907190].
8. M. A. Shifman and A. I. Vainshtein, "Operator Product Expansion And Calculation Of The Two Loop Gell-Mann-Low Function," Sov. J. Nucl. Phys. **44**, 321 (1986).
9. J. P. Bornsen and A. E. van de Ven, "Three-loop Yang-Mills beta-function via the covariant background field method," Nucl. Phys. B **657**, 257 (2003). [arXiv:hep-th/0211246].
10. G. V. Dunne and C. Schubert, "Closed-form two-loop Euler-Heisenberg Lagrangian in a self-dual background," Phys. Lett. B **526**, 55 (2002) [arXiv:hep-th/0111134].
11. G. V. Dunne and C. Schubert, "Two-loop self-dual Euler-Heisenberg Lagrangians. I: Real part and helicity amplitudes," JHEP **0208**, 053 (2002) [arXiv:hep-th/0205004].
12. G. V. Dunne and C. Schubert, "Two-loop self-dual Euler-Heisenberg La-

grangians. II: Imaginary part and Borel analysis," JHEP **0206**, 042 (2002) [arXiv:hep-th/0205005].

13. M. L. Mangano and S. J. Parke, "Multiparton Amplitudes In Gauge Theories," Phys. Rept. **200**, 301 (1991); G. Mahlon, "One loop multiphoton helicity amplitudes", Phys. Rev. D **49** (1994) 2197; Z. Bern, L. J. Dixon, M. Perelstein and J. S. Rozowsky, "Multi-leg one-loop gravity amplitudes from gauge theory," Nucl. Phys. B **546**, 423 (1999); [arXiv:hep-th/9811140]; Z. Bern, A. De Freitas and L. Dixon, "Two-loop helicity amplitudes for gluon gluon scattering in QCD and supersymmetric Yang-Mills theory," JHEP **0203**, 018 (2002); [arXiv:hep-ph/0201161]; T. Binoth, E. W. N. Glover, P. Marquard and J. J. van der Bij, "Two-loop corrections to light-by-light scattering in supersymmetric QED," JHEP **0205**, 060 (2002). [arXiv:hep-ph/0202266].
14. W. Bardeen, "Self-dual Yang-Mills theory, integrability and multiparton amplitudes", Prog. Theor. Phys. Suppl. **123** (1996) 1; D. Cangemi, "Self-duality and maximally helicity violating QCD amplitudes", Int. J. Mod. Phys. **A12** (1997) 1215. [arXiv:hep-th/9610021].
15. L. C. Martin, C. Schubert and V. M. Villanueva Sandoval, "On the low-energy limit of the QED N - photon amplitudes," Nucl. Phys. B **668**, 335 (2003). [arXiv:hep-th/0301022].
16. G. 't Hooft, "Computation Of The Quantum Effects Due To A Four-Dimensional Pseudoparticle," Phys. Rev. D **14**, 3432 (1976). [Erratum-ibid. D **18**, 2199 (1978)].
17. R. Jackiw and C. Rebbi, "Degrees Of Freedom In Pseudoparticle Systems," Phys. Lett. B **67**, 189 (1977); "Spinor Analysis Of Yang-Mills Theory," Phys. Rev. D **16**, 1052 (1977).
18. G. V. Dunne, H. Gies and C. Schubert, "Zero modes, beta functions and IR/UV interplay in higher-loop QED," JHEP **0211**, 032 (2002) [arXiv:hep-th/0210240].
19. E. W. N. Glover, "Progress in NNLO calculations for scattering processes," Nucl. Phys. Proc. Suppl. **116**, 3 (2003); Z. Bern, "Recent progress in perturbative quantum field theory," Nucl. Phys. Proc. Suppl. **117**, 260 (2003). [arXiv:hep-ph/0212406].
20. D. Kreimer, "On the Hopf algebra structure of perturbative quantum field theories," Adv. Theor. Math. Phys. **2**, 303 (1998); [arXiv:q-alg/9707029]; A. Connes and D. Kreimer, "Renormalization in quantum field theory and the Riemann-Hilbert problem. I: The Hopf algebra structure of graphs and the main theorem," Commun. Math. Phys. **210**, 249 (2000); [arXiv:hep-th/9912092]; D. Kreimer, *Knots and Feynman Diagrams*, (Cambridge Univ Press, 2000).
21. G. Dunne, "Two-Loop Diagrammatics in a Self-Dual Background", hep-th/0311167.
22. C. Anastasiou, Z. Bern, L. Dixon and D. A. Kosower, "Planar amplitudes in maximally supersymmetric Yang-Mills theory," arXiv:hep-th/0309040.

A NEW APPROACH TO SCALE SYMMETRY BREAKING AND CONFINEMENT*

P. GAETE
Departamento de Física, Universidad Técnica F. Santa María, Casilla 110-V, Valparaíso, Chile.
patricio.gaete@fis.utfsm.cl

E. GUENDELMAN
Physics Department, Ben Gurion University, Beer Sheva 84105, Israel.
guendel@bgumail.bgu.ac.il

Scale invariant theories which contain (in $4-D$) a four index field strength are considered. The integration of the equations of motion of these $4-index$ field strength gives rise to scale symmetry breaking. The phenomena of mass generation and confinement are possible consequences of this.

1. Introduction

The idea that among the fundamental laws of physics, we have scale invariance (s.i.) as one of the fundamental principles appears as a attractive possibility. In its most naive realizations, such a symmetry is not a viable symmetry, however, since nature seems to have chosen some typical scales, so s.i. has to be broken somehow.

While one may look at quantum effects for the origin of scale symmetry breaking, there is also a way to achieve this at the classical level in certain models. In the first scale invariant model of this kind[1], a metric, a dilaton field and a "measure field" Φ are introduced. The unusual modified measure[2] is an object that has the same transformation properties under general coordinate transformations as $\sqrt{-g}d^4x\,(g = \det(g_{\mu\nu}))$. Φ is a density built out of degrees of freedom independent of the metric. For example, given a four index field strength independent of the metric. for example, given a four index field strength (in four dimensions) $F_{\mu\nu\alpha\beta} = \partial_{[\mu}A_{\nu\alpha\beta]}$,

*Talk presented by E.G.

the measure field Φ is defined as $\Phi = \varepsilon^{\mu\nu\alpha\beta} F_{\mu\nu\alpha\beta}$. One may consider the field strength to be composed of elementary scalars[3]. Four in the case of a four index field strength, as it was done in Refs.[1,2].

The scale symmetry breaking in Ref.[1] comes when considering the integration of the equation of motion of the four index field strength, which introduces an arbitrary scale in the equations of motion. The modified measure idea has been applied to dynamical generation of string and brane tensions[4], to cosmological[5] and to the fermion family problem[6] and to the cosmic coincidences in cosmology[7].

2. Spontaneous Generation of Confining Behavior

Let us now turn our attention to gauge theories and consider first the standard pure Yang-Mills action

$$S = \int d^4x \left(-\frac{1}{4} F^a_{\mu\nu} F^{a\mu\nu} \right), \tag{1}$$

where $F^a_{\mu\nu} = \partial_\mu A^a_\nu - \partial_\nu A^a_\mu + g f^{abc} A^b_\mu A^c_\nu$. This theory is invariant under the scale symmetry $A^a_\mu(x) \to A^a_\mu(x)' = \lambda A^a_\mu(\lambda x)$.

Let us rewrite (1) with the help of an auxiliary field ω

$$S = \int d^4x \left(-\frac{1}{4}\omega^2 + \frac{1}{2}\omega\sqrt{-F^a_{\mu\nu}F^{a\mu\nu}} \right), \tag{2}$$

upon solving the equation of motion obtained from the variation of ω and then replacing into the action (2) we get back (1).

Let us consider now[8] the replacement $\omega \to \varepsilon^{\mu\nu\alpha\beta}\partial_{[\mu} A_{\nu\alpha\beta]}$ and consider now the equation of motion obtained from the variation of $A_{\nu\alpha\beta}$, which is

$$\varepsilon^{\gamma\delta\alpha\beta}\partial_\beta \left(\omega - \sqrt{-F^a_{\mu\nu}F^{a\mu\nu}} \right) = 0, \tag{3}$$

which is solved by

$$\omega = \sqrt{-F^a_{\mu\nu}F^{a\mu\nu}} + M, \tag{4}$$

M being once again a space-time constant which produces s.s.b. of s.i. and in this case it is furthermore associated with the spontaneous generation of confining behavior. Other examples of mass generation were studied also in Ref.[8] with similar technique. Indeed the equations of motion obtained from (2) in the case ω is replaced by $\varepsilon^{\mu\nu\alpha\beta}\partial_{[\mu}A_{\nu\alpha\beta]}$, have the form

$$\nabla_{\mu}\left(\left(\sqrt{-F^{b}_{\alpha\beta}F^{b\alpha\beta}}+M\right)\frac{F^{a\mu\nu}}{\sqrt{-F^{b}_{\alpha\beta}F^{b\alpha\beta}}}\right)=0. \tag{5}$$

Now study the equation (5) for the case of a Abelian theory and for a spherically symmetric electric field $F_{0i} = -E_i$ and $F_{ij} = 0$, where $\mathbf{E} = E(r)\hat{\mathbf{r}}$. Then (5) gives

$$\nabla\cdot\left(\mathbf{E}+\frac{M}{\sqrt{2}}\hat{\mathbf{r}}\right)=0, \tag{6}$$

which is solved by[9]

$$\mathbf{E}=-\frac{M}{\sqrt{2}}\hat{\mathbf{r}}+\frac{q}{r^2}\hat{\mathbf{r}}. \tag{7}$$

The scalar potential V that gives rise to such electric field is

$$V=-\frac{M}{\sqrt{2}}r+\frac{q}{r}, \tag{8}$$

which is indeed resembles very much the Cornell confining potential. Notice that so far (8) refers to the field of one charge and not yet to the interaction energy between two charges. We will see that such interaction energy also has the Cornell form, even at the quantum level. Since Abelian solutions are solutions of the non-Abelian theory, these solutions are also relevant for the non-Abelian generalization.

Before approaching the quantum theory (which will be treated in some approximations) we want to define effective actions that give the equations of motion (5). Indeed one can easily see that [9]

$$L_{eff}=-\frac{1}{4}F_{\mu\nu}F^{\mu\nu}-\frac{M}{4}\sqrt{-F_{\mu\nu}F^{\mu\nu}}, \tag{9}$$

reproduces Eqs. (5).

Since the full treatment of the quantum theory is rather difficult, instead of using (9) we restrict ourselves to a "truncated" phase space model where we consider spherical coordinates (r,θ,φ) in addition to time, but where we set $F_{ij} = 0 = F_{0\varphi} = F_{0\theta}$ and consider only (t,r) dependence of F_{0r}. Then instead of (9), we consider

$$S=4\pi\int drr^2L_{eff}, \tag{10}$$

where

$$L_{eff} = \frac{1}{2}\left(F_{0r}\right)^2 - \frac{M\sqrt{2}}{4}F_{0r}. \tag{11}$$

Similar kind of "reduced phase space" which take into account only the spherical degrees of freedom have been used elsewhere in other examples, see for example Ref.[10].

3. Interaction Energy

As already mentioned, our aim now is to calculate the interaction energy between external probe sources in the model (10). To do this, we will compute the expectation value of the energy operator H in the physical state $|\Phi\rangle$, which we will denote by $\langle H\rangle_\Phi$. The starting point is the two-dimensional space-time Lagrangian (11):

$$L = 4\pi r^2\left\{-\frac{1}{4}F_{\mu\nu}F^{\mu\nu} - \frac{M\sqrt{2}}{8}\varepsilon_{\mu\nu}F^{\mu\nu}\right\} - A_0J^0, \tag{12}$$

where J^0 is the external current. A notation remark, in (12), $\mu,\nu = 0,1$, also, $x^1 \equiv r \equiv x$ and $\varepsilon^{01} = 1$.

We now proceed to obtain the Hamiltonian. For this we restrict our attention to the Hamiltonian framework of this theory. The canonical momenta read $\Pi^\mu = -4\pi x^2\left(F^{0\mu} + \frac{M\sqrt{2}}{8}\varepsilon^{0\mu}\right)$, which results in the usual primary constraint $\Pi^0 = 0$, and $\Pi^i = -4\pi x^2\left(F^{0i} + \frac{M\sqrt{2}}{8}\varepsilon^{0i}\right)$. The canonical Hamiltonian following from the above Lagrangian is:

$$H_C = \int dx\left(\Pi_1\partial^1A^0 - \frac{1}{8\pi x^2}\Pi_1\Pi^1 - \frac{M\sqrt{2}}{4}\varepsilon^{01}\Pi_1 + A_0J^0\right). \tag{13}$$

The consistency condition $\dot{\Pi}_0 = 0$ leads to the secondary constraint $\Gamma_1(x) \equiv \partial_1\Pi^1 - J^0 = 0$. It is straightforward to check that there are no further constraints in the theory, and that the above constraints are first class. The extended Hamiltonian that generates translations in time then reads $H = H_C + \int dx\,(c_0(x)\Pi_0(x) + c_1(x)\Gamma_1(x))$, where $c_0(x)$ and $c_1(x)$ are the Lagrange multipliers. Moreover, it follows from this Hamiltonian that $\dot{A}_0(x) = [A_0(x), H] = c_0(x)$, which is an arbitrary function. Since $\Pi_0 = 0$, neither A^0 nor Π^0 are of interest in describing the system and may be discarded from the theory. The Hamiltonian then takes the form

$$H = \int dx\left(-\frac{1}{8\pi x^2}\Pi_1\Pi^1 - \frac{M\sqrt{2}}{4}\varepsilon^{01}\Pi_1 + c'\left(\partial_1\Pi^1 - J^0\right)\right), \tag{14}$$

where $c'(x) = c_1(x) - A_0(x)$.

According to the usual procedure we introduce a supplementary condition on the vector potential such that the full set of constraints becomes second class. A convenient choice is found to be [11,12,13,14]

$$\Gamma_2(x) \equiv \int\limits_{C_{\xi x}} dz^\nu A_\nu(z) \equiv \int\limits_0^1 d\lambda x^1 A_1(\lambda x) = 0, \tag{15}$$

where λ $(0 \leq \lambda \leq 1)$ is the parameter describing the spacelike straight path $x^1 = \xi^1 + \lambda(x-\xi)^1$, and ξ is a fixed point (reference point). There is no essential loss of generality if we restrict our considerations to $\xi^1 = 0$. In this case, the only nontrivial Dirac bracket is

$$\left\{A_1(x), \Pi^1(y)\right\}^* = \delta^{(1)}(x-y) - \partial_1^x \int\limits_0^1 d\lambda x^1 \delta^{(1)}(\lambda x - y). \tag{16}$$

We are now equipped to compute the interaction energy between point-like sources in the model (14), where a fermion is localized at the origin $\mathbf{0}$ and an antifermion at $\mathbf{y}$. As we have already mentioned, we will calculate the expectation value of the energy operator H in the physical state $|\Phi\rangle$. From our above discussion, we see that $\langle H\rangle_\Phi$ reads

$$\langle H\rangle_\Phi = \langle\Phi| \int dx \left(-\frac{1}{8\pi x^2}\Pi_1\Pi^1 - \frac{M\sqrt{2}}{4}\varepsilon^{01}\Pi_1\right)|\Phi\rangle. \tag{17}$$

Next, as remarked by Dirac[15], the physical state can be written as

$$|\Phi\rangle \equiv \left|\overline{\Psi}(\mathbf{y})\Psi(\mathbf{0})\right\rangle = \overline{\psi}(\mathbf{y}) \exp\left(ie\int\limits_{\mathbf{0}}^{\mathbf{y}} dz^i A_i(z)\right)\psi(\mathbf{0})|0\rangle, \tag{18}$$

where $|0\rangle$ is the physical vacuum state. As we have already indicated, the line integral appearing in the above expression is along a spacelike path starting at $\mathbf{0}$ and ending $\mathbf{y}$, on a fixed time slice.

Taking into account the above Hamiltonian structure, we observe that

$$\Pi_1(x)\left|\overline{\Psi}(y)\Psi(0)\right\rangle = \overline{\Psi}(y)\Psi(0)\Pi_1(x)|0\rangle - e\int_0^y dz_1 \delta^{(1)}(z_1 - x)|\Phi\rangle. \tag{19}$$

Inserting this back into (17), we get

$$\begin{aligned}\langle H\rangle_\Phi = \langle H\rangle_0 &+ \tfrac{e^2}{8\pi}\int dx \tfrac{1}{x^2}\left(\int_0^y dz_1\delta^{(1)}(z_1-x)\right)^2 \\ &+ \tfrac{M\sqrt{2}e}{4}\int dx\left(\int_0^y dz_1 \delta^{(1)}(z_1-x)\right),\end{aligned} \tag{20}$$

where $\langle H \rangle_0 = \langle 0 | H | 0 \rangle$. We further note that

$$\frac{e^2}{2} \int dx \left(\int_0^y dz \delta^{(1)} (z_1 - x) \right)^2 = \frac{e^2}{2} L, \tag{21}$$

with $|y| \equiv L$. Inserting this into Eq.(20), the interaction energy in the presence of the static charges will be given by

$$V = -\frac{e^2}{8\pi} \frac{1}{L} + \frac{M\sqrt{2}e}{4} L, \tag{22}$$

which has the Cornell form. In this way the static interaction between fermions arises only because of the requirement that the $|\overline{\Psi}\Psi\rangle$ states be gauge invariant.

References

1. E. I. Guendelman, *Mod. Phys. Lett.* **A14**, 1043 (1999); *Mod. Phys. Lett.* **A14**, 1397 (1999); gr-qc/9901067.
2. For a review on non scale invariant versions of these models see E. I. Guendelman and A. B. Kaganovich, *Phys. Rev.* **D60**, 065004 (1999). Early papers on the subject: E. I. Guendelman *Phys. Rev.* **D53**, 7020 (1996); *Phys. Rev.* **D55**, 5970 (1997).
3. E. I. Guendelman, E. Nissimov and S. Pacheva, *Phys. Lett.* **B360**, 57 (1995); hep-th/9903245; C. Castro *Int. J. Mod. Phys.* **A13**, 1263 (1998).
4. E. I. Guendelman, *Class Quantum Grav.* **17**, 3673 (2000); E. I. Guendelman, *Phys. Rev.* **D63**, 046006 (2001); E. I. Guendelman, A. B. Kaganovich, E. Nissimov and S. Pacheva, *Phys. Rev.* **D66**, 046003 (2002); hep-th/0304269; hep-th/0210062.
5. E. I. Guendelman and O. Katz, *Class. Quantum Grav* **20**, 1715 (2003) and references there.
6. E. I. Guendelman and A. B. Kaganovich, *Int. J. Mod. Phys.* **A17**, 417 (2002); *Mod. Phys. Lett.* **A17**, 1227 (2002); *Int. J. Mod. Phys.* **A17**, 4419 (2002); *Int. J. Mod. Phys.* **D11**, 1595 (2002).
7. E. I. Guendelman and A. B. Kaganovich, gr-qc/0312006.
8. E. I. Guendelman, hep-th/0306162.
9. P. Gaete and E. I. Guendelman, hep-th/0308025.
10. R. Benguria, P. Cordero and C. Teitelboim, *Nucl.Phys.* **B122**, 61 (1977).
11. P. Gaete, *Phys. Lett.* **B515**, 32 (2001).
12. P. Gaete, *Phys. Rev.* **D59**, 127702 (1999).
13. P. Gaete and I. Schmidt, *Phys. Rev.* **D61**, 125002 (2000).
14. P. Gaete and I. Schmidt, *Phys. Rev.* **D64**, 027702 (2001).
15. P. A. M. Dirac, *The Principles of Quantum Mechanics (Oxford University Press, Oxford, 1958)*; *Can. J. Phys.* **33**, 650 (1955).

GENERALIZATIONS OF YANG-MILLS THEORY WITH NONLINEAR CONSTITUTIVE EQUATIONS*

GERALD A. GOLDIN

Departments of Mathematics and Physics, Rutgers University, Piscataway, New Jersey 08854

VLADIMIR SHTELEN

Department of Mathematics, Rutgers University, Piscataway, New Jersey 08854

We generalize classical Yang-Mills theory by extending nonlinear constitutive equations for Maxwell fields to non-Abelian gauge groups. Such theories may or may not be Lagrangian. We obtain conditions on the constitutive equations specifying the Lagrangian case, of which recently-discussed non-Abelian Born-Infeld theories are particular examples. Some models in our class possess nontrivial Galilean ($c \to \infty$) limits; we determine when such limits exist, and obtain them explicitly.

1. Introduction

General equations for nonlinear, classical electromagnetic fields in media can be written beginning with Maxwell's equations for **E**, **B**, **D**, and **H**, and replacing the usual, *linear* constitutive equations by more general, *nonlinear* equations respecting Lorentz covariance. A general form for such systems was described by Fuschych, Shtelen, and Serov; familiar special cases include Born-Infeld and Euler-Kockel electrodynamics.[1,2] In earlier work we showed that certain nonlinear constitutive equations have well-defined Galilean-covariant limits as the speed of light $c \to \infty$, so that all four of Maxwell's equations remain valid.[3] This is in sharp contrast to linear electrodynamics, where Maxwell's equations are well-known to be incompatible with Galilean relativity.[4,5] Since classical Yang-Mills theory can be understood as an extension of classical electromagnetism to non-Abelian gauge potentials, it is natural to similarly extend Maxwell fields with nonlinear

*Talk presented by G.A.G.

constitutive equations, and to ask whether such extensions also may have Galilean-covariant limits when $c \to \infty$. Non-Abelian generalizations of the Born-Infeld Lagrangian[6] have been known for some time, and recently have attracted renewed interest.[7,8,9,10,11] In this paper we take a different approach, deriving generalizations of classical Yang-Mills theory as non-Abelian extensions of Maxwell systems together with Lorentz-covariant, nonlinear constitutive equations. Standard Yang-Mills theory is a special case of this class, with linear constitutive equations. Particular nonlinear constitutive equations result in Born-Infeld theories. Our approach has important advantages: first, it is general enough to include Lagrangian and non-Lagrangian theories, and secondly, since we directly generalize *nonlinear* Maxwell systems, we have the possibility of obtaining nontrivial Galilean-covariant (nonrelativistic) limits as $c \to \infty$. In Sec. 2 we review Maxwell's equations for media, and characterize the family of nonlinear constitutive equations that results in a theory obtained from an invariant Lagrangian. In Sec. 3 we generalize appropriately from $U(1)$ to non-Abelian gauge theory, and consider the nonrelativistic limit.

2. Nonlinear Electrodynamics

We use SI units, so that c does not enter the definition of $\mathbf{E}$ or $\mathbf{B}$. We have the metric tensor $g^{\mu\nu} = \mathrm{diag}\,(1,-1,-1,-1)$, $x^\mu = (ct, \mathbf{x})$, $x_\mu = g_{\mu\nu}x^\nu = (ct, -\mathbf{x})$, and $x^\mu x_\mu = c^2t^2 - \mathbf{x}^2$. We have $\partial_\mu \equiv \partial/\partial x^\mu = [(1/c)\partial/\partial t, \nabla]$, and we use the antisymmetric Levi-Civita tensor $\varepsilon^{\alpha\beta\mu\nu}$ with $\varepsilon^{0123} = 1$. The tensor fields constructed from vectors $\mathbf{E}$, $\mathbf{B}$, $\mathbf{D}$, $\mathbf{H}$ are

$$F_{\alpha\beta} = \begin{pmatrix} 0 & \frac{1}{c}E_1 & \frac{1}{c}E_2 & \frac{1}{c}E_3 \\ -\frac{1}{c}E_1 & 0 & -B_3 & B_2 \\ -\frac{1}{c}E_2 & B_3 & 0 & -B_1 \\ -\frac{1}{c}E_3 & -B_2 & B_1 & 0 \end{pmatrix}, \qquad F^{\alpha\beta} = g^{\alpha\mu}g^{\beta\nu}F_{\mu\nu}\,,$$

$$\mathcal{F}^{\alpha\beta} = \frac{1}{2}\varepsilon^{\alpha\beta\mu\nu}F_{\mu\nu} = \begin{pmatrix} 0 & -B_1 & -B_2 & -B_3 \\ B_1 & 0 & \frac{1}{c}E_3 & -\frac{1}{c}E_2 \\ B_2 & -\frac{1}{c}E_3 & 0 & \frac{1}{c}E_1 \\ B_3 & \frac{1}{c}E_2 & -\frac{1}{c}E_1 & 0 \end{pmatrix}, \qquad \mathcal{F}_{\alpha\beta} = g_{\alpha\mu}g_{\beta\nu}\mathcal{F}^{\mu\ nu}\,,$$

$$G^{\alpha\beta} = \begin{pmatrix} 0 & -cD_1 & -cD_2 & -cD_3 \\ cD_1 & 0 & -H_3 & H_2 \\ cD_2 & H_3 & 0 & -H_1 \\ cD_3 & -H_2 & H_1 & 0 \end{pmatrix}, \qquad G_{\alpha\beta} = g_{\alpha\mu}g_{\beta\nu}G^{\mu\nu}\,. \tag{2}$$

Maxwell's equations for media in SI units take the form[2]

$$\nabla \times \mathbf{E} = -\frac{\partial \mathbf{B}}{\partial t}, \quad \nabla \cdot \mathbf{B} = 0, \quad \nabla \times \mathbf{H} = \frac{\partial \mathbf{D}}{\partial t} + \mathbf{j}, \quad \nabla \cdot \mathbf{D} = \rho, \tag{3}$$

or in covariant form

$$\partial_\alpha \mathcal{F}^{\alpha\beta} = 0; \quad \partial_\alpha G^{\alpha\beta} = j^\beta, \quad \text{where } j^\beta = (c\rho, \mathbf{j}). \tag{4}$$

The first equation in (4) allows us to introduce $A_\mu = (\phi, -\mathbf{A})$ so that

$$\mathcal{F}^{\alpha\beta} = \epsilon^{\alpha\beta\mu\nu}\partial_\mu A_\nu, \quad F_{\mu\nu} = \partial_\mu A_\nu - \partial_\nu A_\mu. \tag{5}$$

This system is invariant under the Lorentz group as well as the Galilei group; the choice between these symmetries rests in the constitutive equations.[1,3,4] The Lorentz invariant constitutive equations are

$$\mathbf{D} = M\mathbf{B} + \frac{1}{c^2}N\mathbf{E}, \quad \mathbf{H} = N\mathbf{B} - M\mathbf{E}, \tag{6}$$

or in covariant form

$$G^{\mu\nu} = NF^{\mu\nu} + cM\mathcal{F}^{\mu\nu} \equiv M_1 \frac{\partial I_1}{\partial F^{\mu\nu}} + M_2 \frac{\partial I_2}{\partial F^{\mu\nu}}, \tag{7}$$

where M, N, or equivalently M_1, M_2, are functions of Lorentz invariants

$$I_1 = \mathbf{B}^2 - \frac{1}{c^2}\mathbf{E}^2 = \frac{1}{2}\, F_{\mu\nu}F^{\mu\nu}, \quad I_2 = \mathbf{B}\cdot\mathbf{E} = -\frac{c}{4}\, F_{\mu\nu}\mathcal{F}^{\mu\nu}. \tag{8}$$

The standard Maxwell equations for the vacuum correspond to $M = 0$, $N = \text{constant} = (\mu_0)^{-1}$, with $c^2 = (\mu_0\epsilon_0)^{-1}$. A general form of an invariant Lagrangian for a nonlinear theory given by Eqs. (4) and (7) may be written $\mathcal{L} = \mathcal{L}(I_1,\ I_2)$, where $I_1,\ I_2$ are given by (8). In such a theory, the tensor $G^{\mu\nu}$ becomes

$$G^{\mu\nu} = \frac{\partial \mathcal{L}}{\partial F_{\mu\nu}} = \left(\frac{\partial \mathcal{L}}{\partial I_1}\right) 2\,F^{\mu\nu} + c\left(-\frac{\partial \mathcal{L}}{\partial I_2}\right)\mathcal{F}^{\mu\nu}. \tag{9}$$

Comparison of Eqs. (9) with Eqs. (7) yields the conditions

$$2\frac{\partial \mathcal{L}}{\partial I_1} = N, \quad -\frac{\partial \mathcal{L}}{\partial I_2} = M, \tag{10}$$

from which the compatibility condition for a Lagrangian theory reads,

$$2\frac{\partial M}{\partial I_1} + \frac{\partial N}{\partial I_2} = 0. \tag{11}$$

The well-known Born-Infeld Lagrangian is usually written as

$$\mathcal{L} = 1 - R, \quad R = \frac{b^2}{\mu_0 c^2}\sqrt{1 + \frac{c^2}{b^2}I_1 - \frac{c^2}{b^4}I_2^2}\,. \tag{12}$$

Another example is Euler-Kockel electrodynamics.[2] Here, in the first approximation, one has $M = 7\lambda(\mu_0)^{-1}I_2$ and $N = (\mu_0)^{-1}(1 - 2\lambda I_1)$, where λ is a small parameter. The corresponding Lagrangian takes the form

$$\mathcal{L} = -\frac{1}{2\mu_0}I_1 + 2\lambda I_1^2 + \frac{7\lambda}{2\mu_0}I_2^2\,, \tag{13}$$

which coincides with the "toy model" generalization of the Maxwell Lagrangian discussed by DeLorenci *et al.*[13]

3. Generalization of Yang-Mills Theory

To generalize the nonlinear electrodynamics described in Sec. 2 to non-Abelian gauge theory, we replace as usual the partial derivative ∂_μ by the commutator with the covariant derivative D_μ; i.e. $\partial_\mu \to [D_\mu, \quad]$, where

$$D_\mu = \partial_\mu + igT^\ell W_\mu^\ell\,, \tag{14}$$

g is the YM coupling constant, the T^ℓ are the $N^2 - 1$ generators of $SU(N)$, and summation over ℓ is assumed. Then

$$[D_\mu, D_\nu] = igF_{\mu\nu}\,, \text{ where } F_{\mu\nu} = T^\ell F_{\mu\nu}^\ell\,. \tag{15}$$

The field equations of the non-Abelian theory generalizing Maxwell's equations (4) and (7) take the form

$$[D_\mu, \mathcal{F}^{\mu\nu}] = 0, \quad [D_\mu, G^{\mu\nu}] = J^\nu\,, \tag{16}$$

where J^ν is an external current, and where the constitutive equations are to be written in a new way. Letting u_s $(s = 1, 2, 3, \dots)$ be the independent invariants of the Yang-Mills fields, we write

$$G^{\ell\,\mu\nu} = \sum_s M_s(u_1, u_2, \dots)\,\frac{\partial u_s}{\partial F_{\mu\nu}^\ell}\,, \tag{17}$$

where the M_s are functions of the invariants. For the gauge group $SU(N)$ we have no fewer than $5N^2 - 11$ independent invariants, using a simple argument of Roskies.[12] Since the gauge group $SU(N)$ has $N^2 - 1$ parameters, and the Lorentz group has 6 parameters, the number of components of $F_{\mu\nu}^\ell$ is $6(N^2 - 1)$. One can choose a Lorentz frame and an $O(N)$ frame in which $6 + (N^2 - 1) = 5 + N^2$ components vanish. There will then be $6(N^2 - 1) - (5 + N^2) = 5N^2 - 11$ remaining components. Any invariant could be evaluated in this special frame, and therefore could be a function of these $5N^2 - 11$ components. In particular, there are 9 independent invariants for $SU(2)$, as follows:[12]

$$u_1 = \mathrm{tr}(K) = \mathbf{B}^2 - \frac{1}{c^2}\mathbf{E}^2, \quad u_2 = -\frac{1}{2}\mathrm{tr}(J) = \mathbf{B}\cdot\mathbf{E},$$

$$u_3 = \frac{1}{4}\mathrm{tr}(J^2),\ u_4 = \det(J),\ u_5 = \mathrm{tr}(K^2),\ u_6 = \det(K),\ u_7 = \mathrm{tr}(JK),$$

$$u_8 = \frac{1}{6}\varepsilon_{ijk}F^{i\ \nu}_{\mu}F^{j\ \rho}_{\nu}F^{k\ \mu}_{\rho},\quad u_9 = -\frac{c}{6}\varepsilon_{ijk}\mathcal{F}^{i\ \nu}_{\mu}\mathcal{F}^{j\ \rho}_{\nu}\mathcal{F}^{k\ \mu}_{\rho}, \tag{18}$$

where

$$K_{ij} = \frac{1}{2}\ F^i_{\mu\nu}F^{j\ \nu\mu},\quad J_{ij} = \frac{c}{2}\ F^i_{\mu\nu}\mathcal{F}^{j\ \nu\mu},, \tag{19}$$

with $i, j, k = 1, 2, 3$. The factors of c in Eqs. (18)-(19) have been introduced so that in all cases, the limit $c \to \infty$ results in survival of the leading terms. There are now Lagrangian and non-Lagrangian theories determined by Eqs. (16)-(17). In a Lagrangian theory the constitutive equations are

$$G^{\ell\,\mu\nu} = \frac{\partial\mathcal{L}}{\partial F^{\ell}_{\mu\nu}} = \sum_s \frac{\partial\mathcal{L}}{\partial u_s}\frac{\partial u_s}{\partial F^{\ell}_{\mu\nu}}. \tag{20}$$

Thus Eqs. (16)-(17) determine a Lagrangian theory if and only if the coefficients M_s in (17) can be written as $M_s = \partial\mathcal{L}/\partial u_s$ for some scalar-valued function $\mathcal{L} = \mathcal{L}(u_1, u_2, \ldots)$. The corresponding restrictions on the M_s are the compatibility conditions resulting from equalities of the mixed derivatives of $\mathcal{L}$ with respect to u_r and u_s; i.e., $\partial M_s/\partial u_r = \partial M_r/\partial u_s$. In particular, one obtains non-Abelian versions of Born-Infeld or Euler-Kockel theory by taking various generalizations of the respective Lagrangians discussed in Sec. 2. Finally, let us consider the possibility of a nonrelativistic limit. Galilean symmetry transformations (the $c \to \infty$ limit of Lorentz transformations) have the form

$$t' = t,\ \mathbf{x}' = \mathbf{x} - \mathbf{v}t,\quad (\mathbf{E}^{\ell})' = \mathbf{E}^{\ell} + \mathbf{v}\times\mathbf{B}^{\ell},\quad (\mathbf{B}^{\ell})' = \mathbf{B}^{\ell}.$$

Note that there is no corresponding nonrelativistic limit of the standard Yang-Mills equations, since the constitutive equations $G^{\ell\,\mu\nu} = (\mu_0)^{-1}F^{\ell\,\mu\nu}$ immediately break the Galilean symmetry. But our equations (16)-(17) *can* have a $c \to \infty$ limit, *provided the constitutive equations are nonlinear.* One obtains such a Galilean non-Abelian gauge theory from Eqs. (16)-(17), writing these equations explicitly in terms of $\mathbf{E}^{\ell}, \mathbf{B}^{\ell}, \mathbf{D}^{\ell}$ and $\mathbf{H}^{\ell}$, and then taking the limit as $c \to \infty$. The equations of motion (16) will always be the same as in relativistic theory, as the factors of c cancel; only the constitutive equations (17) will be different. With a little help from Maple, we calculated the explicit form of the invariants $u_1, \ldots, u_9$ in Eqs. (18), as well as their limits $\hat{u}_1, \ldots, \hat{u}_9$ as $c \to \infty$. In the notation that follows, B_{13} is the third component of $\mathbf{B}^{\ell=1}$; thus the $\mathbf{B}^{\ell}$ (gauge components $\ell = 1, 2, 3$)

are vectors with spatial components $B_{\ell 1},\ B_{\ell 2},\ B_{\ell 3}$. Here are some of these Galilean YM gauge invariants:

$$\hat{u}_1 = \mathbf{B}^\ell \cdot \mathbf{B}^\ell = B_{11}^2 + B_{12}^2 + B_{13}^2 + B_{21}^2 + B_{22}^2 + B_{23}^2 + B_{31}^2 + B_{32}^2 + B_{33}^2\,,$$

$$\hat{u}_2 = u_2 = \mathbf{B}^\ell \cdot \mathbf{E}^\ell\,,$$

$$\hat{u}_3 = u_3 = (\mathbf{B}^1 \cdot \mathbf{E}^1)^2 + (\mathbf{B}^2 \cdot \mathbf{E}^2)^2 + (\mathbf{B}^3 \cdot \mathbf{E}^3)^2$$
$$+\frac{1}{2}\,[\,(\mathbf{B}^1 \cdot \mathbf{E}^2 + \mathbf{B}^2 \cdot \mathbf{E}^1)^2 + (\mathbf{B}^2 \cdot \mathbf{E}^3 + \mathbf{B}^3 \cdot \mathbf{E}^2)^2 + (\mathbf{B}^3 \cdot \mathbf{E}^1 + \mathbf{B}^1 \cdot \mathbf{E}^3)^2\,]\,,$$

$$\hat{u}_8 = \frac{1}{6}\,\epsilon_{ijk}(\mathbf{B}^i \times \mathbf{B}^j) \cdot \mathbf{B}^k = (\mathbf{B}^1 \times \mathbf{B}^2) \cdot \mathbf{B}^3\,,$$

$$\hat{u}_9 = -\frac{1}{2}\,\epsilon_{ijk}(\mathbf{B}^i \times \mathbf{B}^j) \cdot \mathbf{E}^k\,. \tag{21}$$

Acknowledgment

GG thanks R. Kerner for interesting discussions and helpful suggestions.

References

1. W. I. Fushchich, V. M. Shtelen, and N. I. Serov, *Symmetry Analysis and Exact Solutions of Equations of Nonlinear Mathematical Physics,* Kluwer Acad. Publ., Dordrecht (1993).
2. J. D. Jackson, *Classical Electrodynamics,* third edition, J. Wiley & Sons, Inc., New York (1999).
3. G. A. Goldin and V. M. Shtelen, *Phys. Lett. A* **279**, 321 (2001)
4. M. Le Bellac and J.-M. Levy-Leblond, *Nuovo Cim.* **14B**, 217 (1973)
5. H. R. Brown, P. R. Holland, *Am. J. Phys.* **67**, 204 (1999)
6. M. Born and L. Infeld, *Proc. R. Soc. London* **144**, 425 (1934)
7. T. Hagiwara, *J. Phys. A* **14**, 3059 (1981)
8. A. A. Tseytlin, theory, *Nuclear Phys. B* **501**, 41 (1997)
9. J.-H. Park, *Phys. Lett. B* **458**, 471 (1999)
10. D. Gal'tsov R. Kerner, *Phys. Rev. Let* **84**, 5955 (2000)
11. E. Serié, T. Masson, and R. Kerner, *arXiv:hep-th/030710* **v2** (1 Sep 2003)
12. R.Roskies, *Phys. Rev. D* **14**, 1722 (1977)
13. V. A. DeLorenci, R. Klippert, M. Novello, and J. M. Salim, *Phys. Rev. D,* **65**, 063501 (2002)

EXPLORING THE 4D SUPERCONFORMAL ZOO*

KEN INTRILIGATOR and BRIAN WECHT
U.C.S.D. Physics Dept., 9500 Gilman Drive, La Jolla, CA 92093

We discuss a new constraint for determining the superconformal $U(1)_R$ symmetry of 4d $N=1$ SCFTs: It is the unique one which locally maximizes $a(R) \equiv 3TrR^3 - TrR$. This constraint comes close to proving the conjectured "a-theorem" for $N=1$ SCFTs. Using this "a-maximization", exact results can now be obtained for previously inaccessible 4d $N=1$ SCFTs. We apply this method to a rich class of examples: 4d $N=1$ SQCD with added matter chiral superfields in the adjoint representation. We classify a zoo of SCFTs, finding that Arnold's ADE singularity classification arises in classifying these theories via all possible relevant Landau-Ginzburg superpotentials. We verify that all RG flows are indeed compatible with the "a-theorem" conjecture, $a_{IR} < a_{UV}$, in every case.

There is an intuitive picture of renormalization group (RG) flows from the UV to the IR as being like the flows of water down mountains, through valleys, and eventually ending up in lakes. The water height corresponds to the number of "degrees of freedom" of the theory, which is decreasing because of the coarse graining removal of UV degrees of freedom in RG flow to the IR. Zamolodchikov [1] made this intuition precise in 2d by defining a c-function which counts the number of degrees of freedom of the theory, and proving that it monotonically decreases along RG flows to the IR. In the far IR, the RG can flow to a fixed point, which is a conformal field theory (CFT). The c-function is stationary at the CFT endpoint of the flow, and in the 2d case it becomes the central charge c of the Virasoro algebra.

The intuition about course graining suggests that such a monotonically decreasing c-function should exist for RG flows in any space-time dimension, but a definitive proof above 2d has been elusive. Cardy[2] has suggested that an appropriate central charge for 4d CFTs (i.e. only the endpoints of the RG flow) is the coefficient, conventionally called "a", of the conformal

*Talk presented by K.I.

anomaly $\langle T^\mu_\mu \rangle$ of the theory on a curved space-time background[a] such as S^4. This proposal is the conjectured 4d "a-theorem", that all 4d RG flows have $a_{IR} < a_{UV}$. All known 4d RG flows are compatible with this conjecture[b]. We'll discuss some new RG flows and fixed points which provide additional striking, non-trivial evidence for its validity.

4d asymptotically free gauge theories have $g(\mu) \to 0$ in the extreme UV, i.e. free quarks and gluons, with g increasing in the flow to the IR. Eventually, non-perturbative phenomena can take over, e.g. confinement and a mass gap, and the theory in the far IR can be a free field theory of composites, e.g. pions, or have no massless degrees of freedom at all. Another possibility is a non-trivial IR fixed point, an interacting CFT (no mass gap). This happens if the RG flow drives the coupling g to a critical value g_*, where the beta functions vanish, $\beta(g_*) = 0$. A scenario for realizing this phase was argued for long ago[5,6]: if the theory is just barely asymptotically free, $\beta(g)$ can have a non-trivial zero, $\beta(g_*) = 0$, for $g_* \neq 0$ which is sufficiently weak to justify the perturbative analysis.

Interacting 4d CFTs were once thought to be very rare and exotic, perhaps only occurring in contrived examples. However, we now know of many such interacting RG fixed points. In this talk, we'll mention some new ones, which we've recently found and explored.

For the case of supersymmetric gauge theories, some exact results can be obtained, giving better insight into their non-perturbative dynamics. These studies have dovetailed with the use of other non-perturbative methods, which do not rely on supersymmetry, to explore more general strongly interacting gauge theories. A lesson learned from these studies is that the non-Abelian Coulomb phase exists somewhat generically, provided that there are sufficiently many massless matter fields (but not too many so as to spoil asymptotic freedom).

For example, in $N = 1$ supersymmetric QCD, Seiberg[7] argued that the theory flows to an interacting 4d $N = 1$ SCFT for N_f in the range

---

[a]On a general background $\langle T^\mu_\mu \rangle \sim a(\text{Euler}) + c(\text{Weyl}^2)$. It is known from counterexamples that the coefficient "c" of the Weyl curvature squared term does *not* obey a 4d analog of the c-theorem: there are RG flows with $c_{IR} > c_{UV}$ [3].

[b]Another conjectured c-function candidate is the coefficient of T^4 in the free-energy per unit volume of the theory at finite temperature[4]. This has some advantages over the conformal anomaly: it is not restricted to even space-time dimensions, and it places tighter bounds on the maximum number of massless flavors where chiral symmetry breaking can occur. But it has the disadvantage that it can only be readily computed in free field theories. In particular, supersymmetry does not help to compute it, so we will not discuss this thermal c-function candidate further here.

$\frac{3}{2}N_c < N_f < 3N_c$. More generally, we suspect that $N = 1$ SCFTs are fairly generic for matter content in the range $T(G) < T(Matter) < 3T(G)$ (there is also the more exotic possibility of the free-magnetic phase, as seen in SQCD[7]; see e.g. [9] for a review). Here $T(Matter)$ and $T(G)$ are the quadratic Casimirs of the matter and adjoint representations of the gauge group G, with $T(Matter) < 3T(G)$ for asymptotic freedom and $T(Matter) > T(G)$ to avoid dynamically generated superpotentials, which would spoil conformal invariance.

Supersymmetry relates the stress tensor $T_{\mu\nu}$ to a $U(1)_R$ current R_μ: they are both in a current supermultiplet $T_{\alpha\dot{\alpha}}$ whose first component is the $U(1)_R$ current (α and $\dot{\alpha}$ are spinor indices). Let's call the charge associated with this $U(1)_R$ current R_T. Away from RG fixed points, R_T isn't really conserved: its non-conservation is related by supersymmetry to non-zero T^μ_μ and lack of scale invariance. As the theory RG flows to a SCFT, $R_T \to R_*$ which is conserved. 4d SCFTs necessarily have a conserved $U(1)_{R_*}$ symmetry, with $U(1)_{R_*} \subset SU(2,2|1)$, the 4d superconformal group. We refer to this $U(1)_{R_*}$ as the superconformal R-symmetry. The supersymmetry relation $R_\mu \leftrightarrow T_{\mu\nu}$ implies many powerful results, for example:

• $\Delta(\mathcal{O}) \geq \frac{3}{2}|R_*(\mathcal{O})|$ for all gauge invariant spin zero operators $\mathcal{O}$, with $\Delta(\mathcal{O})$ the exact operator dimension and $R_*(\mathcal{O})$ the operator's $U(1)_{R_*}$ charge. Chiral primary operators have $\Delta(\mathcal{O}) = \frac{3}{2}R_*(\mathcal{O})$, and additivity of the $U(1)_R$ charge for composite operators implies that they form a closed OPE ring, with additive operator dimensions. A unitarity bound, $\Delta(\mathcal{O}) \geq 1$ for spin zero operators, implies that $R_*(\mathcal{O}) \geq \frac{2}{3}$ for spin zero chiral primary operators, with $R_*(\mathcal{O}) = \frac{2}{3}$ if and only if $\mathcal{O}$ is a decoupled free field.

• R_* is anomaly free precisely if the exact NSVZ[10] beta function vanishes, $\beta_{NSVZ}(g_*) = 0$, as is appropriate for a CFT.

• The conformal anomalies a and c can be exactly related[3] to the superconformal R-symmetry's 't Hooft anomalies:

$$a = \frac{3}{32}(3\mathrm{Tr}R_*^3 - \mathrm{Tr}R_*) \qquad c = \frac{1}{32}(9\mathrm{Tr}R_*^3 - 5\mathrm{Tr}R_*).$$

This is extremely powerful, because 't Hooft anomaly matching [8] implies that such 't Hooft anomalies are constant along the RG flows[c] and can thus be evaluated in the weakly coupled UV limit. In what follows, we will rescale a to eliminate the factor of 3/32.

[c]Here we are supposing the the R_* symmetry is conserved along the RG flow, rather than being an accidental symmetry of the IR fixed point. Away from the IR fixed point, R_* differs from the R-current in the $T_{\alpha\dot{\alpha}}$ supermultiplet.

As an example, consider $SU(N_c)$ SQCD, with N_f flavors in the SCFT range[7] $\frac{3}{2}N_c < N_f < 3N_c$. This theory has a unique anomaly free $U(1)_R$ symmetry which commutes with all of the other flavor symmetries: the R-charges of the fundamentals are $R(Q_i) = R(\widetilde{Q}_i) = 1 - (N_c/N_f)$. This $U(1)_R$ is conserved along the entire RG flow, from $g = 0$ in the UV to the g_* in the IR where $\beta(g_*) = 0$, so it's natural to suppose that the superconformal $U(1)_{R_*}$ of the IR SCFT is this unique candidate (rather than being an accidental symmetry). This yields the exact operator dimensions of chiral primary operators via $\Delta = \frac{3}{2}R$, e.g. the gauge invariant mesons $M_{i\bar{j}} = Q_i\widetilde{Q}_{\bar{j}}$ have $\Delta(M) = \frac{3}{2}R(M) = 3(1 - (N_c/N_f))$. And, using 't Hooft anomaly matching, we exactly compute a and c, even at strongly interacting RG fixed points, using the spectrum of the UV free field theory.

More generally, to make use of the above powerful supersymmetry relations, one must be able to identify precisely *which* possible R-symmetry is the special one, R_*, of the superconformal algebra. It is generally not uniquely fixed by the symmetries, as it was in the SQCD example. Indeed, if the theory has a large group $\mathcal{F}$ of non-R global flavor symmetries, with charges F_I, we can make a general R-symmetry by combining any initial R-symmetry, R_0, with any linear combination of the flavor symmetries:

$$R_t = R_0 + \sum_I s_I F_I,$$

with the s_I real parameters. The question, then, is to determine the particular values of the s_I which produce the special $U(1)_{R_*} \subset SU(2,2|1)$.

We recently found a simple solution for this problem[12]: the superconformal R-symmetry is the unique choice of the trial R-symmetry, R_t as given above, which locally maximizes

$$a_{trial}(s_I) = 3\mathrm{Tr}R_t^3 - \mathrm{Tr}R_t.$$

The value of a_{trial} at this local maximum is then the central charge a of the SCFT, so we refer to this procedure for finding the R-charge as "a-maximization." (Because a_{trial} is a cubic function, there is a unique local maximum, but no global maximum.)

We proved the above by showing that the $T_{\mu\nu} \leftrightarrow R_\mu$ supersymmetry relation implies that $9\mathrm{Tr}R_*^2F_I = \mathrm{Tr}F_I$, from which it follows that the superconformal $U(1)_R$ extremizes a_{trial}, and also $\mathrm{Tr}R_*F_IF_J < 0$, which implies that the extremum is a local maximum. The relation $9\mathrm{Tr}R_*^2F_I = \mathrm{Tr}F_I$ is obtained by relating the $F_IR_*^2$ three-current triangle anomaly to another triangle anomaly, with the R_* currents replaced by stress tensors.

The relation $\mathrm{Tr}R_*F_IF_J < 0$ comes from relating the $F_IF_JR_*$ three-current triangle anomaly to the F_I, F_J, stress tensor three-point function, which is then related to the F_I, F_J current-current two point function by a general relation for 4d CFTs [13]. A unitarity condition, fixing the sign of the current-current two-point function, then implies $\mathrm{Tr}R_*F_IF_J < 0$.

As a simple example to illustrate the a-maximization procedure, consider a free chiral superfield Φ, with trial R-charge r. We compute $a_{trial} = 3(r-1)^3 - (r-1)$, which has a unique local maximum at $r = 2/3$, the correct value for a free field: $\Delta = 3R/2 = 1$. a-maximization for interacting theories is essentially just as simple as this free-field example, the only difference being that some constraints associated with the interactions must be imposed. For example, gauge interactions simply have the effect of imposing the condition that the R-symmetry must be anomaly free.

a-maximization also suggests a simple proof of the conjectured a-theorem, at least in the supersymmetric context[12]. Consider a general RG flow, from some CFT in the UV to another CFT in the IR. Often the flavor symmetry group of the IR theory is a subgroup of that of the UV theory, $\mathcal{F}_{IR} \subset \mathcal{F}_{UV}$, because the relevant deformations of the UV CFT broke some of the flavor symmetries. It then follows from a-maximization that $a_{IR} < a_{UV}$, simply because maximizing over a subset leads to a smaller value. This proof is not fully general, because there are examples where $\mathcal{F}_{IR}$ is not a subset of $\mathcal{F}_{UV}$, because of accidental symmetries. In all such examples, $a_{IR} < a_{UV}$ is still satisfied, but this loophole in the proof needs filling. Another concern is that perhaps maximizing over a subspace need not lead to a smaller value, because it's only a local maximum. This latter caveat can be dispensed with by a recent work by Kutasov[22], where the constraints associated with $\mathcal{F}_{IR} \subset \mathcal{F}_{UV}$ are imposed with Lagrange multipliers λ and it's shown that $a(\lambda)$ is monotonically decreasing between the UV and IR endpoints.

We also note that a-maximization always yields the R-charges as solutions of quadratic equations with rational coefficients. Thus, for any 4d $\mathcal{N} = 1$ SCFT, all of the operator R-charges, chiral primary operator dimensions, and central charges a and c are always algebraic numbers. Specifically, they're solutions of quadratic equations with rational coefficients ("quadratic irrational numbers"). In particular, these quantities can't depend on any continuous moduli.

Finally, a general caution: we must maximize $a_{trial} = 3\mathrm{Tr}R_t^3 - \mathrm{Tr}R_t$ over the complete space of all possible trial R-symmetries, including all accidental symmetries. One situation where accidental symmetries are readily apparent, and required, is when a gauge invariant chiral primary operator, e.g. $M = \widetilde{Q}Q$, hits or appears to violate the unitarity bound, $R(M) \geq 2/3$. M then becomes a free field, with an accidental symmetry, J_M, under which only M is charged. The trial R_t must then include mixing with J_M: $R_{t,new} = R_{t,old} + s_M J_M$, with the new parameter s_M again fixed by maximizing a_{trial}. This has an important effect, leading to an additive correction to the quantity a_{trial} to be maximized w.r.t. the other parameters s_I, as was first discussed by Kutasov, Parnachev, and Sahakyan (KPS) [14].

a-maximization can be used to study previously mysterious theories. As an example, consider $SU(N_c)$ SQCD with $0 < N_f < 2N_c$ fundamental flavors, along with an extra matter chiral superfield, X, in the adjoint representation. This theory, with $W_{tree} = 0$, is believed to flow in the IR to an interacting $\mathcal{N} = 1$ SCFT. This was argued for in [9] by turning on $W_{tree} = \lambda Q X \widetilde{Q}$, finding massless monopoles and dyons on submanifolds of a Coulomb branch, and noting that these non mututally-local fields all become massless at the origin upon taking $\lambda \to 0$. We refer to these IR SCFTs as the $\widehat{A}(N_c, N_f)$ theories, or simply $\widehat{A}$. We can also consider the theory with added superpotential $W_{tree} = \mathrm{Tr}X^{k+1}$ which, for particular ranges of N_f (depending on k), can drive a RG flow to different SCFTs in the IR, which we'll refer to as the A_k SCFTs. Finding the $U(1)_R$ symmetry of the $\widehat{A}$ SCFTs requires a-maximization, while that of the A_k theory is determined by W_{tree} and the symmetries.

Let's outline the salient features of the $\widehat{A}$ and A_k theories, obtained by KPS[14]. Write the trial $U(1)_R$ as $R(Q_i) = R(\widetilde{Q}_i) \equiv y$, $R(X) = (1-y)/x$, with $x \equiv N_c/N_f$. To simplify things, take N_c and N_f large, holding $x > \frac{1}{2}$ fixed. For $x \approx \frac{1}{2}$, the theory is just barely asymptotically free, and the RG fixed point is of the "Banks-Zaks" type [5,6], at weak gauge coupling $g_* \ll 1$. As we increase x, the RG fixed point moves to larger and larger values of the gauge coupling g_*. a-maximization can be used to determine the exact R-charges $y(x)$, for all x. As expected, $y(x)$ and $R(X)$ decrease with x, corresponding to the negative anomalous dimensions of gauge interactions.

As x increases, generalized mesons of the form $M_j = \widetilde{Q}X^j Q$ successively hit, and then appear to violate, the unitarity bound $R(\mathcal{O}) \geq 2/3$. Each time this happens, there is an associated accidental flavor symmetry which must be included in the trial R_t, leading to added contributions to the quantity a_{trial} which is to be maximized. Because of this, it's best to let a computer solve the required a-maximizations, numerically obtaining $y(x)$. The $x \to \infty$ limit can be treated analytically, with the result that $R(Q) \to (\sqrt{3}-1)/3 \approx 0.244$, and $R(X) \to 0$ in the limit [14].

Suppose that we sit at the $\widehat{A}$ RG fixed point, and then perturb the theory by adding a superpotential $W_{A_k} = \lambda \mathrm{Tr} X^{k+1}$. This deformation is relevant if $\Delta(W_{tree}) < 3$, i.e. if $R(X) < 2/(k+1)$. Using the a-maximization results, it is seen that W_{A_k} can be relevant for any value of k, arbitrarily large, provided that x is sufficiently large, $x > x_{A_k}^{min}$, with $x_{A_k}^{min}$ determined by solving $R(X) = (1 - y(x_{A_k}^{min}))/x_{A_k}^{min} = 2/(k+1)$. For any $x > x_{A_k}^{min}$, W_{A_k} is a relevant deformation of the $\widehat{A}$ SCFT, driving a RG flow away from the $\widehat{A}$ SCFT to a new SCFT, which we call A_k. The A_k SCFTs exist for x in the range $x_{A_k}^{min} < x < x_{A_k}^{max} = k$, with $x_{A_k}^{max} = k$ the stability bound [15]. This x range is non-empty, $x_{A_k}^{min} < x_{A_k}^{max}$, for all k.

The results of KPS [14] also clarified the meaning of the duality found in[15,16,17] for the A_k theories, which relates the original $SU(N_c)$ theory to a similar one with gauge group $SU(kN_f - N_c)$ and some additional gauge singlet mesons and superpotential, generalizing the duality of[7]. In particular, there is a non-empty "conformal window" range of x, $x_{A_k}^{min} < x < k - \widetilde{x}_{A_k}^{min}$ for which $W_{tree} = \mathrm{Tr} X^{k+1}$ is relevant in the electric theory, and its analog in the magnetic dual is also relevant. This is where the electric and magnetic dual descriptions are both useful.

There are many non-trivial checks of the a-theorem conjecture which can be made here, and $a_{IR} < a_{UV}$ is indeed satisfied for all RG flows. E.g. we can flow from the UV to the IR as: free UV$\rightarrow \widehat{A} \rightarrow A_k \rightarrow A_{k'}$, with $k' < k$. The a-theorem would then predict $a_{freeUV}(x) > a_{\widehat{A}}(x) > a_{A_k}(x) > a_{A_{k'}}(x)$. Interestingly, $a_{A_k}(x) > a_{A_{k'}}(x)$ had some apparent violations for small x [3]. But using the new a-maximization results for $x_{A_k}^{min}$ it is now seen that there never actually is a violation of $a_{IR} < a_{UV}$ in any of those examples. The apparent violation of $a_{IR} < a_{UV}$ always occurred for x outside of the range $x > x_{A_k}^{min}$ needed for the A_k SCFT to exist, so there's no a-theorem violating flow after all[14].

In our work [21], we generalized these examples by considering SQCD with N_f fundamental flavors, Q_i and $\widetilde{Q}_i$, along with $N_a = 2$ adjoint matter flavors, X and Y. As in [9], considering deformations suggests that this theory, with $W_{tree} = 0$, flows to an interacting $\mathcal{N} = 1$ SCFT for all N_f in the asymptotically free range $0 \leq N_f < N_c$, which we write as $x \equiv N_c/N_f > 1$. We'll call these SCFTs $\widehat{O}(N_c, N_f)$, and again consider N_c and N_f large, with x fixed, to simplify things. For $x \approx 1$, the $\widehat{O}$ IR fixed point is at weak coupling, and as x increases, the $\widehat{O}$ IR fixed point is at stronger coupling. We use a-maximization to solve for the R-charges for all x, taking the anomaly free trial R-symmetry to be: $R_t(Q) = R_t(\widetilde{Q}) \equiv y$ and

$R_t(X) = R_t(Y) = \frac{1}{2}(1+\frac{1-y}{x})$. Plugging this into $a_{trial} = 3\mathrm{Tr}R_t^3 - \mathrm{Tr}R_t$ and maximizing w.r.t. y gives $y(x) = 1 + \left(3(8x^2-1)\right)^{-1}\left(3x - 2x\sqrt{26x^2-1}\right)$. We obtain a relatively simple closed form expression for $y(x)$ for the $\widehat{O}$ SCFTs because no operators hit the unitarity bound, so no additional contributions to a_{trial}, associated with accidental symmetries are needed (though we can't definitively rule them out). We thus find that $R(Q)$ and $R(X)$ and $R(Y)$ are decreasing functions of x, and for $x \to \infty$ we obtain $R(Q) \to 1 - \sqrt{26}/12 \approx 0.575$, and $R(X) = R(Y) \to \frac{1}{2}$.

Using the above result, we can classify the relevant superpotential deformations of the $\widehat{O}$ SCFT, which could drive RG flows from $\widehat{O}$ to various new SCFTs in the IR. Let's consider superpotentials involving only the adjoints (there are additional ones using operators which include the quarks). Using our result for the $\widehat{O}$ SCFT that $R(X) = R(Y) > \frac{1}{2}$ for all x, we see that $W = \mathrm{Tr}X^kY^\ell$ is only a relevant deformation of the $\widehat{O}$ SCFT if $k + \ell \leq 3$. The complete list of such relevant superpotential deformations (modulo field redefinitions) of $\widehat{O}$ is thus $W_{\widehat{A}} = \mathrm{Tr}Y^2$, $W_{\widehat{D}} = \mathrm{Tr}XY^2$, and $W_{\widehat{E}} = \mathrm{Tr}Y^3$. Each of these drives $\widehat{O}$ to new SCFTs, which we name $\widehat{A}$, $\widehat{D}$, and $\widehat{E}$ respectively, for all N_f in the range $0 < N_f < N_c$. The $\widehat{A}$ case is that considered by [14] and reviewed above.

We can now use a-maximization to analyze the new $\widehat{D}$ and $\widehat{E}$ SCFTs. E.g. for the $\widehat{D}$ RG fixed points, we impose $R(XY^2) = 2$, and the trial $U(1)_R$ charges are $R(Q) = R(\widetilde{Q}) \equiv y$, $R(X) = 2(1-y)/x$, $R(Y) = 1 + (y-1)/x$. Here we find that operators do hit the unitarity bound, so their accidental symmetries must be accounted for in a_{trial}, in analogy with the $\widehat{A}$ case of [14]. Because of this, we solved for $y(x)$ numerically. The asymptotic values in the limit $x \to \infty$ are $y \to -1/8$, $R(Y) \to 1$, and $R(X) \to 0$.

Using these results for the $\widehat{D}$ SCFT, we can now classify its relevant superpotential deformations. Since $R(X) \to 0$ for $x \to \infty$, we find that $\Delta W = \mathrm{Tr}X^{k+1}$ can be a relevant deformation of the $\widehat{D}$ SCFT, for arbitrarily large k, provided that x is sufficiently large, $x > x^{min}_{D_{k+2}}$. We solved for $x^{min}_{D_{k+2}}$ by using our results for the $\widehat{D}$ SCFT to determine the x where $R(X) = 2/(k+1)$: we find $x^{min}_{D_5} = 2.09$, $x^{min}_{D_6} = 3.14$, etc., with $x^{min}_{D_{k+2}} \to \frac{9}{8}k$ for $k \to \infty$. When $x > x^{min}_{D_{k+2}}$, the relevant ΔW drives a RG flow from the $\widehat{D}$ SCFTs to new SCFTs, which we name D_{k+2}. More generally, we can get to the D_{k+2} SCFTs, provided that $x > x^{min}_{D_{k+2}}$, by perturbing $\widehat{O}$ by $W_{D_{k+2}} = \lambda_1 \mathrm{Tr}X^{k+1} + \lambda_2 \mathrm{Tr}XY^2$. (The D_{k+2} SCFT also must satisfy $x < x^{max}_{D_{k+2}} = 3k$, to prevent generating a W_{dyn}.)

Continuing this process of using a-maximization to analyze each SCFTs, classifying their relevant superpotential deformations, and then using these deformations to flow to new SCFTs, we obtain a classification of the SCFTs that are obtainable from $\widehat{O}$ via the following superpotential deformations (again, there are additional ones making use of operators involving the quarks): $W_{\widehat{O}} = 0$, $W_{\widehat{A}} = \mathrm{Tr}Y^2$, $W_{\widehat{D}} = \mathrm{Tr}XY^2$, $W_{\widehat{E}} = \mathrm{Tr}Y^3$, $W_{A_k} = \mathrm{Tr}(X^{k+1} + Y^2)$, $W_{D_{k+2}} = \mathrm{Tr}(X^{k+1} + XY^2)$, $W_{E_6} = \mathrm{Tr}(Y^3 + X^4)$, $W_{E_7} = \mathrm{Tr}(Y^3 + YX^3)$, and $W_{E_8} = \mathrm{Tr}(Y^3 + X^5)$. Each of these LG superpotentials drives a RG flow to a corresponding interacting SCFT in the IR, with the A_k, D_{k+2}, E_6, E_7, and E_8 cases existing only if x is sufficiently large.

We note that the above superpotential classification agrees with Arnold's singularity classification. That classification also appeared, among other contexts, in the classification of 2d SCFTs having $c < 3$ [23,24]. The appearance of Arnold's ADE in our 4d case is a new guise, which perhaps has a deeper connection to other occurrences of the ADE series in string theory and mathematics. This is a topic for future exploration.

The map of the possible flows between these SCFTs is:

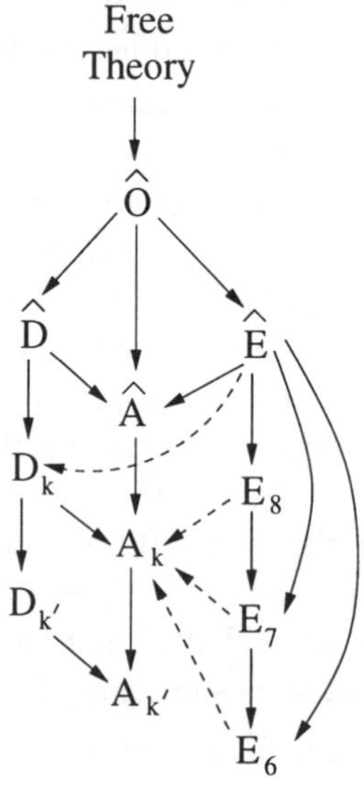

Figure 1: *The map of possible flows between fixed points. Dotted lines indicate flow to a particular value of k.*

Every arrow in the diagram corresponds to a RG flow, from a SCFT in the UV to another SCFT in the IR. For each of these flows, we used a-maximization to exactly compute a_{UV} and a_{IR}, and we verified that $a_{IR} < a_{UV}$ is indeed satisfied for every such RG flow. E.g. $a_{E_6}(x) < a_{E_7}(x) < a_{E_8}(x) < a_{\widehat{E}}(x) < a_{\widehat{O}}(x) < a_{free\ UV}(x)$. There are some apparent violations for small x, e.g. it appears that $a_{E_7}(x) < a_{E_6}(x)$ for $x < 3.16$, but there is actually no violation of the a-theorem conjecture afterall, because we find that that E_7 RG fixed point only exists if $x > x_{E_7}^{min} \approx 4.12$. Likewise, all other such potential violations occur outside of the range where the RG flow can exist.

Our new results also give some insight into a proposed duality [18] for the theory with $W_{D_{k+2}} = Tr(X^{k+1} + XY^2)$, relating the original "electric" $SU(N_c)$ theory to a "magnetic" $SU(3kN_f - N_c)$ dual. We find different phases depending on $x \equiv N_c/N_f$. The duality is most useful in the D_{k+2} conformal window: $x_{D_{k+2}}^{min} < x < 3k - \widetilde{x}_{D_{k+2}}^{min}$, where $W_{D_{k+2}}$ is relevant in both the electric and magnetic theories. It's interesting that this range is non-empty for all k, e.g. for large k we find $x_{D_{k+2}}^{min} \approx \frac{9}{8}k$ and $3k - \widetilde{x}_{D_{k+2}}^{min} \approx 1.8962k$. Both dual descriptions are good in the conformal window. Outside of the conformal window, the theory can instead flow to the electric or magnetic $\widehat{D}$ SCFT, and the other dual description is then not useful.

The a-maximization methods used here can be applied to many other 4d supersymmetric gauge theories, and there are many other superconformal zoos remaining to be explored.

Acknowledgments

KI thanks the organizers for the invitation to this fun conference. This work was partially supported by DOE-FG03-97ER40546.

References

1. A. B. Zamolodchikov, JETP Lett. **43**, 730 (1986) [Pisma Zh. Eksp. Teor. Fiz. **43**, 565 (1986)].
2. J. L. Cardy, Phys. Lett. B **215**, 749 (1988).
3. D. Anselmi, D. Z. Freedman, M. T. Grisaru and A. A. Johansen, Nucl. Phys. B **526**, 543 (1998) [arXiv:hep-th/9708042].
   D. Anselmi, J. Erlich, D. Z. Freedman and A. A. Johansen, Phys. Rev. D **57**, 7570 (1998) [arXiv:hep-th/9711035].
4. T. Appelquist, A. G. Cohen and M. Schmaltz, Phys. Rev. D **60**, 045003 (1999) [arXiv:hep-th/9901109].
5. D. J. Gross and F. Wilczek, Phys. Rev. D **9**, 980 (1974).
6. T. Banks and A. Zaks, Nucl. Phys. B **196**, 189 (1982).
7. N. Seiberg, Nucl. Phys. B **435**, 129 (1995) [arXiv:hep-th/9411149].

8. G. ' t Hooft, *Recent Developments in Gauge Theories* Plenium Press **135** (1980), reprinted in *Unity of Forces in the Universe Vol II* A. Zee ed. World Scientific **1004** (1982)
9. K. A. Intriligator and N. Seiberg, Nucl. Phys. Proc. Suppl. **45BC**, 1 (1996) [arXiv:hep-th/9509066].
10. V. A. Novikov, M. A. Shifman, A. I. Vainshtein and V. I. Zakharov, Nucl. Phys. B **229**, 381 (1983).
11. V.I. Arnold, S.M. Gusein-Zade and A.N. Varchenko, *Singularities of Differentiable Maps* volumes 1 and 2, Birkhauser (1985), and references therein.
12. K. Intriligator and B. Wecht, Nucl. Phys. B **667**, 183 (2003) [arXiv:hep-th/0304128].
13. H. Osborn, Annals Phys. **272**, 243 (1999) [arXiv:hep-th/9808041].
14. D. Kutasov, A. Parnachev and D. A. Sahakyan, JHEP **0311**, 013 (2003) [arXiv:hep-th/0308071].
15. D. Kutasov, Phys. Lett. B **351**, 230 (1995) [arXiv:hep-th/9503086].
16. D. Kutasov and A. Schwimmer, Phys. Lett. B **354**, 315 (1995) [arXiv:hep-th/9505004].
17. D. Kutasov, A. Schwimmer and N. Seiberg, Nucl. Phys. B **459**, 455 (1996) [arXiv:hep-th/9510222].
18. J. H. Brodie, Nucl. Phys. B **478**, 123 (1996) [arXiv:hep-th/9605232].
19. K. A. Intriligator and N. Seiberg, Nucl. Phys. B **431**, 551 (1994) [arXiv:hep-th/9408155].
20. K. A. Intriligator and N. Seiberg, Nucl. Phys. B **444**, 125 (1995) [arXiv:hep-th/9503179].
21. K. Intriligator and B. Wecht, Nucl. Phys. B **677** (2004) 223 [arXiv:hep-th/0309201].
22. D. Kutasov, [arXiv:hep-th/0312098].
23. E. J. Martinec, Phys. Lett. B **217**, 431 (1989).
24. C. Vafa and N. P. Warner, Phys. Lett. B **218**, 51 (1989).

4-DIMENSIONAL EINSTEIN THEORY EXTENDED BY A 3-DIMENSIONAL CHERN-SIMONS TERM

R.JACKIW

Center for Theoretical Physics
Department of Physics
Massachusetts Institute of Technology
Cambridge, Massachusetts 02139
jackiw@lns.mit.edu

When 4-dimensional general relativity is extended by a 3-dimensional gravitational Chern-Simons term an apparent violation of diffeormorphism invariance is extinguished by the dynamical equations of motion for the modified theory. The physical predictions of this recently proposed model show little evidence of symmetry breaking, but require the vanishing of the Pontryagin density.

Introduction

Chern-Simons terms are odd-dimensional entities, interesting to mathematicians [1] and physicists [2]. For physicists they are most useful when they are defined on 3-manifolds, and over the last twenty years they have been widely used to model various physical processes in 3-dimensional space-time, that is, phenomena confined to motion on a plane, like the Hall effect, or gravitational motion in the presence of cosmic strings.

However, these 3-dimensional structures can also be inserted into physical theories in 4-dimensional space-time, and because of the dimensional mismatch this gives rise to kinematical and dynamical violation of Lorentz symmetry, CPT symmetry, etc.

As you have heard earlier in this meeting from Kostelecky, the subject of Lorentz and CPT symmetry violation is interesting these days, mainly due to his initiating and stimulating work by theorists and experimentalists. The former build plausible extensions of standard theories, with small symmetry-breaking terms; the latter perform more and more precise experiments limiting the magnitude of such possible terms. Thus far, no evidence for symmetry breaking has been found; indeed, conventional symmetries are confirmed, with ever-decreasing uncertainty.

Today, I shall describe a Chern-Simons modification of 4-dimensional gravity theory – Einstein's general relativity – and the associated decrease in symmetry.

Actually, more than a decade ago, George Field, Sean Carroll and I investigated a Chern-Simons modification of Maxwell's electromagnetic theory. So in order to set the stage for the gravitational extension, I shall first review the Maxwell story [3].

1. Chern-Simons Modification of Maxwell Theory

The Chern-Simons term for an Abelian gauge theory on an Euclidean 3-space reads

$$CS(A) \equiv \frac{1}{4}\varepsilon^{ijk} F_{ij} A_k = \frac{1}{2}\mathbf{A}\cdot\mathbf{B}. \tag{1.1}$$

The first expression is in tensor notation; the second in vector notation, with $\mathbf{B}$ being the magnetic field, $\mathbf{B} = \nabla \times \mathbf{A}$. All indices are spatial $[i, j, k : x, y, z]$. A related 4-dimensional formula in Minkowski space-time defines the topological Chern-Simons current

$$K^\mu = {}^*F^{\mu\nu} A_\nu, \tag{1.2}$$

where ${}^*F^{\mu\nu}$ is the dual electromagnetic tensor.

$${}^*F^{\mu\nu} = \frac{1}{2}\varepsilon^{\mu\nu\alpha\beta} F_{\alpha\beta} \tag{1.3}$$

It is seen that the Chern-Simons term (1.1) is proportional to the time $t(\mu = 0)$ component of the Chern-Simons current, (1.2) with the time dependence suppressed. Also the divergence of the topological current is the topological Pontryagin density.

$$\partial_\mu K^\mu = \partial_\mu ({}^*F^{\mu\nu} A_\nu) = \frac{1}{2}{}^*F^{\mu\nu} F_{\mu\nu} \tag{1.4}$$

In Chern-Simons modified electromagnetism the Chern-Simons term (1.1) (with field arguments extended to include t) is added to the usual Maxwell Lagrangian.

$$I = \int d^4x \left(-\frac{1}{4} F^{\mu\nu} F_{\mu\nu} + \frac{\mu}{2} \mathbf{A}\cdot\mathbf{B} \right) \tag{1.5}$$

Here μ, with dimension of mass, measures the strength of the extension. Formula (1.5) may be alternatively presented in covariant notation, with the help of an external, constant embedding 4-vector v_μ.

$$I = \int d^4x \left(-\frac{1}{4} F^{\mu\nu} F_{\mu\nu} + \frac{1}{2} v_\mu {}^*F^{\mu\nu} A_\nu \right)$$

$$v_\mu = (\mu, 0) \tag{1.6}$$

In spite of the presence of the vector potential, the action is gauge invariant: Under a gauge transformation it changes by a surface term, since $\partial_\mu {}^*F^{\mu\nu} = 0$. This can be made explicit by recognizing that in (1.6) there occurs the Chern-Simons current K^μ (1.2). Therefore, with the help of (1.4) and an integration by parts the action acquires a gauge invariant form.

$$I = \int d^4x \left(-\frac{1}{4} F^{\mu\nu} F_{\mu\nu} + \frac{1}{4}\theta \; {}^*F^{\mu\nu} F_{\mu\nu} \right)$$

$$\partial_\mu \theta \equiv v_\mu \tag{1.7}$$

The external quantity is now θ, which is taken as $\theta = \mu t$ so that (1.5) and (1.6) are reproduced.

Since the explicitly covariant formulations (1.6), (1.7) involve external, fixed quantities [a fixed constant embedding vector v_μ in (1.6); a fixed function θ, linear in time, in (1.7)], we expect that Lorentz invariance is lost. Also, since $\mathbf{A} \cdot \mathbf{B}$, and ${}^*F^{\mu\nu} F_{\mu\nu}$ are axial quantities, parity is lost; but C and T are preserved, so CPT is also lost. To confirm these statements, we now look to the solutions of the modified equations of motion.

In the electromagnetic equations of motion, which follow from the Chern-Simons extended action, only Ampère's law is modified.

$$-\frac{\partial \mathbf{E}}{\partial t} + \nabla \times \mathbf{B} = \mathbf{J} + \mu \mathbf{B} \tag{1.8}$$

All other Maxwell equations continue to hold. Also the consistency condition on (1.8) remains as in Maxwell theory: the charge density $\rho = \nabla \cdot \mathbf{E}$ and the current $\mathbf{J}$ must satisfy their continuity equation, as is seen by taking the divergence of (1.8) and using $\nabla \cdot \mathbf{B} = 0$.

The modification that we have constructed is particularly felicitous for the following reasons.

(i) Gauge invariance is maintained, so the photon continues to possess just two independent polarizations.

(ii) Eq.(1.8) is not a radical departure; it has played previous roles in physical theory: in plasma physics one frequently replaces the source current $\mathbf{J}$ with a magnetic field $\mathbf{B}$. Of course, we are not working with a collective/phenomenological theory, like plasma physics, rather we are examining the feasibility of (1.8) for fundamental physics.

To assess the actual physical content of the Chern-Simons extended electromagnetism, and its associated symmetry breaking, we have examined some solutions. We found that in the source-free ($\mathbf{J} = \mathbf{0}$) case, plane waves continue to solve the extended equations. The photon posseses two independent polarizations, (as anticipated from gauge invariance) however they travel at velocities which differ from the velocity of light (thus Lorentz boost invariance is lost—as anticipated) and also the two polarizations travel with velocities that differ from each other (thus parity invariance is lost–as anticipated).

The fact that the two photon helicities travel (in vacuum) with different velocities makes empty space behave as a birefringent medium. Consequently linearly polarized light, passing through this birefringent environment, undergoes a Faraday-like rotation, which can be looked for in observations of light from distant galaxies. Much data exists on this phenomenon, and the conclusion is unavoidable: there is no such effect in Nature; $\mu = 0$ is required. This was asserted in our initial investigations [3, 4], and the many other analyses carried out in the intervening years support that conclusion (see e.g. [5]).

2. Chern-Simons Modification of Einstein Theory

A. Gravitational Chern-Simons term in 3-space.

The 3-dimensional, gravitational Chern-Simons term can be presented in terms of the 3-dimensional Christoffel connection ${}^3\Gamma^p_{iq}$, [2]

$$CS(\Gamma) = \varepsilon^{ijk}\left(\frac{1}{2}\,{}^3\Gamma^p_{iq}\,\partial_j{}^3\Gamma^q_{kp} + \frac{1}{3}{}^3\Gamma^p_{iq}\,{}^3\Gamma^q_{jr}\,{}^3\Gamma^r_{kp}\right), \tag{2.1}$$

but it is understood that the Christoffel connection takes the usual expression in terms of the metric tensor, which is the fundamental variable. Variation with respect to the metric tensor of the intergrated Chern-Simons term results in the 3-dimensional "Cotton tensor"; which involves a covariant curl of the 3-dimensional Ricci tensor ${}^3R^i_j$.

$$\frac{\delta}{\delta g_{ij}}\int d^3x CS(\Gamma) = -\sqrt{g}\,{}^3C^{ij} = \frac{1}{2}\varepsilon^{imn}\,{}^3D_m{}^3R^j_n + i \leftrightarrow j \tag{2.2}$$

${}^3C^{ij}$ is symmetric, traceless and covariantly conserved. It vanishes if and only if the 3-dimensional metric tensor is conformally flat. A related formula gives the 4-dimensional Chern-Simons current K^μ,

$$K^\mu = 2\varepsilon^{\mu\alpha\beta\gamma}\left[\frac{1}{2}\,\Gamma^\sigma_{\alpha\tau}\,\partial_\beta\,\Gamma^\tau_{\gamma\sigma} + \frac{1}{3}\,\Gamma^\sigma_{\alpha\tau}\,\Gamma^\tau_{\beta\eta}\Gamma^\eta_{\gamma\sigma}\right], \tag{2.3}$$

whose divergence is the topological Pontryagin density.

$$\partial_\mu K^\mu = \frac{1}{2}{}^*R^\sigma{}_\tau{}^{\mu\nu}\,R^\tau{}_{\sigma\mu\nu} \equiv \frac{1}{2}{}^*RR \tag{2.4}$$

Here $R^\tau{}_{\sigma\mu\nu}$ is the Riemann curvature tensor and ${}^*R^\sigma{}_\tau{}^{\mu\nu}$ is its dual.

$${}^*R^\sigma{}_\tau{}^{\mu\nu} = \frac{1}{2}\,\varepsilon^{\mu\nu\alpha\beta}R^\sigma{}_{\tau\alpha\beta} \tag{2.5}$$

[Notation: $(i, j, ...)$ are 3-dimensional, spatial indices, and 3-dimensional geometric entities are decorated with the superscript "3". Undecorated geometric entities are 4-dimensional, and Greek indices label the 4 space-time coordinates.] Note that unlike in the vector case, the Chern-Simons term (2.1) is not the time component K^0, because the former contains 3-dimensional Christoffel entities, while 4-dimensional ones are present in K^0. This variety allows various extensions general relativity.

B. Gravitational Chern-Simons term in 4-space

In analogy with the electromagnetic formulation (1.7), we choose to extend Einstein theory by adopting the action [6]

$$\begin{aligned} I &= \frac{1}{16\pi G}\int d^4x\left(\sqrt{-g}R + \frac{1}{4}\theta^*RR\right) \\ &= \frac{1}{16\pi G}\int d^4x\left(\sqrt{-g}R - \frac{1}{2}v_\mu K^\mu\right), \qquad v_\mu \equiv \partial_\mu\theta. \end{aligned} \tag{2.6}$$

The first contribution is the usual Einstein-Hilbert term involving the Ricci scalar R. The modification involves an external quantity: θ in the first equality; $\partial_\mu\theta \equiv v_\mu$ in the second equality, which follows from the first by (2.4) and an integration by parts. Eventually we shall take the embedding vector v_μ to possess only a time component, and θ to depend solely on time. So then our modification (2.6) involves the time component of 4-dimensional Chern-Simons current (2.3) [rather than the 3-dimensional Chern-Simons term (2.1)].

The equation of motion that emerges when (2.6) is varied with respect to $g_{\mu\nu}$ is

$$G^{\mu\nu} + C^{\mu\nu} = -8\pi G T^{\mu\nu}. \tag{2.7}$$

Here $G^{\mu\nu}$ is the covariantly conserved (Bianchi identity) Einstein tensor, $G^{\mu\nu} = R^{\mu\nu} - \frac{1}{2}g^{\mu\nu}R$. We have inserted a source with strength G (Newtons constant) consisting of the matter energy-momentum tensor $T^{\mu\nu}$, which also is convariantly conserved, since we assume matter to be conventionally, covariantly coupled to gravity. $C^{\mu\nu}$ is the term with which we are extending the Einstein theory.

$$\sqrt{-g}\; C^{\mu\nu} = \frac{\delta}{\delta g_{\mu\nu}} \frac{1}{4} \int d^4x \; \theta^* RR$$

$$= -\frac{1}{2}\left(v_\sigma \varepsilon^{\sigma\mu\alpha\beta} D_\alpha R^\nu_\beta + v_{\sigma\tau}{}^* R^{\tau\mu\sigma\nu} + \mu \leftrightarrow \nu\right) \qquad (2.8)$$

$C^{\mu\nu}$ is manifestly symmetric; it is traceless because ${}^*RR$ is conformally invariant. $C^{\mu\nu}$'s first term (involving the curl of R^ν_β) is similar to the 3-dimensional $\sqrt{g}^3 C^{ij}$ (2.2). Even the second term can be viewed as a generalization from 3-dimensions: it involves only the Weyl tensor part of Riemann tensor, which vanishes in 3 dimensions. [Cotton defined his tensor in arbitrary dimensions d, and his definition is equivalent to ours in $d = 3$, where it is also given by the variation of the 3-d gravitational Chern-Simons term, as is (2.2). [7] However for $d \neq 3$, Cotton's tensor does not appear to have a variational definition. Our $d = 4$ Cotton-like tensor in (2.8) does possess a variational definition, at the the expense of introducing non-geometrical entities like θ and v_μ.]

Finally we must examine $D_\mu C^{\mu\nu}$, whose vanishing is a consistency requirement on (2.7). However, an explicit evaluation (which involves only geometric identities) shows that, unlike in 3 dimensions, $C^{\mu\nu}$ is not covariantly conserved. Rather

$$D_\mu C^{\mu\nu} = \frac{1}{8\sqrt{-g}}\; v^{\nu *} RR. \qquad (2.9)$$

Thus the vanishing of ${}^*RR$ is a consistency condition of the new dynamics: every solution to (2.7) will necessarily lead to vanishing Pontryagin density.

$$G^{\mu\nu} + C^{\mu\nu} = -8\pi G T^{\mu\nu} \Rightarrow {}^*RR = 0 \qquad (2.10)$$

We may derive and understand the expression for the covariant divergence of $C^{\mu\nu}$ by examining the response of our addition to changes in the coordinates. With the infinitesimal transformation

$$\delta x^\mu = -f^\mu(x) \qquad (2.11)$$

we have

$$\delta g_{\mu\nu} = D_\mu f_\nu + D_\nu f_\mu. \qquad (2.12)$$

The Hilbert Einstein action is of course invariant. To assess the variance properties of our modification, we can proceed in two ways. First observe that ${}^*RR$ is scalar density, so it transforms as $\delta({}^*RR) = \partial_\mu(f^{\mu *}RR)$. θ is an external quantity, therefore we do not transform it.

$$\delta I_{CS} = \frac{1}{4}\int d^4x\ \theta\delta({}^*RR) = \frac{1}{4}\int d^4x\theta\partial_\mu(f^{\mu *}RR)$$

$$= -\frac{1}{4}\int d^4xv_\mu f^{\mu *}RR \qquad (2.13a)$$

Alternatively, we may vary I_{CS}, by varying $g_{\mu\nu}$ according to (2.10), and using the definition (2.8) for $C^{\mu\nu}$.

$$\delta I_{CS} = \int d^4y\ \frac{\delta}{\delta g_{\mu\nu}(y)}\left(\frac{1}{4}\int d^4x\ \theta\ {}^*RR\right) 2D_\mu f_\nu(y)$$

$$= 2\int d^4x\ \sqrt{-g}\ C^{\mu\nu}D_\mu f_\mu = -2\int d^4x\ \sqrt{-g}\ (D_\mu C^{\mu\nu})f_\nu \qquad (2.13b)$$

Equating the two expressions for δI_{CS} establishes (2.9), and also demonstrates that ${}^*RR$ is a measure of the failure of diffeomorphism invariance. But ${}^*RR$ vanishes as a consequence of the equation of motion, so is some sense diffeomorphism invariance is dynamically reinstated.

For another perspective, consider a variant of our model, where θ in (2.6) is a dynamical variable, not an externally fixed quantity. If we postulate that under diffeomorphisms (2.11) θ transforms as a scalar,

$$\delta\theta = f^\mu\partial_\mu\theta = f^\mu v_\mu, \qquad (2.14)$$

then (2.12a) acquires the additional contribution

$$\frac{1}{4}\int d^4x\,\delta\theta\ ({}^*RR) = \frac{1}{4}\int d^4x\ v_\mu\ f^\mu\ {}^*RR, \qquad (2.15)$$

which cancels (2.13a), showing that the Chern-Simons modification with dynamical θ is diffeomorphism invariant. Now let us look at the equations of motion in this variant of modified gravity: varying $g_{\mu\nu}$ still produces (2.7); varying θ, which now acts as a Lagrange multiplier, forces ${}^*RR$ to vanish, but that requirement is already implied by (2.7), (2.10). Thus the equations of the fully dynamical, and diffeomorphism invariant theory coincide with the equations of the non-invariant theory, where θ is a fixed, external quantity.

Formula (2.13a) shows that when v_μ is chosen to have only a time component, $v_\mu = (\frac{1}{\mu}, \mathbf{0})$; equivalently $\theta = t/\mu$, then I_{CS} is invariant under all space-time reparametrizations of the spatial coordinates, and also of shifts in time: f^0 = constant, f^iarbitrary. Henceforth we make this choice for v_μ and θ.

C. Physical effects of the Chern-Simons term in 4-d gravity

We examine some physical processes in the Chern-Simon modified gravity theory.

(i) It is important that the Schwarzschild solution continues to hold; thus our theory passes the "classic" test of general relativity. The result is established in two steps. First we posit a stationary form for the metric tensor

$$g_{\mu\nu} = \begin{pmatrix} N & 0 \\ 0 & g_{ij} \end{pmatrix}, \tag{2.16}$$

with time-independent entries. It follows that C^{00} and $C^{n0} = C^{0n}$ vanish. Also one finds that the spatial components reproduce the 3-dimensional Cotton tensor.

$$\sqrt{-g}\, C^{ij} = \sqrt{g}^3\, C^{ij} \tag{2.17}$$

Next, we make the spherically symmetric *Ansatz*, and find that C^{ij} vanishes. Evidently also ${}^*RR$ must vanish on the Schwarzschild geometry, because the modified equations are satisfied. Since the Kerr geometry, carries non vanishing ${}^*RR$, it will not be a solution to the extended equations. It remains an interesting, open question which deformation of the Kerr geometry satisfies the Chern-Simons modified equations.

(ii) Next we perform a linear analysis by expanding the metric tensor around a flat background $g_{\mu\nu} = \eta_{\mu\nu} + h_{\mu\nu}$. The purpose of the linear analysis is to determine the propagating degrees of freedom, to study the nature of small disturbances (gravity waves) and to illuminate the construction of an energy-momentum (pseudo) tensor, which is symmetric and divergence-free.

Keeping only the linear portions of the Einstein tensor and $C_{\mu\nu}$, we verify that both $G^{linear}_{\mu\nu}$ and $C^{linear}_{\mu\nu}$ are divergence-free.

$$\partial^\mu G^{linear}_{\mu\nu} = 0 = \partial^\mu C^{linear}_{\mu\nu} \tag{2.18}$$

This is seen from the explicit formulas. It also follows from the observation that the exact equation $D^\mu G_{\mu\nu} = 0$ implies the above result for $G^{linear}_{\mu\nu}$; moreover, from (2.9) we see that $D^\mu C_{\mu\nu}$ is of quadratic order, hence the above result for $C^{linear}_{\mu\nu}$ holds also. It is further seen that the linear portions are invariant under the "gauge" transformation

$$h_{\mu\nu} \rightarrow h_{\mu\nu} + \partial_\mu \lambda_\nu + \partial_\nu \lambda_\mu \tag{2.19}$$

In the Einstein theory, one decomposes $h_{\mu\nu}$ into temporal parts, and purely spatial parts h^{ij}. The latter is further decomposed into its trace, its longitudinal part, and its traceless transverse part, denoted by h^{ij}_{TT}. One then finds from the linear equations that, with the exception of h^{ij}_{TT}, all other components of $h_{\mu\nu}$ are either non-propagating or can be eliminated by the gauge transformation (2.19). Only h^{ij}_{TT} survives and it is governed by a d'Alembertian. Since in 4 dimensions a symmetric, transverse and traceless 3×3 matrix possesses two independent components, one concludes that in Einstein's theory small gravitational disturbances are waves, with two polarizations, each moving with the velocity of light (governed by the d'Alembertian).

None of this changes where $C^{linear}_{\mu\nu}$ is included. Again only h^{ij}_{TT} propagates, governed by the d'Alembertian. Explicitly the modified equation for h^{ij}_{TT} reads

$$\begin{aligned}(\delta^{im}\delta^{jn} + \frac{1}{2\mu}\varepsilon^{ipm}\ \delta^{nj}\ \partial_p + \frac{1}{2\mu}\varepsilon^{jpm}\ \delta^{ni}\ \partial_p)\ \Box\ h^{mn}_{TT} \\ = -16\pi\ G\ T^{ij}_{TT}.\end{aligned} \tag{2.20}$$

T^{ij}_{TT} is the transverse traceless part of the stress tensor. The new terms are the (μ^{-1}) contributions; they involve only spatial derivatives. One may consider that the left side of (2.20) involves an operator acting on $\Box\ h^{mn}_{TT}$.

$$\mathcal{O}^{ij}{}_{mn} \Box\ h^{mn}_{TT} = -16\pi G\ T^{ij}_{TT} \tag{2.21a}$$

Acting on this equation with the inverse operator $\mathcal{P} = \mathcal{O}^{-1}$ shows that the effect of the entire extension is to modify the source

$$\begin{aligned}\Box\ h^{mn}_{TT} &= -16\pi G\ \mathcal{P}^{mn}{}_{ij}\ T^{ij}_{TT} \\ &\equiv -16\pi G\ \tilde{T}^{ij}_{TT}\end{aligned} \tag{2.21b}$$

Thus we see that in sharp contrast to the electromagnetic case, the Chern-Simons modification of gravity does not change the velocity of gravity waves and there is no Faraday rotation. It is also noteworthy that the reduction to 2 degrees of freedom (2 polarizations) takes place also in the extended theory. Such a reduction of degrees of freedom is considered to be a consequence of gauge invariance, here diffeomorphism invariance, which evidently continues to hold on our modified theory.

There does exist a physical manifestation of the extension. Although the velocities of the two polarizations are the same, their intensities differ, due to the modification of the source ($T^{ij}_{TT} \to \tilde{T}^{ij}_{TT}$). One finds for a weak modification (large μ) that the ratio of the intensity of waves with negative helicity to those with positive helicity is

$$\frac{-}{+} = \left(1 + \frac{4\omega}{\mu}\right), \tag{2.22}$$

where ω is the frequency. This puts into evidence the parity violation of the modification.

Finally we turn to the topic of the energy-momentum (pseudo) tensor. A straight forward approach to this problem in the Einstein theory is to rewrite the equation of motion by decomposing the Einstein tensor $G_{\mu\nu}$ into its linear and non-linear parts, and moving the non-linear terms to the "right" side, summing it with the matter energy-momentum tensor.

$$G^{linear}_{\mu\nu} = -8\pi\, G \left(T_{\mu\nu} + \frac{1}{8\pi G}\, G^{non-linear}_{\mu\nu}\right) \tag{2.23}$$

Clearly $G^{linear}_{\mu\nu}$ is symmetric and conserved, therefore, so must be the right side, which is now renamed as total (gravity + matter) energy-momentum (pseudo) tensor.

$$\tau_{\mu\nu} = T_{\mu\nu} + \frac{1}{8\pi G}\, G^{non-linear}_{\mu\nu} \tag{2.24}$$

Exactly the same procedure works in the extended theory. We present the equation of motion (2.7) as

$$G^{linear}_{\mu\nu} + C^{linear}_{\mu\nu} = -8\pi\, G \left(T_{\mu\nu} + \frac{1}{8\pi G}(G^{non-linear}_{\mu\nu} + C^{non-linear}_{\mu\nu})\right). \tag{2.25}$$

We have already remarked that the left side is divergenceless. Thus we can identify a symmetric and conserved energy- momentum (pseudo) tensor as

$$\tau_{\mu\nu} = T_{\mu\nu} + \frac{1}{8\pi G}(G^{non-linear}_{\mu\nu} + C^{non-linear}_{\mu\nu}). \tag{2.26}$$

It is striking that this structure is present in a theory that seems to violate Lorentz invariance!

In Ref. [8] there is a survey of other gravitational energy-momentum (pseudo) tensors for Einstein's theory that differ from each other by super potentials. In particular there is described a Noether construction with a

Belinfante improvement, which also yields a symmetric, conserved energy-momentum (pseudo) tensor tied to the Poincaré invariance of the Einstein theory. It would be interesting to reconsider this construction in the extended theory and to compare the result to (2.26).

3. Conclusion

Measuring the intensity of polarized gravity waves is not feasible. Thus for present days, our model is only a theoretical exercise. Nevertheless, it shows interesting and unexpected behavior in that an important symmetry–diffeomorphism invariance–is not present in the action, but is restored by the equations of motion. Correspondingly the physical effects of the symmetry breaking are quite hidden.

An analogy can be made with the Stückelberg formalism for massive, Abelian gauge fields. The action

$$I_m = \int d^4x \left(-\frac{1}{4} F^{\mu\nu} F_{\mu\nu} + \frac{1}{2} m^2 A^\mu A_\mu \right)$$

is not gauge invariant

$$\delta_{\text{gauge}}\, I_m = \int m^2 A^\mu \partial_\mu \lambda = - \int m^2 \partial_\mu A^\mu \lambda,$$

but is seen to be broken by $\partial_\mu A^\mu$. However, the equation of motion

$$\partial_\mu F^{\mu\nu} + m^2 A^\nu = J^\nu$$

has as a consequence (for conserved matter currents) $\partial_\nu A^\nu = 0$, and the equation may be presented in gauge invariant form.

$$\partial_\mu F^{\mu\nu} + m^2 \left(g^{\mu\nu} - \frac{\partial^\mu \partial^\nu}{\Box} \right) A_\mu = J^\nu$$

But an important difference remains. If the the mass is promoted to a field, $m^2 \to m^2(x)$, and this field is varied, then the resulting equation $A^\mu A_\mu = 0$, is not consequent to the original equation of motion. Indeed it is an unacceptable equation, because it prevents finding non trivial solutions. (One recognizes the Higgs mechanism, in unitary gauge, as providing a kinetic term and a potential for the field "$m^2(x)$", so that its equation of motion becomes dynamically acceptable.)

References

1. S. Chern, *Complex Manifolds without Potential Theory*, 2nd ed. (Springer, Berlin, 1979).

2. S. Deser, R.Jackiw and S. Templeton, Ann. Phys. (NY) **140**, 372 (1982), (E) **185**, 406 (1988).
3. S. Carroll, G. Field and R. Jackiw, Phys. Rev. D **41**, 1231 (1990).
4. S. Carroll and G. Field, Phys. Rev. Lett. **79**, 2394 (1997).
5. M. Goldhaber and V. Trimble, J. Astrophys. Astron. **17**, 17 (1996);
   T. Jacobson, S. Liberati, D. Mattingly and F. Stecker, astro-ph/0309681.
6. R. Jackiw and S.-Y. Pi, Phys. Rev. D (in press);
   A. Lue, L. Wang and M. Kamionkowski, Phys. Rev. Lett. **83**, 1506 (1999).
7. A. Garcia, F. Hehl, C. Heinicke and A. Macias, gr-qc/0309008.
8. D. Bak, D. Cangemi and R. Jackiw, Phys. Rev. D **49**, 5173 (1994).

SMALL BLACK HOLES IN RANDALL-SUNDRUM I SCENARIO*

D. KARASIK, C. SAHABANDU, P. SURANYI, L.C.R. WIJEWARDHANA

Department of Physics, University of Cincinnati
Cincinnati, Ohio 45221-0011
karasik@physics.uc.edu

An approximation method to study the properties of a small black hole located on the TeV brane in Randall-Sundrum I scenario is presented. The method enables us to find the form of the metric close to the matter distribution when its asymptotic form is given. The short range solution is found as an expansion in the ratio between the Schwarzschild radius of the black hole and the curvature length of the bulk. Long range properties are introduced using the linearized gravity solution as an asymptotic boundary condition. The solution is found up to first order. It is valid in the region close to the horizon but is not valid on the horizon. The regularity of the horizon is still under study.

1. Introduction

Within the Randall-Sundrum brane world scenarios[1,2] the four dimensional universe is viewed as a three-brane with brane tension σ, embedded in a AdS five dimensional bulk, with cosmological constant $8\pi G_5 \Lambda = -6\ell^{-2}$ where ℓ is the curvature length of the AdS bulk. A flat brane is achieved by fine tuning the brane tension $8\pi G_5 \sigma = \pm 6/\ell$

RSI was postulated [1] to solve the hierarchy problem by making gravity strong at the weak scale. In this scenario the world consists of two opposite tension branes embedded in an ADS bulk with an orbifold structure. The negative tension brane is called the TeV brane, while the positive tension brane is called the Planck brane. Z_2 orbifold symmetry is assumed about each brane. In this scenario the standard model particles are assumed to propagate only on the TeV brane. Gravity, on the other hand, is five dimensional and propagates in the bulk. The five dimensional gravitational coupling is of order 1TeV $G_5 = (M_5)^{-3}$, $M_5 \sim 1TeV$. The hierarchy

*Talk presented by D.K.

appears on the TeV brane at distances much larger than the curvature length ℓ. The gravitational attraction becomes ordinary four dimensional attraction with Newton's constant $G_4 = (M_{\rm Pl})^{-2}$, $M_{\rm Pl} = 10^{16}{\rm Te}V$.

Black holes in theories with extra dimensions have been studied widely. Myers and Perry [3] found Schwarzschild type solutions (MPS) in D-dimensional asymptotically flat space. Black hole solutions were also found in asymptotically AdS space [4,5]. No non-trivial closed form black hole solutions, other than the black string solution [6] have been found in 3-brane theories of the Randall Sundrum type. Given that there is considerable interest surrounding the production of black holes at accelerators [7], and in collisions of cosmic rays [8], it is important to develop approximate methods to find black hole solutions in Randall-Sundrum brane world theories.

A TeV black hole solution in RSI scenario is characterized by two scales: The curvature length of the bulk ℓ, generated by the bulk cosmological constant and the five dimensional Schwarzschild radius, μ, related to the mass of the black hole as $\mu = \sqrt{\frac{8}{3\pi}G_5 M}$. For a reasonable setup [9] the curvature length in RSI is of order $\ell \sim 10({\rm TeV})^{-1}$, while the 'mass' of a TeV black hole is $\mu \sim 1({\rm TeV})^{-1}$.

In this work we use the following method: we expand the solution in the dimensionless parameter $\epsilon = \mu/\ell << 1$. The coordinates are rescaled by μ, therefore, an expansion in ϵ actually means an expansion in inverse powers of ℓ. Zeroth order in ϵ should be nothing but the five dimensional flat background MPS black hole [3]. First order in ϵ "turns on" the brane tension, while second order in ϵ "turns on" the bulk cosmological constant.

Up to first order in ϵ (ϵ-solution) the metric is found in section 2. The ϵ-solution is required to be regular everywhere. Especially it should not include black strings off the brane. The ϵ-solution is valid only in the region $r << \ell$. It is not valid in the asymptotic region $r \to \infty$ and one is unable to identify the mass or satisfy the junction conditions on the second brane. In order to overcome these subtleties we use the linearized solution as an asymptotic boundary condition. The detailed calculations appear in Ref. 10.

In section 3 we analyze the properties of the horizon. Clearly, the MPS solution, (zero order in ϵ), includes a regular horizon with a well defined temperature. The horizon is modified in the ϵ solution. Although the ϵ solution is valid in the region close to the horizon, it is not obvious that the new horizon is regular.

2. An Expansion Close to the Horizon

The ansatz for the metric, up to first order in ϵ, is

$$ds^2 = a(\rho,\psi)\left[-\left(1-\frac{1}{\rho^2}-\epsilon B_1\right)dt^2+\left(\frac{\rho^2}{\rho^2-1}+\epsilon A_1\right)d\rho^2 \right.$$
$$\left. +2\epsilon V_1 d\rho d\psi+\rho^2\left(1+\epsilon U_1\right)d\psi^2+\rho^2\sin^2\psi d\Omega_2^2\right]\ . \qquad (1)$$

The conformal factor in front of the metric is the Randall-Sundrum conformal factor $a(\rho,\psi)=(1-\epsilon\rho\cos\psi)^{-2}$ and was introduced for convenience. For convenience we have rescaled the coordinates ρ and t by μ thus making them dimensionless. The TeV brane is located at $\psi=\pi/2$ and the direction perpendicular to the brane is $\psi=0$. The rescaled equations of motion include the bulk Einstein equations with a cosmological constant $\sim\epsilon^2$, and the Israel junction conditions [11] on the TeV brane with brane tension $\sim\epsilon$. The Planck brane is located at a distance $\sim\ell$ from the point mass and therefore the junction condition on the Planck brane can not be evaluated.

The bulk equations can be solved in terms of two independent functions; a gauge function $F(\rho,\psi)$ and a wave function $H(\rho,\psi)$ (Where $H_0(\rho)=H(\rho,\frac{\pi}{2})$). The solution is

$$B_1=-\frac{2}{\rho^3}\left[6\rho H_0(\rho)-(2\rho^2-1)H_0'(\rho)+\rho F(\rho,\psi)\right], \qquad (2a)$$

$$A_1=2\rho^2\left[\frac{(2\rho^4-1)H_0'(\rho)}{\rho(\rho^2-1)^2}-\frac{6H(\rho,\psi)+2\tan\psi H_{,\psi}(\rho,\psi)}{\rho^2-1}+\frac{6(\rho^2-2)H_0(\rho)}{(\rho^2-1)^2}\right.$$
$$\left.+4\rho\cos\psi-\frac{(2\rho^2-1)H_0''(\rho)}{\rho^2-1}+\frac{(\rho^2-2)F(\rho,\psi)}{(\rho^2-1)^2}+\frac{\rho F_{,\rho}(\rho,\psi)}{\rho^2-1}\right], \qquad (2b)$$

$$U_1=2\tan^2\psi\left[6H(\rho,\psi)-6H_0(\rho)-(2\rho-\frac{1}{\rho})\left(H_{,\rho}(\rho,\psi)-H_0'(\rho)\right)\right.$$
$$\left.-\ F(\rho,\psi)-\cot\psi F_{,\psi}(\rho,\psi)\right], \qquad (2c)$$

$$V_1=\rho\tan\psi\left[-2\rho(2\rho^2-1)\cos\psi+\frac{(2\rho^2-1)\tan\psi H_{,\psi}(\rho,\psi)}{\rho^2-1}+2\rho H_{,\rho}(\rho,\psi)\right.$$
$$\left.-(4\rho-\frac{1}{\rho})H_0'(\rho)+(2\rho^2-1)H_0''(\rho)-\rho F_{,\rho}(\rho,\psi)+\frac{\rho^2F_{,\psi}(\rho,\psi)}{(\rho^2-1)\tan\psi}\right]. \qquad (2d)$$

The gauge function $F(\rho,\psi)$ was introduced by a coordinates transformation $\rho\to\rho(1-\epsilon F)$, $\psi\to\psi+\epsilon F\tan\psi$. The bulk equations are indeed gauge invariant, however, the junction conditions are not invariant and the function $F(\rho,\psi)$ is subject to junction conditions on the brane $F(\rho,\frac{\pi}{2})=0$, $F_{,\psi}(\rho,\frac{\pi}{2})=0$.

The wave function $H(\rho,\psi)$ satisfies the differential equation

$$(\rho^2-1)\left(H_{,\rho\rho}-\frac{1}{\rho}H_{,\rho}\right)+H_{,\psi\psi}+2\frac{\cos^2\psi+1}{\sin\psi\cos\psi}H_{,\psi}=0\ , \tag{3}$$

and is subject to the following boundary conditions:

No black strings - The only source is the (point) mass on the brane at $\psi=\pi/2$. Therefore, the solution must be regular everywhere. In particular one should eliminate the black string which lies on a line perpendicular to the brane at $\psi=0$. The Kretchmann scalar ($R_{ABCD}R^{ABCD}$) depends only on the combination $3H+\tan\psi H_{,\psi}$, which must be regular at $\psi=0$.

Compatibility with Linearized Gravity - At large distances, $\rho\to\infty$, the solution is given by Linearized Gravity which is an expansion in the mass, $M\sim\mu^2$, to first order. Iterating the linearized solution to include higher orders of M one finds that only integer powers of M (even powers of μ) enter the metric. Recalling that $\epsilon=\mu/\ell\sim M^{1/2}$ and $\rho\sim M^{-1/2}$, the functions $H(\rho,\psi)$ and $F(\rho,\psi)$ must be odd functions of ρ.

Junction conditions on the brane - If we expand the function $H(\rho,\psi)$ in powers of $\cos\psi$, $H=\sum H_n(\rho)\cos^n\psi$, then Israel junction conditions on the TeV brane imply that

$$H_1(\rho)=0\ ;\ \ H_3(\rho)=-\frac{2}{3}(\rho^3-\rho)\ . \tag{4}$$

Asymptotic boundary conditions - One would like to impose boundary conditions at $\rho\to\infty$. The problem is that the ϵ expansion is valid only for $1\le\rho<<1/\epsilon$. However, the asymptotic form of the metric is given by the linearized solution. The later is valid for $1<<\rho$. As long as $\epsilon<<1$, there is an intermediate region $1<<\rho<<1/\epsilon$ where both the ϵ-solution and the linearized solution are valid. Therefore, the large ρ limit of the ϵ-solution should coincide with first order in ϵ of the linearized solution.

The solution to the differential equation (3), which satisfies the above boundary conditions, is

$$\begin{aligned}H(\rho,\psi)=\rho\sqrt{\rho^2-1}\Bigg[&Q^1_{1/2}(z)\left(\frac{d_2}{\sin^3\psi}+b_2\frac{9[8\psi+4\psi\cos(4\psi)-3\sin(4\psi)]}{4096\sin^3\psi}\right)\\&+b_2\left(P^1_{1/2}(z)+\frac{2}{\pi^2}\frac{\partial Q^1_{n-1/2}(z)}{\partial n}\Bigg|_{n=1}\right)\frac{3[4\psi-\sin(4\psi)]}{32\sin^3\psi}\\&+\sum_{n=0}a_{2n}Q^1_{n-1/2}(z)\frac{3[n\cos(2n\psi)\sin(2\psi)-\cos(2\psi)\sin(2n\psi)]}{8n(1-n^2)\sin^3\psi}\Bigg]\end{aligned} \tag{5}$$

where $P^1_{n-1/2}$ ($Q^1_{n-1/2}$) are the associated Legendre toroidal functions of the first (second) kind, and $z=2\rho^2-1$. The solution includes a discrete set of parameters; d_2, b_2, and a_{2n} $n = 0, 1...$ These parameters should be fixed by the asymptotic boundary condition and the junction condition (4). The asymptotic condition fixes the value of $a_0 = -2(3\pi)^{-1}$, $b_2 = \pi/3$, and the asymptotic form of the gauge function $F(\rho, \psi) \sim \rho/2(\cos\psi + \psi\csc\psi - \pi/2) + F_2(\psi)/\rho + 0(\rho^{-3})$ where $F_2''(\pi/2) = \pi/32(37 + 72d_2 + 81\pi^2/256)$. The rest of the set a_{2n} can be evaluated by Eq.(4) which together with (5) is an expansion of a given function in a complete set of toroidal functions.[12] The prescription given by Banerjee [12] is quite complicated and we will not present it here. A single parameter d_2 remains undetermined.

3. The Black Hole Horizon

The existence of an event horizon is a delicate issue. It is clear that before turning on the brane ($\epsilon = 0$), there exists a single horizon at $\rho_H = 1$. There is no guarantee that first order in ϵ will not ruin the horizon. The first indication that the solution (5) is not regular on the horizon is that the Legendre functions of the second kind are singular at $\rho = 1$. If we assume that the horizon is located at $\rho = 1 + \epsilon\zeta$, an expansion in ϵ is equivalent to an expansion around $\rho = 1$. The second indication comes from the surface gravity. The surface gravity calculated from the solution (5) is not constant on the horizon. According to Ref. 13 this shows that the horizon is singular.

The regularity of the horizon is still under study. The detailed discussion of this issue is left for a future publication.

4. Summary

The method suggested in this paper is to expand the metric in the dimensionless parameter $\epsilon = \sqrt{G_5 M}/\ell$, and to fix the asymptotic behavior using the weak field approximation (linearized gravity). Zeroth order in ϵ is just a Myers-Perry flat background black hole. First order in ϵ corresponds to the presence of a brane with brane tension which is of order ϵ in a flat bulk. Second order in ϵ includes also the bulk cosmological constant which is of order ϵ^2.

We fix the asymptotic behavior in the following way. The ϵ-solution is valid only in the region $r << \ell$. It is not valid in the asymptotic region $r \to \infty$ and one is unable to identify the mass or satisfy the junction conditions on the second brane (at distance $\sim \ell$). In order to overcome these subtleties we use the linearized solution as an asymptotic boundary condition. Linearized gravity is valid in the region $r >> \mu$. We are interested in the case $\mu << \ell$, therefore, there is an intermediate region $\mu << r << \ell$

where the ϵ-solution and the linearized solution must coincide. As a result, the long range characteristics (mass identification and second brane conditions) are introduced into the ϵ-solution.

We should mention that the method of expanding in ϵ and comparing with the linearized solution as an asymptotic condition is applicable to other scenarios as well. For example: Randall-Sundrum single brane scenario, de Sitter bulk, or flat compact extra dimension.

Although the ϵ solution is valid in the region close to the horizon, it is not obvious that the new horizon is regular. The ϵ solution acquires a null surface defined by $g_{tt} = 0$. The normal vector is null and the static Killing vector is null i.e. the surface is a Killing Horizon. But, the surface gravity of the horizon is not a constant which indicates that the horizon is singular [13]. This leaves us with the following possibilities; (1) The horizon is singular. Which means that there are no small black holes in Randall-Sundrum scenario. (2) The ϵ expansion does not converge on the horizon. It is valid only at a distance ϵ away from the horizon, and therefore one cannot use it to calculate the surface gravity. (3) The boundary conditions should be changed thus changing the entire solution. The issue of the regularity of the horizon is under study and the detailed discussion is left for a future publication.

Acknowledgments

This work is supported in part by the U.S. Department of Energy Grant No. DE-FG02-84ER40153. We thank Philip Argyres and Cenalo Vaz for fruitful discussions.

References

1. L. Randall and R. Sundrum, *Phys. Rev. Lett.* **83**:3370-3 (1999).
2. L. Randall and R. Sundrum, *Phys. Rev. Lett.* **83**:4690-3 (1999).
3. R.C. Myers and M.J. Perry, *Annals Phys.* **172**:304 (1986).
4. S.W. Hawking and D.N. Page, *Commun. Math. Phys.* **87**:577 (1983).
5. D. Birmingham, *Class. Quant. Grav.* **16**:1197-1205 (1999).
6. A. Chamblin, S.W. Hawking, H.S. Reall, *Phys. Rev.* D**61**:065007 (2000).
7. S. Giddings and E. Katz *J. Math. Phys.* **42**:3082-3102 (2001); T. Banks and W. Fischler, *Preprint* hep-th/9906038; S. Dimopoulos and G. Landsberg, *Phys. Rev. Lett.* **87**:161602 (2001); S.B. Giddings and S. Thomas, *Phys. Rev.* D**65**:056010 (2002); D.M. Eardley and S.B. Giddings, *Phys. Rev.* D**66**:044011 (2002) gr-qc/0201034.
8. L.A. Anchordoqui, J.L. Feng, H. Goldberg and A.D. Shapere, *Phys. Rev.* D**65**:124027 (2002) hep-ph/0112247.

9. L. Anchordoqui, H. Goldberg and A. D. Shapere, *Phys. Rev.* D**64**:024033 (2002) hep-ph/0204228.
10. D. Karasik, C. Sahabandu, P. Suranyi, and L.C.R. Wijewardhana, *preprint* gr-qc/0309076.
11. W. Israel, *Nuovo Cimento* B**44**:1 (1966); B**48**:463 (1966).
12. D. P. Banerjee, *Amer. J. Math* **64**:72-80 (1940); D. P. Banerjee, *Proceedings of the Cambridge Philosophical Society*, vol. **34**: 30-32 (1938).
13. I. Rácz and R. M. Wald, *Class. Quant. Grav.* **9**:2643-2656 (1992).

MATRIX MODELS AND $\mathcal{N} = 2$ GAUGE THEORY*

S. G. NACULICH†
Department of Physics
Bowdoin College
Brunswick, ME 04011 USA
E-mail: naculich@bowdoin.edu

H. J. SCHNITZER‡ and N. WYLLARD§
Martin Fisher School of Physics
Brandeis University
Waltham, MA 02454 USA
E-mail: schnitzer@brandeis.edu, wyllard@cern.ch

We describe how the ingredients and results of the Seiberg-Witten solution to $\mathcal{N} = 2$ supersymmetric U(N) gauge theory may be obtained from a matrix model.

Dijkgraaf and Vafa discovered that the non-perturbative effective superpotential for certain $d = 4$ $\mathcal{N} = 1$ supersymmetric gauge theories can be obtained by calculating planar diagrams in a related gauged matrix model[1,2,3] (for a more complete list of references, see Ref. 4). In this talk, we will show that matrix models can also be used to obtain all the ingredients and results of the Seiberg-Witten solution[5] of certain $\mathcal{N} = 2$ supersymmetric gauge theories, specifically U(N) theories without matter, or with matter in fundamental, symmetric, or antisymmetric representations[1,2,6,7,8,9].

1. $\mathcal{N} = 2$ supersymmetric gauge theory

We will focus on the $\mathcal{N} = 2$ U(N) gauge theory with N_f hypermultiplets in the fundamental representation to illustrate the matrix model approach,

*Talk given by S.G.N.

†Research partially supported by the NSF under grant PHY-0140281.
‡Research partially supported by the DOE under grant DE–FG02–92ER40706.
§Research supported by the DOE under grant DE–FG02–92ER40706. Current address: Department of Physics, CERN, Theory Division, CH-1211, Geneva 23, Switzerland.

indicating where differences occur in theories with symmetric or antisymmetric hypermultiplets. To apply the insights of Dijkgraaf-Vafa to an $\mathcal{N}=2$ gauge theory, one begins by expressing its field content in terms of $\mathcal{N}=1$ superfields. Let ϕ denote the adjoint $\mathcal{N}=1$ chiral superfield belonging to the $\mathcal{N}=2$ vector multiplet, and q^I and $\tilde{q}_I$ the $\mathcal{N}=1$ chiral superfields that comprise the $\mathcal{N}=2$ hypermultiplets transforming in the fundamental representation. The $\mathcal{N}=2$ theory has the superpotential

$$W_{\mathcal{N}=2}(\phi,q,\tilde{q})=\sum_{I=1}^{N_f}\left[\tilde{q}_I\phi q^I+m_I\tilde{q}_I q^I\right] \tag{1.1}$$

where m_I are the masses of the fundamentals. The Coulomb branch of the moduli space of vacua is characterized by an arbitrary diagonal vev for the scalar field in ϕ, but we may select a specific (but generic) point $\phi=\mathrm{diag}(e_1,\cdots,e_N)$, at which the U($N$) gauge group is broken to U(1)^N, by adding a perturbation to the superpotential

$$W(\phi,q,\tilde{q})=W_{\mathcal{N}=2}(\phi,q,\tilde{q})+W_0(\phi) \tag{1.2}$$

where $W_0'(x)=\alpha\prod_{i=1}^N(x-e_i)$. The perturbation breaks the supersymmetry to $\mathcal{N}=1$, but the full $\mathcal{N}=2$ supersymmetry will be restored by sending $\alpha\to0$ at the end of the matrix model calculation.

2. The perturbative matrix model

Each $\mathcal{N}=1$ chiral superfield described in the previous section has an analog (denoted by a capital letter) in the corresponding matrix model; specifically, an $M\times M$ hermitian matrix Φ, and M-dimensional vectors Q^I and $\tilde{Q}_I$. (The analog of a symmetric or antisymmetric matter hypermultiplet would be an $M\times M$ symmetric or antisymmetric matrix.) The superpotential (1.2) is reinterpreted[1] as the potential of the matrix model, whose partition function is thus

$$Z=\frac{1}{\mathrm{vol}(G)}\int \mathrm{d}\Phi\,\mathrm{d}Q^I\mathrm{d}\tilde{Q}_I\exp\left(-\frac{W(\Phi,Q,\tilde{Q})}{g_s}\right) \tag{2.1}$$

where G is the unbroken matrix model gauge group, and g_s is a parameter that will be taken to zero in the planar limit $M\to\infty$. The matrix integral (2.1) can be evaluated perturbatively about an extremum

$$\Phi_0=\begin{pmatrix} e_1\mathbf{1}_{M_1} & 0 & \cdots & 0\\ 0 & e_2\mathbf{1}_{M_2} & \cdots & 0\\ \vdots & \vdots & \ddots & \vdots\\ 0 & 0 & \cdots & e_N\mathbf{1}_{M_N}\end{pmatrix},\qquad (Q^I)_0=0,\qquad (\tilde{Q}_I)_0=0 \tag{2.2}$$

where the M_i are arbitrary, subject to $\sum_{i=1}^{N} M_i = M$. The U(M) symmetry of the matrix model is broken by Φ_0 to $G = \prod_{i=1}^{N} \mathrm{U}(M_i)$. The residual gauge symmetry must be gauge-fixed, and ghosts introduced; for details, see Refs. 2, 6. For the U(N) gauge theory with an antisymmetric representation[7,8,9] of mass m, Φ_0 must include an additional diagonal block $m\mathbf{1}_{M_0}$ and the antisymmetric matrix a corresponding block J (the symplectic unit), breaking the symmetry of the matrix model to $\mathrm{Sp}(M_0) \times \prod_{i=1}^{N} \mathrm{U}(M_i)$. The inclusion of the extra block for the antisymmetric case has been put in a broader context in Ref. 10.

2.1. *Topological expansion*

The connected diagrams of the perturbative expansion of (2.1) may be organized, using standard 't Hooft double-line notation, in a topological expansion characterized by the Euler characteristic χ of the surface in which the diagram is embedded

$$\log Z = \sum_{\chi \le 2} g_s^{-\chi} F_\chi(S), \qquad S_i = g_s M_i, \qquad \chi = 2-2g-h-q \tag{2.3}$$

with g the number of handles, h the number of boundary components, and q the number of crosscaps. We now take the large M limit, letting $M_i \to \infty$, $g_s \to 0$ with S_i held fixed. In this limit, the dominant contribution $F_{\mathrm{s}}(S) \equiv g_s^2 \log Z\big|_{\mathrm{sphere}}$ arises from planar diagrams that can be drawn on a sphere. Theories with fundamental representations contain surfaces with boundaries; the dominant such contribution $F_{\mathrm{d}}(S) \equiv g_s \log Z\big|_{\mathrm{disk}}$ comes from planar diagrams on a disk. Theories with symmetric or antisymmetric representations contain nonorientable surfaces; the dominant nonorientable contribution $F_{\mathrm{rp}}(S) \equiv g_s \log Z\big|_{\mathbb{RP}^2}$ comes from planar diagrams on $\mathbb{RP}^2$, a sphere with one crosscap.

2.2. *The effective superpotential*

The values of S_i in the matrix model, hitherto arbitrary, are determined by the extremization of the effective superpotential, given by[1,11,12]

$$W_{\mathrm{eff}}(S) = -\left[\sum_{i=1}^{N} \frac{\partial}{\partial S_i} F_{\mathrm{s}}(S) + F_{\mathrm{d}}(S) + 4F_{\mathrm{rp}}(S)\right] \tag{2.4}$$

in the case where the gauge group U(N) of the gauge theory is broken to $\mathrm{U}(1)^N$. The resulting vevs $\langle S_i \rangle$ may be computed in an expansion in Λ, the scale in the matrix model. For the U(N) theory with N_f fundamentals, the leading term is[6]

$$\langle S_i \rangle = \alpha \frac{\prod_{I=1}^{N_f}(e_i + m_I)}{\prod_{j\neq i}(e_i - e_j)} \Lambda^{2N-N_f} + \mathcal{O}(\Lambda^{4N-2N_f}) \tag{2.5}$$

and the Λ^{4N-2N_f} term is also computed in Ref. 6.

2.3. *Tadpole diagrams*

The Seiberg-Witten solution of the $\mathcal{N} = 2$ gauge theory is expressed, not in terms of the parameters e_i, but in terms of the renormalized order parameters a_i, defined as the periods of the Seiberg-Witten differential[5]. The matrix model prescription for computing a_i was presented and motivated in Ref. 6:

$$a_i = e_i + \left[\sum_{j=1}^{N} \frac{\partial}{\partial S_j} g_s \langle \mathrm{tr}\, \Psi_{ii} \rangle_{\mathrm{sphere}} + \langle \mathrm{tr}\, \Psi_{ii} \rangle_{\mathrm{disk}} + 4 \langle \mathrm{tr}\, \Psi_{ii} \rangle_{\mathrm{rp}} \right] \Bigg|_{\langle S \rangle} \tag{2.6}$$

where Ψ_{ii} is the ith diagonal block of $\Phi - \Phi_0$, and the vevs in Eq. (2.6) represent tadpole diagrams with the specified topology. Equation (2.6) may be computed in an expansion in Λ; the Λ^{2N-N_f} contribution agrees[6] with the one-instanton relation between a_i and e_i computed in SW theory[13].

2.4. *Period matrix and prepotential*

In Seiberg-Witten theory, the matrix τ_{ij} of gauge couplings of the unbroken $\mathrm{U}(1)^N$ gauge theory is given by[5,13]

$$\tau_{ij}(a) = \frac{\partial^2 \mathcal{F}(a)}{\partial a_i \partial a_j} \tag{2.7}$$

$$\mathcal{F}(a) = \mathcal{F}_{\mathrm{pert}}(a) + \frac{\Lambda^{2N-N_f}}{2\pi i} \sum_i \frac{\prod_{I=1}^{N_f}(a_i + m_I)}{\prod_{j\neq i}(a_i - a_j)^2} + \mathcal{O}(\Lambda^{4N-2N_f})$$

where $\mathcal{F}_{\mathrm{pert}}(a)$ is the perturbative prepotential and the one-instanton prepotential is shown explicitly.

The matrix model prescription for the gauge coupling matrix is[1]

$$\tau_{ij} = \frac{1}{2\pi i} \frac{\partial^2 F_s(S)}{\partial S_i \partial S_j} \bigg|_{S=\langle S \rangle} . \tag{2.8}$$

In Ref. 6, this quantity was computed and re-expressed in terms of a_i using Eq. (2.6), and was shown to agree with Eq. (2.7) to one-instanton accuracy.

For theories with symmetric or antisymmetric hypermultiplets, the prescription (2.8) must be modified by including relative signs among the various contributions to $F_s(S)$. The justification for these signs, together with a prescription for computing τ_{ij}, was given in Refs. 8, 9. With this modification, and also the inclusion of the extra block for the case with antisymmetric matter discussed above, the matrix model calculation of τ_{ij} agrees with the SW calculation to one-instanton accuracy[9].

3. SW curve and differential from the matrix model

In this section, we will indicate how the usual ingredients of the Seiberg-Witten approach, the SW curve and SW differential, may be obtained from the matrix model using saddle point methods. In this approach, one introduces the trace of the resolvent

$$\omega(z) = g_s \left\langle \operatorname{tr} \left(\frac{1}{z - \Phi} \right) \right\rangle \tag{3.1}$$

which, like the free energy (2.3), may be expressed in terms of a topological expansion, with $\omega_s(z)$ the leading term in the large M limit. The saddle-point approximation to (2.1) implies[1]

$$\omega_s^2(z) - W_0'(z)\,\omega_s(z) + \tfrac{1}{4} f(z) = 0 \tag{3.2}$$

where $f(z)$ is a polynomial, given by

$$f(z) = 4 g_s \left\langle \operatorname{tr} \left(\frac{W_0'(z) - W_0'(\Phi)}{z - \Phi} \right) \right\rangle. \tag{3.3}$$

The polynomial $f(z)$ is determined by extremizing the effective superpotential (2.4); this may be done exactly[6] using Abel's theorem, or perturbatively[9] using (2.5), as follows:

$$\begin{aligned} f(z) &= 4 \sum_i \frac{W_0'(z)}{z - e_i} \langle S_i \rangle + \mathcal{O}(S^2) \\ &= 4\,\alpha^2 \Lambda^{2N-N_f} \left[\prod_{I=1}^{N_f} (z + m_I) - \tilde{T}(z) \prod_{i=1}^{N} (z - e_i) \right] + \mathcal{O}(\Lambda^{4N-2N_f}) \end{aligned} \tag{3.4}$$

where $\tilde{T}(z)$ is the polynomial part of $\prod_{I=1}^{N_f}(z+m_I)/\prod_{i=1}^{N}(z-e_i)$. Defining $y(z) = -2\omega_s(z) + W_0'(z)$, one obtains

$$y^2 = W_0'(z)^2 - f(z) \tag{3.5}$$

precisely the Seiberg-Witten curve for this theory[13] for the choice of moduli e_i consistent with Eq. (2.6). See Refs. 6, 9 for details.

The Seiberg-Witten differential may also be obtained in the matrix model approach as[14,6,3,8]

$$\lambda_{SW} = z\left[\sum_{i=1}^{N}\frac{\partial}{\partial S_i}\omega_{\rm s} + \omega_{\rm d} + 4\,\omega_{\rm rp}\right]{\rm d}z. \tag{3.6}$$

The cubic Seiberg-Witten curve (and associated SW differential) for the gauge theory with symmetric or antisymmetric matter hypermultiplets may also be obtained[7,8,9] from the matrix model approach, using saddle-point methods, together with extremization of the effective superpotential.

Acknowledgments

SGN would like to thank the organizers of the QTS3 conference for the opportunity to present this work.

References

1. R. Dijkgraaf and C. Vafa, *Nucl. Phys.* **B644**, 3 (2002), hep-th/0206255; *Nucl. Phys.* **B644**, 21 (2002), hep-th/0207106, hep-th/0208048; R. Dijkgraaf, et. al., *Phys. Lett.* **B573**, 138 (2003), hep-th/0211017.
2. R. Dijkgraaf, S. Gukov, V. A. Kazakov, and C. Vafa, *Phys. Rev.* **D68**, 045007 (2003), hep-th/0210238.
3. F. Cachazo, M. Douglas, N. Seiberg, and E. Witten, *JHEP* **0212**, 071 (2002), hep-th/0211170; N. Seiberg, *JHEP* **0301**, 061 (2003), hep-th/0212225.
4. R. Argurio, G. Ferretti and R. Heise, hep-th/0311066.
5. N. Seiberg and E. Witten, *Nucl. Phys.* **B426**, 19 (1994), hep-th/9407087; *Nucl. Phys.* **B431**, 484 (1994), hep-th/9408099.
6. S. G. Naculich, H. J. Schnitzer, and N. Wyllard, *Nucl. Phys.* **B651**, 106 (2003), hep-th/0211123; *JHEP* **0301**, 015 (2003), hep-th/0211254v3.
7. A. Klemm, K. Landsteiner, C. I. Lazaroiu, and I. Runkel, *JHEP* **0305**, 066 (2003), hep-th/0303032.
8. S. G. Naculich, H. J. Schnitzer, and N. Wyllard, *JHEP* **0308**, 021 (2003), hep-th/0303268.
9. S. G. Naculich, H. J. Schnitzer, and N. Wyllard, *Nucl. Phys.* **B674**, 37 (2003), hep-th/0305263.
10. K. Intriligator, et. al., hep-th/0311181.
11. R. Argurio, V. L. Campos, G. Ferretti, and R. Heise, *Phys. Rev.* **D67**, 065005 (2003), hep-th/0210291.
12. H. Ita, H. Nieder, and Y. Oz, *JHEP* **0301**, 018 (2003), hep-th/0211261; R. A. Janik and N. A. Obers, *Phys. Lett.* **B553**, 309 (2003), hep-th/0212069; S. K. Ashok, et. al., *Phys. Rev.* **D67**, 086004 (2003), hep-th/0211291.
13. E. D'Hoker, I. M. Krichever, and D. H. Phong, *Nucl. Phys.* **B489**, 179 (1997), hep-th/9609041.
14. R. Gopakumar, *JHEP* **0305**, 033 (2003), hep-th/0211100.

ON A POSSIBLE NONCOMMUTATIVE GENERALIZATION OF THE SIGNATURE AND EULER TOPOLOGICAL INVARIANTS*†

O. OBREGÓN and M. SABIDO

Instituto de Física de la Universidad de Guanajuato,
P.O. Box E-143,
37150 León, Gto., México
E-mail: octavio@ifug3.ugto.mx, msabido@fisica.ugto.mx

C. RAMÍREZ

Facultad de Ciencias Físico Matemáticas, Universidad Autónoma de Puebla
P.O. Box 1364,
72000 Puebla, México
E-mail: cramirez@fcfm.buap.mx

H. GARCÍA-COMPEÁN

Departamento de Física, Centro de Investigación y de Estudios Avanzados del I.P.N.
P.O. Box 14-740,
07000 México D.F., México
E-mail: compean@fis.cinvestav.mx

Noncommutative topological gravity arises in a similar manner as noncommutative Yang-Mills theories. We use the Seiberg-Witten map to show how such a theory based on a $SL(2,\mathbf{C})$ complex connection can be constructed, and from which generalized noncommutative Euler and the signature invariants can be obtained. Finally, in the discussion, we speculate on the description of noncommutative gravitational instantons, as well as noncommutative local gravitational anomalies.

*Talk given by O.O.

†This work was in part supported by grants numbers 37851E and 33951E of CONACYT and projects of PROMEP and the University of Guanajuato.

1. Introduction

Noncommutativity among the space time coordinates is an old proposal[1]. It has been studied from the mathematical[2], as well as the field theoretical point of view[3]. Noncommutative gauge theory has recently attracted attention in connection with string theory, Seiberg and Witten[4] have shown that one can construct noncommutative gauge fields in terms of the usual commutative ones. This result has been generalized for any symmetry Lie algebra[5].

Given these results, one is tempted to look for a formulation of gravity as a gauge theory and to study its noncommutative generalization. One natural candidate is the MacDowell-Mansouri action[6]. This, however, does not fit so nicely with noncommutativity, in particular because it involves a non-standard Lorentz group trace which is not compatible with noncommutativity, and the resulting noncommutative torsion constraint does not seem to have a solution[7]. We have shown that if we start with the usual self-dual gravity, which is equivalent to the MacDowell-Mansouri action, a noncommutative formulation which circumvents the trace problem can be given, and moreover the noncommutative torsion constraint can be solved[8]. Other proposals for noncommutative gravity have been made, they are mentioned in Ref. 8.

In Ref. 9 the self-dual version of the MacDowell-Mansouri action was proposed. It has been shown that this action contains a complex topological term that results in the signature plus the Euler topological term. We have generalized this topological action to its corresponding noncommutative topological quadratic theory of gravity.

We first present the main features of the Seiberg-Witten map[4] for any Lie group[5]. Then topological gravity is reviewed. We then construct the corresponding noncommutative topological gravity[10] and show that in a power expansion in the noncommutative parameters $\theta^{\rho\sigma}$ of the action, the resulting even terms do not seem to vanish identically, and as the action does not depend on the metric, and term to term is given by a divergence, each one of these terms is topological. At the end of this contribution we briefly discuss these results and speculate on the description of noncommutative gravitational instantons and local gravitational anomalies.

2. The Seiberg-Witten Map

Let us consider a gauge theory with a hermitian connection, invariant under a symmetry Lie group G, with gauge fields A_μ,

$$\delta_\lambda A_\mu = \partial_\mu \lambda + i\,[\lambda, A_\mu]\,, \tag{1}$$

where $\lambda = \lambda^i T_i$, and T_i are the generators of the Lie algebra G of the group G, in the adjoint representation. These transformations are generalized for the noncommutative theory as,

$$\delta_{\widehat{\Lambda}} \widehat{A}_\mu = \partial_\mu \widehat{\Lambda} + i\left[\widehat{\Lambda} \stackrel{*}{,} \widehat{A}_\mu\right], \tag{2}$$

with an associative and noncommutative $*$-product, which is a combination of the matrix product with the $\star$-Moyal function product, which is given by

$$f(x) \star g(x) \equiv \left[\exp\left(\frac{i}{2}\theta^{\mu\nu}\frac{\partial}{\partial\varepsilon^\mu}\frac{\partial}{\partial\eta^\nu}\right) f(x+\varepsilon) g(x+\eta)\right]_{\varepsilon=\eta=0} \,. \tag{3}$$

Inside integrals, this product has the property $\mathrm{Tr}\int A_1 * A_2 * A_3 * \cdots * A_n = \mathrm{Tr}\int A_n * A_1 * A_2 * A_3 * \cdots * A_{n-1}$. In particular, $\mathrm{Tr}\int A_1 * A_2 = \mathrm{Tr}\int A_1 A_2$.

The noncommutative parameters $\widehat{\Lambda}$ have some dependence on λ and the connection A. The commutators $[A \stackrel{*}{,} B] \equiv A * B - B * A$ have the correct derivative properties when acting on products of noncommutative fields.

Due to noncommutativity, commutators like $\left[\widehat{\Lambda} \stackrel{*}{,} \widehat{A}_\mu\right]$ take values in the enveloping algebra of G in the adjoint representation, $U(G, \mathrm{ad})$. Therefore, $\widehat{\Lambda}$ and the gauge fields $\widehat{A}_\mu$ will also take values in this algebra. In general, for some representation R of G, we will denote $U(G, R)$ the corresponding enveloping algebra. Let us write for instance $\widehat{\Lambda} = \widehat{\Lambda}^I T_I$ and $\widehat{A} = \widehat{A}^I T_I$, then,

$$\left[\widehat{\Lambda} \stackrel{*}{,} \widehat{A}_\mu\right] = \left\{\widehat{\Lambda}^I \stackrel{*}{,} \widehat{A}^J_\mu\right\}[T_I, T_J] + \left[\widehat{\Lambda}^I \stackrel{*}{,} \widehat{A}^J_\mu\right]\{T_I, T_J\}\,, \tag{4}$$

where $\{A \stackrel{*}{,} B\} \equiv A * B + B * A$ is the noncommutative anticommutator. Thus all the products of the generators T_I will be needed in order to close the algebra $U(G, \mathrm{ad})$. Its structure can be obtained by successive computation of commutators and anticommutators starting from the generators of G, until it closes,

$$[T_I, T_J] = i f_{IJ}{}^K T_K, \quad \{T^I, T^J\} = d_{IJ}{}^K T_K. \tag{5}$$

The noncommutative field strength is defined as $\widehat{F}_{\mu\nu} = \partial_\mu \widehat{A}_\nu - \partial_\nu \widehat{A}_\mu - i[\widehat{A}_\mu \stackrel{*}{,} \widehat{A}_\nu]$, hence it takes also values in $U(G, \mathrm{ad})$. From Eq. (2) it turns out that,

$$\delta_{\widehat{\Lambda}}\widehat{F}_{\mu\nu} = i\left(\widehat{\Lambda} * \widehat{F}_{\mu\nu} - \widehat{F}_{\mu\nu} * \widehat{\Lambda}\right). \tag{6}$$

We see that these transformation rules can be obtained from the commutative ones, just by replacing the ordinary product of smooth functions by the Moyal product, with a suitable product ordering. This allows the construction of invariant quantities in a simple way.

The fact that the observed world is commutative, means that there must be possible to obtain it from the noncommutative one by taking the limit $\theta \to 0$. Thus the noncommutative fields $\widehat{A}$ are given by a power series expansion on θ, starting from the commutative ones A,

$$\widehat{A} = A + \theta^{\mu\nu}A^{(1)}_{\mu\nu} + \theta^{\mu\nu}\theta^{\rho\sigma}A^{(2)}_{\mu\nu\rho\sigma} + \cdots \quad . \tag{7}$$

The coefficients of this expansion are determined by the Seiberg-Witten map, which states that the symmetry transformations of (7), given by (2), are induced by the symmetry transformations of the commutative fields (1). In order that these transformations be consistent, the transformation parameter $\widehat{\Lambda}$ must satisfy[5],

$$\delta_\lambda\widehat{\Lambda}(\eta) - \delta_\eta\widehat{\Lambda}(\lambda) - i[\widehat{\Lambda}(\lambda) \stackrel{*}{,} \widehat{\Lambda}(\eta)] = \widehat{\Lambda}(-i[\lambda, \eta]). \tag{8}$$

Similarly, the terms in Eq. (7) are functions of the commutative fields and their derivatives, and are determined by the requirement that $\widehat{A}$ transforms as (2) [5].

The fact that the noncommutative gauge fields take values in the enveloping algebra, has the consequence that they have a bigger number of components than the commutative ones, unless the enveloping algebra coincides with the Lie algebra of the commutative theory, as is the case of $U(N)$. However, the physical degrees of freedom of the noncommutative fields can be related one to one to the physical degrees of freedom of the commutative fields by the Seiberg-Witten map, fact used in [5] to construct noncommutative gauge theories, in principle for any Lie group.

In order to obtain the Seiberg-Witten map to first order, the noncommutative parameters are first obtained from Eq. (8) [4,5],

$$\widehat{\Lambda}\left(\lambda, A\right) = \lambda + \frac{1}{4}\theta^{\mu\nu}\left\{\partial_\mu\lambda, A_\nu\right\} + O\left(\theta^2\right). \tag{9}$$

Then, from Eqs. (2) and (7), the following solution is given

$$\widehat{A}_\mu(A) = A_\mu - \frac{1}{4}\theta^{\rho\sigma}\{A_\rho, \partial_\sigma A_\mu + F_{\sigma\mu}\} + O(\theta^2), \tag{10}$$

and then for the field strength it turns out that,

$$\widehat{F}_{\mu\nu} = F_{\mu\nu} + \frac{1}{4}\theta^{\rho\sigma}\Big(2\{F_{\mu\rho}, F_{\nu\sigma}\} - \{A_\rho, D_\sigma F_{\mu\nu} + \partial_\sigma F_{\mu\nu}\}\Big) + O(\theta^2). \tag{11}$$

The higher coefficients in Eq. (7) can be obtained from the observation that the Seiberg-Witten map preserves the operations of the commutative function algebra, hence the following differential equation can be written[4],

$$\delta\theta^{\mu\nu}\frac{\partial}{\partial\theta^{\mu\nu}}\widehat{A}(\theta) = \delta\theta^{\mu\nu}\widehat{A^{(1)}_{\mu\nu}}(\theta), \tag{12}$$

where $\widehat{A^{(1)}_{\mu\nu}}$ is obtained from $A^{(1)}_{\mu\nu}$ in Eq. (7), by substituting the commutative fields by the noncommutative ones under the $*$-product.

3. Topological Gravity

We briefly review four-dimensional topological gravity. We start from the following $SO(3,1)$ invariant action

$$I_{TOP} = \frac{\Theta_G^P}{2\pi}\mathrm{Tr}\int_X R\wedge R + i\frac{\Theta_G^E}{2\pi}\mathrm{Tr}\int_X R\wedge\widetilde{R}, \tag{13}$$

where R is the field strength, corresponding to a $SO(3,1)$ connection ω

$$R_{\mu\nu}{}^{ab} = \partial_\mu\omega_\nu{}^{ab} - \partial_\nu\omega_\mu{}^{ab} + \omega_\mu{}^{ac}\omega_\nu{}_c{}^b - \omega_\mu{}^{bc}\omega_\nu{}_c{}^a, \tag{14}$$

X is a four dimensional closed pseudo-Riemannian manifold and $\widetilde{R}_{\mu\nu}{}^{ab} = -\frac{i}{2}\varepsilon^{ab}{}_{cd}R_{\mu\nu}{}^{cd}$ is the dual with respect to the group. Here the coefficients are the gravitational analogs of the $\Theta-$vacuum in QCD.

In this action, the connection satisfies the first Cartan structure equation, which relates it to a given tetrad. This action can be written as the integral of a divergence, and a variation of it with respect to the tetrad vanishes, hence it is metric independent, and therefore topological.

The action (13) arises naturally from the MacDowell-Mansouri type action[6,9]. Keeping this philosophy in mind, action (13) can be rewritten in terms of the self-dual and anti-self-dual parts, $R^{\pm} = \frac{1}{2}(R \pm \widetilde{R})$, of the Riemann tensor as follows:

$$I_{TOP} = \mathrm{Tr}\int_X \left(\tau R^+\wedge R^+ + \overline{\tau}R^-\wedge R^-\right) = \mathrm{Tr}\int_X \left(\tau R^+\wedge R^+ + \overline{\tau}\,\overline{R^+}\wedge\overline{R^+}\right), \tag{15}$$

where $\tau = \left(\frac{1}{2\pi}\right)\left(\Theta_G^E + i\Theta_G^P\right)$, and the bar denotes complex conjugation. In local coordinates on X, this action can be rewritten as

$$I_{TOP} = 2\text{Re}\left(\tau \int_X d^4x\ \varepsilon^{\mu\nu\rho\sigma} R^{+\ ab}_{\mu\nu} R^{+}_{\rho\sigma ab}\right). \tag{16}$$

Therefore, it is enough to study the complex action,

$$I = \int_X d^4x\ \varepsilon^{\mu\nu\rho\sigma} R^{+\ ab}_{\mu\nu} R^{+}_{\rho\sigma ab}. \tag{17}$$

Further, the self-dual Riemann tensor satisfies, $\varepsilon^{ab}{}_{cd} R^{+\ cd}_{\mu\nu} = 2iR^{+\ ab}_{\mu\nu}$. This tensor has the useful property that it can be written as a usual Riemann tensor, but in terms of the self-dual components of the spin connection, $\omega^{+\ ab}_{\mu} = \frac{1}{2}\left(\omega^{ab}_{\mu} - \frac{i}{2}\varepsilon^{ab}{}_{cd}\omega^{cd}_{\mu}\right)$, as

$$R^{+\ ab}_{\mu\nu} = \partial_\mu \omega^{+\ ab}_{\nu} - \partial_\nu \omega^{+\ ab}_{\mu} + \omega^{+\ ac}_{\mu}\omega^{+\ b}_{\nu\ c} - \omega^{+\ bc}_{\mu}\omega^{+\ a}_{\nu\ c}. \tag{18}$$

In this case, the action (16) can be rewritten as,

$$I = \int_X d^4x\ \varepsilon^{\mu\nu\rho\sigma}\left[2R_{\mu\nu}{}^{0i}(\omega^+)R_{\rho\sigma 0i}(\omega^+) + R_{\mu\nu}{}^{ij}(\omega^+)R_{\rho\sigma ij}(\omega^+)\right]. \tag{19}$$

Now, we define $\omega_\mu{}^i = i\omega^{+0i}_\mu$, from which we obtain, by means of the self-duality properties, $\omega^{+ij}_\mu = -\varepsilon^{ij}{}_k\omega_\mu{}^k$. Then it turns out that

$$R_{\mu\nu}{}^{oi}(\omega^+) = -i(\partial_\mu\omega^i_\nu - \partial_\nu\omega^i_\mu + 2\varepsilon^i_{jk}\omega^j_\mu\omega^k_\nu) = -i\mathcal{R}_{\mu\nu}{}^i(\omega), \tag{20}$$

$$R_{\mu\nu}{}^{ij}(\omega^+) = \partial_\mu\omega^{+ij}_\nu - \partial_\nu\omega^{+ij}_\mu - 2(\omega_\mu{}^i\omega_\nu{}^j - \omega_\nu{}^i\omega_\mu{}^j) = -\varepsilon^{ij}{}_k\mathcal{R}_{\mu\nu}{}^k(\omega). \tag{21}$$

This amounts to the decomposition $SO(3,1) = SL(2,\mathbf{C}) \times SL(2,\mathbf{C})$, such that $\omega_\mu{}^i$ is a complex $SL(2,\mathbf{C})$ connection. If we choose the algebra $s\ell(2,\mathbf{C})$ to satisfy $[T_i, T_j] = 2i\varepsilon_{ij}{}^k T_k$ and $\text{Tr}(T_iT_j) = 2\delta_{ij}$, then we can write

$$I = \text{Tr}\int_X d^4x\ \varepsilon^{\mu\nu\rho\sigma}\mathcal{R}_{\mu\nu}(\omega)\mathcal{R}_{\rho\sigma}(\omega), \tag{22}$$

where, $\mathcal{R}_{\mu\nu} = \partial_\mu\omega_\nu - \partial_\nu\omega_\mu - i[\omega_\mu, \omega_\nu]$ is the field strength. This action is invariant under the $SL(2,\mathbf{C})$ transformations, $\delta_\lambda\omega_\mu = \partial_\mu\lambda + i[\lambda, \omega_\mu]$.

In the case of a Riemannian manifold X, the signature and the Euler topological invariants of X, are the real and imaginary parts of (22)

$$\sigma(X) = -\frac{1}{24\pi^2}\text{Re}\left(\text{Tr}\int_X d^4x\ \varepsilon^{\mu\nu\rho\sigma}\mathcal{R}_{\mu\nu}(\omega)\mathcal{R}_{\rho\sigma}(\omega)\right), \tag{23}$$

$$\chi(X) = \frac{1}{32\pi^2}\text{Im}\left(\text{Tr}\int_X d^4x\ \varepsilon^{\mu\nu\rho\sigma}\mathcal{R}_{\mu\nu}(\omega)\mathcal{R}_{\rho\sigma}(\omega)\right). \tag{24}$$

4. Noncommutative Topological Gravity

We wish to have a noncommutative formulation of the $SO(3,1)$ action (13). Its first term, can be straightforwardly made noncommutative, in the same way as for usual Yang-Mills theory,

$$\mathrm{Tr} \int_X \widehat{R} \wedge \widehat{R}. \tag{25}$$

Instead, for the second term of (13) such an action cannot be written, because it involves the Levi-Civita symbol, an invariant Lorentz tensor, but which is not invariant under the full enveloping algebra. However, as mentioned at the end of the preceding section, this term can be obtained from Eq. (22).

Thus, in general we will consider as the noncommutative topological action of gravity, the $SL(2, \mathbf{C})$ invariant action,

$$\widehat{I} = \mathrm{Tr} \int_X d^4x \; \varepsilon^{\mu\nu\rho\sigma} \widehat{\mathcal{R}}_{\mu\nu} \widehat{\mathcal{R}}_{\rho\sigma}, \tag{26}$$

where $\widehat{\mathcal{R}}_{\mu\nu} = \partial_\mu \widehat{\omega}_\nu - \partial_\nu \widehat{\omega}_\mu - i[\widehat{\omega}_\mu \stackrel{*}{,} \widehat{\omega}_\nu]$, is the $SL(2, \mathbf{C})$ noncommutative field strength. This action does not depend on the metric of X. Indeed, as well as the commutative one, it is given by a divergence,

$$\widehat{I} = \mathrm{Tr} \int_X d^4x \; \varepsilon^{\mu\nu\rho\sigma} \partial_\mu \left(\widehat{\omega}_\nu * \partial_\rho \widehat{\omega}_\sigma + \frac{2}{3} \widehat{\omega}_\nu * \widehat{\omega}_\rho * \widehat{\omega}_\sigma \right). \tag{27}$$

Thus, a variation of (26) with respect to the noncommutative connection, will vanish identically because of the noncommutative Bianchi identities,

$$\delta_{\widehat{\omega}} \widehat{I} = 8 \mathrm{Tr} \int \varepsilon^{\mu\nu\rho\sigma} \delta \widehat{\omega}_\mu * \widehat{D}_\mu \widehat{\mathcal{R}}_{\rho\sigma} \equiv 0, \tag{28}$$

where $\widehat{D}_\mu$ is the noncommutative covariant derivative.

Further, from the first Cartan structure equation, the SO(3,1) connection, and thus its $SL(2, \mathbf{C})$ projection $\omega_\mu{}^i$, can be written in terms of the tetrad and the torsion. Furthermore, from the Seiberg-Witten map, the noncommutative connection can be written as well as $\widehat{\omega}(e)$. Therefore, a variation of the action (26) with respect to the tetrad of the action, can be written as

$$\delta_e \widehat{I} = 8 \mathrm{Tr} \int \varepsilon^{\mu\nu\rho\sigma} \delta_e \widehat{\omega}_\mu(e) * \widehat{D}_\mu \widehat{\mathcal{R}}_{\rho\sigma} \equiv 0, \tag{29}$$

hence it is topological, as the commutative one.

Thus, we see from (26) that, in a $\theta-$power expansion of the action, each one of the resulting terms will be independent of the metric, as well as they will be given by a divergence. Thus, unless these terms vanish identically, they will be topological. Furthermore, the whole noncommutative action, expressed in terms of the commutative fields by the Seiberg-Witten map, is invariant under the SO(3,1) transformations. Thus, each term of the expansion will be also invariant. Thus these terms will be topological invariants.

The action (26) is not real, as well as the limiting commutative action. Hence, it is not obvious that the signature (25) will be precisely its real part. In this case we could neither say that $\widehat{\chi}(X)$ is given by its imaginary part. In fact we can only say that $\widehat{\chi}(X)$ could be obtained from the difference of (26) and (25). However, the real and the imaginary parts of (28) are invariant under SL(2,**C**) and consequently under SO(3,1), and thus they are the natural candidates for $\widehat{\sigma}(X)$ and $\widehat{\chi}(X)$, as in (23) and (24).

In order to write down the θ expansion of these noncommutative actions, we will take the fundamental representation of $SL(2,\mathbf{C})$, with generators given by the Pauli matrices. In this case, to second order in θ, the Seiberg-Witten map for the Lie algebra valued commutative field strength $\mathcal{R}_{\mu\nu} = \mathcal{R}^i_{\mu\nu}(\omega)\sigma_i$, is given by

$$\widehat{\mathcal{R}}_{\mu\nu} = \mathcal{R}_{\mu\nu} + \theta^{\alpha\beta}\mathcal{R}^{(1)}_{\mu\nu\alpha\beta} + \theta^{\alpha\beta}\theta^{\gamma\delta}\mathcal{R}^{(2)}_{\mu\nu\alpha\beta\gamma\delta} + \cdots \,, \tag{30}$$

where, from Eq. (11) we get,

$$\theta^{\rho\sigma}\mathcal{R}^{(1)}_{\mu\nu\rho\sigma} = \frac{1}{2}\theta^{\rho\sigma}\left[2\mathcal{R}_{\mu\rho}{}^{i}\mathcal{R}_{\nu\sigma i} - \omega_\rho{}^{i}\left(\partial_\sigma\mathcal{R}_{\mu\nu i} + D_\sigma\mathcal{R}_{\mu\nu i}\right)\right]\mathbf{1}, \tag{31}$$

with $\mathbf{1}$ being the unity 2×2 matrix. Further, by means of Eq. (12), we get,

$$\begin{aligned}\theta^{\rho\sigma}\theta^{\tau\theta}\mathcal{R}^{(2)}_{\mu\nu\rho\sigma\tau\theta} &= \frac{1}{4}\theta^{\rho\sigma}\theta^{\tau\theta}\Big(\varepsilon^i_{jk}\left[i\partial_\tau\mathcal{R}^j_{\mu\rho}\partial_\theta\mathcal{R}^k_{\nu\sigma} + \partial_\tau\omega^j_\rho\partial_\theta(\partial_\sigma\right.\\ &\quad \left.+ D_\sigma)\mathcal{R}^k_{\mu\nu}\right] - \omega^i_\rho\partial_\tau\omega^j_\sigma\partial_\theta\mathcal{R}_{\mu\nu j} + \mathcal{R}^i_{\mu\rho}[2\mathcal{R}^j_{\nu\tau}\mathcal{R}_{\sigma\theta j}\\ &\quad -\omega^j_\tau(\partial_\theta + D_\theta)\mathcal{R}_{\nu\sigma j}] - \mathcal{R}^i_{\nu\rho}\left[2\mathcal{R}^j_{\mu\tau}\mathcal{R}_{\sigma\theta j} - \omega^j_\tau(\partial_\theta\right.\\ &\quad \left.+D_\theta)\mathcal{R}_{\mu\sigma j}\right] + \frac{1}{2}\omega^j_\tau(\partial_\theta\omega_{\rho j} + \mathcal{R}_{\theta\rho j})(\partial_\sigma + D_\sigma)\mathcal{R}^i_{\mu\nu}\\ &\quad -2\omega^i_\rho\left\{2\partial_\sigma\mathcal{R}^j_{\mu\tau}\mathcal{R}_{\nu\theta j} - \partial_\sigma[\omega^j_\tau(\partial_\theta + D_\theta)\mathcal{R}_{\mu\nu j}]\right\}\Big)\sigma_i.\end{aligned} \tag{32}$$

Therefore the action (26) will be given by,

$$\begin{aligned}\widehat{I} &= \mathrm{Tr}\int_X d^4x\, \varepsilon^{\mu\nu\rho\sigma}\Big[\mathcal{R}_{\mu\nu}\mathcal{R}_{\rho\sigma} + 2\theta^{\tau\vartheta}\mathcal{R}_{\mu\nu}\mathcal{R}^{(1)}_{\rho\sigma\tau\vartheta}\\ &\quad + \theta^{\tau\theta}\theta^{\vartheta\varsigma}\Big(2\mathcal{R}_{\mu\nu}\mathcal{R}^{(2)}_{\rho\sigma\tau\theta\vartheta\varsigma} + \mathcal{R}^{(1)}_{\mu\nu\tau\theta}\mathcal{R}^{(1)}_{\rho\sigma\vartheta\varsigma}\Big)\Big] + \mathcal{O}(\theta^3).\end{aligned} \tag{33}$$

Taking into account (31), we get that the first order term is proportional to $\mathrm{Tr}(\sigma_i)$ and thus vanishes identically. Similarly to the second order term (32), the third order term for $\widehat{\mathcal{R}}$ can be computed by means of Eq. (12). The result is given by a rather long expression, which however is proportional to the unity matrix $\mathbf{1}$, like (31), as a consequence the third order term in (33) also vanishes identically. Further using (32), we finally get,

$$
\begin{aligned}
\widehat{I} = \int_X d^4x \ \varepsilon^{\mu\nu\rho\sigma} \Big\{ & 2\mathcal{R}^i_{\mu\nu}\mathcal{R}_{\rho\sigma i} + \frac{1}{4}\theta^{\tau\theta}\theta^{\vartheta\zeta}\Big[- \varepsilon_{ijk}\mathcal{R}^i_{\mu\nu} \ [\partial_\vartheta \mathcal{R}^j_{\rho\tau}\partial_\zeta \mathcal{R}^k_{\sigma\theta} \\
& -\partial_\vartheta \omega^j_\tau \partial_\zeta(\partial_\theta + D_\theta)\mathcal{R}^k_{\rho\sigma}] + [\mathcal{R}^i_{\mu\tau}\mathcal{R}_{\nu\theta i} - \frac{1}{2}\omega^i_\tau(\partial_\theta + D_\theta)\mathcal{R}_{i\mu\nu}][\mathcal{R}^j_{\rho\vartheta}\mathcal{R}_{\sigma\zeta j} \\
& -\frac{1}{2}\omega^j_\vartheta(\partial_\zeta + D_\zeta)\mathcal{R}_{\rho\sigma j}] + \mathcal{R}^i_{\mu\nu}\{\mathcal{R}_{i\sigma\theta}[2\mathcal{R}^j_{\rho\vartheta}\mathcal{R}_{\tau\zeta j} - \omega^j_\vartheta(\partial_\zeta + D_\zeta)\mathcal{R}_{\rho\tau j}] \\
& +\frac{1}{4}(\partial_\theta + D_\theta)\mathcal{R}_{\rho\sigma i}\omega^j_\vartheta(\partial_\zeta \omega_{\tau j} + \mathcal{R}_{\zeta\tau j}) + \omega_{\theta i}[\partial_\tau(\mathcal{R}^j_{\rho\vartheta}\mathcal{R}_{\sigma\zeta j}) \\
& -\frac{1}{2}\partial_\tau \omega^j_\vartheta(\partial_\zeta + D_\zeta)\mathcal{R}_{\rho\sigma j}]\} - \frac{1}{2}\mathcal{R}^i_{\mu\nu}\omega_{\tau i}\partial_\vartheta \omega^j_\theta \partial_\zeta \mathcal{R}_{\rho\sigma j}\Big]\Big\} + \mathcal{O}(\theta^4),
\end{aligned}
\tag{34}
$$

where the second order correction does not identically vanish.

5. Discussion

Here we show our proposal of a noncommutative version of topological gravity with quadratic actions. This proposal is based on the complex action (26) in terms of the self-dual and anti-self-dual connections, and from which the noncommutative natural generalization of the signature (23) and Euler (24) topological invariants can be extracted. We show this action up to third θ order, and we obtain that the first and the third order vanish, but the second order is not identically zero. The action to this order is given by Eq. (34). It seems that all odd θ orders vanish identically. Thus we show how these topological invariants are modified nontrivially by the noncommutativity deformation.

On a (commutative) Riemannian manifold, the signature and Euler topological invariants characterize gravitational instantons. Thus the study of noncommutative topological invariants should allow us, through the Seiberg-Witten map, to deform gravitational instantons into noncommutative versions of them. In order to make explicit computations, specific gravitational (noncommutative) metrics have to be chosen. In this context, the noncommutative formulation for dynamical gravity[8] would be helpful.

Similarly to the gauge theory case, one could speculate on a definition of noncommutative local gravitational Lorentz anomaly, by a suitable definition of the noncommutative Lorentz group $\widehat{SO(4)}$ in the compact space-time of an Euclidean signature. The application of these ideas to the diffeomorphism transformations connected with the identity might predict new nontrivial noncommutative gravitational effects, which should be computed explicitly as a noncommutative correction to the gravitational contribution to the chiral anomaly[11].

Finally, the ALE gravitational instantons are an important case of gravitational instantons. They can be obtained as smooth resolutions of **A-D-E** orbifold singularities $\mathbf{C}^2/\Gamma$, with Γ being an **A-D-E** finite subgroup of $SU(2)$. These gravitational instantons are classified through the Kronheimer construction[12], which is the analogue of Atiyah, Drinfeld, Hitchin, and Manin construction of Yang-Mills instantons. There is a proposal to extend this construction to the noncommutative case[13]. Thus, it would be interesting to give the noncommutative analogue of the Kronheimer construction of ALE instantons.

References

1. H. Snyder, *Phys. Rev.* **71**, 38 (1947).
2. A. Connes, *Noncommutative Geometry* (Academic Press, London, 1994).
3. M.R. Douglas and N.A. Nekrasov, *Rev. Mod. Phys.* **73**, 977 (2002).
4. N. Seiberg and E. Witten *J. High Energy Phys.* **09**, 032 (1999).
5. See X. Calmet, B. Jurco, P. Schupp, J. Wess and M. Wohlgenannt, *Eur. Phys. J. C.* **23**, 363 (2002), and references therein.
6. S.W. MacDowell and F. Mansouri, *Phys. Rev. Lett.* **38**, 739 (1977)and F. Mansouri, *Phys. Rev.* **D16**, 2456 (1977)
7. A.H. Chamseddine *Phys. Lett.* **B504**, 33 (2001).
8. H. García-Compéan, O. Obregón, C. Ramírez and M. Sabido, *Phys. Rev.* **D68**, 440151 (2003).
9. J.A. Nieto, O. Obregón, and J. Socorro, *Phys. Rev* **D50**, R3583 (1994).
10. Most of the results presented here have been already published in H. García-Compéan, O. Obregón, C. Ramírez and M. Sabido, *Phys. Rev.* **D68**, 450101 (2003).
11. R. Delbourgo and A. Salam, *Phys. Lett.* **40B**, 381 (1972); T. Eguchi and P.G.O. Freund, *Phys. Rev. Lett.* **37**, 1251 (1976).
12. P.B. Kronheimer, *J. Diff. Geom.* **29**, 665 (1989).
13. N. Nekrasov and A. Schwarz, *Commun. Math. Phys.* **198**, 689 (1998).

SUSY GUTS IN FOUR AND FIVE DIMENSIONS

STUART RABY

Department of Physics, The Ohio State University,
174 W. 18th Ave., Columbus, OH 43210, USA
raby@pacific.mps.ohio-state.edu

In this talk we consider SUSY GUTs in four and five dimensions. We focus on the issues of gauge coupling unification and nucleon decay.

1. Gauge coupling unification and nucleon decay in SUSY GUTs

The apparent unification of the three gauge couplings at a scale of order 3×10^{16} GeV is, at the moment, the only experimental evidence for low energy supersymmetry [1,2,3,4,5]. In this talk we consider the status of gauge coupling unification and the demise of minimal SUSY SU(5).

Due to the accuracy of the experimental data, the theoretical analysis of unification requires two loop renormalization group running from M_G to M_Z. Consistency then requires including one loop threshold corrections at both the GUT and weak scales. Once GUT threshold corrections are considered, a precise definition of the GUT scale (M_G) is needed. The three gauge couplings no longer meet at one scale, since

$$\alpha_i(\mu)^{-1} = \alpha_G^{-1} + \Delta_i(\mu) \tag{1}$$

where the corrections $\Delta_i(\mu)$ are logarithmic functions of mass for all states with GUT scale mass. In principle, the GUT scale can now be defined as the mass M_X of the $X, \bar{X}$ gauge bosons mediating proton decay or as the scale where any two couplings meet [6]. We define M_G as the value of μ where

$$\alpha_1(M_G) = \alpha_2(M_G) \equiv \tilde{\alpha}_G. \tag{2}$$

Using two loop RGE from M_G to M_Z, we find

$$\begin{aligned} M_G &\approx 3 \times 10^{16} \text{ GeV} \\ \alpha_G^{-1} &\approx 24. \end{aligned} \tag{3}$$

In addition, good fits to the low energy data require

$$\epsilon_3 \equiv \frac{(\alpha_3(M_G) - \tilde{\alpha}_G)}{\tilde{\alpha}_G} \sim -3\% \text{ to } -4\%. \tag{4}$$

Note the exact value of the threshold correction ϵ_3, needed to fit the data, depends on the weak scale threshold corrections and in particular on the SUSY particle spectrum. We shall return to this later. On the other hand, significant contributions to the GUT threshold correction ϵ_3 typically arise from the Higgs and GUT breaking sectors of the theory. Above M_G there is a single coupling constant $\alpha_G \approx \tilde{\alpha}_G$ which then runs up to some fundamental scale M_*, such as the string scale, where the running is cut off. The GUT symmetry, in concert with supersymmetry, regulates the radiative corrections. Without the GUT, ϵ_3 would naturally take on a value of order one.

Let's consider nucleon decay in the Minimal SU(5) SUSY GUT in more detail. In minimal SUSY SU(5), we have the following gauge and Higgs sectors. The gauge sector includes the gauge bosons for SU(5) which decompose to the SM [SU(3) × SU(2) × U(1)] gauge sector plus the massive gauge bosons $\{X,\ \bar{X}\}$. The $\{X,\ \bar{X}\}$ bosons with charges $\{(3, \bar{2}, -5/3),\ (\bar{3}, 2, 5/3)\}$ are responsible for nucleon decay. The minimal SU(5) theory has, by definition, the minimal Higgs sector. It includes a single adjoint of SU(5), $\Sigma = 24$, used for GUT breaking and for doublet-triplet splitting for the electroweak Higgs sector and the Higgs fields $H = 5,\ \bar{H} = \bar{5}$ given by

$$\begin{pmatrix} H_u \\ T \end{pmatrix}, \begin{pmatrix} H_d \\ \bar{T} \end{pmatrix} \subset H,\ \bar{H}.$$

The superpotential for the GUT breaking and Higgs sectors of the model is given by [7,2,3]

$$W = \frac{\lambda}{3}\mathrm{Tr}(\Sigma^3) + \frac{\lambda V}{2}\mathrm{Tr}(\Sigma^2) + \bar{H}(\Sigma + 3V)H. \tag{5}$$

In general, nucleon decay can have contributions from operators with dimensions 4, 5 and 6. Dimension 4 operators can be excluded by R parity.

Dimension 5 operators [?,9] are obtained when integrating out heavy color triplet Higgs fields.

$$\begin{aligned} W_{Yukawa} \supset\ & H_u\ QY_u\overline{U} + H_d(QY_d\overline{D} + LY_e\overline{E}) \\ & + T(Q\frac{1}{2}c_{qq}Q + \overline{U}c_{ue}\overline{E}) + \bar{T}(Qc_{ql}L + \overline{U}c_{ud}\overline{D}) \end{aligned} \tag{6}$$

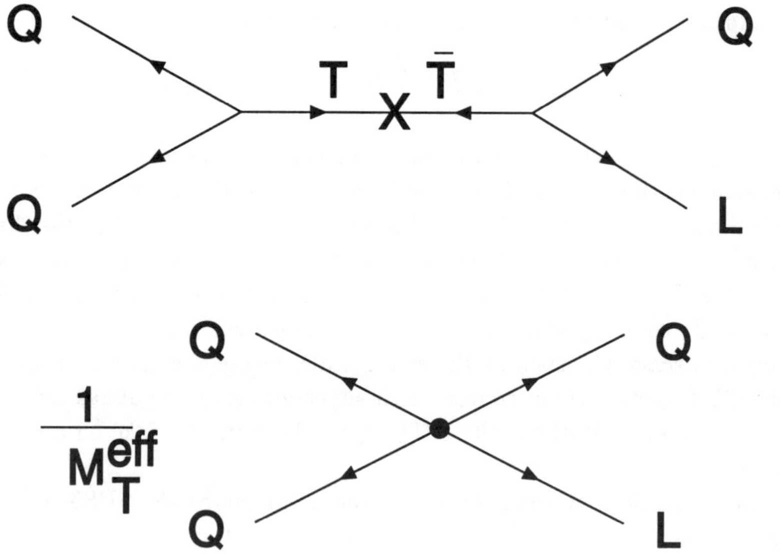

Figure 1. Color triplet Higgs exchange diagram giving the dimension 5 superpotential operator for proton and neutron decay.

If the color triplet Higgs fields $\{T,\ \bar{T}\}$ in Eqn. 6 have an effective mass term $M_T^{eff}\ \bar{T}\ T$ we obtain the dimension 5 operators

$$(1/M_T^{eff})\ \left[\ Q\frac{1}{2}c_{qq}Q\ Qc_{ql}L + \overline{U}c_{ud}\overline{D}\ \overline{U}c_{ue}\overline{E}\ \right], \tag{7}$$

denoted LLLL and $RRRR$ operators, respectively. Nucleon decay via dimension 5 operators was considered in Refs. [9,10]. Consider the contribution to the process $p \to K^+ + \bar{\nu}_\tau$ in Fig. 2. This graph is due to the RRRR operators and gives the dominant contribution at large $\tan\beta$ and a significant contribution for all values of $\tan\beta$ [11,12,13,14].

The largest uncertainty in the nucleon decay rate is due to the color triplet Higgs mass parameter M_T^{eff}. As M_T^{eff} increases, the nucleon lifetime increases. Thus it is useful to obtain an upper bound on the value of M_T^{eff}. This constraint comes from imposing perturbative gauge coupling unification [11,13,15,16,17,14]. Recall, in order to fit the low energy data a GUT scale threshold correction

$$\epsilon_3 \equiv \frac{(\alpha_3(M_G) - \tilde{\alpha}_G)}{\tilde{\alpha}_G} \sim -3\% \ \text{to} \ -4\% \tag{8}$$

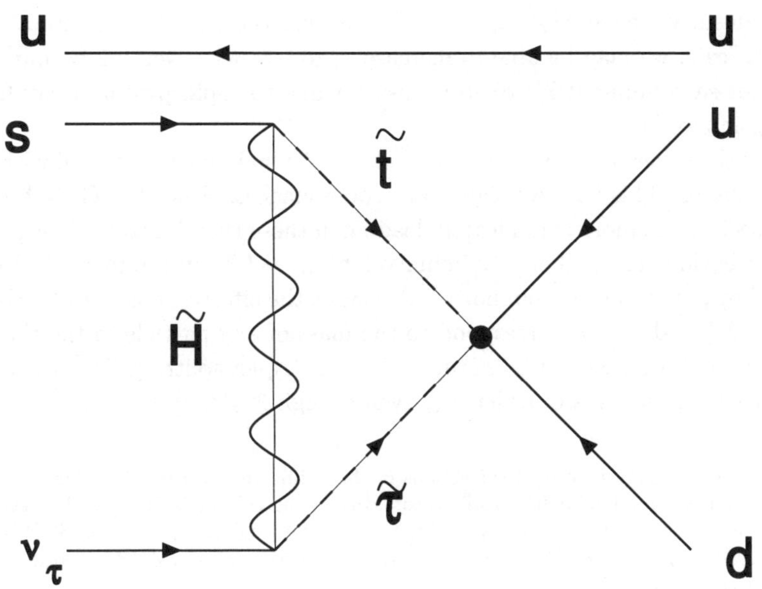

Figure 2. Higgsino and third generation stop - stau loop giving the dominant loop contribution to dimension 5 nucleon decay.

is needed. ϵ_3 is a logarithmic function of particle masses of order M_G, with contributions from the electroweak Higgs and GUT breaking sectors of the theory.

$$\epsilon_3 = \epsilon_3^{\text{Higgs}} + \epsilon_3^{\text{GUT breaking}} + \cdots , \tag{9}$$

$$\epsilon_3^{\text{Higgs}} = \frac{3\alpha_G}{5\pi} \ln\left(\frac{M_T^{eff}}{M_G}\right). \tag{10}$$

In Table 1 we have analyzed three different GUT theories – the minimal SU(5) model, an SU(5) model with natural doublet-triplet splitting [16] and minimal SO(10) [17] (which also has natural doublet-triplet splitting). We have assumed that the low energy data, including weak scale threshold corrections, requires $\epsilon_3 = -4\%$. We have then calculated the contribution to ϵ_3 from the GUT breaking sector of the theory in each case.

In minimal SU(5), the one loop contribution to ϵ_3 from the GUT breaking sector vanishes. Hence, the -4 % must come from the Higgs sector alone, requiring the color triplet Higgs mass $M_T^{eff} \sim 2 \times 10^{14}$ GeV. Note since the Higgs sector is also minimal, with the doublet masses fine-tuned to vanish, we have $M_T^{eff} \equiv M_T$. By varying the SUSY spectrum at the weak scale, we may be able to increase ϵ_3 to -3 % or even -2 %, but this cannot save minimal SU(5) from disaster due to rapid proton decay from Higgsino exchange.

In the other theories, Higgs doublet - triplet splitting is obtained without fine-tuning. This has two significant consequences. First, the GUT breaking sectors are more complicated, leading in these theories to large negative contributions to ϵ_3. The maximum value $|\epsilon_3| \sim 10\%$ in minimal SO(10) is fixed by perturbativity bounds [17]. Secondly, the effective color triplet Higgs mass M_T^{eff} does not correspond to the mass of any particle in the theory. In fact, in both cases with "natural" doublet-triplet splitting [16,17], the color triplet Higgs mass is of order M_G even though $M_T^{eff} \gg M_G$.

Table 1. GUT threshold corrections in three different theories. The first column is the minimal SU(5) SUSY theory, the second column is SU(5) with "natural" Higgs doublet-triplet splitting, and the third column is minimal SO(10) SUSY model. The values for M_T^{eff} are fixed by the value of $\epsilon_3^{\rm Higgs}$ needed to obtain $\epsilon_3 = -4\%$.

	Minimal SU(5)	SU(5) "Natural" D/T	Minimal SO(10)
$\epsilon_3^{\rm GUTbreaking}$	0	- 7.7 %	- 10 %
$\epsilon_3^{\rm Higgs}$	- 4 %	+ 3.7 %	+ 6 %
M_T^{eff} [GeV]	2×10^{14}	3×10^{18}	6×10^{19}

Before discussing the upper bounds on the proton lifetime due to the exchange of color triplet Higgsinos, in generic 4D SUSY GUTs, let us elaborate on the meaning of M_T^{eff}. Consider a simple case with two pairs of SU(5) Higgs multiplets, $\{ \bar{5}_H^i, \ 5_H^i \}$ with $i = 1, 2$. In addition, also assume that only $\{ \bar{5}_H^1, \ 5_H^1 \}$ couples to quarks and leptons. Then M_T^{eff} is defined by the expression

$$\frac{1}{M_T^{eff}} = (M_T^{-1})_{11} \tag{11}$$

where M_T is the color triplet Higgs mass matrix. In the cases with "natural" doublet - triplet splitting, we have

$$M_T = \begin{pmatrix} 0 & M_G \\ M_G & X \end{pmatrix} \tag{12}$$

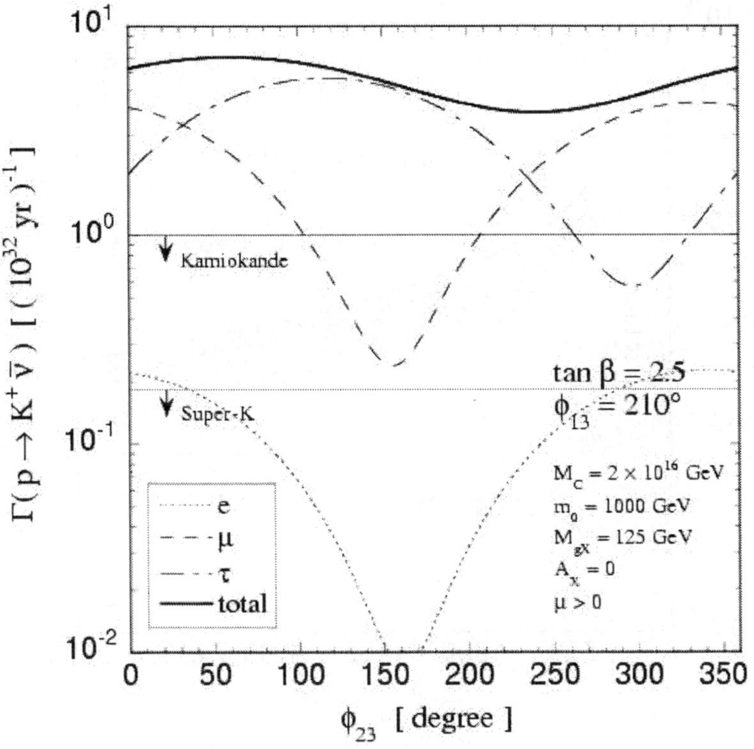

Figure 3. [Fig. 2: Goto and Nihei (1999)] Decay rates $\Gamma(p \to K^+\bar{\nu}_i)$ ($i = e, \mu$ and τ) as functions of the phase ϕ_{23} for $\tan\beta = 2.5$. The soft SUSY breaking parameters are fixed at $m_0 = 1$ TeV, $M_{\tilde{g}} = 125$ GeV and $A_0 = 0$ with $\mu > 0$. M_T and M_Σ are taken as $M_T = M_\Sigma = 2 \times 10^{16}$ GeV. The horizontal lower line corresponds to the Super-Kamiokande limit $\tau(p \to K^+\bar{\nu}) > 5.5 \times 10^{32}$ years, and the horizontal upper line corresponds to the Kamiokande limit $\tau(p \to K^+\bar{\nu}) > 1.0 \times 10^{32}$ years. The new Super-Kamiokande bound is $\tau(p \to K^+\bar{\nu}) > 1.9 \times 10^{33}$ years.

with

$$\frac{1}{M_T^{eff}} \equiv \frac{X}{M_G^2}. \tag{13}$$

Thus for $X \ll M_G$ we have $M_T^{eff} \gg M_G$ and *no* particle has mass greater than M_G [18]. The large Higgs contribution to the GUT threshold correction is in fact due to an extra pair of light Higgs doublets with mass of order X.

Due to the light color triplet Higgsino, it has been shown that minimal SUSY SU(5) is ruled out by the combination of proton decay constrained

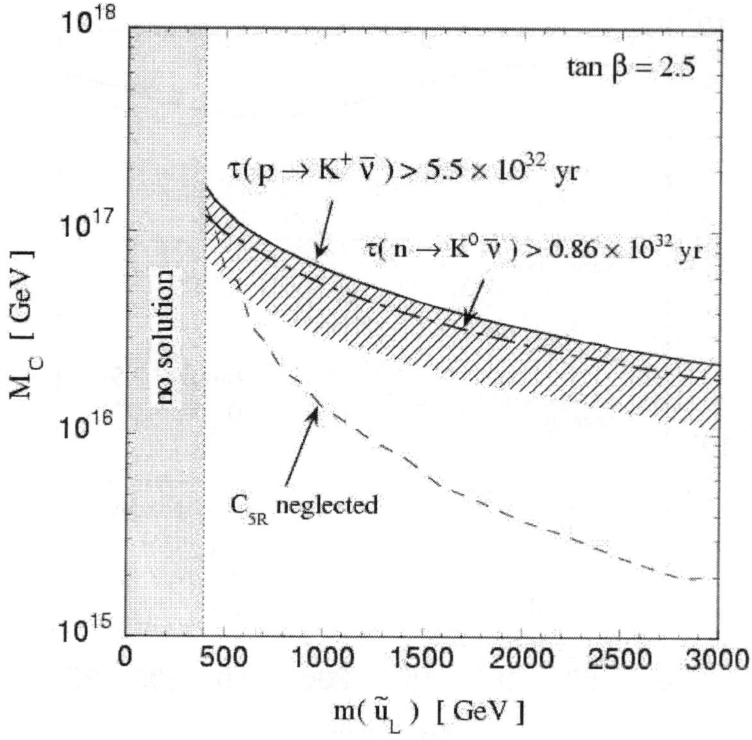

Figure 4. [Fig. 4: Goto and Nihei (1999)] Lower bound on the colored Higgs mass M_T as a function of the left-handed scalar up quark mass $m_{\tilde{u}_L}$. The soft SUSY breaking parameters m_0, $M_{\tilde{g}}$, A_0 are scanned within the range $0 < m_0 < 3$ TeV, $0 < M_{\tilde{g}} < 1$ TeV and $-5 < A_0 < 5$ with $\tan\beta$ fixed at 2.5. Both signs of μ are considered. The whole parameter region of the two phases ϕ_{13} and ϕ_{23} is examined. The solid curve represents the bound derived from the Super-Kamiokande limit $\tau(p \to K^+\bar{\nu}) > 5.5 \times 10^{32}$ years and the dashed curve represents the corresponding result without the RRRR effect. The left-hand side of the vertical dotted line is excluded by other experimental constraints. The dash-dotted curve represents the bound derived from the Kamiokande limit on the neutron partial lifetime $\tau(n \to K^0\bar{\nu}) > 0.86 \times 10^{32}$ years.

by gauge coupling unification [13,14] !! In Figs. 3 and 4 we reprint the figures from the paper by Goto and Nihei [13]. In Fig. 3 the decay rate for $p \to K^+\bar{\nu}_i$ for any one of the three neutrinos ($i = e, \mu$ and τ) is plotted for fixed soft SUSY breaking parameters as a function of the relative phase ϕ_{23} between two LLLL contributions to the decay amplitude. The phase ϕ_{13} is the relative phase between one of the LLLL contributions and the

ϕ_{13} is the relative phase between one of the LLLL contributions and the RRRR contribution. The latter contributes predominantly to the $\bar{\nu}_\tau$ final state, since it is proportional to the up quark and charged lepton Yukawa couplings. As noted in Ref. [13], the partial cancellation between LLLL contributions to the decay rate is completely filled by the RRRR contribution. It is this result which provides the stringent limit on minimal SUSY SU(5). As one sees from Fig. 3, for the color triplet Higgs mass $M_T = 2\times10^{16}$ GeV ($\equiv M_C$ in the notation of Ref. [13]), the universal scalar mass $m_0 = 1$ TeV and $\tan\beta = 2.5$, there is no value of the phase ϕ_{23} which is consistent with Super-Kamiokande bounds. Note, the proton decay rate scales as $\tan\beta^2$; hence the disagreement with data only gets worse as $\tan\beta$ increases. In Fig. 4 the contour of constant proton lifetime is plotted in the $M_T(\equiv M_C)$ – $m(\tilde{u}_L)$ plane, where $m(\tilde{u}_L)$ is the mass of the left-handed up squark for $\tan\beta = 2.5$. Again, there is no value of $m(\tilde{u}_L) < 3$ TeV for which the color triplet Higgs mass is consistent with gauge coupling unification. In Ref. [13] the up squark mass was increased by increasing m_0. Hence all squarks and slepton masses increased.

One may ask whether one can suppress the proton decay rate by increasing the mass of the squarks and sleptons of the first and second generation, while keeping the third generation squarks and sleptons light (in order to preserve "naturalness"). This is the question addressed in Ref. [14]. They took the first and second generation scalar masses of order 10 TeV, with the third generation scalar masses less than 1 TeV. They showed that since the RRRR contribution does not decouple in this limit, and moreover since any possible cancellation between the LLLL and RRRR diagrams vanishes in this limit, one finds that minimal SUSY SU(5) cannot be saved by decoupling the first two generations of squarks and sleptons.

Minimal SUSY SU(5) is excluded by the concordance of experimental bounds on proton decay and gauge coupling unification. The *bottom line* we find for dimension 6 operators [11,14]

$$\tau(p \to \pi^0 + e^+) \approx 5\times10^{36} \left(\frac{M_X}{3\times10^{16}\text{ GeV}}\right)^4 \left(\frac{0.015\text{ GeV}^3}{\beta_{lattice}}\right)^2 \text{ years.} \quad (14)$$

On the other hand for dimension 5 operators in realistic SUSY GUTs we obtain rough upper bounds on the proton lifetime coming from gauge coupling unification and perturbativity [15,16,17]

$$\tau(p \to K^+ + \bar{\nu}) < \left(\frac{1}{3} - 3\right) \times 10^{34} \left(\frac{0.015\text{ GeV}^3}{\beta_{lattice}}\right)^2 \text{ years.} \quad (15)$$

These results should be compared to the latest Super-Kamiokande data [19].

mode	τ/B limit [10^{33} yrs]
$p \to \pi^0 + e^+$	5.7 @ 90%C.L.
$p \to K^+ + \bar{\nu}$	1.9 @ 90%C.L.

2. SUSY GUTs in more than four dimensions

We should mention that there has been a recent flurry of activity on SUSY GUTs in extra dimensions beginning with the work of Kawamura [20]. However the study of extra dimensions on orbifolds goes back to the original work of Dixon et al. [21] in string theory. Although this interesting topic would require another review, let me just mention some pertinent features here. In these scenarios, grand unification is only a symmetry in extra dimensions which are then compactified at scales of order M_G. The effective four dimensional theory, obtained by orbifolding the extra dimensions, has only the standard model gauge symmetry or at most a Pati-Salam symmetry which is then broken by the standard Higgs mechanism. In these theories, it is possible to completely eliminate the contribution of dimension 5 operators to nucleon decay. This may be a consequence of global symmetries as shown in Ref. [22] or a continuous $U(1)_R$ symmetry (with R parity a discrete subgroup) [23]. [Note, it is also possible to eliminate the contribution of dimension 5 operators in 4 dimensional theories with extra symmetries [24], but these 4 dimensional theories are quite convoluted. Thus it is difficult to imagine that nature takes this route. On the other hand, in one or more small extra dimensions the elimination of dimension 5 operators is very natural.] Thus at first glance, nucleon decay in these theories may be extremely difficult to see. However this is not necessarily the case. Once again we must consider the consequences of grand unification in extra dimensions and gauge coupling unification.

Extra dimensional theories are non-renormalizable and therefore require an explicit cutoff scale M_*, assumed to be larger than the compactification scale M_c. The Kałuza-Klein excitations above M_c contribute to threshold corrections to gauge coupling unification evaluated at the compactification scale. The one loop renormalization of gauge couplings is given by

$$\frac{2\pi}{\alpha_i(\mu)} = \frac{2\pi}{\alpha(M_*)} + b_i \log\left(\frac{M_c}{\mu}\right) + \Delta_i \tag{16}$$

where Δ_i are the threshold corrections due to all the KK modes from M_c to M_* and can be expressed as $\Delta_i = b_i^{eff} \log(\frac{M_*}{M_c})$ [25,26,27,28,23]. In a 5D SO(10) model, broken to Pati-Salam by orbifolding and to the MSSM via

Higgs VEVs on the brane, it was shown [29] that the KK threshold corrections take a particularly simple form

$$\frac{2\pi}{\alpha_i(\mu)} = \frac{2\pi}{\alpha(M_*)} + b_i^{MSSM} \log\left(\frac{\hat{M}_c}{\mu}\right) + \hat{\Delta}_i \tag{17}$$

$$= \frac{2\pi}{\alpha(M_*)} + b_i^{MSSM} \log\left(\frac{\hat{M}_c}{\mu}\right) + \frac{2}{3} b_i^{SM}(V) \log\left(\frac{M_*}{\hat{M}_c}\right). \tag{18}$$

with

$$\hat{\Delta}_{\text{gauge}} = \frac{2}{3} b_i^{SM}(V) \log\left(\frac{M_*}{\hat{M}_c}\right) \tag{19}$$

$$\hat{\Delta}_{Higgs} = 0, \tag{20}$$

and $\hat{M}_c \equiv (\pi R/2)^{-1}$. Here b^{MSSM} includes the gauge sector d the Higgs sector together and $b^{SM}(V)$ includes the gauge sector only. The running equation is very simple and permits us to directly compare with well known 4D SUSY GUTs. In the minimal 4D SU(5) SUSY model, the running equation is given by

$$\frac{2\pi}{\alpha_i(\mu)} = \frac{2\pi}{\alpha(M_{GUT})} + b_i^{MSSM} \log\left(\frac{M_T}{\mu}\right) + b_i^{SM}(V) \log\left(\frac{M_{GUT}}{M_T}\right) \tag{21}$$

where M_T is the color triplet Higgs mass. As discussed earlier, we can achieve unification by adjusting the color triplet Higgs mass $M_T = 2 \times 10^{14}$ GeV (see Table 1).

Comparing Eqn. (18) and (21), we observe that if

$$\hat{M}_c = M_T, \quad \frac{M_*}{\hat{M}_c} = \left(\frac{M_{GUT}}{M_T}\right)^{\frac{3}{2}},$$

the same unification is achieved here. Therefore we get $\hat{M}_c = 2 \times 10^{14}$ GeV and $M_* = 3.7 \times 10^{17}$ GeV by using $M_{GUT} = 3 \times 10^{16}$ GeV. A few remarks are in order. There is no problem with proton decay due to dimension 5 operators, even though the color triplet Higgs mass is of order 10^{14} GeV, since these operators are excluded by R symmetry. In a 5D model, the 4D GUT scale has no fundamental significance. The couplings unify at the cutoff scale and there is no scale, above which we have a perturbative SO(10) GUT.

In Fig. 5 we show the running in the difference of couplings for two independent cases.

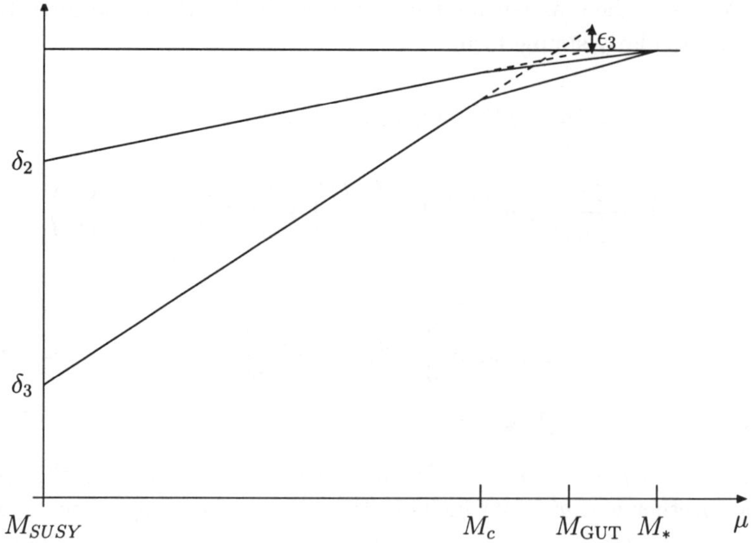

Figure 5. Differential running of $\delta_2 = 2\pi(\frac{1}{\alpha_2} - \frac{1}{\alpha_1})$ and $\delta_3 = 2\pi(\frac{1}{\alpha_3} - \frac{1}{\alpha_1})$.

(1) We show the couplings for four dimensional gauge theories with GUT scale thresholds, in which the GUT scale is defined as the point where α_1 and α_2 meet and ϵ_3 is the relative shift in α_3 due to threshold corrections, and
(2) for a five dimensional SO(10) model where the three couplings meet at the cutoff scale and the threshold corrections due to the KK tower is defined at the compactification scale.

In both cases, the running of the gauge couplings below the compactification scale must be the same. Thus we can use the low energy fits from 4D theories to constrain a 5D theory.

Note, since the the KK modes of the baryon number violating $\{X, \bar{X}\}$ gauge bosons have mass starting at the compactification scale $M_c \approx 10^{14}$ GeV we must worry whether proton decay due to dimension 6 operators is safe. It has been shown [30,23,31] that this depends on where in the extra dimensions the quarks and leptons reside. If they are on symmetric orbifold fixed points, i.e. symmetric under the GUT symmetry, then this leads to the standard dimension 6 proton decay operators. This is ruled out for the first and second families of quarks and leptons. Hence the first two families must be either in the bulk or on broken symmetry fixed points. If they are in the bulk, then the $\{X, \bar{X}\}$ mediated processes take massless modes to KK excitations which is not a problem. Otherwise, if the first two families are on the broken symmetry fixed point, the wave functions for the $\{X, \bar{X}\}$ bosons vanish there. However, certain effective higher dimensional operators on the broken symmetry orbifold fixed points can allow the $\{X, \bar{X}\}$ bosons to couple to the first two families. These operators are allowed by symmetries and they naturally lead to proton lifetimes for $p \to e^+ \pi^0$ of order $10^{34\pm2}$ years. The large uncertainty is due to the order of magnitude uncertainty in the coefficient of these new effective operators.

3. Conclusions

We have considered SUSY GUTs: gauge coupling unification and nucleon decay. We have discussed both four and five dimensional versions. In four dimensions we argue that minimal SU(5) is excluded by the combination of gauge coupling unification and nucleon decay constraints. In five dimensions the dimension 5 operator contribution to nucleon decay is naturally removed. Hence, a minimal SU(5) is not excluded in 5D. There are many topics we did not touch on, in particular fermion masses and mixing angles. Below we summarize some of the virtues and problems for SUSY GUTs in four and five dimensions. In any number of dimensions, SUSY GUTs have the following virtues.

- $M_Z << M_{GUT}$ "Natural"
- Explain charge quantization & family structure
- Gauge coupling unification
- Yukawa coupling unification for the third generation in SO(10) or $b - \tau$ unification in SU(5)
- LSP - dark matter candidate
- Baryogenesis via leptogenesis
- Neutrino mass (See–Saw mechanism) The right-handed neutrino mass, consistent with atmospheric neutrino oscillations, is of order $10^{-1} - 10^{-2}\ M_G$. In 4D, this may be obtained via a See-Saw of the form M_G^2/M_{Pl}. On the other hand, in 5D a right-handed neutrino mass of the same magnitude can be obtained given by the compactification scale M_c. It is however only natural if the right-handed components of quarks and leptons are located in the bulk [32].
- SUSY GUTs plus family symmetry can naturally generate a hierarchy of fermion masses. In 5D there are additional suppression

mechanisms. For example, bulk suppression factors are obtained for bulk modes. On the other hand, orbifold GUT breaking typically destroys Yukawa relations among fermions in the same family, since fermions in the same family come from different GUT multiplets.

We list the major features distinguishing 4 and 5 dimensional SUSY GUTs below.

- Gauge symmetry breaking in 4D requires a complicated Higgs sector, while in 5D GUT breaking is accomplished by orbifold or Wilson line breaking. The latter does not require a complicated Higgs sector. On the other hand it is ad hoc.
- Higgs doublet–triplet splitting in 4D requires a complicated Higgs sector, while in 5D it is accomplished via orbifold/Wilson line breaking.
- Proton decay ($p \to K^+ \bar{\nu}$) due to dimension 5 operators dominates in 4D. In 5D, however, dimension 5 baryon and lepton number violating operators are excluded by R symmetry.

  In 4D proton decay ($p \to e^+ \pi^0$) via dimension 6 operators *is negligible.* On the other hand, in 5D *one is now sensitive to physics at the cutoff and the effects are incalculable (and perhaps even observable ?)*

Acknowledgments

This talk is dedicated to the memory of Freydoon Mansouri.

References

1. S. Dimopoulos, S. Raby and F. Wilczek, Phys. Rev. D **24**, 1681 (1981).
2. S. Dimopoulos and H. Georgi, Nucl. Phys. B **193**, 150 (1981).
3. N. Sakai, Z. Phys. C **11**, 153 (1981).
4. L. E. Ibanez and G. G. Ross, Phys. Lett. B **105**, 439 (1981); M. B. Einhorn and D. R. T. Jones, Nucl. Phys. B **196**, 475 (1982); W. J. Marciano and G. Senjanovic, Phys. Rev. D **25**, 3092 (1982).
5. U. Amaldi, W. de Boer and H. Furstenau, Phys. Lett. B **260**, 447 (1991); J. R. Ellis, S. Kelley and D. V. Nanopoulos, Phys. Lett. B **260**, 131 (1991); P. Langacker and M. x. Luo, Phys. Rev. D **44**, 817 (1991).
6. V. Lucas and S. Raby, Phys. Rev. D **54**, 2261 (1996) [arXiv:hep-ph/9601303].
7. E. Witten, Nucl. Phys. B **188**, 513 (1981).
8. S. Weinberg, Phys. Rev. D **26**, 287 (1982).
9. N. Sakai and T. Yanagida, Nucl. Phys. B **197**, 533 (1982).

10. S. Dimopoulos, S. Raby and F. Wilczek, Phys. Lett. B **112**, 133 (1982); J. R. Ellis, D. V. Nanopoulos and S. Rudaz, Nucl. Phys. B **202**, 43 (1982).
11. V. Lucas and S. Raby, Phys. Rev. D **55**, 6986 (1997) [arXiv:hep-ph/9610293].
12. K. S. Babu and M. J. Strassler, arXiv:hep-ph/9808447.
13. T. Goto and T. Nihei, Phys. Rev. D **59**, 115009 (1999) [arXiv:hep-ph/9808255].
14. H. Murayama and A. Pierce, Phys. Rev. D **65**, 055009 (2002) [arXiv:hep-ph/0108104].
15. K. S. Babu, J. C. Pati and F. Wilczek, Nucl. Phys. B **566**, 33 (2000) [arXiv:hep-ph/9812538].
16. G. Altarelli, F. Feruglio and I. Masina, JHEP **0011**, 040 (2000) [arXiv:hep-ph/0007254].
17. R. Dermisek, A. Mafi and S. Raby, Phys. Rev. D **63**, 035001 (2001) [arXiv:hep-ph/0007213].
18. K. S. Babu and S. M. Barr, Phys. Rev. D **48**, 5354 (1993) [arXiv:hep-ph/9306242].
19. Jung C K, Brookhaven National Lab preprint, UNO 02-BNL (2002).
20. Y. Kawamura, Prog. Theor. Phys. **105**, 999 (2001) [arXiv:hep-ph/0012125]; Y. Kawamura, Prog. Theor. Phys. **105**, 691 (2001) [arXiv:hep-ph/0012352].
21. L. J. Dixon, J. A. Harvey, C. Vafa and E. Witten, Nucl. Phys. B **261**, 678 (1985); L. J. Dixon, J. A. Harvey, C. Vafa and E. Witten, Nucl. Phys. B **274**, 285 (1986); L. E. Ibanez, H. P. Nilles and F. Quevedo, Phys. Lett. B **187**, 25 (1987).
22. E. Witten, arXiv:hep-ph/0201018; M. Dine, Y. Nir and Y. Shadmi, Phys. Rev. D **66**, 115001 (2002) [arXiv:hep-ph/0206268].
23. L. J. Hall and Y. Nomura, Phys. Rev. D **66**, 075004 (2002) [arXiv:hep-ph/0205067].
24. K. S. Babu and S. M. Barr, Phys. Rev. D **65**, 095009 (2002) [arXiv:hep-ph/0201130].
25. L. J. Hall and Y. Nomura, Phys. Rev. D **64**, 055003 (2001) [arXiv:hep-ph/0103125]; L. J. Hall and Y. Nomura, Phys. Rev. D **65**, 125012 (2002) [arXiv:hep-ph/0111068].
26. Y. Nomura, D. R. Smith and N. Weiner, Nucl. Phys. B **613**, 147 (2001) [arXiv:hep-ph/0104041].
27. R. Contino, L. Pilo, R. Rattazzi and E. Trincherini, Nucl. Phys. B **622**, 227 (2002) [arXiv:hep-ph/0108102].
28. Y. Nomura, Phys. Rev. D **65**, 085036 (2002) [arXiv:hep-ph/0108170].
29. H. D. Kim and S. Raby, JHEP **0301**, 056 (2003) [arXiv:hep-ph/0212348].
30. G. Altarelli and F. Feruglio, Phys. Lett. B **511**, 257 (2001) [arXiv:hep-ph/0102301].
31. A. Hebecker and J. March-Russell, Phys. Lett. B **539**, 119 (2002) [arXiv:hep-ph/0204037].
32. H. D. Kim and S. Raby, JHEP **0307**, 014 (2003) [arXiv:hep-ph/0304104].

NORMAL ORDER: COMBINATORIAL GRAPHS*

ALLAN I. SOLOMON[†§], PAWEL BLASIAK[†‡], ANDRZEJ HORZELA[‡], KAROL A. PENSON[†], and GERARD DUCHAMP [§§]

[†] *Université Pierre et Marie Curie*
Laboratoire de Physique Théorique des Liquides, CNRS UMR 7600
Tour 16, 5ième étage, 4, place Jussieu, F 75252 Paris, Cedex 05, France
e-mail: penson@lptl.jussieu.fr

[§] *The Open University*
Physics and Astronomy Department
Milton Keynes MK7 6AA
e-mail: a.i.solomon@open.ac.uk

[‡] *H. Niewodniczański Institute of Nuclear Physics, Polish Academy of Sciences*
Department of Theoretical Physics
ul. Radzikowskiego 152, PL 31-342 Kraków, Poland
e-mail: pawel.blasiak@ifj.edu.pl, andrzej.horzela@ifj.edu.pl

[§§] *LIFAR, Université de Rouen*
76821 Mont-Saint Aignan Cedex, France
e-mail: gduchamp2@free.fr

In this note we consider the normal ordering problem for a string of creation and annihilation operators. In the general case, upon which we touch briefly, this problem leads to combinatorial numbers, the so-called *Rook numbers.* Since we assume that the two species, bosons and fermions, commute, we subsequently restrict ourselves to consideration of a single species, single-mode boson monomials. This problem leads to elegant generalisations of well-known combinatorial numbers, specifically Bell and Stirling numbers. We explicitly give the generating functions for some classes of these numbers. In this note we concentrate on the combinatorial graph approach, showing how some important classical results of graph theory lead to transparent representations of the combinatorial numbers associated with the boson normal ordering problem.

*Talks by K.A.P. and A.I.S.

1. Normal Ordering

In this note we give a brief review of some combinatorial graphs associated with the normal ordering of creation and annihilation operators.

The process of normally ordering a string of creation and annihilation operators simply means reordering the elements of the string so that all the annihilation operators appear on the right, taking into account the commutation (or anti-commutation) relations. The value of such a procedure is that, for example, the expectation value in a coherent state of such a normally ordered string may be immediately seen. Performing this operation on strings of bosons or fermions leads to classical combinatorial numbers and, in the analogous *quon* case, to q-variants of these numbers. In a "supersymmetric" context one may well have to consider a general string of bosons and fermions; but as the bosons and fermions are assumed to commute, the ordering reduces trivially to ordering the two species separately. We thus assume the basic commutation relations

- Bosons $[a_i, {a_j}^\dagger] = \delta_{ij}$
- Fermions $\{f_i, {f_j}^\dagger\} = \delta_{ij}$
- Mixed $[a_i, {f_j}^\dagger] = 0$

We note that the boson ordering problem is equivalent to ordering a string of operators $a \equiv d/dx$ and $a^\dagger \equiv x$.

For both bosons and fermions, the ordering process of a general string results in combinatorial numbers called *Rook numbers.* We shall merely give a brief account of this approach here; further details, especially with respect to the fermion case, may be found in Ref. [1]. In this note we *illustrate* rather than *prove* the methods described.

2. Rook Numbers and Generalized Stirling Numbers

The normal form of a bosonic string or "word", $w = w(a, a^\dagger)$, satisfies $\mathcal{N}(w) = w$. Normal forms of boson strings are connected to the so-called *Rook numbers* [2], [3] in the following way [4]. Draw a North-East ($\nearrow$) line over each creation operator $a^\dagger$ and a South-East ($\searrow$) line over each annihilation operator a so as to produce a continuous line. This gives the diagram below.

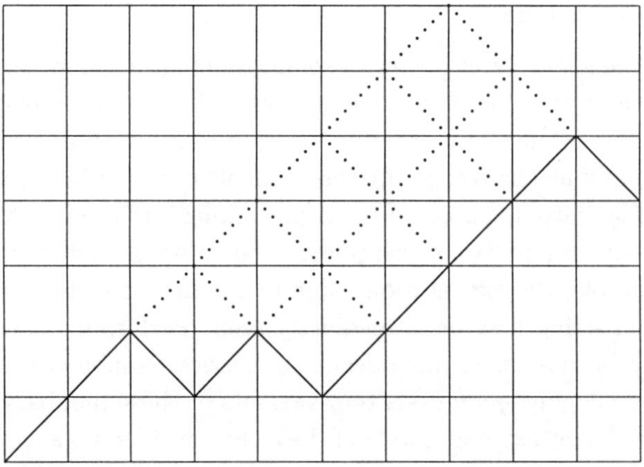

The picture is completed by the dotted lines as shown. The result of this process is a *Ferrers diagram* or *board* [3] representing, in this case, the partition $\{5, 4\}$ of 9 represented by the following diagram.

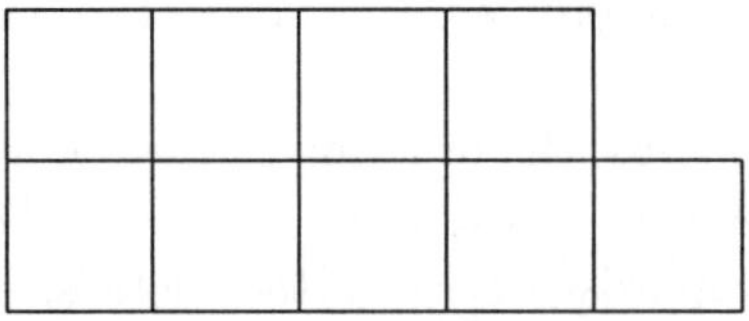

The k-th *rook number* $r_k(B)$ of a Ferrers board B is the number of ways of placing k non-capturing rooks on the board. For our example of the partition $\{5, 4\}$ one has

k	0	1	2	$k > 2$
$r_k(B)$	1	9	16	0

and thus the normal form of $w = a^\dagger a^\dagger a a^\dagger a a^\dagger a^\dagger a^\dagger a^\dagger a$ is given by

$$\mathcal{N}(w) = a^\dagger a^\dagger a^\dagger a^\dagger a^\dagger a^\dagger a^\dagger aaa + 9\, a^\dagger a^\dagger a^\dagger a^\dagger a^\dagger a^\dagger aa + 16\, a^\dagger a^\dagger a^\dagger a^\dagger a^\dagger a. \quad (1)$$

For $w = aa^\dagger a^2(a^\dagger)^2$, the board represents the partition $[3, 2, 2]$ and

$$\begin{aligned}\mathcal{N}(aa^\dagger a^2(a^\dagger)^2) &= a^\dagger a^\dagger a^\dagger aaa + 7\ a^\dagger a^\dagger aa + 10\ a^\dagger a + 2 \\ &= r_0(B)(a^\dagger)^3 a^3 + r_1(B)(a^\dagger)^2 a^2 + r_2(B)(a^\dagger)a + r_3(B). \quad (2)\end{aligned}$$

In fact Eqs.(1) and (2) are illustrations of the general formula expressing the normal form $\mathcal{N}(w)$ with the help of the rook numbers as

$$\mathcal{N}(w) = \sum_{k=0}^{\infty} r_k(B) : w^{(k)} : \quad (3)$$

In Eq.(3) $: w^{(k)} :$ means that in the word $w = w(a, a^\dagger)$ we cross out k a's and k $a^\dagger$'s and then order normally the result *without* taking into account the commutation relations. In fact for finite word w the sum in Eq.(3) has a finite number of non-vanishing terms. Note that coefficients of the normal monomials are the rook numbers of the board; it is possible to give a simple algorithm which computes these numbers.

Similar observations apply to the normal ordering of *quons* a_q (q-bosons) satisfying

$$[a_q, a_q^\dagger]_q \equiv a_q a_q^\dagger - q a_q^\dagger a_q = 1,$$

the q-Weyl algebra[a].

If we restrict ourselves to simple recurring strings in one boson mode, which we shall henceforth do, we see that classical combinatorial numbers appear naturally [5,6].

The normal ordering problem for canonical bosons $[a, a^\dagger] = 1$ is related to certain combinatorial numbers $S(n, k)$ called *Stirling numbers of the second kind* through [7]

$$(a^\dagger a)^n = \sum_{k=1}^{n} S(n, k)(a^\dagger)^k a^k, \quad (4)$$

with corresponding numbers $B(n) = \sum_{k=1}^{n} S(n, k)$ called *Bell numbers.* In fact, for physicists, these equations may be taken as the *definitions* of the Stirling and Bell numbers. For the quons a_q (q-bosons) mentioned above, a natural q-generalisation [8], [9] of these numbers is

$$(a_q^\dagger a_q)^n = \sum_{k=1}^{n} S_q(n, k)(a_q^\dagger)^k a_q^k. \quad (5)$$

[a] An interesting, if as yet experimental, observation is that the resulting polynomials in q obtained by setting $a_q = 1 = a_q^\dagger$ are *unimodal.*

In the *canonical* boson case, for integers $n, r, s > 0$ we define generalized Stirling numbers of the second kind $S_{r,s}(n,k)$ through ($r \geq s$):

$$[(a^\dagger)^r a^s]^n = (a^\dagger)^{n(r-s)} \sum_{k=s}^{ns} S_{r,s}(n,k)(a^\dagger)^k a^k, \tag{6}$$

as well as generalized Bell numbers $B_{r,s}(n)$

$$B_{r,s}(n) = \sum_{k=s}^{ns} S_{r,s}(n,k). \tag{7}$$

For both $S_{r,s}(n,k)$ and $B_{r,s}(n)$ exact and explicit formulas have been found [5,6]. We refer the interested reader to these sources for further information on those extensions. However, in this note we shall mainly deal with the classical Bell and Stirling numbers, corresponding to $B_{1,1}(n)$ and $S_{1,1}(n)$ in our notation, and the extension to $B_{2,1}(n)$ and $S_{2,1}(n)$.

The conventional and picturesque description of the classical Bell and Stirling Numbers of the second kind is in terms of the distribution of differently coloured balls among identical containers or, equivalently, the number of partitions of an n-element set [10,11]. The relation of this classical definition to the normal order expansion of $(a^\dagger a)^n$ is via the contractions induced by the application of Wick's Theorem. In general, $S(n,m)$ gives the coefficient of the term $:(a^\dagger a)^m :\equiv (a^\dagger)^m a^m$ in the normal ordering expansion, directly from the (physicist's) definition of $S(n,m)$.

Taking the concrete example $(a^\dagger a)^3$,

- $S(3,1) = 1$ is the coefficient of the term $: a^\dagger a :\equiv a^\dagger a$
- $S(3,2) = 3$ is the coefficient of the term $: (a^\dagger a)^2 :\equiv (a^\dagger)^2 a^2$
- $S(3,3) = 1$ is the coefficient of the term $: (a^\dagger a)^3 :\equiv (a^\dagger)^3 a^3$

In the next section we relate these combinatorial numbers to certain graphs. We shall elaborate on this graphical approach elsewhere [12].

3. Generating Functions and Graphs

In general, for combinatorial numbers $g(n)$ we may define an *exponential generating function* $G(x)$ through [13]

$$G(x) = \sum_{n=0}^{\infty} g(n) \frac{x^n}{n!}. \tag{8}$$

For the Bell numbers, this generating function takes the particularly nice form [11]

$$G(x) = \sum_{n=0}^{\infty} B(n)\frac{x^n}{n!} = \exp(\exp(x) - 1). \tag{9}$$

We shall derive this exponential generating function by the use of a simple graph theory technique. Some initial terms of the sequence $\{B(n)\}$ are $\{1, 2, 5, 15, 52, 203, 877, \ldots\}$.

Another convenient way of representing combinatorial numbers is by means of *graphs*. To illustrate this, we now consider a graphical method for describing the combinatorial numbers $B(n)$ associated with the normal order expansion of $(a^\dagger a)^n$. For concreteness, take the cases $n = 1, 2, 3$ (Figure 1). In this diagram the filled dots on the left represent the differently coloured balls, while the empty dots represent the identical containers. This pictorial approach gives the values for $S(n, m)$. By convention, $S(n, 0) = \delta_{n,0}$.

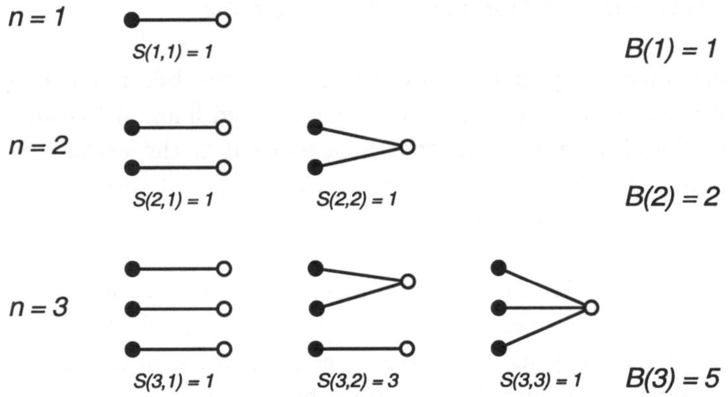

Figure 1. Graphs for $B(n)$, $n = 1, 2, 3$.

One reason the graphical representation is useful is that there exist some rather powerful results which apply to graphs and their associated generating functions. We list and apply two such results, labelled A and B, in what follows.

A. Connected graph theorem

This states that if $C(x) = \sum_{n=1}^{\infty} c(n)x^n/n!$ is the exponential generating function of *labelled connected* graphs, *viz.* $c(n)$ counts the number of connected graphs of order n, then

$$A(x) = \exp(C(x)) \tag{10}$$

is the exponential generating function for *all* graphs.

We may apply this very simply to the case of the $B(n)$ graphs in Figure 1. For each order n, the *connected* graphs clearly consist of a single graph. Therefore for each n we have $c(n) = 1$; whence, $C(x) = exp(x) - 1$. It follows that the generating function for *all* the graphs $A(x)$ is given by

$$A(x) = \exp(\exp(x) - 1) \tag{11}$$

which is the generating function for the Bell numbers.

Such graphs may be generalised to give graphical representations for the extensions $B_{r,s}(n)$ [12]. We illustrate this by using the following powerful result on certain classes of graphs:

B. Generating function for a class of graphs

We generalize the graphical representation for the Bell numbers given in Figure 1. As before, we shall be counting labelled lines. A line starts from a black dot, the *origin*, and ends at a white dot, the *vertex*. What we refer to as *origin* and *vertex* is, of course, arbitrary. At this point there are no other rules, although we are at liberty to impose further restrictions; a black dot may be the origin of 1,2,3,... lines, and a white dot the vertex for 1,2,3,... lines. We may further associate *strengths* V_s with each vertex receiving s lines, and multipliers L_m with a black dot which is the origin of m lines. Again $\{V_s\}$ and $\{L_m\}$ play symmetric roles; in this note we shall only consider cases where the L_m are either 0 or 1.

There is an exponential generating function $G(x, V, L)$ which counts the number $g(n)$ of graphs with n lines arising from the above rules [14,15]:

$$\begin{aligned} G(x, V, L) &= \exp\left(\sum_{m=1}^{\infty} L_m \frac{x^m}{m!} \frac{d^m}{dy^m}\right) \exp\left(\sum_{s=1}^{\infty} V_s \frac{y^s}{s!}\right)\Bigg|_{y=0} \\ &\equiv \sum_{n=0}^{\infty} g(n) \frac{x^n}{n!} \end{aligned} \tag{12}$$

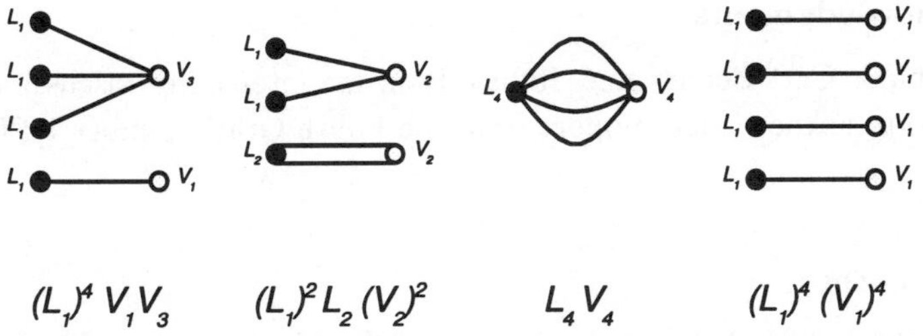

Figure 2. Some examples of 4-line graphs.

Example 1: The exponential generating function corresponding to Figure 1 is obtained by putting $L_m = 0$ for $m \neq 1$ and $V_s = 1$ for all s. This immediately gives the exponential generating function for the Bell numbers $B_{1,1}(n) \equiv B(n)$ as $\exp(\exp(x - 1))$, a result we have already obtained through use of the Connected Graph Theorem above.
Example 2: The diagrams of Figure 1, that is corresponding to $L_m = 0$ for $m \neq 1$, are in a sense generic for the Wick contractions occurring in expressions such as $((a^\dagger)^r a)^n$ [12]. However, the coefficients V_s depend on the exponent r. We obtain the exponential generating function for $B_{2,1}(n)$, corresponding to the normal ordering of $((a^\dagger)^2 a)^n$, by putting $V_s = s!$. This immediately leads to the formal expression

$$
\begin{aligned}
G(x) &= \exp\left(x\frac{d}{dy}\right)\exp\left(\sum_{s=1}^{\infty} y^s\right)\bigg|_{y=0} \\
&\equiv \exp\left(\frac{x}{1-x}\right)
\end{aligned}
\tag{13}
$$

which is the exponential generating function for $B_{2,1}(n)$ [5,6].

Many other applications and extensions of the ideas sketched in this note will be found in [12].

Acknowledgments

We thank Carl Bender and Itzhak Bars for interesting discussions. PB wishes to acknowledge support from the Polish Grant number 1 P03B 051 26.

References

1. Navon, A.M. 1973 Combinatorics and Fermion Algebra, *Nuovo Cimento* **16B**, 324
2. Bryant, V. 1993 *Aspects of combinatorics*, Cambridge Univ. Press
3. Stanley, R. 1999 *Enumerative Combinatorics, Vol 1*, Cambridge Univ. Press
4. van Leeuwen, M. 2003 personal communication.
5. Blasiak, P., Penson, K.A. and Solomon, A.I. 2003 The general boson normal ordering problem, *Phys. Lett. A* **309** 198.
6. Blasiak, P., Penson, K.A. and Solomon, A.I. 2003 The boson normal ordering problem and generalized Bell numbers, *Ann. Comb.* **7** 127.
7. Katriel, J. 1974 Combinatorial aspects of boson algebra, *Lett. Nuovo Cimento* **10** 565.
8. Katriel, J. 2000 Bell numbers and coherent states, *Phys. Lett. A.* **273** 159.
9. Schork, M. 2003 On the combinatorics of normal ordering of bosonic operators and deformations of it, *J. Phys. A: Math. Gen.* **36** 4651
10. Sloane, N.J.A. 2004 On-Line Encyclopedia of Integer Sequences, available online at http://www.research.att.com/~njas/sequences/
11. Comtet, L. 1974 *Advanced Combinatorics*, Reidel, Dordrecht.
12. Blasiak, P., Duchamp, G., Horzela, A., Penson, K.A. and Solomon, A.I.: Model combinatorial field theories, to be published.
13. Wilf, H.S. 1994 *Generatingfunctionology*, Academic Press, New York.
14. Vasiliev, N.A. 1998 *Functional Methods in Quantum Field Theory and Statistical Physics*, Gordon and Breach Publishers, Amsterdam.
15. Bender, C.M, Brody, D.C. and Meister, B.K. 1999 Quantum field theory of partitions, *J.Math. Phys.* **40** 3239; Bender, C.M., Brody, D.C., and Meister B.K. 2000 Combinatorics and field theory, *Twistor Newsletter* **45** 36.

CONSERVATION LAWS IN FLUID DYNAMICS AND SELF-DUALITY OF RANKINE-HUGONIOT SHOCK CONDITIONS

V. V. SREEDHAR

Department of Physics, Indian Institute of Technology, Kanpur 208016, INDIA

It has recently been shown that the maximal invariance group of fluid dynamics is larger than the Galilei group and was determined to be the semidirect product of $SL(2,R)$ and the nine-parameter static Galilei group. In this paper we construct the associated conservation laws and use them to construct a self-dual set of Rankine-Hugoniot jump conditions that describe shocks.

1. Introduction

It has recently been observed that the density profiles of a supernova explosion and an inertial confinement plasma implosion[1,2,3] are strikingly similar. An empirical basis for this intriguing duality between explosion and implosion was given by Drury and Mendonça [4] who pointed out that Euler's equations of fluid dynamics, which describe both the systems, are form–invariant under a set of nonlinear coordinate transformations *viz.* $\vec{x} \to \vec{x}/t, \quad t \to -1/t$. These non-linear transformations suggest that the maximal kinematical invariance group $\mathcal{G}$ of fluid dynamics is larger than the standard Galilei group. It is now known that this larger group is a twelve-parameter semidirect product, $\mathcal{G} = SL(2,R) \wedge G$ [5], where G is the nine-parameter connected static Galilei group:

$$\vec{x} \to \mathbf{R}\vec{x} + \vec{v}t + \vec{a}, \qquad t \to t \tag{1}$$

and $SL(2,R)$ is the group consisting of the transformations:

$$t \to \frac{\alpha t + \beta}{\gamma t + \delta}, \qquad \vec{x} \to \frac{\vec{x}}{\gamma t + \delta} \qquad \text{with} \qquad \alpha\delta - \beta\gamma = 1\,. \tag{2}$$

Physically, the three-parameter $SL(2,R)$ group consists of time translations, scale transformations, and a one-parameter set of time-dependent scale transformations called expansions. The transformations proposed by Drury and Mendonça are a special case of the $SL(2,R)$ transformations with $(\alpha,\ \beta,\ \gamma,\ \delta) = (0,\ -1,\ 1,\ 0)$. The minus sign in the time transformation maps an explosion to an implosion and the inversion allows large time scales to be mapped to small time scales and *vice versa.* So it is the $SL(2,R)$ part of $\mathcal{G}$ which is of interest for understanding the explosion–implosion map.

It should be pointed out that the naive expectation of using time-reversal invariance, to explain the similarity between explosion and implosion, is untenable here, since the length and time scales involved in the two systems are drastically different. Invoking scaling arguments, in addition to time-reversal invariance, is not of much help; since although suitable scalings leave the equations of fluid mechanics and the Reynolds number invariant, they have the property of reversing the direction of time's arrow and thereby violate the second law of thermodynamics. As a consequence, such transformations, when applied to a shock wave, violate entropy conditions that define the physicality of the shock. As is well-known, however, both the supernova explosion and the plasma implosion are driven by the formation and propagation of a shock wave. It is therefore important to examine whether the $SL(2,R)$ group which seems to explain the observed duality upholds the physicality of a shock wave. With this in view, we extend the study of [5], in which the explanation of explosion-implosion duality based on the symmetry group $\mathcal{G}$ was restricted to smooth, subsonic flows in ideal polytropic fluids, to examine shock waves. It will be shown that shock wave solutions are consistent with the symmetries of the maximal kinematical invariance group $\mathcal{G}$ in the following precise sense.

A shock in a fluid is described mathematically by the well-known Rankine-Hugoniot jump conditions [6]. So the natural question to ask is: What happens to these conditions under the action of the $SL(2,R)$ group? As a first step towards answering this question, we construct the Noether charges corresponding to the $SL(2,R)$ symmetry group and use the attendant conservation laws to establish a dual set of Rankine-Hugoniot jump conditions. The dual set is identically satisfied if the standard Rankine-Hugoniot conditions are satisfied and *vice versa.* Although seemingly redundant because of this reason, the dual set holds independently; following, as it does, from the symmetries of the fluid equations. Thus to each physical system governed by the fluid dynamics equations, we can associate two independent sets of conditions which describe the same shock. In this article we will present the new conservation laws that arise because of the $SL(2,R)$ group which will in turn enable us to write down the dual Rankine-Hugoniot conditions. The details will be published elsewhere.

2. Symmetries of Fluid Dynamics

In this section we briefly recapitulate the results of Ref. [5]. The general fluid equations in n-dimensional space are[7]

$$D\rho = -\rho\vec{\nabla}\cdot\vec{u} \tag{3}$$

$$\rho D\vec{u} = -\vec{\nabla}p + \vec{V} \tag{4}$$

$$D\varepsilon = -(\varepsilon + p)\vec{\nabla}\cdot\vec{u} \tag{5}$$

where

$$D = \frac{\partial}{\partial t} + \vec{u}\cdot\vec{\nabla}$$

and

$$V_i = \nabla_j\left(\eta(\nabla_j u_i + \nabla_i u_j - \frac{2}{n}\delta_{ij}\nabla_k u_k)\right) + \nabla_i(\zeta\nabla_k u_k)$$

In the above equations $\rho, \vec{u}, p, \varepsilon$ stand for the density, velocity, pressure and energy density of the fluid respectively and η, ζ are the bulk and shear viscosity fields.

The above differential equations are usually augmented by an algebraic condition called the polytropic equation of state which relates the pressure to the energy density as

$$p = (\gamma_0 - 1)\varepsilon \tag{6}$$

where γ_0 is a constant called the polytropic exponent. As shown in [5], the maximal invariance group of the above set of equations is $\mathcal{G}$ provided the polytropic exponent takes the standard value for an ideal fluid *viz.* $\gamma_0 = 1 + \frac{2}{n}$. For this value, the fluid equations are invariant under the following transformations[5]:

Connected, static Galilei transformations

Let g denote a general element of this sub-group then

$$g: \quad t' = t\,, \qquad \vec{x}' = \mathbf{R}\vec{x} + \vec{v}t + \vec{a} \tag{7}$$

with $\mathbf{R}$ an orthogonal matrix. Under the action of g the fields ρ and $\vec{u}$ transform as

$$\rho' = \rho \quad \text{and} \qquad \vec{u}' = \vec{u} + \vec{v} \tag{8}$$

SL(2,R) transformations

Let σ denote a general element of the $SL(2,R)$ part of the group $\mathcal{G}$ then

$$\sigma: \quad t' = \frac{\alpha t + \beta}{\gamma t + \delta}, \quad \vec{x}' = \frac{\vec{x}}{\gamma t + \delta} \quad \text{where} \quad \alpha\delta - \beta\gamma = 1 \tag{9}$$

Under the action of σ, the fields transform as

$$\rho' = (\gamma t + \delta)^n \rho \quad \text{and} \quad \vec{u}' = (\gamma t + \delta)\vec{u} - \gamma(\vec{x} + \vec{a}) + \delta\vec{v} \tag{10}$$

For both g and σ, the transformations of ε and p can be worked out once the transformation properties of ρ are known since

$$\varepsilon = \rho^{\gamma_0}, \qquad p = (\gamma_0 - 1)\varepsilon, \tag{11}$$

while the viscosity fields transform like tensors *i.e.*

$$\eta' = \Big(\frac{1}{\gamma t + \delta}\Big)^n \eta \quad \text{and} \quad \zeta' = \Big(\frac{1}{\gamma t + \delta}\Big)^n \zeta \tag{12}$$

The above results were derived in [5] by requiring the invariance of the Action for the simple case of an inviscid and isentropic fluid. The symmetry of the equations followed by subsequently relaxing the simplifications to arrive at the general fluid equations. It should be noted that the requirement of the invariance of the Action is sufficient, but not necessary, for the form invariance of the equations that follow from it. Any transformation that leaves the Action invariant upto a multiplicative factor suffices to produce equations of motion which have the same form. If this is taken into account, the determinantal condition $\alpha\delta - \beta\gamma = 1$ is no longer required and $SL(2,R)$ gets replaced by $GL(2,R)$ in the maximal invariance group of the general fluid equations. However, since this extra one-parameter set of transformations does not leave the underlying Action invariant, it does not give rise to any conservation laws. In view of this we will henceforth concentrate on the Noetherian symmetries of the fluid equations and for this purpose $\mathcal{G} = SL(2,R) \wedge G$. Finally, it should also be pointed out that the equations are also invariant under a discrete $\mathbf{Z}_2$ group that reverses the sign of all the $SL(2,R)$ parameters. Although seemingly trivial, this discrete group has an important role to play in the discussion of the transformation properties of the viscosity fields under the $SL(2,R)$ group.

3. Conservation Laws

In this section we construct the conservation laws corresponding to the symmetries outlined in the previous section. In order to do this it is useful to revert back to the Action formalism and obtain the results for the subclass of inviscid, isentropic and irrotational flows. The corresponding expressions for a general fluid can then be worked out along the lines of [5].

For inviscid, isentropic, irrotational flows the Lagrangian density is given by

$$\mathcal{L} = \rho(\dot{\phi} - \frac{1}{2}(\vec{\nabla}\phi)^2) - \rho^{\gamma_0} \tag{13}$$

where $\vec{\nabla}\phi$ stands for the curl-free part of the velocity vector field $\vec{u}$. Let $\mu = 0, 1, 2, 3$ and x^μ be a four-vector under the transformations of the previous section *i.e.* x^i with $i = 1, 2, 3$ are the components of $\vec{x}$ and $x^0 = t$. Let the variations in the coordinates and fields be defined as

$$\delta x^\mu = x^{\mu\prime} - x^\mu \quad \text{and} \quad \delta\phi(x) = \phi'(x') - \phi(x) \tag{14}$$

Then the variations for translations, rotations, boosts, dilatations, and expansions respectively are given by

$$\delta x^\mu = a^\mu, \;\; \delta x^i = \omega^{ij} x^j, \;\; \delta x^i = v^i t, \;\; \delta x^i = \lambda x^i, \;\; \delta t = 2\lambda t \text{ and } \delta x^\mu = -\mu t x^\mu \tag{15}$$

where the parameters λ, a^0, μ are expressible in terms of the $SL(2, R)$ parameters α, β, γ, δ. The field variation is given by

$$\delta\phi = \Lambda = \frac{[\gamma(x + a) - \delta v]^2}{2\gamma(\gamma t + \delta)} \tag{16}$$

The variation in ρ is not important since no derivatives of ρ appear in the Lagrangian density. Using these results we find, by a straightforward application of Noether's theorem [8], that the following quantities integrated over all space are constants of motion:

Temporal Translations

$$\mathcal{H} = \frac{\rho}{2}(\vec{\nabla}\phi)^2 + \rho^{\gamma_0} \tag{17}$$

Spatial Translations

$$\vec{P} = \rho\vec{\nabla}\phi \tag{18}$$

Rotations

$$\vec{L} = \rho\vec{\nabla}\phi \times \vec{x} \tag{19}$$

Boosts

$$\vec{K} = \rho[(\vec{\nabla}\phi)t - \vec{x}] \tag{20}$$

Dilatations

$$D = 2t[\frac{\rho}{2}(\vec{\nabla}\phi)^2 + \rho^{\gamma_0}] + \rho\vec{x}\cdot\vec{\nabla}\phi \tag{21}$$

Expansions

$$A = t^2[\frac{\rho}{2}(\vec{\nabla}\phi)^2 + \rho^{\gamma_0}] + t\rho\vec{x}\cdot\vec{\nabla}\phi + \frac{\rho}{2}\vec{x}^2 \tag{22}$$

The conditions of irrotationality and isentropicity can be relaxed easily and one sees that Euler's equations

$$\dot{\rho} = -\vec{\nabla}\cdot(\rho\vec{u}) \tag{23}$$

$$\rho\dot{\vec{u}} = -\rho(\vec{u}\cdot\vec{\nabla})\vec{u} - \vec{\nabla}p \tag{24}$$

$$\dot{\varepsilon} = -\vec{\nabla}\cdot(\varepsilon\vec{u}) - p\vec{\nabla}\cdot\vec{u} \tag{25}$$

can be re-expressed in the form of conservation laws,

$$\frac{\partial}{\partial t}\rho = -\frac{\partial}{\partial x_j}(\rho u_j) \tag{26}$$

$$\frac{\partial}{\partial t}(\rho u_i) = -\frac{\partial}{\partial x_j}(\rho u_i u_j + \delta_{ij}p) \tag{27}$$

$$\frac{\partial}{\partial t}(\frac{1}{2}\rho\vec{u}^2 + \varepsilon) = -\frac{\partial}{\partial x_j}[(\frac{1}{2}\rho\vec{u}^2 + \varepsilon + p)u_j] \tag{28}$$

for mass and translation generators found above. These can be expressed succinctly as follows:

$$\partial_\mu J^\mu_{(\rho)} = 0, \quad \partial_\mu J^\mu_{(\vec{P})} = 0, \quad \text{and} \quad \partial_\mu J^\mu_{(\mathcal{H})} = 0 \tag{29}$$

The zeroeth components of the above currents, namely ρ, $\rho\vec{u}$, and $\frac{1}{2}\rho\vec{u}^2 + \varepsilon$, give the charge densities which when integrated over all space give the conserved charges. As is well-known these are merely statements of mass, momentum flux, and total energy conservation. The corresponding current densities are

$$J^j_\rho = \rho u_j \tag{30}$$

$$J^j_{P_i} = \rho u_i u_j + \delta_{ij} p \tag{31}$$

$$J^j_{\mathcal{H}} = \left(\frac{1}{2}\rho \vec{u}^2 + \varepsilon + p\right) u_j \tag{32}$$

These are used to set up the standard Rankine-Hugoniot jump conditions. The conservation laws corresponding to rotations, boosts, dilatations and expansions can be stated similarly

$$\partial_\mu J^\mu_{(\vec{L})} = 0, \quad \partial_\mu J^\mu_{(\vec{K})} = 0, \quad \partial_\mu J^\mu_{(D)}, \quad \text{and} \quad \partial_\mu J^\mu_{(A)} = 0 \tag{33}$$

The charge densities are shown in (19) -(22) respectively, and the corresponding currents are

$$\vec{J}_{L_i} = \epsilon_{ikl} x_k \vec{J}_{P_l} \tag{34}$$

$$\vec{J}_{K_i} = t\vec{J}_{P_i} - x_i \vec{J}_\rho \tag{35}$$

$$\vec{J}_D = x_i \vec{J}_{P_i} - 2t\vec{J}_{\mathcal{H}} \tag{36}$$

$$\vec{J}_A = \frac{1}{2}\vec{x}^2 \vec{J}_\rho - t x_i \vec{J}_{P_i} + t^2 \vec{J}_{\mathcal{H}} \tag{37}$$

These may be used to set up the dual Rankine-Hugoniot jump conditions. The multiplet structure with respect to $SL(2,R)$ is

$$\rho, \quad \vec{L}, \quad (\vec{P}, \vec{K}), \quad (\mathcal{H}, D, A) \tag{38}$$

i.e. all conservation laws except that for angular momentum $\vec{L}$ can be obtained from Euler's equations (in conservation law form) by $SL(2,R)$ transformations. The linear relations between the currents is noteworthy and will play a crucial role when we consider flows with singularities. In particular, they can be used to show that the dual Rankine-Hugoniot conditions are identically satisfied because of the the standard Rankine-Hugoniot conditions. As mentioned already, the details of this proof will be given elsewhere.

It is well-known that Rankine-Hugoniot conditions describe not only shocks, but other discontinuities like rarefactions, for example. Therefore, the map between the dual sets of Rankine-Hugoniot conditions would be relevant to explosion-implosion duality only if both the sets refer to shocks. In other words, only those Rankine-Hugoniot conditions that describe a shock are of interest, and only those $SL(2,R)$ transformations which map a shock to a shock are physical for explosion-implosion duality. Physical shocks are distinguished from other discontinuities because their entropy always increases across the shock front. This requirement is unaffected by the $SL(2, R)$ transformations. The physicality of the map may also be established in the following way: First, Euler equations ought to be thought of as a special case with vanishingly small viscosity of a more general set of fluid equations. Requiring the sign of the viscosity to remain unchanged under the transformations establishes the physicality. It may be recalled that although Navier-Stokes equations are only Galilean invariant, the larger group $\mathcal{G}$ is the maximal kinematical invariance group for the general fluid equations with the viscosity fields transforming like tensors under the $SL(2, R)$ transformations. We therefore use these equations for our purpose of examining the limit of vanishing viscosity. The detailed proofs of these results are beyond the scope of this article and will be presented elsewhere.

References

1. B. Remington, "Supernova Hydrodynamics Up Close," Science and Technology Review Jan'/Feb' 2000, Lawrence Livermore National Laboratory; `http://www.llnl.gov/str`
2. I. Hachisu *et al.*, "Rayleigh-Taylor instabilities and mixing in the helium star models for Type Ib/Ic supernovae," Astrophysical Journal **368** (1991) L27–30;
3. H. Sakagami and K. Nishihara, "Rayleigh-Taylor instability on the pusher-fuel contact surface of stagnating targets," Physics of Fluids **B 2** (1990) 2715–2730.
4. L. O'C Drury and J. T. Medonça, "Explosion Implosion Duality and the Laboratory Simulation of Astrophysical Systems", Physics of Plasma **7** (2000) 5148-5152
5. L. O'Raifeartaigh and V. V. Sreedhar, "The maximal kinematical invariance group of fluid dynamics and explosion-implosion duality," Annals Phys.293:215-227,2001 hep-th/0007199.
6. Robert D. Richtmyer, Principles of Advanced Mathematical Analysis, Vol. 1 1978 Springer-Verlag, See section 2.9, 17.1 –17.6
7. L. D. Landau and E. M. Lifshitz, Fluid Mechanics, Pergamon Press (1959).
8. Herbert Goldstein, Classical Mechanics, Second Edition 1980, Addison Wesley, See section 12-7.
9. O. Jahn and V. V. Sreedhar, Am. J. Phys. 69, 1039 (2001) math-ph/0102011
10. R. Jackiw, A.P. Polychronakos, Phys.Rev. D62 (2000) 085019, hep-th/0004083
11. Tulsi Dass, Mathematical Method in Classical and Quantum Physics. 1998, University Press, New Delhi.

B FIELDS FROM A LUDDITE PERSPECTIVE

MARK STERN
Department of Mathematics, Duke University,
Durham, NC, 27708, USA
E-mail: stern@math.duke.edu

In this talk, we discuss the geometric realization of B fields and higher p-form potentials on a manifold M as connections on affine bundles over M. We realize D branes on M as special submanifolds of these affine bundles. As an application of this geometric understanding of the B field, we give a simple geometric explanation for the Chern-Simons modification of the field strength of the heterotic B field.

1. Introduction

The mathematical description of $B-$fields in terms of gerbes (see for example [4]) is too abstract to be useful for many basic computations. In this talk we discuss some of our recent work [10,9] developing a simple, geometric representation of B fields and higher $p-$form potentials on a manifold M as connections on various affine bundles over M.

Our geometric representation has three ingredients:

(1) the representation of the B field on a manifold M as a connection on an affine bundle E over M,
(2) a dictionary between string sigma model fields and differential operators on E, and
(3) the representation of D branes as submanifolds of E (not M).

The choice of E depends on the sector of string theory under consideration. Choosing E to be an affine bundle is probably useful only in a low energy approximation. In fact, heterotic string theory already requires more complicated (principal affine) bundles in order to see the full Green-Schwarz mechanism.

Our geometric model of B fields and D branes closely agrees with our stringy expectations. Some easy consequences of this model include

- A derivation of noncommutative Yang-Mills associated to a B field which suggests generalizations to deformations associated to higher p-form potentials.

- A geometric representation of topological B type D branes corresponding to coherent sheaves that need not be locally free and that may be "twisted".
- A simple geometric explanation for the introduction of the Yang-Mills gauge transformations of B fields in heterotic string theory.

We will not discuss noncommutative deformations here, but give a simple example in section 3 of the geometric realization of a topological B type D brane corresponding to a coherent sheaf which is not locally free. In section 4, we show how the geometry of affine bundles leads to the introduction of Yang-Mills gauge transformations of B fields in heterotic string theory.

2. The model

In [9], T duality considerations suggested that B fields should be realized as connections on a space of connections. Here we consider a finite dimensional analog: B fields as connections on affine bundles locally modelled on T^*M. (See [5] for related ideas.) First we review some basic facts about connections on T^*M.

Local coordinates (x^i) on M define vector fields $\frac{\partial}{\partial x^i}$ on M. These coordinates also determine local coordinates (x^i, p_s) on T^*M which define vector fields $\frac{\partial}{\partial x^i}$ and $\frac{\partial}{\partial p_s}$ on T^*M. Here we are abusing notation in a standard way, using the same symbol, $\frac{\partial}{\partial x^i}$, to denote two different vector fields, one on M and the other on T^*M. The coordinates thus define a lifting of vector fields on M to vectorfields on T^*M given in a notationally confusing manner as $\frac{\partial}{\partial x^i} \to \frac{\partial}{\partial x^i}$. This lift is obviously coordinate dependent. The coordinate dependence may be removed by introducing a connection. For example, with the Levi Civita connection, we have the globally well defined lift given by

$$\frac{\partial}{\partial x^i} \to \frac{\partial}{\partial x^i} + p_n \Gamma^n_{is} \frac{\partial}{\partial p_s}. \tag{1}$$

We now break the vector space structure on T^*M to an affine structure by introducing new allowed local coordinate transformations

$$(x, p) \to (x, p + \lambda(x)),$$

where λ is a locally defined 1 form on M. Then (1) is no longer coordinate independent. To fix this we introduce a 1 form μ and a 2 form b and define

$$\frac{\partial}{\partial x^i} \to \frac{\partial}{\partial x^i} + (\mu_{i;s} + b_{is} + p_n \Gamma^n_{is}) \frac{\partial}{\partial p_s}. \tag{2}$$

This lift is coordinate independent if

$$\mu \to \mu + \lambda, \text{ and } b \to b + d\lambda, \text{ when } p \to p + \lambda.$$

We denote the new bundle equipped with this affine structure, $T^*_B M$. Here the b field is the component of the connection on an affine bundle corresponding to the translation subspace of the affine transformations.

Our new b field immediately runs into a problem. The cohomology class of db vanishes. To realize B fields with cohomologically nontrivial field strength, we pass to quotients of $T^*_B M$. There are two obvious ways to do this. The first is to consider discrete quotients; then our fibers become products of tori and affine spaces. This cure allows the field strength of b to lie in a discrete subgroup of $H^3(M)$ but restricts the geometry of M.

A second solution is given by quotienting by "gauge equivalence class". In other words, consider sections p mod exact sections df. This does not give a finite dimensional bundle; so, to stay in the geometric regime, we consider the finite dimensional approximation to this equivalence given by 1 jets of sections quotiented by 1 jets of exact sections. The resulting quotient space is an affine space locally modelled on $\bigwedge^2 T^* M$. If we allow affine changes of coordinates $(x, p) \to (x, p + \lambda)$, where now p and λ are 2 forms, then we may define lifts of vector fields of the form

$$\frac{\partial}{\partial x^i} \to \frac{\partial}{\partial x^i} + (b_{ji;k} + b_{ik;j} + c_{ijk} + p_{nk} \Gamma^n_{ij} + p_{jn} \Gamma^n_{ik}) \frac{\partial}{\partial p_{jk}}, \tag{3}$$

where b and c are locally defined 2 and 3 forms respectively. This is well defined if

$$b \to b - \lambda, \text{ and } c \to c + d\lambda, \text{ when } p \to p + \lambda.$$

If we restrict to closed translations, λ, then we may choose $c = 0$ and obtain b fields with arbitrary field strength without restricting the topology of M. For nonclosed λ it is necessary to include the 3 form potential c. Once again, its field strength is cohomologically trivial in this formulation, and this may be remedied by passing to quotients. Discrete quotients constrain the geometry of M and lead to field strengths of c lying in discrete subgroups of $H^4(M)$. Gauge quotients lead to higher p-form potentials. We see that there is an analog of our p-form potential construction for all $p < dim M$.

We now see how these bundles are related to string theory. Let s, t denote coordinates for a string with t a timelike parameter and s the position along the string. Let $X^\mu(s,t)$ denote coordinates for the string world sheet. We would like to study a quantum mechanical system which reflects some low energy information about the string sigma model. The simplest method to do this is to consider only the average value $x^\mu(t)$ of $X^\mu(s,t)$ and its conjugate momentum. This leads to the quantum mechanics of a point particle moving on the target space M. This system loses too much string data. Motivated by the quantum mechanics of the (affine) 1 jet approximation of the string maps, we include the average value of X^μ_s in our system. The average value of X^μ_s is just $(X^\mu(\pi,t) - X^\mu(0,t))/\pi$. Jet space constructions suggest that we represent $X^\mu(\pi,t) - X^\mu(0,t)$ as a differential operator $\frac{1}{i}\frac{\partial}{\partial p_\mu}$ tangent to the fiber of an affine bundle, whose coordinates p_μ we think of as velocities, or conjugate momenta to the $(X^\mu(\pi,t) - X^\mu(0,t))/\pi$.

Assuming the p_μ are coordinates for an affine fiber is clearly at best a low energy approximation. For example, if X^1 wraps a small circle then $X(s,t)$ will not lie in a single coordinate chart for all s and the average value of X_s will also encode winding number (and only winding number for the closed strings). This dictates that in the p_1 direction, the noncompact affine fiber be replaced by a circle fiber dual to the wrapped circle, thus leading to discrete quotients of the affine fibers as required for cohomologically nontrivial B field field strengths. Then the commutator

$$\left[\frac{1}{i}\frac{\partial}{\partial p_1}, e^{2\pi i w p_1}\right] = 2\pi w e^{2\pi i w p_1}$$

with $\frac{1}{i}\frac{\partial}{\partial p_1}$ corresponding to $X^1(\pi,t) - X^1(0,t)$ suggests we interpret p_1 as the infinitesimal generator of motion along the circle. (Further deformations of the affine geometry of the fibers which are suggested by supersymmetry are considered in [10] .)

These considerations lead to the rough dictionary

$$\text{average } X^\mu_s \to \frac{\partial}{\partial p_\mu}, \text{ and}$$

$$\text{average } X^j_t \text{ (or better - } \pi^j) \to g^{ij}\left(\frac{\partial}{\partial x^i} + (\mu_{i;\nu} + b_{i\nu} + p_n \Gamma^n_{i\nu})\frac{\partial}{\partial p_\mu}\right).$$

Here π^j denotes the total momentum in the jth direction. The interpretation of p_1 as a generator of motion in the X^1 direction further suggests we associate

$$X^\mu_s(\pi,t) \to p_\mu.$$

3. D branes

We now use the dictionary of the preceding section to see what form D branes must take in our model. An n-brane (in M^n) is given by fully Neumann boundary conditions: $X_s^\mu(0) = X_s^\mu(\pi) = 0$. From our dictionary, we see that this corresponds to a zero section (in a choice of local affine coordinates) of T_B^*M. A change in time parameter for the string world sheet $(s,t) \to (s,\tau(s,t))$ induces an affine change

$$\frac{\partial}{\partial s} \to \frac{\partial}{\partial s} + \frac{\partial \tau}{\partial s}\frac{\partial}{\partial \tau},$$

and correspondingly an affine change of coordinates in T_B^*M. Thus the B field reflects the nonuniqueness of the time coordinate.

Consider next a p brane corresponding to a p dimensional submanifold S of M. Choose local coordinates so that S is given by $x^\nu = 0$ for $\nu > p$. Then the $p-$brane boundary conditions $X_t^\nu = 0$, X_s^ν free, for $\nu > p$ and $X_s^\nu = 0$, X_t^ν free, for $\nu \le p$ translate under our dictionary to $x^\nu = 0$ and p_ν free for $\nu > p$ for some choice of local affine coordinates. This is an n dimensional submanifold Z of T_B^*M which, in a choice of local affine coordinates, is the conormal bundle of S. In particular, a zero brane is just an affine fiber over a point.

Fixing coordinates so that Z is identified with a conormal bundle removes the gauge freedom to vary μ_ν arbitrarily for $\nu \le p$. Hence the D brane comes equipped with a locally defined 1 form μ on S. When the B field is trivial, this defines the gauge field for a line bundle on S (equipped with a local frame). On the overlap of two coordinate neighborhoods, U_α and U_β, we have $b^\alpha - d\mu^\alpha = b^\beta - d\mu^\beta$. Thus, μ determines a connection on a line bundle only if the field strength of b vanishes in $H^3(S)$. This is, of course, to be expected. In the presence of a B field with nontrivial field strength, the gauge field of a D brane is not that of a vector bundle, but of a "twisted bundle" (see for example [11,6,2]) or more generally perhaps of an infinite dimensional C^* algebra [1]. Interpreting μ as a coordinate of the brane in a bundle is, perhaps, geometrically simpler than working with infinite dimensional C^* algebra bundles.

If we have two distinct D branes Z_1 and Z_2 which correspond after distinct affine transformations to the normal bundle of a single submanifold S of M, then we see that we have a gauge enhancement. In addition to the previously identified gauge fields, we also have the 1 form measuring the relative displacement in the fiber between the 2 branes.

In general, we would like to define a D brane to be an n dimensional submanifold, Z of T_B^*M which corresponds to a choice of boundary condition, Dirichlet or Neumann, for each coordinate. If the Neumann boundary condition were well defined, then for Z to represent a choice of boundary condition would imply that the symplectic form $\omega := dx^\mu \wedge dp_\mu$ vanishes when pulled back to Z. I.e., Z is Lagrangian. Neither the form ω nor the Neumann condition is well defined. One can define a connection dependent analog of each, but we will instead use the provisional definition:

Provisional Definition 3.1. A D brane is an n dimensional submanifold Z of T_B^*M such that for every point $p \in Z$ there exists a neighborhood of p in Z which, after an affine choice of coordinates can be identified with an open set in the conormal bundle of some submanifold S of M.

BPS conditions will restrict the possible Z which occur. This definition may be too broad. In particular, the distinction between p brane and p' brane for $p \neq p'$ becomes somewhat fuzzy. Nonetheless, it does provide a geometrical framework which includes sheaves. For example, assume now that M is a complex manifold, and consider the ideal sheaf $\mathcal{I}_D$ of a divisor D. Let z be a local defining function for D. The connection form $\frac{dz}{z}$ of $\mathcal{I}_D$ is singular along D. Hence the n brane Z_D given by the graph of $\frac{dz}{z}$ becomes vertical as it approaches D. This singularity cannot be removed by a (finite) affine change of coordinates and reflects the fact that $\mathcal{I}_D$ is not the sheaf of sections of a vector bundle. Assuming in this example that $T_B^*M = T^*M$, we may try to deform Z_D to the zero section $Z_\mathcal{O}$, which is the n brane corresponding to the trivial sheaf $\mathcal{O}$. Allowing only bounded affine shifts, we are left in the limit with

$$Z_D \to Z_\mathcal{O} \cup Z_{\mathcal{O}_D},$$

where $Z_{\mathcal{O}_D}$ is the brane corresponding to the sheaf $\mathcal{O}_D$. This gives a geometrical analog of the exact sequence of sheaves

$$0 \to \mathcal{I}_D \to \mathcal{O} \to \mathcal{O}_D \to 0.$$

4. Chern-Simons augmentation of the field strength

In this section we show for heterotic strings how our treatment of B fields leads to the modification of the B field field strength by the addition of Chern-Simons terms. We will only treat the Yang-Mills term in detail. Some of the material in this section postdate the conference and were elaborated at the suggestion of Eric Sharpe.

Let $\mathfrak{g}$ be the Lie algebra of $SO(32)$ or $E_8 \times E_8$. Let $\mathfrak{h}$ be a Cartan subalgebra. Let $P \subset \mathfrak{h}^*$ denote the set of roots of $\mathfrak{g}$ with respect to $\mathfrak{h}$, and let $\{\alpha_i\}_{1 \leq i \leq 16}$ be a basis of simple roots. Let Γ denote the lattice in $\mathfrak{h}^*$ generated by P. With respect to an inner product $(\cdot, \cdot)$ on $\mathfrak{h}$ determined by the Killing form, Γ is self dual and every element of P has square length 2.

We use the bosonized description of the heterotic string. So, let X be a 26 dimensional manifold which is a torus bundle with 10 dimensional base M and fiber isometric to $\mathfrak{h}/\Gamma$. Consider an affine cotangent bundle over X. Quotient the affine fibers by the action generated by 1 forms dual to Killing vectors generating the torus lattice to obtain an affine bundle $T^*_{hetB}M$.

For $\alpha \in P$, let τ_α denote the vector field on $\mathfrak{h}/\Gamma$ which satisfies

$$\tau_\alpha \beta = (\alpha, \beta), \text{ for all } \beta \in P.$$

Up to the familiar cocycle difficulty [3] (most easily corrected by adjoining gamma matrix prefactors), the vector fields $e^{2\pi i \alpha}\tau_\alpha, \alpha \in P$ and $\tau_{\alpha_i}, 1 \leq i \leq 16$ generate an algebra isomorphic to $\mathfrak{g}$, when endowed with the Lie bracket given by their commutator composed with projection onto their span.

Let B denote a $B-$field on X. We modify our prior assumption that B is locally a 2 form pulled back from X with the (chiral) assumption that it is allowed to vary in the $\mathfrak{h}^*/\Gamma$ fiber but only (in an appropriate frame) as $B(x, t, s) = B(x, t + s)$, where x is a local coordinate on M, and t and s denote coordinates in $\mathfrak{h}/\Gamma$ and $\mathfrak{h}^*/\Gamma \simeq \mathfrak{h}/\Gamma$ respectively. More precisely, we assume that B has the form

$$B = b_{\nu\rho} dx^\nu \wedge dx^\rho + A_{\nu\alpha} e^{2\pi i \alpha(t+s)} dx^\nu \wedge d\alpha + A_{\nu i} dx^\nu \wedge d\alpha_i + \text{massive},$$

where other terms may occur but will be thought of as massive and ignored at this level. Only those terms in B which are invariant in the fiber and which are annihilated by interior multiplication by vectors tangent to the torus fiber descend to a 2 form on M. Thus, these lead to the "B field" b on M.

We also allow affine transformations $p \to p + \lambda$ to vary in the torus fiber; we assume that λ has the form

$$\lambda = \lambda^{-\alpha}(x) e^{2\pi i \alpha(t+s)} d\alpha + \lambda^i(x) d\alpha_i + \lambda^0(x) + \text{massive}.$$

Here $\lambda^0(x)$ is the pull back of a 1 form on M and λ^α and λ^i are the pull back of functions on M. Under this transformation

$$b \to b + d\lambda^0 - A_{\mu\alpha} dx^\mu \wedge d\lambda^{-\alpha} - A^i_\mu dx^\mu \wedge d\alpha_i(\lambda^0)/2,$$

which under a natural identification can be written as

$$b \to b + d\lambda^0 + TrA \wedge d\lambda.$$

Similarly we obtain a variation in A which can be written as

$$A \to A + d(\lambda - \lambda^0) + [A, \lambda].$$

The new term in the variation of the b (not B) field differs by an exact term from the usual Yang-Mills modification of the heterotic B field. This leads as usual to the Yang-Mills Chern-Simons modification of the field strength of b:

$$H = db - \omega_{YM},$$

but now H arises simply as the component of dB which descends to a form on M. Here ω_{YM} denotes the Yang-Mills Chern-Simons form.

To obtain the gravitational Chern-Simons form, ω_L, requires a bit more work, but seems likely to cast more light on the relation of heterotic string theory to M theory. The affine bundle must be replaced by a principal affine bundle. Then the modified field strength

$$H = db + \omega_L - \omega_{YM}$$

again arises by taking the component of dB which descends to M.

Acknowledgments

I would like to thank Ilarion Melnikov, Ronen Plesser, and Eric Sharpe for helpful discussions. This work was supported in part by NSF grant DMS 0204188.

References

1. P. Bouwknegt, V. Mathai, "D-branes, B fields, and twisted K-theory" JHEP 0003 (2000) 007 hep-th/0002023.
2. A. Caldararu, S. Katz, and E. Sharpe, "D-branes, B fields and Ext groups" hep-th/0302099.
3. M. Green, J. Schwarz, and E. Witten, Superstring theory Volumes 1 and 2, Cambridge, Cambridge University Press (1987).
4. N. Hitchin, Lectures on special Lagrangian submanifolds, *Winter School on Mirror Symmetry, Vector Bundles, and Lagrangian Submanifolds*, 151-182, AMS/IP Stud Adv. Math. 23, Amer. Math. Soc., Providence, RI, 2001 DG/9907034.
5. N. Hitchin, Generalized Calabi-Yau manifolds, DG/0209099.
6. A. Kapustin, D-Branes in a topologically nontrivial B-field, Adv. Theor. Math. Phys. 4 (2001) 127 hep-th/9909089.
7. H. Ooguri, Y. Oz, and Z. Yin, D-branes on Calabi-Yau Spaces and Their Mirrors, hep-th/9606112, Nuc. Phys. B 477 (1996) pp. 407-430.

8. E. Sharpe, D-Branes, Derived Categories, and Grothendieck Groups, hep-th/9902116, Nucl. Phys. B561 (1999) pp.433-450.
9. M. Stern, Quantum mechanical mirror symmetry, D branes, and B fields, hep-th/0209192.
10. M. Stern, Mechanical D branes and B fields, hep-th/0310020.
11. E. Witten, D branes and K theory, hep-th/9810188, JHEP 9812:025 (1998) 032.

GEOMETRIC TRANSITIONS, MATRIX MODELS AND EFFECTIVE FIELD THEORIES

RADU TATAR

Theoretical Physics Group
Lawrence Berkeley National Laboratory
Berkeley, CA 94720, USA
rtatar@socrates.berkeley.edu

We discuss the geometric engineering and large N transition for a supersymmetric $U(N)$ chiral gauge theory with one adjoint, one conjugate symmetric, one anti-symmetric and eight fundamental chiral multiplets. The IIB realization involves an orientifold of a non-compact Calabi-Yau.

1. Introduction

String theory provides nontrivial information about supersymmetric gauge theories by means of geometric engineering. An interesting class of models is based on local conifold geometries [1,2,5,6,7]. These arise in IIB string theory on a resolved ADE fibration over the complex plane, with D5-branes partially wrapped on the exceptional divisors. On the common non-compact part of the world volumes, one obtains a supersymmetric field theory in four dimensions, whose precise nature is dictated by the Calabi-Yau geometry. The gauge group of such models is a product of $U(N)$ factors, with N fixed by the number of partially wrapped D5-branes. It has been argued that at large N but fixed $g_{string}N$ such models undergo a transition in which the resolved geometry is replaced by the background obtained through a conifold transition, realized by blowing down all exceptional curves and smoothing out the resulting singular fibration. The exact effective superpotential of the gauge theory can then be calculated by evaluating period integrals of the deformed geometry.

The approach described above leads to theories whose gauge group is a product of $U(N)$ factors, with matter in adjoint or bifundamental chiral multiplets. To engineer more general models, one can consider orientifolds of such backgrounds, possibly supplemented by the introduction of further D-branes.

Our model contains chiral multiplets Φ, A, S in the adjoint, antisymmetric and conjugate symmetric representations of $U(N)$ as well as eight quarks $Q_1 \ldots Q_8$ in the fundamental representation. We consider the tree-level superpotential:

$$W_{tree} = \text{tr}\ [W(\Phi) + S\Phi A] + \sum_{f=1}^{8} Q_f^T S Q_f \quad , \tag{1}$$

where:

$$W(z) = \sum_{j=1}^{d+1} \frac{t_j}{j} z^j \tag{2}$$

is a complex polynomial of degree $d+1$.

2. Chiral field theories from geometric engineering

2.1. *The IIB background*

Our starting point is a Calabi-Yau A_2 fibration, chosen to admit a certain holomorphic $\mathbf{Z_4}$ action. We shall first consider the singular limit of this fibration, then provide global and local models of its resolution following [10].

2.1.1. *The singular limit X_0*

Consider the Calabi-Yau hypersurface X_0 defined by the equation:

$$xy = u(u - W'(z))(u + W'(z)) \quad , \tag{3}$$

where z, u, x, y are the affine coordinates of C^4 and W is a polynomial of degree $d+1$ with complex coefficients. Throughout this section, we restrict to the generic case when the derivative W' has only simple roots, which we denote by z_j. The index j runs from 1 to d.

The space (3) is a singular A_2 fibration over the z-plane, which admits a 3-section Σ_0 obtained by requiring $x = y = 0$. Together with (3) this gives the planar curve with equation:

$$u(u - W'(z))(u + W'(z)) = 0 \quad . \tag{4}$$

Thus Σ_0 has three rational components, which we index as follows:

$$C_0 : \ u = -W'(z) \quad , \quad C_1 : \ u = 0 \quad , \quad C_2 : \ u = +W'(z) \quad . \tag{5}$$

One easily checks that the components meet only when $u = 0$ and z coincides with a critical point of W. In fact Σ_0 has triple points sitting at $u = 0$, $z = z_j$ and no other singular points.

The total space (3) also has d singular points, which coincide with the triple points of Σ_0. The fiber sitting above z_j is a standard A_2 singularity:

$$xy = u^3 \quad . \tag{6}$$

When z is away from all z_j, the fiber $X_0(z)$ is a smooth deformation of (6).

2.1.2. *The resolution*

Consider the space $\hat{X}$ obtained by resolving each of the singular A_2 fibers $X_0(z_j)$. Following [10], we describe it as the complete intersection in $P^1 \times P^1 \times C^4$ cut by the equations:

$$\begin{aligned} \beta_1(u + W'(z)) &= \alpha_1 x \\ \alpha_2(u - W'(z)) &= \beta_2 y \\ \alpha_1\beta_2 u &= \beta_1\alpha_2 \quad . \\ u(u - W'(z))(u + W'(z)) &= xy \end{aligned} \tag{7}$$

We let $[\alpha_i, \beta_i]$ be the homogeneous coordinates on the two P^1 factors ($i = 1, 2$). Forgetting these gives the resolution map $\tau : \hat{X} \to X_0$. The exceptional fibers of τ are given by $z = z_j, u = x = y = 0$ and $\alpha_2 = 0$ or $\beta_1 = 0$. The case $\alpha_2 = 0$ defines a rational curve $D_j^{(1)}$ which can be identified with $P^1[\alpha_1, \beta_1]$, while $\beta_1 = 0$ gives the curve $D_j^{(2)}$, identified with $P^1[\alpha_2, \beta_2]$.

The orientifold action takes the form:

$$\hat{\kappa} : \quad (z_1, x_1, u_1) \to (z_1, u_1, -x_1) \tag{8}$$

with fixed point set $\hat{O}$ given by $\hat{C}_1$.

As in [10], the brane construction arises by performing T-duality with respect to the $U(1)$ action on $\hat{X}$, which has the following form in local coordinates:

$$\hat{\rho}(\theta) : \quad (z_j, x_j, u_j) \longrightarrow (z_j, \ \ e^{i\theta} x_j, \ \ e^{-i\theta} u_j) \quad . \tag{9}$$

The fixed point locus of (9) consists of the three components $\hat{C}_j$, with local equations:

$$\hat{C}_j : \quad u_j = x_j = 0 \quad , \tag{10}$$

which take the following form in the coordinates of the patch:

$$\begin{aligned} \hat{C}_0 &: \quad x_1 u_1 = -W'(z) \quad , \quad x_1 = \infty \\ \hat{C}_1 &: \quad x_1 = u_1 = 0 \\ \hat{C}_2 &: \quad x_1 u_1 = +W'(z) \quad , \quad u_1 = \infty \ . \end{aligned} \tag{11}$$

The $U(1)$ action stabilizes the exceptional curves $D_j^{(\alpha)}$.

2.2. *Geometric description of supersymmetric vacua*

Let us recall the relevant part of the classical moduli space of supersymmetric vacua, which was discussed in [9]. The vacua of interest are solutions of the D- and F-flatness constraints, with the supplementary assumption that the VEV of Φ is a normal matrix, i.e. $[\Phi, \Phi^\dagger] = 0$. This condition allows us to eliminate possible baryonic branches, which are irrelevant for our purpose.

The condition that Φ is normal means that it can be diagonalized through a unitary gauge transformation. Hence one can take:

$$\Phi = \mathrm{diag}(z_1 1_{N_1} \dots z_d 1_{N_d}) \ , \tag{12}$$

where N_j are nonnegative integers such that $\sum_{j=1}^d N_j = N$ (we use the convention that if some N_j vanishes, then the corresponding eigenvalue does not appear in (12)). Then it was showed in [9] that the D- and F-flatness constraints imply that the VEVs of S, A and Q_f must vanish in this family of vacua. The unbroken gauge group in such a vacuum is given by $\prod_{j=1}^d U(N_j)$, with the convention that $U(0)$ is the trivial group.

3. The geometric transition and effective superpotential

To describe the geometric transition of our system, we follow [3] by considering the most general log-normalizable deformation of X_0

$$xy = (u - t_0(z))(u - t_1(z))(u - t_2(z)) \tag{13}$$

where:

$$\begin{aligned} t_0(z) &= -\frac{2W_1'(z) + W_2'(z)}{3} \\ t_1(z) &= \frac{2W_2'(z) + W_1'(z)}{3} \\ t_2(z) &= \frac{W_1'(z) - W_2'(z)}{3} \end{aligned} \tag{14}$$

and W_1, W_2 are two polynomials (note that $t_0(z) + t_1(z) + t_2(z) = 0$).

After the geometric transition $\hat{X} \longrightarrow X_0 \longrightarrow X$ of [3,4], the D5-branes wrapping the exceptional fibers of $\hat{X}$ are replaced by fluxes through the S^3 cycles created by smoothing. This is given by $x = y = u = 0$, which is the curve C_1. The D5-branes wrapped over the orientifold fixed locus also survive the transition. Thus we end up with a compactification with NS-NS and R-R fluxes and a $\mathbf{Z_4}$ orientifold which fixes an O5 plane with worldvolume $R^{1,3} \times C_1$, together with 8 fractional D5-branes wrapping the orientifold fixed 'plane'. As above, the RR charge of the D5-branes cancels the RR charge of the orientifold. The worldvolume of these D5-branes carries an $SO(8)$ symmetry, which is frozen to a global symmetry of the system because the internal part of the D5-brane worldvolume is non-compact. This gives the geometric realization of the $SO(8)$ flavor symmetry, which is unbroken after confinement.

The common effective superpotential of these theories receives contributions from both orientable and unorientable planar diagrams. In the associated matrix models, the former correspond to the strict planar limit, while the latter give the subleading contribution in the $1/\hat{N}$ expansion. Recall from [9] that the gaugino superpotential can be expressed as:

$$W_{eff} = \sum_{j=1}^{d} \left[N_j \frac{\partial F_0}{\partial S_j} + 4F_1 + \alpha_j S_j \right] \ , \tag{15}$$

where S_j are the gaugino condensates after confinement in the $U(N_j)$ factors and α_j are related to the effective gauge couplings in these factors. Here F_0 and F_1 are the leading and subleading contributions to the free energy of the holomorphic matrix model associated to our chiral theory:

$$F = F_0 + \frac{g}{\hat{N}} F_1 + O\left(\left(\frac{g}{\hat{N}}\right)^2\right) \ , \tag{16}$$

where $\hat{N}$ describes the size of the various matrices involved. One has the relation [9]:

$$F = F_{SO} - \frac{g}{\hat{N}} \ln 2 \ . \tag{17}$$

In the chiral theory, the effective superpotential arises from planar diagrams with or without boundaries associated to the fundamentals Q_f. The former correspond to the closed string contribution to the free energy after the geometric transition, while the latter arise from unoriented open strings ending on the eight fractional D5-branes in the presence of the orientifold.

4. Konishi anomaly

Let us now discuss the generalized Konishi anomalies [8] for this model. The structure of the tree level superpotential implies:

$$j\frac{\partial W_{eff}}{\partial t_j} = \langle \mathrm{tr}\,(\Phi^j)\rangle \ . \tag{18}$$

The strategy is to extract a set of chiral ring relations which determine the generating function $T(z) = \langle \mathrm{tr}\,(\frac{1}{z-\Phi})\rangle$ of the correlators $\langle \mathrm{tr}\,(\Phi^j)\rangle$. Then integration of (18) allows one to compute the effective superpotential up to a piece independent of t_j.

We will project onto the chiral ring, i.e. the equivalence class of operators which are annihilated by the anti-chiral supercharge $\bar{Q}_{\dot{\alpha}}$. As is well-known, these are in one-to-one correspondence with the lowest components of chiral superfields.

We define:

$$R(z) := -\frac{1}{32\pi^2}\left\langle \mathrm{tr}\left(\frac{W^2}{z-\Phi}\right)\right\rangle, T(z) := \left\langle \mathrm{tr}\left(\frac{1}{z-\Phi}\right)\right\rangle \ , \tag{19}$$

$$M(z) := \left\langle \mathrm{tr}\left(S\frac{1}{z-\Phi}A\right)\right\rangle,$$

$$M_Q(z) := \sum_f \left\langle Q_f^T \frac{1}{z-\Phi^T} S \frac{1}{z-\Phi} Q_f \right\rangle \ , \tag{20}$$

$$K(z) := -\frac{1}{32\pi^2}\left\langle \mathrm{tr}\left(S\frac{W^2}{z-\Phi}A\right)\right\rangle, L(z) := \sum_f \left\langle Q_f^T S \frac{1}{z-\Phi} Q_f \right\rangle, \tag{21}$$

and introduce the degree $d-1$ polynomials:

$$f(z) := -\frac{1}{32\pi^2}\left\langle \mathrm{tr}\left(\frac{W'(z)-W'(\Phi)}{z-\Phi}W^2\right)\right\rangle \ , \tag{22}$$

$$c(z) := \left\langle \mathrm{tr}\left(\frac{W'(z)-W'(\Phi)}{z-\Phi}\right)\right\rangle \ . \tag{23}$$

Assuming that vevs of spinor fields vanish due to Lorentz invariance, we find the following Ward identities for the generating functions (19-21):

$$R(z)^2 - 2W'(z)R(z) + 2f(z) = 0 \ , \tag{24}$$

$$T(z)R(z) - W'(z)T(z) - 2R'(z) + c(z) = 0 \ , \tag{25}$$

$$K(z) - \frac{1}{2}R(z)^2 = 0 \ , \tag{26}$$

$$M(z) - R(z)T(z) - 2R'(z) = 0 \ , \tag{27}$$

$$M_Q(z) + M(z) - R(z)T(z) + 2R'(z) = 0 \ , \tag{28}$$

$$2L(z) - N_F R(z) = 0 \ . \tag{29}$$

Given a solution $(R(z), T(z))$ of these constraints, the quantities K, M, M_Q and L are determined by $R(z)$ and $T(z)$. Hence all solutions are parameterized by the $2d$ complex coefficients of the polynomials $f(z)$ and $c(z)$.

References

1. C. Vafa, "Superstrings and topological strings at large N," J. Math. Phys. **42**, 2798 (2001) [arXiv:hep-th/0008142].
2. F. Cachazo, K. A. Intriligator and C. Vafa, "A large N duality via a geometric transition," Nucl. Phys. B **603**, 3 (2001) [arXiv:hep-th/0103067].
3. F. Cachazo, S. Katz and C. Vafa, "Geometric transitions and N = 1 quiver theories," [arXiv:hep-th/0108120].
4. F. Cachazo, B. Fiol, K. A. Intriligator, S. Katz and C. Vafa, "A geometric unification of dualities," Nucl. Phys. B **628**, 3 (2002)[arXiv:hep-th/0110028].
5. R. Dijkgraaf and C. Vafa, "Matrix models, topological strings, and supersymmetric gauge theories," Nucl. Phys. B **644**, 3 (2002) [arXiv:hep-th/0206255].
6. R. Dijkgraaf and C. Vafa, "On geometry and matrix models," Nucl. Phys. B **644**, 21 (2002) [arXiv:hep-th/0207106].
7. R. Dijkgraaf and C. Vafa, "A perturbative window into non-perturbative physics," arXiv:hep-th/0208048.
8. F. Cachazo, M. R. Douglas, N. Seiberg and E. Witten, "Chiral rings and anomalies in supersymmetric gauge theory," JHEP **0212**, 071 (2002)[arXiv:hep-th/0211170].
9. K. Landsteiner, C. I. Lazaroiu and R. Tatar, "Chiral field theories, Konishi anomalies and matrix models," arXiv:hep-th/0307182.
10. K. h. Oh and R. Tatar, "Duality and confinement in N = 1 supersymmetric theories from geometric transitions,", Adv. Theor. Math. Phys. **6**, 141 (2003) [arXiv:hep-th/0112040].

THE SOLUTION OF THE DIRAC EQUATION WITH THE INTERACTION TERM*

YEONG-SHYEONG TSAI

Dept. of Applied Math., National Chung Hsing University, Taichung, Taiwan
ystsai@dragon.nchu.edu.tw

HUNG-MING TSAI

Dept. of Physics, National Taiwan University, Taipei, Taiwan
r91222016@ntu.edu.tw

PO-YU TSAI

Dept. of Chemistry, Tunghai University, Taichung, Taiwan
byt@pchome.com.tw

LU-HSING TSAI

Dept. of Physics, National Taiwan Normal University, Taipei, Taiwan

With the best knowledge of the authors, it is very difficult, if it is not impossible, to find the exact solution of the Dirac equation with the interaction term [1,2]. In this article, we carefully analyze the mathematical theory of the quantum field and try to find the exact solution of the field equation with the interaction term. For simplicity, we take the Dirac field equation with the interaction term as a template of the gauge field equation. We will show how to obtain the exact solution of the Dirac equation with the interaction term. The results can be applied to other fields as well, such as QCD.

1. Overview of mathematical theory of quantum physics

Quantum mechanics began playing its role in Physics since 1900. Now the quantum theory is a well-developed theory because many geniuses have made their contribution in this subject. Though the whole system of quantum physics is complex, the central part of the quantum theory is simple. It consists two parts, the Lagrangian density and its related quantization.

*Talk given by H.-M.T.

With these two parts, we are able to infer or obtain meaningful quantities of physics [3,4,5].

We take QED as an example. The Lagrangian density of the Dirac field with interaction is typically written as $L = (i\hbar c\overline{\psi}\gamma^{\mu}\partial_{\mu}\psi - mc^2\overline{\psi}\psi) - (\frac{1}{4}F^{\mu v}F_{\mu v}) - q(\overline{\psi}\gamma^{\mu}\psi)A_{\mu}$. The quantization is written as anti-commutation relations and commutation relations.

$$\begin{cases} \psi_\alpha(\mathbf{x},t)\psi_\beta^\dagger(\mathbf{x}',t) + \psi_\beta^\dagger(\mathbf{x}',t)\psi_\alpha(\mathbf{x},t) = \delta_{\alpha\beta}\delta^{(3)}(\mathbf{x}-\mathbf{x}'). \\ \psi_\alpha(\mathbf{x},t)\psi_\beta(\mathbf{x}',t) + \psi_\beta(\mathbf{x}',t)\psi_\alpha(\mathbf{x},t) = 0. \\ \psi_\alpha^\dagger(\mathbf{x},t)\psi_\beta^\dagger(\mathbf{x}',t) + \psi_\beta^\dagger(\mathbf{x}',t)\psi_\alpha^\dagger(\mathbf{x},t) = 0. \end{cases}$$

$$\begin{cases} A_\mu(\mathbf{x},t)\dot{A}_\nu(\mathbf{x}',t) - \dot{A}_\nu(\mathbf{x}',t)A_\mu(\mathbf{x},t) = i\delta_{\mu\nu}\delta^{(3)}(\mathbf{x}-\mathbf{x}'). \\ A_\mu(\mathbf{x},t)A_\nu(\mathbf{x}',t) - A_\nu(\mathbf{x}',t)A_\mu(\mathbf{x},t) = 0. \\ \dot{A}_\mu(\mathbf{x},t)\dot{A}_\nu(\mathbf{x}',t) - \dot{A}_\nu(\mathbf{x}',t)\dot{A}_\mu(\mathbf{x},t) = 0. \\ \psi_\alpha(\mathbf{x},t)A_\mu(\mathbf{x}',t) - A_\mu(\mathbf{x}',t)\psi_\alpha(\mathbf{x},t) = 0. \\ \psi_\alpha(\mathbf{x},t)\dot{A}_\mu(\mathbf{x}',t) - \dot{A}_\mu(\mathbf{x}',t)\psi_\alpha(\mathbf{x},t) = 0. \end{cases} \tag{1}$$

From the Lagrangian density of a specific field, it is quite easy to obtain the field equations of this field. Usually, there are two approaches, Lagrange approach or Hamilton approach. Now, there are the field equations and the quantization. We might ask a question: Is the whole system of quantum theory compatible? By careful analysis, we find that we are not able to find any affirmative answer because the commutation relations and the anti-commutation relations make the field variables dependent. Therefore, the system is subjected to some constraints. It is inappropriate to deduce the field equations from the Lagrangian density without applying the method of Lagrange multipliers. By the method of Lagrange multiplier, the Lagrangian becomes

$$\begin{aligned} \tilde{L}(t) &= \int_{\mathbf{x}^3} L(\psi, \psi^\dagger, \partial_\mu\psi, \partial_\mu\psi^\dagger, t)d\mathbf{x} \\ &\quad + \sum_i \int_{\mathbf{x}^3}\int_{\mathbf{x}'^3} \Lambda^i(\mathbf{x},\mathbf{x}',t)\cdot \text{ constraint } d\mathbf{x}'d\mathbf{x} \\ &= \int_{\mathbf{x}^3} (L + \sum_i \int_{\mathbf{x}'^3} \Lambda^i(\mathbf{x},\mathbf{x}',t)\cdot \text{ constraint } d\mathbf{x}')d\mathbf{x}, \end{aligned}$$

where the summation are taken all the constraints listed in (1), because all quantization relations are nothings but constraints, and where $\Lambda^i(\mathbf{x},\mathbf{x}',t)$ are the Lagrangian multipliers. The redundant constraints must be excluded.

Let the Lagrangian density $\tilde{L} = L + \sum_i \int_{\mathbf{x}'^3} \Lambda^i(\mathbf{x}, \mathbf{x}', t) \cdot$ constraint $d\mathbf{x}'$. Before we find the Euler-Lagrange equation, some work must be done first. Given two field variable ϕ_1 and ϕ_2. Define two Lagrangians

$$\ell_c = [\phi_1, \phi_2] = \phi_1\phi_2 - \phi_2\phi_1 \text{ and } \ell_a = \{\phi_1, \phi_2\} = \phi_1\phi_2 + \phi_2\phi_1.$$

It is obvious that

$$\frac{\partial \ell_a}{\partial \phi_1} = 2\phi_2, \quad \frac{\partial \ell_a}{\partial \phi_2} = 2\phi_1, \quad \frac{\partial \ell_c}{\partial \phi_1} = 0, \quad \frac{\partial \ell_c}{\partial \phi_2} = 0.$$

2. The Dirac equation

By the above example, it is understood that the commutation relations will yield zeroes when they are substituted in the Euler-Lagrange equation. Only the anti-commutation relations are required to be put in the Lagrangian density. Therefore, the Lagrangian density is written as the following form.

$$\begin{aligned}\tilde{L} &= i\hbar c(\overline{\psi}\gamma^\mu \partial_\mu \psi) - mc^2(\overline{\psi}\psi) - \left(\frac{1}{4}F^{\mu\nu}F_{\mu\nu}\right) - q(\overline{\psi}\gamma^\mu\psi A_\mu) \\ &+ \int_{\mathbf{x}'^3} \sum_{\beta=1}^{4}\sum_{\alpha=1}^{4} \overline{\Lambda}_{\beta\alpha}(\mathbf{x}, \mathbf{x}', t)(\{\psi_\alpha(\mathbf{x}, t), \psi_\beta^\dagger(\mathbf{x}', t)\} - \delta_{\alpha\beta}\delta^{(3)}(\mathbf{x} - \mathbf{x}'))d\mathbf{x}' \\ &+ \int_{\mathbf{x}'^3} \sum_{\beta\geq\alpha}\sum_{\alpha=1}^{4} \overline{\overline{\Lambda}}_{\beta\alpha}(\mathbf{x}, \mathbf{x}', t)\{\psi_\alpha(\mathbf{x}, t), \psi_\beta(\mathbf{x}', t)\}d\mathbf{x}'\end{aligned}$$

In order to make it comprehensive, the integrations are translated into summations during the process of finding the field equations. By applying the Euler-Lagrange equation, we are able to obtain the Dirac equation with some parameters, the Lagrange multipliers.

$$i\hbar c\gamma^\mu \partial_\mu \psi - mc^2\psi + \Lambda\psi = q\gamma^\mu \psi A_\mu,$$

where Λ is 4×4 matrix of which the entries are the Lagrange multipliers with some coefficients.

From the relations in (1), there are many useful identities obtained, such as $\int_{\mathbf{x}'^3}\{\psi_\alpha^\dagger(\mathbf{x}', t), \psi_\beta(\mathbf{x}, t)\}d\mathbf{x}' = \delta_{\alpha\beta}$. With these identities, the parameters Λ_{ij} are eliminated. The Dirac equation becomes

$$\gamma^\mu \partial_\mu \psi + U\psi = 0, \tag{2}$$

where U is 4×4 matrix, $U_{ij} = \int_{\mathbf{x}'^3}\{\psi_j^\dagger, (\gamma^\mu \partial_\mu \psi)_i\}d\mathbf{x}'$. Here, $\{A, B\} = AB + BA$. If ψ satisfies

$$\gamma^\mu \partial_\mu \psi = 0 \tag{3}$$

then ψ is also a solution of (2). The solution set of the system of equation (2) might be larger than the solution space of equation (3). Here, we assume that the two solution spaces are the same. It is obvious that the solutions of (3) are also solutions of the wave equation, $\Box\psi = 0$.

3. The solution of the Dirac equation

The basis of the solution space is $\bigcup_s \left(\bigcup_{\mathbf{p}} \left(\left\{u^{(s)}(\mathbf{p}) \exp\left(\frac{i\mathbf{p}\cdot\mathbf{x}}{\hbar} - \frac{i|\mathbf{p}|ct}{\hbar}\right)\right\} \cup \left\{v^{(s)}(\mathbf{p}) \exp\left(\frac{-i\mathbf{p}\cdot\mathbf{x}}{\hbar} + \frac{i|\mathbf{p}|ct}{\hbar}\right)\right\}\right)\right)$. Therefore, $\psi(\mathbf{x}, t)$ is a linear combination of the basis.

$$\psi(\mathbf{x},t) = N \sum_{\mathbf{p}} \sum_{s=1,2} \left(b_{\mathbf{p}}^{(s)} u^{(s)}(\mathbf{p}) \exp\left(\frac{i\mathbf{p}\cdot\mathbf{x}}{\hbar} - \frac{i|\mathbf{p}|ct}{\hbar}\right) \right.$$
$$\left. + d_{\mathbf{p}}^{(s)\dagger} v^{(s)}(\mathbf{p}) \exp\left(\frac{-i\mathbf{p}\cdot\mathbf{x}}{\hbar} + \frac{i|\mathbf{p}|ct}{\hbar}\right)\right),$$

where N is a normalization constant. It is not necessary to discuss the anti-commutation relations of the operators $b_{\mathbf{p}}^{(s)}$ and $d_{\mathbf{p}}^{(s)\dagger}$ here. Let $\mathbf{p} = (p_1, p_2, p_3)$, we have
$u^{(1)}(\mathbf{p}) = \sqrt{\frac{1}{2}}\left(1, 0, \frac{p_3}{|\mathbf{p}|}, \frac{(p_1+ip_2)}{|\mathbf{p}|}\right)^T$, $u^{(2)}(\mathbf{p}) = \sqrt{\frac{1}{2}}\left(0, 1, \frac{(p_1-ip_2)}{|\mathbf{p}|}, \frac{-p_3}{|\mathbf{p}|}\right)^T$,
$v^{(1)}(\mathbf{p}) = \sqrt{\frac{1}{2}}\left(\frac{(p_1-ip_2)}{|\mathbf{p}|}, \frac{-p_3}{|\mathbf{p}|}, 0, 1\right)^T$, $v^{(2)}(\mathbf{p}) = -\sqrt{\frac{1}{2}}\left(\frac{p_3}{|\mathbf{p}|}, \frac{(p_1+ip_2)}{|\mathbf{p}|}, 1, 0\right)^T$.

4. The Maxwell field equation

It is also obvious that the Maxwell equation is the same as before, $\Box A^\mu = \frac{1}{c} j^\mu = q\overline{\psi}\gamma^\mu\psi$. Let the solution of the Maxwell equation to be $A^\mu = A_h^\mu + A_{par}^\mu$, where $\Box A_h^\mu = 0$, it will not be discussed here because it is well-known, and A_{par}^μ is the particular solution. Let $A_{par}^\mu = A_{par1}^\mu + A_{par2}^\mu + A_{par3}^\mu + A_{par4}^\mu$. From differential calculus,

$$\Box(\phi_2\gamma^\mu\phi_1) = (\Box\phi_2)\gamma^\mu\phi_1 + \phi_2\gamma^\mu(\Box\phi_2)$$
$$+2\left[\sum_{i=1}^{3}\left(\frac{\partial\phi_2}{\partial x^i}\gamma^\mu\frac{\partial\phi_1}{\partial x^i}\right) - \left(\frac{\partial\phi_2}{\partial(ct)}\gamma^\mu\frac{\partial\phi_1}{\partial(ct)}\right)\right].$$

By the identity listed above, it is obvious that

$$A_{par1}^\mu = \frac{\hbar^2 q}{2}\sum_{\mathbf{p}}\sum_{\mathbf{p}'}\frac{1}{w(\mathbf{p},\mathbf{p}')}\left(\sum_{s=1}^{2} b_{\mathbf{p}'}^{(s)\dagger}\overline{u}^{(s)}(\mathbf{p}')\right)\gamma^\mu\left(\sum_{s=1}^{2} b_{\mathbf{p}}^{(s)}u^{(s)}(\mathbf{p})\right)$$
$$\times \exp\left(\frac{i}{\hbar}((\mathbf{p}-\mathbf{p}')\cdot\mathbf{x} - (|\mathbf{p}|-|\mathbf{p}'|)ct)\right)$$

$$A^{\mu}_{par2} = \frac{\hbar^2 q}{2} \sum_{\mathbf{p}} \sum_{\mathbf{p}'} \frac{1}{w(\mathbf{p}, \mathbf{p}')} \left(\sum_{s=1}^{2} d^{(s)}_{\mathbf{p}'} \overline{v}^{(s)}(\mathbf{p}') \right) \gamma^{\mu} \left(\sum_{s=1}^{2} d^{(s)\dagger}_{\mathbf{p}} v^{(s)}(\mathbf{p}) \right)$$
$$\exp\left(\frac{-i}{\hbar}((\mathbf{p} - \mathbf{p}') \cdot \mathbf{x} - (|\mathbf{p}| - |\mathbf{p}'|)ct) \right)$$

$$A^{\mu}_{par3} = -\frac{\hbar^2 q}{2} \sum_{\mathbf{p}} \sum_{\mathbf{p}'} \frac{1}{w(\mathbf{p}, \mathbf{p}')} \left(\sum_{s=1}^{2} d^{(s)}_{\mathbf{p}'} \overline{v}^{(s)}(\mathbf{p}') \right) \gamma^{\mu} \left(\sum_{s=1}^{2} b^{(s)}_{\mathbf{p}} u^{(s)}(\mathbf{p}) \right)$$
$$\exp\left(\frac{i}{\hbar}((\mathbf{p} + \mathbf{p}') \cdot \mathbf{x} - (|\mathbf{p}| + |\mathbf{p}'|)ct) \right)$$

$$A^{\mu}_{par4} = -\frac{\hbar^2 q}{2} \sum_{\mathbf{p}} \sum_{\mathbf{p}'} \frac{1}{w(\mathbf{p}, \mathbf{p}')} \left(\sum_{s=1}^{2} b^{(s)\dagger}_{\mathbf{p}'} \overline{u}^{(s)}(\mathbf{p}') \right) \gamma^{\mu} \left(\sum_{s=1}^{2} d^{(s)\dagger}_{\mathbf{p}} v^{(s)}(\mathbf{p}) \right)$$
$$\exp\left(\frac{-i}{\hbar}((\mathbf{p} + \mathbf{p}') \cdot \mathbf{x} - (|\mathbf{p}| + |\mathbf{p}'|)ct) \right)$$

where $w(\mathbf{p}, \mathbf{p}') = p_1 p_1' + p_2 p_2' + p_3 p_3' - |\mathbf{p}||\mathbf{p}'|$.
There are singularities when $w(\mathbf{p}, \mathbf{p}') = 0$, which can be handled easily by the method of elementary differential equation. We will not show it here.

5. Conclusion and Discussion

This article opens another door to get into the quantum field theory. Hung-Ming Tsai , Po-Yu Tsai and Lu-Hsing Tsai have organized two papers. One of these papers suggests that[6]

$$\psi(\mathbf{x}, t) = N \sum_{\mathbf{p}} \sum_{s=1,2} f(\mu, E, T) \left(b^{(s)}_{\mathbf{p}} u^{(s)}(\mathbf{p}) \exp\left(\frac{i\mathbf{p} \cdot \mathbf{x}}{\hbar} - \frac{i|\mathbf{p}|ct}{\hbar} \right) \right.$$
$$\left. + \, d^{(s)\dagger}_{\mathbf{p}} v^{(s)}(\mathbf{p}) \exp\left(\frac{-i\mathbf{p} \cdot \mathbf{x}}{\hbar} + \frac{i|\mathbf{p}|ct}{\hbar} \right) \right)$$

where $f(\mu, E, T)$ are the thermal factors, such as $\left(\sqrt{\frac{1}{1+\exp\left(\frac{E-\mu}{k_B T}\right)}} \right)^{1/2}$ and $\left(\sqrt{\frac{1}{1-\exp\left(\frac{E-\mu}{k_B T}\right)}} \right)^{1/2}$, where E is the energy of a particle, μ is the chemical potential, k_B is the Boltzman constant and T is the temperature. The arguments are: ψ is a member of a Hilbert space whose basis is $\bigcup_s \left(\bigcup_{\mathbf{p}} \left(\left\{ u^{(s)}(\mathbf{p}) \exp\left(\frac{i\mathbf{p}\cdot\mathbf{x}}{\hbar} - \frac{i|\mathbf{p}|ct}{\hbar} \right) \right\} \cup \left\{ v^{(s)}(\mathbf{p}) \exp\left(\frac{-i\mathbf{p}\cdot\mathbf{x}}{\hbar} + \frac{i|\mathbf{p}|ct}{\hbar} \right) \right\} \right) \right)$. Therefore, ψ is a linear combination of the basis. Since ψ is associated with a system of fermions and A^μ is associated with a system of bosons, the collection of particles must obey their statistical laws respectively. There is no argument to support that all $b_{\mathbf{p}}^{(s)}$ and $d_{\mathbf{p}}^{(s)\dagger}$ operators are equally likely appearing in ψ. Similarly, all operators $a_{\mathbf{k}}$ and $a_{\mathbf{k}}^\dagger$ in A^μ are not equally likely either. After introducing the thermal factors, all anti-commutation relations are unchanged except $\{\psi_\beta^\dagger(x'), \psi_\alpha(x)\} = \delta_{\alpha\beta}F(x,x')$, where F is a eight variables regular function with very nice mathematical properties rather than a singular function.

Though it is very difficult to simplify the Dirac equation with the new quantization, it has been proved that the solution set of the new equation is the Hilbert space whose basis is $\bigcup_s \left(\bigcup_{\mathbf{p}} \left(\left\{u^{(s)}(\mathbf{p}) \exp\left(\frac{i\mathbf{p}\cdot\mathbf{x}}{\hbar} - \frac{i|\mathbf{p}|ct}{\hbar}\right)\right\} \cup \left\{v^{(s)}(\mathbf{p}) \exp\left(\frac{-i\mathbf{p}\cdot\mathbf{x}}{\hbar} + \frac{i|\mathbf{p}|ct}{\hbar}\right)\right\}\right)\right)$. The details will not be provided here.

Hopefully, with new quantization, all the divergences of quantum fields shall be removed without applying the renormalization theory. The other paper proves that there is a lower bound of the energy of a photon[7], that is, the energy of a photon must be greater than $B, B > 0$. From the results of these two papers, the catastrophes of QED are removed automatically. Since the thermal factor is a function of the temperature T, quantities derived from quantum field shall depend on the temperature T. It is expected that the experiment of the energy levels of Hydrogen atom might verify this prediction.

People will be surprised when he find that the famous Dirac equation is changed. Here, it should be stressed that nothing has been changed except the vector space in which ψ is a member. It becomes a new Hilbert space with new basis. All the meaningful operators, the momentum operators, the energy operators etc., are not changed because the Lagrangian of the field is not changed.

References

1. D. Lurie, *Particles and Fields*, Wiley, (1968).
2. J.J. Sakurai, *Advanced Quantum Mechanics*, Adison Wesley, (1967).
3. D. Griffiths, *Introduction to Elementary Particles*, Wiley, (1987).
4. J. Schwinger, *Phys. Rev.* **82**, 914.
5. J. Schwinger, *Phys. Rev.* **91**, 713.
6. Hung-Ming Tsai, Po-Yu Tsai and Lu-Hsing Tsai, "The recursive relation beteween a field eqaution and its quantization", will appear soon.
7. Hung-Ming Tsai, Po-Yu Tsai and Lu-Hsing Tsai, "Estimation of the lower bound of photon's energy emited by electrons", will appear soon.

PERTURBATIVE AND NONPERTURBATIVE RENORMALIZATION OF ANOMALOUS QUARK TRIANGLES

ARKADY VAINSHTEIN

William I. Theoretical Physics Institute, University of Minnesota,
116 Church St. SE, Minneapolis, MN 55455
vainshte@umn.edu

Anomalous quark triangles with one axial and two vector currents are studied in special kinematics when one of the vector currents carries a soft momentum. The anomalous longitudinal part of the triangle is not renormalized in the chiral limit according to the Adler-Bardeen theorem. We derive a new nonrenormalization theorem for the transversal part of the triangle. In difference with the longitudinal part this nonrenormalization holds on only perturbatively, nonperturbative effects break it.

1. Introduction

In this talk I report a derivation of new nonrenormalization theorem for the anomalous quark triangles with one axial and two vector currents.[1] In the special kinematics when one of the vector currents carries a soft momentum we find a perturbatively exact relation between transversal and longitudinal part of the triangle. A particular physical situation where the triangle diagrams enter in such kinematics occurs for the two-loop electroweak corrections to the muon anomalous magnetic moment. In Ref. 2 — the study which stimulated the present work — the nonrenormalization theorem for the transversal part was used to show an absence of gluon corrections to light quark loops.

Study of fermion triangle diagrams with one axial and two vector currents represents a remarkable story. The Adler-Bell-Jackiw anomaly in the divergence of axial current, the Adler-Bardeen nonrenormalization theorem, the Wess-Zumino effective action, calculation of the $\pi^0 \to \gamma\gamma$ amplitude, the 't Hooft consistency condition, and the solution of the U(1) problem give an incomplete list of acts where these triangles played a major role.

The famous Adler-Bardeen theorem[3] proves nonrenormalization for the longitudinal part of triangles associated with the divergence of the axial current. What can be said about the transversal part? We argue that in special kinematics when one of the vector currents carries a soft momentum the transversal part is unambiguously fixed by the longitudinal one in the chiral limit of perturbation theory. Such relation immediately proves an absence of perturbative corrections to the transversal part of fermion triangles in the kinematics considered.

The difference between longitudinal and transversal parts shows up at a nonperturbative level. The Operator Product Expansion (OPE) demonstrates this explicitly: in the chiral limit only the transversal part in the OPE contains nonleading operators.[2] Nonrenormalization of the longitudinal part, both perturbatively and nonperturbatively, constitutes the 't Hooft consistency condition,[4] i.e. the exact quark-hadron duality. In QCD this duality is realized as a correspondence between the infrared singularity of the quark triangle and the massless pion pole in terms of hadrons.[a] It is clear that for the transversal part this kind of duality cannot be exact in QCD: there is no massless particle contributing to the transversal part. Thus, the transversal part of the triangle with a soft momentum in one of the vector currents provides us with an interesting object: no perturbative corrections but nonperturbatively it is modified, this is an analogous to the quark condensate $\langle\bar{q}q\rangle$.

2. Hadronic corrections to quark triangles

Let us start with a definition of vector, j_μ, and axial, j^5_ν, currents,

$$j_\mu = \bar{q}\,V\gamma_\mu q, \qquad j^5_\nu = \bar{q}\,A\gamma_\nu\gamma_5 q\,, \tag{1}$$

where the quark field q^i_f has color (i) and flavor (f) indices and the matrices V and A are diagonal matrices of vector and axial couplings acting on flavor indexes. To avoid dealing with the U(1) anomaly in respect to gluon interactions we assume that $\mathrm{Tr}\,A = 0$. In the case of electroweak corrections one can view the vector current as an electromagnetic one with V being the matrix of electric charges and j^5_ν as the axial part of the Z boson current with matrix A given by the weak isospin projection.

The amplitude for the triangle diagram in Fig. 1 involving the axial current j^5_ν and two vector currents j_μ and $\tilde{j}_\gamma = \bar{q}\,\widetilde{V}\gamma_\mu q$ (for generality we

[a] A pioneering effort to analyze the axial current anomaly in terms of infrared singularity was made by Dolgov and Zakharov.[5]

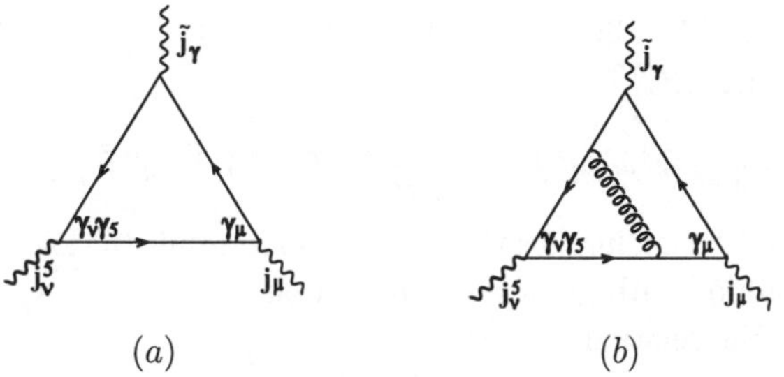

Figure 1. Quark triangle, diagram (a), and a gluon correction to it, diagram (b).

use a different matrix $\widetilde{V}$ for the soft momentum current) can be written as

$$T_{\mu\gamma\nu} = -\int \mathrm{d}^4x \mathrm{d}^4y\, \mathrm{e}^{iqx-iky} \langle 0|\, T\{j_\mu(x)\, \tilde{j}_\gamma(y)\, j^5_\nu(0)\}|0\rangle \,. \qquad (2)$$

We can view the current $\tilde{j}_\gamma$ as a source of a soft photon with the momentum k. Introducing a polarization vector of a soft photon $e^\gamma(k)$ we come to the amplitude $T_{\mu\nu}$

$$T_{\mu\nu} = T_{\mu\gamma\nu} e^\gamma(k) = i \int \mathrm{d}^4x\, \mathrm{e}^{iqx} \langle 0|\, T\{j_\mu(x)\, j^5_\nu(0)\}|\, \gamma(k)\rangle\,, \qquad (3)$$

which can be viewed as a mixing between the axial and vector currents in the external electromagnetic field.

It is clear that the expansion of $T_{\mu\nu}$ in the small momentum k starts with linear terms and we neglect quadratic and higher powers of k. There are only two Lorentz structures for $T_{\mu\nu}$ which are linear in k and consistent with the conservation of electromagnetic current,

$$T_{\mu\nu} = -\frac{i}{4\pi^2}\Big[w_T(q^2)\Big(-q^2 \tilde{f}_{\mu\nu} + q_\mu q^\sigma \tilde{f}_{\sigma\nu} - q_\nu q^\sigma \tilde{f}_{\sigma\mu}\Big) + w_L(q^2)\, q_\nu q^\sigma \tilde{f}_{\sigma\mu}\Big], \quad (4)$$

$$\tilde{f}_{\mu\nu} = \frac{1}{2}\,\epsilon_{\mu\nu\gamma\delta} f^{\gamma\delta}\,, \qquad f_{\mu\nu} = k_\mu e_\nu - k_\nu e_\mu\,.$$

Both structures are transversal with respect to vector current, $q^\mu T_{\mu\nu} = 0$. As for the axial current, the first structure is transversal with respect to q^ν while the second is longitudinal.

The one-loop result for the invariant functions w_T and w_L can be taken from the classic papers by Bell and Jackiw,[6] Adler,[7] and Rosenberg.[8] In the chiral limit

$$w_L^{1-\mathrm{loop}}[m=0] = 2\, w_T^{1-\mathrm{loop}}[m=0] = \frac{2N_c \,\mathrm{Tr}\,(A\, V\, \widetilde{V})}{Q^2}\,. \qquad (5)$$

Nonvanishing in the chiral limit, $m = 0$, the longitudinal part $q^\nu T_{\mu\nu}$ represents the axial anomaly,[6,7]

$$q^\nu T_{\mu\nu} = \frac{i}{4\pi^2}\, Q^2 w_L\, q^\sigma \tilde{f}_{\sigma\mu} = \frac{i}{2\pi^2}\, N_c \,\mathrm{Tr}\,(A\, V\, \widetilde{V})\, q^\sigma \tilde{f}_{\sigma\mu}\,, \tag{6}$$

and its nonrenormalization implies that the one-loop result (5) for w_L stays intact when interaction with gluons is switched on.

We claim that the relation

$$w_L[m=0] = 2\, w_T[m=0] \tag{7}$$

which holds at the one-loop level, see Eq. (5), gets no perturbative corrections from gluon exchanges. This follows from the following line of argumentation.

In the chosen kinematics the fermion triangle with $m = 0$ possesses a special feature: namely, a symmetry under permutation of indexes of axial and vector currents, $\mu \leftrightarrow \nu$. Indeed, in the triangle diagrams (a) and (b) in Fig. 1 one can move γ_5 from the axial vertex $\gamma_\nu\gamma_5$ to the vector vertex γ_μ. In the chiral limit it moves via even number of gamma matrices in any order of perturbation theory. Together with the change of the momentum $q \to -q$ (which does not affect $T_{\mu\nu}$) it shows the symmetry of the amplitude $T_{\mu\nu}$. The same can be derived from the $\mathrm{SU}(N_f)_L{\times}\mathrm{SU}(N_f)_R$ features of the amplitude.

At first glance the symmetry under the $\mu \leftrightarrow \nu$ permutation seems to be in contradiction with the general decomposition (4): the transversal part of $T_{\mu\nu}$ is antisymmetric, the longitudinal part has no symmetry, and there is no way to choose w_T and w_L which makes the $T_{\mu\nu}$ symmetric. Note, however, that the term $q^2 \tilde{f}_{\mu\nu}$ in the transversal structure in Eq. (4) actually produces a term in $T_{\mu\nu}$ which does not depend on q. It is because $w_T \propto 1/q^2$. The $\mu \leftrightarrow \nu$ symmetry holds for a singular in q part of $T_{\mu\nu}$ when the condition (7) relating w_T to w_L is fulfilled. The constant in q part is then fixed by the conservation of the vector current, $q^\mu T_{\mu\nu} = 0$. An independence on q for this part of the amplitude provides, in fact, an alternative proof of the Adler-Bardeen theorem. Indeed, gluon corrections would lead to logarithmic dependence on q instead of the constant.

Another way to be automatically consistent with the vector current conservation is to use the Pauli-Vilars regulators. Technically it reduces to subtraction from the triangles with massless quarks similar triangles with the heavy regulator fermions propagating on the loops. The regulator triangles produce terms which are polynomial in momenta, in our case terms linear in k and independent on q. Moreover, it is simple to see that these terms are antisymmetric under the $\mu \leftrightarrow \nu$ permutation. Indeed, in the propagator of the heavy regulator the leading term contains no gamma-matrix that leads to the sign change when γ_5 from the axial vertex $\gamma_\nu\gamma_5$ is moved to the vector vertex γ_μ.

Thus, we see that the crossing symmetry of the singular part in the triangle amplitude $T_{\mu\nu}$ leads to the relation (7) in perturbation theory. Nonrenormalization of w_L implies the same for w_T.

2.1. *Nonperturbative effects and OPE*

To study a nonperturbative effect in the triangle amplitude $T_{\mu\nu}$ one can use the OPE methods.[2] The OPE for the T-product of electromagnetic and axial currents at large Euclidean q^2 has the form

$$\hat{T}_{\mu\nu} \equiv i \int \mathrm{d}^4x \, \mathrm{e}^{iqx} \, T\{j_\mu(x) \, j_\nu^5(0)\} = \sum_i c^i_{\mu\nu\gamma_1\ldots\gamma_i}(q) \, \mathcal{O}_i^{\gamma_1\ldots\gamma_i} , \qquad (8)$$

where the local operators $\mathcal{O}_i^{\gamma_1\ldots\gamma_i}$ are constructed from the light fields and supplied by a normalization point μ separating short distances (accounted in the coefficients c^i) and large distances (in matrix elements of $\mathcal{O}_i$). The field can be viewed as light if its mass is less than μ. In the problem under consideration besides quark and gluon light fields include also the soft electromagnetic field A_μ which could enter in a form of the gauge invariant field strength $F_{\mu\nu} = \partial_\mu A_\nu - \partial_\nu A_\mu$.

The amplitude $T_{\mu\nu}$ is given by the matrix element of the operator $\hat{T}_{\mu\nu}$ between the photon and vacuum states.

$$T_{\mu\nu} = \langle 0 | \, \hat{T}_{\mu\nu} \, | \gamma(k) \rangle = \sum_i c^i_{\mu\nu\alpha_1\ldots\alpha_i}(q) \, \langle 0 | \, \mathcal{O}_i^{\alpha_1\ldots\alpha_i} \, | \gamma(k) \rangle . \qquad (9)$$

The leading (by a minimal dimension) is the $d = 2$ operator

$$\mathcal{O}_F^{\alpha\beta} = \frac{1}{4\pi^2} \, \tilde{F}^{\alpha\beta} = \frac{1}{4\pi^2} \, \epsilon^{\alpha\beta\rho\delta} \partial_\rho A_\delta , \qquad (10)$$

where $\tilde{F}^{\alpha\beta}$ is the dual of the electromagnetic field strength. The OPE coefficients $c^F_{L,T}$ for $\mathcal{O}_F^{\alpha\beta}$ in one loop coincide with expressions (5) for $w_{L,T}$,

In the chiral limit the next after leading operators are the d=6 operators of the type

$$\bar{q}\sigma^{\alpha\beta}\gamma_5 q \, \bar{q}q . \qquad (11)$$

Their OPE coefficients $c_T^{(6)} \propto 1/q^6$, $c_L^{(6)} = 0$ lead to $(\Lambda_{\rm QCD}/q)^4$ corrections to w_T but no corrections to w_L. The distinction between longitudinal and transversal parts showing up at the $d = 6$ level of four-fermion operators was firstly demonstrated in Ref. 9. Power corrections in w_T reflect nonvanishing masses of meson resonances contributing to the transversal part. This was used in Ref. 2 for construction of a resonance model for w_T consistent with the OPE constraints.

3. Conclusions

The quark triangles in special kinematics with one soft photon are similar to polarization operators: in this case it is a nondiagonal mixing of the axial and vector currents in the background of a soft vector field. In this kinematics we find that perturbative gluon corrections are absent in the chiral limit not only for the longitudinal part but for the transversal part as well. At the nonperturbative level the transversal part is corrected in contrast with the longitudinal one. In this respect the hadronic shift in the transversal part represents an object similar to the quark condensate: it appears only at nonperturbative level. This feature was emphasized recently in Ref. 10 where derivation of nonrenormalization theorems for the transversal part was extended to more general kinematics.

Acknowledgments

I am grateful to S. L. Adler, A. Czarnecki, G. Gabadadze, E. D'Hoker, W. J. Marciano, A. Schwimmer, M. Shifman and M. Voloshin for helpful discussions.

The work was supported in part by DOE grant DE-FG02-94ER408.

References

1. A. Vainshtein, Phys. Lett. B **569**, 187 (2003) [hep-ph/0212231].
2. A. Czarnecki, W. J. Marciano and A. Vainshtein, Phys. Rev. D **67**, 073006 (2003) [hep-ph/0212229].
3. S. L. Adler and W. A. Bardeen, Phys. Rev. **182**, 1517 (1969).
4. G. 't Hooft, in *Recent Developments In Gauge Theories,* Eds. G. 't Hooft *et al.*, (Plenum Press, New York, 1980).
5. A.D. Dolgov and V.I. Zakharov, Nucl. Phys. B **27**, 525 (1971).
6. J. S. Bell and R. Jackiw, Nuovo Cim. A **60**, 47 (1969).
7. S. L. Adler, Phys. Rev. **177**, 2426 (1969).
8. L. Rosenberg, Phys. Rev. **129**, 2786 (1963).
9. M. Knecht, S. Peris, M. Perrottet and E. De Rafael, JHEP **0211**, 003 (2002) [hep-ph/0205102].
10. M. Knecht, S. Peris, M. Perrottet and E. de Rafael, *New nonrenormalization theorems for anomalous three point functions,* hep-ph/0311100.

BICONFORMAL SUPERGRAVITY

JAMES T. WHEELER

Utah State University Department of Physics, 4415 Old Main Hill, Logan, UT 84322, jwheeler@cc.usu.edu

Biconformal supergravity is a new gauging of the superconformal group relevant to the Maldacena conjecture. Using the group quotient method to biconformally gauge SU(2,2—N), we generate a 16-dim superspace. We write the most general even-parity action linear in the curvatures. The bosonic sector is known to descend to general relativity on a 4-dim manifold.

1. Biconformal supergravity

For the past half decade, studies of M/string theory have been dominated by interest in the relationships between string on specified backgrounds and Yang-Mills gauge theories in lower dimensions. These investigations, triggered by Maldacena [1], include the conjecture that type IIB string on an $AdS_5 \times S^5$ background is dual to $N = 4$, $d = 4$ supersymmetric Yang-Mills theory. Since the isometry group of the manifold $AdS_5 \times S^5$ is the superconformal group, the corresponding IIB string theory could have a ghost-free, conformal supergravity theory as its low energy limit. Therefore, it is interesting to revisit – and extend – the set of conformal supergravity theories. Here we find an alternative to the classical conformal supergravity theories. The new theory has an action linear in superconformal curvatures, and therefore could in fact be ghost-free.

The first conformal supergravity theories were developed by Freund, Ferber and Crispim-Romao in one series of articles [2] and by Kaku, Townsend, van Nieuwenhuizen and Ferrara [3] in another series of articles. We use the group quotient menthod developed by Cartan [5] to construct a new class of models.

The group quotient method provides a systematic way to implement a given local symmetry as a gravity theory. One begins with a (graded) Lie group, G, containing the local symmetry (super)group of interest, H, as a sub-(super)group. Then the quotient G/H is a manifold with local H

symmetry. The cosets of H in G provide a projection from G to G/M, so the resulting structure is a principal fiber bundle with fibers isomorphic to H.

We begin with the superconformal group and require H to contain the homothetic group (Lorentz transformations and dilatations). We also demand that H contain the full $U(n)$ internal symmetry. The group and subgroup symmetries then determine the dimension of the physical spacetime to be $d = \dim(G/H)$. The physical space may be any manifold consistent with the local structure of the principal fiber bundle $(G, G/H)$. The action may be any H-invariant d-form. Using this method, the group structure alone gives a general class of geometries with the desired symmetries, including dimension of physical space, the form of the curvatures, and the relevant fields of the theory. From the curvatures it is straightforward to write the most general linear action functional.

A distinguishing feature of our work arises from our use of the homothetic group (Lorentz transformations and dilatations) for H. This is not new in itself. However, applying the Cartan method leads uniquely to an 8-dim symplectic space. This *biconformal gauging* of the conformal group, which now permits actions linear in the curvatures, gives a new class of models.

In the first use of an 8-dim conformal gauging, Ivanov and Niederle [7] resticted the 4 new dimensions to a submanifold generated by conformal transformations, effectively.making these directions gauge degrees of freedom. Biconformal gauging [8] treats all 8-dim equally. The resulting space is symplectic, so we employ a Hamiltonian interpretation. Wehner and Wheeler showed that because the biconformal volume form is scale invariant, it is possible to write a gravity action which is linear in the curvatures [9]. The first order Hamiltonian formulation, the 8-dim space and the linear action combine to preserve the correct number of degrees of freedom for the gravity problem. Assuming minimal [9] or vanishing [8] torsion, the field equations reduce in a particular set of gauges to the Einstein equation on an 4-dim submanifold. To clarify the meaning of the symplectic structure, biconformal gauging was applied to a globally-conformally-invariant generalization of Newton's second law [10]. The result, when the integral of the dilatational gauge field is identified with the action, is Hamiltonian dynamics. This strengthens our confidence and understanding of the interpretation of relativistic biconformal spaces as relatives of phase space.

The goal of the present work is to generalize the biconformal construction to the superconformal group.

The structure of the superconformal group, $SU(2,2|N)$ is well known [2,11]. To construct $SU(2,2|N)$ we demand that the generators of the graded

Lie algebra preserve a complex super-metric, $\mathcal{H}$, diagonally composed of a 4-dim Hermitian matix, Q and an N-dim, anti-hermitian, (symmetric) matrix P. For a super-transformation of the form $T = \begin{pmatrix} A & B \\ C & D \end{pmatrix}$, we obtain conditions on A, B, C and D. Using a spinor representation for conformal Lie algebra, and choosing $Q = -i\gamma^5$, the invariance condition for the A-sector bosonic generators selects a basis for the Lie algebra $su\,(2,2)$. The D-sector generators preserve the anti-Hermitian form P, which we take to be symmetric, $P^t = P$. Therefore, the internal symmetry is $U\,(N)$. Among the bosonic symmetries are two commuting generators: $i1_4$ in the A-sector and $i1_N$ in the D-sector, where 1_4 and 1_N denote the 4- and N-dim identies, respectively. By demanding vanishing superdeterminant for the superconformal group, we eliminate these in favor of the supertraceless combination $E = -\frac{i}{4} diag\left(1_4, \frac{4}{N} 1_N\right)$. The generator E functions as a central charge in the conformal subalgebra and appears in one fermionic anticommutator. There are therefore a total of $N^2 + 16$ bosonic generators and $8N$ fermionic generators. The B sector provides the fermionic generators, with C determined in terms of B.

A central issue in gauging the group is to find a suitable generalization of the final local symmetry subgroup, H. This choice determines the dimension and local character of the physical superspace. In the bosonic case of biconformal gauging [7,8] H is the homogeneous homothetic group, consisting of Lorentz transformations and dilatations only. We therefore require a sub-supergroup H of the entire supergroup such that (1) the A-sector of the bosonic part of H is homothetic and (2) the D-sector of the bosonic part of H must be $U\,(N)$, thereby retaining the entire D-sector as a local internal symmetry. There are three possible choices for H satisfying these two conditions. Two of these are mathematically equivalent, so there are two distinct group quotients that give rise to the desired homothetic bosonic fiber symmetry:

(1) No nonzero spinor B survives in the local symmetry. All fermionic degrees of freedom are then coordinate degrees of freedom, so we have a 16-dim superspace (spanned by ω^a, ω_a, χ and ψ) with local homothetic and $U(N)$ symmetry. The supervolume form is dimensionless.

(2) Only left handed spinors B provide local symmetries. Half of the fermionic degrees of freedom lie on the fiber and half become coordinate degrees of freedom. The volume form has nonzero scaling

dimension and there is no evident real-valued action linear in the curvatures.

The third case is like case 2, but with right handed spinors. The first case, in which all fermionic degrees of freedom become superspace coordinates, has the dimensionless volume form characteristic of biconformal gauging. We examine this case. The local symmetry includes Lorentz transformations, dilatations, the central charge E, and $U(N)$ transformations. The quotient G/H is a 16-dim superspace.

With the relevant groups in hand, we turn to the construction of the geometry. Introducing the dual basis of connection one-forms,

$$\omega^{\Pi} \in \left\{\omega_b^a, \omega_a, \omega^a, \omega, \psi_\beta^B, \chi_\beta^B, \alpha, \pi_R^{\rho\sigma}, \pi_I^{\rho\sigma}\right\}$$

we proceed to write the expressions for the curvatures. We find expressions in agreement with other authors for nine independent curvature 2-forms: the Lorentz curvature, $\mathbf{\Omega}_b^a$, the torsion and co-torsion, $\mathbf{\Omega}^a$ and $\mathbf{\Omega}_a$, the dilatational curvature $\mathbf{\Omega}$, the fermionic curvature $\mathbf{\Theta}_\beta^B$ and its conjugate, and the curvatures of the internal symmetries, $\mathbf{\Pi}_{\rho\sigma}^R, \mathbf{\Pi}_{\rho\sigma}^I$, and $\mathbf{A}$. Each curvature is a 2-form in the horizontal basis forms $\left\{\omega_a, \omega^a, \psi_\beta^B, \chi_\beta^B\right\}$.

The most general, even-parity, homothetic gauge-invariant action linear in the curvatures for the case of a $U(1)$ internal symmetry follows directly:

$$\begin{aligned} S = \int \Big\{ & \left(\alpha_1 \mathbf{\Omega}_b^a \left[\sigma_a^b\right]_{AB} + \left(\alpha_2 \mathbf{\Omega} + \alpha_3 \mathbf{A}\right) Q_{AB}\right) \chi_\alpha^A \psi_\beta^B P^{\alpha\beta} \left(\chi\psi\right)^2 \phi_b \\ & + \left(\alpha_4 \mathbf{\Omega}_b^a + \left(\alpha_5\, \mathbf{\Omega} + \alpha_6 \mathbf{A}\right) \delta_b^a\right) \varepsilon_{acde} \varepsilon^{bfgh} \omega_{fgh} \omega^{cde} \phi_f \\ & + \alpha_7 (\mathbf{\Omega}^m \left[\gamma_m\right]_{BD}\ \chi_\alpha^B \chi_\beta^D - \mathbf{\Omega}_m \left[\gamma^m\right]_{BD} \psi_\alpha^B \psi_\beta^D) P^{\alpha\beta} \left(\chi\psi\right)^2 \phi_b + \alpha_8 \Phi \\ & + \beta \left(\mathbf{\Theta}_\alpha^M \left[\gamma^a\right]^A{}_M \omega_b \psi_\beta^B - \overline{\mathbf{\Theta}}_\alpha^M \left[\gamma_b\right]^A{}_M \omega^a \chi_\beta^B\right) \\ & \times Q_{AB} P^{\alpha\beta} \left(\chi\psi\right)^3 \varepsilon_{acde} \varepsilon^{bfgh} \omega_{fgh} \omega^{cde} \\ & + \lambda_1 \left(\mathbf{\Pi}_{\alpha\lambda}^R + \mathbf{\Pi}_{\alpha\lambda}^I\right) \chi_\beta^A \psi_\rho^B P^{\lambda\rho} P^{\alpha\beta} Q_{AB} \left(\chi\psi\right)^3 \delta_b^a \varepsilon_{acde} \varepsilon^{bfgh} \omega_{fgh} \omega^{cde} \phi_f \\ & + \lambda_2 \left(\mathbf{\Pi}_{\alpha\lambda}^R + \mathbf{\Pi}_{\alpha\lambda}^I\right) \left[\gamma_5\right]_{BD} \psi_\rho^B \chi_\beta^D P^{\rho\lambda} P^{\alpha\beta} \left(\chi\psi\right)^2 \phi_b \Big\} \end{aligned} \tag{1}$$

where

$$\left(\chi\psi\right)^n = \left(\chi_\alpha^A Q_{AB} P^{\alpha\beta} \psi_\beta^B\right)^n$$

for any integer, n and where $\alpha_1, \alpha_2 ... \alpha_8, \beta, \lambda_1$ and λ_2 are arbitrary constant coefficients. The most general, linear, odd parity action may be found in [6]. This action is constructed purely from the geometry without auxiliary fields.

Gauge transformations do not mix the bosonic and fermionic curvatures. This fact gives raises the question: where is the supersymmetry? The answer is in the field equations. The field equations relate components of the bosonic curvatures, $\mathbf{\Omega}_a, \mathbf{\Omega}^a, \mathbf{\Omega}, \mathbf{\Omega}^a_b, \mathbf{A}$, to those of the fermionic curvatures $\mathbf{\Theta}^A, \overline{\mathbf{\Theta}}^A$. Under coordinate transformations the components of the two types of curvatures mix. Thus, supersymmetry is carried by supercoordinate transformations in the 16-dim superspace.

Among the equations of motion following from eq.(1), one contains the expected Rarita-Schwiinger term:

$$[\gamma^a]^M{}_L \Theta^L{}_{an} \sim \gamma^a \left(\partial_a \psi_n - \partial_n \psi_a\right) \sim \partial\!\!\!/ \psi_n$$

Thus, cases with $\alpha_1 = \alpha_2 = \alpha_8 = 0$ give the massless Dirac equation for a spin-3/2 particle.

These results are important for several reasons:

(1) The $N = 5$ case is a gauging of the $AdS_5 \times S^5$ background of the Maldacena conjecture, and may provide a low energy limit of string theory.
(2) The large class of actions linear in the supercurvatures directly imply (1) GR-type gravity theory and (2) Dirac-type and Rarita-Schwinger-type spinor equations.
(3) Supersymmetrization introduces matter systematically into biconformal space.
(4) Our use of Cartan's group manifold methods gives superspace automatically.

Numerous papers [2,3] examine conformal supergravity in four dimensions. Most use actions quadratic in the curvatures, typically making use of the MacDowell-Mansouri approach. All of these papers except Crispim Romão et al. [2], use 4-dim integrals, so the local symmetry is the Poincaré group. The gauge transformations then mix all of the curvatures. This is in sharp constrast to our action, which requires only local Lorentz and dilatational invariance.

Our action displays other important differences from previous ones. Previous actions involve curvature 2-forms over 4-dim spacetime, or over 8-dim superspace. By contrast, the biconformal action is expanded in terms of eight *bosonic* coordinates and *sixteen* fermionic coordinates. When the curvatures 2-forms are expanded, each of the nine curvatures contains ten H-invariant tensors among its components.

Only Crispim Romão et al. and Castellani et al. [3] claim actions linear in the curvatures. This is accomplished using auxiliary H-invariant tensors that arise by reducing the local symmetry from co-Poincaré to homothetic, combined with the demand that $R^A = 0$ solve the field equations. It seems likely that these theories follow from the biconformal approach by integrating over the extra 4 coordinates.

References

1. Maldacena, J., Adv. Theor. Math. Phys. 2 (1998) 231-252; Int. J. Theor. Phys. 38 (1999) 1113-1133.
2. Freund, Peter G. O., J. Math. Phys., Vol. 17, No. 3 (1976) 424; Ferber, Crispim Romão, J., A. Ferber and Peter G. O. Freund, Nuc. Phys. B**126** (1977) 429-435; A. and P. G. O. Freund, Nuc. Phys. B**122** (1977) 170-176; Ferber, A., Nuc. Phys. B132 (1978) 55.
3. Kaku, M., P.K. Townsend and P. van Nieuwenhuizen, Phys. Lett. **69**B (1977) 304; Ferrara, S., M. Kaku, P.K. Townsend and P. van Nieuwenhuizen, Nuc. Phys. B**129** (1977) 125-134; Kaku, M., P.K.Townsend and P. van Nieuwenhuizen, Phys. Rev. Lett. 39, No. 18 (1977) 1109; Kaku, M., P.K.Townsend and P. van Nieuwenhuizen, Phys. Rev. D**17** (15 June, 1978) 853; Townsend, P. K. and P. Van Nieuwenhuizen, Phys. Rev. D, Vol 19, No. 10 (1979) 3166; Castellani, L., P. Fré and P. van Nieuwenhuizen, Annals of Physics 136 (1981) 398-434.
4. Ne'eman, Yuval and Tuillio Regge, Phys. Lett. 74B, no. 1,2 (1978) 54; Ne'eman, Y. and T. Regge, Riv. Nuovo Cimento **1** (1980), 1.
5. Kobayashi and Nomizu, *Foundations of Differential Geometry* (Wiley, New York, 1963).
6. Anderson, Lara B. and James T. Wheeler, hep-th/0309111.
7. Ivanov, E. A. and J. Niederle, Phys. Rev. D25, No 4 (1982) 988; Ivanov, E. A. and J. Niederle, Phys. Rev. D25, No 1 (1982) 976.
8. Wheeler, James T., J. Math. Phys., 39 (1) (1998) 299-328.
9. Wehner, A. and James T. Wheeler, Nuc. Phys. B557 (1999) 380-406.
10. Wheeler, James T., ArXiv hep-th/0305017.
11. Freund, P. G. O. and I. Kaplansky, J. Math. Phys., Vol 17, No. 2 (1976) 228-231; Kac, V. G., Commun. Math. Phys. 53 (1977) 31-64; Kac, V. G., Adv. Math. 26 (1977) 8-96.

VI. STRING THEORY

ON D-BRANE TOPOLOGY CHANGE*

GARNIK G. ALEXANIAN

Physics Department, Syracuse University, Syracuse, New York 13244

A.P. BALACHANDRAN

Physics Department, Syracuse University, Syracuse, New York 13244

PEDRO J. SILVA

Physics Department, Syracuse University, Syracuse, New York 13244

A study of the relation between topology change, energy and Lie algebra representations for fuzzy geometry in connection to M-theory is presented. We encounter two different types of topology change, related to the different features of the Lie algebra representation appearing in the matrix models of M-theory. From these studies, we propose a new method of obtaining non-commutative solutions for the non-Abelian D-brane action found by Myers. This mechanism excludes one of the two topology changing processes previously found in other non-commutative solutions of many matrix-based models in M-theory i.e. in M(atrix) theory, Matrix string theory and non-Abelian D-brane physics.

1. Introduction

During the last few years we have seen how non-commutative geometry has come to play an important role in string theory. It appears not only at the fundamental Planck distances where a smooth geometry can not be trusted, but also at the level of effective theories in D-brane physics, where the Chan-Paton factors result in matrix degrees of freedom. In these cases the effective Lagrangian of N-Dp-branes comes with built-in non-commutative features in the form of matrix models. Myers [1] proposed an effective action for these non-Abelian D-branes by demanding consistency with t-duality among the different Dp-branes. In particular, he started from the well known D9-brane action and proceeded by t-dualizing. The agreement with the weak background actions of Taylor and Van Raamsdonk [2] was used as

*Talk presented by G.G.A.

a consistency check. Note that these linearized actions came from a very different theoretical framework associated with the BFSS M(atrix) theory proposal [3].

This remarkable characteristic of the built-in non-commutativity is not new in the framework of M-theory. We already have at least two other examples where this type of construction is found, namely M(atrix) theory [3] and Matrix string theory [4,5,6,7,8].

An essential characteristic of the matrix actions is their capability to describe non-commutative geometries that correspond to different extended objects of the theory. For example, higher dimensional Dq-branes may be formed by smaller Dp-branes ($q > p$). To be more precise, consider the dielectric effect [1] where N Dp-branes form a single $D(p+2)$-brane generating a configuration corresponding to a non-commutative two-sphere (fuzzy sphere [9,10]). Another kind of construction corresponds to N D1-branes forming n D5-branes ($n < N$), this time using a fuzzy four-sphere [a,11]. Also it is worth mentioning that there are other non-commutative manifolds (apart from the fuzzy n-spheres) which are relevant to matrix models, e.g. tori, CP(n), RP(n), etc.

In this talk we will report on the main results of [12], where we have studied the relations between the discrete partonic picture (the non-commutative picture) and the smooth geometry that is obtained from it in the limit of large number of partons, i.e. the "reconstruction" of the geometry.

2. Quantum geometry and new solutions with the obstractions to topology change

In the original calculation [1], Myers considered N D0-branes in a constant four-form Ramond-Ramond (RR) field strength background of the form $F^{(4)}_{t123} = -2f\epsilon_{123}$. The relevant D0-brane action is given by

$$S_{Do} = -\mu_0\lambda^2 \int dt \left(\frac{1}{2}\partial_t\Phi^i\partial_t\Phi_i + \frac{1}{4}[\Phi,\Phi]^2 + \frac{i}{3}F^{(4)}_{tijk}\Phi^i\Phi^j\Phi^k\right), \qquad (1)$$

where Φ^i are matrix-valued scalars in the adjoint representation of the $U(N)$; they represent the nine directions transverse to the D0-brane, $i = (1..9)$, μ_0 is the charge of the D0-brane and $\lambda = 2\pi\alpha'$ (α' is the string length squared). For static configurations the kinetic term vanishes and center of

[a]This particular type of quantum geometry has been used extensively, see for example [13,14,7].

mass degrees of freedom decouple. Hence, variation of the action gives the following polynomial equation in Φ,

$$[\Phi^j, [\Phi_j, \Phi_i]] + if\epsilon_{ijk}[\Phi^j, \Phi^k] = 0. \tag{2}$$

This is solved by setting to zero all but the first three scalars Φ^i ($i = (1,2,3)$) which are replaced by Lie algebra generators T^i (in this case $su(2)$) times a scalar r. Then the above equation becomes an equation for r,

$$r - \frac{f}{2} = 0. \tag{3}$$

Therefore, three of the scalars Φ^i are of the form $\Phi^i = f/2T^i$. The non-commutative geometry appears since these Φ also correspond to the first three cartesian coordinates transverse to the brane.

However, as it is explained in appendix B of [12], the above equations do not fully define the quantum geometry. Still, it is necessary to fix the representation of T^i as different representations will give different solutions and topologies. Each of these solutions has a characteristic energy that is a function of the quadratic Casimir given by,

$$V_N = -\frac{\mu_0 \lambda^2 f^2}{6} \sum_{i=1}^{3} \mathrm{Tr}[\Phi^i \Phi_i]. \tag{4}$$

Given a fixed size of the representation N, the irreducible representation corresponds to the lower bound (strictly speaking this is the fuzzy sphere), while the reducible representations have higher energies corresponding to more complicated topologies of the products of fuzzy spheres. An important characteristic of the above construction is that in the large N limit the algebra of functions defined on these solutions becomes the algebra of functions on the classical manifold (see appendix B of [12]).

This example contains all of the ingredients that define the process of finding the non-commutative solutions. Generalizations of this program have appeared, but the underlying structure is the same. These solutions are usually called "fuzzy spaces", although not all of them are properly well defined, the best known example being the so called fuzzy four-sphere[b].

We would like to emphasize two important aspects of these solutions. First, the representation in which the different scalars Φ^i act is not fixed by the equations of motion. Second, the existence of the possibility of the

---

[b] In this case it is known that the algebra of functions defined on the "fuzzy S^4" does not close and some extra structure will be needed to properly define the quantum geometry [16].

topology change process suggested by the natural decay of higher energy solutions into lower energy ones. This cascade process has already received some attention in [17], the main result being the discovery of unstable modes that trigger the topology change. These modes are related to the relative position of the different fuzzy manifolds that appear in the product of higher energy solutions. Hence, given a solution A we can always construct another solution (of higher energy and more complicated topology) by considering larger matrices of two or more copies of A. The different topology of the above type of solutions comes from the fact that they correspond to reducible representations.

Nevertheless, it is important to note that this is not the only way of obtaining topology change. There is also the possibility of having different quantum geometries defined in the same group structure, which are not related by reducible-irreducible relations. For example, $SU(4)$ contains many different fuzzy geometries among which we have $(CP(3), CP(2), S^2 \times S^2)$ (see appendix B of [12]).

Regarding the first aspect, it is obvious that if the representation is fixed by the equations of motion there is no room for topology change. The fact that we can choose the representation signals the existence of a degeneracy in the set of quantum geometries, a symmetry that allows the topology change. What we have seen in this section is that from the point of view of the D-brane physics, the equations of motion only define the group structure, while energy is (in all of the solutions found) the only quantity that differentiates between quantum geometries and therefore "chooses" the classical geometry in the large N limit via the decay processes.

Therefore, we have a clear relation between topology change and the fact that the equations of motion do not fix the representation of the group. Although all of the solutions that currently exist share this behavior, there is no reason to believe that there are no circumstances in which the equations of motion can fix the representation and hence preclude topology changing processes from happening. To investigate this matter we need to understand in greater depth the definition of fuzzy geometry and the relations between fuzzy coordinates, representations and invariants of the algebraic structure.

An intuitive way to understand a fuzzy geometry [18] is to define it by a modification of the algebra of functions on classical geometry. For example, take the functions on the sphere. It will be enough to consider the spherical harmonics. In a classical case there is an infinite number of these. Then truncate the basis at a given angular momentum j by projecting out all the higher angular momentum modes. The resulting algebra will represent basically the fuzzy sphere.

An important mathematical point is that this algebra of functions can be obtained from the symmetric irreducible representations of size $2j + 1 = N$ of $su(2)$ using coherent-state techniques (see appendix B of [12]). Actually, there exist a precise relation between cartesian coordinates in R^3 (which defines the embedding of the sphere in R^3) and the $su(2)$ Lie algebra generators in the corresponding representation. These are the three matrices T^i appearing in the dielectric effect of Myers (note that $\mathrm{Tr}(T^i T_i) = NR^2$, with R equal to the radius of the fuzzy sphere).

An interesting generalization of the fuzzy sphere with rich enough algebraic structure is the 2n-dimensional fuzzy complex projective planes (fuzzy $CP(n)$) [15]. These quantum geometries are strongly related to $su(n)$. It turns out that to define fuzzy $CP(n)$[c] one needs to consider only the *symmetric* representations of $SU(3)$ for which the following equation is sufficient,

$$d^{ijk} T_j T_k = w(N) T^i, \tag{5}$$

where d_{ijk} is the only invariant rank three symmetric tensor in $su(n+1)$, and $w(N)$ is related to the representation used. We can now show how to find a concrete example where the D-brane equations of motion include an invariant tensorial equation (like equation 5) that determines the representation and hence enforces an obstruction to the topology change. In doing so we will use equation (5) as a hint and will search for a simple configuration where fuzzy $CP(2)$[d] could appear.

In order to do this, consider N D1-branes with a constant world-volume electric field $F_{[2]}$ in the presence of a five-form RR field strength F_{ijklm}, flat metric constant dilaton and zero B-field. The energy density for this system is obtained by expanding the non-Abelian action proposed by Myers[e] followed by the Hamiltonian transformation. Here we show the final form, once we have restricted our study to static and constant configurations.

$$\begin{aligned} E(\Phi, F_{[2]}, F_{[5]}) = \mu_1 \lambda^2 & \left[\frac{1}{4} \Phi^{ij} \Phi_{ji} - \frac{1}{4} F_{[2]}^2 \right] \\ & + \mu_1 \lambda^4 \left[\frac{1}{16} (\Phi^{ij} \Phi_{ji})^2 - \frac{1}{8} \Phi^{ij} \Phi_{jk} \Phi^{kl} \Phi_{li} \right. \\ & \left. - \frac{1}{16} F_{[2]}^2 \Phi^{ij} \Phi_{ji} + \frac{1}{10} \Phi^i \Phi^j \Phi^k \Phi^l \Phi^m F_{mlkji} F_{\tau\sigma} \right], \end{aligned} \tag{6}$$

[c] Construction of fuzzy $CP(n)$ is presented in appendix B of [12].
[d] This is smallest non-trivial $CP(n)$ geometry, $CP(1)$ is the sphere.
[e] see appendix A of [12] for a detail derivation and conventions.

where we have used the convention $\Phi^{ij} = [\Phi^i, \Phi^j]$ and $i = (1,..,8)$. We also set the the five-form to be

$$F_{ijklm} = h\, d^{pq}{}_{[i} f^p{}_{jk} f^p{}_{lm]} \,, \tag{7}$$

where "h" represents the strength of the RR field. Note that this is the only invariant tensor with five indices in $su(3)$ ($F_{[5]}$ is related to $(U^{-1}\, dU)^5$, the only closed 5-form in $SU(3)$).

To find the extremal points of this potential we will use configurations such that each Φ^i is proportional to the generator of $su(3)$, T^i i.e.

$$\Phi^i = \rho T^i, \tag{8}$$

The detailed form of the equations of motion can be found in appendix A of [12]. These equations are complicated and do not shed extra light on the discussion. For our purposes it is enough to show the general structure. After some algebra, one can show (see appendix A of [12]) that they are similar to (5)

$$T^i = w\, d^i_{\ jk} T^j T^k, \tag{9}$$

where w is a function of $(N, h, F_{[2]}, \rho)$. These equations can only be solved if the $su(3)$ generators are in the specific representation of appendix B of [12]. Then, once this representation is chosen, the above expression becomes an algebraic equation defining ρ as a function of $(F_{[2]}, h, N)$. Also, all the matrix products in the field equations simplify. To study the stability of these solutions in terms of the variable ρ, it is sufficient to substitute the ansatz back into the expression for the energy density

$$E = \mu_1 \lambda^2 N \Bigg[-\tfrac{1}{4}F^2 + \tfrac{3}{16}(4 - \lambda^2 F^2) c_2 \rho^4 + \tfrac{9\lambda^2}{32}(c_2 - 1) c_2 \rho^8 + \tfrac{\lambda^2}{40} F_{\tau\sigma} h (2n+3) c_2 \rho^5 \Bigg],$$

where $N = (n+1)(n+2)/2$, $c_2 = (n^2+3n)/3$ is the quadratic Casimir of the symmetric representation of T^i and n is an arbitrary positive integer.

Clearly, this potential has a global minimum at some value of ρ that we will call ρ_0. It depends on the value of the electric field $F_{[2]}$, the strength of the background RR field h and N. In particular, for $F_{\tau\sigma} = \sqrt{2}$ and $\lambda = 1$ the expressions for ρ_0 and radius of S^7 in which it is embedded simplify and can are given by

$$\rho_0 = \left(\frac{h(2n-3)}{9\sqrt{2}(c_2-1)}\right)^{1/3},$$

$$R = \lambda\sqrt{Tr(\Phi^i\Phi_i)/N} = \left(\frac{h(2n-3)c_2^{3/2}}{9\sqrt{2}(c_2-1)}\right)^{1/3} \tag{10}$$

that in the large N limit gives $R \sim \frac{\sqrt{2}h}{27}n^{2/3}$. In the above expression we can see how the radius increases with the number of D1-branes and the strength of the RR field. Also, note that in order to achieve this effect we needed a nontrivial electric field on the D1-brane, so that we have fundamental strings diluted into the D1-brane as a requirement.

Therefore, we have found new type of solutions corresponding to N D1-branes forming a fuzzy $CP(2)$, with an obstruction for the topology change. In the large N limit this configuration goes over a D5-brane with topology $R^{(1,1)} \times CP(2)$.

In fact, we can check this correspondence by looking at the different couplings of the non-abelian D1-branes to various RR fields. For example, using solution (8) one can show that in the large N limit it takes the form of the coupling of a single D5-brane [12].

There exist other types of examples of non-Abelian D-branes forming higher dimensional D-branes, where the resulting geometry is a fuzzy CP(2) manifold. These cases however, are different in nature from the one presented here since the resulting fuzzy geometry is determined by the lowest energy condition and not by the equations of motion. The dual picture corresponding to D-branes with CP(2) topology has already been studied, we refer the reader to [14] for further information.

Acknowledgments

We would like to thanks V.P. Nair for useful discussions. The work of PJS was supported in part by NSF grant PHY-0098747 to Syracuse University and by funds from Syracuse University. GA and APB were supported in part by the DOE and NSF under contract numbers DE-FG02-85ER40231 and INT-9908763 respectively.

References

1. R.C. Myers, JHEP 9912 (1999) 022, hep-th/9910053.

2. W. Taylor, M. Van Raamsdonk, Nucl.Phys. B558 (1999) 63-95, hep-th/9904095. W. Taylor, M. Van Raamsdonk, Nucl.Phys. B573 (2000) 703-734, hep-th/9910052.
3. T. Banks, W. Fischler, S.H. Shenker and L. Susskind, Phys.Rev. D55 (1997) 5112-5128, hep-th/9610043. L. susskind, hep-th/9704080.
4. R. Dijkgraaf, E. Verlinde, H. Verlinde, Nucl.Phys. B500 (1997) 43-61, hep-th/9703030.
5. Tom Banks, Nathan Seiberg, Nucl.Phys. B497 (1997) 41-55, hep-th/9702187. R. Dijkgraaf, E. Verlinde, H. Verlinde, Nucl.Phys.Proc.Suppl. 62 (1998) 348-362, hep-th/9709107. Lubos Motl, hep-th/9701025.
6. Ricardo Schiappa, Nucl.Phys. B608 (2001) 3-50, hep-th0005145.
7. Pedro J. Silva, JHEP 0202 (2002) 004,hep-th/0111121.
8. D Brecher, B Janssen and Y Lozano, hep-th/0112180.
9. J. Madore,*An Introduction to Noncommutative Differential Geometry and its Applications*,Cambridge University Press,
10. H. Grosse and P. Prešnajder, *Lett.Math.Phys.* **33**, 171 (1995) and references therein; H. Grosse, C. Klimčík and P. Prešnajder, *Commun.Math.Phys.* **178**,507 (1996); **185**, 155 (1997); H. Grosse and P. Prešnajder, *Lett.Math.Phys.* **46**, 61 (1998) and ESI preprint, 1999; H. Grosse, C. Klimčík, and P. Prešnajder, hep-th/9602115 and *Commun.Math.Phys.* **180**, 429 (1996); H. Grosse, C. Klimčík, and P. Prešnajder, in *Les Houches Summer School on Theoretical Physics*, 1995, hep-th/9603071; P. Prešnajder, hep-th/9912050 and *J.Math.Phys.* **41** (2000) 2789-2804; Cambridge, 1995; gr-qc/9906059.
11. Neil R. Constable, Robert C. Myers, Oyvind Tafjord, JHEP 0106 (2001) 023, hep-th/0102080.
12. G.Alexanian, A.P.Balachandran and P.Silva, hep-th/0207052.
13. J. Castelino, S. Lee, W. Taylor, Nucl.Phys. B526 (1998) 334-350, hep-th/9712105.
14. S. P. Trivedi, S. Vaidya, JHEP 0009 (2000) 041, hep-th/0007011.
15. G.Alexanian, A.P.Balachandran, G.Immirzi, B.Ydri, J.Geom.Phys. 42 (2002) 28-53, hep-th/0103023.
16. P. Ming Ho and S Ramgoolam, Nucl.Phys. B627 (2002) 266-288, hep-th/0111278.
17. D. P. Jatkar, G. Mandal, S. R. Wadia, K.P. Yogendran, JHEP 0201 (2002) 039, hep-th/0110172. K. Hashimoto, hep-th/0204203. J. Madore, L.A. Saeger, Class.Quant.Grav. 15 (1998) 811-826, gr-qc/9708053.
18. M. R. Douglas and N. A. Nekrasov, "Noncommutative field theory," Rev. Mod. Phys. **73**, 977 (2002) hep-th/0106048; A. Connes, "Noncommutative geometry: Year 2000," math.qa/0011193; A. P. Balachandran, "Quantum spacetimes in the year 1," hep-th/0203259.

KÄHLER VERSUS NON-KÄHLER COMPACTIFICATIONS*

MELANIE BECKER
Department of Physics, University of Maryland, College Park, MD 20742

KESHAV DASGUPTA
Department of Physics, Stanford University, Stanford CA 94305

We review our present understanding of heterotic compactifications on non-Kähler complex manifolds with torsion. Most of these manifolds can be obtained by duality chasing a consistent F-theory compactification in the presence of fluxes. We show that the duality map generically leads to non-Kähler spaces on the heterotic side, although under some special conditions we recover Kähler compactifications. The dynamics of the heterotic theory is governed by a new superpotential and minimizing this superpotential reproduces all the torsional constraints. This superpotential also fixes most of the moduli, including the radial modulus. We discuss some new connections between Kähler and non-Kähler compactifications, including some phenomenological aspects of the latter compactifications.

The Calabi-Yau (CY) compactifications of Candelas et al. [1] have led to some major progress in our understanding of string theory vacua. Compactifying the heterotic string on such manifolds results in four-dimensional models with minimal supersymmetry (susy). In terms of the corresponding two dimensional non-linear sigma model, we demand conformal invariance so that all the tadpoles vanish and the string equations of motion are satisfied. In this way we recover again CY spaces. For the bosonic case this will give us the model studied in [2]. By definition CY manifolds are Kähler and have a vanishing first Chern class. By Yau's theorem therefore, for a *given* complex structure and a *given* cohomology class of the Kähler form there is a *unique* Ricci-flat metric with $SU(3)$ holonomy[a].

*Talks presented by M.B. and K.D.

[a]Ricci flatness is not an essential property of the compactifying manifold as has been demonstrated in [3]. We can restore Kählerity without having a Ricci flat metric (a field redefination relates them). For the non-Kähler manifolds, however, we can never have a Ricci flat metric.

Generically, when considering ordinary CY compactifications of the heterotic string theory the three-form background fluxes (at weak coupling and constant dilaton) are equal to zero. The four-dimensional spacetime is Minkowski and therefore has zero cosmological constant. Although susy would also allow anti de-Sitter solutions, here only the Minkowski solution is realized. The cancellation of the two-loop sigma model beta function puts a strong constraint on the vector bundle, namely it has to be identified with the tangent bundle, implying that the three form is in the cohomology classes of the manifold [b]. Further, by the Uhlenbeck-Yau theorem, there is an essentially unique choice of vector field for any *given* holomorphic stable vector bundle satisfying the Donaldson-Uhlenbeck-Yau (DUY) equations.

This attractive scenario however is clouded by some inherent problems which are related to the degeneracy of string vacua. Essentially there are two different degeneracies appearing in string theory compactifications. First, there are thousands of CY manifolds that could be potential solutions to the low energy effective theory. Second, once we choose a *particular* CY manifold, there are many different moduli associated with the complex structure and the Kähler structure deformations of the manifold. All these moduli are unstable at tree level and thus lead to a situation that is unattractive for phenomenology. It turns out that the radial modulus of the CY is one of the Kähler moduli. Therefore, when this field is not stabilized the CY will runaway to infinite size. This ruins the whole consistency of the compactification scenario [4].

One way to remove the degeneracy for a given CY is to relax the restriction on the fundamental two-form J by allowing manifolds that have $dJ \neq 0$, i.e spaces that are non-Kähler. These more general compactifications were first discussed in detail in [5], [6] and [7]. At first sight it may not be apparent at all how one could remove degeneracies by going to non-Kähler manifolds. This will be explained below. But first observe that breaking the Kähler condition is not straightforward, as one can show that in the absence of background fluxes *and* warped metrics, the generic solution is always Kähler. Therefore for a non-Kähler manifold to be a solution of the equations of motion one has to switch on non trivial three form fluxes H, which will essentially play the role of a torsion. The torsion is not closed because an embedding of the spin-connection into the gauge-connection is not allowed, as this would lead back to CY manifolds. Furthermore, the

[b]This is a sufficient condition, but not a necessary one. We shall discuss this in more detail as we go along.

dilaton is generically non constant and related to the warp factor of the underlying manifold.

The first concrete example of such manifolds was constructed in [8] and [9] by duality chasing a particular model of the general class of M-theory compactifications with non vanishing fluxes considered in [10]. The manifold constructed is a T^2 bundle over a four dimensional $K3$ base. In [9] and [11] many properties of this manifold were explicitly found by going to the orbifold limit of the $K3$ base. So for example, the manifold is compact, complex, has a vanishing first Chern class and $SU(3)$ holonomy. It was observed in [12], [11] that the Betti numbers of this manifold are different from the Betti numbers of a simple product $K3 \times T^2$ (appearing for vanishing fluxes), implying a topology change. The topology change is achieved by considering an additional *twist* (along with the flux) so that a consistent solution that preserves minimal susy in four dimensions is obtained. Due to this fact it is *not* possible to construct this manifold directly in the heterotic theory using a supergravity analysis. This has been explained in [13]. However the duality chasing that we performed miraculously takes the topology change into account, so that a consistent non-Kähler manifold appears on the heterotic side. See [9] and [13] for a more detailed analysis of this. In fact, the *twist* that we expect on the heterotic side is actually one component of the G-flux in M-theory.

An immediate advantage that compactifications on manifolds with torsion have is moduli stabilization at tree level. The fluxes give rise to a potential that stabilizes *all* the complex structure moduli [9], the radial modulus and some of the remaining Kähler moduli [13], [11]. This stabilization can be understood in terms of a superpotential first constructed in [13] and verified later by dimensional reduction in [14]. Contrary to popular belief, the superpotential is complex and is given by $W = \int (H + idJ) \wedge \Omega$. Here J is the fundamental two form which may not necessarily be integrable for an arbitrary choice of fluxes and Ω is the unique holomorphic (3,0) form wrt the almost complex structure that characterizes these manifolds. The form of this superpotential implies that the dynamics of the heterotic theory compactified on these non-Kähler manifolds can be described by a *complex* three form G. This three form is anomaly free and gauge invariant (see [11] and [13] for a derivation of this) and therefore can be used to construct the scalar potential for all the moduli from its kinetic term $\int |G|^2$. This potential incorporates terms that are of higher order in α' (see [15] for details). Notice, that the no-scale structure of the potential is broken in a rather interesting way. The anomaly free three form can be shown to depend secretly on the radial modulus (and some of the Kähler moduli) by solving the anomaly condition [13]. The non-trivial radial dependence comes from the fact that the Bianchi identity incorporates the three form on both sides of the equation and therefore can be solved iteratively order by order in α' [11,13]. In [13] a simple analysis was performed to evaluate the radius of the non-Kähler manifold. It was shown that the $K3$ base can be made large enough by choosing large flux densities, but the fiber generically has a size of order α'. This again implies that a simple supergravity analysis cannot be performed directly in the heterotic theory and duality chasing needs to be performed [c].

There are a couple of questions one might ask at this point. One immediate one would be whether we can choose *any* fourfold in M-theory, or whether this choice is restricted. As in the usual relation between compactifications of F-theory on a fourfold and the heterotic string on a threefold there is, of course, the restriction that the fourfold should be an elliptic fibration. Observe that the particular fourfold that was chosen for duality chasing i.e $K3 \times K3$, has an orientifold description on the type IIB side, which makes the description easier. The orientifold operation actually involves three actions: world sheet parity (the usual orientifold), fermion number reversal and space reversal (the orbifold action). The presence of these three actions guarantees that a set of U-dualities will take us to the heterotic string [8,9] and not bring us back to the strongly coupled type IIB theory. This aspect was used in [15] to construct new examples of non-Kähler manifolds that have non zero Euler characteristics. The original non-Kähler manifolds constructed in [8,9] have zero Euler characteristics [12,11].

The next question is the choice of fluxes. Anomaly cancellation in M-theory implies that fluxes are necessary, if compactifications on manifolds with non-vanishing Euler characteristics are being considered [10,16]. An alternative picture in which fluxes are traded with space filling branes and their consequent effect on the geometry is discussed in [15]. From this discussion one could naively conclude that one should *always* get a torsional compactification on the heterotic side and the simpler CY compactifications are ruled out. Interestingly, it turns out that for some special choice of fluxes, we will get back an ordinary CY compactification. Let us elab-

[c]Although the duality chasing works for most cases, sometimes it may not give the complete answer. For example, the full heterotic Bianchi identity, or the scalar potential do not follow from simple duality chasing. The derivation of these require higher order α' corrections of the action and the T-duality rules, that do not appear in the supergravity approximation. These subtleties have been explained in [11] and [15].

orate this a little bit. More details will appear elsewhere. There are two different types of fluxes that we can choose on the M-theory side: one that is positioned over the full fourfold and the other that is *localized* at the fixed points of the fourfold. These fixed points are the points where the T^2 fibration degenerates. As discussed in detail in [8,9], the non-localized fluxes appear on the type IIB side as the NS and RR three form fields. If these fluxes are not vanishing we eventually obtain the non-Kähler manifolds with torsion three form on the heterotic side after further U-dualities. The localized fluxes, on the other hand, become the seven brane gauge fields on the type IIB side [9,11], which under further U-dualities become the heterotic gauge fields. These gauge fields originate in M-theory by decomposing the G-fluxes in terms of the localized (normalizable) harmonic (1,1) forms near each singularity [17]. In the presence of non-localized G-fluxes the localized (1,1) forms themselves change by the backreaction of the fluxes on the geometry. This can be worked out with some effort [17], but will be ignored in the following. Generically in the presence of both fluxes we would get non-Kähler spaces with torsion. Furthermore, the background G-fluxes *warp* the geometry in some special way [8,9,11]. The equation for the warp factor was given in [8,9]. What happens if we choose only the localized fluxes? It is easy to see that in type IIB theory we will get gauge fields on the seven branes and no three form background fluxes. The anomaly cancellation condition will put some restriction on the total instanton numbers of these gauge fields. This is of course the usual restriction that we expect also on the heterotic side. Therefore we seem to recover ordinary CY compactifications, except for the fact that in the presence of fluxes in M-theory we will typically get a warped metric. This would naively ruin the $dJ = 0$ property. However a careful analysis reveals that the warp factor equation e.g. on the type IIB side, is in fact proportional to $\Omega_3(\omega) - \Omega_3(A)$, where Ω_3 is the Chern-Simons three form (see [8] and [9] for the derivation of this). Therefore if we *embed* the spin connection ω into the gauge connection A we will recover a trivial warp factor and the manifold will become Kähler! This is precisely the reason behind embedding the spin-connection into the gauge connection. Here we have rederived this property from M-theory by demanding the consistency of CY compactifications.

There is another rather interesting aspect that comes to mind at this point. Imagine that we do not turn on the non-localized gauge fluxes and at the same time do not allow the standard embedding. Then the naive expectation would be that we should get a non-Kähler manifold with the non-Kählerity coming precisely from the difference $\Omega_3(\omega) - \Omega_3(A)$. This would seem to contradict the result of [18] where it was found that a fractional gauge Chern-Simons term can appear in an ordinary CY compactification. However this apparent puzzle can be resolved by taking the background *gaugino condensate* into account[d]. To see how this helps we need to back up a little for more generality.

The complex three form that appears in the heterotic theory in the presence of torsion has to be imaginary self-dual (ISD) to preserve supersymmetry in four dimensions. This implies the background equation $dJ = *H$, where $*$ is the Hodge duality in six dimensions (over the non-Kähler manifold). This equation is more general than the constraint derived in [5,6,7] and it reduces to the known form when the manifold is complex (see [13] for a derivation of this fact). This equation makes the non-Kähler nature of the manifold manifest. Observe that if we scale the metric then the three form H scales linearly too. On the other hand from the Bianchi identity we observe, that the three form does not scale (at least to the lowest order in α'). This implies that the radial modulus should get stabilized, as we saw earlier. This argument, although correct, is rather naive at this point. The fact that the Bianchi identity does not scale is only true for the Kähler case. In the non-Kähler case since the three form appears on both sides of the identity, the issue is more subtle. Therefore the correct way to study the potential for the radial modulus would be to evaluate the three form flux order by order in α' and use the kinetic term to calculate the potential. This was done in [13,15].

Coming back to the torsional constraint, we see that in the absence of three form fluxes the manifold can still become non-Kähler via the relation $dJ = *\, \alpha'[\Omega_3(\omega) - \Omega_3(A)]$. In [18] the $\Omega_3(\omega)$ term was cancelled by one of the Chern-Simons terms of the gauge fields. It turns out that one can make $dJ = 0$ using the gaugino condensate contribution to the superpotential. The gaugino condensate will change the torsional equation by an additional term (for a derivation of this see for example [19]). This additional term can be used to cancel the remaining gauge Chern-Simons part and we recover the Kähler property[e].

---

[d]The gaugino condensate contributes to the (3,0) and the (0,3) part of the threeform, and therefore breaks susy.

[e]A more "dynamical" way to achieve this would be to take our proposed superpotential and solve for dJ using the various contributions (tree level, perturbative and non-perturbative). If $dJ = 0$ with integrable J we get Kähler CY compactifications. All other cases, i.e when $dJ \neq 0$ with J integrable or non-integrable, will correspond to non-Kähler compactifications. As we saw earlier, for this case the radius is stabilized at tree level. For the Kähler case, the radius is fixed non-perturbatively [20,18]. In both cases the σ-model conformal invariance is restored at *this* particular radius.

Therefore we see that Kähler compactifications are a special case of the more general non-Kähler compactifications with torsion (at least those non-Kähler compactifications that could be constructed from F-theory using duality chasing). To summarize: the most generic superpotential governing these backgrounds is complex and only for the special case $dJ = 0$ we recover the superpotential proposed in [21].

Our next goal is to understand the fibration structure of our non-Kähler manifold. The precise metric of the fiber torus has been worked out in [9,11]. As we discussed earlier, there is a change in Betti numbers when we go from $K3 \times T^2$ to the non-Kähler space. It turns out that this change of "topology" can be understood from a *brane-box* configuration in the type IIB theory. The brane-box divides the region in two parts: one that is inside the box and the other outside. The walls of the box are made out of NS5-branes which are actually sources of NS flux. The RR three form flux can be obtained from the T-dual twist of a type IIA configuration (see [15] for a detailed analysis of this scenario)[f]. Under a set of U-dualities each side of the brane box transforms into a Taub-NUT space which reproduces the fibration structure. The fact that the metric of the system works out correctly has been checked in [15]. To see how the superpotential works out from the brane box configuration we observe that the RR three form sources can also be replaced by D5 branes (forming, say, the two other sides of the box). These D5 branes become, under a set of U-dualities, NS5 branes wrapping some two cycles of the non-Kähler space whose fibration structure is determined from the U-dual brane box. The NS5 branes show a jump of the H charge precisely as $H = - * dJ$ [22], and therefore contribute the complex dJ part of the superpotential. More details on this will appear in a future publication. Thus we can reproduce the full structure of the non-Kähler space using a brane configuration. This should come as no surprise because far away from the brane box configuration there is no distinction from the geometrical picture and the brane picture. There are some subtleties in this construction primarily related to the hidden orientifold nature of the system. The box configuration survives this projection. But one also has to take into account the F-theory monodromies to reproduce the complete gauge bundles. These monodromies appear as a stringy cosmic string in our model (again for more details the readers are advised to look in [15]).

Before moving ahead, we still have to clarify the reason of why these manifolds have an $SU(3)$ holonomy (wrt the torsional connection). The

[f] As an additional advantage we get non constant three form fluxes, as opposed to the constant fluxes of [8,9].

existence of minimal susy in four dimensions gives us a necessary condition that is, however, not sufficient. A slightly more stronger argument[g] will be to observe that if we choose the metric so that the cube of J is the product of the holomorphic three form Ω and its dual, then the holomorphic connection that respects the metric will also respect the holomorphic three form and therefore will have SU(3) holonomy. The profound aspect of Yau's theorem is that, in the CY case, the metric can be chosen to be Kähler in addition to the above property. Since there is no Kähler metric for the more general non-Kähler compactifications, we have nothing comparable to prove. We do, however, need another principle for choosing the particular metric we have chosen out of an infinite dimensional space of metrics. The above criterium is satisfied by our choice of metric.

We can also explain this using the torsion classes W_i (with $i = 1, ..., 5$) of [23] and [24]. In this classification, a non-Kähler manifold that preserves an $SU(3)$ holonomy will have to necessarily satisfy the equation $2W_4 + W_5 = 0$, with an additional condition $W_1 = W_2 = 0$, so that the complex structure is integrable. For more details on the physics aspects of this see for example [23]. For our non-Kähler manifold the torsion classes have been worked out in [11] and for the Iwasawa manifold, in [23]. Using this analysis, it can also be shown that one *cannot* define any Kähler metric on these manifolds (see [12,11] for more details). Therefore these manifolds are explicitly non-Kähler.

So far our discussion has been mostly restricted to non-Kähler manifolds that have zero Euler characteristics. Having zero Euler characteristics is not much of a problem, of course, because we are not embedding the spin connection into the gauge connection. The reason why we would like to look for more general manifolds with non-zero Euler characteristics is purely to extend our understanding of these manifolds. This may serve as a new and interesting direction in mathematics and may also turn out to be phenomenologically more attractive.

The first interesting example of such a manifold is four dimensional and can be described in terms of a $K3$ manifold with torsion. The metric has been worked out in [5], where it was shown to pick up an overall conformal factor from the back reaction of the fluxes on the geometry. One may wonder if it is possible to obtain this manifold from duality chasing in F-theory. Indeed this has been achieved in [15] (at least for the non-compact case). Therein it was shown that the metric is related to a Gimon-Polchinski

[g]We thank Paul Green for providing the following argument.

kind of model with the axion-dilaton τ fixed at a *particular* value. In this sense this is different from our earlier examples where we had a vanishing axion-dilaton and therefore F-theory was at *constant* coupling [25]. Now since the axion-dilaton is non-trivial, there would be sizable non-perturbative corrections to the model. These corrections actually convert the intersecting orientifold planes and branes to smooth hyperbolas. For details on this see [26] for the case of zero torsion and [15] for the case with torsion. An alternative way to appreciate this would be to observe that the metric of a torsional $K3$ is precisely the metric of a NS5 brane at a point on the K3. Therefore under a set of U-dualities the system maps to a configuration of T^4 orientifolds with seven branes. These seven branes are the sources of non trivial τ in this framework (see [15] for more details).

There are many subtleties that we have ignored here, which have been discussed in [15] though. The generic orientifold action in the type IIB case is in fact *ambiguous*. This ambiguity gives different heterotic duals and is present in four- as well as six-dimensional models. Of course, once we fix the heterotic compactification, we also fix the type IIB ambiguity. Nevertheless, the six-dimensional compactification becomes rather involved because of this subtlety. For a more detailed discussion of this issue in the generic context see [27], and for the torsional case, see [15].

This brings us to the next interesting case of a six-dimensional compactification with non zero Euler characteristics. The duality chains from F-theory to the heterotic theory were given in [15]. If we keep the orientifold action Γ as generic, then the type IIB manifold will be of the form N_6/Γ, where N_6 is a six-dimensional compact manifold (in the absence of fluxes) that has non zero Euler characteristics. In the presence of fluxes, the heterotic dual of this manifold will look like a Z_2 action of a torus fibration (locally) over a compact base. If we choose another action of Γ that gives a P^1 fibration over a $P^1 \times P^1$ base on the type IIB side (in the absense of fluxes), then the heterotic manifold will be a non-trivial six manifold that is a Z_2 action of a torus fibration over a $P^1 \times P^1$ base. The generic fibration structure is not difficult to work out (though the analysis may get very tedious). The result, in a compact form, is presented in [15] and therefore we refer the reader to this paper. However, the full topological data have not yet been worked out. This will be presented elsewhere.

The examples discussed above are all of orbifold nature but in principle smooth examples do exist. Some of these have been discussed in [15]. They include the connected sums of $S^3 \times S^3$ (first discussed in [28]) and some examples of flops of an elliptically fibered CY space. The flops used herein are the ones that break the Kähler condition. In [15] it was discussed in detail how the bundles follow the manifold through the flop. But whether these smooth examples are solutions of string equations of motion have not been discussed yet. This is relegated to a future publication.

The question of finding stable vector bundles for our manifolds is a very important one especially because we are no longer allowed to embed the spin connection into the gauge connection (see [29,11] for a discussion on this). The DUY equations take the same form as in the CY case, i.e the (2,0) and the (0,2) part of the curvature vanishes and the (1,1) part is traceless. However, there are two additional conditions. First, notice that J is no longer closed (and may not be integrable either) and second, there is now a constraint on $\mathrm{tr}(F \wedge F)$ coming from the $i\partial\bar{\partial}J$ part (see [15] for details). These conditions look very restrictive and one might wonder if there exists *any* solution at all to these equations. Again the duality chasing comes to the rescue here. The Weierstrass equation governing the F-theory background allows an D_4^4 bundle to propagate to the heterotic side taking the orientifold action into account. This bundle does satisfy all the conditions as has been explicitly demonstrated in [11]. One has to carefully take the orbifold singularities, localized fluxes *and* higher order corrections in F-theory into account. Failing to do so will not reproduce the correct result [11]. In this way we obtain one particular consistent example, for which the above equations can be solved.

There are several open questions that might be raised at this point. First is the question of the *stability* of the bundle[h]. For Kähler compactifications the holomorphic gauge fields that satisfy the DUY equations are equivalent to Mumford-Takemoto stable holomorphic bundles. For non-Kähler compactifications, assuming the metric to be (approximately) Gauduchon, a similar statement can also be made (though under some special circumstances). For more details see [15]. Unfortunately a full understanding of the bundles has not yet been achieved and we hope to address this question in a near future.

Another open question is to understand the full non-abelian nature of the bundle. From the duality chasing one expects the full non-abelian nature to show up. But in practice, it is only the abelian part that is manifest in this scenario (the localized G-fluxes form the Cartan subalgebra of the D_4 algebra). The full non-abelian part can be seen, if we consider the M2 branes to be wrapping the degenerating cycles of Taub-NUT.

[h]Earlier work on this appeared in [30].

In [11] the intersection matrices of these two cycles were shown to reproduce exactly the Cartan matrix of the D_4 algebra (at least near one of the four orbifold singularties).

Finally, one might ask about the number of generations for these non-Kähler models. Since we do not allow the standard embedding, the number of generations is *not* equal to the Euler number of the non-Kähler manifold. However we need to evaluate the number of generations when (a) the non-trivial warp factor is taken into consideration, and (b) the spin connection is the torsional connection (in addition to not having standard embedding). Somewhat surprisingly, the number of generations is still given by the third Chern class c_3 of the bundle. For a proof of this see [15]. In [15] the number of generations for a rather simple example with a $U(1)$ bundle has been worked out. For a more realistic non-Kähler compactification one would have to compute the number of generations for a phenomenologically relevant group (say for example $SU(5)$). This would mean that we need an $SU(5)$ bundle on our non-Kähler space satisfying the DUY equations. Whether this is indeed possible remains to be seen. More details on this will be reported elsewhere.

Having a detailed analysis of non-Kähler manifolds, it is now time to ask whether we can formulate a non-linear sigma model description of these manifolds. Some details of this have appeared in the early works of [5,6], where the basic constraints were shown to follow from a (0,2) sigma model description (for a more recent exposure see [11] and [15]). One would expect a reasonable non-linear sigma model description behind our construction but the calculation of massless spectra is rather subtle here. See [15] for a discussion on this and the issue of H-twisted sheaf cohomology for the counting of states. However at this point, it is not clear how to formulate a simple *linear* sigma model that flows to the conformally invariant background that we have.

The non-linear sigma model, on the other hand, can be efficiently used to describe various interesting aspects of the torsional backgrounds. In particular one can explicitly derive the preferred connection for this background. The preferred connection is chosen in such a way that with this choice the heterotic σ-model action resembles the Green-Schwarz superstring action in a background configuration with 24 free fields. The identification becomes precise as soon as we embed the gauge connection in the modified spin-connection. However, as we discussed above, the standard embedding is not allowed here; and therefore only to the lowest order in α' the two actions may be identified. On the other hand, with this choice of connection one can in fact use the Green-Schwarz superstring action to compute $O(\alpha')$ corrections to the heterotic beta functions. All these corrections are suppressed by the size of the six-manifold, and conformal invariance is restored when the manifold attains the size that is dictated by the minima of our superpotential. For this and other details regarding σ-model description see [6] for an early exposition and [11] for a more recent discussion related to the superpotential. The fact that three form appears on both sides of the Bianchi identity can also be easily shown. The torsional equation and the DUY relations follow from demanding world sheet susy [6,5,11]. One can also show that the fundamental form should be H-covariantly constant, so that the manifold allows an integrable complex structure.

The above discussion, hopefully, summarizes our present understanding of heterotic compactifications on non-Kähler complex manifolds with torsion. We have shown that it is possible to get an almost *rigid* vacua by using background fluxes. To finish this summary let us remark that there is another important issue that has not been addressed so far. This has to do with the *number* of possible string vacua. Recall that at the beginning of this note we discussed two kind of degeneracies: one of them originated from the fact that many different manifolds can be solutions to the string equations of motion and the other one resulted from the deformations of a *given* vacuum. What we discussed so far concerns only the lifting of the second type of degeneracy i.e. the fixing of the moduli for a *given* vacuum. But we have, at present, no understanding on how to fix the other degeneracy. A recent counting of flux vacua in a different context has revealed that this number could be finite in some cases [31]. It would be rather interesting to understand if such a calculation can be performed in the present scenario.

Acknowledgments

It is our pleasure to thank K. Becker, E. Goldstein, P. S. Green, S. Prokushkin and E. Sharpe for their collaboration leading to some of the results presented herein. Furthermore we would like to thank the organizers of QTS3 and SUSY 03 for organizing a stimulating conference. The work of M.B. is supported by NSF grant PHY-01-5-23911 and an Alfred Sloan Fellowship. The work of K.D is supported by a David and Lucile Packard Foundation Fellowship 2000-13856.

References

1. P. Candelas, G. T. Horowitz, A. Strominger and E. Witten, "Vacuum configurations for superstrings," Nucl. Phys. B **258** (1985) 46-74.
2. D. Nemeschansky and S. Yankielowicz, "Critical d imension of string theories in curved space," Phys. Rev. Lett. **54**, 620 (1985) [Erratum-ibid. **54**, 1736 (1985)]; D. Friedan, Z. a. Qiu and S. H. Shenker, "Superconformal invariance in two dimensions and the tricritical Ising model," Phys. Lett. B **151**, 37 (1985).
3. D. Nemeschansky and A. Sen, "Conformal invariance of supersymmetric sigma models on Calabi-Yau manifolds," Phys. Lett. B **178**, 365 (1986); P. Candelas, M. D. Freeman, C. N. Pope, M. F. Sohnius and K. S. Stelle, "Higher order corrections to supersymmetry and compactifications of the heterotic string," Phys. Lett. B **177**, 341 (1986).
4. M. Dine and N. Seiberg, "Couplings and scales in superstring models," Phys. Rev. Lett. **55** (1985) 366-369.
5. A. Strominger, "Superstrings with torsion", Nucl. Phys. **B274** (1986) 253-284.
6. C. M. Hull, "Superstring compactifications with torsion and space-time supersymmetry," *In Turin 1985, Proceedings, Superunification and Extra Dimensions*, 347-375.
7. B. de Wit, D. J. Smit and N. D. Hari Dass, "Residual supersymmetry of compactified D = 10 supergravity," Nucl. Phys. B **283** (1987) 165-191.
8. K. Dasgupta, G. Rajesh and S. Sethi, "M theory, orientifolds and G-flux," JHEP **9908** (1999) 023, hep-th/9908088.
9. K. Becker and K. Dasgupta, "Heterotic strings with torsion," JHEP **0211** (2002) 006, hep-th/0209077.
10. K. Becker and M. Becker, "M-Theory on eight-manifolds,", Nucl. Phys. **B477** (1996) 155-167, hep-th/9605053.
11. K. Becker, M. Becker, K. Dasgupta and P. S. Green, "Compactifications of heterotic theory on non-Kähler complex manifolds. I," JHEP **0304** (2003) 007, hep-th/0301161.
12. E. Goldstein and S. Prokushkin, "Geometric model for complex non-Kähler manifolds with SU(3) structure," hep-th/0212307.
13. K. Becker, M. Becker, K. Dasgupta and S. Prokushkin, "Properties of heterotic vacua from superpotentials," Nucl. Phys. B **666**, 144 (2003), hep-th/0304001.
14. G. L. Cardoso, G. Curio, G. Dall'Agata and D. Lust, "BPS action and superpotential for heterotic string compactifications with fluxes," JHEP **0310**, 004 (2003), hep-th/0306088.
15. K. Becker, M. Becker, P. S. Green, K. Dasgupta and E. Sharpe, "Compactifications of heterotic strings on non-Kähler complex manifolds. II," Nucl. Phys. B **678**, 19 (2004), hep-th/0310058.
16. S. Sethi, C. Vafa and E. Witten, "Constraints on low-dimensional string compactifications," Nucl. Phys. B **480** (1996) 213-224, hep-th/9606122.

17. K. Dasgupta, G. Rajesh, D. Robbins and S. Sethi, "Time-dependent warping, fluxes, and NCYM," JHEP **0303** (2003) 041, hep-th/0302049; K. Dasgupta and M. Shmakova, "On branes and oriented B-fields," Nucl. Phys. B **675**, 205 (2003), hep-th/0306030.
18. S. Gukov, S. Kachru, X. Liu and L. McAllister, "Heterotic moduli stabilization with fractional Chern-Simons invariants," hep-th/0310159.
19. G. L. Cardoso, G. Curio, G. Dall'Agata and D. Lust, "Heterotic string theory on non-Kähler manifolds with H-Flux and gaugino condensate," hep-th/0310021.
20. E. I. Buchbinder and B. A. Ovrut, "Vacuum stability in heterotic M-theory," hep-th/0310112.
21. E. Witten, "New issues in manifolds of $SU(3)$ holonomy," Nucl. Phys. B **268** (1986) 79-112.
22. J. P. Gauntlett, N. w. Kim, D. Martelli and D. Waldram, "Fivebranes wrapped on SLAG three-cycles and related geometry," JHEP **0111** (2001) 018, hep-th/0110034. J. P. Gauntlett, D. Martelli, S. Pakis and D. Waldram, "G-structures and wrapped NS5-branes," hep-th/0205050; J. P. Gauntlett, D. Martelli and D. Waldram, "Superstrings with intrinsic torsion," hep-th/0302158.
23. G. L. Cardoso, G. Curio, G. Dall'Agata, D. Lust, P. Manousselis and G. Zoupanos, "Non-Kähler string backgrounds and their five torsion classes," Nucl. Phys. B **652**, 5 (2003), hep-th/0211118.
24. S. Gurrieri, J. Louis, A. Micu and D. Waldram, "Mirror symmetry in generalized Calabi-Yau compactifications," Nucl. Phys. B **654**, 61 (2003), hep-th/0211102.
25. K. Dasgupta and S. Mukhi, "F-theory at constant coupling," Phys. Lett. B **385** (1996) 125-131, hep-th/9606044.
26. A. Sen, "A non-perturbative description of the Gimon-Polchinski orientifold," Nucl. Phys. B **489** (1997) 139-159, hep-th/9611186; A. Sen, "F-theory and the Gimon-Polchinski orientifold," Nucl. Phys. B **498** (1997) 135-155, hep-th/9702061.
27. R. Gopakumar and S. Mukhi, "Orbifold and orientifold compactifications of F-theory and M-theory to six and four dimensions," Nucl. Phys. B **479** (1996) 260-284, hep-th/9607057.
28. J. Gutowski, S. Ivanov and G. Papadopoulos, "Deformations of generalized calibrations and compact non-Kähler manifolds with vanishing first Chern class," math.dg/0205012.
29. S. Ivanov and G. Papadopoulos, "A no-go theorem for string warped compactifications," Phys. Lett. B **497** (2001) 309-316, hep-th/0008232.
30. J. Li and S.-T. Yau, "Hermitian Yang-Mills connections on non-Kähler manifolds," in *Mathematical Aspects of String Theory*, World Scientific, 1987.
31. M. R. Douglas, "The statistics of string/M theory vacua" JHEP **0305** (2003) 046, hep-th/0303194; S. Ashok and M. R. Douglas," Counting flux vacua", hep-th/0307049.

ON EFFECTIVE ACTION OF STRING THEORY FLUX COMPACTIFICATIONS

ALEX BUCHEL
Department of Applied Mathematics
University of Western Ontario
London, Ontario N6A 5B7, Canada
Perimeter Institute for Theoretical Physics
Waterloo, Ontario N2J 2W9, Canada
abuchel@uwo.ca

We discuss four dimensional effective actions of string theory flux compactifications. These effective actions describe four dimensional gravity coupled to overall Kähler modulus of the compactification manifold. We demonstrate the agreement between ten dimensional equations of motion of supergravity with localized branes, and equations of motion derived from the effective action. The agreement is lost however if one evaluates the full effective action on the equations of motion for a subset of the supergravity modes, provided these modes depend on-shell on the Kähler modulus.

1. Introduction

Probably the biggest surprise of recent cosmological measurements is the evidence that the cosmological constant of the Universe is positive. It is thus important to understand how (and if) four dimensional de-Sitter vacua arise in string theory compactifications. The problem is complicated by the well known no-go theorem in supergravity [1,2] which forbids non-singular (warped) de-Sitter compactifications with finite four dimensional Newton's constant. This no-go theorem can be phrased in the statement on the energy condition on the matter stress tensor upon consistent Kaluza-Klein (KK) reductions: if the matter of the higher dimensional theory obeys the strong energy condition, the matter of the lower dimensional effective description would satisfy the strong energy condition as well, implying that lower dimensional de Sitter (or more generally accelerating Universe) is not possible. In ten dimensional supergravities matter satisfies the strong energy condition. The situation is different with positive tension extended objects, D-branes, in string theory. In ten dimensions, whether or not a

Dp-brane satisfies the strong energy condition depends on its codimension. Specifically, codimension three and higher branes ($p < 7$) satisfy strong energy condition. In consistent supergravity KK reductions the effective lower dimensional action can be obtained by integrating out compact directions of a ten dimensional supergravity effective action. If one assumes that the same prescription is true for a D-brane effective action, one immediately concludes that upon KK reduction D-branes can be sources for de Sitter. The point simply is that KK reduction transverse to a brane *decreases* its codimension, thus making lower dimensional accelerating Universe possible. In explicit string construction this idea has been realized for the first time in [3], KKLT. An obvious question of KKLT construction is whether effective description of $\overline{D3}$ is indeed given by simple dimensional reduction of its ten dimensional effective action. The question is nontrivial, as KKLT model does not allow for a consistent ten dimensional lift: the Kähler modulus stabilization involves string instanton corrections, which are highly nonlocal from the ten dimensional perspective (assuming their ten dimensional description exists at all). Thus strictly speaking, there is no starting point for the derivation of the $\overline{D3}$ effective action in KKLT model. To the best of our knowledge, there is no consistent derivation of the effective brane description in the Kaluza-Klein reduction for any compactification in the literature as well.

In the next section we discuss a simple exactly soluble D-dimensional toy model in which we confirm that contrary to a supergravity mode, in a consistent KK reduction to $d < D - 2$ dimensions, the energy condition of the extended positive tension d-dimensional defect indeed changes. The defect contribution to the d-dimensional vacuum energy Λ_d is positive. The apparent conflict with the energy condition upon KK reduction is resolved because the defect introduction 'shifts' the stabilization value of the overall Kähler modulus of the compactification manifold. This is precisely what is observed in KKLT construction. However, though the vacuum energy in the toy model compactification increases, it can never become positive. The latter point is just a special case of the no-go theorem for supergravity with localized sources satisfying a BPS-like condition of Ref. [4]. Additionally, we explain the dangers of using incomplete set of D-dimensional equations of motion to deduce lower dimensional effective action.

The same arguments can be applied to type IIB string theory flux compactifications with localized sources, as discussed in details in Ref. [5].

2. Defect effective action in consistent KK reductions

Consider the following $D = d + q$ dimensional (we assume $d > 2$) model of gravity coupled to the q-form field strength $F_{[q]}$

$$S_D = \frac{1}{2k_D^2} \int_{M_D} d^D\xi \sqrt{-g} \Big(R^D - \frac{1}{2q!} F_{[q]}^2 \Big) \,, \tag{1}$$

where k_D is D-dimensional gravitational coupling. We will use effective action (1) to study Kaluza-Klein reductions of direct warped product compactifications $M_D = M_d \times S^q$ on S^q,

$$\begin{aligned} dS_D^2 &= g_{MN} d\xi^M d\xi^N \\ &= \sigma^{-2} \, \tilde{g}_{\mu\nu} dx^\mu dx^\nu + \sigma^{\frac{2(d-2)}{q}} \, (dS^q)^2 \,, \end{aligned} \tag{2}$$

where $\sigma = \sigma(x)$ is a q-sphere Kähler modulus, $\tilde{g}_{\mu\nu} = \tilde{g}_{\mu\nu}(x)$ is the metric on M_d, and $(dS^q)^2$ is the metric on a round S^q. The metric ansatz (2) is supplemented with

$$F_{[q]} = b \; vol(S^q) \,. \tag{3}$$

Additionally, we will consider $N \gg 1$, d-dimensional neutral domain walls of tension $T_d > 0$, uniformly distributed on S^q. Each domain wall has a DBI-type action

$$S_d = -T_d \int_{M_d} d^d x \sqrt{-\tilde{g}} \; \sigma^{-d} \,. \tag{4}$$

The total effective action is

$$S_{tot} = S_D + S_{(loc)} = S_D + \sum_{i=1}^{N} S_d \,, \tag{5}$$

where the summation is over all localized defects on S^q.

Note that the stress tensor of the localized defects is

$$T_{MN}^{(loc)} = -\frac{2}{\sqrt{-g}} \frac{\delta S_{(loc)}}{\delta g^{MN}} \,, \tag{6}$$

so

$$T_{\mu\nu}^{(loc)} = -\frac{T_d N}{vol(S^q) \; \sigma^d} \; \tilde{g}_{\mu\nu} \,, \qquad T_{mn}^{(loc)} = 0 \,. \tag{7}$$

In D-dimensions, the stress tensor T_{MN}^{loc} satisfies the strong energy condition provided

$$T_{tt}^{(loc)} - \frac{1}{D-2} \, g_{tt} \, T_K^{K\ (loc)} \; > \; 0 \,, \tag{8}$$

which using (7) evaluates to

$$-\frac{T_d N}{vol(S^q)\ \sigma^d}\,\frac{q-2}{D-2}\,\tilde{g}_{tt}\ >\ 0\,. \tag{9}$$

Clearly, codimension higher than two ($q > 2$) defects satisfy the strong energy condition, while defects of lower codimension violate it.

In what follows we assume that σ is constant. We consider first D-dimensional solutions of (5), and show that they are equivalent to the solutions of the effective d-dimensional action, obtained by simply integrating out the S^q coordinates in (5).

2.1. *D-dimensional equations of motion*

With the flux ansatz (3), the q-form Bianchi identity are trivially satisfied. We are left with Einstein equations

$$R_{MN} = \frac{1}{2(q-1)!}\left(F_{M\alpha_2\cdots\alpha_q}F_N{}^{\alpha_2\cdots\alpha_n} - \frac{q-1}{q(d+q-2)}F^2_{[q]}g_{MN}\right) + k_D^2\left(T^{loc}_{MN} - \frac{1}{d+q-2}T^{loc}g_{MN}\right)\,. \tag{10}$$

Eq. (10) for the components along the defect implies that M_d is an Einstein manifold

$$r^{(d)}_{\mu\nu} = \Lambda_d \tilde{g}_{\mu\nu}\,, \tag{11}$$

with

$$\Lambda_d = -\frac{b^2(q-1)}{2(d+q-2)}\sigma^{2(1-d)} - k_D^2\frac{T_d N}{vol(S^q)\ \sigma^d}\ \frac{q-2}{d+q-2}\,. \tag{12}$$

Note from (12) that as long as defects satisfy the strong energy condition in D dimensions, *i.e.*, $q > 2$, in a consistent KK reductions they *can not* produce d-dimensional accelerating Universe. This is a special case of a more general result that follows from a no-go theorem for supergravity compactifications with branes satisfying BPS-like condition of Ref. [4]. From (12) is appears that a positive tension defect satisfying a strong energy condition provides a *negative* contribution to d-dimensional cosmological constant. This conclusion would be correct, if before and after defect introduction the Kähler modulus of the compactification manifold (in this case σ) would not change. In fact, the Kähler modulus is shifted so that after all, the contribution of defect tension to Λ_d is positive. Indeed, from the Einstein equations along the S^q directions

$$(q-1)\sigma^{2(2-d)/q} = \frac{b^2}{2}\,\frac{d-1}{(d+q-2)}\,\sigma^{2(2-d)} + k_D^2 \frac{T_d N}{vol(S^q)\;\sigma^{(d-2)}}\,\frac{d}{d+q-2}\,. \tag{13}$$

From (13) it follows that defect introduction always results in the expansion of S^q, *i.e.*, σ increases. Using (12), (13),

$$\Lambda_d = -\frac{b^2}{2d}\,\frac{1}{\sigma^{2d-2}} - \frac{(q-2)(q-1)}{d}\,\frac{1}{\sigma^{2(d+q-2)/q}}\,, \tag{14}$$

so, it is monotonically increasing function of σ for $q > 2$. Thus, introduction of localized sources (4) will always increase Λ_d.

2.2. *d-dimensional effective description*

We would like to reproduce the results of the previous section from the lower dimensional effective action, obtained by integrating out S^q coordinates. For constant σ and the metric and flux ansatze (2) and (3) correspondingly, we find

$$\begin{aligned}\sqrt{-g}&\left(R^D - \frac{1}{2q!}F_{[q]}^2\right)\\ &= \sqrt{-\tilde{g}}\sqrt{g_{S^q}}\;\sigma^{-2}\left(r^d\sigma^2 + (q-1)q\sigma^{2(2-d)/q} - \frac{1}{2}b^2\sigma^{4-2d}\right).\end{aligned} \tag{15}$$

Thus Kaluza-Klein reduction of (1) yields

$$S_D^{eff} = \frac{vol(S^q)}{2k_D^2}\int_{M_d} d^dx\sqrt{-\tilde{g}}\Big(r^d - V_{bulk}(\sigma)\Big)\,, \tag{16}$$

where

$$V_{bulk} = -(q-1)\;q\;\sigma^{2(2-d-q)/q} + \frac{1}{2}b^2\sigma^{2-2d}\,. \tag{17}$$

The KK reduction of the defect action $S_{(loc)}$ is simply

$$S_{(loc)}^{eff} = -NT_d\int_{M_d} d^dx\sqrt{-\tilde{g}}\;\sigma^{-d} \equiv \int_{M_d} d^dx\sqrt{-\tilde{g}}\;\Big(-V_{(loc)}(\sigma)\Big)\,. \tag{18}$$

Notice that the stress tensor computed from $S_{(loc)}^{eff}$ violates the strong energy condition in d-dimensions. This is so because extended defects are now codimension zero objects. The complete effective action reads

$$\begin{aligned}S^{eff} &= S_D^{eff} + S_{(loc)}^{eff}\\ &= \frac{vol(S^q)}{2k_D^2}\int_{M_d} d^dx\sqrt{-\tilde{g}}\Big(r^d - V(\sigma)\Big)\,,\end{aligned} \tag{19}$$

with

$$V = V_{bulk} + V_{(loc)}$$
$$= -(q-1)\; q\; \sigma^{2(2-d-q)/q} + \frac{1}{2} b^2 \sigma^{2-2d} + k_D^2 \frac{2T_d N}{vol(S^q)\; \sigma^d} \,. \tag{20}$$

It is easy to verify that effective action (19) equations of motion

$$r^d_{\mu\nu} - \frac{1}{2} r^d\; \tilde{g}_{\mu\nu} + \frac{1}{2} V\; \tilde{g}_{\mu\nu} = 0\,, \tag{21}$$

$$\frac{\partial V}{\partial \sigma} = 0\,, \tag{22}$$

are equivalent to D-dimensional equations of motion (12), (13).

2.3. *Incorrect d-dimensional effective description*

In this section we illustrate the pitfalls of using incomplete D-dimensional equations of motion to deduce d-dimensional effective action. This approach was used in existing literature, and has led to quite unexpected conclusions concerning consistency of KK reductions in the type IIB flux compactifications.

For a constant modulus σ we expect the effective action to be of the type (19), which we rewrite one more time as

$$S^{eff} = \frac{vol(S^q)}{2k_D^2} \int_{M_d} d^d x \sqrt{-\tilde{g}} \Big(r^d - \tilde{V}(\sigma) \Big) \,. \tag{23}$$

Einstein equations from (23) imply that

$$r^d_{\mu\nu} - \frac{1}{2} r^d\; \tilde{g}_{\mu\nu} + \frac{1}{2} \tilde{V}\; \tilde{g}_{\mu\nu} = 0\,,$$
$$\frac{1}{2}\; \tilde{g}_{\mu\nu} \Big(\tilde{V} - (d-2)\Lambda_d \Big) = 0\,, \tag{24}$$

where in the second line we used the fact that M_d must be an Einstein manifold. With (24), it is tempting to use D-dimensional equation of motion (12) to conclude

$$\tilde{V} = (d-2)\Lambda_d$$
$$= -\frac{b^2(q-1)(d-2)}{2(d+q-2)} \sigma^{2(1-d)} - k_D^2 \frac{T_d N}{vol(S^q)\; \sigma^d} \frac{(d-2)(q-2)}{d+q-2} \,. \tag{25}$$

Potential $\tilde{V}$ (25) derived in this way disagrees with (20), and is manifestly incorrect. For one reason, in the absence of any defects ($N = 0$), $\tilde{V}$ implies that σ can not be stabilized. This conclusion differs from exact D-dimensional analysis.

Acknowledgments

I would like to thank Shanta de Alwis, Steve Giddings, Shamit Kachru, Renata Kallosh, Jim Liu, Rob Myers, Joe Polchinski, Radu Roiban, Christian Romelsberger, Sav Sethi, Henry Tye, and Johannes Walcher for numerous discussions, correspondence and comments. I would like to thank Aspen Center for Physics for hospitality where part of this work was done, and acknowledge the partial support by the U.S. Department of Energy while at the University of Michigan, where this work started. Research at the Perimeter Institute is supported in part by funds from NSERC of Canada.

References

1. G.W. Gibbons, in *Supersymmetry, Supergravity and Related Topics*, edited by F. de Aguila, J.A. de Azcárraga and L. Ibañez, 346 (World Scientific, Singapore, 1985).
2. J. M. Maldacena and C. Nunez, *Int. J. Mod. Phys. A* **16**, 822 (2001). [arXiv:hep-th/0007018].
3. S. Kachru, R. Kallosh, A. Linde and S. P. Trivedi, *Phys. Rev. D* **68**, 046005 (2003).
4. S. B. Giddings, S. Kachru and J. Polchinski, *Phys. Rev. D* **66**, 106006 (2002).
5. A. Buchel, arXiv:hep-th/0312076.

IS THERE A STRING THEORY LANDSCAPE? SOME CAUTIONARY REMARKS

MICHAEL DINE

Santa Cruz Institute for Particle Physics, Santa Cruz CA 95064
dine@scipp.ucsc.edu

There is evidence that string theory possesses a large discretuum of stable and/or metastable ground states, with zero or four supersymmetries in four dimensions. We discuss critically the nature of this evidence. Assuming this "landscape" exists, anthropic explanations of some quantities are almost inevitable. We explain that this landscape is likely to lead to a prediction of low energy supersymmetry. But we argue that many features of low energy physics are not anthropic and, as currently understood, the landscape picture will get them wrong. This indicates that this viewpoint is potentially falsifiable. Moreover, if it is correct, many questions must be answered through more conventional scientific explanations.

1. Introduction: The "Vacuum Selection" Problem

From the beginning, the seemingly vast array of possible ground states has made string theory both attractive and problematic. Ground states with more than four supersymmetries have the virtue that they are theoretically tractable, but they are also totally unrealistic. It has long been clear that no potential for the moduli exists, and the duality revolution spoiled any remaining hope that some sort of non-perturbative inconsistency might permit us to discard these states. It also strongly suggested that this proliferation of possible ground states is an inherent feature of any sort of quantum general relativity. Apart from anthropic arguments (to be discussed below), we have no inkling why nature doesn't select one of these states. With four or less supersymmetries there is a vast proliferation of candidate ground states, revealed in various approximations. Some of these have features which resemble those of the real world. Unlike the case of more supersymmetries, there are potentials for the moduli, tadpoles (either at the perturbative or non-perturbative level), and some possibility of non-perturbative anomalies. Faced with this plethora of states, I, for a long time, comforted myself that not a single example of a (meta)stable ground state of this sort had been exhibited in a controlled approximation, and so perhaps there might be some unique or at least limited set of sensible states.

One of the most exciting – and troubling – developments in string theory in the last few years has been the suggestion that there is a vast array of stable or highly metastable states of string theory with four or less supersymmetries. Crucial to the emerging picture is the role of compactification with fluxes.[1,2,3,4,5,6] The most persuasive elaboration of this possibility to date is due to Kachru, Kallosh, Linde and Trivedi (KKLT),[7] who argue for the existence of a *discretuum* or *landscape*,[8] both supersymmetric with $N = 1$ supersymmetry, as well as non-supersymmetric, with supersymmetry softly broken. The existence of a landscape, if established, raises questions about the very nature of scientific explanation. Most importantly, this assertion places the anthropic principle at center stage. There has been strong reaction to this fact, ranging from near celebration by advocates of the anthropic principle to a great deal of handwringing and even denunciation from those who find the anthropic principle objectionable.

In this talk, I would like to give an overview of some of the issues raised by the possible existence of a landscape. I will explain why, even before we accept the landscape, some element of anthropic explanation is probably inevitable in quantum general relativity. Understanding the number of supersymmetries and the dimension of space-time may well require invoking some *extremely* weak anthropic considerations (what we might call the Minimalist Anthropic Principle, or MAP).

But the landscape requires a much broader application of the anthropic principle. I will stress in this talk is that the questions of the applicability and validity, of the anthropic principle are *scientific* ones. As to applicability, I will discuss what I believe has been reliably established by KKLT, and what has not (in this context I will present a few, admittedly very tenuous reasons, to hope that some sort of unique or nearly unique set of predictions emerge from string theory). I will mention various proposals for anthropic explanations of the cosmological constant, and show that, apart from the landscape, our experience with string theory renders the others (almost) implausible.

Then I will turn to the question of whether the anthropic principle is predictive or falsifiable. Michael Douglas[9] has put forward, most clearly and persuasively, the question of anthropic prediction in this context, though he prefers not to use this language. In particular, he stresses that the important question is determining the probability distributions for various physical quantities in the flux discretuum. One could imagine that, if these distributions were well understood, the anthropic principal could become predictive – and falsifiable. The typical vacuum consistent with anthropic constraints might not look at all like the Standard Model, or it might make a prediction, say of supersymmetry or some large dimension. Indeed, while these distributions are not yet sufficiently well known to make definite statements, I will argue that it is almost inevitable that this framework will predict low energy supersymmetry. But I will also explain why it is also likely to make a number of incorrect predictions. More precisely, if the flux discretuum is to describe nature, there are many questions whose explanations cannot be anthropic, but must emerge as a result of physical principles. These principles are not apparent in the landscape as envisioned by KKLT. This may be a sign that the discretuum is the *wrong* direction to understand fundamental issues in physics. Alternatively, it may mean that the business of particle physicists will be extracting these principles. In either case, we should neither *despair nor give up!*

2. How Many Supersymmetries?

Why don't we live in a universe with more than four supersymmetries? As I indicated above, there is almost certainly nothing wrong with these vacua. One could imagine that these are somehow disjoint theories from that which describes what we observe. This viewpoint has been put forward by Banks.[10] It has long seemed to me that some very mild anthropic considerations might rule these out (no conventional stars? no inflation? no structure?) I don't think that most of us would find this Minimalist Anthropic Principle particularly disturbing. It would not render the theory unpredictive or unfalsibiable.

What about theories with less supersymmetry? In some sense, theories with $N = 1$ supersymmetry are not, at first sight, obviously different than those with $N = 0$ supersymmetry. After all, generically one expects potentials on the moduli spaces of both albeit perturbatively in one case and non-perturbatively in the other. In my view, there is some tentative evidence that there is a real distinction. First, however, since we are speaking, at best, of approximate supersymmetry, we need some definitions. I will refer to $N = 0$ theories as theories for which there is no limit of the moduli space with only a finite number of supersymmetries. Examples include (compactifications of) the $O(16) \times O(16)$ heterotic theory, the non-supersymmetric heterotic M-theories studied by Fabinger and Horava, and others. $N = 1$ theories are those with only four supersymmetries in *some* limit of the moduli space. Familiar example include the heterotic and Type I theories on Calabi-Yau spaces, orientifolds of Type II theories on Calabi-Yau, and many more.

The $N = 0$ theories have a number of properties, which suggest that they might be qualitatively different from the supersymmetric ones. Conceivably many (most?) of them suffer from various diseases. One potential problem is the appearance of tachyons appear in the moduli space, but it is not clear that this signals any kind of consistency. Fabinger and Horava[11] have argued that $N = 0$ theories may suffer, generically, from other instabilities, particularly Witten's "decay to nothing".[12]. With Fox and Gorbatov,[13] I have been following up this latter issue. We have sharpened, but nor resolved, the question of the extent to which bubbles of nothing are generic to these spaces. There are, we believe, good reasons to think these instabilities *are* generic. We have also understood how these instabilities arise even in the presence of a potential on the moduli space, in $N = 0$ theories (but not in $N = 1$ theories). But while the prospect of decay to nothing is frightening, it is not clear whether this means that these states don't make sense as quantum theories. Still, it seems possible that with some work we can show that large classes of $N = 0$ theories don't make sense.

The discretuum suggests another approach to this question. As we will see, it is plausible (*but certainly not established*) that vacua with low energy supersymmetry, consistent with modest anthropic constraints, are by far the most numerous.

Before exploring anthropic arguments, it is worth noting that in controlled approximations, we don't presently know how to make sense of most (any?) string solutions with four or less supersymmetries. Generally, because there are potentials on on the moduli space, they are cosmological with singularities in their past or future, which we don't know how to treat.[14] Perhaps this is cause for optimism. After all, for space-like singularities, we know that string theory sometimes provides a resolution, and we know (in weakly coupled strings and sometimes elsewhere) the criterion (modular invariance). Perhaps there is an analogous condition for time-like singularities. If we were lucky, perhaps this condition would yield only a small set of consistent theories.

3. The Dreaded Anthropic Principle

Linde was probably the first to realize that inflation leads to a framework in which one might sensibly implement the anthropic principle.[15] Perhaps in a very vast universe, the fundamental parameters take different values in different regions. The most promising application of this idea has long been to the problem of the cosmological constant, for which it is fair to say that we currently have no equally successful candidate solution.[16]

The response to this suggestion has been enthusiasm in some quarters and revulsion in others. I have been more agnostic, and I strongly believe that the question itself is a scientific one. If string theory provides a means to implement the anthropic principle, then we *have no choice but to consider it seriously.* Rather than wasting time arguing about philosophical issues, I think the smart money will be on trying to figure out whether one can make predictions or falsify the theory. Some ideas for how predictions might emerge will be discussed below. If string theory does not provide a setting for the anthropic principle, while we can't, perhaps, "disprove" it, I think we can justifiably, *and smugly*, ignore it.

There are, two my knowledge, two principle scenarios for implementing the anthropic principle:

(1) Extremely light scalars:[17,18,19] Here, if one is to solve the cosmological constant problem, one requires a scalar with a mass smaller than the present Hubble constant, but which varies over such a large region of field space that it can cancel an energy density at least of order TeV^4.
(2) A discretuum or landscape [20,1,2,7] (as for the references, the first provided a simple model based on an "irrational axion", but no example of this type has been exhibited in string theory; the second and third used fluxes in a manner not too different than that currently of interest; the last is the most fully developed of these ideas).

Each of these scenarios is quite radical, and their plausibility cannot be assessed without a theory which is fundamental (in the sense of including gravity and other interactions, and finite) like string theory.

3.1. *Extremely Light Scalars*

In string theory, it appears unlikely that there are scalars with the requisite properties to solve the cosmological constant problem.[a] Ordinary scalars in string theory have masses generally have masses consistent with dimensional analysis, e.g. set by the scale of supersymmetry breaking. String theory also has periodic scalars, for which shift symmetries can suppress masses. So these might be candidates, but one needs decay constants far greater (exponentially greater) than the Planck mass. Searches in string/M theory have not yielded any candidates.[24] (Such fields would also be of interest as candidate inflatons.[22])

While this cannot yet be considered a theorem, it seems unlikely that this sort of implementation of the anthropic principle is realized in string/M theory. What is perhaps most interesting is that it might be possible, with a finite amount of effort, to rule out this implementation completely.

3.2. *Discretuum or Landscape*

There have been various proposals for a discretuum. That of 20 does not seem to be realized in string theory. That of 1, while interesting, made assumptions about the stabilization of moduli for which there was no support. The proposal of 7 is the most promising to date, including, as it does, a detailed picture of the stabilization of moduli. This proposal is the subject of the next section.

4. The Proposal of KKLT

KKLT have put forth a quite detailed proposal for how a landscape might emerge in string theory. There is not space here to fully review this set of ideas, but we should note, first, that they are based on effective field theory. Banks has argued cogently that one cannot use effective field theory to study multiple vacua in theories of gravity.[23] For example, in many circumstances there are no transitions between the different states, and an observer in one can not do experiments which will indicate the existence of others. So it is not clear that the multiplicity of states has any meaning. These arguments have been reviewed and elaborated in 25.

[a] Dimopoulos and Thomas[21] have recently suggested a possibility which could conceivably be realized in string theory: they have argued that a conformal field theory might yield a huge Z factor for some scalar field. This possibility is worthy of further exploration. This discussion itself is an example of how the anthropic principle is an issue whose validity and relevance can be decided by scientific means.

Setting aside these larger questions of principle, I want to focus here on questions of self-consistency with the effective action analysis. I will not be able to fully answer whether KKLT have actually demonstrated the existence of a discretuum. My tentative conclusion will be that there is a discretuum of states with $N = 1$ supersymmetry, but that only a limited set of quantities can actually be computed in these states. Because of the latter problem, studies of supersymmetry breaking may be more problematic, but the existence of a supersymmetric discretuum is at least plausible. Some of these questions can probably be answered through further investigation.

Compactifications of string theory (IIB on a CY, X, or F-theory on a Calabi-Yau four-fold, for definiteness) permit many possible quantized fluxes, F_{IJK}, H_{IJK}. The number of possible three cycles (b_3) can be of order 100's. These fluxes are not highly constrained. Tadpole cancellation gives one condition on many fluxes. Because there are so many possible choices of fluxes, there are potentially an exponentially large number of states.

KKLT, following earlier work[5,3,6] noted that the presence of flux tends to stabilize moduli. For example, 5 considered orientifolds of Type II theory on a Calabi-Yau space near a conifold point. If z is the modulus describing the distance from the point, then in the presence of flux on the collapsing three cycles one finds both stabilization and warping. There is a superpotential for z and the dilaton, $\tau = \frac{i}{g} + a$,

$$W = (2\pi)^3 \alpha' (M\mathcal{G}(z) - K\tau z) \tag{1}$$

where M and K denote the flux quanta, and

$$\mathcal{G}(z) = \frac{z}{2\pi i} \ln(z) + \text{ holomorphic.} \tag{2}$$

This effective action has a supersymmetric minimum where

$$D_z W = \frac{\partial W}{\partial z} + \frac{\partial K}{\partial z} W = 0 \tag{3}$$

This is solved by:

$$z \approx \exp\left(-\frac{2\pi K}{M g_s}\right) \tag{4}$$

If the ratio N/M is large, then z is very small. The corresponding space can be shown to be highly warped. In addidtion,

$$W_o = < W > \tag{5}$$

is exponentially small.

Including additional fluxes, it is possible to fix other complex structure moduli and also the dilaton, τ.

$$W = (2\pi)^3\alpha'[M\mathcal{G}(z) - \tau(Kz + K'f(z))] \tag{6}$$

$$D_\tau W = \frac{\partial W}{\partial \tau} + \frac{\partial K}{\partial \tau} W = 0$$

for

$$\bar{\tau} = \frac{M\mathcal{G}(0)}{K'f(0)} \qquad W = 2(2\pi)^3\alpha' M\mathcal{G}(0)$$

z is still exponentially small, and the space is highly warped, but *now* W_o *is no longer exponentially small.* This is crucial to the KKLT picture. In the limit that R is very large, there is no potential for R; the compactification radius is not fixed.

4.1. *Fixing the Remaining Moduli?*

KKLT noted that in flux vacua, W_o is generically large (of order some typical flux integer), but argued that among the vast number of possible fluxes, W_o will sometimes be small, simply by chance. Other effects (e.g. gluino condensation) will generate a superpotential for $\rho = R^4 + ib$:

$$W = W_o + e^{-\rho/c} \tag{7}$$

This has a supersymmetric minimum, with

$$\rho \sim -c\ln(W_o) \tag{8}$$

Consistency of the analysis requires that ρ be large, but this will be the case only in an exponentially small fraction of states.

If there is a systematic approximation, it consists of integrating out the string modes followed by the KK modes, the complex structure and dilaton, and finally the radial mode. Consistency requires a hierarchy of masses:

$$M^2_{string} >> M^2_{kk} = 1/R^2 >> M_{\tau^2, z^2} >> m^2_\rho. \tag{9}$$

Before making detailed calculations, these conditions seem to hold at large radius. The masses of the τ and z fields are suppressed by ρ relative to the Kaluza-Klein modes, and by $\rho^{3/2}$ relative to the string scale. But in the presence of large fluxes or large b_3, there are a number of sources of enhancement, which can be inferred from the expression for the superpotential:

(1) Factors of fluxes, $(N, M)^2$

(2) Factors of τ, τ^2

(3) Factors due to the large size of the mass matrix, b_3. In the absence of any detailed understanding of the mass matrix, we might worry that a large, random matrix has eigenvalues which grow with the size of the matrix.

This list suggest that one might expect that

$$\frac{m_{z,\tau}^2}{M_{string}^2} = \frac{b_3 \tau^2 N^2}{\rho^{3/2}}. \tag{10}$$

If we require, say $\tau = 5$, and suppose that a typical N is of order 3, this requires that ρ, be quite large, and therefore that W_o is *extremely* small. Indeed, if we take for the number of states that suggested by KKLT and by Douglas, $e^{b_3 \ln(\frac{\chi}{24}/2)}$, there might well not be enough states that one could imagine doing a self-consistent computation in *any* of them.

But this may not call the KKLT discretuum into question.[b] Imagine studying the theory at extremely large ρ. Here, one can compute W_o reliably. One can also compute the leading ρ-dependent terms in the superpotential. Because of non-renormalization theorems, this form of the superpotential will remain valid until the point where higher order exponentials become important. This requires only that e^{ρ} be small, which would seem to be a much weaker requirement than equation 9.

Supersymmetry breaking, from this point of view, could be more problematic. One will not have the same level of control of the Kahler potential, for example.

It is worth stressing what serves as the small parameter which justifies analysis in the landscape. It is not large flux numbers or large b_3, per se, but the fact that in the vast discretuum of states, there is a small fraction – but large number – of states with small W_o.

5. Supersymmetry and Supersymmetry Breaking

For now, let's accept that a discretuum exists, and that the universe samples all of these states in its cosmic history. KKLT argued that anti D3-branes, located well down the throat can give exponentially small effects in a warped geometry. An alternative possibility, which is perhaps somewhat easier to think about, is to note that in some fraction of this vast array of states, the

---

[b] I thank the participants at the KITP workshop on string cosmology for discussions of these issues, and especially Joe Polchinski.

low energy dynamics presumably breaks supersymmetry. (By assumption, the discretuum contains states with complicated gauge groups and chiral fermions.) Then

$$V = \exp(-8\pi^2/b_o g^2) \qquad (11)$$

If g^{-2} is distributed more or less uniformly, V will be distributed roughly uniformly on a log scale. Correspondingly, there will be a substantial number of states where the cosmological constant,

$$V = [\exp(-8\pi^2/b_o g^2) - 3|W_o|^2] \qquad (12)$$

is small compared to $|W_o|^2$.

6. Anthropics

So we have a picture in which there are many, many states. Among these states, quantities relevant to low energy physics vary:

- Low energy gauge groups
- The matter content
- The values of the parameters of the low energy lagrangian.

If the universe samples all of these states, there will only be observers in a subset with suitable properties.

6.1. *The Cosmological Constant*

The cosmological constant provides the most compelling application of the anthropic principle.[16] Holding all other fundamental and cosmological parameters fixed, one finds that suitable structure forms only if Λ is less then or of order 10 times its observed value[c].

6.2. *Does the Anthropic Landscape Predict Low Energy Supersymmetry*

We cannot prove at present that the landscape predicts low energy supersymmetry, but it seems likely that it would. As Douglas has explained (see also 25), for small W_o, the distribution of W_o is likely to be uniform. Suppose, also, that the origin of supersymmetry breaking is dynamical, as

[c]Allowing other quantities to vary permits a far broader range of Λ. See, e.g., 26.

described above, and that $\frac{8\pi^2}{g^2}$ roughly uniformly distributed. Then, e.g., if $W_o = 10^{-28}$ (susy breaking $\sim 10^4$ GeV), in about 10^{-3} of states,

$$V_o = e^{-\frac{8\pi^2}{bg^2}} - 3|W_o|^2 < |W_o|^2.$$

In this subset of states one has the possibility of obtaining a small cosmological constant.

Suppose that the anthropic argument for Λ is correct. We can compare the number of states with sufficiently small Λ with and without supersymmetry. Of course, since we have not done a reliable counting, we are describing here only a program which might lead to such a prediction, and its hypothetical results.

Without supersymmetry, we might expect simply:

$$P(\Lambda) \approx \frac{\Lambda}{M_p^4}.$$

(here P is the probability of cosmological constant less than Λ).

With supersymmetry, we start with the probability of small W_o:

$$P(W_o) \approx \frac{W_o}{M_p^3}.$$

Now small Λ will favor small supersymmetry breaking. This is the usual argument about the connection of the cosmological constant and the scale of supersymmetry breaking, and it remains valid in this framework. This means that we will require more anthropic input. In particular, we might imagine that the ratio: M_w/M_p is fixed (note that we would also require something like this for non-supersymmetric theories, presumably, to account for the hierarchy). So we might estimate the fraction of suitable states along these lines:

(1) 10^{-10} have suitable susy breaking (i.e. low energy gauge groups with suitable properties)
(2) 10^{-2} have susy breaking comparable to W_o
(3) 10^{-13} have suitable W_o.
(4) 10^{-60} of these have small Λ.

So we guess that for supersymmetric states, we pay a factor of 10^{-85} to realize this set of anthropic constraints, vs. $10^{-120} \times 10^{-32}$ for non-susy states. So SUSY wins unless there are an overwhelmingly large number of non-susy states relative to supersymmetric ones. Note that this picture favors susy breaking at the lowest possible scale (gauge mediation?)

So one sees here how the anthropic principle, coupled with knowledge about the distribution of states, might lead to a real prediction. But before getting too excited, there are other issues to face in the flux discretuum.

6.3. *Anthropic Pitfalls*

A program of implementing the anthropic principle within the landscape faces several hurdles, and at least at first sight, seems likely to fail. Within the landscape, as we currently imagine it, there are numerous states with different gauge groups, particle content, and couplings. These features must either be fixed anthropically, or are otherwise random. But there are many features of the Standard Model which seem neither anthropically constrained *nor random.*[27]

In thinking about this sort of anthropic selection, it is useful to organize the problem by considering physics first at very large distance scales, and then at progressively shorter scales, using the language of effective actions and the renormalization group.

At the very largest distances, we face the problem of the cosmological constant, which we have already discussed. At shorter distances, we face the question of the existence of an unbroken $U(1)$ symmetry. This is plausibly anthropic. The question of whether the strong group is $SU(3)$ or something else might be anthropic, but this is more difficult to decide. E.g. for groups other than $SU(3)$, by varying m_u and m_d we can probably reproduce many features of nuclear physics. Conceivably deuterium is essential for stellar nucleosynthesis, and this might single out $SU(3)$. The ratio m_e/Λ_{qcd}, another important parameter of the low energy lagrangian, might be fixed, for example, by molecular physics and/or by astrophysics. The relative size of m_u and m_d might also be determined by the details of nuclear physics; at the grossest level, the fact that $m_d > m_u$ is necessary for proton stability (as opposed to neutron stability).

While plausibly constrained by anthropic arguments, making persuasive arguments for these quantities will be challenging, at the very least. But at higher energies, we encounter couplings whose prediction is more problematic: m_s, m_c, m_b, V_{km}. It is hard to see how these could be anthropic (if there were obvious anthropic arguments, one might have predicted the values of these quantities prior to their discovery). One of the most puzzling is θ_{qcd}. It is not at all clear what sort of anthropic argument might require $\theta_{qcd} << 1$.

In the landscape, it would appear that all of these quantities are simply random numbers. θ_{qcd}, for example, is presumably not small in a typical vacuum. The fluxes, generically, break CP and contribute to θ_{qcd}. Moreover, the assumption is that all of the moduli are fixed would seem to preclude axions. E.g. cross terms in the potential for ρ involving W_o will give a large potential to the pseudoscalar component.

At still high energies, one encounters further questions. The weak gauge group may be hard to understand anthropically. Dark matter, cosmological parameters (the size of inflationary fluctuations, the number of e-foldings, and the like) will pose deeper challenges.

In the flux discretuum, the parameters of low energy physics seem to be random numbers. If this is really true, the landscape is not a correct description of physics. Alternatively, there are some set of principles in the landscape which explain those laws of nature which do not seem to be anthropically constrained. Within the flux discretuum, it is not obvious what these might be. As another example, anthropically, the proton lifetime is probably not required to be much larger than 10^{16} years. So one might to hope to understand the length of the proton lifetime from symmetries. But most states of the flux discretuum don't have symmetries.[d] For θ_{qcd}, we might want to find a reason why some modulus is not fixed at high energies. One could imagine that this is cosmological.

7. Conclusions

Whether one likes it or not, it is quite possible that quantum theories of gravity predict a landscape. It is plausible that the flux vacua explored by KKLT exist, but it is by no means certain. Perhaps the strongest reason for doubt is Banks' general critique of effective action methods in gravity theories, but I have mentioned some concerns within the conventional framework of effective actions. I have also mentioned some curious issues of scales, within the framework of effective field theories. One interesting feature of the proposed discretuum which we have noted is that the small parameter is essentially a random variable.

If there is a discretuum, it seems quite likely that we will be able to falsify this whole picture. Typical states, even subject to anthropic constraints, will disagree violently with observation. To avoid this conclusions,

[d]For example, in the case of the T_6/Z_2 orbifold, one has a $Z_2^5 \times S_6$ symmetry at some points in the moduli space. But half of the fluxes must vanish to preserve even one Z_2. In more realistic models, this is likely to correspond to a drastic reduction of the number of states in the discretuum.

we would need to show that in some subset of states – which are picked out by other considerations – there is some rational explanation of the many features of the Standard Model which don't seem susceptible to anthropic explanation. We have seen, for example, how low energy supersymmetry could conceivably emerge from the requirement of small cosmological constant and suitable electroweak symmetry breaking. Perhaps some considerations might lead to approximate flavor symmetries, small θ, and the like. On a pessimistic note, we have argued that symmetries are unlikely to play this role.

Mention of the anthropic principle brings out strong reactions from most physicists, who ask what can be the role of science in such a situation. But the lesson of the KKLT proposal is not so pessimistic. First, the existence of a landscape within string theory is a question we should be able to decide. If we decide that there is such a discretuum, we will probably be forced to contemplate the anthropic principle; if not, we can dismiss it. But even if we do adopt the anthropic principle, it will at best explain only a few quantities: either we will falsify string theory, or we will uncover principles which explain most of the features of the Standard Model. We will likely make additional predictions for accelerators and cosmology as well. So surrender to the anthropic principle will not be necessary or possible; we won't have to give up.

Acknowledgments

I wish to thank my collaborators on this work, Tom Banks and Elie Gorbatov. I also wish to thank Michael Douglas, Patrick Fox, Renata Kallosh, Shamit Kachru, Andrei Linde, Joe Polchinski, Eva Silverstein, and Sandip Trivedi for conversations which have shaped my thinking on these matters. I particularly want to thank the hospitality of the KITP and the participants in the string cosmology workshop for helpful comments and criticisms.

References

1. R. Bousso and J. Polchinski, JHEP **0006**, 006 (2000) [arXiv:hep-th/0004134].
2. J. L. Feng, J. March-Russell, S. Sethi and F. Wilczek, Nucl. Phys. B **602**, 307 (2001) [arXiv:hep-th/0005276].
3. K. Becker and M. Becker, Nucl. Phys. B **477**, 155 (1996) [arXiv:hep-th/9605053].
4. J. de Boer, R. Dijkgraaf, K. Hori, A. Keurentjes, J. Morgan, D. R. Morrison and S. Sethi, Adv. Theor. Math. Phys. **4**, 995 (2002) [arXiv:hep-th/0103170].

5. S. B. Giddings, S. Kachru and J. Polchinski, Phys. Rev. D **66**, 106006 (2002) [arXiv:hep-th/0105097].
6. B. S. Acharya, arXiv:hep-th/0212294 and arXiv:hep-th/0303234.
7. S. Kachru, R. Kallosh, A. Linde and S. P. Trivedi, arXiv:hep-th/0301240.
8. L. Susskind, arXiv:hep-th/0302219.
9. M. R. Douglas, JHEP **0305**, 046 (2003) [arXiv:hep-th/0303194]; S. Ashok and M. R. Douglas, arXiv:hep-th/0307049.
10. T. Banks, arXiv:hep-th/0306074.
11. M. Fabinger and P. Horava, Nucl. Phys. B **580**, 243 (2000) [arXiv:hep-th/0002073].
12. E. Witten, Nucl. Phys. B **195**, 481 (1982).
13. M. Dine, P.J. Fox, and E. Gorbatov, to appear.
14. T. Banks and M. Dine, JHEP **0110**, 012 (2001) [arXiv:hep-th/0106276].
15. Linde was apparently the first to appreciate that inflation provides a natural setting to implement the anthropic principle: A.D. Linde, "Nonsingular Regenerating Inflationary Universe" Print-82-0554 (Cambridge); "The New Inflationary Universe Scenario" in The Very Early Universe, eds. G.W. Gibbons, S. W. Hawking and S. Siklos (Cambridge University Press, Cambridge 1983) 205-249. It is also discussed in A.D. Linde, *Partile Physics and Inflationary Cosmology*, (Chur, Switzerland, Harwood, 1990)
16. S. Weinberg, arXiv:astro-ph/0005265.
17. T. Banks, Phys. Rev. Lett. **52**, 1461 (1984).
18. A.D. Linde, "Inflation And Quantum Cosmology," Print-86-0888 (June 1986), in: Three hundred years of gravitation, (Eds.: Hawking, S.W. and Israel, W., Cambridge Univ. Press, 1987), 604-630.
19. J. Garriga and A. Vilenkin, Phys. Rev. D **61**, 083502 (2000) [arXiv:astro-ph/9908115].
20. T. Banks, M. Dine and N. Seiberg, Phys. Lett. B **273**, 105 (1991) [arXiv:hep-th/9109040].
21. S. Dimopoulos and S. Thomas, Phys. Lett. B **573**, 13 (2003) [arXiv:hep-th/0307004].
22. N. Arkani-Hamed, H. C. Cheng, P. Creminelli and L. Randall, Phys. Rev. Lett. **90**, 221302 (2003) [arXiv:hep-th/0301218].
23. T. Banks, arXiv:hep-th/0011255; T. Banks, arXiv:hep-th/0211160.
24. T. Banks, M. Dine, P. J. Fox and E. Gorbatov, arXiv:hep-th/0303252.
25. T. Banks, M. Dine and E. Gorbatov, arXiv:hep-th/0309170.
26. A. Aguirre, Phys. Rev. D **64**, 083508 (2001) [arXiv:astro-ph/0106143].
27. Related observations have been made by J. Donaghue, in J. F. Donoghue, arXiv:hep-th/0310203.

CHIRALITY CHANGE IN HETEROTIC STRING THEORY*

MICHAEL DOUGLAS

Department of Physics and Astronomy
Rutgers University,
Piscataway, NJ 08855-0849, USA
and
I.H.E.S.,
Le Bois-Marie, Bures-sur-Yvette, 91440 France
E-mail: mrd@physics.rutgers.edu

CHEN-GANG ZHOU

Department of Physics and Astronomy,
University of Kentucky,
Lexington, KY 40517
E-mail: czhou@pa.uky.edu

It is known that string theory compactifications leading to low energy effective theories with different chiral matter content (*e.g.* different numbers of standard model generations) are connected through phase transitions, described by non-trivial quantum fixed point theories. We point out that, contrary to widespread belief, such compactifications are also connected on a purely classical level, through transitions that can be described using standard effective field theory. Explicit examples are given in heterotic string theory compactifications.

1. Introduction

One of the early successes of string theory was a natural mechanism producing a multiplicity of generations minus antigenerations of standard model like spectrum, in the context of heterotic string compactification on a Calabi-Yau threefold (CY_3) M, using simple topological properties of the CY_3 [1].

More generally, one starts with $d = 10$, $\mathcal{N} = 1$ supergravity/super Yang-Mills with $E_8 \times E_8$ or $Spin(32)/\mathbb{Z}_2$ gauge group and with the stringy corrections needed for anomaly cancellation, and still compactify on a CY_3,

*Talks by M.D. and C.-G.Z.

only generalizing by taking a general background gauge connection for a vector bundle V satisfying anomaly cancellation constraints [2,3,4]. These are often referred to as "(0, 2)" models from the world-sheet supersymmetry of the heterotic sigma model; in nonperturbative heterotic string theory [5] this class of models is further generalized, but the number of generations minus antigenerations, which we will denote N_{gen}, is still determined by the choice of V.

In this talk we discuss whether heterotic string theory vacua with different chiral matter spectrum, including standard models with different numbers of generations, are connected. Previous work on this question includes [6] who discuss a mechanism which connects such vacua through a non-trivial phase transition, whose IR behavior is governed by the dimensional reduction of the phase transition associated to the $(1,0)$ superconformal theory in six dimensions. Further work supporting this idea appears in [7].

These works are based on a (still) common belief that the number of generations provides an "index" which in classical physics is conserved under continuous variations of the background, so that such transitions could only proceed through a phase transition. We will give simple arguments that this is false and provides explicit examples. Before studying more complicated case of CY_3, we begin with a toy model of gauge theory compactification on S^2, which illustrates the phenomena of chirality change nicely.

2. Toy Model: Gauge Theory Compactification on S^2

Consider KK reduction of 6d gauge theory over S^2 to R^4. Early work on KK reduction of Einstein-Maxwell system from 6d to 4d on S^2 can be found in [8,9]. On a two dimensional manifold, line bundles are classified topologically by the first Chern class $c_1(L)$, taking value in integer valued cohomology class $H^2(M,\mathbb{Z}) \cong \mathbb{Z}$. By index theorem, the number of chiral fermion zero modes for a fermion coupling to the $U(1)$ connection associated to the line bundle L is determined by this number. For example, a charge $+1$ fermion coupled to connection on $\mathcal{O}(n)$ will have n net positive chirality zero modes. Upon KK reduction, these fermion zero modes correspond to massless chiral fermions in the low energy theory. Obviously, there is no transition between connections corresponding to line bundles $\mathcal{O}(m)$ and $\mathcal{O}(n)$ for $m \neq n$: topological obstruction denoted by c_1 prevents it.

However, such obstruction disappears when the line bundle is embedded in a $SU(2)$ bundle V. The topological type of V is also uniquely classified by the first Chern class, which is trivially zero. But there are infinitely many ways to split V into direct sum of two line bundles[10]:

$$V \cong \mathcal{O} \oplus \mathcal{O} \cong \mathcal{O}(1) \oplus \mathcal{O}(-1) \cong \mathcal{O}(1) \oplus \mathcal{O}(-1) \cong \cdots . \qquad (1)$$

Each splitting type produces $U(1)$ gauge theory in R^4 with different chiral fermion content upon KK reduction. Because they are embedded in the same unique bundle V, one can easily construct smooth paths in the space of $SU(2)$ connections on S^2 to connect these reduced $U(1)$ connections. Although in this case, the path does not lie entirely within the extremal configurations of the Yang-Mills connections, which is equivalent to say that it passes out of the moduli space of vacua, it illustrates the principle of smooth chirality change. This phenomena can be generalized to any Riemann surface, by appealing to the analysis of Atiyah and Bott[11] on general solutions to the Yang-Mills equations of Riemann surfaces and how they sit in the total space of connections. More realistic compactification can be found in the case of high dimension (CY_3) which we will turn to in the next section.

3. Chirality Change on Compact Calabi-Yau Manifold in Heterotic Theory

Similar to the case that different $U(1)$ bundles on S^2 can be connected inside $SU(2)$ bundle, in heterotic string compactification on CY_3 to 4d, all vacua obtained by making different choices of V must be connected already on the classical level, without going through non-classical phase transition. The argument is essentially the following. The low energy gauge group is the commutant in the ten dimensional gauge group (say) $E_8 \times E_8$ of the structure group G of the bundle V. To get a realistic grand unified gauge group, we take V to have structure group $SU(k)$ (for example $k = 4$ leads to unbroken $SO(10)$). The number of generations is then $N_{gen} = c_3(V)/2$, as follows from the index theorem[12]. The only constraint on V is from anomaly cancellation condition, $c_2(V) = c_2(M)$, as required by the well-definedness of the three form H, which satisfies $dH = \frac{1}{4\pi}(trF \wedge F - trR \wedge R)$.

Now for bundles with structure group $SU(k)$, $c_3(V)$ is a topological invariant. On the other hand, if we allow general motions in configuration space (possibly through a potential barrier), nothing prevents us from deforming the connection on V into the full ten dimensional gauge group. Now, for E_8 and $Spin(32)/Z_2$ bundles, $c_3(V)$ is not a topological invariant; in fact it cannot even be defined: its definition uses the third order symmetric invariant d^{abc}, which does not exist for these Lie algebras. (An equivalent way is to say c_3 of a E_8 bundle is zero, whose similarity to $SU(2)$ bundle on S^2 is obvious. Indeed, as is familiar from other discussions [13], all E_8 vector bundles on a manifold M of (real) dimension less than 15 with a specified rank and second Chern class $c_2(V)$ are connected by smooth deformations. Such a deformation will correspond to a path in the configuration space of the full string theory, connecting vacua with two possibly different values of N_{gen}.

Naively, this might be thought paradoxical as chiral matter cannot be lifted by varying uncharged fields or parameters in a low energy effective Lagrangian. The obvious loophole which this suggests is that chiral matter must be lifted by giving vevs to charged fields, as is familiar even in the standard model. When the gauge connection in V deforms into E_8 (or any larger group containing the original $SU(k)$), some or all of the low energy gauge symmetry is broken. If the chiral matter is nonchiral under the remaining unbroken gauge symmetry, there is no barrier to giving it a mass. As an example, after a deformation breaking $SO(10)$ to $SO(9)$, a chiral spinor in the 16 of $SO(10)$ becomes a nonchiral spinor of $SO(9)$, which can be lifted.

The remaining question is how the chiral matter can remain massive when the original gauge symmetry is restored with different c_3. The answer is that the final result is a new model in which the original gauge symmetry is *not* restored. Rather, an isomorphic but different copy of the gauge group becomes unbroken. One should think of the low energy gauge group as a small unbroken subgroup of a large (even infinite dimensional) gauge group, the group of all gauge transformations of E_8 gauge theory in six dimensions, almost all of which is broken by the background. Even if the unbroken gauge groups G are isomorphic on the two sides of the transition, they can be *two different embeddings* of this gauge group in the large gauge group.

The minimal effective field theory realization of this phenomenon is in terms of $G \times G$ gauge theory, with matter chosen to always break at least one copy of G, and along the interpolating path break to a subgroup of G.

One expects the mechanism of [6] to involve no potential barrier, because it goes through a superconformal theory, with no preferred energy scale. As for the classical mechanism we discuss, on general grounds, one might expect variations of the gauge connection on a compact manifold of "size" R to go through a potential barrier of height $V \sim 1/g_{YM4}^2 R^4$. This can be made small at large radius or strong coupling but would be non-zero.

After clearing up the conceptual issues, we return to the actual construction of vector bundles realizing such transition. On the quintic Calabi-Yau

3-fold, monad construction[14,15,16,17,18] is employed to find vector bundle V with fixed second class, which is defined by the exact sequence

$$0 \to V \to \oplus_{i=1}^{n+M} \mathcal{O}(n_i) \stackrel{\phi=(Q_j^i(X))}{\longrightarrow} \oplus_{j=1}^{M} \mathcal{O}(m_j) \to 0. \quad (2)$$

All the topological data of the bundle V are encoded in the sets m_i and n_j. To give some examples, we list the rank-4 bundles here.

Table 1. Rank-four vector bundles on quintic Calabi-Yau manifold with second Chern class satisfying anomaly cancellation condition $c_2(V) = c_2(M)$.

rank (n)	(n_i, m_j)	N_{gen}
4	(1122222222;333333)	80
4	(11122222;3334)	85
4	(111122;44)	90
4	(11111;5)	100

These vector bundles generically has no singularity. One can also (partially) prove that they are generically stable, which give string theory vacua with effective $d = 4$ $\mathcal{N} = 1$ $SO(10)$ gauge theory. In this case, one does not need to go to the full E_8 gauge group to realize the transition, a $SO(8)$ subgroup of the E_8 group is enough. Explicitly, $V \oplus V^*$ (where V^* is the complex conjugate of the bundle V) can be regarded as a $SO(8)$ bundle with zero c_3, and all the $SU(4)$ bundle V with different c_3 can be connected within this bundle, and realizing chirality change classically. The low energy effective theory would be two isomorphic $SO(10)$ gauge theories with different number of chiral matters being connected through a $SO(8)$ gauge theory without chiral matter, and can be described by Higgs mechanism.

4. Conclusion

More examples of chirality change can be found. In particular, the brane construction in type IIA orientifold theory gives example of chirality change proceeds entirely through supersymmetric configuration[19].

The basic loophole in the naive argument that classical chirality change is not possible is that the true effective field theory description of the situation is in terms of an infinite gauge group, which in the gauge theory examples is the group of all maps from the internal space to the (finite dimensional) gauge group, which contains many copies of any given low energy gauge group. Chirality change takes place by breaking one copy of the low energy gauge group, and restoring a different copy with different chiral matter.

If one desires to think of "all" or large sets of string theory compactifications as described by a single effective field theory, the "effective field theory of everything," with all quantum effects already summarized in the effective Lagrangian, then it would have to contain an infinite number of fields, almost all of which are massive in any small region of configuration space, and would probably not be the best way to study any one compactification. Nevertheless it might be very useful as a unifying and organizing concept behind the general discussion. This is reminiscent of work by Giveon, Porrati and Rabinovici[20] on $\mathcal{N} = 4$ supergravity-super Yang-Mills theories arising from toroidal compactification of the heterotic string.

References

1. P. Candelas, G. T. Horowitz, A. Strominger and E. Witten, *Nucl.Phys.* **B258**:46-74, 1985.
2. C. Hull and E. Witten, *Phys.Lett.***B160**(1985)398.
3. E. Witten, *Nucl. Phys.***B268**(1986)79.
4. J. Distler and B. Greene, *Nucl. Phys.***B304**(1988)1-62.
5. P. Horava and E. Witten, *Nucl.Phys.***B475**(1996)94-114, hep-th/9603142.
6. S. Kachru, E. Silverstein, *Nucl.Phys.***B504**(1997)272-284, hep-th/9704185; E. Silverstein, hep-th/9709209.
7. B. Ovrut, T. Pantev and J. Park, **JHEP 0005**:045, 2000, hep-th/0001133.
8. A. Salam, J. Strathdee, *Annals Phys.***141**(1982) 316-352.
9. S. Randjbar-Daemi, A. Salam, J. Strathdee, *Nucl.Phys.***B214**(1983) 491-512.
10. A. Grothendieck, *Am.J.Math.* **79**(1957), 121-38.
11. M. Atiyah, R. Bott, *Phil. Trans. R. Soc. Lond.***A308** 523-615 (1982).
12. M.B. Green, J.H. Schwarz, E. Witten, *Superstring Theory*, Volume II, Cambridge: Cambridge University Press.
13. D.-E. Diaconescu, G. Moore and E. Witten, *Adv.Theor.Math.Phys.***6**:1031-1134,2003, hep-th/0005090.
14. J. Distler, B. Green, *Nucl. Phys. B* **304** (1988) 1.
15. S. Kachru, *Phys.Lett. B* **349**(1995) 76.
16. P.Candelas, M.Lynker and R. Schimmrigk, *Nucl. Phys. B* **341** (1990) 383.
17. T. Hübsch, Calabi-Yau manifolds, A Bestiary for Physicists, World Scientific, Singapore 1991.
18. A. Lukas and B.Ovrut, hep-th/9908100.
19. M. Douglas and C-G Zhou, in preparation.
20. A. Giveon, M. Porrati and E. Rabinovici, *Phys.Rept.***244**(1994)77-202, hep-th/9401139.

SUPERSTRINGS AND WZNW MODELS*

P. A. GRASSI[A,B,†], G. POLICASTRO[C,‡] and P. VAN NIEUWENHUIZEN[A,§]

[(A)] *C.N. Yang Institute for Theoretical Physics, State University of New York at Stony Brook, NY 11794-3840, USA*

[(B)] *Dipartimento di Scienze e Tecnologie Avanzate, Università del Piemonte Orientale C.so Borsalino, 54, Alessandria, 15100 Italy*

[(C)] *DAMTP, Center for Mathematical Science, Wilberforce Road, Cambridge CB3 0WA, UK*

We give a brief review of our approach to the quantization of superstrings. New is a derivation of the measure at tree level and a path integral formula for this measure.

1. A brief review

In a series of papers [1] we have presented a new approach to the old problem of the covariant quantization of the superstring. The fields of the classical superstring are $x^m(z,\bar{z})$ and two (i=1,2) spacetime spinors $\theta^{\alpha i}(z,\bar{z})$ where $m = 0,\ldots,9$ is a vector index and $\alpha = 1,\ldots,16$ the index of a real chiral spinor. A natural choice as action might seem to be the supersymmetric line element

$$S_1 = \int d^2x \frac{1}{2} \Pi^m_\mu \Pi^\mu_m\,, \tag{1}$$

where $\Pi^m_\mu = \partial_\mu x^m - i\sum_i \theta^{\alpha i}\gamma^m_{\alpha\beta}\partial_\mu\theta^{\beta i}$ with $\gamma^m_{\alpha\beta} = \gamma^m_{\beta\alpha}$ real Dirac matrices.

It is invariant under the rigid supersymmetry transformations $\delta_\epsilon x^m = i\epsilon^i\gamma^m\theta^i$ and $\delta_\epsilon\theta^{\alpha i} = \epsilon^{\alpha i}$. However, there is no kinetic term for $\theta^{\alpha i}$. One

*Talk given by P.v.N.

†pgrassi@insti.physics.sunysb.edu

‡g.policastro@damtp.cam.ac.uk

§vannieu@insti.physics.sunysb.edu

can add a Wess-Zumino term [2]

$$S_{WZ} = \int d^3x \sum_i (-)^i d\theta^i \gamma^m d\theta^i \Pi_m \,, \tag{2}$$

It is separately supersymmetrically invariant, and $S_1 + S_{WZ}$ has the (in)famous local κ symmetry which allows one to choose the light-cone gauge in target space. Manifest Lorentz covariance is then lost. So far the covariant quantization of this action has remained an open problem.

Our approach begins with an idea of Berkovits [3] who started from the BRST charge

$$Q_B = \oint dz \lambda^\alpha d_\alpha \,, \qquad d_\alpha = p_\alpha + i\partial x^m (\gamma_m \theta)_\alpha + \frac{1}{2}(\gamma^m\theta)_\alpha(\theta\gamma_m\partial\theta) \,. \tag{3}$$

We are now on the Euclidean worldsheet with light-cone derivatives $\partial = \partial_\sigma - i\partial_\tau$ and restrict our attention to only one θ^α. The OPE's $x^m(z)x^n(w) \sim -\eta^{mn} ln(z-w)$ and $p_\alpha(z)\theta^b(w) \sim \delta_\alpha^{\ \beta}(z-w)^{-1}$ can be used to show that

$$d_\alpha(z)d_\beta(w) \sim 2i\frac{\gamma^m_{\alpha\beta}\Pi_m}{(z-w)} . \tag{4}$$

The operator d_α arises naturally from the action $S_1 + S_{WZ}$ [1], and is the only ingredient we retain of the Green-Schwarz model [2]. We build upon this d_α a whole new structure, and we expect that the results we obtain in the end will be equivalent to whatever one would call the covariant quantum superstring.

From (4) it follows that $Q_B^2 = 2i \oint \lambda\gamma^m\lambda\Pi_m$. Clearly, if λ are so-called pure spinors, which satisfy by definition $\lambda\gamma^m\lambda = 0$, we obtain a nilpotent BRST charge [3]. These λ^α must be complex in order to have a nontrivial solution of $\lambda\gamma^m\lambda = 0$.

In [1] we have relaxed the constraint of pure spinors by adding a Lagrange multiplier term to Q_B of the form $-i\oint \beta_m\lambda\gamma^m\lambda$. Now λ^α need not longer be complex, and we take it real, just like θ^α. In fact we first determined the superalgebra generated by d_α; it contains in addition to Π_m also $\partial\theta^\alpha$, and in the OPE's one finds terms with single poles and double poles. Denoting these three operators by $J^g_M = \{\Pi_m, -id_\alpha, \partial\theta^\alpha\}$, the OPE has the form

$$J^g_M(z)J^g_N(w) \sim \frac{J^g_P f^P{}_{MN}}{(z-w)} - \frac{\mathcal{H}_{MN}}{(z-w)^2} \,. \tag{5}$$

The BRST charge is then given

$$Q = -\oint dz \left(J^g_M + \frac{1}{2}J^{gh}_M\right) c^M \tag{6}$$

where $c^M = \{\xi^m, \lambda^\alpha, \chi_\alpha\}$ are the ghosts, and the BRST ghost current J^{gh}_M depends also on b_M, where $b_M = \{\beta_m, w_\alpha, \kappa^\alpha\}$ are the antighosts. The susy generator $q_\alpha = p_\alpha - i\partial x^m(\gamma_m\theta)_\alpha - \frac{1}{6}(\gamma_m\theta)_\alpha(\theta\gamma^m\partial\theta)$ transforms θ^α into ϵ^α and x^m into $i\,\epsilon\gamma^m\theta$.

The three operators J^g_M define a non-semisimple super Lie algebra [4], but the invariant metric $\mathcal{H}_{MN}$ is still nonsingular. The generators of this algebra are $(P_m, Q_\alpha, K^\alpha)$ where K^α is a fermionic central charge. In a coset approach K^α corresponds to new anticommuting coordinates ϕ_α which form with the usual θ^α a bigger, but little understood superspace [4].

We have shown in [1] that a WZNW action based on the currents J^g_M, J^{gh}_M and J^h_M, but in terms of ϕ_α instead of p_α, reduces to the free field action which in turn determines the propagators, provided one expresses $\partial_z\phi_\alpha$ in terms of $p_{z\alpha}$ by $i\partial_z\phi_\alpha = q_{z\alpha}$. Thus all our previous work turns out to have been based on a WZNW model. We work from now on in the formulation without ϕ_α but with p_α.

The ghost currents $J^{gh}_M = \{2\kappa\gamma_m\lambda, 2\xi^m(\gamma_m\kappa) + 2i\beta^m(\gamma_m\lambda, 0\}$ satisfy the same OPE's as J^g_M but without double poles. It then follows as usual that the BRST charge Q is classically (using single contractions) nilpotent, but at the quantum level it ceases to be nilpotent due to the double poles. For that reason we have added, in the spirit of gauged WZNW models [5], a new triplet of currents J^h_M which have the same OPE's as J^g_M except that the sign of the double poles is opposite. The BRST charge is obtained by replacing J^g_M by $J^g_M + J^h_M$. Fatal double poles in contractions now cancel and this extended Q is nilpotent. However, we have doubled all coordinates: in addition to x^m, θ^α and p_α we also have x^m_h, θ^α_h and p^h_α.

This model has a very interesting conformal field theory. The energy-momentum tensor T_{zz} is

$$T_{zz} = -\frac{1}{2}\partial_z x^m \partial_z x_m - p_{z\alpha}\partial_z\theta^\alpha - \beta_{zm}\partial_z\xi^m - \kappa^\alpha_z\partial_z\chi_\alpha - w_\alpha\partial_z\lambda^\alpha + \frac{1}{2}J^h_{zm}J^{h,m}_z + i\,J^h_{z\beta}J^{h,\beta}_z\,. \tag{7}$$

and the total conformal charge vanishes: $c = 10-32-20+32+32+10-32 = 0$. The ghost current is

$$j^{gh}_z = -\beta_{zm}\xi^m - \kappa^\alpha_z\chi_\alpha - w_{z\alpha}\lambda^\alpha\,. \tag{8}$$

Since the anomaly in the OPE $j^{gh}_z(z)j^{gh}_w(w) = c_j/(z-w)^2$ of the ghost current with itself is not zero but given by $c_j = -22$, while $T_{zz}(z)j^{gh}_w(w) = 22/(z-w)^3 + j^{gh}_z(z)/(z-w)^2$, this superconformal algebra seems to be a twisted version of an $N=2$ superconformal algebra. The BRST current is nilpotent $j_z^B(z)j_w^B(w) \sim 0$. There is also a fermionic current B_{zz} which is the dual of the BRST current

$$B_{zz} = b_M(J_N^g - J_N^{gh})\mathcal{H}^{NM}\,. \tag{9}$$

It has spin 2 and squares to $B_{zz}(z)B_{ww}(w) \sim \frac{F_{www}(w)}{(z-w)}$. The current F_{zzz} is not only BRST closed but also BRST exact

$$j_z^B(z)\Phi_{www}(w) \sim \frac{F_{www}}{(z-w)}\,. \tag{10}$$

The six currents $j_z^B, B_{zz}, j_z^{gh}, T_{zz}, F_{zzz}$ and ϕ_{zzz} generate a closed algebra. However, it is not an ordinarey N=2 superalgebra, but rather a deformation called a Kazama algebra [7].

Following [8] one can recover an ordinary N=2 superconformal algebra by adding a topological gravity quartet, also called a Koszul quartet [6]. It contains the usual anticommuting ghost pair (b_{zz}, c^z) and a commuting spin $(2,-1)$ counterpart (β_{zz}, γ^z). The ghost numbers are $(-1,1)$ and $(-2,2)$ respectively. The currents of this model form an ordinary $N = 2$ superconformal algebra, so without Φ_{zzz} and F_{zzz} currents. However, if one adds the currents of the Koszul model to the currents of the WZNW model, and modifies B_{zz} appropiately, one ends up with an ordinary $N = 2$ superconformal algebra for the combined system.

2. Tree level measure

In this section, we give a derivation of the measure in our formalism. For that purpose, we follow the idea that the tree level action for the target space theory (the string field theory action) is determined by the so-called BV measure. Assuming that the target space action is of Chern-Simons type [9], one has

$$S_{D=10} = \int d\mu \left(\Phi Q \Phi + \frac{2}{3}\Phi \star \Phi \star \Phi\right), \tag{11}$$

where Φ is the string field (see (13)) and the product $\star$ denotes the usual matrix multiplication and tracing over $U(n)$ matrices. The integral in the above equation is the path integral over the zero modes of $x^m, \theta^\alpha, \lambda^\alpha, \xi^m$ and χ_α (and ξ, see below) which remains after one has performed the path over all nonzero modes. In order to perform the integration, we have to establish the measure $d\mu$. Since Φ has ghost number one, the ghost number of the Lagrangian is three. Therefore, we have to impose that the measure has ghost number -3 (notice that, in constrast to Berkovits' work we have more ghosts present, therefore there are several possible combinations).

A second requirement is the BRST invariance of the measure. To construct the measure it is convenient to consider a ghost number 3 polynomial $\Omega^{(3)}$ which is invariant under super-Poincaré transformations (up to Q-exact terms) such that

$$\langle \Omega^{(3)} \rangle \equiv \int d\mu \, \Omega^{(3)} = 1 \,. \tag{12}$$

The operator $\Omega^{(3)}$ represents the Poincaré dual to the measure (see for example Bott and Tu [10]). The condition of BRST invariance of $\Omega^{(3)}$ is therefore given by $\int d\mu \{Q, \Omega^{(3)}\} = 0$.

The measure $d\mu$ itself should also be BRST invariant. We implement this requirement by imposing the condition that if $\Omega^{(3)} = \{Q, \mathcal{K}\}$ where $\mathcal{K}$ is any polynomial with ghost number 2, then

$$\int d\mu \{Q, \mathcal{K}\} = 0 \,. \tag{13}$$

Since the action in (12) should reproduce the super-Yang-Mills action in $d = (9,1)$ dimensions, we can choose a given coupling to normalize the integral in (12). For example the gluon-gluino-gluino coupling $f_{abc} v^a_\mu(x) \psi^{b\alpha} \gamma^m_{\alpha\beta} \psi^{c\beta}$ may provide a convenient normalization. A generic vertex reads

$$\begin{aligned} \Phi = {} & \lambda^\alpha A_\alpha + \xi^m A_m + \chi_\alpha W^\alpha + \\ & + \xi \left(\xi^m \xi^n F_{mn} + \xi^m \chi_\alpha F_m{}^\alpha + \lambda^\alpha \chi_\beta F_\alpha{}^\beta + \chi_\alpha \chi_\beta F^{\alpha\beta} \right) , \end{aligned} \tag{14}$$

where the superfields $A_\alpha, \ldots, F^{\alpha\beta}$ depend only the combinations $x^m - x^m_h$ and so on. (The other combinations are excluded by imposing the condition $B_0 \Phi = 0$ on the vertex operator.) We also assume that all superfields are Lie algebra valued. The field ξ is obtained by the bosonization of the superghosts $\gamma^z = \eta e^\phi$ and $\beta_{zz} = \partial \xi e^{-\phi}$ of the twisted topological Koszul quartet.

By expanding the superfields A_m and W^α one finds that the first coefficients coincide with the gluon field v^a_m and with the gluino $\psi^{a\alpha}$, respectively. Therefore, a candidate for $\Omega^{(3)}$ is given by

$$\Omega^{(3)} \sim \gamma^{m,\alpha\beta} \chi_\alpha \xi_m \chi_\beta \,. \tag{15}$$

because substitution into (11) and (10) reproduces the gluon-gluino-gluino coupling. Notice that this candidate for $\Omega^{(3)}$ has the correct ghost number, and also from a conformal point of view it has the right properties, namely it is a scalar, just like the corresponding element $\Omega^{(3)}_{bos} = c\partial c \partial^2 c$ of the bosonic string. The form of $\Omega^{(3)}$ and equation (12) imply that

$$\begin{aligned} &\langle (\xi^m A_m)(z_1)(\chi_\alpha W^\alpha)(z_2)(\chi_\alpha W^\alpha)(z_3)\rangle \qquad (16)\\ &\to \int d^{10}x\, f_{abc} v^a_m \psi^{b\alpha}\psi^{c\beta}\langle \xi_m \chi_\alpha \chi_\beta\rangle = \int d^{10}x f_{abc} v^a_m \psi^{b\alpha}\psi^{c\beta}\, \gamma^m_{\alpha\beta}\,. \end{aligned}$$

However, $\Omega^{(3)}$ in (15) fails to be BRST invariant. To repair it, we consider the following two combinations of ghosts

$$\hat{\xi}^m = \xi^m + i\lambda^\alpha \gamma^m_{\alpha\beta}\theta^b\,, \qquad \hat{\chi}_\alpha = \chi_\alpha + 2i\xi^m \gamma_{m\alpha\beta}\theta^\beta - \frac{4i}{3}\gamma^m_{\alpha\beta}\theta^\beta\lambda^\gamma\gamma_{m\gamma\delta}\theta^\delta \quad (17)$$

which are invariant under BRST transformations. Then we have finally

$$\begin{aligned} \Omega^{(3)} &= \gamma^m_{\alpha\beta}\hat{\chi}_\alpha\hat{\xi}_m\hat{\chi}_\beta \qquad (18)\\ &= (\chi\gamma^m\chi)\xi_m + 2\xi^m\xi^n\chi\gamma_{mn}\theta + \xi^m(\chi\gamma_m\gamma_n\theta)(\lambda\gamma_n\theta) + (\lambda\gamma^n\theta)(\chi\gamma_n\chi)\\ &+ (\lambda\gamma^m\theta)(\lambda\gamma^m\theta)(\chi\gamma_{mn}\theta) + \xi^m\xi^n\xi^r\theta\gamma_{mnr}\theta + \xi^m\xi^n(\lambda\gamma^r\theta)\theta\gamma_{mnr}\theta\\ &+ \xi^m(\lambda\gamma^n\theta)(\lambda\gamma^r\theta)\theta\gamma_{mnr}\theta + (\lambda\gamma^m\theta)(\lambda\gamma^n\theta)(\lambda\gamma^r\theta)\theta\gamma_{mnr}\theta \end{aligned}$$

where the first term reproduces the combination needed to fix the coupling of the gluon-gluino-gluino vertex. More important is the observation that the last term in $\Omega^{(3)}$ is exactly the combination discussed in [3]. Substitution of this term with pure spinors into (12) and (11) reproduces the complete super-Yang-Mills theory, and the measure is in this case also BRST invariant. In our covariant approach the λ's are not constrained to be pure spinors, and then we need all terms in (18) for BRST invariance.

In order, to fix the integration, one has to assume that a particular monomial $\Omega^{(3)}_i$ of $\Omega^{(3)}$ has the property $\int d\mu \Omega^{(3)}_i = 1$. One choice is obviously the last term, as we have learnt from [3]. Another choice is to fix the first term. However, we can also impose that another combination of terms gives the expected result.

3. Path integral formula for the measure

In order to show that the ghosts and antighosts of the present formulation are emerging from the Berkovits formulation, we derive here a path integral formula for the measure (18) assuming that only the last piece in (18) (the so-called Berkovits term) $\Omega^{(3)} = (\lambda\gamma^m\theta)(\lambda\gamma^n\theta)(\lambda\gamma^r\theta)\theta\gamma_{mnr}\theta$ contributes to (12). Therefore, we write

$$1 = \int d^{16}\theta\, d^{11}\lambda\, \mu(\lambda,\theta)\, [(\lambda\gamma^m\theta)(\lambda\gamma^n\theta)(\lambda\gamma^r\theta)\theta\gamma_{mnr}\theta]\,. \qquad (19)$$

and we want to compute $\mu(\lambda, \theta)$. At the first step, we notice that in order that the integral over the pure spinor fields λ is well defined the measure should contain a Dirac delta function to render the integration over the λ's finite: $\mu(\lambda, \theta) = \delta^{11}(\lambda)\mu(\lambda, \theta)'$. The delta function in the measure is also needed for compensating the ghost number of $d^{11}\lambda$. In order that the measure $\mu'(\lambda, \theta)$ projects out three components of the pure spinor, it should have the form

$$\mu(\lambda, \theta) = \delta^{11}(\lambda)\mu^{(\alpha\beta\gamma)}(\theta)\partial_{\lambda^\alpha}\partial_{\lambda^\beta}\partial_{\lambda^\gamma}\,. \tag{20}$$

One needs exactly three λ-derivatives to take into account the ghost number.

The next step is to represent the delta function by an integral over variables w_α which we identity with our antighosts.[a] The three derivatives become then three components of w

$$1 = \int d^{16}\theta\, d^{11}\lambda\, d^{11}w\, e^{iw_\alpha\lambda^\alpha}\, [\mu^{(\alpha\beta\gamma)}(\theta)w_\alpha w_\beta w_\gamma]\, \Omega^{(3)}\,. \tag{21}$$

The integration $d^{16}\theta$ should be saturated by the θ's present in $\mu^{(\alpha\beta\gamma)}(\theta)$ and those present in $\Omega^{(3)}$. Knowing that in the latter there are 5 θ's, we obtain

$$\mu^{\alpha\beta\gamma}(\theta) = \mu^{(\alpha\beta\gamma)}_{[\rho_1\ldots\rho_{11}]}\theta^{\rho_1}\ldots\theta^{\rho_{11}}\,, \tag{22}$$

where $\mu^{(\alpha\beta\gamma)}_{[\rho_1\ldots\rho_{11}]}$ are numerical constants. Using the following formula for Berezin integrals

$$\theta^{\alpha_1}\ldots\theta^{\alpha_{n-l}} = \frac{(n-l)!(-i)^l}{l!}\epsilon^{\alpha_1\ldots\alpha_{n-l}\rho_{n-l+1}\ldots\rho_n}\int d^n p\, (p_{\rho_{n-l+1}}\cdots p_{\rho_n})\, e^{ip_\alpha\theta^\alpha}\,, \tag{23}$$

where p_α are 16 fermionic new variables and $\epsilon^{\alpha_1\ldots\alpha_{16}}$ is the invariant tensor of $Spin(9,1)$, we obtain

$$1 = \int d^{16}\theta\, d^{11}\lambda\, d^{11}w e^{i\lambda^\alpha w_\alpha + ip_\alpha\theta^\alpha}\, [(p\gamma^m w)\,(p\gamma^n w)\,(p\gamma^r w_\gamma)\,(p\gamma_{mnr}p)]\, \Omega^{(3)}\,. \tag{24}$$

The contractions in (24) with the invariant tensors are dictated by Lorentz invariance. In particular we have

$$\mu^{(\alpha\beta\gamma)}_{[\rho_1\ldots\rho_{11}]} = \epsilon_{\alpha_1\ldots\alpha_5\,\rho_1\ldots\rho_{11}}\gamma^{m,\alpha\alpha_1}\gamma^{n,\beta\alpha_2}\gamma^{r,\gamma\alpha_3}\gamma^{\alpha_4\alpha_5}_{mnr}\,. \tag{25}$$

---

[a]This needs further study because the antighosts $w_{z\alpha}$ are vectors on the worldsheet which do not have zero modes on a genus zero Riemann surface.

To simplify further the integral, we introduce 10 anticommuting variables ξ^m which correspond exactly to our ghost fields $\xi^m(z)$ discussed in the previous sections. We can then rewrite (24) as follows

$$1 = \int d^{16}\theta\, d^{11}\lambda\, d^{11}w d^{10}\xi e^{i\lambda^\alpha w_\alpha + ip_\alpha\theta^\alpha + \xi^m p\gamma_m w} \times \left[\epsilon_{m_0 \dots m_9}\xi^{m_0} \dots \xi^{m_6}(p\gamma^{m_7 m_8 m_9}p)\right] \Omega^{(3)} . \quad (26)$$

Again it is convenient to use (23) to introduce variables β_m which correspond to our antighosts $\beta^m_z(z)$ and which simplify the expression for the measure further. This leads to

$$1 = \int d^{16}\theta\, d^{16}p\, d^{11}\lambda\, d^{11}w\, d^{10}\xi\, d^{10}\beta \times e^{(i\lambda^\alpha w_\alpha + ip_\alpha\theta^\alpha + i\xi^m\beta_m + \xi^m p\gamma_m w)}\, (p\, \not{\beta}\, \not{\beta}\, \not{\beta}\, p)\, \Omega^{(3)} .$$

where $\beta_{\alpha\beta} = \gamma^m_{\alpha\beta}\beta_m$. As a last step we introduce commuting variables χ_α and κ^α which are the final pairs of (anti)ghosts in our work. The operator $(p\, \not{\beta}\, \not{\beta}\, \not{\beta}\, p)$ can be written as follows

$$(p\, \not{\beta}\, \not{\beta}\, \not{\beta}\, p) = \partial_{\chi_\alpha}(p\, \not{\beta}\chi)\, \gamma^m_{\alpha\beta}\beta_m\, \partial_{\chi_\alpha}(p\, \not{\beta}\chi) , \quad (27)$$

The integration over χ_α requires the introduction of another delta function $\delta^{16}(\chi)$ in order to make these integrals finite. So we have

$$1 = \int d^{16}\theta\, d^{16}p\, d^{11}\lambda\, d^{11}w\, d^{10}\xi\, d^{10}\beta\, d^{16}\chi\, d^{16}\kappa \times e^{(i\lambda^\alpha w_\alpha + ip_\alpha\theta^\alpha + i\xi^m\beta_m + i\kappa^\alpha\chi_\alpha + \xi^m p\gamma_m w + \beta_m p\gamma^m\chi)} \left(\kappa^\alpha\gamma^m_{\alpha\beta}\kappa^\beta\beta_m\right) \Omega^{(3)} .$$

We have thus constructed a formula for the measure

$$\mu(\theta, \lambda; \xi, \beta, \chi, \kappa, w, p) = e^{(i\lambda^\alpha w_\alpha + ip_\alpha\theta^\alpha + i\xi^m\beta_m + i\kappa^\alpha\chi_\alpha + \xi^m p\gamma_m w + \beta_m p\gamma^m\chi)} \left(\kappa^\alpha\gamma^m_{\alpha\beta}\kappa^\beta\beta_m\right) , \quad (28)$$

which contains the Fourier transform of all independent pairs $(\lambda, w), \dots (\kappa, \chi)$. The remaining piece $\left(\kappa^\alpha\gamma^m_{\alpha\beta}\kappa^\beta\beta_m\right)$ has a very interesting property. It can be identified with the spin 3, ghost number -3 field $\Phi_{zzz}(z) = \left(\kappa^\alpha_z\gamma^m_{\alpha\beta}\kappa^b_z\beta_{zm}\right)$ appearing in the previous section. Notice that we can also make a redefinition of the fields $\lambda^\alpha \to \lambda^\alpha - i\xi^m(\gamma_m p)^\alpha$ and $\kappa^\alpha \to \kappa^\alpha - i\beta^m(\gamma_m p)^\alpha$ to recast the measure in a more elegant form

$$\mu(\theta, \lambda; \xi, \beta, \chi, \kappa, w, p) = e^{(i\hat{\lambda}^\alpha w_\alpha + ip_\alpha\theta^\alpha + i\xi^m\beta_m + i\hat{\kappa}^\alpha\chi_\alpha)} \left(\kappa^\alpha\gamma^m_{\alpha\beta}\kappa^\beta\beta_m\right) . \quad (29)$$

We conclude that we have provided a path integral formula for the tree level measure. We need all ghosts of our formalism to be able to construct the formula (29). We have recovered the operator Φ_{zzz} from a completely different point of view. The exponent in (29) suggests that the ghost current should be changed to $j^{gh} \to j^{gh} + p_{z\alpha}\theta^\alpha$. Notice that the action $\int d^2z p_{z\alpha}\bar{\partial}\theta^\alpha$ is invariant under the rigid transformations generated by $\oint p_{z\alpha}\theta^\alpha$. The new ghost current j^{gh} has an anomaly equal to $c_j = 6$.

4. Summary and open problems

We have constructed a tree level measure for the string field theory formulation of our approach to the superstring. If one projects it onto the last term (how that should be done is to be worked out), one obtains Berkovits' measure. Then only the terms

$$A_\alpha \sim a_m(\gamma^m\theta)_\alpha + (\gamma_m\theta)_\alpha(\theta\gamma^m\psi) + F_{mn}(\theta\gamma^{mnt}\theta)(\gamma_t\theta)_\alpha + \mathcal{O}(\theta^5) \quad (30)$$

remain. (These are the only structures which remain in the WZ gauge $\theta^\alpha A_\alpha = 0$.) The action in (10) then yields the complete SYM action in 10 dimensions.

So the following picture emerges. Our completly covariant approach with many more ghosts then only λ^α has linearized the theory to a point where no constraints are any longer present, and no non-covariant canonical approach is needed. Yet, at the end we get rid of all excess ghosts, and recover the formulation with only λ^α. Thus this approach combines the best of the covariant approach with the maximal number of ghosts and the minimal approach with only λ^α. We hope that these ideas will also be effective at the loop level.

There are (of course) many problem left:

- To define physical states we had to restrict vertex operators such that they have "non-negative grading". The grading is a quantum number one can assign to the ghosts because the underlying algebra is non-semisimple. In [1] we have shown that it is related to homological perturbation theory, but it is desirable to find an extra BRST operator which selects non-negative grading.
- We should calculate the cohomology of the model with J^g_M, J^{gh}_M and J^h_M together with the Koszul quartet. We have found two BRST charges which anticommute, and further conditions involving $B_{zz}(z)$ which remove the dependence on the differences $x^m - x^m_h$, etc, but a complete analysis is in preparation.
- So far we focused only on the left-moving sector *i.e.* the heterotic string, but a complete treatment containing both left-moving and

right-moving sectors for all fields, in particular for the WZNW model, is still lacking. Also this we intend to construct.

- The big problem is, of course, the calculation of amplitudes: path integrals, measures and vertex operators. We hope that here our covariant methods will be helpful.

References

1. P. A. Grassi, G. Policastro, M. Porrati and P. van Nieuwenhuizen, *Covariant quantization of superstrings without pure spinor constraints*, JHEP **10** (2002) 054, [hep-th/0112162]; P. A. Grassi, G. Policastro, and P. van Nieuwenhuizen, *The massless spectrum of covariant superstrings*, JHEP **11** (2002) 004, [hep-th/0202123]; P. A. Grassi, G. Policastro and P. van Nieuwenhuizen, *On the BRST cohomology of superstrings with / without pure spinors,* [hep-th/0206216]; P. A. Grassi, G. Policastro and P. van Nieuwenhuizen, *The covariant quantum superstring and superparticle from their classical actions,* Phys. Lett. **B 553** (2003) 96, [hep-th/0209026]; P. A. Grassi, G. Policastro and P. van Nieuwenhuizen, *Yang-Mills theory as an illustration of the covariant quantization of superstrings*, Proceedings 3rd Sacharov Conference, 2002 Moscow, [hep-th/0211095]; P. A. Grassi, G. Policastro and P. van Nieuwenhuizen, *Introduction to the covariant quantization of superstrings*, Proceedings Leuven Conference, 2002, Leuven, [hep-th/0302147]; P.A. Grassi, G. Policastro and P. van Nieuwenhuizen, *Quantum Superstrings as a WZNW and N=2 superconformal algebra,* Nucl. Phys. B **676** 43, 2003.
2. M.B. Green, J.H. Schwarz, and E. Witten, *Superstring Theory,* vol. 1, chapter 5 (Cambridge U. Press, 1987).
3. N. Berkovits, *Super-Poincaré covariant quantization of the superstring,* JHEP 0004, 018 (2000); N. Berkovits and B. C. Vallilo, *Consistency of super-Poincaré covariant superstring tree amplitudes,* JHEP 0007, 015 (2000) [hep-th/0004171]; N. Berkovits, *Cohomology in the pure spinor formalism for the superstring,* JHEP 0009, 046 (2000) [hep-th/0006003]; N. Berkovits, *Covariant quantization of the superstring,* Int. J. Mod. Phys. A 16, 801 (2001) [hep-th/0008145]; N. Berkovits and O. Chandia, *Superstring vertex operators in an AdS(5) x S(5) background,* Nucl. Phys. B **596**, 185 (2001) [hep-th/0009168].
4. M.B. Green, *Supertranslations, Superstrings and Chern-Simons Forms*, Phys. Lett. B **223** (1989) 157; W. Siegel, *Classical Superstring Mechanics*, Nucl. Phys. **B 263** (1986) 93; W. Siegel, *Randomizing the Superstring*, Phys. Rev. D **50** (1994) 2799.
5. H. J. Schnitzer, *A Path Integral Construction of Superconformal Field Theories from a Gauged Supersymmetric Wess-Zumino-Witten action*, Nucl. Phys. B **324**, 572 (1989); D. Karabali and H. J. Schnitzer, *Brst Quantization Of The Gauged WZW Action And Coset Conformal Field Theories,* Nucl. Phys. B **329**, 649 (1990).
6. R. Dijkgraaf, H. Verlinde and E. Verlinde, *Notes On Topological String The-*

ory And 2-D Quantum Gravity, PUPT-1217 *Based on lectures given at Spring School on Strings and Quantum Gravity, Trieste, Italy, Apr 24 - May 2, 1990 and at Cargese Workshop on Random Surfaces, Quantum Gravity and Strings, Cargese, France, May 28 - Jun 1, 1990*

7. Y. Kazama, *A Novel Topological Conformal Algebra,* Mod. Phys. Lett. A **6** (1991), 1321; E. Getzler, *Manin Pairs And Topological Field Theory,* Annals Phys. **237**, 161 (1995) [hep-th/9309057].
8. J. M. Figueroa-O'Farrill and S. Stanciu, *Nonreductive WZW models and their CFTs, II: N=1 and N=2 cosets,* Nucl. Phys. B **484**, 583 (1997) [hep-th/9605111]; J. M. Figueroa-O'Farrill and S. Stanciu, *N=1 and N=2 cosets from gauged supersymmetric WZW models,* [hep-th/9511229]; . M. Figueroa-O'Farrill and S. Stanciu, *Nonreductive WZW models and their CFTs,* Nucl. Phys. B **458**, 137 (1996) [hep-th/9506151].
9. E. Witten, *Non-commutative Geometry and String Field Theory,* Nucl. Phys. B **268**, 253 (1986)
10. R. Bott, L.W. Tu *Differential Forms in Algebraic Topology,* Springer-Verlag Berlin and Heidelberg, 1994.

STRING CREATION AND COSMOLOGY

STEVEN S. GUBSER
Joseph Henry Laboratories
Princeton University
Princeton, NJ 08544
E-mail: ssgubser@Princeton.EDU

I argue that string creation may have played a role in reheating the universe after inflation. For strings in four dimensions that arise from branes wrapping cycles in the extra dimensions, estimates from effective field theory show that the string tension need only fall a couple of orders of magnitude below the Planck scale in order for string creation to extract a significant fraction of the energy in coherent motion of the inflaton field. I also comment on a special four-dimensional background which involves only Neveu-Schwarz fields and offers the possibility of studying closed string creation on the worldsheet.

1. Introduction

As the experimental data on the large-scale structure and early history of the universe improve, there has been increased interest in seeing how ideas from string theory may fit in to understanding cosmology at the earliest of times. On the whole, efforts in this direction have focused on either finding a home for inflation in string theory (for instance, Refs. 1, 2) or replacing it with something quite different but string-inspired (for instance, Refs. 3, 4, 5). Important as it is to establish in a fundamental theory of gravity either a sound underpinning for inflation or a solid alternative to it, my focus here will be instead on the possibility of string physics playing an important role just after inflation has ended. The characteristic scale of string theory is usually closely to the Planck mass, considerably higher than a typical inflaton mass scale. However, it turns out that objects known as "tensionless strings"[6,7,8,9] can be produced copiously from coherent motion of the inflaton field. Without significant fine-tuning, it is possible that creation of such strings can extract a significant fraction of energy from the inflaton, thus providing an alternative to preheating.[10,11,12]

String creation is interesting in its own right, but on the whole it is ill-understood. The best studied examples of it are decaying D-brane solutions,[13,14,15,16] where boundary CFT methods are available. We will rely instead on estimates from effective field theory, employing a method of steepest descent to extract Bogliubov coefficients from approximate wave equations for massive string modes. This method was developed in Refs. 17, 18, and its application to preheating was considered in Ref. 18. It would be highly desirable to improve our ability to compute string creation amplitudes: this is an important feature of time-dependent backgrounds in string theory, and it is relevant to almost all the stringy proposals for the physics of the early universe.

The organization of this paper is as follows. In section 2 we will briefly summarize the results of the steepest descent analysis of string creation. In section 3 we will explain in some detail how this analysis may be applied to provide an alternative to preheating. In section 4 we will remark on a special background in which a worldsheet treatment of closed string creation may be possible.

The results summarized here are based mainly on Ref. 18, except that section 4 is based on ongoing work with J. Friess and I. Mitra.

2. Estimating string creation

The density of states in tree-level string theory is exponential:

$$\frac{dN}{dE} \propto E^{\gamma} e^{E/T_H} \qquad T_H = \frac{1}{2\pi\sqrt{\alpha' c_\perp/6}}\,, \tag{1}$$

where T_H is the Hagedorn temperature and γ depends on details of the string theory in question. The quantity $c_\perp$ is the central charge for transverse modes: $c_\perp = 12$ for the type II superstring. The exponential density of states (1) is in contrast with the power-law behavior one encounters for point-particle theories. The states that dominate (1) are highly excited strings. String interactions are not taken into account in (1). T_H appears to be a temperature at which some phase transition occurs. Above the phase transition strings themselves may not be a good description of the physics. There is an extensive literature on this so-called Hagedorn transition (see for example Refs. 19, 20, 21), and string interactions do enter into the discussion in important ways. For the purposes of the present discussion, we will not need to consider interactions explicitly.

The on-shell constraints for quantum states of a string include $L_0|\psi_{\rm phys}\rangle = 0$, which leads to an equation roughly of the form

$$\ddot{\chi} + \omega(t)^2 \chi = 0\,, \tag{2}$$

where $\omega(t)^2$ depends on the excitation level and the momentum: in a flat background, $\omega^2 = k^2 + N/\alpha'$, where N is the level (up to factors of 2 that depend somewhat on details). In a cosmological background, a reasonable first guess is

$$\omega(t)^2 = k^2 + a(t)^2 N/\alpha' \,, \tag{3}$$

where $a(t)$ is the scale factor and t is conformal time. One would of course like to improve upon (2): this is one place where input from a worldsheet treatment is needed. The form (3) is predicated on working with the string frame metric, where the parameter α' is a constant length squared.

The steepest descent method, developed in this context in Refs. 17, 18, allows us to crudely estimate the Bogliubov coefficients for (2): if $|\beta| \ll 1$, then

$$\beta \approx \frac{i\pi}{3} \exp\left(-2i \int_{-\infty}^{4} dt\, \omega(t)\right) \exp\left(-2i \int_{r}^{t_*} dt\, \omega(t)\right) , \tag{4}$$

where $t_* = r - i\mu$ is the location of the zero of $\omega(t)^2$ in the lower half plane which is closest to the real axis, and r and μ are real. The last of the three factors in (4) is the most important: it leads to exponential suppression of $|\beta|^2$ when ω is large. Assuming that $\omega(t)$ is well-approximated by an elliptical arc over the contour from r to t_*, one has

$$|\beta|^2 \approx \left(\frac{\pi}{3}\right)^2 e^{-\pi\mu\omega(r)} \,. \tag{5}$$

From this one can estimate the total number of strings created:

$$N_{\text{tot}} = \int^{\infty} d\omega \, \frac{dN}{d\omega} |\beta|^2 \sim \int^{\infty} d\omega \, e^{\omega/T_H} e^{-\pi\mu\omega} \,. \tag{6}$$

Evidently, this total number *converges* precisely if

$$\pi\mu T_H > 1 \,. \tag{7}$$

If this bound is violated, it's somewhat like passing to a temperature above the Hagedorn temperature: clearly, interactions and/or back-reaction must then be involved in rendering the total energy passing into strings finite.

The punch-line of the next section will be that, for non-perturbative strings, it is possible without appreciable fine-tuning of parameters to violate (7) just after inflation. If this happens, then strings must indeed play an important role in reheating. Moreover, it would be highly excited strings that would make the dominant contribution, so the effects of strings would be quite unlike anything in a point-particle theory—hence characteristic of string theory.

Needless to say, it would be very desirable to learn what happens in time-dependent backgrounds when the number or energy of strings created diverges. Decaying D-branes may be the simplest laboratory in which to attack this question: see for example Ref. 22.

3. Extracting energy from inflaton oscillations

Let us start with a brief overview of the theory of preheating,[10,11,12] which will be a useful point of comparison when considering post-inflationary string creation.

- At the end of inflation, the universe is cold and empty.
- The inflaton ϕ is oscillating around its minimum with an amplitude assumed to be comparable to the Planck scale.
- These coherent oscillations may die off via perturbative production of particles. This is ordinary reheating.
- Preheating can occur if another *boson* has a coupling $g\phi^2\chi^2$ to the inflaton.
- The effective mass of χ varies: $m^2_{\text{eff}} = m_0^2 + g\phi^2$. This means parametric resonance can occur and give exponentially growing occupation numbers for χ.
- Preheating is "efficient" iff m^2_{eff} varies by a large factor from its minimum to its maximum: m_0^2 needs to be a small contribution to the mass.

The last point appears contrived, especially when it is stated more precisely, as we will do shortly. "Efficient" in this context means that a significant fraction of the energy in the coherent inflaton oscillations should be extracted via the parametric resonance. We may then ask, if string states could be produced in a similar way, could they play a role in reheating?

We have already noted in the introduction that the string scale is usually only slightly smaller than the four-dimensional Planck scale, $M_{\text{Pl},4} = (8\pi G_N)^{-1/2} \sim 10^{18}\,\text{GeV}$, while a typical value for the inflaton mass (in power-law inflation) is $m_{\text{inflaton}} \sim 10^{13}\,\text{GeV}$. It will not come as a surprise that this prevents copious production of fundamental strings, unless we make somewhat unnatural assumptions. Although our eventual aim is to consider production of non-perturbatively constructed strings which are tensionless at some point in moduli space, it will be instructive first to consider the production of fundamental strings in slightly more detail; also, this discussion will tie in with the special background discussed in section 4.

The first goal in assessing a model of preheating is to find the regime of parameters where preheating is efficient. The analogous goal here is to find where the bound (7) is first violated. When this happens, the particle occupation numbers of the individual string modes should still be small: if they are not, then according to (6), the number of strings created will diverge exponentially. Thus we have no need of parametric resonance and can continue to use the steepest descent method.

The onshell condition in string metric is, approximately, $\omega^2 = k^2 + m^2|g_{tt}^{str}|$, where t is conformal time. The string metric is related to the Einstein metric by

$$ds_{4E}^2 = e^{\gamma\varphi/M_{\rm Pl,4}} ds_{str}^2 = a(t)^2(-dt^2 + d\vec{x}^2)\,, \tag{8}$$

where φ is the dilaton and γ is a constant of order unity. Thus

$$\omega(t)^2 = k^2 + e^{\gamma\varphi(t)/M_{\rm Pl,4}} a(t)^2 \frac{N}{\alpha'}\,. \tag{9}$$

The dilaton is probably not the same field as the inflaton; however, it seems reasonable to assume some finite overlap, so when the inflaton is oscillating at the end of inflation, the dilaton is too. Thus $\omega(t)^2$ oscillates between slowly varying extremes:

$$\begin{aligned}
\omega(t)^2 &= \frac{\omega_{\rm max}^2 + \omega_{\rm min}^2}{2} + \frac{\omega_{\rm max}^2 - \omega_{\rm min}^2}{2}\cos\Omega t \\
\omega_{\rm min}^2 &= k^2 a(t)^2 + e^{\gamma\varphi_{\rm min}/M_{\rm Pl,4}}\frac{N}{\alpha'} \\
\omega_{\rm max}^2 &= k^2 a(t)^2 + e^{\gamma\varphi_{\rm max}/M_{\rm Pl,4}}\frac{N}{\alpha'}\,.
\end{aligned} \tag{10}$$

The zeroes of $\omega(t)^2$ are

$$t_{*,n}^{\pm} = \frac{2\pi n}{\Omega} \pm \frac{i}{\Omega}\log\frac{\omega_{\rm max} + \omega_{\rm min}}{\omega_{\rm max} - \omega_{\rm min}}\,, \tag{11}$$

and it is plausible to assume that $t_* = t_{*,0}^- = r - i\mu$ makes the biggest contribution to string production. String production is finite if

$$\pi\mu T_H > 1\,, \tag{12}$$

which translates approximately to

$$\sqrt{\tau_{\rm min}} \gtrsim m_{\rm inflaton}\,, \tag{13}$$

where $\tau_{\rm min}$ is the minimum string tension as measured in Einstein frame.

If the bound (12) is violated, then what must happen is that a significant fraction of the energy of the inflaton oscillations passes into excited string states, and the back-reaction slows down the inflaton. This would correspond to the idea of efficient preheating. Note however that $\omega_{\rm min} \sim \sqrt{\tau_{\rm min}}$ and $\Omega \sim m_{\rm inflaton}$, so for (13) to occur, the string tension as measured in Einstein frame must dip temporarily below $m_{\rm inflaton}$. This is possible, but it seems like asking a lot in conventional string compactifications. The upshot, then, is that dimensional analysis didn't fail, and we have proposed a contrived scenario.

Let's compare with conventional preheating, where it is useful to define "Mathieu parameters" A and q as follows:

$$A + 2q = \left(\frac{\omega_{\rm max}}{2\Omega}\right)^2 , \quad A - 2q = \left(\frac{\omega_{\rm min}}{2\Omega}\right)^2 . \tag{14}$$

A "broad resonance" occurs in the standard theory when $2q < A < 2q+\sqrt{bq}$ for some order one constant b: this leads to efficient particle production. On the other hand, efficient string production occurs when $2q < A < 2xq$ for some $x > 1$. So string production occurs in a qualitatively bigger wedge of parameter space, albeit still a small one.

Another problem with proposing that fundamental strings might have been created after inflation is that they can wind around the extra dimensions. In phenomenologically interesting string compactifications, there usually is a non-trivial first homotopy group on the compactification manifold, so such windings can be topologically stable. Moreover, the density of states of wound strings is smaller only by roughly a constant factor compared to unwound strings, and this constant factor is at most a couple of orders of magnitude in conventional setups. Wound strings would act as dark matter, or perhaps they could even have visible sector effects like fractional charge. Just the dark matter consideration is discouraging: a naive estimate suggests that wound strings would overclose the universe today by something close to a factor of 10^{17}.

Considering non-perturbatively constructed strings in place of fundamental strings alleviates all the difficulties so far discussed. These strings would arise as branes wrapped on a shrinking cycle in the extra dimensions. Consider for example D3-branes on a shrinking S^2. The S^2 may become small only at a point on the compactification manifold. This avoids the difficulty of having wound strings, because there is a strong energy barrier against wrapped branes wandering away from the location of the shrinking cycle. The parameter controlling the size of the S^2 is in fact complex: it is

$$\varphi = \int_{S^2} (J_2 + iB_2) , \tag{15}$$

where J_2 is the Kahler form on the extra dimensions (assumed to form a complex manifold) and B_2 is the Neveu-Schwarz two-form potential of string theory. Starting from the DBI action for D-branes, one arrives at

$$S = -\tau_{D3} \int d^4\xi \sqrt{G_{\mu\nu} + B_{\mu\nu}} + \ldots \tag{16}$$

for a D3-brane wrapped on the S^2. Using $\tau_{D3} \sim 1/(\alpha'^2 g_s)$, one obtains

$$\tau_{\text{eff}} \approx M|\varphi| \qquad \text{with} \qquad M \sim \frac{M_{\text{Pl},4}}{g_s}\,, \tag{17}$$

where φ has been rescaled to be a canonically normalized complex scalar field. We have derived (17) using a particular construction of light strings in four dimensions, but it is in fact typical for a variety of non-perturbative constructions, many of which are related to one another by string dualities.

Let us now consider the situation where after inflation, various moduli of string theory are rolling around, presumably before stabilizing in a preferred vacuum. The scalar φ in particular rolls around the complex plane, and if it comes close to the origin, then the light strings we have discussed become nearly tensionless. Based on our earlier discussion, we may expect appreciable string creation at such a moment. To estimate it, suppose $\varphi(t) \approx \varphi_0 + \dot{\varphi}t$, where φ_0 is the closest approach to the origin on the trajectory $\varphi(t)$. Then

$$\tau_{\text{eff}} \approx M|\varphi_0 + \dot{\varphi}t| = M\sqrt{|\varphi_0|^2 + (|\dot{\varphi}|t)^2}\,. \tag{18}$$

To estimate the amount of string creation, we want to find a complex solution t_* to $\omega(t)^2 = 0$. For highly excited strings, it is equivalent to solve $\tau_{\text{eff}} = 0$, which gives

$$t_* \equiv r - i\mu = -i|\varphi_0|/|\dot{\varphi}|\,. \tag{19}$$

Thus the convergence criterion on the total number of strings produced is

$$\pi\mu T_H \sim \frac{|\varphi_0|}{|\dot{\varphi}|}\sqrt{M|\varphi_0|} \gtrsim 1\,. \tag{20}$$

The minimum tension for the light strings is $\tau_{\text{min}} \approx M|\varphi_0|$. We can estimate that $\dot{\varphi} \sim M_{\text{Pl},4} m_{\text{inflaton}}$: the φ oscillations are of amplitude $M_{\text{Pl},4}$ and of frequency m_{inflaton}. The condition (20) becomes

$$1 \lesssim \frac{\sqrt{M\varphi_0^3}}{\dot{\varphi}} = \frac{\tau_{\text{min}}^{3/2}}{M\dot{\varphi}} \sim \frac{\tau_{\text{min}}^{3/2}}{M M_{\text{Pl},4} m_{\text{inflaton}}} \sim 10^4 \left(\frac{\tau_{\text{min}}}{M_{\text{Pl},4}^2}\right)^{3/2}, \tag{21}$$

or equivalently,

$$\sqrt{\tau_{\rm min}} \gtrsim \frac{1}{20} M_{\rm Pl,4}\,, \tag{22}$$

where we have assumed $M \sim 10 M_{\rm Pl,4}$ and $m_{\rm inflaton} \sim 10^{-5} M_{\rm Pl,4}$.

The estimates leading to (22) are inevitably fairly crude as long as we have not specified what the inflaton is and how much it overlaps with the scalar φ controlling the effective string tension. The result (22) is rather different from what we had for fundamental strings, and the reason is that we almost have a discontinuity in the first derivative of $\omega(t)^2 \propto \tau_{\rm eff} \sim M|\varphi| \approx M|\dot\varphi t|$.

Evidently, the non-perturbatively constructed strings do not have to dip very far below the Planck scale in order to violate this convergence criterion! So their participation in reheating is quite plausible. One would still have to answer the question of how dense in moduli space the loci of tensionless strings are. Quantifying this is technically difficult, but the fact that the loci are typically real codimension two is encouraging.

4. Backgrounds involving only Neveu-Schwarz fields

Let us now consider a background which might be more amenable to a worldsheet treatment.[a] As noted earlier, it is desirable to explore the possibilities for such vacua, since one would eventually want to go beyond the effective field theory method of estimating string creation that we described in section 2. Essentially all string theory compactifications give rise to the following terms in the effective action (written in four-dimensional string frame):

$$\begin{aligned} S &= \frac{1}{2\kappa_4^2} \int d^4x \sqrt{G} e^{-\phi} \left[R + (\partial\phi)^2 - \frac{1}{2} H_3^2 \right] \\ e^{-\phi} &= \frac{{\rm Vol}\, M_6}{(2\pi\sqrt{\alpha'})^6} \frac{1}{g_s^2} \qquad 2\kappa_4^2 = 2\pi\alpha'\,. \end{aligned} \tag{23}$$

In four-dimensional Einstein frame, defined by $ds_E^2 = e^{-\phi} ds_{str}^2$, we have instead

$$S = \frac{1}{2\pi\alpha'} \int d^4x \sqrt{g} \left[R - \frac{1}{2}(\partial\phi)^2 - \frac{1}{2} e^{2\phi} (\partial\chi)^2 \right]\,, \tag{24}$$

and the fundamental string tension as measured in Einstein frame is $\tau_{\rm eff} = e^{\phi}/2\pi\alpha'$. Except in type I string theory, H_3 is the four-dimensional part of

[a]This section represents work in progress with J. Friess and I. Mitra.

the field strength of the Neveu-Schwarz two-form, and χ is defined via the equation $d\chi = e^{-\phi} * H_3$.

The scalar ϕ and the axion χ parametrize the upper half plane. As a first guess, we might suppose that there are FRW solutions where $\chi + ie^{-\phi}$ approximately describes a semi-circular geodesic in the upper half plane (with the usual metric), and that by making g_s small, we can avoid significant back-reaction of the scalars on the geometry. Solving the scalar equations of motion in a background assumed to be flat Minkowski space, one finds

$$\chi + ie^{-\phi} = e^{-\phi_0}\left(\tanh\frac{t}{t_0} + i\,\mathrm{sech}\,\frac{t}{t_0}\right) \qquad \tau_{\text{eff}} = \frac{e^{\phi_0}\cosh t/t_0}{2\pi\alpha'}, \tag{25}$$

where ϕ_0 and t_0 are integration constants. The usual convergence criterion for the total number of strings created is $t_0\sqrt{\tau_{\text{eff}}(0)} \gtrsim 1$, or, more explicitly,

$$e^{\phi_0} \gtrsim \frac{2\pi\alpha'}{t_0^2}. \tag{26}$$

It would appear that by making ϕ_0 very negative (corresponding to extremely weak string coupling), with t_0 large compared to $\sqrt{\alpha'}$ but fixed, we can get the bound in (26) to fail. This is a first indication that this type of solution is an interesting laboratory for string creation. It is nice that the wave equation for string modes is approximately a Mathieu equation, and that in the early and late time limits it is possible to construct infinite order adiabatic in and out vacua.

In fact, it is not difficult to promote the approximate solution described above to an exact solution of the equations of motion following from the two-derivative action:

$$\begin{aligned} H_3 &= e^{\phi} * d\chi = h dx^1 \wedge dx^2 \wedge dx^3 \qquad e^{\phi} = e^{\phi_0}\sec ht \\ ds_{4E}^2 &= -e^{\phi_0}(\sec^2 ht)\left(\frac{\cos\frac{ht}{2} + \sin\frac{ht}{2}}{\cos\frac{ht}{2} - \sin\frac{ht}{2}}\right)^{\sqrt{3}} dt^2 \\ &\quad + e^{\frac{-\phi_0}{3}}\left(\frac{\cos\frac{ht}{2} + \sin\frac{ht}{2}}{\cos\frac{ht}{2} - \sin\frac{ht}{2}}\right)^{1/\sqrt{3}} d\vec{x}^2. \end{aligned} \tag{27}$$

Here t runs from $-\pi/2h$, where there is a curvature singularity, to $\pi/2h$, which is actually a non-singular infinite future when expressed in terms of proper time. This solution is not new: to my knowledge it first appeared in Ref. 23.

The main qualitative difference between the full solution (27) and the no-back-reaction approximation (25) is that the in vacuum is more difficult to define in (27), due to the singularity. Two possibly interesting approaches would be to replace the early time evolution by an appropriate instanton, or to find a CFT formulation in which α' corrections are under control.

The solution (27) can be lifted to ten dimensions, and in the cases where it comes from IIA strings or $E_8 \times E_8$ heterotic strings, there is a further lift to eleven dimensions, where the size of the eleventh dimension is controlled by the dilaton. When the lift is done to Horava-Witten theory, it describes two E_8 that start far apart at $t = -\pi/2h$, move swiftly together to a minimum distance at $t = 0$, and then move apart again. This is reminiscent of ekpyrosis, but actually it is quite different: in four-dimensional Einstein frame, there is never any contraction of the scale factor. There is also no singularity when the branes are at their distance of closest approach. The reason they don't actually collide is that the four-form field G_4 is turned on in such a way as to make the E_8 branes repel. It is notable that in ekpyrosis as envisioned in Ref. 5, strings become tensionless at the instant that the branes collide. It seems extremely likely that string creation is important for understanding how the universe might pass through this singular event.

The point of involving only Neveu-Schwarz fields is that it is more likely that one can perform a worldsheet analysis of such backgrounds. In fact, the background we have considered resembles a limit of NS5-branes, doubly Wick rotated so that radius is replaced by time. The main stumbling block to finding a CFT description is that the geometry does not factorize in the way that the NS5-brane geometry does, into a three-sphere part, a linear dilaton background, and flat space.

5. Conclusions

Part of the appeal of the discussion in section 3 is that nearly tensionless strings arise in a variety of contexts, on loci in moduli space of real codimension two. Thus, independent of detailed ideas about how string theory connects to the Standard Model, it is plausible to have non-perturbatively constructed strings become temporarily light enough to play a role in post-inflationary cosmology. Because we remain ignorant of what string vacuum we might really be living in, it's highly desirable to have some ideas about characteristic features of string theory which may become testable, which are difficult to contrive without invoking strings, and which are common to many backgrounds. Creation of light strings seems to qualify on all counts. On the other hand, the experimental constraints on precisely how the universe was reheated seem rather weak, so to really make the idea of post-inflationary string creation testable, one must gain a better idea of how such string creation would affect quantities that will be measured in the near future, like properties of the primordial spectrum of perturbations.

On the formal side, a better understanding of string creation from a worldsheet point of view is needed, and it would also be useful to learn more about the decay of highly excited strings. One reason that it is particularly important to have some examples of closed string creation described on the string worldsheet is that for bounds like (7) to be violated, one is usually forced to have string scale curvatures in the background.[b] So a worldsheet treatment is needed to be sure one has α' corrections under control, even before quantum creation processes are contemplated. I hope to report on some of these issues in the future.

Acknowledgments

I thank the organizers of QTS3 for putting together a stimulating workshop. This work was supported in part by the Department of Energy under Grant No. DE-FG02-91ER40671, and also by a Sloan Fellowship.

References

1. S. H. S. Alexander, "Inflation from D - anti-D brane annihilation," *Phys. Rev.* **D65** (2002) 023507, `hep-th/0105032`.
2. S. Kachru, R. Kallosh, A. Linde, J. Maldacena, L. McAllister, and S. P. Trivedi, "Towards inflation in string theory," *JCAP* **0310** (2003) 013, `hep-th/0308055`.
3. R. H. Brandenberger and C. Vafa, "Superstrings in the early universe," *Nucl. Phys.* **B316** (1989) 391.
4. G. Veneziano, "Scale factor duality for classical and quantum strings," *Phys. Lett.* **B265** (1991) 287–294.
5. J. Khoury, B. A. Ovrut, N. Seiberg, P. J. Steinhardt, and N. Turok, "From big crunch to big bang," *Phys. Rev.* **D65** (2002) 086007, `hep-th/0108187`.
6. E. Witten, "Some comments on string dynamics," `hep-th/9507121`.
7. A. Strominger, "Open p-branes," *Phys. Lett.* **B383** (1996) 44–47, `hep-th/9512059`.
8. O. J. Ganor and A. Hanany, "Small E_8 Instantons and Tensionless Non-critical Strings," *Nucl. Phys.* **B474** (1996) 122–140, `hep-th/9602120`.
9. N. Seiberg and E. Witten, "Comments on String Dynamics in Six Dimensions," *Nucl. Phys.* **B471** (1996) 121–134, `hep-th/9603003`.
10. J. H. Traschen and R. H. Brandenberger, "Particle production during out-of-equilibrium phase transitions," *Phys. Rev.* **D42** (1990) 2491–2504.

[b]The treatment of nearly tensionless strings in section 3 may be an exception to this.

11. L. Kofman, A. D. Linde, and A. A. Starobinsky, "Towards the theory of reheating after inflation," *Phys. Rev.* **D56** (1997) 3258–3295, hep-ph/9704452.
12. I. Zlatev, G. Huey, and P. J. Steinhardt, "Parametric resonance in an expanding universe," *Phys. Rev.* **D57** (1998) 2152–2157, astro-ph/9709006.
13. A. Sen, "Rolling tachyon," *JHEP* **04** (2002) 048, hep-th/0203211.
14. A. Strominger, "Open string creation by S-branes," hep-th/0209090.
15. A. Maloney, A. Strominger, and X. Yin, "S-brane thermodynamics," hep-th/0302146.
16. N. Lambert, H. Liu, and J. Maldacena, "Closed strings from decaying D-branes," hep-th/0303139.
17. A. E. Lawrence and E. J. Martinec, "String field theory in curved spacetime and the resolution of spacelike singularities," *Class. Quant. Grav.* **13** (1996) 63–96, hep-th/9509149.
18. S. S. Gubser, "String production at the level of effective field theory," hep-th/0305099.
19. Y. I. Kogan, "Vortices on the world sheet and string's critical dynamics," *JETP Lett.* **45** (1987) 709–712.
20. J. J. Atick and E. Witten, "The Hagedorn transition and the number of degrees of freedom of string theory," *Nucl. Phys.* **B310** (1988) 291–334.
21. S. S. Gubser, S. Gukov, I. R. Klebanov, M. Rangamani, and E. Witten, "The Hagedorn transition in non-commutative open string theory," *J. Math. Phys.* **42** (2001) 2749–2764, hep-th/0009140.
22. D. Gaiotto, N. Itzhaki, and L. Rastelli, "Closed strings as imaginary D-branes," hep-th/0304192.
23. E. J. Copeland, A. Lahiri, and D. Wands, "String cosmology with a time dependent antisymmetric tensor potential," *Phys. Rev.* **D51** (1995) 1569–1576, hep-th/9410136.

BLACK HOLE SINGULARITY IN ADS/CFT

VERONIKA HUBENY

Department of Physics, Stanford University, Stanford, CA 94305, USA
E-mail: veronika@itp.stanford.edu

We present a short review of hep-th/0306170. In the context of AdS/CFT correspondence, we explore what information from behind the horizon of the bulk black hole geometry can be found in boundary CFT correlators. In particular, we argue that the CFT correlators contain distinct, albeit subtle, signals of the black hole singularity.

It is a well-founded expectation that black holes should provide a key, or at least a window, into quantum gravity. Typically, black holes contain curvature singularities, where classical laws of general relativity break down, to be replaced by more fundamental quantum physics. Unfortunately, an asymptotic observer cannot see this quantum singularity resolution in action, because the near-singularity physics is cloaked by an event horizon, the defining feature of a black hole. The fact that black holes are causally nontrivial, signaled by an event horizon, has also associated with it long-standing puzzles, such as the information paradox.

To unravel the mysteries of black holes, we call to aid string theory, which has already proved very successful in resolving many timelike singularities. Spacelike singularities appear to be more subtle, and we expect that a fully nonperturbative formulation of string theory will be essential for this endeavor. While there are several remarkable formulations, the one best suited for our purposes is the well-known AdS/CFT correspondence[1]. In particular, we will use the duality between 10-dimensional IIB string theory on $AdS_5 \times S^5$ and $\mathcal{N} = 4$, 4-dimensional super Yang Mills gauge theory, living on the boundary of AdS.

More specifically, to study a singularity and an event horizon in the bulk gravity theory, we will consider the Schwarzschild-AdS geometry, describing a large black hole in AdS, whose gauge dual corresponds to an approximately thermal state. By analyzing the properties of this state, such as the expectation values of various operators, we can hope to extract information about the black hole. Indeed, due to the powerful nature of the AdS/CFT duality, we expect to be able to decode the full quantum near-singularity bulk physics from the gauge dual.

However, since such decoding requires that the gauge theory encodes behind-the-horizon physics in the first place, we must first ask, *how much of the bulk physics does the gauge dual encode?* Naive considerations might lead one to think that the CFT encodes only the region of the bulk which is causally connected to the boundary, *i.e.* outside the event horizon. On the other hand, an event horizon is a global object (defined as the boundary of the past of the future infinity), which means that we cannot determine the presence or position of the event horizon without knowing the entire future evolution of the spacetime. Hence if the horizon were to bound what the CFT can encode, then the AdS/CFT correspondence would have to be very nonlocal in time.

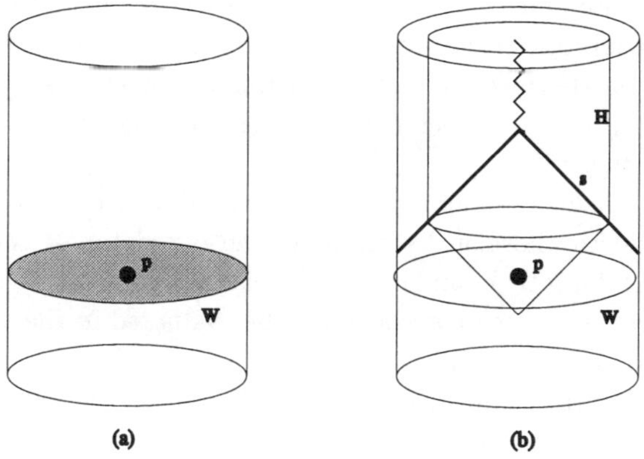

Figure 1. Sketch of the proposed process of abstracting information from inside of the horizon: a) An event p, in locally pure AdS space is measured by W. b) After this measurement is performed, a shell s collapses, forming a black hole with a horizon H which encompasses p.

We illustrate this point by the following gedanken-experiment[2] (*cf.* Fig.1): Start with "empty" AdS space, and consider some event, labeled p in Fig.1a. As a boundary observer, one can obtain instantaneous information about the event p, for instance by measuring appropriately decorated Wilson loop[3] W. *After* this measurement has been made, i.e. entirely in the future of W, one can send in a shell s of radiation with sufficient energy such that when it implodes in the center of AdS, it forms a large black hole, as sketched in Fig.1b. The main observation is that the global event horizon for this spacetime, denoted by H in Fig.1b, can originate at the center of AdS prior to p. In particular, provided the shell is sent in within time of order the AdS radius after the measurement of p is performed, one can always make the resulting black hole large enough for its horizon to encompass p. This implies that one has in fact succeeded in "measuring" an event, p, which is inside a black hole horizon.

While the above argument indicates that the CFT should encode at least some physics inside a horizon, it does not lend itself to detailed computational analysis. Instead, it is more fruitful to turn to a simpler set-up, namely that of the eternal black hole, which is static outside the horizon and has both future and past singularities, as well as *two* boundaries. This geometry was analyzed previously in 3 dimensions by a number of authors[4,5,6]; here we study the higher-dimensional analog[7].

The real-time or thermofield formalism for thermal field theory[8] is especially well-suited for this analysis. One copy of the CFT resides on each of the two asymptotic regions; the CFTs are noninteracting but entangled through the Hartle-Hawking state[5]. In this approach, the one boundary thermal description is recovered by tracing over the Hilbert space of the other boundary CFT.

Moreover, as demonstrated previously for 3 dimensions[6], one can probe physics behind the horizon by studying the correlator of two operators, one on each asymptotic boundary, each creating a large mass bulk particle. As the mass $m \to \infty$, the correlator can be evaluated in the semi-classical geodesic approximation and is given by $\exp(-m\mathcal{L})$, where $\mathcal{L}$ is the (regularized) proper length of the spacelike geodesic joining the boundary points. Because the geodesic passes through spacetime regions inside the horizon, this boundary correlator reveals information about the geometry behind the horizon.

Unfortunately, the three dimensional case, which is simple enough to study analytically, is rather special. The geometry is locally that of pure AdS_3, and the black hole singularity is merely a result of the orbifold nature of this geometry[9]. Consequently, the geodesics are not sensitive to the position of the singularity, and correspondingly the correlation function is relatively structureless.

On the other hand, this situation changes drastically in higher dimensions. In particular, the singularity in $d > 3$ dimensions approaches that of the d-dimensional Schwarzschild black hole singularity, which has a dramatic effect on the spacelike geodesics: pictorially, spacelike radial geodesics "bounce off" from the singularity. For sake of simplicity, we focus on $d = 5$ where the boundary CFT is four dimensional $\mathcal{N} = 4$ SYM; however, our results are qualitatively similar to those in other higher dimensions as well.

The first surprise, underlining the marked difference from the previously-studied 3-D case, is that the Penrose diagram[a] of the Schw-AdS_5 spacetime is different. While the Penrose diagram for the Schw-AdS_3 geometry has the shape of a square, this can no longer be true in higher dimensions[7]. One reason, apparent from Fig.2, is that outgoing radial null geodesics starting at the past singularity do not reach the boundary at a time-symmetric point. Apart from the differences in the Penrose diagrams, there are two important differences in the symmetric radial spacelike geodesics: Unlike the 3-D case, in higher dimensions there exists a particular time t_c beyond which there are no geodesics connecting the two boundaries, and the geodesics cross each other.

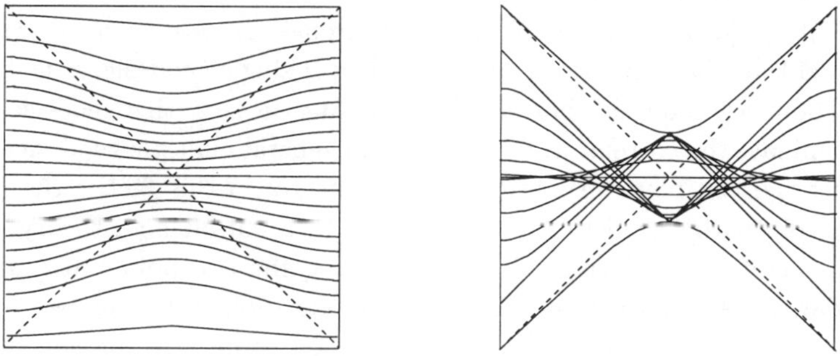

(a) (b)

Figure 2. Penrose diagrams of Schw-AdS in (a) 3 and (b) 5 dimensions. The top and bottom curves are the future and past singularities, the vertical lines are the boundaries, the dashed diagonal lines are the event horizons, and the remaining curves are spacelike radial geodesics.

This has an important consequence for our present goal of seeing a signature of the singularity in the dual field theory. Consider the boundary to boundary correlator $\langle\phi\phi\rangle(t)$ between two high-dimension operators inserted in a symmetric fashion on the two boundaries at time t. Since this

[a]Penrose diagram is a conformal diagram of a spacetime including its boundary, used to describe its causal properties; time runs up and space runs sideways, such that null rays lie at 45 degree angles.

correlation function is related to the proper length of a spacelike geodesic connecting the two points, we expect that it reflects the special behaviour of the geodesics as $t \to t_c$. In fact, direct evaluation suggests that we should see a pole in the correlation function:

$$\langle \phi\phi \rangle(t) \sim e^{-m\mathcal{L}(t)} \sim \frac{1}{(t-t_c)^{2m}} \qquad \text{as} \qquad t \to t_c \ . \tag{1}$$

This corresponds to a light-cone singularity in the field theory, since the geodesics are becoming almost null.

Had this been the full story, we would have a glaring signature of the black hole singularity in the gauge theory dual. However, general considerations of the boundary field theory rule this out; one can easily show that $|\langle \phi\phi \rangle(t)| \leq |\langle \phi\phi \rangle(0)| < \infty$. What went wrong? In evaluating Eq.(1) we assumed that the correlator is dominated by a single geodesic, namely the real, "bounce" geodesic shown in Fig.2b. But, in fact, this geodesic does *not* dominate the correlator, for there are in general multiple geodesics that connect the two boundary points. One indicator of this fact is the intersection of nearby geodesics around $t = 0$; in a cut-off field theory there would therefore be *three* geodesics connecting the two points at (cut-off) boundaries. Alternately, in a Euclidean picture, one real and two purely imaginary geodesics contribute. At $t = 0$ their proper distances coincide, creating a branch point[b] in the correlator which behaves as $t^{4/3}$ for small t. (The 3 in the denominator of the exponent, implying a 3-sheeted Riemann surface, corresponds to the 3 geodesics.)

By studying various resolutions of this branch point, we show that as t increases from 0, the correlator defined by the boundary CFT is given by a symmetric sum of the two complex branches of this expression. Each of these can be attributed to a complex geodesic in complexified spacetime. But the correlator is an analytic function of t and can be continued onto the real sheet. This is the essential feature of the CFT which enables us to extract information which is not directly measurable. On the real sheet, the "light cone singularity" does appear. So the boundary correlator *does* contain information about the singularity, albeit in a subtle way.

---

[b]The full analytic structure of the correlator is given by the 3-sheeted Riemann surface of $\mathcal{L}(t)$, which can be derived implicitly in terms of "energy" E of the real geodesics;

$$t = \frac{1}{2}\ln\left(\frac{\frac{1}{2}E^2 - E + 1}{\sqrt{1+\frac{1}{4}E^4}}\right) - \frac{1}{2}\,i\,\ln\left(\frac{-\frac{1}{2}E^2 + iE + 1}{\sqrt{1+\frac{1}{4}E^4}}\right)\ , \qquad \mathcal{L} = \ln\left(\frac{2}{\sqrt{1+\frac{1}{4}E^4}}\right)$$

So far, our discussion dealt with infinitely massive field ϕ, as well as classical spacetime with vanishing string length l_s and coupling g_s. We can extract further information about the singularity by deviating away from this limit. Although the branch point at $t = 0$ is smoothed out for any finite m, preventing analytic continuation to the bouncing geodesic, at each order in the $1/m$ expansion the branch point persists and one can follow this geodesic and the accompanying fluctuation corrections into the bouncing domain. Moreover, since the $1/m$ corrections are largest near the singularity, we can effectively isolate the near-singularity behaviour.

One may similarly make use of the l_s and g_s corrections, thereby extracting stringy and quantum near-singularity behaviour. The main corrections at finite but small l_s (with g_s kept equal to zero) can be expressed as modifications of the supergravity field equations, which does not change the basic picture outlined above: There still exists a branch point, allowing analytic continuation onto the real sheet. The g_s expansion is more analogous to the $1/m$ expansion; finite g_s smooths out the singularity. But by taking appropriate $g_s \to 0$ limits we can extract behavior around the singularity to all orders in g_s, as well as study certain leading nonperturbative effects.

In summary, we have seen that the CFT encodes physics behind the horizon, including the near-singularity region. Unlike the previously-studied three dimensional case, in higher dimensions there is a genuine curvature singularity which is bowed in on the Penrose diagram, leading to a special behaviour of the geodesics as a critical time t_c is approached. Despite subtleties related to the nontrivial analytic structure of $\mathcal{L}(t)$, the CFT correlators reveal distinct signals of the black hole singularity. In fact, given[c] the CFT data $\langle \phi\phi \rangle$, the properties of the singularity are computationally accessible[7]. Since the "t_c" singularity persists to all orders in $\frac{1}{m}$ and g_s, as well as for small l_s, we can also extract the stringy and quantum behaviour near the black hole singularity. Thus, using analyticity, we have demonstrated that a significant amount of information from behind the horizon, and in particular from near the singularity, is encoded in boundary theory correlators. Since analyticity was essential for our arguments, it would be worthwhile to understand its implications in the context of AdS/CFT at a much deeper level.

[c]Of course, it is not presently feasible to obtain the CFT "data" $\langle \phi\phi \rangle$ in the relevant regime by a direct computation: the CFT is strongly coupled, and $m \to \infty$ implies infinitely large dimension operators. Nevertheless, it is still of use to ask the matter-of-principle question: with given information about the CFT, what can one learn about the bulk?

Acknowledgments

I wish to thank my collaborators, Lukasz Fidkowski, Matthew Kleban, and Stephen Shenker. This work is supported in part by NSF grant PHY-9870115 and by the Stanford Institute for Theoretical Physics.

References

1. J. M. Maldacena, *Adv. Theor. Math. Phys.* **2**, 231 (1998) [arXiv:hep-th/9711200]. E. Witten, *Adv. Theor. Math. Phys.* **2**, 253 (1998) [arXiv:hep-th/9802150]. O. Aharony, S. S. Gubser, J. M. Maldacena, H. Ooguri and Y. Oz, *Phys. Rept.* **323**, 183 (2000) [arXiv:hep-th/9905111].
2. V. E. Hubeny, *Int. J. Mod. Phys. D* **12**, 1693 (2003) [arXiv:hep-th/0208047].
3. L. Susskind and N. Toumbas, *Phys. Rev. D* **61**, 044001 (2000) [arXiv:hep-th/9909013]. B. Freivogel, S. B. Giddings and M. Lippert, *Phys. Rev. D* **66**, 106002 (2002) [arXiv:hep-th/0207083].
4. V. Balasubramanian and S. F. Ross, *Phys. Rev. D* **61**, 044007 (2000) [arXiv:hep-th/9906226]. J. Louko, D. Marolf and S. F. Ross, *Phys. Rev. D* **62**, 044041 (2000) [arXiv:hep-th/0002111].
5. J. M. Maldacena, *JHEP* **0304**, 021 (2003) [arXiv:hep-th/0106112].
6. P. Kraus, H. Ooguri and S. Shenker, *Phys. Rev. D* **67**, 124022 (2003) [arXiv:hep-th/0212277].
7. L. Fidkowski, V. Hubeny, M. Kleban and S. Shenker, arXiv:hep-th/0306170.
8. M. Le Bellac, *Thermal Field Theory*, Cambridge University Press (1966).
9. M. Banados, C. Teitelboim and J. Zanelli, *Phys. Rev. Lett.* **69**, 1849 (1992) [arXiv:hep-th/9204099].

Due to space restrictions, only a very limited set of references is listed here. Please refer to Ref.[7] for a more complete list.

D-BRANE STABILITY, GEOMETRIC ENGINEERING, AND MONODROMY IN THE DERIVED CATEGORY

ROBERT L. KARP

Center for Geometry and Theoretical Physics
Duke University, Durham, NC 27708-0320
E-mail: karp@cgtp.duke.edu

1. Background

The original idea that a D-brane is a subspace in target space where open strings can *end* is too simple. In the case of closed strings we know that even at $g_s = 0$ there are α'-corrections which modify the classical geometry. Accordingly, one expects that the picture of D-branes as vector bundles over submanifolds is to be altered as well.

On a Calabi–Yau n-fold X, there are two types of boundary conditions [1], and hence D-branes: A-type and B-type. To gain additional control one can consider a suitable topologically twisted version of the boundary non-linear sigma model. The topological twisting can itself be of A-type or B-type. Moreover, A-type branes are compatible with the A-twist, while B-type branes are compatible with the B-twist. For topological D-branes the following picture emerges:

- A-type – wrapping Lagrangian real n-cycles, and giving rise to objects in the derived Fukaya category of Lagrangian submanifolds
- B-type – wrapping holomorphic cycles, and giving rise to objects in the bounded derived category of coherent sheaves $\mathbf{D}(X)$ of X.

The homological mirror symmetry conjecture of Kontsevich gives a categorical equivalence of the above triangulated A_∞-categories. The conjecture has already been proved for elliptic curves, and very recently for quartic K3-surfaces [2].

In this paper we focus on **B-branes**, where using mirror symmetry, all the α'-corrections can be understood exactly [3]. But now, instead of physical B-branes, living in the boundary conformal field theory, we have *topological B-branes*, and a priori it is not clear what is relationship between them. This is where the notion of Π-*stability* pioneered by Douglas et al. [4,5] comes into play: a topological B-brane is physical if it is Π-stable. We will have more to say about Π-stability, but first we have to review some terminology. For the precise definition of Π-stability the reader should consult for example [3].

Given two D-branes, A and B, the totality of open strings stretching between them is represented by $\mathrm{Hom}(\mathsf{A}[n], \mathsf{B})$, for any integer n. Note that for a D-brane A, A[1] represents its anti-brane.

The most interesting open strings are in $\mathrm{Hom}(\mathsf{A}[-1], \mathsf{B})$. Giving **vev** to such a string forms a potential bound state, depending on whether the open string is tachyonic or not. Mathematically this gives us a cone, $\mathrm{Cone}(f : \mathsf{A}[-1] \to \mathsf{B})$, and a distinguished triangle:

$$\mathsf{A}[-1] \to \mathsf{B} \to \mathrm{Cone}(f) \to \mathsf{A}. \tag{1}$$

This construction brings democracy to bound states: any vertex of this triangle is a potential bound state of the other two vertices.

Each stable D-brane is given a grade $\varphi \in \mathbb{R}$ which varies continuously over the moduli space, although not in a single-valued way [4,5]: $\varphi(\mathsf{A}) = -\arg(Z(\mathsf{A}))/\pi$. Near the large radius limit the central charge is

$$Z(\mathrm{F}) = \int_X \mathrm{e}^{B+iJ}\, ch(\mathrm{F})\sqrt{td(X)} + \ldots (\text{quantum corrections}). \tag{2}$$

Using Mirror symmetry we can evaluate all the quantum corrections [3]. Then the mass of an open string becomes exactly computable as well: $M^2_{\mathsf{A}[-1]\to\mathsf{B}} \sim \varphi(\mathsf{B}) - \varphi(\mathsf{A})$. It is $M^2_{\mathsf{A}[-1]\to\mathsf{B}}$ that determines the stability of $\mathrm{Cone}(f)$ in (1), whether it is a bound state or not.

2. Monodromy

To learn more about the behaviour of D-branes in non-trivial backgrounds, we will subject them to monodromy transformations. Henceforth we follow the set of stable D-branes as we traverse a loop in the moduli space of complexified Kahler forms. The interesting loops are those that cross a line of marginal stability for some D-brane B, causing it to decay, and those that encircle points in moduli space where some D-brane A becomes massless. To restore the physics we had before looping we are required to relabel the D-branes. This relabeling is the monodromy action on D-branes. Note that monodromy is not a statement about a certain D-brane manifestly becoming another D-brane, but rather it is the relabeling process.

A good place to start looping is the discriminant locus Δ. This is the subvariety in the moduli space of complexified Kahler forms where the associated conformal field theory is singular. Alternatively, it is the locus in the moduli space of complex structures of the mirror Calabi–Yau where this latter one becomes singular.

The paradigmatic example of monodromy is given by the following

Conjecture 1. *(Kontsevich, Horja, Morrison) The monodromy action for looping around a component of the discriminant locus where a single D-brane* A *becomes massless is:* $\mathsf{B} \longmapsto \mathrm{Cone}(\hom(\mathsf{A},\mathsf{B}) \otimes \mathsf{A} \to \mathsf{B})$.

This monodromy can be understood physically, at least for some branes B, as B splitting off $\mathsf{A}[1]$'s a certain number of times [6]. Naturally the question arises: what could A be? An instance is given by

Conjecture 2. *At any point on the primary component of* Δ *the 6-brane* O_X *and its translates become massless. At a generic point no other D branes become massless.*

But can we do better, deal with more interesting degenerations? Maybe several D-branes becoming massless? Possibly infinitely many becoming massless? The typical situation as we approach the wall of the Kähler cone is that some subspace E of the Calabi–Yau X shrinks to a subspace Z:

$$\begin{array}{ccc} E & \overset{i}{\hookrightarrow} & X \\ \downarrow{\scriptstyle q} & & \\ Z & & \end{array} \tag{3}$$

Call it an EZ-degeneration. In this case the following conjecture is known:

Conjecture 3. *[6] Any D-Brane which becomes massless at a point on the discriminant associated with an EZ-degeneration is generated by objects of the form* $i_* q^* \mathsf{z}$ *for* $\mathsf{z} \in \mathbf{D}(Z)$.

In the moduli space there is a rational curve $\mathbb{P}^1$ connecting the two limit points: the large radius limit point with the EZ-point, corresponding to the phase reached by following the EZ-degeneration through the wall of the Kähler cone. We have the following

Conjecture 4. *[6, 7] The monodromy around the discriminant inside the above* $\mathbb{P}^1$ *is the autoequivalence of* $\mathbf{D}(X)$*:* $\mathsf{B} \mapsto \mathrm{Cone}(i_* q^* q_* i^! \mathsf{B} \to \mathsf{B})$.

In what follows, we present two different ways of checking the consistency of these conjectures. The first example is based upon the fact that a certain monodromy can be computed in two different ways. Firstly by making use of the Conjectures, and secondly by exploring the simple structure of loops on marked $\mathbb{P}^1$'s. The second example, based on geometric engineering, will also probe aspects of Π-stability. Nothing we say here is new, most of what follows can be found in [6] and [8]. The reader is also referred to these papers for many of the details.

3. Consistency check

3.1. *The model*

As a starting point we consider X to be the well known degree 8 hypersurface in the resolution of $\mathbb{P}^4_{\{2,2,2,1,1\}}$ (see e.g. [9]). X is then a K3-fibration $\pi: X \to C$ over a base $C \cong \mathbb{P}^1$, with a section. Call the generic K3 fiber F.

What we want is the quantum-corrected moduli space of complexified Kähler forms on X. This can be understood via mirror symmetry as the complex structure moduli space of the mirror Y. Choosing algebraic coordinates on the moduli space, the "primary" component of the discriminant is given by $\Delta_0 = (1 - 2^8 x)^2 - 2^{18} x^2 y$. The Kähler cone is two dimensional, where x controls the size of the K3-fiber, and y the size of the base $C \cong \mathbb{P}^1$.

3.2. *Consistency of different monodromies*

We can make the base $\mathbb{P}^1$ very large by $y \to 0$. But the primary component Δ_0 has second order intersection with $y = 0$, call this point P. The $y = 0$ curve is rational, and has three distinguished points: the large Calabi–Yau limit point ($x = y = 0$), the hybrid "$\mathbb{P}^1$-phase" point, and P.

We want to analyze the monodromy around P. The transition associated to P is of EZ-type, and consists of collapsing X onto the $\mathbb{P}^1$ base: $E = X$, $Z = \mathbb{P}^1$, $i = id_X$ and $q = \pi$. Using Conjecture 4 we can immediately deduce the monodromy action: $\mathsf{B} \mapsto \boldsymbol{H}(\mathsf{B}) = \mathrm{Cone}(\pi^*\pi_*\mathsf{B} \to \mathsf{B})$.

On the other hand singularity theory allows us to write the monodromy around P in a completely different way (see e.g. [10]), as $\boldsymbol{L}^{-1}\boldsymbol{KLK}$, where $\boldsymbol{L}$ is the large radius monodromy $\boldsymbol{L}(\mathsf{B}) = \mathsf{B} \otimes \mathcal{O}_X(F)$, while $\boldsymbol{K}$ is the Kontsevich et al. monodromy of Conjecture 1: $\boldsymbol{K}(\mathsf{B}) = \mathrm{Cone}(\hom(\mathcal{O}_X, \mathsf{B}) \otimes \mathcal{O}_X \to \mathsf{B})$.

By making extensive use of the structures in the derived categories of X and $\mathbb{P}^1$, it was shown in [6] that $\boldsymbol{H} = \boldsymbol{L}^{-1}\boldsymbol{KLK}$ indeed holds, providing a strong consistency check for the monodromy conjectures.

3.3. *Geometric engineering*

We can use the techniques of [11] to engineer an $N = 2$ pure SU(2) gauge theory using the above mentioned Calabi–Yau X [8]. The goal is to understand the stable BPS solitons spectrum of the Yang–Mills theory [12, 13] from the Π-stable BPS D-branes spectrum of the Type IIA theory.

The geometric engineering limit requires again the base C to be very large, $y \to 0$, and we must also lie very close to the discriminant locus. Note that the second order intersection point $(2^{-8}, 0)$ splits for $y = \epsilon$. Introducing the variable u by $u^{-2} = 4y/(2^8 x - 1)^2$ places the two intersection points at $u = \pm 1$. In this limit the Picard–Fuchs equations satisfied by the periods on X reduce to the one satisfied by the Seiberg-Witten periods: $(1 - u^2)\partial_u^2 - 2u\partial_u - 1/4 = 0$ [8]. But it is precisely the periods that together with the D-brane charges determine the central charges, and hence the stability of solitons.

The D-brane charge is measured by K-theory: $K^*(\mathbb{P}^1) = H^0(C) \oplus H^2(C) = \mathbb{Z}^2$. For convenience take $\mathcal{O}_C$ and $\mathcal{O}_p$, which also generate $\mathbf{D}(C)$, as the generators of $K^*(\mathbb{P}^1)$. Therefore, we require only two basic central charges: $a = Z(\pi^*\mathcal{O}_p)$, $a_D = Z(\pi^*\mathcal{O}_C)$.

Conjecture 3 and and the same knot theoretic computation as in the previous subsection gives the following assignments:

- magnetic monopole: $\mathcal{O}_X = \pi^*\mathcal{O}_C$, i.e., the 6-brane wrapping X
- dyon: $\mathcal{O}_X(F) = \pi^*\mathcal{O}_C(1)$, i.e., a 6-brane with 4-brane charge
- W-boson: $\pi^*\mathcal{O}_p$, i.e., a 4-brane wrapped around a K3 fiber.

The charge assignment is in agreement with the short exact sequence of sheaves $0 \to \mathcal{O}_C(-1) \to \mathcal{O}_C \to \mathcal{O}_p \to 0$, and it's pullback to X. This is in itself a non-trivial fact. We can also read off the following identifications: magnetic charge = *rank*, electric charge = *degree*.

3.3.1. *Weak coupling region*

For very large $|u|$ we are at the large radius limit of X, where all the α'-corrections are very small. In this limit B-type D-branes correspond to stable holomorphic vector bundles over subvarieties of X. Conjecture 3 allows us to work on $C \cong \mathbb{P}^1$ [8]. For brevity we consider only the stability of rank 2 bundles.

By Grothendieck's theorem, a rank r vector bundle V over $C \cong \mathbb{P}^1$ is a direct sum of line bundles: $V \cong \mathcal{O}_C(s_1) \oplus \mathcal{O}_C(s_2) \oplus \ldots \oplus \mathcal{O}_C(s_r)$. Thus the general rank 2 bundles is of the form $\mathcal{O}_C(s) \oplus \mathcal{O}_C(t)$. But $\mathcal{O}_C(s) \oplus \mathcal{O}_C(t)$

fits into the triangle $\mathcal{O}_C(s) \to \mathcal{O}_C(s) \oplus \mathcal{O}_C(t) \to \mathcal{O}_C(t) \to \mathcal{O}_C(s)[1]$ where $f \in \mathrm{Hom}(\mathcal{O}_C(s), \mathcal{O}_C(t)[1]) = \mathrm{Ext}^1(\mathcal{O}_C(s), \mathcal{O}_C(t))$.

There are two cases to consider, depending on whether $f = 0$ or not, resulting in the following [8]: *the rank two bundle* $\mathcal{O}(s) \oplus \mathcal{O}(t)$ *will decay into* $\mathcal{O}(s') \oplus \mathcal{O}(t')$ *for* $s' + t' = s + t$ *and* $|s' - t'| \leq 1$. Note that for $\mathcal{O}(s) \oplus \mathcal{O}(t)$ with $|s-t| \leq 1$, $\mathrm{Ext}^1(\mathcal{O}_C(s), \mathcal{O}_C(t)) = \mathrm{Ext}^1(\mathcal{O}_C(t), \mathcal{O}_C(s)) = 0$, so there are no open strings to either stabilize or destabilize, and the two constituents are free to drift apart.

Similarly, one proves that any dyon (r, m) for $r > 1$ is unstable to decay into dyons with $r = 1$, and there are no stable states of charge $(0, m)$ for $m > 1$ either [8], in line with [12,13].

3.3.2. *Strong coupling region*

By considering the distinguished triangle $\mathcal{O}_C(m) \to \mathcal{O}_C(m+1) \to \mathcal{O}_p \to \mathcal{O}_C(m)[1]$ we can show that the only stable solitons are: $\mathcal{O}_C$ and $\mathcal{O}_C(1)$, i.e., the monopole and the dyon, again in line with [12,13]. In particular, the W-boson $\mathcal{O}_p$ decays into $\mathcal{O}_C(1)$ and $\mathcal{O}_C[1]$ [8].

3.3.3. *Returning to weak coupling*

Suppose now that we continue our journey, back to the weakly coupled region. We would expect to recover the original set of stable solitons. But they may not be the same elements of the derived category, as they might undergo monodromy.

One would expect that as we cross the MS-line, $\mathcal{O}_p$ would become stable again. Unfortunately $M^2_{\mathcal{O}_C \to \mathcal{O}_C(1)} \sim \varphi(\pi^*\mathcal{O}_C(1)) - \varphi(\pi^*\mathcal{O}_C[-1])$ gets larger and larger, so $\mathcal{O}_p$ never stabilizes. To rescue come some new bound states! Define X by the triangle $\pi^*\mathcal{O}_C(1)[-1] \to \pi^*\mathcal{O}_C[2] \to \mathrm{X} \to \pi^*\mathcal{O}_C(1)$, then iterate it: $\mathrm{Y}_m \to \mathrm{Y}_{m+1} \to \mathrm{X} \to \mathrm{Y}_m[1]$. It turns out that *all* of the Y_m's become stable as we pass the MS-line back into the weak-coupling regime.

This infinite tower of new states Y_m is the "replacement" for the old states $\mathcal{O}_C(m)$, of charge $(1, m)$. It is also true that under monodromy around $u = 1$ (we can use Conjecture 1 for this) the set of states $\mathcal{O}_C(m)$ turn into the set of states Y_m.

But there is still an aesthetic discrepancy: the original set of weak coupling states have clear geometric interpretation as wrapped branes, while after monodromy we have rather exotic objects. However, from the derived category point of view they look quite symmetric (see the sketch below): in the strong-coupling region we have only two stable D-branes: A and B. As we move across the MS-line into the weak coupling regime they can form a bound state W which plays the rôle of the W-boson. Crossing upwards this is done by $\mathsf{A} \to \mathsf{B}$, while crossing downwards by $\mathsf{B} \to \mathsf{A}[3]$. This is essentially Serre duality at work. Observe also the **3** appearing, as if the solitons in Seiberg–Witten theory knew that they should be associated with a 3-fold!

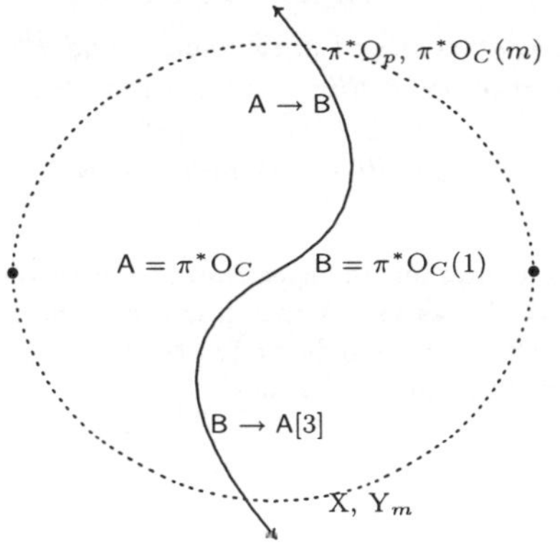

Acknowledgments

I would like to thank my advisor Paul Aspinwall, for his careful guidance, and Paul Horja for enjoyable collaboration. This work was supported in part by the NSF grant DMS-0074072.

References

1. K. Becker, M. Becker, and A. Strominger, Nucl. Phys. **B456** (1995) 130–152, hep-th/9507158. H. Ooguri, Y. Oz, and Z. Yin, Nucl. Phys. **B477** (1996) 407–430, hep-th/9606112.
2. P. Seidel, math.SG/0310414.
3. P. S. Aspinwall and M. R. Douglas, JHEP **05** (2002) 031, hep-th/0110071.
4. M. R. Douglas, B. Fiol, and C. Römelsberger, hep-th/0002037.
5. M. R. Douglas, J. Math. Phys. **42** (2001) 2818–2843, hep-th/0011017.
6. P. S. Aspinwall, R. P. Horja, and R. L. Karp, hep-th/0209161.
7. P. Horja, math.AG/0103231.
8. P. S. Aspinwall, R. L. Karp, JHEP **0304**, 049 (2003) [arXiv:hep-th/0211121].
9. P. Candelas et al., Nucl. Phys. B **416**, 481 (1994) [arXiv:hep-th/9308083].
10. P. S. Aspinwall, J. Math. Phys. **42** (2001) 5534–5552, hep-th/0102198.
11. S. Kachru et al., Nucl. Phys. **B459** (1996) 537–558, hep-th/9508155. S. Katz, A. Klemm, and C. Vafa, Nucl. Phys. **B497** (1997) 173–195, hep-th/9609239.
12. N. Seiberg and E. Witten, Nucl. Phys. **B426** (1994) 19–52, hep-th/9407087.
13. F. Ferrari and A. Bilal, Nucl. Phys. **B469** (1996) 387–402, hep-th/9602082.

B-BRANES AND THE DERIVED CATEGORY

SHELDON KATZ

Departments of Mathematics and Physics
University of Illinois at Urbana-Champaign
Urbana, IL 61801
Email: katz@math.uiuc.edu

By a direct CFT computation, the spectrum of the topological B-model is compared to Ext groups of sheaves. A match can only be made if abstract vector bundles on holomorphic submanifolds are twisted by the canonical Spin^c structure of its support in describing physical branes.

1. Introduction

In recent years, there has been much fruitful interplay between physics and geometry through string theory. Many questions in physics correspond to questions in pure geometry. The motivations in physics and geometry are often of a completely different nature, so that finding the precise dictionary between the two areas is key to making fundamental progress.

The spectrum of the closed topological string is well understood[15], and the correlation functions can be computed. In this talk, I outline a similar description for the open string sector of the topological B-model on a compact Kähler manifold X. The first task is to understand the boundary conditions, which are to be identified with the derived category in well-known ways[10,12,5,1]. The simplest case arises from a holomorphic submanifold $i : S \hookrightarrow X$ and holomorphic vector bundle E on X, so that a D-brane is wrapped on S with gauge bundle E. This data can be identified with the sheaf i_*E of $\mathcal{O}_X$ modules, which is not a bundle unless $S = X$. In any case, i_*E is an object of the derived category of X. More general objects of the derived category can be identified with boundary conditions as well.

This pleasing identification of geometric objects with boundary conditions is complicated by worldsheet anomalies, which can be eliminated by twisting the gauge bundle by a Spin^c structure[6]. A canonical choice is given by the canonical Spin^c structure on X. This corresponds to modifying the gauge bundle E to become, somewhat imprecisely, $E \otimes K_S^{-1/2}$. With this choice, the open string spectrum matches Ext groups exactly[9]. With any other twisting or no twisting at all, the spectrum can fail to match Ext groups. So the identification of objects of the derived category with branes is a bit more subtle than one would like.

In this talk, I give the basic examples illustrating how to compute the spectrum directly in BCFT and how the results are only sensible when the anomaly is included in the analysis. More details are available elsewhere[9,13], as are generalizations to orbifolds[8], flat B-field backgrounds[3], and D-branes wrapping nonreduced schemes[4,14].

2. Closed string B model

I start by reviewing the topological B-model on a Calabi-Yau threefold X. Locally X has three complex coordinates (ϕ^1, ϕ^2, ϕ^3) which can be identified with bosonic fields on the worldsheet. A nowhere vanishing antisymmetric holomorphic tensor $\Omega = \Omega_{ijk}$ is required to soak up fermion zero modes. This condition is satisfied for a Calabi-Yau threefold. The topological B-model can be compared to physical strings in 10D $M^4 \times X$; I do not do that here but make some brief comments at the end.

Tensors on X have holomorphic indices i and antiholomorphic indices $\bar{i}$. These are associated to the holomorphic and antiholomorphic tangent bundles of X which are global objects, i.e.

$$\rho^i \in T^{1,0}(X)$$
$$\rho^{\bar{i}} \in T^{0,1}(X)$$

with $\det(T^{1,0}(X))$ a trivial bundle, required by fermion zero modes coupling to $T^{1,0}(X)$ as mentioned above.

The worldsheet is a Riemann surface Σ with local complex coordinate z. Covariant tensors split into holomorphic and antiholomorphic pieces. The holomorphic part is called the canonical bundle and is denoted by $K_\Sigma = (T^{1,0}(\Sigma))^*$. The antiholomorphic part is denoted by $\bar{K}_\Sigma = (T^{0,1}(\Sigma))^*$

The fields in the topological string combine the tensor structures on Σ and X. So the fields in the theory are:

$$\begin{array}{lcl} & & \phi^i(z) \\ \psi^{\bar{i}}_\pm(z) & \in & \Gamma\left(\phi^* T^{0,1} X\right), \\ \psi^i_+(z) & \in & \Gamma\left(K \otimes \phi^* T^{1,0} X\right), \\ \psi^i_-(z) & \in & \Gamma\left(\overline{K} \otimes \phi^* T^{1,0} X\right) \end{array}$$

The ψ fields anticommute with each other, and the ϕ are bosonic. The bulk B model Lagrangian is

$$\tfrac{1}{2}g_{i\bar{j}}\partial\phi^i\bar{\partial}\phi^{\bar{j}} + \tfrac{1}{2}g_{i\bar{j}}\partial\phi^{\bar{j}}\bar{\partial}\phi^i + ig_{i\bar{j}}\psi_-^{\bar{j}}D_z\psi_-^i + ig_{i\bar{j}}\psi_+^{\bar{j}}D_{\bar{z}}\psi_+^i + R_{i\bar{i}j\bar{j}}\psi_+^i\psi_+^{\bar{i}}\psi_-^j\psi_-^{\bar{j}}$$

with BRST transformations

$$\begin{aligned}
\delta\phi^i &= 0,\\
\delta\phi^{\bar{\imath}} &= i\alpha\left(\psi_+^{\bar{\imath}} + \psi_-^{\bar{\imath}}\right),\\
\delta\psi_+^i &= -\alpha\partial\phi^i,\\
\delta\psi_+^{\bar{\imath}} &= -i\alpha\psi_-^{\bar{j}}\Gamma_{\bar{j}\bar{m}}^{\bar{\imath}}\psi_+^{\bar{m}},\\
\delta\psi_-^i &= -\alpha\bar{\partial}\phi^i,\\
\delta\psi_-^{\bar{\imath}} &= -i\alpha\psi_+^{\bar{j}}\Gamma_{\bar{j}\bar{m}}^{\bar{\imath}}\psi_-^{\bar{m}}.
\end{aligned}$$

The BRST symmetry satisfies $\delta^2 = 0$. Consideration is restricted to BRST-closed operators, $\delta F = 0$. Addition of a BRST-exact operator does not alter correlation functions

$$F \sim F + \delta G,$$

so the spectrum is described by BRST cohomology

$$H^*_{\mathrm{BRST}} = \frac{\{\mathrm{BRST}-\text{closed operators}\}}{\{\mathrm{BRST}-\text{exact operators}\}}.$$

It is convenient to change variables and define

$$\begin{aligned}
\eta^{\bar{\imath}} &= \psi_+^{\bar{\imath}} + \psi_-^{\bar{\imath}},\\
\theta_i &= g_{i\bar{j}}\left(\psi_+^{\bar{j}} - \psi_-^{\bar{j}}\right),\\
\rho_z^i &= \psi_+^i,\\
\rho_{\bar{z}}^i &= \psi_-^i,
\end{aligned}$$

A vertex operator is of the form

$$b_{\bar{\imath}_1\cdots\bar{\imath}_n}^{j_1\cdots j_m}(\phi_0)\,\eta^{\bar{\imath}_1}\cdots\eta^{\bar{\imath}_n}\theta_{j_1}\cdots\theta_{j_m}$$

This operator can be identified with a tensor field on X:

$$b_{\bar{\imath}_1\cdots\bar{\imath}_n}^{j_1\cdots j_m}(\phi)d\phi^{\bar{\imath}_1}\wedge\ldots\wedge d\phi^{\bar{\imath}_n}\otimes\frac{\partial}{\partial\phi^{j_1}}\wedge\ldots\wedge\frac{\partial}{\partial\phi^{j_m}}$$

where $\wedge$ denotes a completely antisymmetrized product.

Under this identification, BRST-cohomology is identified with the Dolbeault cohomology of $\Lambda T^{1,0}(X)$, an important geometric invariant of the holomorphic structure of X.

$$H^q\left(\Lambda^p T^{1,0}(X)\right) = \frac{\{\bar\partial - \text{closed } (0,q) \text{ forms}\}}{\{\bar\partial - \text{exact } (0,q) \text{ forms}\}}$$

where the forms take values in $\Lambda^p T^{1,0}(X)$. Here $\bar\partial$ is the antiholomorphic differential, e.g.

$$\bar\partial f = \frac{\partial f}{\partial \bar\phi^i} d\bar\phi^i.$$

Among these Dolbeault cohomology classes, $H^1(T^{1,0}(X))$ parametrizes the holomorphic structure.

Correlation functions of the associated vertex operators are important. For example, three point correlators of operators associated to $H^1(T^{1,0}(X))$

$$\langle A^{i_1}_{\bar j_1}\theta_{i_1}\eta^{\bar j_1}, A^{i_2}_{\bar j_2}\theta_{i_2}\eta^{\bar j_2}, A^{i_3}_{\bar j_3}\theta_{i_3}\eta^{\bar j_3}\rangle$$

are identified with the Yukawa couplings in the gauge sector of the heterotic string.

This is the kind of description to be extended to the open string.

3. Open string B-model

In extending the previous analysis to open string the first step is to find the boundary conditions. The simplest supersymmetric B-branes are described by a holomorphic submanifold $S \subset X$ together with a gauge field A_i on S. The coordinates on S are Neumann directions while coordinates transverse to S are Dirichlet directions. Using the gauge field, introduce the usual boundary term in the action

$$\int_{\partial\Sigma} A_i \frac{\partial\phi^i}{\partial t} + \dots.$$

Along Neumann directions, the boundary conditions are

$$\psi^i_+|_{\partial\Sigma} = \psi^i_-|_{\partial\Sigma}, \qquad \theta_i = (\mathrm{Tr}\ F_{i\bar j})\,\eta^{\bar j},$$

F being the field strength, while along Dirichlet directions, the boundary conditions are

$$\psi^i_+|_{\partial\Sigma} = -\psi^i_-|_{\partial\Sigma}, \qquad \eta^{\bar\imath} = 0.$$

Some care is needed in interpreting the intrinsic meaning of a Dirichlet direction, since the tangent bundle of X along S has a C^∞ splitting into the sum of the tangent bundle to S and the normal bundle to S in X, i.e. $TX|_S \simeq TS \oplus N_{S/X}$, but there need not be a holomorphic splitting.

The vertex operators are

$$b^{\alpha\beta j_1 \cdots j_m}_{\bar{i}_1 \cdots \bar{i}_n}(\phi_0)\, \eta^{\bar{i}_1} \cdots \eta^{\bar{i}_n} \theta_{j_1} \cdots \theta_{j_m}$$

where the $\bar{i}$ are indices on S and the θ_j are constrained by the boundary conditions. Also, α, β are gauge degrees of freedom (Chan-Paton indices).

The immediate goal is to evaluate the BRST cohomology and identify it with a geometric object. Mathematically, a novel complex of differential forms has been defined, the novelty being the boundary conditions, and the problem is to compute its cohomology. The answer is expected to match Ext groups of objects in the derived category of algebraic geometry. This does check, with some surprises along the way.

Now $U(n)$ gauge fields live in rank n vector bundles E on S. Locally on S, E takes the form $U_\alpha \times \mathbf{C}^n$ for $U_\alpha \subset S$. Vector bundles are glued together over intersections $U_\alpha \cap U_\beta$ via $r \times r$ matrices $g_{\alpha\beta}$ acting on $\mathbf{C}^r$ by multiplication. Sections of bundles are identified with local $\mathbf{C}^r$-valued functions s_α on each U_α satisfying

$$s_\alpha = g_{\alpha\beta} s_\beta. \tag{1}$$

The sections of E form an important object of geometry called a *coherent sheaf* on X. This means that sections s_α of E can be multiplied by holomorphic functions f on X, and that the result only depends on the restriction of such functions to S (coherence is a technical finiteness condition which is automatically satisfied here). Explicitly the multiplication is simply the identity $f|_S \cdot s_\alpha = g_{\alpha\beta}((f|_S) \cdot s_\beta)$ obtained from (1) after multiplication by $f|_S$. The resulting sheaf is denoted $i_* E$ where $i : S \hookrightarrow X$ is the inclusion.

Coherent sheaves form the heart of a t-structure on the derived category of X. From this heart, all objects of the derived category can be built by shifts and extensions. The heart, shifts, and extensions model branes, antibranes, and tachyon condensation. The notion of stability of bundles has a corresponding physical interpretation (*pi-stability*)[5] which is being developed mathematically as well.[2,7]. The surprises which are described in this talk will need to be incorporated into these notions.

Naively, the brane described by E is to be identified with the sheaf $i_* E$ of the derived category. But this is not quite correct due to anomalies;

the standard correspondence between objects of the derived category and branes has to be modified.

Now if two branes S, T are considered, with strings from S to T, there are various combinations of Dirichlet and Neumann conditions at the respective ends of the string. The fermion zero modes couple to directions in S and T as well as to directions not in either S nor T. Geometrically, these correspond to

$$T(S \cap T), \qquad \mathcal{N} = \frac{T(X)}{T(S) \oplus T(T)}$$

respectively, where the notational simplification of writing e.g. $T(S)$ in place of $T^{1,0}(S)$ has been adopted. The triviality of

$$\det(T(S \cap T)) \otimes \det(\mathcal{N})$$

is required to soak up fermion zero modes. This is equivalent to the triviality of the bundle

$$\det\left(N_{S\cap T,T}\right) \otimes \det\left(N_{S\cap T,S}\right),$$

where $N_{S\cap T,T}$ is the bundle of normal vector fields to $S \cap T$ in T.

4. The Freed-Witten anomaly

An important anomaly in the path integral measure was described by Freed and Witten[6]. There are two interesting factors in the path integral:

$$\mathrm{Pfaff}(D),$$

the Pfaffian of the worldsheet Dirac operator, and

$$\exp\left(i \int_{\partial\Sigma} A\right)$$

They found that the Pfaffian has an anomaly, characterized by the second Stiefel-Whitney class $w_2(S) \in H^2(S, \mathbf{Z}_2)$: that in traversing a family of images of the boundary of the worldsheet $\partial\Sigma$ parametrized by a curve C,

$$\mathrm{Pfaff}(D) \mapsto (-1)^{\langle w_2(S), C\times\partial\Sigma\rangle} \mathrm{Pfaff}(D).$$

So $\exp\left(i \int_{\partial\Sigma} A\right)$ must transform in the same way to cancel the anomaly:

$$\exp\left(i \int_{\partial\Sigma} A\right) \mapsto (-1)^{\langle w_2(S), C\times\partial\Sigma\rangle} \exp\left(i \int_{\partial\Sigma} A\right)$$

This says that the gauge field does *not* live in a bundle, but rather in a *twisted* bundle. In fact, the gauge field must live in a twisted bundle E of the form

$$E = B \otimes (L)^{1/2},$$

where B is an ordinary bundle, and L is a line bundle with

$$c_1(L) \equiv w_2(S) \pmod 2. \tag{2}$$

This suggests that the map

$$\text{Bundles on } S \to \text{ physical branes}$$

should not be the identity map, but should be modified to

$$E \mapsto E \otimes L^{1/2}.$$

for some L. The line bundle L corresponds to a choice of Spin^c structure on X. Any bundle L satisfying (2) will give an identification, but only one choice will identify the open string spectrum with Ext groups of sheaves.

A complex manifold X need not be spin, but is always Spin^c. In fact, there is a canonical Spin^c structure[11], and it is not a surprise that this is the one that gives the twisting of the gauge bundle that relates the spectrum to Ext groups. The result is that the correct identification of bundles with twisted sheaves is given by

$$E \mapsto E \otimes (K_S^*)^{1/2}.$$

Said differently, this corresponds to a trivialization of a twisting of K-theory dictated by the anomaly.

5. Computation of the spectrum

With all the machinery in place, the method for computing the spectrum can now be outlined in the case of parallel coincident branes on $S \hookrightarrow X$. Denote by E, F the gauge bundles on the respective ends of the string. The boundary vertex operators are of the form

$$b^{\alpha\beta j_1 \cdots j_m}_{\bar{\imath}_1 \cdots \bar{\imath}_n}(\phi_0)\, \eta^{\bar{\imath}_1} \cdots \eta^{\bar{\imath}_n} \theta_{j_1} \cdots \theta_{j_m}.$$

Recall the effect of the boundary conditions: the θ indices are determined by their values in directions normal to S, the η indices are constrained to only live along directions tangent to S, and ϕ zero modes ϕ_0 are constrained to only map out S.

By explicit computation[9], the BRST cohomology classes are in one-to-one correspondence with ext groups

$$\mathrm{Ext}^n_X(i_*(E), i_*F)$$

In particular, if $E = F$ then $\mathrm{Ext}^1_X(i_*E, i_*E)$ describes first order deformations of E as a sheaf or object of the derived category.

Here is a small part of the computation. Restricting attention to θ indices tangent to S, the BRST condition determines bundle-valued differential forms living on S, i.e.

$$b^{\alpha\beta j_1\cdots j_m}_{\bar{\imath}_1\cdots\bar{\imath}_n}(\phi_0)\,\eta^{\bar{\imath}_1}\cdots\eta^{\bar{\imath}_n}\theta_{j_1}\cdots\theta_{j_m} \in H^n\left(S, E^*\otimes F\otimes \Lambda^m N_{S/X}\right)$$

where $N_{S/X}$ is the normal bundle to S in X.

Putting the boundary conditions

$$\theta_i = (\mathrm{Tr}\ F_{i\bar{\jmath}})\,\eta^{\bar{\jmath}}$$

back in iteratively, the *local to global spectral sequence*

$$E_2^{p,q} : H^p\left(S, \mathcal{E}^\vee\otimes\mathcal{F}\otimes\Lambda^q N_{S/X}\right) \Longrightarrow \mathrm{Ext}^{p+q}_X\left(i_*\mathcal{E}, i_*\mathcal{F}\right)$$

is reproduced exactly, each iteration corresponding to an iteration of the spectral sequence. This shows that the BRST cohomology is the abutment of the spectral sequence, i.e. the anticipated groups $\mathrm{Ext}^n_X\left(i_*\mathcal{E}, i_*\mathcal{F}\right)$.

The Freed-Witten anomaly is not manifest here — it cancels out of the opposite ends of the string, so has been suppressed from the notation. But this is not the case in general. Consider branes

$$i : S \hookrightarrow X,\ \text{gauge field in } E\otimes (K_S^*)^{1/2}$$

$$j : T \hookrightarrow X,\ \text{gauge field in } F\otimes (K_T^*)^{1/2}$$

Now the BRST cohomology is on one-to-one correspondence with elements of

$$\mathrm{Ext}^n_X(i_*E, j_*F))$$

The computation matches the local to global spectral sequence with E_2 term

$$E_2^{p,q} = H^p\left(S\cap T, E^*|_{S\cap T}\otimes F|_{S\cap T}\otimes \Lambda^{q-m}\mathcal{N}\otimes \det N_{S\cap T/T}\right)$$

$$\Longrightarrow \mathrm{Ext}^{p+q}_X\left(i_*E, j_*F\right)$$

where m is the rank of $N_{S\cap T/T}$ and

$$\mathcal{N} = \frac{T(X)}{T(S) \oplus T(T)}.$$

as before. Note that the term $\det N_{S\cap T/T}$ is physically absent if the anomaly is ignored. The computation works only because the B-model anomaly cancels the Freed-Witten anomaly. To see this, use the identity

$$\sqrt{\frac{K_S|_{S\cap T}}{K_T|_{S\cap T}}} \cong \det N_{S\cap T/T}$$

This may seem a bit odd that the B-model anomaly is required, since for physical strings, the condition on the triviality of

$$\det\left(N_{S\cap T,T}\right) \otimes \det\left(N_{S\cap T,S}\right) \tag{3}$$

is absent, while there is general confidence that the derived category and ext groups gives the correct answer. This suggests that for physical strings (3) is a new supersymmetry selection rule.

I would like to use this description of the spectrum to compute correlation functions, but there are a number of serious obstacles. All that can be said at present is that three-point functions can be computed via the Yoneda pairing.

6. Generalizations

The analysis above extends to a wide range of string models, with similar results. The extension to orbifolds[8] shows that this analysis is not constrained to large radius and shows how Ext groups in (quotient) stacks arise in string theory. The extension to flat B-field backgrounds[3] shows how Ext groups in the derived category of twisted sheaves arises in string theory. Finally, the boundary conditions can correspond to sheaves on nonreduced schemes, which can sometimes be described physically by turning on nilpotent Higgs fields. The resulting spectrum coincides with Ext groups of sheaves associated to these nonreduced schemes[4].

7. Conclusions

Open string vertex operators can be explicitly and completely described by complex geometry, in the situation where the boundary conditions correspond to gauge fields in (twisted) bundles on holomorphic submanifolds and beyond. It is desirable to describe the twisting needed to cancel the anomaly intrinsically in terms of the derived category. Thus boundary branes can be described directly in CFT. It is desirable to extend existing methods to compute higher correlation functions.

Acknowledgments

I would like to thank Eric Sharpe for many valuable discussions and for collaboration on this topic, and Robert Karp for comments on an earlier draft.

References

1. P. Aspinwall and A. Lawrence, *Derived categories and zero-brane stability*, *JHEP* **0108** 004 (2001), hep-th/0104147.
2. T. Bridgeland, *Stability conditions on triangulated categories*, math.AG/0212237.
3. A. Caldararu, S. Katz, and E. Sharpe, *D-branes, B fields, and Ext groups*, ATMP **7** 382–403 (2003), hep-th/0302099.
4. R. Donagi, S. Katz, and E. Sharpe, *Spectra of D-branes with Higgs vevs*, hep-th/0309270.
5. M. Douglas, *D-branes, categories, and $\mathcal{N} = 1$ supersymmetry*, *J. Math. Phys.* **42** 2818–2843 (2001), hep-th/0011017.
6. D. Freed and E. Witten, *Anomalies in string theory with D-branes*, hep-th/9907189.
7. A. Gorodentscev, S. Kuleshov, A. Rudakov, *Stability data and t-structures on a triangulated category*, math.AG/0312442.
8. S. Katz, T. Pantev, and E. Sharpe, *D-branes, orbifolds, and Ext groups*, *Nucl. Phys.* **B673** 263–300 (2003), hep-th/0212218.
9. S. Katz and E. Sharpe, *D-branes, open string vertex operators, and Ext groups*, *Adv. Theor. Math. Phys.* **6** 979–1030 (2003), hep-th/0208104.
10. M. Kontsevich, *Homological algebra of mirror symmetry*, Proceedings of the 1994 International Congress of Mathematicians, Birkäuser, Zürich, 1995, p. 120, alg-geom/9411018.
11. H. B. Lawson and M.-L. Michelson, Spin Geometry, Princeton University Press, Princeton 1989.
12. E. Sharpe, *D-branes, derived categories, and Grothendieck groups*, *Nucl. Phys.* **B561** 433–450 (1999), hep-th/9902116.
13. E. Sharpe, *Lectures on D-branes and sheaves*, hep-th/0307245.
14. E. Sharpe, *Mathematical Aspects of D-branes*, these proceedings.
15. E. Witten, *Mirror Manifolds and Topological Field Theory*, in Essays on Mirror Manifolds, S.T. Yau (ed), International Press, Hong Kong 1992.

REMARKS ON BRANES, FLUXES, AND SOFT SUSY BREAKING*

ALBION LAWRENCE

Physics Dept., Brandeis University, Waltham, MA 02454
albion@brandeis.edu

JOHN McGREEVY

Physics Dept., Princeton University, Princeton, NJ 08544
mcgreevy@princeton.edu

We review recent work identifying soft SUSY-breaking terms in local type II string models with branes and magnetic fluxes. We then make a new observation about the space of D-brane vacua in Calabi-Yau backgrounds, and identify vevs for nonperturbative charged hypermultiplets in Calabi-Yau backgrounds with $N = 2$ Fayet-Iliopoulos terms.

1. Introduction

A wide class of phenomenologically attractive string theory backgrounds with low-energy $\mathcal{N} = 1$ SUSY are described combinations of D-branes, orientifold planes, and magnetic fluxes. Nontrivial gauge dynamics is typically localized in regions of the compactification manifold, and a fairly generic scenario for SUSY breaking will have supersymmetry broken in one region of the manifold, with the standard model dynamics localized somewhere else. Supersymmetry breaking will be communicated via 10d supergravity effects at tree level, and via radiative corrections as in anomaly or gaugino mediation. However, the detailed appearance and origins of such terms in the low energy effective action of specific models is understood only in a few very specific examples.

In these proceedings we review and slightly extend recent work [1] studying the appearance of tree-level soft SUSY-breaking terms for local models of Calabi-Yau threefold backgrounds with D-branes. The closed string

*Talk given by A.L.

modes live in multiplets of $\mathcal{N} = 2$, $d = 4$ SUSY. D-branes and magnetic fluxes break SUSY to $\mathcal{N} = 1$ or $\mathcal{N} = 0$. The D-brane modes, and some closed string modes controlling the local geometry, have finite 4d kinetic terms even when the CY is noncompact. The other closed string modes appear as "spurions", as their dynamics decouples from the low energy physics of the local model. The auxiliary fields for these modes appear as soft SUSY-breaking couplings in the local model. This provides a set of building blocks for more complete models, and in a given model should allow one to address questions such as whether the squark masses are aligned with the quark masses. Therefore, in Section 2 we review the identification of auxiliary fields for light closed string modes with magnetic fluxes, and review how these appear as soft SUSY-breaking couplings on D-branes. This section is based on the talk given by the first author at the Quantum Theory and Symmetries 3 conference at the University of Cincinnati.

Magnetic fluxes and D-branes are crucial aspects of type I, type II, and F-theory models. The interplay between these two aspects of string theory makes apparent some features of the string theory models which are highly nontrivial from the point of view of the low energy field theory. In section 3 we discuss the impact of fluxes on the space of D-branes in Calabi-Yau compactifications. In section 4 we discuss the degrees of freedom responsible for tuning $\mathcal{N} = 2$ Fayet-Iliopoulos terms in type II models on CY threefolds.

Because of the limited space allowed for these proceedings, we will be minimalist about referencing. A more complete bibliography appears in [1]. We apologize to those who are not referenced here.

2. Closed string modes and soft SUSY breaking

2.1. *Auxiliary fields for $\mathcal{N} = 2$ vector multiplets*

We begin with type II string theory on a local (*e.g.* noncompact) Calabi-Yau threefold X times $\mathcal{R}^4$, together with D-branes filling $\mathcal{R}^4 \times \mathcal{C}$, $\mathcal{C} \subset X$. Although the D-branes preserve at most $\mathcal{N} = 1$ supersymmetry at most, the closed string modes lie in $\mathcal{N} = 2$ supermultiplets. The $\mathcal{N} = 2$ properties of the latter constrain their couplings to the D-branes [3]. Therefore, it is important to understand the underlying $\mathcal{N} = 2$ SUSY structure. Vevs for auxiliary fields in the closed string supermultiplets can break SUSY to $\mathcal{N} = 1$ or $\mathcal{N} = 0$ via computable operators at tree level.

Closed string vector multiplets arise from complex structure deformations of X. We can write them as chiral superfields in terms of the $\mathcal{N} = 2$ superspace variables $\theta, \hat{\theta}$ which are a doublet of Weyl spinors under the

$SU(2)_R$ symmetry of $\mathcal{N} = 2$ theories. Translations in these superspace directions are generated by spacetime supercharges built from left- and right-moving worldsheet sectors, respectively.

A vectormultiplet can be described by a superfield V which solves the chiral constraints

$$\begin{aligned}\bar{\nabla}_{\dot{\alpha}} V &\equiv \left(-\frac{\partial}{\partial\bar{\theta}^{\dot{\alpha}}} - i\sigma^{\mu}_{\beta\dot{\alpha}}\theta^{\beta}\partial_{\mu}\right) V = 0 \\ \hat{\bar{\nabla}}_{\dot{\alpha}} V &\equiv \left(-\frac{\partial}{\partial\hat{\bar{\theta}}^{\dot{\alpha}}} - i\sigma^{\mu}_{\beta\dot{\alpha}}\hat{\theta}^{\beta}\partial_{\mu}\right) V = 0 \; . \end{aligned} \quad (1)$$

The superspace expansion for such a field is:

$$\begin{aligned} V \;=\; & w^a + \theta^{\alpha}\zeta^a_{\alpha} + \hat{\theta}^{\alpha}\hat{\zeta}^a_{\alpha} + \theta^2 D^a_{++} + \theta^{\alpha}\hat{\theta}^{\beta}\left(\epsilon_{\alpha\beta}D^a_{+-} + F^a_{\alpha\beta}\right) \\ & + \hat{\theta}^2 D^a_{--} + \theta^{\alpha}\hat{\theta}^2\chi^a_{\alpha} + \hat{\theta}^{\beta}\theta^2\hat{\chi}^a_{\beta} \\ & + \theta^2\hat{\theta}^2 C^a \; . \end{aligned} \quad (2)$$

One may impose the additional constraints: [1,4]

$$(\epsilon_{ij}\nabla^i\sigma_{\mu\nu}\nabla^j)(\epsilon_{kl}\nabla^k\sigma^{\mu\nu}\nabla^l)V = -96\partial^2\bar{V} \quad (3)$$

which render C,χ as dependent variables; impose the constraint that $\sigma^{\mu\nu}_{\alpha\beta}F^{\alpha\beta}$ be an anti-delf-dual tensor; and impose the "reality constraints" $\Box D_{++} = \Box D^*_{--}$, $\Box D_{+-}$ real.

In type IIB string theory, we can identify the bosonic degrees of freedom as follows. If we choose "CFT coordinates" on the moduli space [1], the scalar component w^a can be associated to the perturbation

$$\delta^m (ds)^2 = \delta g^m_{\bar{i}\bar{j}} d\bar{z}^{\bar{i}} d\bar{z}^{\bar{j}} \; . \quad (4)$$

The label m denotes a direction in the complex structure moduli space. Factoring out reparameterizations, each such deformation can be associated to an elements of $H^{(2,1)}(X)$:

$$\omega^m_{\bar{i}jk} = \delta g^m_{\bar{i}\bar{j}} g^{i\bar{j}} \Omega_{ijk} \quad (5)$$

where Ω is the holomorphic $(3,0)$ form on X and g is the metric. Choose a basis ω^a of harmonic representatives of $H^{(2,1)}(X)$. Auxiliary fields correspond to deformations of the NS-NS 3-form $H = \sum_m h^m\omega^m$ + h.c.; of the RR 3-form $F = \sum_m f^m\omega^m$ + h.c.; and of $T = i(\partial - \bar{\partial})J = \sum_m \tau^m\omega^m$ + h.c., where $J = g_{i\bar{j}} dz^i d\bar{z}^{\bar{j}}$. In terms of components h, f, τ:

$$\begin{aligned} D^m_{++} &= (\tau^m + h^m) \\ D^m_{+-} &= g_s(f^m - C^{(0)}h^m) \\ D^m_{--} &= (\tau^m - h^m) \; , \end{aligned} \quad (6)$$

where $C^{(0)}$ is the type IIB RR axion.

The results above can be proven using RNS worldsheet techniques. Using these, one may also find the auxiliary fields for the "special geometry" coordinates on complex structure moduli space. That is, choose a symplectic basis A^a, B_a of $H_3(X)$, such that $A^a \cup B_b = \delta^a_b$, $A^a \cup A^b = B_a \cup B_b = 0$. A good set of coordinates on moduli space is $t^a = \int_{A^a} \Omega$. The "dual periods" $F_a = \int_{B_a} \Omega$ can be written as functions of t. The auxiliary fields corresponding to t^a can be written as:

$$\begin{aligned} D^a_{++} &= \int_{A^a} \left(\tilde{T} + \tilde{H}\right) \\ D^a_{+-} &= g_s \int_{A^a} \left(\tilde{F} - C^{(0)} \tilde{H}\right) \\ D^a_{--} &= \int_{A^a} \left(\tilde{T} - \tilde{H}\right) , \end{aligned} \tag{7}$$

where the tildes denote the projection of the forms into $H^{(2,1)}(X)$. The auxiliary fields for F_a are as above, only with A^a replaced by B_a. It is posssible to combine these statements into a 'supermultiplet of three-forms' which incorporates all of the complex structure multiplets, of the form

$$V = \Omega + \theta^i \theta^j D_{ij} + ... \tag{8}$$

where (i, j) run over $SU(2)$ doublet indices $\pm$.

A similar story holds for hypermultiplets in type IIA compactifications. [1]. The identification of auxiliary fields for vector multiplets in IIA and hypermultiplets in IIB is not yet completely understood.

2.2. *Soft SUSY breaking*

Vevs for auxiliary fields D_{ij} break supersymmetry to $\mathcal{N} = 1$ or $\mathcal{N} = 0$. For example, let $D^m_{--} \neq 0$. The SUSY transformations related to $\hat{\theta}$ are broken, as

$$\delta \hat{\zeta}_\alpha = \hat{\epsilon}_\alpha D_{--} \ . \tag{9}$$

If in addition $D_{+-} = D_{++} = 0$, an $\mathcal{N} = 1$ SUSY is still unbroken.

When D_{ij} is related by SUSY to the nonpropagating complex structure deformations of X, one may fix its value by hand. In this case, SUSY is explicitly broken by couplings of these nondynamical fields to the propagating modes. One can show explicitly [5,1] that nontrivial vevs for $G = F - \tau H$, where $\tau = C^{(0)} + i/g_s$, breaks the supersymmetry generating translations along $\theta - i\hat{\theta}$, and leads to a superpotential for complex structure moduli [6,2].

We can also use fluxes to introduce soft SUSY breaking terms in $\mathcal{N}=1$ models with D-branes placed in CY backgrounds. For example, let us study a D5-brane in type IIB wrapping a rational curve $\mathcal{C}$ inside X, which preserves $\mathcal{N}=1$ SUSY. Holomorphic deformations of $\mathcal{C}$ correspond to open string chiral multiplets, [3] with superfield description $\Phi = \phi + \theta\psi + \theta^2 F_\phi$. To all orders in string perturbation theory, the superpotential

$$W = W(t^a, \Phi^i) = \sum_n g_n(t^a)\mathrm{tr}\Phi^n \qquad (10)$$

for these modes depends only on the complex structure moduli of X, and not on the Kähler class. If $D_{ij} \neq 0$ is chosen so that the $\mathcal{N}=1$ SUSY preserved by the D5-brane is broken, one induces explicit, computable SUSY-violating operators. For example, expand t^a in the superspace direction for which the D-brane preserves translation invariance: $V^a = t^a + \tilde{\theta}^2 F^a + ...$ where F is the corresponding auxiliary field. The couplings G_k should be written as superfields, so that:

$$g_k \longrightarrow g_k(t^a) + \tilde{\theta}^2 F^a \partial_a g_k \equiv g_k + \theta^2 \Delta_k \ , \qquad (11)$$

leading to soft SUSY-violating terms of the form

$$\int d^2\theta W + h.c. = \Delta_2 \mathrm{tr}\phi^2 + \Delta_3 \mathrm{tr}\phi^3 + \mathrm{h.c.} + ... \qquad (12)$$

such terms are induced in the presence of RR flux through cycles whose periods appear in the functions $g_k(t^a)$. [1]

3. Connecting closed-string vacua by paths in open-string field space

In this section, we will show that by moving in open-string configuration space, it is possible to connect vacua with different values of closed-srting three-form flux. This amplifies and applies some remarks made in Ref. [7].

Consider type IIB on a CY X, with a D5-brane wrapped on a holomorphic curve $\mathcal{C} \subset X$ that is a member of a family $\mathcal{M}$ of holomorphic curves such that $\pi_1(\mathcal{M})$ is nontrivial. Examples arise [10,11] when $\mathcal{C}$ is an exceptional curve in the resolution of an A_1 singularity over a Riemann surface $\mathcal{M} = \mathcal{S}_g$ of genus $g > 0$.

The moduli space of D5-branes is lifted by deforming the complex structure of X in such a way that the family $\mathcal{M}$ becomes obstructed; such deformations are in correspondence [11] with sections dW_0 of the canonical bundle $T^\star\mathcal{S}_g$. After a generic such deformation, the moduli space is reduced to a collection of isolated points where $0 = dW_0$. We will be discussing motion in off-shell configuration space where W_0' is not necessarily zero. We may choose W_0 to be proportional to some small control parameter ϵ. The potential hills between vacua are then parametrically small compared to the string scale, and the field space for low-energy excitations of the D5-brane is still well-described by $\mathcal{M}$.

This correspondence between one-forms on the moduli space and complex-structure moduli of the CY implies a map from one-cycles of the moduli space $\mathcal{M}$ to three-cycles of the CY. [12,11]. A path γ maps to the three-cycle $\pi^{-1}\gamma$ obtained by fibering the exceptional curve $\mathcal{C}_x$ over each point in γ. Moving the D-brane around a loop γ in the moduli space $\mathcal{M}$ generates a quantum of RR flux though the cycle $\pi^{-1}\gamma$. [7] This follows from the fact that the D5-brane is magnetically charged under the three-form flux.[a] We can see this fact further manifest itself in the superpotential. [9]

The effective superpotential governing the open-string moduli and complex structure moduli is: [8,6,9]

$$W = W_{GVW} + W_{obstruction} = \int_X \Omega \wedge G + \int_\Xi \Omega \tag{13}$$

where G is the three-form associated to a linear combination of D_{ij} preserving the same SUSY as the D5-brane, and includes the RR flux. [1] For deformations of a D5-brane from a rational curve $\mathcal{C}_0$ to a curve $\mathcal{C}$, the three-chain $\Xi = \mathcal{C} - \mathcal{C}_0$. There are two ambiguities in defining this obstruction contribution to the superpotential. [8]

(1) $\mathcal{C}_0$ is a base point on the closed string moduli space; changing $\mathcal{C}_0$ changes Ξ and so changes W by an additive constant.
(2) Since $H_3(X)$ is nontrivial, a 3-chain Ξ such that $\partial\Xi = \mathcal{C} - \mathcal{C}_0$ is only determined up to the addition of an element of $H_3(X, \mathbf{Z})$. This also additively changes the superpotential. We will give a concrete example below.

Moving the brane in a loop γ in $\mathcal{M}$ shifts the chain Ξ by $\Xi \mapsto \Xi + \pi^{-1}\gamma$. This in turn shifts the superpotential by $\int_{\pi^{-1}\gamma} \Omega$. By Poincaré duality, this can be identified with

$$\delta W = \int_X \Omega \wedge [\pi^{-1}\gamma]. \tag{14}$$

[a] An illustrative analogy arises in Maxwell theory on $R^3 \times S^1$: start with a magnetic monopole-antimonopole pair, and move the magnetic monopole around the circle. This causes the magnetic flux through the transverse plane to jump.

But this can be absorbed in W_{GVW}, if the the RR flux shifts by $[\pi^{-1}\gamma]$. This possibility of interchanging contributions between the two terms in W is made clearest by writing

$$W = \int_{G+\Xi} \Omega. \tag{15}$$

where G is the 3-cycle Poincaré dual to the flux.

Let us consider an explicit example of a patch of this model. Consider the hypersurface in $\mathbf{C}^4$ given by

$$y^2 + u^2 + v^2 = W_0'(x)^2 + f(x). \tag{16}$$

In this example, we take $\mathcal{C}$ to be an S^2 which can be resolved out of any double roots of the RHS of (16) at a point x, and Ξ is this S^2 times a curve in the x-plane ending at x. There is no reason for Ξ to be special Lagrangian, and it is not. The superpotential is $W = \int_\Xi \Omega$. All of the information about the threefold and its complex structure, including this integral, can be represented in terms of information on the Riemann surface Σ at $u = v = 0$, defined by $y^2 = (W_0')^2 + f$. Σ is a double cover of the x-plane, each of whose fibers represent a two-sphere homologous to $\mathcal{C}$. The superpotential integral can be represented as

$$W(x) = \int_{x_0}^{x} y(\tilde{x})d\tilde{x} \tag{17}$$

where $\mathcal{C}$ is the S^2 over the point x, and x_0 specifies the base-point curve.

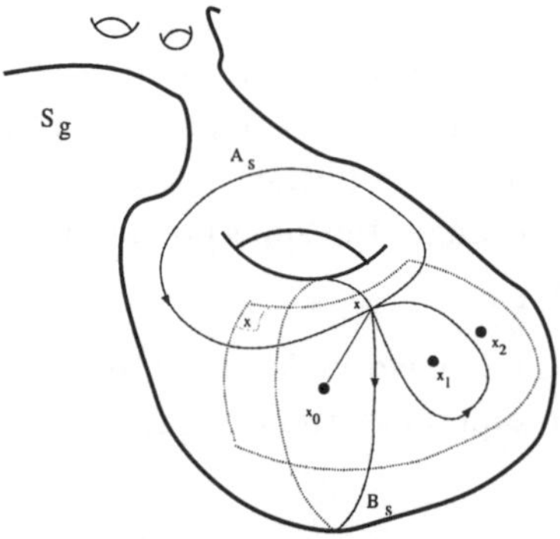

Fig. 1: *Each point in this picture represents a hemisphere of the exceptional P^1 in the ALE singularity $y^2 + u^2 + v^2 = 0$.*

It is important to distinguish Σ from $\mathcal{S}_g$. The geometry of the x-plane embeds into $\mathcal{S}_g$ as shown in Fig. 1. In this example, the moduli space of the curve $\mathcal{C}$ when $W_0' = 0, f = 0$ is $\mathcal{S}_g$; for generic W_0 at $f = 0$, holomorphic curves appear only at the critical points of W_0.

The shift in the superpotential we have described is effected by changing the flux through the cycle dual to the 3-cycle the D5-brane sweeps out in moving through a loop in $H_1(\mathcal{M})$. We can see further that this is consistent with rules for tadpole cancellation when one turns on N units of NS-NS three-form flux H through the three-cycle Ξ that the D5-brane is sweeping out.

In this case, if the D5-brane sweeps out a cycle, we have stated that it induces a jump $\delta F = [\pi^{-1}\gamma]$ in the RR flux. Since $\int H \wedge F = M$ induces an RR 4-form tadpole that must be cancelled by adding M three-branes. But this tadpole is precisely cancelled by D3-brane charge on the D5-brane which sweeps out the cycle Ξ with H-flux. H-flux through Ξ means that there is a gradient for the B-field through the sphere the D5-brane is wrapping, with respect to the direction on $\mathcal{M}$ it is moving. H-flux quantization means that the B-field will shift by $2\pi N$ upon traversing the loop γ, in units where $B = B + 2\pi$ when there is no brane. Because of this, B induces N units of D3-charge via the worldvolume Chern-Simons coupling $\int_{D5} B \wedge C_{(4)}$. This phenomenon is essentially identical to the phenomenon described in [13]: domain walls in $\mathcal{R}^4$ achieved by wrapping D5-branes around cycles Ξ with NS-flux $\int_\Xi H = N$ interpolate between vacua with D3-brane number differing by N.

This result implies that that in going around what was apparently a loop in this open-string moduli space $\mathcal{M}$, the string theory does not come back to itself. Rather, the closed-string background is changed. We conclude that the open-string moduli space in fact has no π_1. For example, stable cosmic string solutions corresponding to the putative loop do not exist.

Said another way, the moduli space or low-energy field space is in fact a *multiple cover* – with infinitely many sheets – of the Riemann surface $\mathcal{S}_g$, and the RR flux labels the sheets. Note that this generically does involve going off-shell, since in the presence of the obstruction superpotential there is not in fact a moduli space.

A similar discussion implies that it is possible to interpolate between values of NSNS flux quanta by moving wrapped NS5-branes.

4. $\mathcal{N} = 2$ Fayet-Iliopoulos terms in type II models

Taylor and Vafa [2] showed that in local models of type IIB Calabi-Yau compactifications, the superpotential [6] for complex structure moduli can arise from electric and magnetic Fayet-Iliopoulos (FI) terms [14] which spontaneously break the global $\mathcal{N} = 2$ SUSY to $\mathcal{N} = 1$. This is consistent with the identification of magnetic flux [1,5] with auxiliary fields, as the auiliary fields will be equated to the FI terms on-shell. [14]

One might ask whether the FI terms should be identified as separate degrees of freedom, equated to the magnetic flux via on-shell equations of motion. This appears to be the case. Study a deformed conifold in type IIB with vanishing 3-cycle A and dual cycle B. D3-branes wrapping A are light hypermultiplets charged with respect to the vector multiplets associated with the period t_A. The hypermultiplet can be written in terms of two $\mathcal{N} = 1$ chiral multiplets $Q, \tilde{Q}$ with opposite $U(1)$ charge. If we write the $\mathcal{N} = 2$ supermultiplet $V = t_A + \ldots$ in terms of an $\mathcal{N} = 1$ chiral multiplet A and a $\mathcal{N} = 1$ vector multiplet, the coupling of $Q, \tilde{Q}$ to A includes the following superpotential term:

$$W = \int d^2\theta A Q \tilde{Q} \tag{18}$$

If the scalar component of $< Q\tilde{Q} >$ gets a vev, this will appear as an electric FI term, or a magnetic FI term with respect to the vector multuplet associated to t_B.

It would be nice to show microscopically that such a vev induces magnetic 3-form flux, and that the potential for $< Q\tilde{Q} >$ has discrete minima associated to different values of NS-NS and RR flux through B. We can, however, note that when $\int_A H = -K \gg 1$, $\int_B F = N \gg 1$, this result is consistent with the conjectured field theory dual. [15,13] This geometry is described by an $\mathcal{N} = 1$ $SU(NK + N) \times SU(N)$ gauge theory with bifundamentals in $(NK + N, \bar{N})$ and $(\overline{NK + N}, N)$. At low energies, the gauge invariant degrees of freedom include "meson" and "baryon" degrees of freedom, constructed from the bifundamentals. The mesons are dual to motions of D3-branes in the Klebanov-Strassler geometry. The baryons correspond to D3-branes wrapping the 3-cycles of this geometry. [15,16] The space of vacua contains branches where either the mesons or baryons have vevs. Domain walls connecting the meson and baryon branches were argued [13] to be dual to D5-branes wrapping A. The disappearance of the D3-branes is consistent with the tadpole cancellation arguments reviewed above.

Acknowledgments

A.L. would like to thank the organizers of QTS3 for organizing an exciting conference, and for inviting him to speak there. A.L. is supported in part by the NSF grant PHY-0331516; J.M. is supported in part by a Princeton University Dicke Fellowship and in part by the DOE under Grant No. DE-FG03-92ER40701.

References

1. A. Lawrence and J. McGreevy, arXiv:hep-th/0401034.
2. T. R. Taylor and C. Vafa, *Phys. Lett.* **B474**, 130 (2000) [arXiv:hep-th/9912152].
3. I. Brunner, M. R. Douglas, A. E. Lawrence and C. Romelsberger, *J.High Energy Phys.* **0008**, 015 (2000) [arXiv:hep-th/9906200].
4. M. de Roo, J. W. van Holten, B. de Wit and A. Van Proeyen, *Nucl. Phys.* **173**, 175 (1980).
5. C. Vafa, *J. Math. Phys.* **42**, 2798 (2001) [arXiv:hep-th/0008142].
6. S. Gukov, C. Vafa and E. Witten, *Nucl. Phys.* **B584**, 69 (2000) [Erratum-ibid. **B608**, 477 (2001)] [arXiv:hep-th/9906070].
7. M. Aganagic, A. Klemm and C. Vafa, *Z. Naturforsch.* A **57**, 1 (2002) [arXiv:hep-th/0105045].
8. E. Witten, *Nucl. Phys.* **B507**, 658 (1997) [arXiv:hep-th/9706109].
9. W. Lerche, P. Mayr and N. Warner, arXiv:hep-th/0208039; W. Lerche, arXiv:hep-th/0312326.
10. S. Kachru, S. Katz, A. E. Lawrence and J. McGreevy, *Phys. Rev.* **D62**, 126005 (2000) [arXiv:hep-th/0006047].
11. S. Katz, D. R. Morrison and M. Ronen Plesser, *Nucl. Phys.* **B477**, 105 (1996) [arXiv:hep-th/9601108].
12. C. H. Clemens and P. A. Griffiths, *Ann. Math.* **95**, 281 (1972).
13. S. Kachru, J. Pearson and H. Verlinde, *J. High Energy Phys.* **0206**, 021 (2002) [arXiv:hep-th/0112197].
14. I. Antoniadis, H. Partouche and T. R. Taylor, *Phys. Lett.* **B372**, 83 (1996) [arXiv:hep-th/9512006]; I. Antoniadis and T. R. Taylor, *Fortsch. Phys.* **44**, 487 (1996) [arXiv:hep-th/9604062]; H. Partouche and B. Pioline, *Nucl. Phys. Proc. Suppl.* **56B**, 322 (1997) [arXiv:hep-th/9702115].
15. I. R. Klebanov and M. J. Strassler, *J. High Energy Phys.* **0008**, 052 (2000) [arXiv:hep-th/0007191].
16. S. S. Gubser and I. R. Klebanov, *Phys. Rev.* **D58**, 125025 (1998) [arXiv:hep-th/9808075].

THE DYNAMICS OF THE Dp + D($p-2$) SYSTEM

LOUIS LEBLOND

Laboratory for Elementary Particle Physics, Cornell University
Ithaca, NY 14853
lleblond@mail.lns.cornell.edu

We study the dynamics of the non-BPS Dp + D($p-2$) system. We shall approach this problem with the help of Boundary Superstring Field Theory (BSFT). It is possible to obtain the Dp+D($p-2$) action in BSFT as a particular vortex solution of two D-branes and one anti-D-brane (DD$\bar{\text{D}}$). In this framework, we were able to show that the DD$\bar{\text{D}}$ system has codimension 2 defects after tachyon condensation, in direct opposition with the idea that any codimension 2 defect (D($p-2$)-brane) should dissolve into an constant flux spreading throughout the entire volume of the Dp-brane. This is an important step in understanding the properties and stability of a D1-string inside a D3-brane which has many applications to the study of cosmic strings in brane inflationary scenarios.

1. Introduction

There is currently a major theoretical effort to embed the inflationary paradigm into superstring theory and to determine if there could be any distinct signatures coming from these models. One of the more serious proposal is the brane inflationary scenario [1–4], in which the interaction between D3-branes and anti-D3-branes naturally generates a potential for inflation. The exit of inflation occurs when the two branes (or stack thereof) collide. As the branes approach each other, a tachyonic mode appears, and the tachyon rolling triggers the exit of inflation. For the $\mathbb{KLMT}$ scenario, the branes will annihilate, and codimension 2 defects are expected to form. These defects will appear as cosmic strings (never as monopoles or domain walls [5, 6]). Their tensions can be predicted and are within reach of the next series of experiments on the Cosmic Microwave Background. Since they provide an important signature of the inflation era, it is very important to understand their dynamics in the presence of the D3-brane which spans our four dimensional universe. For the rest of this proceeding we will treat the more general Dp + D($p-2$) system although we are really interested in the D3-D1 system.

It is well known that the Dp + D$(p-2)$ system is non-BPS and has a tachyon mode in its spectrum. There exists a BPS state with the same quantum numbers to which we will refer as the DpD$(p-2)$ bound state. This bound state was previously thought of as a Dp brane with a constant flux throughout its entire volume [7,8]; the D$(p-2)$ is 'smeared' inside the Dp-brane. This state has the correct mass and charge to be interpreted as the bound state of the Dp+D$(p-2)$ system. If this dissolution process is generic and happens early enough, there may be no cosmic strings signatures left to be detected.

Recently, it was shown [9] that the coupling between the bulk RR C_{p-1} field and the gauge field completely changes this picture. This coupling effectively generates a mass to the gauge field, and a constant flux spread throughout the entire volume of the Dp brane is therefore impossible to achieve. In this proceeding, we argue that the cosmic strings are stable when we turn on the coupling to C_{p-1}. We set up an action for the DD$\bar{\text{D}}$ that enables us to study the dynamics of the formation and dissolution (or non-dissolution) of the Dp + D$(p-2)$ system. We argue that to calculate the tension of the final vortex, two paths can be taken [see Fig.1].

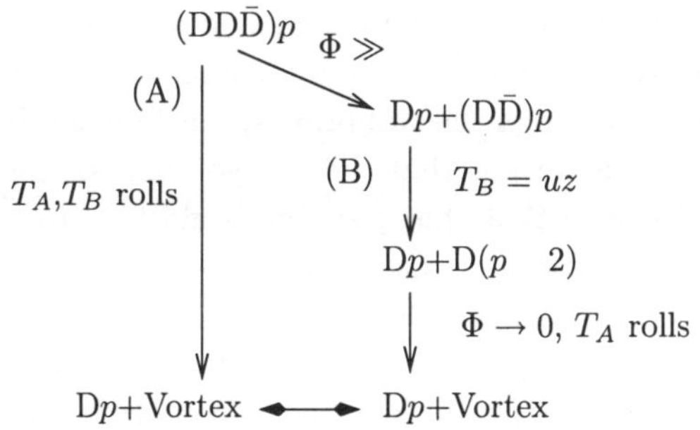

Figure 1. The two possible paths to get a vortex on a Dp-brane. Φ is the distance between the D-brane and the D$\bar{\text{D}}$ system, T_A and T_B are the two complex tachyons in the DD$\bar{\text{D}}$ system.

Much of this work can be found in [10]. A good introduction and reference to the subject of BSFT can be found in [11] and references therein.

2. The D$\bar{\text{D}}$ system and its extension to DD$\bar{\text{D}}$

The action we obtain for the (D$\bar{\text{D}}$)p system in BSFT is (the B field has been put to zero and we have dimensionally reduce to p+1 dimensions putting all the extra scalar fields to zero):

$$S_{(\mathrm{D\bar{D}})p} = -2\tau_p \int d^{p+1}x\sqrt{-g}\, e^{-2\pi\alpha' T\bar{T}}\, \mathcal{F}(\mathcal{X} + \sqrt{\mathcal{Y}})\, \mathcal{F}(\mathcal{X} - \sqrt{\mathcal{Y}}) \quad (1)$$

where we use the covariant notation of [12],

$$\mathcal{X} \equiv 2\pi\alpha'^2 g^{\mu\nu}\partial_\mu T\partial_\nu \bar{T}, \quad \mathcal{Y} \equiv \left(2\pi\alpha'^2\right)^2 \left(g^{\mu\nu}\partial_\mu T\partial_\nu T\right)\left(g^{\alpha\beta}\partial_\alpha \bar{T}\partial_\beta \bar{T}\right).$$

In this proceeding, the exact form of $\mathcal{F}(x)$ will not be relevant. The RR part of the action can also be calculated in this framework for N branes and M anti-branes and the coupling to the C_{p-1} form is the one relevant to vortex formation.

$$S_{\mathrm{RR}} = \tau_p g_s \int \sum_{\text{odd } p} C_p \wedge \mathrm{Tr}\left[\begin{pmatrix} \mathbb{1}_{N\times N} & 0 \\ 0 & -\mathbb{1}_{M\times M} \end{pmatrix} O\right] e^{2\pi\alpha' i\mathcal{F}}, \quad (2)$$

$$\mathcal{F} = \begin{pmatrix} F^1 + iT^\dagger T & -i(DT)^\dagger \\ -iDT & F^2 + iTT^\dagger \end{pmatrix},$$

where F^1 and F^2 are the gauge field strengths on the N D-branes and M anti-D-branes respectively. The gauge symmetry on the D$\bar{\text{D}}$ system is $U(1)\times U(1)$, and the tachyon couples only to a linear combination of the two $U(1)$'s. After annihilation of the two branes, there should be no gauge symmetry remaining. This means that the tachyon breaks the $U(1)$ it couples to, and it is well known that this breaking could lead to vortex formation ($\Pi_1(U(1)) = \mathbb{Z}$). The action (1) has very simple solitonic solutions given by $T = uz$ and $A = 0$ (in complex coordinates). The parameter u should be taken to be very large ($u \to \infty$). It was shown in [12] that these exact solutions gives the correct charge and tension to be interpreted as codimension 2 branes! To extend this action to the case of DD$\bar{\text{D}}$, some approximations [13] are needed since the path integral is no longer gaussian. The DD$\bar{\text{D}}$ system is invariant under $U(2)\times U(1) = SU(2)\times U(1)_- \times U(1)_+$. The tachyon field is a doublet under this $SU(2)$ and, as with the D$\bar{\text{D}}$ system, it couples to a linear combination of the two $U(1)$'s, namely $U(1)_-$ in the notation above. A priori, the pattern of symmetry breaking ($SU(2) \times U(1)_- \to U(1)_\gamma$) is exactly the same as the electroweak symmetry breaking, and as in the standard model, no defects are expected. We can show that this conclusion is wrong by looking at the DD$\bar{\text{D}}$ system. We were able to get an approximate action for the DD$\bar{\text{D}}$ system by neglecting the mixing terms in the path integral coming from the non-abelian nature of the stack of branes. With a linear profile for the tachyon field ($T_i = u_i x_i$) we get:

$$S_{(\mathrm{DD\bar{D}})p} = -\tau_p \int d^{p+1}x \left(1 + 2e^{-2\pi\alpha'(u_i x_i)^2} \prod_{i=1}^{4} \mathfrak{F}\left(\frac{(2\pi\alpha')^2}{\pi} u_i\right)\right). \quad (3)$$

The RR part of the action, being topological, is believed to be exact. We are again interested only in the coupling to the C_{p-1} form. First, lets define correctly all the gauge fields.

$$SU(2) \to A_3, W^{\pm}, \qquad U(1)_{DD} \to B_1, \qquad U(1)_{\bar{D}} \to B_2$$

where B_1 is the U(1) gauge field in U(2) and B_2 is the gauge field associated with the U(1) on the $\bar{D}$. The various field strength that we will need are:

$$\mathcal{A}^1 = \begin{pmatrix} A_3 + \frac{1}{2}B_1 & W^+ \\ W^- & -A_3 + \frac{1}{2}B_1 \end{pmatrix},$$
$$F^1 = d\mathcal{A}^1 + \mathcal{A}^1 \wedge \mathcal{A}^1, \; F^2 = dB_2,$$
$$F^- = d\mathcal{A}^1 + \mathcal{A}^1 \wedge \mathcal{A}^1 - dB_2.$$

The relevant coupling terms (between C_{p-1} and the gauge field) in the RR action for the DD$\bar{\text{D}}$ system are [13] (this is an exact result)

$$S_{RR} \propto \int_{M_{p+1}} C_{p-1} \wedge (F^- + \frac{1}{t}(e^{-t} - 1)(\mathrm{Tr}\,(\mathcal{T}F^1) - tF^2)), \quad (4)$$

where

$$t = T^\dagger T = T_A\bar{T}_A + T_B\bar{T}_B,$$
$$\mathcal{T} = TT^\dagger = \begin{pmatrix} T_A\bar{T}_A & T_A\bar{T}_B \\ T_A\bar{T}_A & T_B\bar{T}_B \end{pmatrix}.$$

With those two results in hand (3,4), we can study the dynamics of dissolution of a D$(p-2)$ inside a Dp.

3. Vortex inside and outside the Dp brane

There are two possible routes we can take to study the dynamics of dissolution [see Fig 1]. The first one (A) is to try to go directly to the bound state from the DD$\bar{\text{D}}$ action. For this path, the mixing terms between the two tachyons might qualitatively change the answer so (3) is not really reliable though (4) is. For the second route (B), we first dimensionally reduce and then take a D$\bar{\text{D}}$ pair away from the other brane by turning on the corresponding scalar field (Φ) representing the distance between the branes. Then, we form a metastable D$(p-2)$-brane using one of the two

This D($p-2$)-brane has an NS-NS attraction toward the other brane and when it gets close to it, the other tachyonic field will roll down its potential, triggering the final decay to the bound state [see fig 2.]. For this method, the mixing terms are well under control.

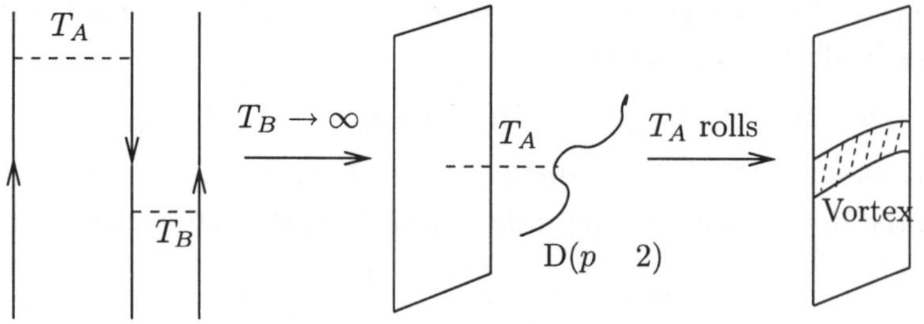

Figure 2. Route B) Creation of a vortex of flux by first forming the D($p-2$)-brane outside the Dp-brane and then bringing it back. Arrows down represents D-brane and arrows up represents anti-D-brane.

Irrespective of the path chosen, it is possible to prove the existence of vortices using topological arguments. The important factor to consider is that for a coupling of the form $\int C_{p-1} \wedge dA$, the gauge field become massive (this is the Green-Schwarz mechanism) [9]. If we start with the action (4) and set T_B to zero while T_A rolls, then we have the possibility of vortex formation.

$$S_{RR} \propto \int_{M_{p+1}} C_{p-1} \wedge \left(d\left(-A_3 + \frac{1}{2}B_1\right) + W^+ \wedge W^- \right) \tag{5}$$
$$+ C_{p-1} \wedge \left(e^{-T_1 \bar{T}_1} \left(d\left(A_3 + \frac{1}{2}B_1\right) + W^+ \wedge W^- - dB_2 \right) \right).$$

This equation is easily interpreted. All the gauge fields that will be broken are the ones following the exponential. The $U(1)_\gamma$ remaining is just $-A_3 + \frac{1}{2}B_1$ but this gauge symmetry is broken via its coupling with C_{p-1}. Hence the moduli space of the tachyon after breaking is trivial, there is no U(1) left over. The fundamental group is now

$$\Pi_1\left(SU(2) \times U(1)_-\right) = \mathbb{Z}. \tag{6}$$

We conclude that vortex solutions are topologically allowed and stable. There is an extra cubic interaction term ($C_{p-1} \wedge W^+ \wedge W^-$) which plays no role in this discussion.

4. Conclusion

Though we were able to prove that vortices are topologically allowed, it still remains to use our approximate action (3) to calculate the tension of this vortex. If the final state is to be the BPS bound state then we know what the tension of our vortex should be. Just from SUSY arguments, one can obtain the mass of the BPS bound state:

$$M_{\mathrm{D}p\mathrm{D}(p-2)} = \sqrt{\tau_p^2 V_p^2 - \tau_{p-2}^2 V_{p-2}^2}$$

and the mass of the vortex is therefore expected to be

$$M_{\mathrm{vortex}} = \sqrt{\tau_p^2 V_p^2 - \tau_{p-2}^2 V_{p-2}^2} - \tau_p V_p \approx \frac{\tau_{p-2}^2 V_{p-2}^2}{2\tau_p V_p}.$$

To calculate the tension, we will need to find the correct profile for the tachyons and calculate the energy from the action (3). A variational principle could be used to find the solution that minimizes the energy.

Acknowledgments

This work was done in collaboration with S.-H. Henry Tye. This material is based upon work supported by the National Science Foundation under Grant No. PHY-0098631.

References

1. G. R. Dvali and S.-H. Henry Tye. *Phys. Lett.*, **B450:72–82**, 1999.
2. Burgess, C. P. and Majumdar, M. and Nolte, D. and Quevedo, F. and Rajesh, G. and Zhangand, R.J. *JHEP*, **07:047**, 2001.
3. Kachru, Shamit and others *JCAP*, **0310:013**, 2003.
4. Firouzjahi, Hassan and Tye, S. H. Henry, 2003.
5. Jones, Nicholas and Stoica, Horace and Tye, S.-H. Henry. *JHEP*, **07:051**, 2002.
6. Sarangi, Saswat and Tye, S.-H. Henry *Phys. Lett.*, **B36:185-192**, 2002.
7. E. Gava, K. S. Narain, and M. H. Sarmadi. *Nucl. Phys.*, **B504:214–238**, 1997.
8. Akikazu Hashimoto and IV Taylor, Washington. *Nucl. Phys.*, **B503:193–219**, 1997.
9. Edmund J. Copeland, Robert C. Myers, and Joseph Polchinski. 2003.
10. Louis Leblond and S.-H. Henry Tye. work in progress.
11. Tadaoki Uesugi. 2002.
12. Nicholas T. Jones and S. H. Henry Tye. *JHEP*, **01:012**, 2003.
13. Nicholas T. Jones, Louis Leblond, and S. H. Henry Tye. *JHEP*, **10:002**, 2003.

SMOOTH COSMOLOGIES FROM M-THEORY*

H. LÜ

George P. and Cynthia W. Mitchell Institute for Fundamental Physics
Texas A&M University, College Station TX 77843-4242

J.F. VÁZQUEZ-PORITZ

Department of Physics and Astronomy
University of Kentucky, Lexington, KY 40506

Department of Physics
University of Cincinnati, Cincinnati OH 45221-0011

We review two ways in which smooth cosmological evolution between two de Sitter phases can be obtained from M/string-theory. Firstly, we perform a hyperbolic reduction of massive type IIA* theory to $D = 6$ $\mathcal{N} = (1,1)$ $SU(2) \times U(1)$ gauged de Sitter supergravity, which supports smooth cosmological evolution between $\mathrm{dS}_4 \times S^2$ and a dS_6-type geometry. Secondly, we obtain four-dimensional de Sitter gravity with $SU(2)$ Yang-Mills gauge fields from a hyperbolic reduction of standard eleven-dimensional supergravity. The four-dimensional theory supports smooth cosmological evolution between $\mathrm{dS}_2 \times S^2$ and a dS_4-type geometry. Although time-dependent, these solutions arise from a first-order system via a superpotential construction. For appropriate choices of charges, these solutions describe an expanding universe whose expansion rate is significantly larger in the past than in the future, as required for an inflationary model.

1. $D = 6$ $\mathcal{N} = (1,1)$ de Sitter supergravity from massive type IIA*

The advantage of embedding de Sitter solutions in $*$ theories [1,2] is that they can be viewed as "supersymmetric," even though the anti-commutators of the super-charges are no longer positive definite [a]. The time-like T-dualization of type IIB theory can yield massive type IIA* supergravity,

*Talk by J.F.V.-P.

[a]The first de Sitter solution within the context of an extended supergravity theory was found in [3].

whose bosonic Lagrangian is given by [4]

$$\begin{aligned}\mathcal{L}_{10} = & \hat{R}\,\hat{*}\mathbb{1} - \tfrac{1}{2}\hat{*}d\hat{\phi}\wedge d\hat{\phi} + \tfrac{1}{2}e^{\frac{3}{2}\hat{\phi}}\,\hat{*}\hat{F}_{(2)}\wedge\hat{F}_{(2)} - \tfrac{1}{2}e^{-\hat{\phi}}\,\hat{*}\hat{F}_{(3)}\wedge\hat{F}_{(3)} \\ & + \tfrac{1}{2}e^{\frac{1}{2}\hat{\phi}}\,\hat{*}\hat{F}_{(4)}\wedge\hat{F}_{(4)} - \tfrac{1}{2}d\hat{A}_{(3)}\wedge d\hat{A}_{(3)}\wedge\hat{A}_{(2)} - \tfrac{1}{6}m\,d\hat{A}_{(3)}\wedge(\hat{A}_{(2)})^3 \\ & - \tfrac{1}{40}m^2(\hat{A}_{(2)})^5 + \tfrac{1}{2}m^2 e^{\frac{5}{2}\hat{\phi}}\,\hat{*}\mathbb{1}, \end{aligned} \qquad (1)$$

where $\hat{F}_{(2)} = d\hat{A}_{(1)} + m\,\hat{A}_{(2)}$, $\hat{F}_{(3)} = d\hat{A}_{(2)}$ and $\hat{F}_{(4)} = d\hat{A}_{(3)} + \hat{A}_{(1)}\wedge d\hat{A}_{(2)} + \frac{1}{2}m\,\hat{A}_{(2)}\wedge\hat{A}_{(2)}$. The hyperbolic reduction[b] of massive type IIA* theory can be obtained as an analytical continuation of the S^4 reduction [8] of massive type IIA supergravity [9]:

$$\begin{aligned} d\hat{s}_{10}^2 &= c^{\frac{1}{12}}\,X^{\frac{1}{8}}\Big[\Delta^{\frac{3}{8}}\,ds_6^2 + 2g^{-2}\,\Delta^{\frac{3}{8}}\,X^2\,d\xi^2 + \tfrac{1}{2}g^{-2}\,\Delta^{-\frac{5}{8}}\,X^{-1}\,s^2\xi\sum_{i=1}^{3}(h^i)^2\Big], \\ \hat{F}_{(4)} &= -\tfrac{\sqrt{2}}{6}\,g^{-3}\,c^{1/3}\,s^3\,\Delta^{-2}\,U\,d\xi\wedge\epsilon_{(3)} - \sqrt{2}g^{-3}\,c^{4/3}\,s^4\,\Delta^{-2}\,X^{-3}\,dX\wedge\epsilon_{(3)} \\ &\quad -\sqrt{2}g^{-1}\,c^{1/3}\,s\,X^4 *F_{(3)}\wedge d\xi - \tfrac{1}{\sqrt{2}}c^{4/3}\,X^{-2} *F_{(2)} \\ &\quad +\tfrac{1}{\sqrt{2}}g^{-2}\,c^{1/3}\,s\,F^i_{(2)}\,h^i\wedge d\xi \\ &\quad -\tfrac{1}{4\sqrt{2}}g^{-2}\,c^{4/3}\,s^2\,\Delta^{-1}\,X^{-3}\,F^i_{(2)}\wedge h^j\wedge h^k\,\epsilon_{ijk}\,, \qquad (2) \\ \hat{F}_{(3)} &= c^{2/3}\,F_{(3)} + g^{-1}\,c^{-1/3}\,s\,F_{(2)}\wedge d\xi\,, \\ \hat{F}_{(2)} &= \tfrac{1}{\sqrt{2}}\,c^{2/3}\,F_{(2)}\,, \qquad e^{\hat{\phi}} = c^{-5/6}\,\Delta^{1/4}\,X^{-5/4}\,, \end{aligned}$$

where $X = e^{-\frac{1}{2\sqrt{2}}\phi}$, $\Delta \equiv -X\,s^2\xi + X^{-3}\,c^2\xi$ and $U \equiv X^{-6}\,c^2 + 3X^2\,s^2 - 4X^{-2}\,s^2 - 6X^{-2}$. We have defined $h^i \equiv \sigma^i - g\,A^i_{(1)}$, $\epsilon_{(3)} \equiv h^1\wedge h^2\wedge h^3$, $s = \sinh\xi$, $c = \cosh\xi$ and $m = \frac{\sqrt{2}}{3}g$. σ_i are $SU(2)$ left-invariant 1-forms which satisfy $d\sigma^i = -\frac{1}{2}\epsilon_{ijk}\,\sigma^j\wedge\sigma^k$. $*$ is the six-dimensional Hodge dual. The resulting six-dimensional theory[c] is given by

$$\begin{aligned}\mathcal{L}_6 = & R*\mathbb{1} - \tfrac{1}{2}*d\phi\wedge d\phi + g^2\Big(\tfrac{2}{9}e^{\frac{3}{\sqrt{2}}\phi} - \tfrac{8}{3}e^{\frac{1}{\sqrt{2}}\phi} - 2e^{-\frac{1}{\sqrt{2}}\phi}\Big)*\mathbb{1} \\ & -\tfrac{1}{2}e^{-\sqrt{2}\phi}*F_{(3)}\wedge F_{(3)} + \tfrac{1}{2}e^{\frac{1}{\sqrt{2}}\phi}\Big(*F_{(2)}\wedge F_{(2)} + *F^i_{(2)}\wedge F^i_{(2)}\Big) \qquad (3) \\ & +A_{(2)}\wedge\big(\tfrac{1}{2}dA_{(1)}\wedge dA_{(1)} + \tfrac{1}{3}g\,A_{(2)}\wedge dA_{(1)} \\ & +\tfrac{2}{27}g^2\,A_{(2)}\wedge A_{(2)} + \tfrac{1}{2}F^i_{(2)}\wedge F^i_{(2)}\big)\,, \end{aligned}$$

where $F_{(3)} = dA_{(2)}$, $F_{(2)} = dA_{(1)} + \frac{2}{3}g\,A_{(2)}$ and $F^i_{(2)} = dA^i_{(1)} + \frac{1}{2}g\,\epsilon_{ijk}A^j_{(1)}\wedge A^k_{(1)}$.

---

[b] No-go theorems prohibiting de Sitter compactifications [5,6] are surmounted by reducing over a non-compact space [7].

[c] De Sitter supergravities as hyperbolic reductions of type IIB* and M*-theories have recently been obtained in [10].

Since the theory allows us to truncate to $U(1)^2$, we can include in our theory a vector-tensor multiplet with the Lagrangian given by[d]

$$\hat{e}^{-1}\mathcal{L}_6 = \hat{R} - \tfrac{1}{2}(\partial\phi_1)^2 - \tfrac{1}{2}(\partial\phi_2)^2 - \hat{V} + \tfrac{1}{4}\sum_{i=1}^{2} X_i^{-2}\,(\hat{F}^i_{(2)})^2\,, \tag{4}$$

where $X_i = e^{\frac{1}{2}\vec{a}_i\cdot\vec{\phi}}$ with $\vec{a}_1 = (\sqrt{2}\,,\frac{1}{\sqrt{2}})$ and $\vec{a}_2 = (-\sqrt{2}\,,\frac{1}{\sqrt{2}})$. The scalar potential is given by

$$\hat{V} = -\tfrac{4}{9}g^2\,(X_0^2 - 9X_1\,X_2 - 6X_0\,X_1 - 6X_0\,X_2)\,, \tag{5}$$

where $X_0 = (X_1\,X_2)^{-3/2}$. We consider the ansatz

$$\begin{aligned} ds_6^2 &= -d\tau^2 + a^2\,dx_j^2 + b^2\,d\Omega_2^2\,,\\ F^i_{(2)} &= \lambda_i\,\Omega_{(2)}\,, \end{aligned} \tag{6}$$

where the functions a and b depend only on the co-moving time coordinate τ and $d\Omega_2^2$ is the metric for the unit 2-sphere S^2. A cosmological solution can be obtained from the following first-order equations [4]

$$\begin{aligned} \dot{\vec{\phi}} &= \sqrt{2}\Big(-\frac{1}{2\sqrt{2}\,g}\,(q_1\,\vec{a}_1\,X_1^{-1} + q_2\,\vec{a}_2\,X_2^{-1})\,b^{-2} + \frac{dW}{d\vec{\phi}}\Big)\,,\\ \frac{\dot{b}}{b} &= -\frac{1}{4\sqrt{2}}\Big(\frac{3}{\sqrt{2}\,g}\,(q_1\,X_1^{-1} + q_2\,X_2^{-1})\,b^{-2} + W\Big)\,,\\ \frac{\dot{a}}{a} &= \frac{1}{4\sqrt{2}}\Big(\frac{1}{\sqrt{2}\,g}\,(q_1\,X_1^{-1} + q_2\,X_2^{-1})\,b^{-2} - W\Big)\,, \end{aligned} \tag{7}$$

provided that the two $U(1)$ charges satisfy $g\,(\lambda_1 + \lambda_2) = 1$. The superpotential is given by $W = \frac{g}{\sqrt{2}}\,(\frac{4}{3}X_0 + 2X_1 + 2X_2)$. In particular, if $\lambda_1 = g^{-1}$ and $\lambda_2 = 0$, then we can consistently set $\phi_1 = 2\,\phi_2$ and (7) can be solved explicitly. After the coordinate transformation $d\tau = \frac{3}{2}\,e^{\frac{3}{\sqrt{8}}\phi_2}(g\,t)^{-1}\,dt$, the metric of the solution can be expressed as

$$ds_6^2 = \widetilde{H}^{-\frac{1}{4}}\,H^{\frac{1}{4}}\Big[-\tfrac{9}{4}\,\widetilde{H}\,H^{-1}\,\frac{dt^2}{(g\,t)^2} + (g\,t)^2\,(dx_i^2 + \widetilde{H}\,g^{-2}\,d\Omega_2^2)\Big]\,. \tag{8}$$

where

$$e^{\sqrt{8}\,\phi_2} = \frac{\widetilde{H}}{H} = \frac{1 + \frac{3}{4}\,\frac{1}{(g\,t)^2}}{1 + \frac{9}{4}\,\frac{1}{(g\,t)^2}}\,. \tag{9}$$

[d]Though in general not the case, the truncation is consistent for the present purpose of constructing cosmological solutions.

The solution can be viewed as an intersection of a spatial domain wall wrapped on Ω_2, characterized by the function H, and an S2-brane characterized by the function $\widetilde{H}$. This is a smooth cosmological solution in which the co-moving time runs from an infinite past, which is $dS_4 \times S^2$, to an infinite future, which is a dS_6-type geometry with the boundary $R^3 \times S^2$.

For general values of q_i, the first-order equations cannot be solved analytically. However, it is straightforward to find the fixed-point solution of $dS_4 \times S^2$ where b and $\vec{\phi}$ are constants. Using a numerical method, we have verified that there are cosmological solutions which run from $dS_4 \times S^2$ in the infinite past and are of dS_6-type in the future. The ratio of the Hubble constant in the infinite past and future can be straightforwardly calculated. In particular, for $-\lambda_1 \gg \lambda_2$, the ratio is given by

$$\frac{H_{\rm past}}{H_{\rm future}} \frac{\sqrt{2}}{16}\Big(\frac{3}{2}\sqrt{-3\frac{\lambda_1}{\lambda_2}}\Big)^{3/4}, \tag{10}$$

which can be arbitrarily large. It is surprising that we can get a somewhat realistic cosmological model from a first-order system, which implies supersymmetry from the point of view of " theories [1].

2. $D = 4$ Yang-Mills de Sitter gravity from M-theory

Four-dimensional de Sitter spacetime arises in standard eleven-dimensional supergravity as a warped product with a hyperbolic 7-space [11]. The latter can be expressed as a foliation of two 3-spheres, on which $SU(2)$ Yang-Mills fields can reside. This enables us to use the following reduction ansatz for $D = 11$ supergravity [13]:

$$\begin{aligned} ds_{11}^2 &= \Delta^{\frac{2}{3}}\, ds_4^2 + g^{-2}\,\Delta^{\frac{2}{3}}\, d\theta^2 \\ &\quad + \tfrac{1}{4} g^{-2}\,\Delta^{-\frac{1}{3}}\Big[c^2 \sum_i (h^i)^2 + s^2 \sum_i (\tilde{h}^i)^2\Big], \end{aligned} \tag{11}$$

$$\begin{aligned} F_{(4)} &= 2g\,\epsilon_{(4)} - \tfrac{1}{4} g^{-2}\Big(s\,c\, d\theta \wedge h^i \wedge {*F^i_{(2)}} - s\,c\, d\theta \wedge \tilde{h}^i \wedge {*F^i_{(2)}} \\ &\quad - \tfrac{1}{4} c^2\, \epsilon_{ijk}\, h^i \wedge h^j \wedge {*F^k_{(2)}} + \tfrac{1}{4} s^2\, \epsilon_{ijk}\, \tilde{h}^i \wedge \tilde{h}^j \wedge {*F^k_{(2)}}\Big), \end{aligned} \tag{12}$$

where $c = \cosh\theta$, $s = \sinh\theta$, $\Delta = \cosh(2\theta)$, and $*$ denotes the four-dimensional Hodge dual. σ_i and $\tilde{\sigma}_i$ are $SU(2)$ left-invariant 1-forms and the vielbein h^i and $SU(2)$ Yang-Mills field strengths $F^i_{(2)}$ have been defined in the previous section. It is straightforward to verify that the Bianchi identity $dF_{(4)} = 0$ and the equation of motion $d\hat{*}F_{(4)} = \frac{1}{2} F_4 \wedge F_{(4)}$ are satisfied provided that the $SU(2)$ Yang-Mills fields $A^i_{(2)}$ satisfy the lower-dimensional equations of motion $D{*}F^i_{(2)}=0$, where the covariant derivative D is defined by $DV^i = dV^i + g\,\epsilon_{ijk}\,A^j\wedge V^k$, for any vector V^i.

The evaluation of the $D=11$ Einstein equations of motion is substantially more complicated, and we have not performed the calculation in full detail. However, we have verified that the equations of motion work for the $U(1)$ subsector of the $SU(2)$ gauge fields. Combining the result, the lower-dimensional equations of motion can be obtained from the Lagrangian

$$e^{-1}\mathcal{L}_4 = R - \tfrac{1}{4}(F^i_{(2)})^2 - 8g^2\,. \tag{13}$$

Thus, we have obtained four-dimensional Einstein $SU(2)$ Yang-Mills de Sitter gauged gravity from $D=11$ by consistent Kaluza-Klein reduction on a hyperbolic 7-space [13].

A regular cosmological solution of (13) is given by [12]

$$\begin{aligned} ds_4^2 &= H^2\,(-f^{-1}\,dt^2 + f\,dx^2 + t^2\,d\Omega_2^2)\\ F^3_{(2)} &= \frac{2\ell}{(t\,H)^2}\,dt\wedge dx\,, \qquad {*F^3_{(2)}} = 2\ell\,\Omega_{(2)}\,,\\ H &= 1+\frac{\ell}{t}\,, \qquad f = \tfrac{4}{3}g^2t^2\,H^4 - 1\,. \end{aligned} \tag{14}$$

For $g^2\,\ell^2 = \frac{3}{64}$, this is smooth cosmological evolution between $\mathrm{dS}_2\times S^2$ and a dS_4-type geometry with the boundary of $S^2\times S^1$. It is straightforward to lift the solution back to $D=11$ and obtain a regular cosmological solution in M-theory. The corresponding metric is given by

$$\begin{aligned} ds_{11}^2 &= \Delta^{\frac{2}{3}}\,H^2\Big(-f^{-1}\,dt^2 + f\,dx^2 + t^2\,d\Omega_2^2\Big) + g^{-2}\,\Delta^{\frac{2}{3}}\,d\theta^2\\ &+\tfrac{1}{4}g^{-2}\,\Delta^{-\frac{1}{3}}\Big[c^2\Big(d\omega_2^2 + (\sigma_3 - \frac{2g}{H}\,dx)^2\Big)\\ &+s^2\Big(d\tilde\omega_2^2 + (\tilde\sigma_3 - \frac{2g}{H}\,dx)^2\Big)\Big]\,. \end{aligned} \tag{15}$$

3. Discussion

The crucial issue regarding both of the above cosmological solutions is their stability. Although the * theories are necessary from the point of view of time-like T-duality, they suffer from an instability due to the ghost-like nature of the supergravity fields. It is of interest, therefore, to further study whether the stability is protected by time-like T-duality or "supersymmetry." Our second example might be more appealing since it is embedded within standard eleven-dimensional supergravity. The optimistic view is that the time scale of the instability is large enough to, nevertheless, validate the cosmological solutions.

Metastable de Sitter vacua have recently been constructed in type IIB theory [14]. All moduli are frozen due to fluxes as well as corrections to the superpotential from Euclidean D-brane instantons or gaugino condensation. However, these techniques can only yield inflationary models under restrictive conditions [15]. If such moduli-freezing techniques can be incorporated into our smooth inflationary models, then this might greatly enhance the cosmological landscape of M-theory [16]. In particular, we would have cosmologies which smoothly evolve (without topological transitions) between de Sitter-type spacetimes of different dimensionalities.

Acknowledgments

H.L. is supported in part by DOE grant DE-FG03-95ER40917. J.F.V.P. is supported in part by DOE grant DE-FG01-00ER45832.

References

1. C.M. Hull, *JHEP* **9807**, 021 (1998), hep-th/9806146.
2. C.M. Hull, *JHEP* **9811**, 017 (1998), hep-th/9807127.
3. S.J. Gates, Jr. and B. Zwiebach, *Phys. Lett.* **B123**, 200 (1983).
4. H. Lü and J.F. Vazquez-Poritz, hep-th/0305250.
5. G.W. Gibbons, *GIFT seminar on supersymmetry, supergravity and related topics*, edited by F. del Aguila, J.A. de Azcarraga and L.E. Ibanez, World Scientific (1984).
6. J. Maldacena and C. Nunez, *Int. J. Mod. Phys.* **A16**, 822 (2001), hep-th/0007018.
7. C.M. Hull and N.P. Warner, *Class. Quant. Grav.* **5**, 1517 (1988).
8. M. Cvetič, H. Lü and C.N. Pope, *Phys. Rev. Lett.* **83**, 5226 (1999), hep-th/9906221.
9. L.J. Romans, *Phys. Lett.* **B169**, 374 (1986).
10. J.T. Liu, W.A. Sabra and W.Y. Wen, *JHEP* **0401**, 007 (2004), hep-th/0304253.
11. G.W. Gibbons and C.M. Hull, hep-th/0111072.
12. H. Lü, C.N. Pope and J.F. Vazquez-Poritz, hep-th/0307001.
13. H. Lü and J.F. Vazquez-Poritz, hep-th/0308104.
14. S. Kachru, R. Kallosh, A. Linde and S.P. Trivedi, *Phys. Rev.* **D68**, 046005 (2003), hep-th/0301240.
15. S. Kachru, R. Kallosh, A. Linde, J. Maldacena, L. McAllister and S.P. Trivedi, *JCAP* **0310**, 013 (2003), hep-th/0308055.
16. L. Susskind, hep-th/0302219.

MORE ON SUPERSYMMETRIC TENSIONLESS ROTATING STRINGS IN $AdS_5 \times S^5$*

DAVID MATEOS

Perimeter Institute for Theoretical Physics
Waterloo, Ontario N2J 2W9, Canada
dmateos@perimeterinstitute.ca

TONI MATEOS

Departament ECM, Facultat de Física
Universitat de Barcelona, Institut de Física d'Altes Energies and
CER for Astrophysics, Particle Physics and Cosmology
Diagonal 647, E-08028 Barcelona, Spain
tonim@ecm.ub.es

PAUL K. TOWNSEND†

Institució Catalana de Recerca i Estudis Avançats.
Departament ECM, Facultat de Física
Universitat de Barcelona
Diagonal 647, E-08028 Barcelona, Spain
p.k.townsend@damtp.cam.ac.uk

Rotating IIB strings in $AdS_5 \times S^5$ become ultra-relativistic, and hence effectively tensionless, in the limit of large angular momentum on S^5. We have shown previously that such tensionless strings may preserve supersymmetry. Here we extend this result to include a class of supersymmetric tensionless strings with *arbitrary* $SO(6)$ angular momentum. We close with some general comments on tensionless strings in $AdS_5 \times S^5$.

1. Introduction

A recent advance in our understanding of the AdS/CFT correspondence has come about via consideration of classical, closed IIB strings in $AdS_5 \times S^5$ that rotate within S^5 and thus carry $SO(6)$ momenta [1, 2]. In the limit

*Talk presented by D.M.

†On leave from DAMTP, University of Cambridge, UK.

of large angular momentum J (precisely, for $J \gg 1, \sqrt{\lambda}$, where $\sqrt{\lambda}$ is the string tension) the energy of the string is given by the classical result, and this admits an expansion in positive integer powers of λ/J^2. This yields a prediction for the dimension of the dual operator, with corresponding large $SO(6)$ R-charge, in *perturbative* $N = 4$ super Yang-Mills (SYM) theory. In the planar limit, this dimension can be computed by spin chain techniques [3] (or by means of an equivalent sigma model [4]) and a precise agreement with the string theory predictions has been found; remarkably, both routes lead to the same integrable model [5, 6, 7, 8].

The cases that were first studied, and which are still best understood, are those for which the $SO(6)$ angular momentum two-form has only two *independent* skew-eigenvalues. It was argued in [1, 2] that the agreement between the string and SYM calculations was, in contrast to previous quantitative tests, evidence for the AdS/CFT conjecture in a 'non-supersymmetric sector'. However, as we pointed out in a previous paper [9], the rotating strings under consideration are effectively tensionless for large angular momentum because almost all of the energy is due to the rotation, and the states of tensionless strings in the same $SO(6)$ representation preserve either 1/4 or 1/8 supersymmetry. Thus, the limit in which the AdS/CFT prediction can be checked by semiclassical, perturbative techniques on both sides of the correspondence is a limit in which supersymmetry is partially restored.

More recently, predictions of the AdS/CFT correspondence have been verified for a larger class of rotating strings for which the $SO(6)$ angular momentum has three independent skew-eigenvalues [10]. One finds that the energies, like the corresponding anomalous dimensions on the $N = 4$ SYM side of the correspondence, are again determined by the solution to an integrable model, but of a more general type. Here we shall show, extending our previous result, that these rotating strings are again supersymmetric in the ultra-relativistic, and hence tensionless, limit. In fact, the tensionless strings are supersymmetric for *arbitrary* $SO(6)$ angular momentum, generically preserving 1/8 supersymmetry.

This paper is written with the intention that it be read in conjunction with [9]. We will begin by presenting the tensionless version of the general 'three-spin' rotating string solution of [10]. We then present our main new result, which is a demonstration that these tensionless strings generically preserve 1/8 supersymmetry. We close with some further comments on tensionless strings in $AdS_5 \times S^5$.

2. Null rotating strings on S^5

Since the strings of interest here lie at the origin of AdS_5, the relevant part of the $AdS_5 \times S^5$ metric is

$$ds^2 = -dt^2 + d\theta^2 + \sin^2\theta d\phi^2 + \sum_{i=1}^{3} a_i^2 (d\alpha_i)^2 , \tag{1}$$

where $a_1 = \cos\theta$, $a_2 = \sin\theta\cos\phi$, $a_3 = \sin\theta\sin\phi$, and α_i are polar angles in three orthogonal planes. In order to avoid coordinate singularities, we will assume here that $a_1a_2a_3$ is non-zero, which corresponds to the assumption that the angular momentum two-form has maximal rank.

Let (τ, σ) be the worldsheet coordinates. In the gauge $t = \tau$ the phase-space form of the Lagrangian density for a tensionless string in the above background is

$$L = p_\theta \dot{\theta} + p_\phi \dot{\phi} + \sum_i p_i \dot{\alpha}_i - H - s \left(p_\theta \theta' + p_\phi \phi' + \sum_i p_i \alpha_i' \right) , \tag{2}$$

where s is the Lagrange multiplier for the string reparametrization constraint, and H is the Hamiltonian density

$$H = \sqrt{p_\theta^2 + \frac{p_\phi^2}{\sin\theta^2} + \sum_i \frac{p_i^2}{a_i^2}} . \tag{3}$$

Note that the α_i equation of motion is $\dot{p}_i = (sp_i)'$, from which it follows that the angular momentum, $J_i = \oint d\sigma\, p_i$, is a constant of motion.

We seek solutions for which θ and ϕ are constant and $p_\theta = p_\phi = 0$. This requires, in order to solve the corresponding equations of motion, that we set

$$a_i^2 = |p_i| \Big/ \sum_j |p_j| , \tag{4}$$

from which we see that

$$H = \sum_i |p_i| . \tag{5}$$

Given (4), the p_i equations of motion reduce to $\dot{\alpha}_i - s\alpha_i' = 1$, while the constraint imposed by s is $\sum_i p_i \alpha_i' = 0$. We may choose a gauge for which $s = 0$, in which case the equations above have the solution[a]

$$\alpha_i = t + m_i \sigma , \qquad p_i = J_i / 2\pi , \tag{6}$$

[a] This is not the unique solution, but it is the one of relevance when one considers the tensionless string as a limit of the tensionful string.

for integers m_i satisfying $\sum_i m_i J_i = 0$. Because each p_i has a definite sign, integration of (5) yields the total energy

$$E = |J_1| + |J_2| + |J_3| \,. \qquad (7)$$

An argument analogous to that of [9] shows that this energy saturates a BPS bound implied by the $PSU(2,2|4)$ supersymmetry algebra of $AdS_5 \times S^5$. Because of this, the rotating strings are supersymmetric. Since we have assumed that all J_i are non-zero, the fraction of supersymmetry preserved is 1/8, as we will confirm in the following section. If only two J's are non-zero, then 1/4 supersymmetry is preserved. And if only one J is non-zero, then 1/2 supersymmetry is preserved.

3. Supersymmetry

The condition for a IIB superstring to be supersymmetric in the ultra-relativistic limit in which it becomes null is

$$p_M \, e^M{}_A \, \Gamma^A \, \epsilon = 0 \,, \qquad (8)$$

where p_M is the ten-momentum, $e^M{}_A$ is the obvious orthonormal frame associated to the metric (1), Γ^A are the corresponding tangent-space (constant) ten-dimensional Dirac matrices, and ϵ is a chiral, complex Killing spinor of the $AdS_5 \times S^5$ background. This spinor can be written in terms of a constant spinor ϵ_0 as

$$\epsilon = e^{i\frac{t}{2}\tilde{\Gamma}} e^{i\frac{\theta}{2}\Gamma_{\phi 123}} e^{\frac{\phi}{2}\Gamma_{\theta\phi}} e^{\frac{i}{2}(\alpha_1 \Gamma_{\theta\phi 23} + \alpha_2 \, i\Gamma_{2\theta} + \alpha_3 \, i\Gamma_{3\phi})} \, \epsilon_0 \,, \qquad (9)$$

where $\tilde{\Gamma}$ commutes with all other matrices in the problem and $\Gamma_i \equiv \Gamma_{\alpha_i}$. It is clear from this expression that the matrices occurring inside the brackets in the last exponential generate, when acting on ϵ_0, rotations in each of the three orthogonal two-planes parametrised by α_i. Note that they all square to unity (so they have eigenvalues ± 1), they are mutually-commuting, and the product of any two of them yields the third one, up to a sign.

For the rotating string solutions of the previous section, the supersymmetry condition (8) reduces to

$$a_i \, \Gamma_{ti} \, \epsilon = \epsilon \,, \qquad (10)$$

where we have assumed, for definiteness, that all J_i are positive. If the angular momentum two-form has maximum rank then all a_i are non-zero, and equation (10) is equivalent to

$$i\Gamma_{2\theta} \, \epsilon_0 = \epsilon_0 \,, \qquad i\Gamma_{3\phi} \, \epsilon_0 = \epsilon_0 \,, \qquad \Gamma_{t1} \, \epsilon_0 = \epsilon_0 \,. \qquad (11)$$

Let us first show that (11) implies (10). We see from the first two conditions in (11) that ϵ_0 is an eigen-spinor of the last exponential in equation (9), so this exponential cancels on both sides of (10). The first exponential also cancels because $\tilde{\Gamma}$ commutes with all other matrices in equation (10), so the latter may be rewritten as

$$a_i \, \Gamma_{ti} \, \epsilon_0 = e^{-\frac{\phi}{2}\Gamma_{\theta\phi}} e^{i\theta\Gamma_{\phi 123}} e^{\frac{\phi}{2}\Gamma_{\theta\phi}} \, \epsilon_0 = (a_1 + a_2 \, i\Gamma_{\phi 123} + a_3 \, i\Gamma_{123\theta}) \, \epsilon_0 \,, \quad (12)$$

which is identically satisfied by virtue of (11).

With some further algebra, it can also be shown that (10) implies (11), but we will not do this here. Instead, we note that, being mutually-commuting projections, the conditions (11) imply that the fraction of preserved supersymmetry is 1/8, as expected for a string carrying three independent angular momenta. We also recall that, as observed above, they imply that ϵ_0 is invariant (up to a phase) under rotations in the α_i-planes associated to the non-zero components of the angular momentum two-form.

4. Discussion

There are now many string solutions in $AdS_5 \times S^5$ whose classical energies have been matched, in the large-energy limit, to anomalous dimensions of composite operators, with a large number of fields, in perturbative $N = 4$ SYM theory. *All* these strings become ultra-relativistic in this limit, and hence effectively tensionless, in the sense that their kinetic energy, K, is much larger than the energy associated to their tension; their classical energy then admits an expansion in positive powers of λ/K^2.[b]

This property is, of course, necessary for a comparison to *perturbative* SYM to be possible, and is a typical property of large-energy strings moving on S^5, as opposed to AdS_5. The reason is that the size of a generic string moving on S^5 is bounded above by the radius of the sphere, so in this case the kinetic energy always dominates in the large-energy limit. On the contrary, the size of a string moving in AdS_5, and hence the contribution of its tension to its energy, can grow without bound. In other words, the large-energy limit for generic strings moving in AdS_5 is not an ultra-relativistic limit. It is therefore not surprising that the energies of these strings cannot be precisely matched to dimensions of SYM operators computed within perturbation theory.

Almost all known strings that become tensionless in the large-energy limit also become supersymmetric, as we have shown in this article for the

[b]Some aspects of these ultra-relativistic strings have been studied in [11].

class of rotating string solutions of [10] with generic $SO(6)$ angular momenta. Of course, in all these cases the energy saturates a BPS bound in the large-energy limit, $E = K = |J_1|+|J_2|+|J_3|$, and so this limit is equivalent to a large-angular-momentum limit. The only apparent exception that we are aware of is a pulsating string solution, for which the classical energy matches an anomalous dimension of a SYM operator computed in perturbation theory despite the fact that this energy is not close to saturating a BPS bound [8]. All string solutions for which it has been checked that the sigma-model quantum corrections to the classical energy vanish in the large-energy limit [1] are also those that happen to be supersymmetric in this limit [9]. The match for the pulsating string is therefore mysterious, since it relies on the neglect of these corrections in the absence of supersymmetry. This might suggest that the key property behind the success of these tests of the AdS/CFT conjecture is the ultra-relativistic nature of the corresponding strings, rather than the nearly-supersymmetric property emphasised in our previous paper.

Acknowledgments

We are grateful to Jorge Russo for useful discussions. DM is partially supported by funds from NSERC of Canada.

References

1. S. Frolov and A.A. Tseytlin, *Nucl. Phys.* **B668** 77 (2003); *JHEP* **0307** 016 (2003) 016;
2. S. Frolov and A.A. Tseytlin, *Phys. Lett.* **B570** 96 (2003).
3. N. Beisert, J.A. Minahan, M. Staudacher and K. Zarembo, *JHEP* **0309** 010 (2003).
4. M. Kruczenski, *Spin chains and string theory*, hep-th/0311203.
5. G. Arutyunov, S. Frolov, J. Russo and A.A. Tseytlin, *Nucl. Phys.* **B671** 3 (2003).
6. N. Beisert, S. Frolov, M. Staudacher and A.A. Tseytlin, *JHEP* **0310** 037 (2003).
7. G. Arutyunov and M. Staudacher, *Matching higher conserved charges for strings and spins*, hep-th/0310182
8. J. Engquist, J.A. Minahan and K. Zarembo, *JHEP* **0311** 063 (2003).
9. D. Mateos, T. Mateos and P.K. Townsend, *JHEP* **0312** 017 (2003).
10. G. Arutyunov, J. Russo and A.A. Tseytlin, *Spinning strings in $AdS_5 \times S^5$: new integrable system relations*, hep-th/0311004.
11. A. Mikhailov, *JHEP* **0312** 058 (2003).

MATRIX THEORY OF pp WAVES

JEREMY MICHELSON
Department of Physics and Astronomy
University of Kentucky
Lexington, KY 40506 USA
E-mail: jeremy@pa.uky.edu

The Matrix Theory that has been proposed for various pp wave backgrounds is discussed. Particular emphasis is on the existence of novel nontrivial supersymmetric solutions of the Matrix Theory. These correspond to branes of various shapes (ellipsoidal, paraboloidal, and possibly hyperboloidal) that are unexpected from previous studies of branes in pp wave geometries.

1. Introduction

The BMN[1] Matrix Theory, which describes the maximally supersymmetric pp wave of M-theory[2], is a very nice arena for learning about M-theory. It is simple enough to be reasonably tractable, yet describes a curved background. Moreover, it has a dimensionless parameter $\frac{\mu\alpha'}{R}$—the ratio, in string units, between the strength of the four-form flux and the DLCQ radius—which, when large, permits a perturbative treatment of the Yang-Mills theory[3]. Additionally, the four-form flux supports fuzzy sphere solutions to the Matrix Theory[1]. The fuzzy spheres are the general vacua of the Matrix Theory, which preserve *all* the SUSYs. These fuzzy spheres have very interesting properties upon compactification[4] of the theory to a IIA Matrix String Theory. These results have already appeared in the literature[5,6] and space constraints prevent us from summarizing them here.

There are many pp wave solutions to M-theory[7,8,9,10] and a Matrix Theory has been proposed[8,11] for each. The original argument noted that the form reproduces BMN and has the right number of SUSYs[8]. A derivation from membrane quantization has since been given[11]. One can also show, beyond counting fermionic generators, that the full SUSY algebras match[12].

Having proposed a Matrix Theory for each M-theory pp wave, one should now study them. The zeroth question to answer is: what are the vacua which preserve all of the SUSYs? We will study this for three *pp* waves: the Penrose limit of $\mathrm{AdS}_3 \times S^3$ [7,8,13,14]; the T-dual[4] of the maximally supersymmetric *pp* wave of the IIB theory[15]; and the 26 supercharge *pp* wave[10]. For the first two, we find, at infinite N, brane solutions which do not follow from treating g_{++} as a superpotential[16].[a]

2. Matrix Theories

The general 11-dimensional *pp* wave considered is[17,8,9]

$$ds^2 = 2dx^+dx^- - \left[\sum_{i=1}^{9} \mu_i^2 (x^i)^2\right] (dx^+)^2 + (dx^i)^2, \qquad F = dx^+ \wedge \Theta, \quad (1)$$

with constant μ_i, Θ, and also Θ is a "spatial" (support in the x^i-directions)[b] three-form, satisfying the equation of motion (e.o.m.) $\Theta_{ijk}^2 = 12 \sum_i \mu_i^2$.

This geometry admits $16 + 2n$ Killing spinors. To present them—and define n—set, (with all sums explicit here, and $U_{(i)} \equiv 3\Gamma^i \not{\Theta} \Gamma^i + \not{\Theta}$)

$$\Omega_+ \equiv -\frac{1}{12}\not{\Theta}(\Gamma^+\Gamma^- + 1\!\!1), \quad \Omega_- \equiv 0, \quad \Omega_i \equiv \frac{1}{24}\Gamma^i U_{(i)} \Gamma^+. \qquad (2)$$

The solutions of the Killing spinor equation, $\mathcal{D}_A \epsilon = \nabla_A \epsilon - \Omega_A \epsilon = 0$, are

$$\epsilon(x, \epsilon_0) = \left[1 + \sum_i x^i \Omega_i\right] e^{-\frac{1}{12}(\Gamma^+\Gamma^- + 1\!\!1)\not{\Theta} x^+} \epsilon_0, \qquad (3)$$

where the constant spinor ϵ_0 obeys[17,8,9]

$$U_{(i)}^2 \Gamma^+ \epsilon_0 = -144 \mu_i^2 \Gamma^+ \epsilon_0. \qquad (4)$$

So there are 16 "standard" SUSYs which obey $\Gamma^+ \epsilon_0 = 0$ and an even number, $2n$, of "supernumerary"[7] SUSYs obeying

$$U_{(i)}^2 \epsilon_0 = -144 \mu_i^2 \epsilon_0, \qquad (5)$$

for each i. Equation (5) need not have any solutions, and never has 12 or 14 solutions[10]. Existence of "supernumerary" SUSYs guarantees a solution to the e.o.m.s, by summing Eq. (5) over i and tracing over Killing spinors.

---

[a] Unfortunately there is no room to list the myriad papers on branes in *pp* waves.

[b] The index $i, j, \ldots$ runs from 1 to 9, and $A, B, \ldots$ runs over all coordinates. Γ-matrices are defined with respect to the elfbein $e^{\hat{-}} = dx^- - \frac{1}{2}\sum_{i=1}^{9} \mu_i^2 (x^i)^2 dx^+$, $e^{\hat{+}} = dx^+$, $e^{\hat{i}} = dx^i$. The Feynman slash, *e.g.* $\not{\Theta} = \frac{1}{3!}\Gamma^{ijk}\Theta_{ijk}$, is used extensively.

In units for which the 11-dimensional Planck length, $\ell_P = 1$, and employing Majorana $\mathfrak{so}(9)$ spinors Ψ, the Matrix Theory action[8,11] is[c]

$$S = R\int d\tau\,\mathrm{Tr}\left\{\frac{1}{2R^2}(D_\tau X^i)^2 + \frac{i}{R}\Psi^\mathsf{T} D_\tau\Psi + \Psi^\mathsf{T}\Gamma^i\left[X^i,\Psi\right] + \frac{1}{4}\left[X^i,X^j\right]^2 \right.$$
$$\left. -\frac{1}{2}\sum_i \frac{\mu_i^2}{R^2}(X^i)^2 - \frac{i}{4R}\Psi^\mathsf{T}\not{\Theta}\Psi - \frac{i}{3R}\Theta_{ijk}X^iX^jX^k\right\}. \quad (6)$$

The first line is the familiar BFSS Matrix Theory[18] action; the second line contains mass terms for the bosons and fermions, and the Myers term[19,20].

The Matrix Theory (6) is invariant under the nonlinearly realized SUSY $\delta\Psi = e^{\frac{1}{4}\not{\Theta}\tau}\epsilon_+$ corresponding to the 16 "standard" SUSYs of the *pp* wave. Consider also the transformations,

$$\delta X^i = i\Psi^\mathsf{T}\Gamma^i\epsilon(\tau), \qquad \delta A_\tau = iR\Psi^\mathsf{T}\epsilon(\tau), \qquad \epsilon(\tau) \equiv e^{-\frac{1}{12}\not{\Theta}\tau}\epsilon_0, \quad (7a)$$

$$\delta\Psi = \frac{1}{2R}D_\tau X^i\Gamma^i\epsilon(\tau) + \frac{1}{R}X^i\hat{\Omega}_i\epsilon(\tau) + \frac{i}{4}\left[X^i,X^j\right]\Gamma^{ij}\epsilon(\tau), \quad (7b)$$

where $\hat{\Omega}_i \equiv \frac{1}{24}\Gamma^i U_{(i)}$ and ϵ_0 is constant (*cf.* Bonelli[21]). These transformations preserve the action precisely when ϵ_0 obeys (5). So the Matrix quantum mechanics (6) preserves exactly the right number of SUSYs to describe the *pp* wave (1)[7,11]. Indeed, the SUSY algebras match[12].

3. The AdS$_3 \times S^3$ *pp* Wave

The Penrose limit of $AdS_3 \times S^3$ has been considered in[7,8,13,14]. Lifting the geometry supported with NS-NS flux to 11-dimensions,

$$ds^2 = 2dx^+dx^- - \mu^2\sum_{a=1}^{4}(x^a)^2(dx^+)^2 + \sum_{i=1}^{9}(dx^i)^2,$$
$${}^{(4)}F = 2\mu dx^+\wedge dx^1\wedge dx^2\wedge dx^9 + 2\mu dx^+\wedge dx^3\wedge dx^4\wedge dx^9. \quad (8)$$

The M-theory direction is x^9, and the directions x^a, a=$1\cdots 4$ are the spacelike $AdS_3\times S^3$ directions transverse to the null geodesic of the Penrose limit.

Vanishing of the fermionic variation (7b) gives the conditions

$$\left[X^1,X^2\right] = \left[X^3,X^4\right], \quad \left[X^2,X^3\right] = \left[X^1,X^4\right], \quad \left[X^1,X^3\right] = -\left[X^2,X^4\right],$$
$$\left[X^2,X^9\right] = i\frac{\mu}{R}X^1, \quad \left[X^1,X^9\right] = -i\frac{\mu}{R}X^2,$$
$$\left[X^4,X^9\right] = i\frac{\mu}{R}X^3, \quad \left[X^3,X^9\right] = -i\frac{\mu}{R}X^4. \quad (9)$$

[c] $D_\tau X^i = \partial_\tau X^i + i\left[A_\tau, X^i\right]$ and $(\Psi_\alpha\Psi_\beta)^* = \Psi_\beta^*\Psi_\alpha^*$. R is the DLCQ radius.

All other commutators vanish, and the vacua are static.

For finite N, all vacua preserving all eight linearly-realized SUSYs are trivial: $X^1 = X^2 = X^3 = X^4 = 0$. However, there is a family of solutions for infinite N. Specifically, take X^1, X^2 and X^3, X^4 to form two noncommutative planes, with equal noncommutativity parameter,

$$[X^1, X^2] = i\vartheta = [X^3, X^4]; \qquad [X^1, X^3] = 0 = [X^2, X^3], \tag{10}$$

and take

$$X^9 = X^9_0 - \frac{\mu}{2R\vartheta}\left[(X^1)^2 + (X^2)^2 + (X^3)^2 + (X^4)^2\right], \tag{11}$$

where X^9_0 commutes with all the matrices. Then it is straightforward to see that equations (9) are satisfied. This solution describes a longitudinal fivebrane of a (fuzzy) paraboloidal shape (11). The fivebrane wraps the entire "$AdS_3 \times S^3$ directions" but it extends into the M-theory direction through the paraboloid equation.

4. The 26 Supercharge Matrix Theory Vacua

The *pp* wave with 26 supercharges was presented in[10]. Supersymmetric vacua obey ($I, J, \cdots = 1 \ldots 7$, $I'', J'', \cdots = 1, 2, 3$, with the conventions[10])

$$\begin{gathered} D_\tau X^I = 0, I = 1 \ldots 7, \quad D_\tau X^8 = \frac{\mu}{2} X^9, \quad D_\tau X^9 = -\frac{\mu}{2} X^8, \\ X^4 = X^5 = X^6 = X^7 = 0, \qquad \left[X^{I''}, X^{J''}\right] = -i\frac{\mu}{R}\epsilon_{I''J''K''}X^{K''}, \\ \left[X^8, X^{I''}\right] = 0 = \left[X^9, X^{I''}\right]. \end{gathered} \tag{12}$$

That is, states preserving all the SUSYs are fuzzy spheres at the origin of the $(4,5,6,7)$-hyperplane, and orbit in the $(8,9)$-plane with frequency $\frac{\mu}{2}$.

Recall[10] that the 26 supercharge *pp* wave can be compactified to a 26 supercharge IIA *pp* wave. Such a reduction occurs along a Killing vector which rotates in the (8,9)-plane. Thus, in the IIA theory, the fully supersymmetric solutions (12) are static. This is reminsicent of the maximally supersymmetric *pp* wave, for which[5] the orbiting fuzzy spheres broke the same half of the (supernumerary) SUSYs as were broken by the reduction to the IIA theory, yielding fully (24 supercharge) supersymmetric IIA fuzzy spheres located at any static value of $X^8_{\rm IIA}$. Here, the solutions have identical form, but with two additional supercharges.

5. IIB T-dual Matrix Theory Vacua

The T-dual, lifted to M-theory, of the IIB maximally supersymmetric *pp* wave is[4] (the coordinates have been reindexed relative to the reference)

$$ds^2 = 2dx^+dx^- - \left[4\mu^2\sum_{I=1}^{6}(x^I)^2 + 16\mu^2(x^7)^2\right](dx^+)^2 + \sum_{i=1}^{9}(dx^i)^2, \quad (13a)$$

$$F = -4\mu dx^+ \wedge dx^7 \wedge dx^8 \wedge dx^9 + 8\mu dx^+ \wedge dx^5 \wedge dx^6 \wedge dx^7. \quad (13b)$$

The linear SUSYs of the Matrix Theory are parametrized by $\Gamma^{5689}\epsilon_0 = \epsilon_0$.

Fully supersymmetric vacua must be static, at $X^{1\ldots4} = 0$ and satisfy

$$-2\frac{\mu}{R}X^7 + \frac{i}{2}\left[X^8, X^9\right] - \frac{i}{2}\left[X^5, X^6\right] = 0, \quad (14a)$$

$$\left[X^6, X^7\right] = 2i\frac{\mu}{R}X^5, \quad \left[X^7, X^5\right] = 2i\frac{\mu}{R}X^6, \quad (14b)$$

$$\left[X^5, X^8\right] = -\left[X^6, X^9\right], \quad \left[X^6, X^8\right] = \left[X^5, X^9\right], \quad (14c)$$

$\left[X^7, X^8\right] = 0 = \left[X^7, X^9\right]$. These equations are similar to those for non-maximally SUSic vacua of the BMN theory[22]. Solutions to equation (14) satisfy the e.o.m.s.

For finite N, the only solutions are fuzzy ellipsoids, ($\left[J^a, J^b\right] = i\epsilon_{abc}J^c$)

$$X^5 = 2\sqrt{2}\frac{\mu}{R}J^1, \qquad X^6 = 2\sqrt{2}\frac{\mu}{R}J^2, \qquad X^7 = 2\frac{\mu}{R}J^3. \quad (15)$$

Constant values of $X^{8,9}$, that are diagonal and proportional to the identity in each irreducible SU(2) block, give the positions of the fuzzy ellipsoids.

At strictly infinite N there are many more solutions. For example,

$$X^{5,6} = 2\sqrt{2}\frac{\mu}{R}J^{1,2}\otimes \mathbb{1}, \qquad X^7 = 2\frac{\mu}{R}\left(J^3 - \frac{1}{8}\vartheta\mathbb{1}\right)\otimes\mathbb{1}, \quad (16a)$$

$$X^8 = \mathbb{1}\otimes\hat{x}^8, \qquad X^9 = \mathbb{1}\otimes\hat{x}^9, \quad (16b)$$

where $\left[\hat{x}^8, \hat{x}^9\right] = i\vartheta$ is a c-number. That is, $X^{5,6,7}$ form a fuzzy ellipsoid and X^8 and X^9 parameterize a noncommutative plane orthogonal to the ellipsoid and translated, along X^7, by an amount $-\frac{\mu}{4R}\vartheta$ from the origin. This is the longitudinal M5-brane, formed as a stack of M2-branes[18,23], with one M2-brane blown up into a fuzzy ellipsoid. Thus, the solution (16) describes a longitudinal M5-brane of topology $\mathbb{R}^{1,3}\times S^2$.

One can also find a solution for every complex simple Lie algebra of rank 2. If the algebra is not simply-laced—*i.e.* except for $\mathfrak{su}(3)$—the algebra gives two inequivalent solutions. Explicitly, if the two roots of the algebra are $\alpha_1 = (a_1, -b_1)$ and $\alpha_2 = (0, a_2)$, with Cartan subalgebra h_1, h_2, ($[h_1, \alpha_1] = a_1\alpha_1$, $\left[\alpha_2, \alpha_2^\dagger\right] = a_2 h_2$, *etc.*), then

$$X^7 = \frac{2\mu}{a_1 R} h_1, \qquad X^5 = \frac{2\mu}{a_1 R}(e_1 + e_1^\dagger), \qquad X^6 = \frac{2\mu}{a_1 Ri}(e_1 - e_1^\dagger),$$
$$X^8 = -2i\sqrt{\frac{b_1}{a_1^2 a_2}}\frac{\mu}{R}(e_2 + e_2^\dagger), \qquad X^9 = -2\sqrt{\frac{b_1}{a_1^2 a_2}}\frac{\mu}{R}(e_2 - e_2^\dagger), \tag{17}$$

is a solution. Demanding hermitian matrices then restricts to a particular noncompact realification of each algebra. The resulting realifications are $\mathfrak{su}(2,1)$, $\mathfrak{sp}(1,1) \cong \mathfrak{so}(1,4)$, $\mathfrak{sp}(2,\mathbb{R}) \cong \mathfrak{so}(2,3)$ and $\mathfrak{g}_{2(2)}$. Two inequivalent solutions are obtained from the last algebra. As these are all noncompact, their nontrivial unitary representations (hermitian matrices) are all infinite dimensional.

For example, explicit[24] $\mathfrak{su}(2,1)$ solutions can be written in terms of three commuting sets of annihilation operators a_1, a_2, b, and their hermitian conjugate creation operators,

$$X^5 = 2\sqrt{\frac{2}{3}}\frac{\mu}{R}\left(a_1^\dagger a_2 + a_2^\dagger a_1\right), \qquad X^6 = -i2\sqrt{\frac{2}{3}}\frac{\mu}{R}\left(a_1^\dagger a_2 - a_2^\dagger a_1\right),$$
$$X^8 = \frac{2}{\sqrt{3}}\frac{\mu}{R}\left(a_1 b + a_1^\dagger b^\dagger\right), \qquad X^9 = -i\frac{2}{\sqrt{3}}\frac{\mu}{R}\left(a_1 b - a_1^\dagger b^\dagger\right),$$
$$X^7 = \frac{2}{3}\frac{\mu}{R}\left(a_1^\dagger a_1 - 2a_2^\dagger a_2 - bb^\dagger\right). \tag{18}$$

The quadratic Casimir of $\mathfrak{s}u(2,1)$ takes the form

$$C_2^{\mathfrak{s}u(2,1)} = -\frac{3}{4}\frac{R^2}{\mu^2}\left((X^8)^2 + (X^9)^2\right) + \frac{3}{8}\frac{R^2}{\mu^2}\left((X^5)^2 + (X^6)^2 + 2(X^7)^2\right)$$
$$- \frac{9}{64}\frac{R^4}{\mu^4}\left[X^8, X^9\right]^2 + \frac{9}{32}\frac{R^4}{\mu^4}\left(\left[X^5, X^8\right]^2 + \left[X^5, X^9\right]^2\right). \tag{19}$$

The first line suggests a hyperboloidal interpretation to the solution, though this needs refinement.

Compactifying the Matrix Theory along X^8 gives[12], at least for weak string coupling, the Green-Schwarz action for the IIB string in the *pp* wave background. The radius of compactification matches precisely. This demonstration requires a field redefinition in the Matrix String Theory that is the inverse of the coordinate transformation that makes the isometry manifest[4] on the IIB side. Unfortunately, there is no space to give details here.

Acknowledgments

This work was supported in part by Department of Energy contract #DE-FG01-00ER45832.

References

1. D. Berenstein, J. Maldacena and H. Nastase, *J. High Energy Phys.* **04**, 013 (2002); [hep-th/0202021].
2. J. Kowalski-Glikman, *Phys. Lett.* **B134**, 194 (1984).
3. K. Dasgupta, M. M. Sheikh-Jabbarri and M. Van Raamsdonk, *J. High Energy Phys.* **05**, 056 (2002); [hep-th/0205185].
4. J. Michelson, *Phys. Rev.* **D66**, 066002 (2002); [hep-th/0203140].
5. S. R. Das, J. Michelson and A. D. Shapere, UK/03-09, hep-th/0306270.
6. Y.-X. Chen and J. Shao, hep-th/0310062.
7. M. Cvetič, H. Lü and C. N. Pope, CTP-TAMU-04/02, UPR-982-T, MCTP-02-15, hep-th/0203082.
8. M. Cvetič, H. Lü and C. N. Pope, *Nucl. Phys.* **B644**, 65–84 (2002); [hep-th/0203229].
9. J. P. Gauntlett and C. M. Hull, *J. High Energy Phys.* **06**, 013 (2002); [hep-th/0203255].
10. J. Michelson, *Class. Quant. Grav.* **19**, 5935–5949 (2002); [hep-th/0206204].
11. N. Iizuka, *Phys. Rev.* **D68**, 126002 (2003); [hep-th/0211138].
12. J. Michelson, unpublished.
13. J. G. Russo and A. A. Tseytlin, *J. High Energy Phys.* **04**, 021 (2002); [hep-th/0202179].
14. J. Gomis, L. Motl and A. Strominger, *J. High Energy Phys.* **11**, 016 (2002); [hep-th/0206166].
15. M. Blau, J. Figueroa-O'Farrill, C. Hull and G. Papadopoulos, *J. High Energy Phys.* **01**, 047 (2002); [hep-th/0110242].
16. Y. Hikida and S. Yamaguchi, *J. High Energy Phys.* **01**, 072 (2003); [hep-th/0210262].
17. J. Figueroa-O'Farrill and G. Papadopoulos, *J. High Energy Phys.* **08**, 036 (2001); [hep-th/0105308].
18. T. Banks, W. Fischler, S. H. Shenker and L. Susskind, *Phys. Rev.* **D55**, 5112–5128 (1997); [hep-th/9610043].
19. R. C. Myers, *J. High Energy Phys.* **12**, 022 (1999); [hep-th/9910053].
20. W. Taylor and M. Van Raamsdonk, *Nucl. Phys.* **B573**, 703–734 (2000); [hep-th/9910052].
21. G. Bonelli, *J. High Energy Phys.* **08**, 022 (2002); [hep-th/0205213].
22. J.-H. Park, *J. High Energy Phys.* **10**, 032 (2002); [hep-th/0208161].
23. T. Banks, N. Seiberg and S. Shenker, *Nucl. Phys.* **B490**, 91–106 (1997); [hep-th/9612157].
24. M. Günaydin and C. Saçlioğlu, *Commun. Math. Phys.* **87**, 159–179 (1982).

BLOCK SPINS AND D-BRANES: COARSE-GRAINING MATRICES AND QUIVERS

K. NARAYAN

Center for Geometry and Theoretical Physics,
Duke University, Durham, NC 27708.
Email: narayan@cgtp.duke.edu

Drawing intuition from critical phenomena, we describe a block-spin-like transformation that acts on the space of D-brane configurations by averaging over nearest-neighbour blocks within their (in general noncommuting) matrix representations. Matrices become "less-off-diagonal" under this blocking transformation, in general given key assumptions involving gauge invariance. We then describe a similar coarse-graining on a simplified subset of supersymmetric quiver gauge theories arising on the worldvolumes of D-brane probes of orbifold geometries, by sequentially Higgsing the gauge symmetry in these theories.

1. Introduction: the basic ideas and motivations

The past few years have seen the emergence of powerful new dualities between gauge theories and gravity [1 2], by taking various decoupling limits of D-brane configurations. The description in terms of an effective spacetime in these cases emerges in an appropriate large-N limit from the dynamics of matrix entries. In general, the worldvolume scalars describing the motion of branes in the transverse space are matrices. Thus various configurations of collections of D-branes are described by matrix representations.

The emergence of a continuum spacetime description from discrete matrix indices is reminiscent of the emergence of a continuum limit in lattice theories of critical phenomena. In these cases, the renormalization group provides a useful conceptual and technical framework for understanding the continuum limit and studying its properties. The continuum limit emerges in an RG approach as a fixed point of a transformation in which the dynamical degrees of freedom are "thinned" by removing some high-energy modes. The renormalization group itself stemmed from Kadanoff's intuition on approximating a lattice spin system near its critical point in terms of block spins: holding these fixed and averaging over the microscopic spins yields a new theory describing the interactions of the block spins with new appropriately scaled parameters. In what follows, it is this intuition from real space renormalization that we will take over to D-brane matrix variables to define a block-spin-like coarse-graining transformation acting on the space of D-brane configurations. This talk is based on the papers [4] [5], which we refer to for a detailed exposition of what follows.

2. Coarse-graining D-brane configurations via matrices

Consider the scalars X^i ($N \times N$ hermitian matrices in the adjoint representation of the gauge group, which we take for concreteness to be $U(N)$ for the present) representing the transverse coordinates of a collection of an infinite number of D-branes (*i.e.* N is essentially infinite) and define blocked up matrices as

$$\tilde{X}^i_{a,b} = \alpha(X^i_{2a-1,2b-1} + X^i_{2a,2b-1} + X^i_{2a-1,2b} + X^i_{2a,2b}), \qquad a, b = 1 \ldots \frac{N}{2}. \quad (1)$$

In other words, define a matrix entry in the blocked up matrix by averaging over the nearest neighbour matrix entries in 2×2 blocks of the original matrix. The overall factor α is a normalization constant for averaging uniformly all over the matrix. In general, one imagines that nonuniform averaging over the matrix implies inhomogeneities in the matrix space and hence perhaps inhomogeneities in spacetime. Also we note that the averaging constant can be different for different matrices, *i.e.* $\alpha = \alpha^i$. Demanding that the identity matrix is preserved under blocking up, *i.e.* $\mathbf{1} \to \mathbf{1}$, fixes $\alpha = \frac{1}{k}$ for $k \times k$ blocks. With intuition from block spins, one would expect that the block size should not matter and in what follows we will verify this. It is clear from the definition of blocking (1) that this preserves hermiticity: the $\tilde{X}^i$s are also hermitian, so that they are also interpretable as scalars in the adjoint representation of a gauge group. While this scheme of coarse-graining a priori seems somewhat ad hoc, we will see in what follows that it leads to interesting patterns of RG-like flows if certain key assumptions are made: in particular, we fix gauge invariance, identifying D-brane locations at the level of matrix variables with D-brane locations in physical space.

Let us illustrate this in an example: consider the algebra representing the noncommutative plane $[X^1, X^2] = i\theta^{12}\mathbf{1}$, which arises, *e.g.* in Matrix theory and on D-brane worldvolumes in constant B_{NSNS} fields. This algebra has translation isometries generated by the identity matrix to preserve which, we fix the averaging constant $\alpha = \frac{1}{2}$ for 2×2 blocks. In terms of an ordered basis of the familiar harmonic oscillator energy eigenstates, the representation of the X^1 matrix coarse-grains under blocking up as

$$X^1 = \sqrt{\frac{\theta^{12}}{2}} \begin{pmatrix} 0 & 1 & 0 & 0 & \dots \\ 1 & 0 & \sqrt{2} & 0 & \dots \\ 0 & \sqrt{2} & 0 & \sqrt{3} & \dots \\ 0 & 0 & \sqrt{3} & 0 & \dots \\ & & & \cdot & \\ & & & \cdot & \end{pmatrix} \to \frac{1}{2}\sqrt{\frac{\theta^{12}}{2}} \begin{pmatrix} 2 & \sqrt{2} & 0 & 0 & \dots \\ \sqrt{2} & 2\sqrt{3} & \sqrt{4} & 0 & \dots \\ 0 & \sqrt{4} & 2\sqrt{5} & \sqrt{6} & \dots \\ 0 & 0 & \sqrt{6} & 2\sqrt{7} & \dots \\ & & & \cdot & \\ & & & \cdot & \end{pmatrix} \quad (2)$$

Subtracting the diagonal part D^1 of the matrix, it is easy to see that the off-diagonal part of the matrix is the same as before, except with a new rescaled θ-parameter. The representation matrix for X^2 exhibits a very similar behaviour, except that the diagonal matrix D^2 generated has zero entries. Thus under blocking up,

$$\tilde{X}^1 = \sqrt{\frac{\theta^{12}}{2}} D^1 + \frac{1}{\sqrt{2}}\sqrt{\frac{\theta_{12}}{2}}\, (a + a^\dagger), \qquad \tilde{X}^2 = \frac{1}{\sqrt{2}}\sqrt{\frac{\theta^{12}}{2}}\, i(-a + a^\dagger), \quad (3)$$

We suggestively call the D^i "counterterm" matrices. The algebra of the new noncommutative coordinates is "self-similar" in form under blocking up, modulo expressions involving the "counterterm" matrices generated under blocking up. Continuing this blocking up procedure, the algebra of the new coordinates after n iterations is

$$[\tilde{X}^1, \tilde{X}^2] = i\frac{\theta^{12}}{2^n}\mathbf{1} + \Delta\theta^{12}. \quad (4)$$

We see from (4) that θ^{12} shrinks exponentially with each iteration while the commutative coordinates are held fixed. Thus the limit of the blocking up transformation is $\theta^{12} = 0$. It is easy to see that the block size does not matter here: indeed under $k \times k$ blocks, we see a periodic repetition of off-diagonal blocks involving factors of $\sqrt{k}$ exhibiting a self-similar structure like the one above with 2×2 blocks and θ^{12} shrinks as $\theta^{12} \to \theta^{12}\ \frac{1}{k^n}$. As currently formulated, the counterterm matrices are in general scheme-dependent, in other words, depend on the gauge-fixing procedure. A similar self-similarity exists for the fuzzy sphere and the fuzzy torus. Since the off-diagonal pieces of the matrices scale downwards, well-behaved functions of the noncommutative plane coordinates (3) that have matrix Taylor expansions $f(\tilde{X}^1, \tilde{X}^2) = \sum_{a_m} f_{a_1 a_2 \dots a_m}\ \tilde{X}^{a_1}\tilde{X}^{a_2}\dots\tilde{X}^{a_m},\ a_i = 1, 2$ asymptote to their commutative restrictions in the limit of a large number $n \to \infty$ of blocking iterations. It is important to note that all of this structure above, using a particular representation of the algebra, is gauge-fixed. Operating on these matrices using basis-changing gauge transformations does

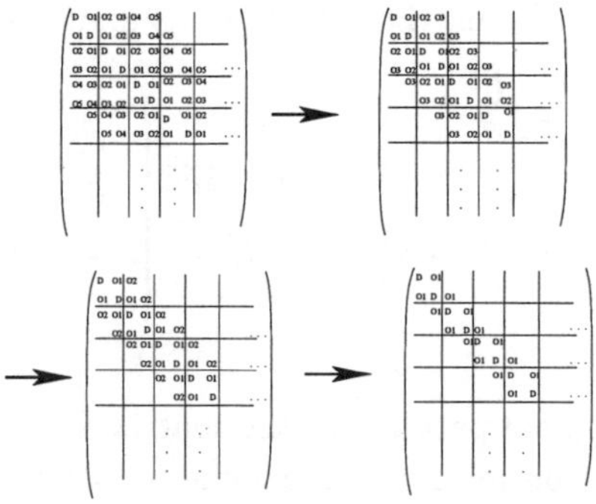

Figure 1. Successive block-transformations make the matrices "less-off-diagonal".

not preserve the precise self-similarity that we have seen above although a general scaling behaviour does in fact prevail. Indeed, consider matrix representations of the general spatially-varying algebra $[X^i, X^j] = i\theta^{ij}(X)$. We assume that these matrices have been put in as near-off-diagonal a form as possible, *e.g.* using basis-changing gauge transformations. Non-linear $\theta^{ij}(X)$ gives rise to matrix representations with nonzero far-off diagonal modes. Under repeated block-transformations (see Figure 1), the "off-diagonality" of the matrix shrinks, similar to the correlation length in the analogy with spin systems. More quantitatively, consider the scaling relation $X^i_{a,b} \sim X^i_{a,a}\ q^{|b-a|}$ for some $|q| < 1$. It is easy to see that the block-matrices then satisfy a similar relation with $\tilde{q} \sim q^2\ (\frac{1+q}{2q})^{[1/|b-a|]}$. It is interesting to note that $\tilde{q} \sim q^2$ for $b \gg a$, *i.e.* far-off-diagonal modes, which thus scale down fast, while $\tilde{q} \sim q$ for, *e.g.* $b = a + 1$, *i.e.* the one-off-the-diagonal modes. Thus this heuristic scaling behaviour shows that under this coarse-graining, the matrices in general become "less off-diagonal". Furthermore, in the case of commuting matrices of D-brane configurations, it is easy to verify that the masses $|X^i_{aa} - X^i_{bb}|$ of the lightest open string states stretched between the block D-branes are on the average less than those stretched between the microscopic D-branes.

3. Coarse-graining quivers

With a view to understanding the physics underlying this transformation, consider a simplified subset of the space of supersymmetric quiver gauge theories arising on the worldvolumes of D-brane probes of orbifold geometries and a similar block-spin-like transformation defined by sequentially Higgsing the gauge symmetry in these theories, thereby performing a sequence of partial blowups of the orbifolds. This is Wilsonian in the region of orbifold moduli space where the closed string blowup modes exhibit a hierarchy of scales which translates to a corresponding hierarchy in the strengths of the Higgs vevs for the bifundamental scalars. As an example, consider the $\mathbb{C}^2/\mathbb{Z}_N$ quiver with a $U(1)^N$ gauge group and bifundamental matter. In the region of moduli space where the D-term parameters satisfy $D_i = \rho_i = (-1)^i\rho$, the bifundamental scalars obtain the vevs $X_{2j} = \sqrt{\rho}$ with all other vevs zero through the F- and D-term equations. This choice of vevs breaks each block $U(1)_{2j} \times U(1)_{2j+1} \to U(1)$, reducing the gauge group in all down to $U(1)^{N/2}$ and further giving masses to some of the scalars and fermions. Defining $Z_j^{\pm} = \frac{1}{2}(Z_{2j,2j} \pm Z_{2j+1,2j+1})$, $j = 1, 2, \dots M = \frac{N}{2}$, we see that X_{2j} is eaten by the super-Higgs mechanism, leaving as the light degrees of freedom $X_{2j+1,2j+2}, Y_{2j+1,2j}$, and Z_j^+. The superpotential W changes as

$$\sum_{i=1}^{N} X_{i,i+1}Y_{i+1,i}(Z_{i,i} - Z_{i+1,i+1}) = \sum_{j=1}^{N/2} X_{2j+1,2j+2}Y_{2j+1,2j}(Z_j^+ - Z_{j+1}^+)$$

$$+ \sqrt{\rho}\sum_{j=1}^{N/2} Y_{2j,2j+1}Z_j^- \quad \to \sum_{j=1}^{N/2} X_{j,j+1}Y_{j+1,j}(Z_{j,j} - Z_{j+1,j+1}), \tag{5}$$

(in the last term, we have relabelled the two branes in each block as $(2j, 2j+1) \to j$ and retained only the light fields) giving an effective $\mathbb{C}^2/\mathbb{Z}_{N/2}$ quiver describing the geometry seen by the block-brane probe. We can realize the RG-like ideas mentioned above by repeating the above process iteratively at a hierarchical sequence of scales. In other words, if M is even, we can now repeat the process with $D_j = (-1)^j\rho'$ for $\rho' \ll \rho$, etc. Figure 2 shows the similar coarse-graining of a slightly more complicated $\mathbb{C}^3/\mathbb{Z}_N$ quiver while the general structure of flows in this simplified supersymmetric subset of the space of quivers obtained under this coarse-graining transformation is shown in Figure 3. This sequential Higgsing calculation makes contact with the matrix coarse-graining defined earlier when lifted to the upstairs $N \times N$ scalar matrices of the N image branes that live on the orbifold covering space: the block matrices are of precisely the same form as those describing the residual quiver if some counter-term matrices are subtracted off. The "counterterm" matrices required to be subtracted here turn out to correspond to fields that become massive under the Higgsing. Thus perhaps such "counterterm" matrices are expected in general to recover recognizable flows under this matrix coarse-graining. Here also, the qualitative structure of the flows is unchanged under changes of the block size.

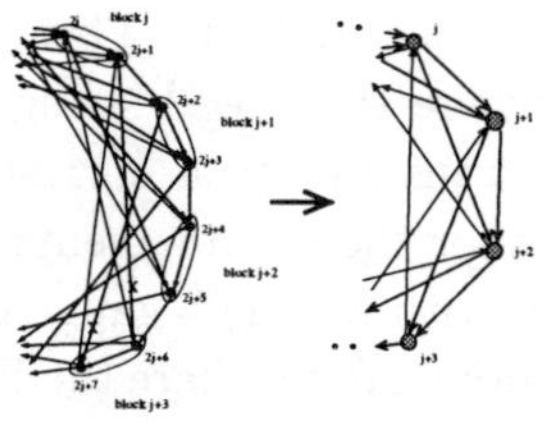

Figure 2. Coarse-graining a more general quiver, *e.g.* $\mathbb{C}^3/\mathbb{Z}_N : (1,5,-6) \to (1,2,-3)$.

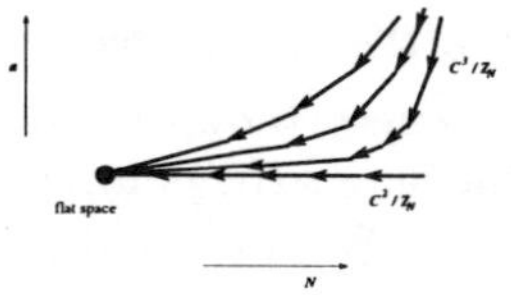

Figure 3. The structure of flows in the space of quivers.

4. Conclusions and open questions

While the notion of "nearest neighbour" is easily defined via the quiver lattices in the calculations here, the notion of locality appears much harder to define in a gauge-invariant way in more general D-brane configurations. It would be interesting to understand if the qualitative structure of flows described above can be obtained in terms of block-spin-like transformations acting on a partition function defined on the space of D-brane configurations and to study notions of universality therein. The matrix coarse-graining described here is then best interpreted as a convenient mnemonic for such RG-like transformations coarse-graining D-branes and spacetime geometry.

Acknowledgments

It is a pleasure to thank the organizers of QTS3, Cincinnati, OH, for a stimulating Symposium. This work is partially supported by NSF grant DMS-0074072.

References

1. T. Banks, W. Fischler, S. Shenker, L. Susskind, Phys. Rev. **D55**, 5112 (1997), [arxiv:hep-th/9610043].
2. See, *e.g.*, the review, O. Aharony, S. .Gubser, J. Maldacena, H. Ooguri, Y. Oz, Phys. Rept. **323**, 183 (2000), [arxiv:hep-th/9905111].
3. See *e.g.*, K. G. Wilson, J. Kogut, Phys. Rep. **12** 75 (1974).
4. K. Narayan, [arxiv:hep-th/0211110].
5. K. Narayan, M. Ronen Plesser, [arxiv:hep-th/0309171].

GLOBAL ASPECTS OF PLANE WAVES

MUKUND RANGAMANI
Dept of Physics, University of California, Berkeley, CA 94720
Theory Group, LBL, Berkeley, CA 94720
mukund@socrates.berkeley.edu

We summarise the recent developments in understanding the global aspects of plane wave and pp-wave geometries. We discuss their causal structure and properties of discrete spacelike quotients of these geometries.

1. Introduction

pp-waves (or "plane-fronted waves with parallel rays") are all spacetimes with covariantly constant null Killing field. In general relativity, they form simple solutions to Einstein's equations with many curious properties. The presence of the covariantly constant null Killing field implies that these spacetimes have vanishing scalar curvature invariants, much the same as flat space. *Plane waves* are a subset of these which have in addition an extra "planar" symmetry along the wavefronts. They can be thought of as arising from the so-called Penrose limit [1] of any spacetime, which essentially consists of zooming in onto any null geodesic in that spacetime. As the notion of Penrose limit makes sense for any Lorentzian manifold, plane waves in a sense are capable of capturing some non-trivial universal features of a given spacetime, much like the trivial information captured by the tangent space.

pp-wave spacetimes are especially important within the context of string theory. This is because they yield exact classical backgrounds for string theory, since all curvature invariants, and therefore all α' corrections, vanish. Hence the pp-wave spacetimes correspond to exact conformal field theories. Because of this fact, they provide much-needed examples of classical solutions in string theory, which can in turn be used as toy models for studying its structure and properties. Plane waves happen to be even simpler, for the action in light-cone gauge is quadratic. Moreover, recently certain plane wave geometries have been shown to be dual to specific subsectors of gauge theroies [2]; this can be ascertained by invoking the AdS/CFT correspondence and taking an appropriate Penrose limit.

Nevertheless, plane waves are distinct from flat spacetime and their structure is much richer. Analysis of causal structure and other global properties while being an interesting endeavour in the general relativistic context, is also capable of shedding light on holography [3]. The exotic causal properties of plane waves make them ideal candidates for these explorations. Moreover, the relative simplicity of world-sheet string theory for these geometries makes them excellent candidates for understanding notions of stringy versions causal structure.

The *pp-wave* spacetimes, which are defined as all spacetimes admitting a covariantly constant null Killing field, ∂_v, and their metric can be written as

$$ds^2 = -2\,du\,dv - F(u,x^i)\,du^2 + dx^i\,dx^i\ . \tag{1}$$

The only non-vanishing component of the Ricci tensor is $R_{uu} = \frac{1}{2}\nabla_x^2\,F(u,x)$ which for a consistent solution to Einstein's equations must be balanced by an appropriate stress-tensor. *Plane wave* spacetimes admit planar symmetry in the transverse directions and the function $F(u,x^i)$ in Eq. (1) is in fact quadratic, $F(u,x^i) = f_{ij}(u)\,x^i x^j$. The maximally symmetric plane wave geometries have in addition a full transverse rotation symmetry and $f_{ij}(u) = \mu^2\delta_{ij}$. One such is the supersymmetric plane wave solution [2] obtained by a Penrose limit of $AdS_5 \times \mathbf{S}^5$.

2. Causal structure of plane and pp-waves

The causal structure of plane wave geometries are one of the most inriguing aspects of their global properties. As such one is usually accustomed to geometries whose causal/conformal boundary is co-dimension one. Many plane waves turn out to have a one-dimensional causal boundary! To understand the rudiments of the causal structure it is helpful to know the behaviour of null geodesics in the spacetimes. The null geodesics for the geometries in Eq. (1) are easily written down. For simplicity we will here just present those for plane waves i.e., $F(u,x^i) = f_{ij}(u)\,x^i\,x^j$;

$$\ddot{x}_i + f_{ij}(u)\,x^j = 0\ , \qquad\qquad 2\,\dot{v} = -f_{ij}(u)\,x^i\,x^j + \dot{x}^i\,\dot{x}^i\ . \tag{2}$$

Here we choose the affine parameter $\lambda = u$ for the null geodesics and $\dot{} \equiv \frac{d}{du}$. Apart from this family of null geodesics note that the null isometry of the geometry implies that $v = \lambda$ at constant (u, x^i) are also null geodesics.

From Eq. (2) we see that the geodesic motion in the transverse directions is just coupled time dependent harmonic motion. If the eigenvalues of the matrix $f_{ij}(u)$ are positive definite then the motion is oscillatory. For the null energy condition to hold we need $\mathrm{Tr}\, f \geq 0$. In this case the oscillatory nature of the geodesics predicates an important property of plane waves: Plane wave spacetimes that are sourced by matter satisfying the null energy condition are not globally hyperbolic [4]. This follows from the oscillations of the null geodesics because it is easy to show that there are congruences of null geodesics that converge to two different null geodesics in the geometry.

The maximally symmetric plane waves are the easiest to discuss causal structure since they can be conformally mapped to the Einstein Static Universe. Performing the requisite conformal transformation one sees for instance that the 10 dimensional maximally supersymmetric plane wave maps into $\mathbf{R} \times \mathbf{S}^9$ and the conformal boundary is read off from the conformal factor to be a one dimensional null curve [5]. To discuss the causal structure of other plane wave geometries requires more work since the trick of mapping them to the ESU is not practicable. In [6], the ideal point construction of [7] was used to show that the causal boundary of some plane waves is one-dimensional. This was later extended in [8] to pp-waves. The upshot is that for a large class of plane wave and pp-wave spacetimes it can be argued that the causal boundary is a one-dimensional null line.

3. Stable causality and discrete quotients

A very interesting property of plane waves is that by applying T-duality transformations to the geometry along a spacelike isometry one reproduces a Gödel-like spacetime, with CTCs through every point. It is easy to see that the pathology of CTCs isn't some mysterious property of string theoretic T-duality, but rather an artifact of quotienting by the particular spacelike isometry in question.

Plane waves are stably causal [9], which implies that not only are they free of causal pathologies such as CTCs, but also neighbourhood geometries related by small fluctuations are also CTC free. It is then perhaps a little surprising that certain spacelike quotients of them are not free of CTCs. The answer to the puzzle lies in realising that stable causality is not a guarantor of CTC freedom; one needs a further simple criterion to achieve this. The criterion [9] is that the spacelike isometry in question should be compatible with atleast some time function that one can define in the geometry (since by definition the time function is one whose gradient is everywhere timelike, it is clear that the existence of one automatically implies the existence of an infinite number). By compatible we mean that there exists some time function τ such that for the isometry ξ^a with $\xi^a \xi_a > 0$, one has $\mathcal{L}_\xi \tau = 0$.

Using the causal properties of plane waves it is easy to argue that the spacelike isometry used to T-dualize to Gödel-like spacetime is not compatible with any time function on the plane wave goemetry. In particular, it is possible to demonstrate that whenever a plane wave has ∂_u as a Killing vector, quotients by any isometry involving ∂_u always lead to spacetimes with closed timelike curves. This is simply related to the causal properties of plane wave geometries – for all plane waves satisfying the null energy condition, there always exist points $p_0 = (u_0, v_0, x_0^i)$ and $p_1 = (u_1, v_1, x_1^i)$ with $u_1 > u_0$ and any values of other coordinates such that p_1 is in the causal future of p_0. Hence when an isometry involves ∂_u one ends up identifying a points which are timelike seperated.

This remarkable phenomenon is peculiar to plane waves in the sense that it simply follows from the causal properties of the spacetime – plane waves simply have too large a light-cone. The situation for pp-waves is even more bizzare; many pp-waves (especially those that satisfy the energy conditions for the matter supporting them) are non-distinguishing. This means that there are distinct points in the spacetime which have identical causal pasts and/or futures.

4. Event horizons

Given the fact that plane waves and pp-waves are exact solutions to string theory world-sheet beta functions, one is led to ask whether among this class of spacetimes there is any chance of finding a black hole spacetime. By this we mean a spacetime with an event horizon which shields part of the spacetime from any asymptotic observer.

Conventionally, notions such as event horizons are applied only for spacetimes whose asymptopia are well-undestood; typically this implies flat space or AdS/dS type of asymptopia. In the case of plane waves we have seen that the asymptopia are quite unconventional. However, it is still possible to characterize what we mean by black holes in these spacetimes [10]. With an appropriate definition of event horizon it is possible to argue based on the causal relations in the geometry that spacetimes that have a covariantly constant null Killing vector cannot admit event horizons [10]. For plane waves this can intuitively be understood from the fact that a local limiting procedure such as the Penrose limit cannnot incorporate global concepts such as event horizons.

It is possible to provide black string geometries which have a null Killing vector [11]. These solutions are typically extremal black holes with a degenrate horizon. It is also possible to find non-extremal black holes which are asymptotically (in the metrical sense) maximally symmetric plane waves, by an algebraic solution generating technique [12].

5. Conclusions

Plane waves provide an exciting opportunity to understand non-trivial causal properties in the context of string theory. While they are extremely simple spacetimes from the point of view of the world-sheet string theory, they nonetheless possess very exotic and intricate causal properties.

We have seen that in many instances these classes of spacetimes possess a one-dimensional causal boundary. In addition, it is possible to exploit the causal relations in these spacetimes to generate upon quotienting spacetime with closed timelike curves. The properties of global structures such as event horizons, etc., have also provided some intriguing results.

To summarise, we have learnt a reasonably good amount of information regarding the global nature of plane wave from a general relativistic viewpoint.

Acknowledgments

I would like to thank the organisers of the QTS conference for a very simulating meeting. I would also like to thank my collaborators Eric Gimon, Akikazu Hashimoto, Oleg Lunin, Veronika Hubeny and Simon Ross for very fruitful collaborations. This work was supported by funds from the Berkeley Center for Theoretical Physics in addition to partial support from the DOE grant DE-AC03-76SF00098 and the NSF grant PHY-0098840.

References

1. R. Penrose, in *Differential geometry and relativity*, pp 271-275, Reidel, Dordrecht, 1976.
2. D. Berenstein, J. M. Maldacena and H. Nastase, *J. High. Energy. Phys.* **0204**, 013 (2002).
3. R. Bousso, *Rev. Mod. Phys.* **74**, 825 (2002).
4. R. Penrose,*Rev. Mod. Phys.* **37**, 215 (1965).
5. D. Berenstein and H. Nastase, [arXiv:hep-th/0205048].

6. D. Marolf and S. F. Ross, *Class. Quant. Grav.* **19**, 6289 (2002).
7. R. Geroch, E. H. Kronheimer, and R.Penrose, *Proc. Roy. Soc. Lond. A.* **327**, 545 (1972).
8. V. Hubeny and M. Rangamani, *J. High. Energy. Phys.* **0212**, 043 (2002).
9. V. Hubeny, M. Rangamani and S. F. Ross, *to appear in Phys. Rev. D.* [arXiv:hep-th/0307257].
10. V. Hubeny and M. Rangamani, *J. High. Energy. Phys.* **0211** 021 (2002).
11. V. Hubeny and M. Rangamani, *J. High. Energy. Phys.* **0301**, 031 (2003).
12. E. G. Gimon, A. Hashimoto, V. Hubeny, O. Lunin and M. Rangamani, *J. High. Energy. Phys.* **0308**, 035 (2003).

THE HIDDEN SYMMETRIES OF THE $ADS_5 \times S^5$ SUPERSTRING

RADU ROIBAN

Department of Physics
University of California, Santa Barbara, CA 93106
radu@vulcan2.physics.ucsb.edu

In this talk I describe the identification of an infinite number of nonlocal conserved charges in the world sheet theory of the Green-Schwarz superstring in $AdS_5 \times S^5$. These integrals of motion are similar to those appearing in well-known two-dimensional integrable field theories. This talk is based on work[1] done in collaboration with Iosif Bena and Joe Polchinski.

1. Introduction

The motivation to study string theory in $AdS_5 \times S^5$ comes largely from the AdS/CFT correspondence. This duality provides perhaps the first explicit realization of the old idea of 't Hooft that there exists a relation between gauge theories and closed string theories. In its original formulation it relates the strongly coupled N=4 SYM theory to string theory in $AdS_5 \times S^5$. To prove the correspondence one needs, however, a regime in which the gauge theory can be handled independently – the weak coupling regime. The AdS/CFT dictionary implies that one is then interested in the strong world sheet coupling limit of the string theory.

It is not immediately obvious that trading a strongly-coupled 4-dimensional theory for a strongly-coupled 2-dimensional one is the most profitable way to proceed. While it is true that there are various techniques to exactly solve 2-dimensional theories, the best known ones are based on the existence of holomorphic current algebras (e.g. WZW models) which do not exist for the superstring on $AdS_5 \times S^5$.

The existence of an infinite set of conserved quantities in a theory is usually a hallmark of its integrability, even if they do not form a current algebra. For example, many integrable 2-dimensional theories exhibit a Yangian algebra at the classical level. The natural expressions for its generators are nonlocal and certain combinations are also integrals of motion.

We will show that such a structure exists in the Green-Schwarz sigma model corresponding to strings in $AdS_5 \times S^5$.

2. Nonlocal charges: review of principal Chiral model

The idea of the construction is based on the existence of world sheet currents which, if treated as connections, have vanishing curvature

$$da + a \wedge a = 0 \ . \tag{1}$$

Whenever such a current exists, a Wilson line constructed out of it will be conserved provided that appropriate boundary conditions are imposed.

$$U(x,y) = Pe^{-\int_x^y d\sigma a_\sigma(z,t)} \Rightarrow \dot{U}(x,y) = a_\tau(x)U(x,y) - U(x,y)a_\tau(y) \tag{2}$$

Thus, if we take the line to be infinite and impose the standard boundary conditions at spatial infinity we find that $U(-\infty,\infty)$ is conserved. Alternatively, we can define U on a circle and impose periodic boundary conditions on fields in which case the trace of any power of U is conserved.

An infinite set of such flat currents are most easily packaged in a single current depending on a spectral parameter λ. Such families of currents exist in all cases of interest such as the principal chiral model (PCM). Let us now review some of the results of Ref. 2. The Lagrangian density is $L \propto Tr[J \wedge *J]$, where $J = g^{-1}dg$ are left-invariant currents of some group G. Using the flatness of $j = -gJg^{-1}$ as well as the fact that it is conserved it is easy to see that the connections

$$a^{\pm}(\lambda) = \alpha_{\pm}(\lambda)j + \beta(\lambda) * j \ ; \quad \alpha_{\pm}(\lambda) = \tfrac{1}{2}(1 \pm \cosh\lambda) \ , \ \beta(\lambda) = \tfrac{1}{2}\sinh\lambda \tag{3}$$

are flat for all λ. Thus, the construction described before applies and we find two generating functionals of conserved charges $U^{\pm}(\lambda)$. To understand their structure, it is instructive to expand U^- in the spectral parameter. The first nontrivial term is the Noether current of right-multiplication. The next term is the bilocal

$$Q_2 = \tfrac{1}{2}\int_{-\infty}^{\infty} d\sigma_1 \int_{-\infty}^{\sigma_1} d\sigma_2 j_\tau(\sigma_1)j_\tau(\sigma_2) - \tfrac{1}{4}\int_{-\infty}^{\infty} d\sigma j_\sigma(\sigma) \ . \tag{4}$$

Obviously, this structure extends to higher charges, Q_n being n-local.

This construction can be applied, with some modifications, to supersymmetric PCM as well as to sigma models on symmetric coset spaces as done, for example, in Ref. 3.

3. The Green-Schwarz superstring on $AdS_5 \times S^5$

The construction described before does not immediately apply for strings in $AdS_5 \times S^5$ because, even though $AdS_5 \times S^5$ is (part of) a coset superspace,

$$AdS_5 \times S^5 \subset \frac{PSU(2,2|4)}{SO(4,1) \times SO(5)} \ , \tag{5}$$

the world sheet theory is not a coset model. This is due to the existence of a Wess-Zumino term as well as to the Siegel symmetry, which prevents the space-time fermions from having standard kinetic terms.

An appropriate complex form of the maximally-extended 4-dimensional superconformal group, $PSL(4|4)$ possesses a $\mathbb{Z}_4$ automorphism which is useful in classifying the components of its left-invariant 1-form J:

$$\begin{array}{cccccccc} J = & H & + & Q_1 & + & P & + & Q_2 \\ \mathbb{Z}_4 & 0 & & i & & -1 & & -i \ . \end{array} \tag{6}$$

Here H is the left-invariant 1-form corresponding to the $SO(4,1) \times SO(5)$ subgroup. This automorphism determines the combinations of currents which can appear in the equations of motion, while the relative coefficients are fixed by the Siegel symmetry.

The construction of flat connections proceeds in two steps. First, to simplify the calculations, we remove the coset connection dependence by conjugating the coset currents:

$$p = gPg^{-1} \qquad q = g(Q_1 + Q_2)g^{-1} \qquad q' = g(Q_1 - Q_2)g^{-1} \tag{7}$$

The equations of motion expressed in terms of these quantities are:

$$\begin{aligned} d{*p} &= p \wedge {*q} + {*q} \wedge p + \tfrac{1}{2}(q \wedge q' + q' \wedge q) \\ 0 &= p \wedge ({*q} - q') + ({*q} - q') \wedge p \quad . \\ 0 &= p \wedge (q - {*q'}) + (q - {*q'}) \wedge p \end{aligned} \tag{8}$$

Second, we notice that, without fixing the Siegel symmetry, we cannot use the Hodge dual of q and q' since their differentials are not fixed by the equations of motion. Combining these two steps, the ansatz for the families of flat connections is:

$$a(\lambda) = \alpha(\lambda)p + \beta(\lambda) * p + \gamma(\lambda)q + \delta(\lambda) * q \ . \tag{9}$$

Requiring that $a(\lambda)$ is flat leads to six equations for the four undetermined coefficients. Three of them are, however, redundant and we find, as in the case of PCM, two 1-parameter families of solutions:

$$\alpha = -2\sinh^2\lambda \ ; \quad \beta = \mp 2 \sinh\lambda \cosh\lambda \ ; \quad \gamma = 1 \pm \cosh\lambda \ ; \quad \delta = \sinh\lambda \ .$$

One might think that fixing the Siegel symmetry will eliminate the redundancy among the constraints following from the flatness conditions. This is, however, not the case. We have found gauges which lead to four instead of six equations but this still leaves one redundant equation. Anyhow, using (9) in equation (2) leads to integrals of motion as previously discussed.

The fact that we constructed the currents (9) in a gauge-invariant theory leaves open the possibility that they are gauge artifacts. Furthermore, the integrals of motion constructed out of them do not commute with the Siegel symmetry generators. It is, however, easy to see that we did not find a complicated way of constructing zero. For example, by setting all space-time fermionic coordinates to zero we recover the flat currents for the bosonic sigma model on $AdS_5 \times S^5$ constructed in Ref. 4, as we should.

In an unitary gauge, as for example the light-cone gauge, the only remaining operators are the physical ones. The computations are quite intractable for strings in $AdS_5 \times S^5$. In the plane wave and flat space limits however, things simplify considerably and it is possible to show[5] that the integrals of motion (2) constructed with the currents (9) are nontrivial. It was also shown that the pure spinor formulation of strings in $AdS_5 \times S^5$ exhibits analogous conserved charges.

4. Questions and discussion

An obvious question is whether there are other integrals of motion. We searched[1] for chiral currents of standard type and found that they are rather trivial. It is possible that they have simple forms only in specific gauges. In this direction it is worth mentioning that there exists a Siegel symmetry gauge which leads to a stress tensor whose AdS_5 and S^5 parts are separately chiral:

$$\partial_- STr[P_+^a P_+^a] = 0 = \partial_- STr[P_+^{a'} P_+^{a'}] \tag{10}$$

where a and a' label AdS_5 and S^5, respectively. In the flat space limit this gauge corresponds to half of the fermions being chiral while the other half being anti-chiral.

The construction described above extends to other string theories of interest like, for example, strings in $AdS_3 \times S^3$. Analyzing this theory using the techniques advocated here might be a good warm-up exercise for using the nonlocal charges for strings in $AdS_5 \times S^5$. It is usually the case that they form the starting point of the inverse scattering method which is related in turn to the Bethe ansatz. In this direction, perhaps the most pressing question is finding the algebra of the generating function of charges.

In ordinary sigma models it is a Yangian algebra. Due to the second class constraints of the Green-Schwarz superstring it is not immediately clear that this structure is unaltered.

There currently seem to be two possible routes toward solving the world sheet theories. Both of them are successful when applied to massive theories. The case of conformal theories is, however, less clear.

– Conformal invariance can be broken by choosing a suitable integrable discretization of the theory[7]. The result would be some integrable spin chain which might be solved using the Algebraic Bethe Ansatz (ABA). In following this route important guidance is provided by the the classical r-matrix which can be thought of, in some sense, as the structure constant of the algebra of the generating function of nonlocal charges. The result of the ABA is the spectrum of the world sheet Hamiltonian which can be used to determine the spectrum of string states.

– According to Ref 8, it is possible to define an analog of the S-matrix even for massless (or conformal) field theories. As usual in exactly solvable theories, this S-matrix is a solution of the Yang-Baxter equation which, in some sense, can be thought of as defining the quantum version of the algebra of the generating function of nonlocal charges. As in the other approach, the classical generating function provides valuable information. The S-matrix is the starting point of the Thermodynamic Bethe Ansatz (TBA). While it is considerably less powerful than the ABA, it suffices to determine the central charge of the theory, as well as the energies of the ground states of certain sectors transforming nontrivially under some discrete subgroups of the 4-dimensional superconformal group. The result is therefore part of the spectrum of the superstring on $AdS_5 \times S^5$.

These issues are subject of ongoing research in collaboration with O. DeWolfe, J. Polchinski and A. Volovich.

From the perspective of the AdS/CFT correspondence an important question is finding the gauge theory duals of the nonlocal charges constructed here. This question was partly answered in Ref. 10 (see also L. Dolan's talk in this volume) where, under the assumption that the expected Yangian algebra is not modified by the presence of the second class constraints, it was shown that they are consistent with an integrable structure exhibited by the 1-loop dilatation operator of the $N = 4$ super-Yang-Mills theory. Other consequences of this extended symmetry algebra on the gauge theory side of the correspondence remain subject to ongoing research.

Acknowledgments

I would like to thank I. Bena and J. Polchinski for collaboration, O. DeWolfe, J. Polchinski and A. Volovich for collaboration on related subjects and the organizers of the QTS3 conference and especially Philip Argyres for the opportunity to present this talk. This work was supported in part by the National Science Foundation under Grant PHY99-07949 (AB) and PHY00-98395 as well as by the Department of Energy under Grant DE-FG02-91ER40618.

References

1. I. Bena, J. Polchinski and R. Roiban, arXiv:hep-th/0305116.
2. M. Luscher and K. Pohlmeyer, *Nucl. Phys.* **B137**, 46 (1978).
3. T. L. Curtright and C. K. Zachos, *Phys. Rev.* **D21**, 411 (1980).
4. G. Mandal, N. V. Suryanarayana and S. R. Wadia, *Phys. Lett.* **B543**, 81 (2002)
5. I. Bena and R. Roiban, unpublished
6. B. C. Vallilo, arXiv:hep-th/0307018.
7. L. D. Faddeev and N. Y. Reshetikhin, *Annals Phys.* **167**, 227 (1986).
8. A. B. Zamolodchikov and A. B. Zamolodchikov, *Nucl. Phys.* **B379**, 602 (1992).
9. P. Fendley, *Nucl. Phys.* **B374**, 667 (1992) [arXiv:hep-th/9109021].
10. L. Dolan, C. R. Nappi and E. Witten, *JHEP* **0310**, 017 (2003)

MATRIX MODEL THERMODYNAMICS

GORDON W. SEMENOFF

Department of Physics and Astronomy, University of British Columbia, Vancouver, British Columbia, Canada V6T 1Z1 Email: semenoff@nbi.dk

We discuss a Hagedorn-like phase transition which can occur in the large N limit of matrix quantum mechanics. We apply the idea to the matrix model of M-theory on a pp-wave background where we present a series of speculative arguments that the phase transition is the Hagedorn transition of little string theory in a certain limit of the 5-brane geometry.

1. Motivation

The BFSS [1] matrix model of M-theory is supersymmetric matrix quantum mechanics with action

$$S = \int dt \mathrm{Tr} \left[\frac{1}{2R}(DX^i)^2 + \frac{R}{4\ell_{pl}^6}[X^i, X^j]^2 + \bar{\psi} D\psi + \frac{R}{\ell_{pl}^3}\bar{\psi}\Gamma^i[X^i, \psi] \right] \quad (1)$$

where $i, j = 1, ..., 9$ and all degrees of freedom are $N \times N$ Hermitian matrices. This is a gauge invariant theory with covariant time derivative $D = \frac{d}{dt} - i[A, ...$. The gauge theory coupling constant is given in terms of the null compactification radius R and the eleven dimensional Planck length ℓ_{pl} (or the ten dimensional IIA string coupling g_s and string length $\ell_s = \sqrt{\alpha'}$) by

$$g_{YM}^2 = \left(\frac{R}{\ell_{pl}^2} \right)^3 = g_s \ell_s^3 \ , \ \ell_{pl} = g_s^{1/3} \ell_s \ , \ R = g_s \ell_s \quad (2)$$

This model has three uses. It is conjectured to describe a discrete light-cone quantization of M-theory[1] where R is the compactification radius of the light-cone, there are N units of light-cone momentum and ℓ_p is the Planck length of 11-dimensional supergravity. Secondly, and historically a little earlier[2], with parameters suitably re-identified, it describes the low energy dynamics of a collection of N D0-branes of type IIA superstring theory. Finally, it is a matrix regularization of the light-cone action for the 11-dimensional supermembrane[3].

One motivation for understanding the behavior of matrix models such as the BFSS model at finite temperature comes from the conjecture that their finite temperature states are related to black hole states of type IIA supergravity. This idea was studied in a series of papers by Kabat and Lowe[4]. They begin with the Beckenstein-Hawking entropy of a black DO-brane solution of IIA supergravity – the area of its even horizon in Planck units. They convert the entropy to the free energy, which they then write in terms of gauge theory parameters to obtain

$$F/T = -2.52N^2 \left(\frac{T^3}{g_{YM}^2 N} \right)^{3/5} \tag{3}$$

A derivation of (3) from the matrix model would be an important result, a first principles computation of non-extremal black hole entropy using string theory. However, one would expect to find (3) in a low temperature and therefore strong coupling limit of the matrix model, making it inaccessible to perturbation theory. A variational technique was applied and claimed agreement with the formula over some range of temperature[4].

The formula (3) is remarkable in three respects. First, it has the correct dependence on N and the 'tHooft coupling $g_{YM}^2 N$ to be the leading order of the 'tHooft limit of the gauge theory. If it could be derived in a perturbative expansion, it would obtain contributions only from planar Feynman diagrams. This means that the 'tHooft limit should be part of the limiting process that would extract the classical physics of black holes from the matrix model.

Secondly, the scaling with N in (3) is as if this 0-dimensional gauge theory were in a de-confined phase. This is particularly true if we assume that we are taking the 'tHooft limit. The use of the word "deconfined" in a theory where there is no spatial extent over which particles can be separated must be justified carefully. The gauge theory has a Gauss law constraint so that quantum states of the Hamiltonian must be singlets under the gauge symmetry. The gauge field in 1) enforces this constraint. The number of singlets at a given energy do not scale like N^2, rather they are of order one. As an example of this, consider a matrix harmonic oscillator with Hamiltonian

$$H = \sum_{i=1}^{d} \omega \mathrm{Tr}(\alpha_i^\dagger \alpha_i) \tag{4}$$

and creation and annihilation operator algebra

$$\left[\alpha^i_{ab}, \alpha^{j\dagger}_{cd}\right] = \delta^{ij}\delta_{ad}\delta_{bc} \tag{5}$$

States are created by $\alpha^\dagger_{ab}$ operating on a vacuum $|0>$. To get a state with energy $E = n\omega$ we must act with n creation operators.

The analog of gauge invariance is to require a physical state condition of invariance under the unitary transformation

$$\alpha^\dagger \to u\alpha^\dagger u^\dagger$$

where

$$u \in U(N)/U(1)$$

Then, physical states are created by operating with invariant combinations of creation operators. In the limit $N \to \infty$ all such combinations are traces

$$\left[\mathrm{Tr}\left(\alpha^\dagger_i\right)\right]^{n_1}\left[\mathrm{Tr}\left(\alpha^\dagger_{j_1}\alpha^\dagger_{j_2}\right)\right]^{n_2}\left[\mathrm{Tr}\left(\alpha^\dagger_{k_1}\alpha^\dagger_{k_2}\alpha^\dagger_{k_3}\right)\right]^{n_3}\ldots|0> \tag{6}$$

where the energy is

$$E = \omega(n_1 + 2n_2 + 3n_3 + \ldots)$$

The number of these traces with a fixed energy, E, does not scale like N^2 as $N \to \infty$, instead it approaches a constant as N is taken large. Thus for normal thermodynamic states, one would not expect the free energy to be of order N^2.

However, the number of independent traces does increase rapidly with E. It has been shown[5] that, in the large N limit, the oscillator has a hagedorn-like density of states at high energy, $\rho(E) \sim \frac{1}{E}e^{E/T_H}$ where the Hagedorn temperature is $T_H = \omega/\ln d$. At temperatures higher than T_H, the thermodynamic canonical ensemble does not exist. It could be made to exist by keeping N large but finite. That would cut off the exponential growth in the asymptotic density of states at some large energy – where ρ reached the the maximum number of independent traces - N. Then we could consider a temperature that is greater than T_H. Both the energy and entropy would be dominated by states at and above the cutoff scale, and the free energy would be negative and would be of order N^2. Then, if we take the limit $N \to \infty$ at constant temperature (noting that the Hagedorn temperature does not depend on N), the free energy would become very large, diverging like N^2. As we shall show in the following, a behavior like this can indeed by found in the matrix model. In more conventional terms, it occurs as a large N Gross-Witten type of phase transition[6,7] which is familiar in unitary matrix models. It is this behavior that we call "deconfinement". It does have the behavior of a finite temperature deconfinement transition in gauge theory in the fact that the free energy jumps from being of order one to being of order N^2.

In an adjoint gauge theory such as (1), there is an order parameter for confinement, the Polyakov loop[8,9]. It occurs when the gauge theory is path integral quantized, there it is the trace of the holonomy of the gauge field around the finite temperature Euclidean time circle,

$$P = \frac{1}{N}\mathrm{Tr}\left(e^{i\oint A}\right) \tag{7}$$

This operator gets a nonzero expectation value when a gauge theory is deconfined. An interesting question is whether it has a nonzero expectation value in the BFSS matrix model. In such a low dimensional theory, it can only have a nonzero expectation value when N is infinite. Indications from weak coupling computations[10] are that it has.

The third remarkable fact about (3) is, though this formula is thought to apply to the black hole only for a range of temperatures[11], the expression at low temperature is reminiscent of critical scaling with a critical temperature $T = 0$ and simple, rational critical exponent.

In spite of the simplicity of these interesting features, there is no analytic derivation of the formula (3) from the matrix model. One of the difficulties in finding a derivation is the intractability of the model itself. These difficulties are well-known from previous attempts to analyze its thermodynamics[10].

2. PP-Wave matrix model

Recently a variant of the matrix model which is conjectured to describe a discrete light cone quantization of M-theory on a pp-wave background has been formulated[12]. It is a 1-parameter deformation of (1),

$$\begin{aligned} S = \int dt\mathrm{Tr}\Bigg[&\frac{1}{2R}(DX^i)^2 + \frac{R}{4\ell_{pl}^6}[X^i, X^j]^2 + \bar{\psi}D\psi + \frac{R}{\ell_{pl}^3}\bar{\psi}\Gamma^i[X^i, \psi] \\ &- \frac{\mu^2}{18R}(X^a)^2 - \frac{\mu^2}{72R}(X^{i'})^2 - \frac{\mu}{4}\bar{\psi}\psi - i\mu\epsilon^{abc}X^aX^bX^c\Bigg] \end{aligned} \tag{8}$$

where the indices $a, b, .. = 1, 2, 3$ and $i' = 4, ..., 9$. The M-theory background is

$$ds^2 = dx^i dx^i - 2dx^+ dx^- - \left(\frac{\mu^2}{9}(x^a)^2 + \frac{\mu^2}{36}(x^{i'})^2\right) dx^+ dx^+$$

$$F_{+123} = \mu$$

This background is a maximally symmetric solution of 11-dimensional supergravity. It is known to support a spherical membrane solution,

$$x^+ = p^+ \tau \ , \ \sqrt{(x^a)^2} = \frac{1}{6} p^+ \mu$$

and a spherical transverse 5-brane

$$x^+ = p^+ \tau \ , \ (x^{i'})^2 = \sqrt{\frac{\mu p^+}{6}}$$

These objects are conjectured to make their appearance in the solutions of the matrix model. The membrane is found immediately in semiclassical quantization. The classical bosonic potential is

$$V = \frac{R}{2} \mathrm{Tr} \left[\left(\frac{\mu}{3R} X^a + i\epsilon^{abc} X^b X^c \right)^2 + \frac{1}{2} \left(i[X^{i'}, X^{j'}] \right)^2 + \right. \\ \left. + \left(i[X^{i'}, X^a] \right)^2 + \left(\frac{\mu}{6R} \right)^2 (X^{i'})^2 \right] \qquad (9)$$

It is minimized by

$$X^{i'}_{\mathrm{cl}} = 0 \ , \ X^a_{\mathrm{cl}} = \frac{\mu}{3R} J^a$$

where J^a is an N-dimensional representation of the SU(2) algebra, $[J^a, J^b] = i\epsilon^{abc} J^c$. The configurations are fuzzy spheres, which are matrix regularizations of the membranes. The 5-branes on the other hand do not seem to appear in the perturbative states of the matrix model. It has been conjectured[13] that the 5-branes indeed appear as the large N limit is taken in a certain way. First of all, J^a need not be an irreducible representation, but can contain a number of irreducible components,

$$J^a = \begin{pmatrix} s_1^a & 0 & 0 & 0 \\ 0 & s_2^a & 0 & 0 \\ 0 & 0 & s_3^a & 0 \\ 0 & 0 & 0 & s_4^a \end{pmatrix}$$

To get a (multi-)membrane state, the large N limit is taken by holding the number of representations fixed and sending the dimension of each of the representations to infinity. To get a five-brane state on the other hand, we hold the dimensions of the representations fixed and repeat them an infinite number of times to get the large N limit. An important difference between these limits is in the realization of the gauge symmetry. In the classical sectors, the gauge symmetry is partially realized in a Higgs phase, with the residual symmetry being that which interchanged representations of the same size. In a membrane state the residual gauge group thus has finite rank, whereas in a 5-brane state its rank always goes to infinity. The single 5-brane state is $X^a = 0$ whereas the state with k coincident 5-branes has the k-dimensional representation repeated $N/k \to \infty$ times. A state with k non-coincident 5-branes has largest representation k-dimensional and a number of smaller representations all repeated an infninite number of times.

The effective coupling constant which governs a semi-classical expansion about one of the classical ground states is

$$\lambda = \left(\frac{3R}{\mu}\right)^3 n$$

where n is the rank of the residual gauge group. The 5-brane limit is where we are required to take the weakest coupling limit. It is the conjecture of ref.[13] that the membranes and all other degrees of freedom decouple in this limit and it isolates the internal dynamics of the 5-brane.

3. Perturbative expansion

Now, let us consider a perturbative expansion of the pp-wave matrix model at finite temperature. The partition function is the path integral

$$Z = \int [dA][dX^i][d\psi] e^{-\int_0^\beta d\tau L[A,X^i,\psi]}$$

where L is the lagrangian with Euclidean time and $\beta = 1/T$ is the inverse temperature. The bosonic and fermionic variables have periodic and antiperiodic boundary conditions, respectively

$$A(\tau+\beta) = A(\tau) \quad , \quad X^i(\tau+\beta) = X^i(\tau) \quad , \quad \psi(\tau+\beta) = -\psi(\tau)$$

We must begin by fixing the gauge. The most convenient gauge fixing is to make the gauge field static and diagonal,

$$\frac{d}{d\tau}A_{ab} = 0 \quad , \quad A_{ab} = A_a \delta_{ab}$$

The Faddeev-Popov determinant for this gauge fixing is

$$\prod_{a\neq b} |A_a - A_b| \cdot \prod_{ab} \det{}' \left(\frac{d}{d\tau} \left(\frac{d}{d\tau} - i(A_a - A_b) \right) \right) = \prod_{a \neq b} \sin \frac{\beta}{2} |A_a - A_b|$$

where the prime on the determinant indicates that the static mode is omitted.

If we expand about the classical vacuum corresponding to the single 5-brane, $X^a_{\rm cl} = 0 = X^i_{\rm cl}$, we find the partition function in the 1-loop approximation is

$$Z = \int dA_a \prod_{a\neq b} \frac{\sin \frac{\beta}{2}|A_a - A_b| \det{}^8 \left(iD_{ab} - \frac{\mu}{4}\right)}{\det{}^{3/2}\left(-D^2_{ab} + \frac{\mu^2}{9}\right) \det{}^3 \left(-D^2_{ab} + \frac{\mu^2}{36}\right)} \tag{10}$$

where $D_{ab} = \frac{d}{d\tau} - i(A_a - A_b)$. Using the formula

$$\det(id/dt + \omega) = \sin \beta\omega/2$$

with periodic boundary conditions and

$$\det(id/dt + \omega) = \cos \beta\omega/2$$

with antiperiodic boundary conditions, we can write

$$Z = \int dA_a \prod_{a\neq b} \frac{[1 - e^{i\beta(A_a - A_b)}][1 + e^{-\beta\mu/4 + i\beta(A_a - A_b)}]^8}{[1 - e^{-\beta\mu/3 + i\beta(A_a - A_b)}]^3 [1 - e^{-\beta\mu/6 + i\beta(A_a - A_b)}]^6} \tag{11}$$

When N is finite, it is straightforward to do this integral using the method of residues. However, to apply to the 5-brane, we require the integral when $N \to \infty$. There are N integration variables A_a and the action, which is the logarithm of the integrand is generically of order N^2 which is large in the large N limit. For this reason, the integral can be done by saddle point integration. This amounts to finding the configuration of the variables A_a which minimize the effective action:

$$\begin{aligned} S_{\rm eff} = \sum_{a\neq b} \Big(&- \ln[1 - e^{i\beta(A_a - A_b)}] - 8\ln[1 + e^{-\beta\mu/4 + i\beta(A_a - A_b)}] + \\ &+ 3\ln[1 - e^{-\beta\mu/3 + i\beta(A_a - A_b)}] + 6\ln[1 - e^{-\beta\mu/6 + i\beta(A_a - A_b)}]\Big) \end{aligned} \tag{12}$$

To study the behavior, it is illuminating to Taylor expand the logarithms in the phases (this requires some assumptions of convergence for the first log)

$$S_{\rm eff} = \sum_{n=1}^{\infty} \frac{1 - 8(-)^{n+1} e^{-n\beta\mu/4} - 3e^{-n\beta\mu/3} - 6e^{-n\beta\mu/6}}{n} |{\rm Tr} e^{in\beta A}|^2 \tag{13}$$

Each term contains the modulus squared of a (multiply would) Polyakov loop operator (7). When a coefficient becomes negative, the loop wants to condense. In fact, as we raise the temperature from zero (and lower β from infinity), the first mode to condense is $n = 1$. This occurs when $T_C = \frac{\mu}{12} \ln 3 \approx .0758533\mu$.

4. A closer look at the phase transition

At temperatures greater than T_C the eigenvalues A_a distribute themselves so that they are clustered near a particular point on the unit circle. To examine the possibility, we consider the equation of motion for the eigenvalues,

$$P\omega(z)+4\omega(-r^3z)+4\omega(-r^{-3}z)=\frac{3}{2}\omega(r^4z)+\frac{3}{2}\omega(r^{-4}z)+ \\ +3\omega(r^2z)+3\omega(r^{-2}z) \qquad (14)$$

where $r=e^{\beta\mu/12}$ and

$$\omega(z)=\frac{1}{N}\sum_{a=1}^{N}\frac{z+e^{i\beta A_a}}{z-e^{i\beta A_a}} \qquad (15)$$

$\omega(z)$ is holomorphic for z away from the unit circle and has asymptotic behavior, $\omega(\infty)=1$ and $\omega(0)\sim -1$. In the large N limit the poles in $\omega(z)$, which occur at the location of the eigenvalues, coalesce to form a cut singularity on a part or perhaps all of the unit circle. $\omega(z)$ remains holomorphic everywhere away from the unit circle.

One subtlety to remember is that equation (14) is valid only when z is one of the eigenvalues. In that case, the sum over eigenvalues in (15) which turns into an integral in the large N limit, must be defined as a principal value. It is straightforward to get this principal value by averaging over approaching the unit circle from the inside and from the outside,

$$P\omega(z)=\frac{1}{2}\lim_{\epsilon\to 0}\left(\omega(e^{\epsilon}z)+\omega(e^{-\epsilon z})\right) \qquad |z|=1\ ,\ z\in \text{cut} \qquad (16)$$

We will take this as the definition of $\omega(z)$ in (14).

It is easy to find one exact solution of (14). If we consider the case where the eigenvalues are uniformly distributed over the unit circle, so that the sum over eigenvalues in (15) is symmetric under $z\to e^{i\theta}z$. In that case, we can average over the symmetry orbit to get

$$\omega(z)=\begin{cases}1 & |z|>1\\ -1 & |z|<1\end{cases} \qquad (17)$$

The result is discontinuous on the entire unit circle. Using the principal value prescription (16) defines its value on the unit circle $\omega(z)=0$ when $|z|=1$.

The resolvent (17) is always a solution of (14) for any value of r. This is the symmetric solution, where the Polyakov loop operator, (whose expectation value is the constant moment of $\omega(z)$ for large z) vanishes. This symmetric solution is only stable if the temperature is low enough. At the critical temperature it becomes an unstable solution and there should be other solutions which have lower free energy.

To examine the equation for the eigenvalues, we consider r large but not infinite. The asymptotic expansions of the resolvent are

$$\omega(z) = 1 + 2\sum_{n=1}^{\infty} \frac{1}{z^n} \langle \eta^n \rangle \tag{18}$$

$$\omega(z) = -1 - 2\sum_{n=1}^{\infty} z^n \left\langle \frac{1}{\eta^n} \right\rangle \tag{19}$$

where $\langle \eta^n \rangle$ are moments of the eigenvalue distribution:

$$\langle \eta^k \rangle \equiv \frac{1}{N} \sum_{a=1}^{N} \eta_a^k$$

If we assume that the distribution is even $< \eta^k > = < \eta^{-k} >$ (because of spontaneous breaking of the phase symmetry such an alignment of the solution always exists), the large r expansion of the equation is

$$P\omega(z) = \left(\frac{6}{r^2} + \frac{3}{r^4} + \frac{8}{r^3}\right)\left(\frac{1}{z} - z\right)\langle \eta \rangle + \frac{6}{r^4}\left(\frac{1}{z^2} - z^2\right)\langle \eta^2 \rangle + \ldots \tag{20}$$

This is similar to the equation for eignevalue distribution of adjoint unitary matrix models[7,14,15]. Remember that this equation is valid only when z is inside the cut discontinuity of $\omega(z)$ which is assumed to occur on a segment of the unit circle (placed symmetrically about $z = 1$).

For example, the semi-circle distribution of Gross and Witten[6] has a cut singularity on the unit circle between branch points $z_{\pm} = -t \pm i\sqrt{1-t^2}$ and

$$\omega(z) = \frac{1}{1+t}\left(\frac{1}{z} - z\right) - \frac{1}{(1+t)}\left(1 + \frac{1}{z}\right)\sqrt{1 + 2tz + z^2} \tag{21}$$

This solution exists for $-1 < t < 1$. When $t \to 1$ then endpoints of the cut touch each other and the cut covers the whole unit circle. This is the where the Gross-Witten phase transition occurs in their unitary matrix model.

The principal value prescription averages the square root to zero along its cut singularity and therefore, for z in the cut, $P\omega(z) = \frac{1}{1+t}\left(\frac{1}{z} - z\right)$. If, for the moment, we ignore the quadratic terms on the right-hand-side of (20), we see that the equation is solved by the semi-circle distribution when

$$\frac{1}{1+t} = \left(8r^{-3} + 3r^{-4} + 6r^{-2}\right)\frac{3-t}{4}$$

Also, remembering that $t < 1$, we get the critical value of r by setting $t = 1$, so this equation can be solved only when $r < r_{\text{crit}}$ where r_{crit} occurs at the same critical temperature as the zero of the $n = 1$ term in the expansion on the right-hand-side of (13). Thus, we see that the semicircle distribution is an approximate solution of the model for temperatures just above the critical temperature (where r is large but $r \leq r_{\text{crit}} = 3$ (not really large, but still larger than one).

To include the quadratic term on the right-hend-side of (20), we can modify the semicircle distribution as:

$$\omega(z) = \frac{1}{(1+t)(1+a/2-3at/2)} \left\{ \left(\frac{1}{z} - z\right) + a\left(\frac{1}{z^2} - z^2\right) + \right.$$
$$\left. + \left((1-at)\left(1+\frac{1}{z}\right) + a\left(z + \frac{1}{z^2}\right)\right)\sqrt{1+2tz+z^2} \right\} \quad (22)$$

where a is assumed to be a small number which controls the deviation from a semi-circle distribution. We have added the correction in such a way as to preserve the property $\omega(1/z) = -\omega(z)$ (where the sign of the square root is changed appropriately) and the asymptotic limits. Plugging this into (14) and asymptotically expanding for large r yields

$$\frac{1}{(1+t)(1+a/2-3at/2)}\left(\left(\frac{1}{z} - z\right) + a\left(\frac{1}{z^2} - z^2\right)\right) =$$
$$\frac{3-t+2a(1-t)^2}{4\,(1+a/2-3at/2)}\left(\frac{1}{z} - 1\right)\left(\frac{8}{r^3} + \frac{6}{r^2} + \frac{3}{r^4}\right) +$$
$$+\frac{(1-t)^2 - \frac{a}{4}\left(9t^3 - 13t^2 + 3t - 7\right)}{4(1+a/2-3at/2)}\left(\frac{1}{z^2} - z^2\right)\frac{6}{r^4} + \text{ order }\left(r^{-6}\right) \quad (23)$$

We will solve this equation nfinitesimally close to the phase transition, where $t = 1 - 2\epsilon$, Equating coefficients, we find

$$\frac{8}{r^3} + \frac{6}{r^2} + \frac{3}{r^4} = 1 + \epsilon^2 \quad (24)$$

and $a \sim \epsilon^2$. This demonstrates that, close to the phase transition, the semi-circle distribution gives an accurate description of the de-confined phase. It would be interesting to explore this further to determine precise thermo-dynamic properties of that phase.

5. A stack of 5-branes

The fluctuation spectrum of the matrices about a stack of k 5-branes is known. If, rather than the trivial vacuum, we had chosen the one where the

k-dimensional representation of SU(2) is repeated N/k times, the residual gauge invariance would be $SU(N/k)$. The gauge field would have a classical solution with $k \times k$ unit matrices times gauge fields A_a representing the whole blocks. The spectrum is known[13] and we can again get an estimate of the Hagedorn temperature[5]

$$T_H(k = \infty) = .073815...$$

We see that the temperature is reduced only slightly in this case. We take this as evidence that the Hagedorn temperature in this weak coupling limit is insensitive to th enumber of 5-branes. This seemingly contradicts the k-dependence of the Hagedorn phase transition of little string theory which has been computed using holography and which behaves like $T_C \sim \sqrt{k}$.

An explanation for this contradiction can be found in the limit that we are using. It is a large N 'tHooft limit and a further expansion in weak 'tHooft coupling. In this limit, if we were to translate the parameters of our model to those that would describe the NS five-brane in type II string theory, the radius of the spherical five-brane would be

$$\frac{r^2}{\alpha'} \sim \lambda^{1/4}$$

This means that we are expanding about a small, highly curved five-brane, whereas the usual holographic result for the Hagedorn temperature is for a large flat 5-brane. This is a similar difficulty as the one which appears in the AdS/CFT correspondence in general. There, the analog of the matrix model, which is maximally supersymmetric Yang-Mills theory, can be readily analyzed only in the weak coupling limit using an asymptotic expansion in λ. Supergravity and holography give tools which compute the strong coupling limit, where λ is large. Thermodynamic quantities like the free energy in particular were computed in both theories and they do not agree with each other for this reason.

6. Discussion

The phase transition in the matrix model is a peculiar one. Closer analysis reveals that it is of first order, but there is no co-existence region where one phase is meta-stable and the other is stable, as is normally the case for first order phase transitions. There is a question as to whether this is an artifact of the approximation that we have done here, i.e. if we expanded the effective action for eigenvalues to higher loop order the phase transition might be more conventional.

Indeed, in other unitary matrix models applied to non-Abelian Coulomb gases[14,15], where the eigenvalues live on a higher dimensional space (a D=1 or even D>1 unitary matrix model) there is a coexistence region.

A similar problem afflicts weakly coupled four-dimensional supersymmetric Yang-Mills theory[16]. In that case, Witten argued using AdS/CFT that, in the strong coupling limit of planar Yang-Mills theory, there should be a de-confining phase transition, identified with the Hawking-Page phase transition of supergravity on asymptotically anti-de Sitter space[17]. Hawking-Page is a normal first order phase transition where there is the possibility of metastable phases. At weak coupling, the analysis looks very similar to what we have done here for the matrix model and the phase transition has the same nature. Aharony et.al. have conjectured that the effect of higher loop corrections (3-loops) to the effective action in Yang-Mills theory would indeed change the phase transition to a more conventional one.

Finally, in the matrix model that we have analyzed, there is the question of whether the phase transition that we have found has anything to do with collapse to black holes. The subject of black holes on pp-wave backgrounds is a murky one which is presently being sorted out. In any case, the classical black hole should appear at a strong coupling limit, rather than the weak coupling limit that we have analyzed. It is tempting to conjecture that, if there were black holes, collapse to black holes at finite temperature is what our phase transition would describe if it persists at strong coupling.

Of course of the conjectured relationship with the 5-brane is valid, then, indirectly, the Hagedorn phase transition of little string theory is related to horizon formation in the 5-brane geometry. In that case, our phase transition and its thermodynamics indeed describes the black hole, again in a limit which is far away from previous analysis of such objects.

Acknowledgments

The work reviewed in this paper was done in collaboration with Ehud Schreiber and Kazayuki Furuuchi. I have benefitted from conversations with Shiraz Minwalla and Mark Van Raamsdonk. Early work on this project was done in collaboration with Jan Ambjorn and Yuri Makeenko.

References

1. T. Banks, W. Fischler, S. H. Shenker and L. Susskind, "M theory as a matrix model: A conjecture," Phys. Rev. D **55**, 5112 (1997) [arXiv:hep-th/9610043].

2. E. Witten, Nucl. Phys. B **443**, 85 (1995) [arXiv:hep-th/9503124].
3. For a review, see W. Taylor, "M(atrix) theory: Matrix quantum mechanics as a fundamental theory," Rev. Mod. Phys. **73**, 419 (2001) [arXiv:hep-th/0101126].
4. D. Kabat, G. Lifschytz and D. A. Lowe, "Black hole thermodynamics from calculations in strongly coupled gauge theory," Phys. Rev. Lett. **86**, 1426 (2001) [Int. J. Mod. Phys. A **16**, 856 (2001)] [arXiv:hep-th/0007051].
5. K. Furuuchi, E. Schreiber and G. W. Semenoff, "Five-brane thermodynamics from the matrix model," arXiv:hep-th/0310286.
6. D. J. Gross and E. Witten, "Superstring Modifications Of Einstein's Equations," Nucl. Phys. B **277**, 1 (1986).
7. For a review, see G. W. Semenoff and R. J. Szabo, "Fermionic Matrix Models," Int. J. Mod. Phys. A **12**, 2135 (1997) [arXiv:hep-th/9605140].
8. A. M. Polyakov, "Thermal Properties Of Gauge Fields And Quark Liberation," Phys. Lett. B **72**, 477 (1978).
9. For some examples of computations of the Polyakov loop operator in other lower dimensional gauge theories, see: G. Grignani, G. W. Semenoff, P. Sodano and O. Tirkkonen, "Charge Screening and Confinement in the Hot 3D-QED," Nucl. Phys. B **473**, 143 (1996) [arXiv:hep-th/9512048]; G. Grignani, G. W. Semenoff, P. Sodano and O. Tirkkonen, "Charge Screening in the Finite Temperature Schwinger Model," Int. J. Mod. Phys. A **11**, 4103 (1996) [arXiv:hep-th/9511110]; G. Grignani, G. W. Semenoff and P. Sodano, "Confinement - deconfinement transition in three-dimensional QED," Phys. Rev. D **53**, 7157 (1996) [arXiv:hep-th/9504105].
10. J. Ambjorn, Y. M. Makeenko and G. W. Semenoff, "Thermodynamics of D0-branes in matrix theory," Phys. Lett. B **445**, 307 (1999) [arXiv:hep-th/9810170].
11. N. Itzhaki, J. M. Maldacena, J. Sonnenschein and S. Yankielowicz, "Supergravity and the large N limit of theories with sixteen supercharges," Phys. Rev. D **58**, 046004 (1998) [arXiv:hep-th/9802042].
12. D. Berenstein, J. M. Maldacena and H. Nastase, "Strings in flat space and pp waves from N = 4 super Yang Mills," JHEP **0204**, 013 (2002) [arXiv:hep-th/0202021].
13. J. Maldacena, M. M. Sheikh-Jabbari and M. Van Raamsdonk, "Transverse fivebranes in matrix theory," JHEP **0301**, 038 (2003) [arXiv:hep-th/0211139].
14. G. W. Semenoff and K. Zarembo, "Adjoint non-Abelian Coulomb gas at large N," Nucl. Phys. B **480**, 317 (1996) [arXiv:hep-th/9606117].
15. G. W. Semenoff, O. Tirkkonen and K. Zarembo, "Exact solution of the one-dimensional non-Abelian Coulomb gas at large N," Phys. Rev. Lett. **77**, 2174 (1996) [arXiv:hep-th/9605172].
16. O. Aharony, J. Marsano, S. Minwalla, K. Papadodimas and M. Van Raamsdonk, "The Hagedorn / deconfinement phase transition in weakly coupled large N gauge theories," arXiv:hep-th/0310285.
17. E. Witten, "Anti-de Sitter space, thermal phase transition, and confinement in gauge theories," Adv. Theor. Math. Phys. **2**, 505 (1998) [arXiv:hep-th/9803131].

MATHEMATICAL ASPECTS OF D-BRANES

ERIC SHARPE

Physics Department
University of Illinois
Urbana, IL 61801
E-mail: ersharpe@uiuc.edu

In this lecture we review recent work on describing D-branes with nonzero Higgs vevs in terms of sheaves, which gives a physical on-shell D-brane interpretation to more sheaves than previously understood as describing D-branes. The mathematical ansatz for this encoding is checked by comparing open string spectra between D-branes with nonzero Higgs vevs to Ext groups between the corresponding sheaves. We illustrate the general methods with a few examples.

1. Introduction

Several years ago, Kontsevich [1] proposed a formulation of mirror symmetry in terms of derived categories. Motivated by Kontsevich's work together with physical pictures of tachyon condensation [2] and K-theory [3], it was originally proposed in [4] that Kontsevich's proposal could be physically realized via off-shell states in the open string B model, in which objects in the derived category (represented by complexes of locally-free sheaves) are represented by D-brane/anti-D-brane configurations, with maps between sheaves represented by tachyons, and localization on quasi-isomorphisms realized by renormalization group flow. Some evidence was given for this proposal by considering T-duality arguments.

This proposal lay dormant until the work of [5,6], which popularized these ideas, and also introduced the notion of "π-stability."

However, to this day, there are still many open problems in identifying (off-shell) open string B model boundary states with objects in the derived category of coherent sheaves.

The most basic problem is that of confirming that the mathematical operation of localizing on quasi-isomorphisms really is realized physically via renormalization-group flow. Since in general cases it is all but impossible to see the entire RG flow directly, one must perform indirect tests; however, until the work of [7,8,9] (see [10,11] for a review) no technology existed for confirming any of the predictions of the derived category program directly in BCFT. Interesting subtleties were also uncovered in [7,8,9]; for example, the Freed-Witten anomaly [12] complicates the dictionary between D-branes and sheaves, and plays a vital role in deriving Ext groups as open string states.

A second issue in the derived categories program has been known for almost eight years, even predating [4]. This issue is the face that the GSO projection tells us that degrees of Ext group elements cannot match $U(1)_R$ charges in general, except in the special case of open strings connected D-branes to themselves, contradicting naive assumptions. This mismatch is manifest in BCFT but very obscure in massive nonconformal models, and as such mismatches can have drastic implications for RG flow arguments, not only is understanding it important, but its existence underlines the importance of checks of RG flow arguments, of the form just mentioned.

Third, the open string analogue of the B model Calabi-Yau condition places a constraint on the possible morphisms of a derived category that can be realized physically by open strings. Although it appears that all objects of a derived category can be represented by off-shell open string states, it seems that not all the morphisms in a derived category can be represented physically.

Finally, we now believe we understand that sheaves of the form $i_* \mathcal{E}$, where i is an inclusion, model on-shell D-branes, but do other coherent sheaves model on-shell D-branes, and if so, which on-shell D-branes? This question was partially answered in [13,14], and this short note is devoted to reviewing the answer to this particular question.

Very briefly, some sheaves not of the form $i_* \mathcal{E}$ have a physical interpretation as D-branes with Higgs vevs. 'Easy' Higgs vevs merely move the support of the sheaf, but *e.g.* nilpotent Higgs vevs deform the sheaf nontrivially to sheaves not of the form $i_* \mathcal{E}$. In a non-topological theory, nilpotent Higgs vevs are often excluded by D-terms (with the prominent exception of orbifolds, where the nilpotents map out exceptional divisors, as seen in [15]), but in a topological theory, there is no such constraint.

2. Description of D-branes with nonzero Higgs vevs in terms of sheaves

A sheaf of the form $i_* \mathcal{E}$, where $i : S \hookrightarrow X$, describes a D-brane on S with gauge bundle $\mathcal{E} \otimes \sqrt{K_S^\vee}$, and with vanishing Higgs vevs, which clearly is not the most general case.

To describe D-branes with nonzero Higgs vevs in terms of sheaves we can use a mathematical prescription that has appeared previously in [16]. In a nutshell, one uses the Higgs vevs to define a deformation of the action of the ring of algebraic functions on the sheaf. Coordinates describing directions normal to the support, that would ordinarily annihilate the module corresponding to the sheaf, instead are taken to act nontrivially on the module, with action defined by the Higgs fields. The result is a new sheaf, living in the total space of the normal bundle to the support of the old sheaf. We shall see this explicitly in examples below.

A mathematical ansatz for encoding Higgs vevs in sheaves is not enough to guarantee that the resulting sheaves have physical content. To verify that this mathematical ansatz is indeed relevant, we compare Ext groups between the sheaves generated by this ansatz to the physical open string spectra, which are calculated using essentially the methods of [7], except that nonzero Higgs vevs deform the worldsheet BRST operator [6,14] to the form

$$Q = \overline{\partial} + \Phi_i^L \theta_i - \Phi_i^R \theta_i.$$

In [14] we prove the

Theorem 2.1. *Ext groups between sheaves generated by the mathematical ansatz above always match open string spectra computed by the BRST operator above.*

3. Example: Single D0 brane on C

The simplest nontrivial example consists of a single D0 brane on the complex line **C**. Such a D0-brane has a single Higgs field, call it Φ.

The sheaf corresponding to that Higgs vev is straightforward to work out. We begin with a skyscraper sheaf supported at the origin, described by the module $\mathbf{C}[x]/(x)$, which we can describe in terms of a single generator α obeying the relation $x \cdot \alpha = 0$. Giving the Higgs field a vev deforms the ring action to become $x \cdot \alpha = \Phi\alpha$ or more simply, $(x - \Phi) \cdot \alpha = 0$. In other words, the new module is $\mathbf{C}[x]/(x - \Phi)$, describing a skyscraper sheaf shifted away from the origin, exactly as one would expect.

As a further check, let us compare open string spectra between the D0-brane above and another D0-brane with no Higgs vevs, to Ext groups between the corresponding sheaves. Since the sheaves corresponding to the two D0-branes are skyscraper sheaves supported at different points, all Ext groups between the sheaves vanish, so there should be no physical open string states.

Let us check that BRST cohomology gives the same result. Degree zero states are defined by constants b. Demanding that such states be annihilated by the BRST operator, which in the present case simplifies to be merely $Q = \Phi\theta$, tells us that $b\Phi = 0$, so $b = 0$ (unless $\Phi = 0$), so there are no BRST-invariant degree zero states.

Degree one states are of the form $b\theta$, for constant b. All these states are annihilated by the BRST operator above; they are all also in the image of the BRST operator, hence there are no degree one states.

Thus, just as predicted, there are indeed no physical open string states between a pair of D0-branes when one has a nonzero Higgs vev.

4. Example: D0 branes on $\mathbf{C}^2$

For our next example, we shall consider a pair of D0 branes at the origin of $\mathbf{C}^2$. There are two Higgs fields; we shall give them vevs

$$\Phi_x = \begin{bmatrix} 0 & 1 \\ 0 & 0 \end{bmatrix}, \quad \Phi_y = \begin{bmatrix} 0 & 0 \\ 0 & 0 \end{bmatrix}.$$

(Note that these Higgs vevs commute, as needed to satisfy F-term conditions.)

We can find the corresponding sheaf as follows. Start with a pair of skyscraper sheaves $\mathcal{O}_0^2$ supported at the origin, describing the original pair of D0 branes before the Higgs vevs. Describe that pair of skyscraper sheaves as a module over $\mathbf{C}[x, y]$ with generators α, β, annihilated by x and y:

$$x \cdot \alpha = x \cdot \beta = y \cdot \alpha = y \cdot \beta = 0.$$

Create a new module by deforming the ring action as follows:

$$x \cdot \begin{bmatrix} \alpha \\ \beta \end{bmatrix} = \Phi_x \begin{bmatrix} \alpha \\ \beta \end{bmatrix} = \begin{bmatrix} \beta \\ 0 \end{bmatrix},$$
$$y \cdot \begin{bmatrix} \alpha \\ \beta \end{bmatrix} = \Phi_y \begin{bmatrix} \alpha \\ \beta \end{bmatrix} = \begin{bmatrix} 0 \\ 0 \end{bmatrix}.$$

This is the module[a] $\mathbf{C}[x, y]/(x^2, y)$, corresponding to a sheaf we shall denote D_x.

[a]Identify α with $1/(x^2, y)$ and β with $x/(x^2, y)$.

We shall check that this sheaf is physically-relevant by comparing open string spectra between D-branes to Ext groups between sheaves. For simplicity, let us compute open string spectra between the D0-branes with Higgs vevs as above, and another pair of D0-branes with vanishing Higgs vevs.

Mathematically, it can be shown (see [14] for details) that

$$\mathrm{Ext}^n_{\mathbf{C}^2}\left(\mathcal{O}_0^2, D_x\right) = \begin{cases} \mathbf{C}^2 & n = 0, 2, \\ \mathbf{C}^4 & n = 1. \end{cases}$$

In general, the mathematical computation is easier than the physical computation – indeed, this is one of the motivations for describing open string states in terms of Ext groups – but in this simple case, it is also straightforward to do the physical computation, as we shall now outline.

Degree zero states are of the form

$$V = \begin{bmatrix} a & b \\ c & d \end{bmatrix}.$$

Demanding that these states be annihilated by the BRST operator

$$Q = \overline{\partial} + \Phi_i^L \theta_i - \Phi_i^R \theta_i$$

implies the constraint $V\Phi_x = 0$, which implies that the BRST-invariant states have $a = c = 0$, and so the space of BRST-invariant states is two-dimensional, matching the Ext group computation.

Degree one states are of the form

$$V = \begin{bmatrix} a_x & b_x \\ c_x & d_x \end{bmatrix} \theta_x + \begin{bmatrix} a_y & b_y \\ c_y & d_y \end{bmatrix} \theta_y.$$

States that are annhilated by the BRST operator have $a_y = c_y = 0$, and the image of the BRST operator has the form

$$\begin{bmatrix} 0 & a \\ 0 & c \end{bmatrix} \theta_x$$

so we see that the space of degree one states has dimension $6 - 2 = 4$, matching the Ext group computation.

Degree two states are of the form

$$V = \begin{bmatrix} a & b \\ c & d \end{bmatrix} \theta_x \theta_y.$$

All of these states are annihilated by the BRST operator, and states in the image of the BRST operator have $a = c = 0$, so we see that the space of degree two states has dimension $4 - 2 = 2$, matching the Ext group computation.

5. Relevance to D-branes in orbifolds

As mentioned in the introduction, nilpotent Higgs fields occur in non-topological theories in orbifolds. In [15] it was shown that classical Higgs moduli spaces of D-branes in orbifolds encode resolution of quotient spaces. It is straightforward to check that the exceptional divisors of those resolutions are described by nilpotent Higgs fields, which correspond via our construction to G-equivariant nonreduced schemes on the covering space (equivalently, nonreduced schemes on the quotient stack). For example, the previous section contained a set of $\mathbf{Z}_2$-invariant Higgs fields relevant to describing D0-branes on $[\mathbf{C}^2/\mathbf{Z}_2]$, which we saw explicitly were described by a nonreduced scheme on the covering space $\mathbf{C}^2$.

The McKay correspondence, in the form [17], also relates exceptional divisors to G-equivariant nonreduced schemes, via a different mechanism, not directly involving D-branes.

In effect, the construction we have outlined here closes this circle of ideas involving the McKay correspondence and the calculations of [15]. See our paper [14] for more details.

Acknowledgments

We would like to thank Sheldon Katz for many useful conversations and collaborations on this material.

References

1. M. Kontsevich, "Homological algebra of mirror symmetry," Proceedings of the 1994 International Congress of Mathematicians, Birkäuser, Zürich, 1995, p. 120, `alg-geom/9411018`.
2. A. Sen, "Stable non-BPS states in string theory," *JHEP* **9806**, 007 (1998), `hep-th/99803194`.
3. E. Witten, "D-branes and K theory," *JHEP* **9812**, 019 (1998), `hep-th/9810188`.
4. E. Sharpe, "D-branes, derived categories, and Grothendieck groups," *Nucl. Phys.* **B561**, 433-450 (1999), `hep-th/9902116`.
5. M. Douglas, "D-branes, categories, and $\mathcal{N} = 1$ supersymmetry," *J. Math. Phys.* **42**, 2818-2843 (2001), `hep-th/0011017`.
6. P. Aspinwall and A. Lawrence, "Derived categories and zero-brane stability," *JHEP* **0108**, 004 (2001), `hep-th/0104147`.
7. S. Katz and E. Sharpe, "D-branes, open string vertex operators, and Ext groups," *Adv. Theor. Math. Phys.* **6**, 979-1030 (2003), `hep-th/0208104`.
8. S. Katz, T. Pantev, and E. Sharpe, "D-branes, orbifolds, and Ext groups," *Nucl. Phys.* **B673**, 263-300 (2003), `hep-th/0212218`.

9. A. Caldararu, S. Katz, and E. Sharpe, "D-branes, B fields, and Ext groups," *Adv. Theor. Math. Phys.* **7**, 381-405 (2003), `hep-th/0302099`.
10. E. Sharpe, "Lectures on D-branes and sheaves," `hep-th/0307245`.
11. S. Katz, "D-branes and the derived category," these proceedings.
12. D. Freed and E. Witten, "Anomalies in string theory with D-branes," `hep-th/9907189`.
13. T. Gomez and E. Sharpe, "D-branes and scheme theory," `hep-th/0008150`.
14. R. Donagi, S. Katz, and E. Sharpe, "Spectra of D-branes with Higgs vevs," `hep-th/0309270`.
15. M. Douglas, B. Greene, and D. Morrison, "Orbifold resolution by D-branes," *Nucl. Phys.* **B506**, 84-106 (1997), `hep-th/9704151`.
16. R. Donagi, L. Ein, and R. Lazarsfeld, "A non-linear deformation of the Hitchin dynamical system," `alg-geom/9504017`, a.k.a. "Nilpotent cones and sheaves on K3 surfaces," pp. 51-61 in *Birational Algebraic Geometry*, Contemp. Math. 207, Amer. Math. Soc., Providence, Rhode Island, 1997.
17. T. Bridgeland, A. King, and M. Reid, "Mukai implies McKay," *J. Amer. Math. Soc.* **14**, 535-554 (2001), `math.AG/9908027`.

A REVIEW OF THE PLANE-WAVE/SYM DUALITY

M.M. SHEIKH-JABBARI

Department of Physics, Stanford University
382 via Pueblo Mall, Stanford CA 94305-4060, USA
E-mail: jabbari@itp.stanford.edu

In this talk which is based on [1] we review the plane-wave/SYM duality. According to this duality type IIB strings on the maximally supersymmetric plane-wave background is dual to a particular large R-charge sector of $\mathcal{N}=4,\ D=4$ superconformal $U(N)$ gauge theory in the large 't Hooft coupling limit.

1. Introduction

In 1974 't Hooft [2] observed a property of $SU(N)$ gauge theories which was very suggestive of a correspondence or "duality" between the gauge dynamics and string theory. The first remarkable observation of 't Hooft was that the true expansion parameter for an $SU(N)$ gauge theory (with or without matter fields) is not the Yang-Mills coupling g^2_{YM}, but rather g^2_{YM} dressed by N, the 't Hooft coupling λ:

$$\lambda = g^2_{YM} N \ . \tag{1.1}$$

The second remarkable observation 't Hooft made was that in addition to the expansion in powers of λ one may also classify the Feynman graphs appearing in the correlation function of generic gauge theory operators in powers of $1/N^2$. This observation is based on the fact that the operators of this gauge theory are built from simple $N \times N$ matrices. One is then led to expand any correlation function in a double expansion, in power of λ as well as $1/N^2$. In the $1/N^2$ expansion, which is a useful one for large N, the terms of lowest order in powers of $1/N^2$ arise from the subclass of Feynman diagrams which can be drawn on a plane (and hence these are called *planar* graphs), once the 't Hooft double line notation is used. In the same spirit one can classify all Feynman graphs according to the lowest genus surface that they may be placed on without any crossings. For genus h surfaces, with $h > 0$, such diagrams are called *non-planar*. The genus h graphs are suppressed by a factor of $\left(1/N^2\right)^h$ with respect to the planar diagrams.

The genus expansion of Feynman diagrams in a gauge theory resembles a similar pattern in string theory: stringy loop diagrams are suppressed by g_s^h where h is now the genus of the string worldsheet and g_s is the string coupling constant.

The 't Hooft's observations was not realized in the context of string theory, until 1997, when a study of the near horizon geometry of D3-branes led to the conjecture that [3]
Strings of type IIB string theory on the $AdS_5 \times S^5$ background are the 't Hooft strings of an $\mathcal{N} = 4$, $D = 4$ supersymmetric Yang-Mills theory.

According to this conjecture any physical object or process in the type IIB theory on $AdS_5 \times S^5$ background can be equivalently described by $\mathcal{N} = 4$, $D = 4$ super Yang-Mills (SYM) theory [3]. In particular, the 't Hooft coupling (1.1) is related to the AdS radius R as

$$\left(\frac{R}{l_s}\right)^4 = g_{YM}^2 N \tag{1.2}$$

where l_s is the string scale. On the string theory side of the duality, l_s/R appears as the worldsheet coupling; hence when the gauge theory is weakly coupled the two dimensional worldsheet theory is strongly coupled and non-perturbative, and vice-versa. In this sense the AdS/CFT duality is a weak/strong duality. Due to technical difficulties our understanding of the string theory side of the duality has been mainly limited to the low energy supergravity limit, and in order for the supergravity expansion about the AdS background to be trustworthy, we generally need to keep the AdS radius large. At the same time we must also ensure the suppression of string loops. As a result, most of the development and checks of the duality from the string theory side have been limited to the regime of large 't Hooft coupling and the $N \to \infty$ limit on the gauge theory side.

The σ-model for strings on $AdS_5 \times S^5$ is difficult to solve. However, there is a specific limit in which $AdS_5 \times S^5$ reduces to a plane wave (for a more detailed discussion see [1]) and in this limit the string theory σ-model becomes solvable [4]. In this special limit we then know the string spectrum, at least for non-interacting strings, and one might ask if we can find the same spectrum from the gauge theory side. For that we first need to understand how this specific limit translates to the gauge theory side. We then need a definite proposal for mapping the operators of the gauge theory to (single) string states. This proposal, following the work of Berenstein-Maldacena-Nastase [5], is known as the BMN conjecture. In section 2 we introduce this conjecture and summarize the supporting evidence for this duality. These pieces of evidence on both sides of the duality are results of detailed and explicit calculations can be found in [1] and references therein.

2. A summary of results in the plane-wave/SYM duality

In this section we recall some basic observations and facts which led BMN to their conjecture [5] as well as a brief summary of the results obtained based on and in support of the conjecture.

• Plane-waves comprise a class of supergravity solutions which have a *globally defined null Killing vector field*, and have the important property that are exact solutions without α' corrections[6].

• The Penrose limit of $AdS_5 \times S^5$ geometry [5] results in another maximally supersymmetric background of type IIB which will be called "the" plane-wave geometry. The plane-wave background possesses a $PSU(2|2) \times PSU(2|2) \times U(1)_- \times U(1)_+$ superalgebra [1].

• Although so far we have not been able to solve the string σ-model in the $AdS_5 \times S^5$ background and obtain the spectrum of (free) strings, string σ-model on its Penrose limit, the plane-wave background, (in the light-cone gauge) is solvable, allowing us to deduce the spectrum of (free) strings on this background. It is worth noting that the solubility of strings σ-model in this background is true for both bosonic and fermionic modes [4].

• Taking the Penrose limit on the gravity side corresponds to restricting the gauge theory to operators with a large charge under one of its global symmetries (more precisely the R-symmetry charge) J, the BMN sector, and simultaneously taking the large N limit.

• The BMN sector of $\mathcal{N} = 4,\ D = 4\ U(N)$ SYM theory is comprised of operators with large conformal dimension Δ and large R-charge J, such that

$$\frac{1}{\mu}p^- \equiv \Delta - J = \text{fixed}\,, \tag{2.1a}$$

$$\alpha'\mu p^+ \equiv \frac{1}{2\sqrt{g_{YM}^2 N}}(\Delta + J) = \text{fixed}\,, \tag{2.1b}$$

together with

$$g_{YM} = \text{fixed}\ , \qquad \frac{J^2}{N} = \text{fixed}, \qquad N \to \infty\ , \qquad J \to \infty\ . \tag{2.2}$$

$\frac{1}{\mu}p^-$ and $\alpha'\mu p^+$ are the corresponding string light-cone Hamiltonian and light-cone momentum, respectively. The parameter μ is a convenient but auxiliary parameter, the role of which will become clear momentarily.

• In (2.1a), p^- should be understood as the *full* plane-wave light-cone string (field) theory Hamiltonian. Explicitly, one can interpret (2.1a) as an equality between two operators, the plane-wave light-cone string field theory Hamiltonian, H_{SFT}, on one side and the difference between the dilatation and the R-charge operators on the other side, i.e.

$$\frac{1}{\mu}H_{SFT} = \mathcal{D} - \mathcal{J} , \tag{2.3}$$

where $\mathcal{D}$ is the dilatation operator and $\mathcal{J}$ is the R-charge generator. Therefore according to the identification (2.3) which is the (improved form of the original) BMN conjecture, the spectrum of strings, which are the eigenvalues of the light-cone Hamiltonian $p^- = H_{SFT}$, should be equal to the spectrum of the dialtation operator, which is the Hamiltonain of the $\mathcal{N} = 4$, gauge theory on $\mathbb{R} \times S^3$, restricted to the operators in the BMN sector of the gauge theory (defined through (2.1) and (2.2)).

• As stated above, equation (2.3) sets an equivalence between two operators. However, the second part of the BMN conjecture is about the correspondence between the Hilbert spaces that these operators act on; on the string theory side, it is the string (field) theory Hilbert space which is comprised of direct sum of the zero string, single string, double string and ... string states, quite similarly to the flat space case. On the gauge theory side it is the so-called BMN operators, the set of $U(N)$ invariant operators of large R-charge J and large dimension in the free gauge theory, subject to (2.1) and (2.2).

• According to the BMN proposal [5], single string states map to certain single trace operators in the gauge theory.[a] In particular, the single string vacuum state in the sector with light-cone momentum p^+, $(\alpha'\mu p^+)^2 = \frac{J^2}{g_{YM}^2 N}$, is identified with the chiral-primary BPS operator

$$|0, p^+\rangle \longleftrightarrow \mathcal{N}_J \mathrm{Tr}(Z^J)|vac\rangle , \tag{2.4}$$

where $\mathcal{N}_J^{-2} = JN^J$ is a normalization constant. In the above, $Z = \frac{1}{\sqrt{2}}(\phi^5 + i\phi^6)$, where ϕ^5 and ϕ^6 are two of the six scalars of the $\mathcal{N} = 4$, $D = 4$ gauge multiplet. The R-charge we want to consider, J, is the eigenvalue for a $U(1)$ generator, $U(1) \subset SU(4)_R$, so that Z carries unit charge and all the other bosonic modes in the vector multiplet, four scalars and four gauge fields, have zero R-charge. Since all scalars have $\Delta = 1$ (classically),

[a]The proposal as stated is only true for free strings. This proposal should be modified once string interactions are included.

for Z and hence Z^J, $\Delta - J = 0$. The advantage of identifying string vacuum states with the chiral-primary operators is two-fold: *i)* they have $\frac{1}{\mu}p^- = \Delta - J = 0$, and *ii)* their anomalous dimension is zero and hence the corresponding p^- remains zero to all orders in the 't Hooft coupling and even non-perturbatively.

• As for stringy excitations above the vacuum, BMN conjectured that we need to work with certain "almost" BPS operators, i.e. certain operators with large J charge and with $\Delta - J \neq 0$, but $\Delta - J \ll J$. In particular, single closed string states were (originally) proposed to be dual to single trace operators with $\Delta - J = 2$. The exact form of these operators and a more detailed discussion regarding them can be found in section 5 of [1]. However, this identification of the single string Hilbert space with the single trace operators, because of the mixing between single trace and multi trace operators, should be modified. Note that this mixing is present both for chiral primaries and "almost" primaries.

• As is clear from (2.2), in the BMN limit the 't Hooft coupling goes to infinity and naively any perturbative calculation in the gauge theory (of course except for chiral-primary two and three point functions) is not trustworthy. However, the fact that we are working with "almost" BPS operators motivates the hope that, although the anomalous dimensions for such operators are non-vanishing, being close to primary, nearly saturating the BPS bound, some of the nice properties of primary operators might be inherited by the "almost" primary operators.

• As a result of explicit gauge theory calculations with the BMN operators, it has been shown that the 't Hooft coupling in the BMN sector is dressed with powers of $1/J^2$. More explicitly, the effective coupling in the BMN sector is λ', rather than the 't Hooft coupling λ, where

$$\lambda' \equiv \frac{\lambda}{J^2} = g_{YM}^2 \frac{N}{J^2} = (\alpha' \mu p^+)^{-2} . \tag{2.5}$$

The last equality is obtained using (2.1) and (2.2).

• Moreover, we will see that the ratio of non-planar to planar graphs is controlled by powers of the genus counting parameter

$$g_2 \equiv \frac{J^2}{N} = 4\pi g_s (\alpha' \mu p^+)^2 , \tag{2.6}$$

which also remains finite in the BMN limit (2.2). Note that in (2.6) $g_s = e^{\phi}$, where ϕ is the value of the dilaton field[b], is *not* the coupling for strings on

[b]Note that for the plane-wave background we are interested in, the dilaton is constant.

the plane-wave, although it is related to it.

• One can do better than simply finding the free string mass spectrum; we can study real interacting strings, their splitting and joining amplitudes and one loop corrections to the mass spectrum. The *one-loop* mass corrections compared to the tree level results are suppressed by powers of the "effective one-loop string coupling"

$$g^{eff}_{one-loop} = \sqrt{\lambda' g_2^2} = g_{YM}\frac{J}{\sqrt{N}} = 4\pi g_s\, \alpha' \mu p^+ . \tag{2.7}$$

It has been argued that all higher genus (higher loop) results replicate the same pattern, i.e. g_2 always appears in the combination $\lambda' g_2^2$. However, *a priori* there is no reason why such a structure should exist and in principle g_2 and λ' can appear in any combination. This point is still an unresolved issue and is currently under study.

• The above observations revive the hope that we might be able to do a full-fledged *interacting* string theory computation using *perturbative* gauge theory with (modified) BMN operators.

Acknowledgments

I would like to thank Darius Sadri for his collaboration in preparing the comprehensive review whose results is summarized here. It is a pleasure to thank organizers of QTS3 for organizing a stimulating conference. Work of the author is supported in part by NSF grant PHY-9870115 and in part by funds from Stanford Institute for Theoretical Physics.

References

1. D. Sadri and M. M. Sheikh-Jabbari, "The plane-wave / super Yang-Mills duality," *To appear in Reviews of Modern Physics*, arXiv:hep-th/0310119.
2. G. 't Hooft, "A Planar Diagram Theory For Strong Interactions," Nucl. Phys. B **72**, 461 (1974); "A Two-Dimensional Model For Mesons," Nucl. Phys. B **75**, 461 (1974).
3. O. Aharony, S. S. Gubser, J. M. Maldacena, H. Ooguri and Y. Oz, "Large N field theories, string theory and gravity," Phys. Rept. **323**, 183 (2000) [arXiv:hep-th/9905111].
4. R. R. Metsaev, "Type IIB Green-Schwarz superstring in plane wave Ramond-Ramond background," Nucl. Phys. B **625**, 70 (2002) [arXiv:hep-th/0112044].
5. D. Berenstein, J. M. Maldacena and H. Nastase, "Strings in flat space and pp waves from N = 4 super Yang Mills," JHEP **0204**, 013 (2002) [arXiv:hep-th/0202021].
6. G. T. Horowitz and A. R. Steif, "Space-Time Singularities In String Theory," Phys. Rev. Lett. **64** (1990) 260.

QUIVERS, DIBARYONS, AND EXCEPTIONAL COLLECTIONS

JOHANNES WALCHER

Kavli Institute for Theoretical Physics
University of California
Santa Barbara, CA 93106, USA

We discuss aspects of the dictionary between brane configurations in del Pezzo geometries and dibaryons in the dual superconformal quiver gauge theories. The basis of fractional branes defining the quiver theory at the singularity has a K-theoretic dual exceptional collection of bundles which can be used to read off the spectrum of dibaryons in the weakly curved dual geometry. Our prescription identifies the R-charge R and all baryonic $U(1)$ charges Q_I with divisors in the del Pezzo surface. As one application of the correspondence, we identify the cubic anomaly $\mathrm{tr}RQ_IQ_J$ as an intersection product for dibaryon charges in large-N superconformal gauge theories.

1. Introduction

I would like to present results of some attempts to understand the connections between various things listed here:
*quivers
*dynamics of gauge theories
*field theory dualities
*(stringy) geometry
*...
In string theory, these connections are obtained using D-branes, and can be perturbative or holographic. Most of what I am presenting here belongs to the holographic category. The examples of interest arise from the so-called del Pezzo geometries ($\mathbb{P}^2$ blown up at n points, with $0 \leq n \leq 8$) and their associated quivers. This talk is based on work done in collaboration with Christopher Herzog and has appeared in [1].

2. Quivers

Let me begin with some definitions. A *quiver* is simply a directed graph, ie, a (finite) set of nodes labelled i, a (finite) collection of arrows labelled $a = (ij)$ with head $h(a) = j$ and tail $t(a) = i$, and multiplicities w_a. The *path algebra* of a quiver is the set of all paths A with product defined by concatenation. A *quiver with relations* is a quiver where some paths have been set to zero in the path algebra. These relations may or may not be integrable to a *superpotential*, which is a closed path W such that removing single arrows generates the relations. A *representation* of a quiver is a representation of the path algebra, ie, a *dimension vector* $d = (d_i)$, a vector space V_i of dimension d_i for each node and a map $X_a : V_i \longrightarrow V_j$ for each arrow $a = (ij)$. The X_a are to satisfy the relations dictated by the superpotential.

3. Gauge theories from quivers

Given a quiver and a dimension vector, we define a *quiver gauge theory* by associating a unitary gauge group of rank d_i with each node and a chiral matter field X_a in the $\overline{d_i} \times d_j$ bifundamental representation with each arrow $a = (ij)$. The relations come from the holomorphic gauge invariant superpotential W. We have a correspondence [2] between *solutions of F-terms of the gauge theory (modulo gauge equivalence)* and *(equivalence classes of) representations of the quiver.*

Let us express various conditions of the gauge theory in the quiver language. Cancellation of *chiral gauge anomalies* becomes the condition that the number of incoming arrows equal the number of outgoing arrows at each node. In other words, $\mathcal{I}d = 0$, where $\mathcal{I}$ is the antisymmetric matrix with entries $w_{ij} = -w_{ji}$. The gauge theory can be *conformal* if there exists an anomaly-free R-symmetry, ie, an assignement of R-charges $R(X_a)$ such that the NSVZ beta-function is zero at each node, and the superpotential has charge 2. These conditions are essentially a system of inhomogeneous linear equations for the $R(X_a)$, which may or may not have a solution, depending on the quiver and the choice of d. If a solution exists, it might not be unique, its ambiguities being parametrized by the solutions of the corresponding homogeneous equation. These ambiguities are identified as *baryonic* $U(1)$ *charges* Q_I. One can show that the number of independent Q_I's is one less that the number of solutions to $\mathcal{I}d = 0$. In other words, the baryoninc $U(1)$ charges are in one-to-one correspondence with non-conformal deformations obtained by changing d.

An interesting result in this context was recently obtained by Intriligator and Wecht [3] and says that the exact superconformal R-charge is distinguished by the fact that it (locally) maximizes the conformal anomaly $a \sim 3\mathrm{tr}R^3 - \mathrm{tr}R$. This is equivalent to the conditions on the cubic 't Hooft anomalies, $\mathrm{tr}R^2Q_I = 0$ for all I and that the matrix $\mathfrak{K}_{IJ} = \mathrm{tr}RQ_IQ_J$ be negative definite. To give away the punchline, let me say that these conditions will be given the interpretation as an intersection form in a holographically dual geometry.

4. Dibaryons

An interesting subset of the chiral operators of the theory are the so-called *dibaryons.* These are generalized determinant operators, and can be associated with paths in the quiver with equal ranks at head and tail. See [1] for an efficient description.

5. Connection with string theory

Quiver gauge theories arise naturally in string theory as low-energy limit of open strings on D-branes (with small g_sN) [2], see [4] for a pedagogical introduction. For example, a stack of N D3-branes at smooth points of a Calabi-Yau three-fold $\mathbb{X}$ gives rise to a $U(N)$ gauge theory. Moving the D3-branes close to an isolated singularity of $\mathbb{X}$ leads to a gauge symmetry enhancement $U(N) \longrightarrow \prod U(Nd_i)$ and additional chiral matter in bifundamentals, ie, a quiver gauge theory. In string theory language, one says that the D3 brane fractionates into the fractional branes E_i associated with the nodes of the quiver.

6. Quivers from exceptional collections

Here, we will be interested in the case where $\mathbb{X}$ is the complex cone over a del Pezzo surface $\mathbb{V}$. In that case, there is a general procedure for constructing the quiver theory, see eg, [5] for a nice summary of this construction: The first step is to blow up the singularity to the total space of the anti-canonical bundle over $\mathbb{V}$. The fractional branes then become branes wrapped around various cycles of $\mathbb{V}$ with gauge bundles. At large volume, the fractional branes turn out to form what is known as an *exceptional collection* of bundles on $\mathbb{V}$. Defining for two bundles E and F on $\mathbb{V}$ the relative Euler character $\chi(E,F) = \sum_i (-1)^i \mathrm{Ext}^i(E,F) = \int \mathrm{ch}(E^*)\mathrm{ch}(F)\mathrm{Td}(\mathbb{V})$, we can (partially) characterize an exceptional collection as an ordered collection

$(E_1, \ldots, E_m)$ such that the matrix $S_{ij} = \chi(E_i, E_j)$ is upper triangular with 1's on the diagonal.

The essential point of this relation between gauge theories and exceptional collections at small $g_s N$ is that holomorphic information, ie, the chiral spectrum and superpotential can (at least in principle) be computed in the blown up geometry at large volume.

7. Holographic duals

Having obtained the gauge theory describing branes on $\mathbb{X}$ from an exceptional collection on $\mathbb{V}$, we now turn to considering the theory in the limit of large 't Hooft coupling. Via AdS/CFT, we have a holographic dual description by string theory involving the near horizon geometry, which here turns out to be the Sasaki-Einstein manifold $\mathbb{Y}$, which is the base of $\mathbb{X}$ as a real cone, and is also an S^1 fibration over $\mathbb{V}$.

The goal for the rest of this presentation is to provide some kinematical tests of this correspondence by finding an exact mapping between states and operators for the case of the dibaryons.

By an argument of Witten [6], dibaryons are "non-perturbative" in the $1/N$ expansion. Therefore, the corresponding states should be D-branes wrapped on cycles in $\mathbb{Y}$. By an arugment due to Mikhailov [7], supersymmetry requires that this cycle project down to a *holomorphic cycle* in $\mathbb{V}$ and that over each point, the brane wraps the $U(1)$ fiber.

The problem we propose to solve is to understand the map between R-charge and baryonic $U(1)$ charges Q_I and geometric information about the corresponding cycles. Previous results on this problem are obtained in [8,9], and include, for example, the realization that the R-charge, which determines the dimension of the operator should correspond to the volume of the wrapped cycle, which can be computed from the degree of the corresponding holomorphic curve. But what about the baryonic charges Q_I?

8. Results

The topological charges of a brane configuration C corresponding to a (di)baryon B obtained from a path A in the quiver Q corresponding to an exceptional collection $\mathcal{E}$ can be read off from the path and a dual exceptional collection $\mathcal{E}^\vee$.

Here, the dual exceptional collection is defined by the property that $\chi(E_i, E_j^\vee) = \delta_{ij}$. For full details of the prescription, we have to refer to [1]. Suffice it to say here that the essential step is to associate with each generator $X_a = X_{ij}$ of the path algebra a fractional divisor $D_a = D_{ij} = c_1(E_j^\vee)/r(E_j^\vee) - c_1(E_i^\vee)/r(E_i^\vee)$ if $\chi(E_i^\vee, E_j^\vee) \neq 0$, (and we add $-K$ otherwise). Moreover, the baryonic $U(1)$ charges Q_I get associated with divisors of $\mathbb{V}$, for example, the R-charge is the canonical divisor.

9. Comments and implications

In the simplest case, we get an actual dibaryon by taking $\mathrm{lcm}(d_i, d_j)$ copies of X_{ij} to form a gauge invariant. This is just right to make $\mathrm{lcm}(d_i, d_j)D_{ij}$ an integral homology class. We point out that we have not been able to establish in full generality that our presription will give an actual curve around which we can wrap a brane, but we have performed a lot of checks in examples.

For example, one can show that this assignement of R-charges satisfies the maximizing-a principle of Intriligator and Wecht. More precisely, as we have mentioned above, this check proceeds by identifying the cubic 't Hooft anomaly $\mathrm{tr}Q_1Q_2R$ with the intersection of the divisors corresponding to the (R or baryonic $U(1)$) charges Q_1 and Q_2). The two conditions then say that the orthogonal complement of the canonical divisor has a definite signature. It is indeed known that the orthogonal complement of the canonical divisor of the n-th del Pezzo is isomorhic to the root lattice of the exceptional Lie group E_n.

10. Concluding remarks

Dibaryons provide interesting tests of AdS/CFT by matching spectra of states and operators. We have here seen that exceptional collections are a useful tool for obtaining and spelling out this correspondence in detail.

Let us also mention that exceptional collections can be used to organize quivers related by (generalized) Seiberg dualities. We also hope that the non-trivial identifications we have provided here prepare the ground for non-conformal deformations of AdS/CFT a la Klebanov-Strassler [10,11].

Acknowledgments

I would like to thank Chris Herzog for collaboration on the results presented here, and I would like to thank the organizers of the QTS03 symposium for giving me the opportunity to talk at this marvelously organized conference. This work was supported in part by the NSF under Grant No. PHY99-07949.

References

1. C. P. Herzog and J. Walcher, JHEP **0309**, 060 (2003) [arXiv:hep-th/0306298].
2. M. Douglas and G. Moore, "D-Branes, Quivers and ALE Instantons," `hep-th/9603167`.
3. K. Intriligator and B. Wecht, "The exact superconformal R-symmetry maximizes a," `hep-th/0304128`.
4. M. R. Douglas, "Lectures on D-branes on Calabi-Yau manifolds," SPIRES entry *Prepared for ICTP Spring School on Superstrings and Related Matters, Trieste, Italy, 2-10 Apr 2001*
5. M. Wijnholt, "Large volume perspective on branes at singularities," `hep-th/0212021`.
6. E. Witten, "Baryons and branes in anti de Sitter space," *JHEP* **9807** (1998) 006, `hep-th/9805112`.
7. A. Mikhailov, "Giant gravitons from holomorphic surfaces," *JHEP* **0011** (2000) 027, `hep-th/0010206`.
8. K. Intriligator and B. Wecht, "Baryon charges in 4D superconformal field theories and their AdS duals," `hep-th/0305046`.
9. C. P. Herzog and J. McKernan, "Dibaryon Spectroscopy," `hep-th/0305048`.
10. I. R. Klebanov and M. J. Strassler, "Supergravity and a confining gauge theory: Duality cascades and chiSB-resolution of naked singularities," *JHEP* **0008** (2000) 052, `hep-th/0007191`.
11. S. Franco, Y. H. He, C. P. Herzog, and J. Walcher, work in progress and to appear

VII. LOOP QUANTUM GRAVITY

LN(3) AND BLACK HOLE ENTROPY

OLAF DREYER
Perimeter Institute for Theoretical Physics
35 King Street North
Waterloo, Ontario
Canada N2J 2W9
E-mail: odreyer@perimeterinstitute.ca

We review an idea that uses details of the quasinormal mode spectrum of a black hole to obtain the Bekenstein-Hawking entropy of $A/4$ in Loop Quantum Gravity. We further comment on a recent proposal concerning the quasinormal mode spectrum of rotating black holes. We conclude by remarking on a recent proposal to include supersymmetry.

1. Introduction

Ever since Bekenstein[4] and Hawking[5] argued that a black hole has an entropy S which is a quarter of its horizon area A it has been a challenge to theoretical physicists to explain what this is an entropy of. In recent years candidate theories of quantum gravity have provided concrete realizations of quantum mechanical microstates. Both String theory[2] and Loop Quantum Gravity[3] have given derivations for the Bekenstein-Hawking entropy.

In this article we will describe how one can fix a free parameter, known as the Immirzi[1] parameter, to obtain the exact Bekenstein-Hawking result for the entropy in Loop Quantum Gravity. We will do this using an input from classical relativity, namely the quasinormal mode spectrum of black holes. The first three sections of this paper deal with this argument. The following sections then deal with rotating black holes and also discuss a new insight into the problem that relies on supersymmetry.

2. Quasinormal Modes in the High Damping Limit

The reaction of a black hole to perturbations is dominated by an infinite discrete set of damped oscillation called quasinormal modes[6,7]. One remarkable feature of the quasinormal mode spectrum for non-rotating black holes is that the real part of the frequency approaches the non-zero value[a] $\omega_\infty = \ln 3/8\pi M$ as the damping increases (see figure 1).

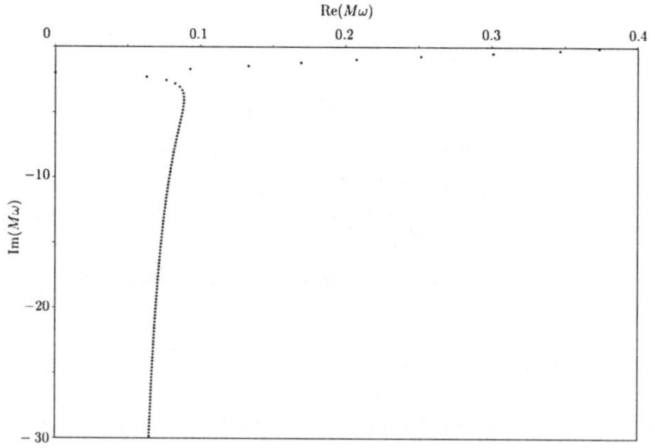

Figure 1. The position of the quasinormal mode frequencies in the complex ω-plane. For higher damping the real part of the frequency becomes approaches the limiting value of $\ln 3/8\pi M$.

A non-rotating black hole has thus associated with it one distinguished frequency. One might then ask if there is a special process in the underlying quantum theory that gives rise to this particular frequency. We will come back to this question after having taken a closer look at Loop Quantum Gravity.

3. Facts from Loop Quantum Gravity

In Loop Quantum Gravity a surface is thought to acquire area through the punctures of spin network edges. Spin networks form the basis of the Hilbert space of LQG. They consist of graphs whose edges are labelled by representations of the gauge group of the theory. In our case this is SU(2) (or SO(3)). The labels are thus (half-)integers. For a puncture with label j the area is

$$A(j) = 8\pi l_P^2 \gamma \sqrt{j(j+1)}. \tag{1}$$

Here l_P^2 is the Planck length. The free parameter γ is the so-called Immirzi[1] parameter.

---

[a]This fact was first observed by Hod[8] using numerical data by Nollert[9] and it has been recently proved rigorously by Motl[10].

One can think of the horizon area of a black hole to be the result of a large number of spin network edges puncturing the horizon surface. The Hilbert space of the theory living on the boundary is increased by each puncture. If the edge has a label j the dimension of the Hilbert space increases by a factor of $2j+1$. The statistically most important contribution comes from the lowest possible spin j_{min}. The dimension of the Hilbert space is thus

$$(2j_{\text{min}} + 1)^N, \tag{2}$$

where N is the number of punctures. Since the area due to one puncture is $A(j_{\text{min}})$ (see formula (1)) this number N is given by $A/A(j_{\text{min}})$. The entropy of the black hole is thus

$$S = \frac{A}{8\pi l_P^2 \gamma \sqrt{j_{\text{min}}(j_{\text{min}} + 1)}} \ln(2j_{\text{min}} + 1). \tag{3}$$

The entropy is proportional to the area but the result so far is not satisfactory since it contains the free parameter γ.

4. Fixing the Immirzi Parameter

When discussing the high damping limit of the quasinormal mode frequencies we asked the question whether there could exist a quantum mechanical process that could give rise to the the limiting frequency ω_∞. Given the picture developed in the last section such a process is easy to imagine. It is the appearance or disappearance of a puncture of the horizon (see figure 2**a**).

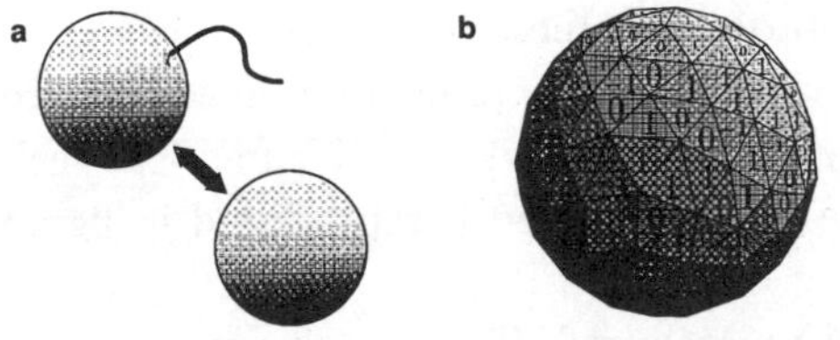

Figure 2. **a** The proposed quantum transition giving rise to the emission of a quantum of energy $\hbar\omega_\infty$. **b** The emerging "It from trit" picture.

As Hod[8] pointed out, the emission of a quanta of frequency ω_∞ would change the area of a non-rotating black hole by $4 \ln 3\ l_P^2$. This has to be equated with the change in area $A(j_{\text{min}})$ due to one puncture. This equation fixes the Immirzi parameter and gives for the entropy

$$S = \frac{A}{4l_P^2}\frac{\ln(2j_{\min}+1)}{\ln 3}. \tag{4}$$

We see that we obtain exactly the Bekenstein-Hawking result if $j_{\min} = 1$. This can be achieved if one chooses SO(3) as the gauge group instead of SU(2)[11]. Wheelers[12] "It from bit" gets thus modified to "It from trit" (see figure 2b)[b].

5. Rotating Black Holes

In the last section we have used quasinormal modes to learn something about the quantum mechanics of non-rotating black holes. A new argument by Hod[13] tries to go the other way. This time the quantum mechanics of black holes is used to gain insight into the quasinormal mode spectrum of a rotating black hole. The starting point is the first law of black hole mechanics

$$\Delta M = T\Delta S + \Omega \Delta J. \tag{5}$$

Using the expressions in the Kerr metric for the temperature $T = \sqrt{M^2-a^2}/4\pi M r_+$ and the angular velocity $\Omega = a/(r_+^2+a^2)$, where $r_\pm = M \pm \sqrt{M^2-a^2}$ are the horizon radii, together with the identification $\Delta M = \hbar\,\mathrm{Re}(\omega)$ and $\Delta S = 4\ln(3)$ one obtains an expression for $\mathrm{Re}(\omega)$ from equation (5):

$$\mathrm{Re}(\omega) = \frac{\ln(3)}{\pi}\frac{(M^2-a^2)^{1/2}}{M(M+(M^2-a^2)^{1/2})} + \frac{2a}{(M+(M^2-a^2)^{1/2})^2+a^2} \tag{6}$$

We thus obtain a prediction for how the frequency $\mathrm{Re}(\omega)$ changes as a function of the rotation parameter $a = J/M$.

To check the validity of this prediction Hod compared it to numerical data obtained by Onozawa[14], who used numerical techniques developed by Leaver[15]. The graph obtained is reproduced in figure 3 using our own numerical data.

The two graphs match well. The largest deviation occurs at the origin. This is to be expected since we are looking at low damping.

It is assumed that the situation improves for larger damping. This is not the case though for the largest damping we have tested. Figure 4**a** shows quasinormal modes for $n = 79$. Figure 4**b** then shows the $\mathrm{Re}(\omega)(a)$

---

[b]I thank Jonathan Oppenheim for this quote.

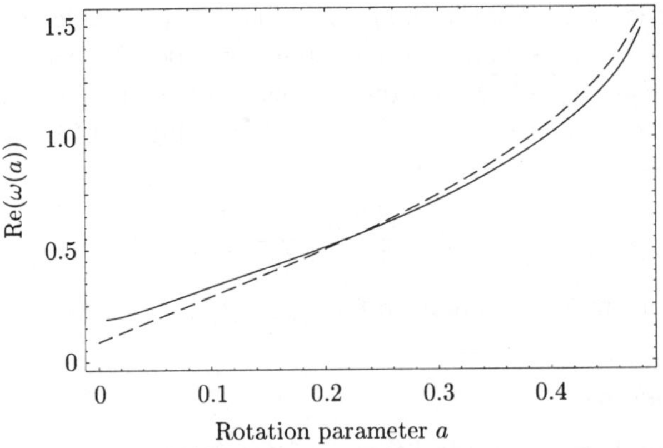

Figure 3. Comparing Hod's prediction with numerical data for the $n = 8$ and $l = m = 2$ modes. The dashed line is the theoretical prediction. The curve is matched well. The biggest discrepancy occurs at the origin, which is to be expected since we are looking at small damping and the ω is assumes its limiting value for large damping.

plot, this time including all the values of m. None of the graphs shows a good matching with the theoretical prediction.

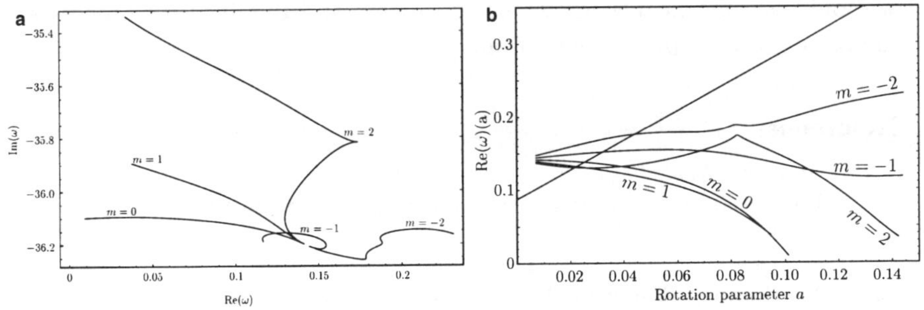

Figure 4. **a** The Kerr quasinormal modes for higher damping. We have $n = 79$, $l = 2$, $m = -2, \ldots, 2$, and $a = 0 \ldots \approx 0.14$.**b** The function $\mathrm{Re}(\omega)(a)$. There is no good agreement for any of the curves.

6. Supersymmetry

In section 4 we saw that we could get perfect agreement of the loop quantum gravity calculation with the Bekenstein-Hawking result provided we change the gauge group from SU(2) to SO(3). This step is problematic since it is not clear how to include fermions in such a theory. Using supersymmetric spin networks Ling and Zhang[16] were able to circumvent this problem. For supersymmetric spin networks the area spectrum is $A(j) = 8\pi\gamma l_P^2\sqrt{j(j+1/2)}$ and the dimension of the representation spaces is $4j+1$. Using the same arguments as above Ling and Zhang obtain the following result for the entropy:

$$S = \frac{A}{4}\frac{\ln(4j_{\text{min}}+1)}{\ln(3)} \tag{7}$$

Exact agreement is thus obtained for $j_{\text{min}} = 1/2$.

7. Conclusions

We have seen how the classical theory of quasinormal modes could be used to fix an ambiguity in Loop Quantum Gravity. A more sophisticated treatment of black holes in Loop Quantum Gravity will be required to see how real this connection is.

The recent results by Hod concerning rotating black holes are interesting but the numerical data is ambiguous. More numerical data on highly damped Kerr modes is required to settle this question.

The appearance of supersymmetry is surprising. The deeper significance will only become clear when we better understand the dynamic processes underlying the proposed connection.

References

1. G. Immirzi, *Nucl. Phys. Proc. Suppl.* **57**, 65 (1997).
2. J. Polchinski, *String Theory I&II*, Cambridge University Press, (1998)
3. T. Thiemann, gr-qc/0110034.
4. J. D. Bekenstein, *Phys. Rev.* **D7**, 2333 (1973).
5. S. W. Hawking, *Commun. Math. Phys.* **43**, 199 (1975).
6. H.-P. Nollert, *Class. Quantum Grav.* **16**, R159 (1999).
7. K. D. Kokkotas and B. G. Schmidt, *Living Reviews in Relativity*, (1999).
8. S. Hod, *Phys. Rev. Lett.* **81** 4293 (1998).
9. H.-P. Nollert, *Phys. Rev.* **D47**, 5253 (1993).
10. L. Motl, *Adv. Theor. Math. Phys.* **6** 1135 (2003). (2003).
11. O. Dreyer, *Phys. Rev. Lett.* **90**, 081301 (2003).
12. J. Wheeler, in *Sakharov Memorial Lectures in Physics*, edited by L. Keldysh and V. Feiberg (Nova Science, New York, 1992), Vol. 2.
13. S. Hod, *Phys. Rev.* **D67**, 081501 (2003).
14. H. Onozawa, *Phys. Rev.* **D55**, 3593 (1997).
15. E. W. Leaver *Proc. R. Soc. Lond.* **A402** 285 (1985).
16. Y. Ling and H. b. Zhang, *Phys. Rev.* **D68**, 101501 (2003).

OBSERVABILITY AND GEOMETRY IN THREE DIMENSIONAL QUANTUM GRAVITY*

KARIM NOUI and ALEJANDRO PEREZ

Center for Gravitational Physics and Geometry, Pennsylvania State University University Park, PA 16802, USA

We consider the coupling between massive and spinning particles and three dimensional gravity. This allows us to construct geometric operators (distances between particles) as Dirac observables. We quantize the system à la loop quantum gravity: we give a description of the kinematical Hilbert space and construct the associated spin-foam model. We quantize the physical distance operator and compute its spectrum.

1. Coupling gravity to point particles

It is well known [5] that three dimensional gravity can be formulated as a Chern-Simons theory whose gauge group G depends on the signature of the metric and on the sign of the cosmological constant. In this article, we will exclusively concentrate on the euclidean case with zero cosmological constant. In that case, the gauge group is the isometry group $G = ISU(2)$ of the three dimensional euclidean space $\mathbb{E}^3$. We will denote by $(P_a)_{a=0,1,2}$ and $(J_a)_{a=0,1,2}$ respectively the translational and rotational generators of the Lie algebra $\mathfrak{g} = isu(2)$.

The coupling between a point particle (mass m and spin s) and the gravitational field can be formulated as a minimal gauge coupling.

1.1. *The coupled system as a minimal gauge coupling*

The formalism we would like to develop is based on the fact that we can identify the degrees of freedom of a massive and spinning relativistic particle evolving in the three dimensional euclidean space $\mathbb{E}^3$ with an element of $G = ISU(2)$. Any element $X \in G$ can be decomposed, by definition, as the

*Talk given by K.N.

semi-direct product $X = (\Lambda, q)$ of a rotation $\Lambda \in SU(2)$ and a translation q: we can clearly identify q with the position of a particle in $\mathbb{E}^3$ and we will show that Λ is related to the momentum of the particle[a]). In that framework, the dynamics between times t_1 and t_2 of a relativistic particle is defined by the following first order action:

$$S_p[X] = \int_{t_1}^{t_2} dt < \chi(m,s) \, , \, X^{-1}\frac{dX}{dt} > \\ \text{where } \chi(m,s) = mJ_0 + sP_0 \in \mathfrak{g} \, . \quad (1)$$

The application $<,>$: $\mathfrak{g} \times \mathfrak{g} \to \mathbb{C}$ is a non-degenerate invariant bilinear form on $\mathfrak{g}$ such that the only non-vanishing product is $< P_a, J_b >= \delta_{ab}$. One can show that the action S_p reproduces the motion of a free massive and spinning relativistic particle: its trajectory is a geodesic of $\mathbb{E}^3$ and its momentum is conserved [4].

Before going to the coupled system, it is interesting to note that the free particle action (1) is invariant under the global transformation $X \mapsto gX$ for any $g \in G$. The coupling to the gravitational field consists precisely of gauging this global symmetry using a G-connection ω. Naturally, the dynamics of the gauge field ω is given by a Chern-Simons action $S_{CS}[\omega]$ of gauge group G defined on a three dimensional manifold $M = \Sigma \times [t_1, t_2]$:

$$S_{CS}[\omega] \;=\; \int_M d^3x \, \epsilon^{\mu\nu\rho} \left(< \omega_\mu, \partial_\nu \omega_\rho > + \frac{1}{3} < \omega_\mu, [\omega_\nu, \omega_\rho] > \right) \, . \quad (2)$$

A naive minimal gauge coupling introduces ambiguities and therefore it is necessary to consider a regularization. The basic idea[1] consists of replacing the point particle by a small loop ℓ whose time evolution draws a cylinder $B = \ell \times [t_1, t_2]$ in the space-time M.

The particle lives on the boundary B and the coupling between the particle and the gravitational field is defined as an integral over B:

$$S_c[\omega, X] \;=\; \frac{1}{2\pi} \int_B dt \, d\varphi < \chi(m,s), X^{-1}\frac{dX}{dt} + X^{-1}\omega_t X > \, . \quad (3)$$

Note that $\varphi \in [0, 2\pi]$ is the angular variable parametrizing the loop ℓ and the origin $\varphi = 0$ is arbitrary. In this integral, X depends only on t and ω_t is the component of the connection in the time direction.

The dynamics of the gauge field ω is, à priori, described by a Chern-Simons action (2). However, because of the regularization, the space-time

---

[a]The momentum p and spin-vector σ of a three-dimensional relativistic particle are proportional and related by the expression $\sigma = \frac{s}{m} p$ where m and s are respectively the mass and the spin.

admits a boundary B which breaks the gauge invariance. To restore the invariance, first we have to ask the question of what type of symmetry we would like at the boundary. As the parameter φ does not have any physical meanning, we impose that gauge transformations are constant along ℓ. To have a theory invariant under these transformations, the Chern-Simons action has to be supplemented with the boundary term $S_B[\omega] = \int_B dtd\varphi < \omega_t, \omega_\varphi >$. It is trivial to verify that the resulting action $S_{CS}[\omega] + S_B[\omega]$ is invariant under the required gauge transformations. Then, the action for the coupled system is given by:

$$S[\omega, X] \;=\; S_{CS}[\omega] \;+\; S_B[\omega] \;+\; S_c[\omega, X]\,. \tag{4}$$

The generalization to an arbitrary number N of particles is straightforward: we call $\wp_i$ the ith particle of mass m_i and spin s_i whose dynamical variable is denoted $X_i = (\Lambda_i, q_i) \in G$.

1.2. *Symmetries and observables*

This section aims to present the main results of the canonical analysis of the action (4). In particular, we obtain the following constraint defined for any $v \in C^\infty(\Sigma, \mathfrak{g})$:

$$\Phi(v) \;\equiv\; \int_\Sigma d^2x \; \epsilon^{ij} < v, F[\omega]_{ij} > +2 \int_\ell d\varphi < v, \omega_\varphi + \frac{1}{2\pi} X\chi(m,s)X^{-1} > \tag{5}$$

As a consequence, the connection is flat on the surface Σ but its holonomy H_ℓ around the loop ℓ is given by $H_\ell = X\exp(-\chi(m,s))X^{-1}$. Note that the connection ω is defined everywhere on Σ and therefore there is no singularity.

When v is constant along ℓ, then the constraint (5) is first class and therefore generates gauge transformations. To this constraint, we have to add the constraints that generate internal symmetries associated to the particle: they are parametrized by an element ε of the Cartan subalgebra of $\mathfrak{g}$. As a result, infinitesimal gauge transformations on the variables X and on the spatial components of the connection $(\omega_i)_{i=1,2}$ read:

$$\delta X \;=\; -X\varepsilon - v(\ell)X \qquad \text{and} \qquad \delta\omega_i \;=\; D_i v \tag{6}$$

where $v(\ell)$ denotes the evaluation of the function v on the loop ℓ. These results are trivially extended to a system $P = \{\wp_1, \cdots, \wp_N\}$ of N particles.

In the absence of particles, the theory reduces to a Chern-Simons theory associated to the group G and the physical phase space is known as the moduli space of flat G-connections. Classical observables are defined in terms of spin-networks. In the presence of particles, we can define new observables which capture the dynamics of the particle and contain, in particular, the relative positions between particles. To understand this point, let us define the following function:

$$O_\gamma[\omega, X_i, X_j] \equiv X_i(t_1)^{-1} U_\gamma[\omega] X_j(t_1) . \tag{7}$$

In this definition, $U_\gamma[\omega]$ denotes the holonomy of the connection along the curve γ whose starting point and end point are respectively the points of coordinates $\varphi_i \in \ell_i$ and $\varphi_j \in \ell_j$. As the connection ω is flat, O_γ depends only on the homotopy class of the path γ. Moreover, it is clearly invariant under the gauge transformations (6) and therefore is an observable. We deal with internal symmetries of the particle by considering initial configuration variables in (7).

To extract physical informations from $O_\gamma[\omega, X_i, X_j]$, we shall write it in the vectorial representation and we obtain:

$$\begin{pmatrix} \Lambda_i(t_1)^{-1} h_\gamma[A] \Lambda_j(t_1) & \Lambda_i(t_1)^{-1}(h_\gamma[A] q_j(t_1) + q_\gamma[e, A] - q_i(t_1)) \\ 0 & 1 \end{pmatrix} . \tag{8}$$

$h_\gamma[A] \equiv P\exp\int_\gamma A$ denotes the holonomy of the spin-connection A along the path γ and we have introduced the notation $q_\gamma[e, A] \equiv \int_\gamma h_{\gamma<x}[A] e_\mu(x) dx^\mu$ where the path $\gamma < x$ (resp. $\gamma > x$) is the part of γ which ends to (resp. starts from) the point $x \in \gamma$. One can see that the translational part $q_\gamma(i,j)$ of O_γ (8) is the analogous of the position of the particle $\wp_j$ in the rest frame of $\wp_i$. Then, the relative distance $D_\gamma(i,j)$ is given by the formula:

$$D^2_\gamma(i,j) = q_\gamma(i,j)^\dagger \, q_\gamma(i,j) . \tag{9}$$

This distance operator is a gauge invariant function which depends only on the homotopy class of γ. As a result, we have constructed a distance operator which is a Dirac observable and therefore should be related to a physical process.

2. Loop quantization

The following aims at exploring loop quantum gravity techniques as an interesting quantization scheme of the coupled system. In particular, we define the notion of kinematical Hilbert space of the coupled system, we quantize the distance observables and compute their action on particular states.

For that purpose, we decompose the first class constraints Φ (5) into its torsion part Φ_T and its curvature part Φ_C defined by:

$$\Phi_T(v) \equiv \Phi(v^a J_a) \quad \text{and} \quad \Phi_C(v) \equiv \Phi(v^a P_a) \tag{10}$$

for any element $v \in C^\infty(\Sigma, \mathbb{R}^3)$ which is constant on the boundary ℓ. These constraints respectively generate the rotational and the translational gauge transformations (6) on the dynamical variables.

2.1. *Kinematical Hilbert space*

The strategy will consist of quantizing the theory before implementing the first class constraints. The kinematical Hilbert space $H_{kin}(\Sigma, P)$ will be defined as the set of states satisfying all first class constraints but not the curvature constraint Φ_C endowed with a suitable scalar product. Then, we define the physical Hilbert space $H_{phys}(\Sigma, P)$ from the kinematical one by introducing a projector which implements the constraint Φ_C. Finally, we make use of this projector to define a spin foam model describing dynamics of massive and spinning self-gravitating particles [2].

In the pure gravitational case, kinematical states are defined as gauge invariant cylindrical functions on Σ. The set of invariant cylindrical functions $\mathrm{Cyl}(\Sigma)$ in naturally endowed with an invariant measure and spin-network states provide it with an orthonormal basis. To generalize this notion to the coupled system, let us note that the action (1) of the free relativistic particle $\wp$ admits a "loop quantization" scheme which leads to the following formulation for its Hilbert space:

$$H(\wp) \equiv \bigoplus_{j-s\in\mathbb{N}} \{\overset{j}{\pi}(\Lambda)^l_s | l \in [-j, +j]\} \,. \tag{11}$$

In that definition, $\overset{j}{\pi}$ denotes the spin j-representation of $SU(2)$ and s is the spin of the particle. Therefore, $H(\wp)$ is a sub-space of polynomial functions of $SU(2)$ whose Hilbert structure is naturally given by the $SU(2)$ Haar measure. We make use of this formulation to generalize the space of cylindrical functions to the coupled system as follows:

$$\mathrm{Cyl}(\Sigma, P) \equiv \mathrm{Cyl}(\Sigma) \otimes H(\wp_1) \otimes \cdots \otimes H(\wp_N) \,. \tag{12}$$

The inner products on $\mathrm{Cyl}(\Sigma)$ and on $H(\wp_i)$ naturally endow $\mathrm{Cyl}(\Sigma, P)$ with a Hilbert structure and we call auxiliary Hilbert space $H_{aux}(\Sigma, P)$ its completion.

The generalized Gauss constraint Φ_T (10) generates $SU(2)$ gauge transformations on the variables of the coupled system. As a result, the kinematical Hilbert space $H_{kin}(\Sigma, P)$ is defined by the Hilbert subspace of gauge invariant functions of $H_{aux}(\Sigma, P)$. At this stage, we can generalize the notion of spin-network states to the coupled system and provide $H_{kin}(\Sigma, P)$ with an orthonormal basis [2]. The construction of the physical Hilbert space is presented in [2] and is a generalization of the techniques developed in the pure gravitational case [3].

2.2. *Distance operators*

Let us consider a classical (square) distance operator (9) between the particles $\wp_i$ and $\wp_j$ and associated to an oriented path γ linking them. One can see that there is no ambiguity to quantize it and one obtains a quantum operator $\hat{D}^2_\gamma(i,j)$ acting on $H_{kin}(\Sigma, P)$. Let us consider a gauge invariant generalized spin-network state $\Psi_S \in H_{kin}(\Sigma, P)$ associated to a graph Γ such that there is one and only one edge $\zeta \in E_\Gamma$ (colored by i_ζ) which intersects the curve γ on only one point $x \notin \ell_1, \ell_2$. The action of $\hat{D}^2_\gamma(i,j)$ on such a state is immediate and the result reads[2]:

$$\hat{D}^2_\gamma(i,j) \triangleright \Psi_S \;=\; i_\zeta(i_\zeta + 1)\, \Psi \,. \tag{13}$$

This result is identical to the usual one obtained in pure loop quantum gravity even if our expression of the distance operator is, à priori, completely different to the usual one. Thus, it seems that we can interpret each holonomy of any spin-network as a flux of geometry. Finally, the action of the distance operator acting on a spin-network whose associated graph intersects γ many times (but not on the boundaries) is still diagonal and is given by a contribution for each edge $\zeta \in E_\Gamma$ intersected.

Aknowledgments

The authors would like to thank Lee Smolin for the kind invitation to the QTS3. This work has been supported in part by NSF grants PHY-0090091 and INT-0307569 and the Eberly Research Funds of Penn State.

References

1. E. Buffenoir and K. Noui, Unfashionable observations about three dimensional gravity, *pre-print* **gr-qc/0305079**.
2. K. Noui and A. Perez, Three dimensional loop quantum gravity: coupling to point particles, *to appear.*

3. K. Noui and A. Perez, Three dimensional loop quantum gravity: physical scalar product and spin-foam models, *to appear.*
4. P. de Sousa Gerbert, On spin and (quantum) gravity in 2+1 dimensions, *Nucl. Phys.* **B346**, 440 (1990).
5. E. Witten, 2+1 dimensional gravity as an exactly soluble system, *Nucl. Phys.* **B311**, 46 (1988).

DYNAMICS OF LOOP QUANTUM GRAVITY AND SPIN FOAM MODELS IN THREE DIMENSIONS*

KARIM NOUI and ALEJANDRO PEREZ

Center for Gravitational Physics and Geometry, Pennsylvania State University, University Park, PA 16802, USA

We present a rigorous regularization of Rovellis's generalized projection operator in canonical 2+1 gravity. This work establishes a clear-cut connection between loop quantum gravity and the spin foam approach in this simplified setting. The point of view adopted here provides a new perspective to tackle the problem of dynamics in the physically relevant 3+1 case.

1. Introduction

The goal of the *spin foam* approach [1] is to construct a mathematically well defined notion of path integral for non-perturbative quantum GR as a devise for computing the '*dynamics*' of the theory. By '*dynamics*' here we mean the characterization of the kernel of the quantum constraints given by the representation of the classical constraints in connection variables, schematically, $G_i = D_a E^a_i$, $V_b = E^a_i F^i_{ab}$, and $S = E^a E^b F_{ab}(A)$ [a].

Spin foam models have been studied as an attempt to give an explicit construction of the generalized projection operator P from the kinematical Hilbert space H_{kin}, where the constraints above are defined as operators, into their kernel H_{phys}. A tentative regularization of the formal projector into the kernel of $S(x)$ and $V^a(x)$

$$P = \text{“} \prod_{x \in \Sigma} \delta(\hat{V}^a(x))\delta(\hat{S}(x)) \text{”} = \int D[N_\mu] \exp(i \int_\Sigma N_0 \hat{S} + N_a \hat{V}^a) \qquad (1)$$

*Talk given by A.P.

[a] The field A is an $SU(2)$ connection and E is its canonical momentum represented by the sensitized triad both defined on a 3-manifold Σ defining space. Space-time is assume to be of the topology $\Sigma \times \mathbb{R}$.

was presented in [2b]. Given two *spin network* states s, s' the physical scalar product $< s, s' >_{phys} := < Ps, s' >$ can be formally defined by

$$< s, s' >_{phys} = \langle Ps, s' \rangle = \int D[N] \sum_{n=0}^{\infty} \frac{i^n}{n!} < \left[\int_{\Sigma} N(x) \hat{S}(x) \right]^n s, s' >, \quad (3)$$

where the exponential in (1) has been expanded in powers. From early on, it was realized that smooth loop states are naturally annihilated (independently of any regularization ambiguity) by $\hat{S}$ [3,4]. Consequently, $\hat{S}$ acts only on *spin network* nodes. Generically, it does so by creating new links and nodes modifying in this way the underlying graph of the *spin network* states. The action of $\hat{S}$ can be visualized as an 'interaction vertex' in the time evolution of the node (see diagram on the right of Figure 1). Therefore, each term in the sum (3) represents a series of transitions–given by the local action of $\hat{S}$ at *spin network* nodes–through different *spin network* states interpolating the boundary states s and s' respectively. They can in fact be expressed as a sum over 'histories' of *spin network* that can be pictured as a system of 'colored' branching surfaces described by a 2-complex labeled by spins. Every such history is called a *spin foam*.

Before even considering the issue of convergence of (3), the problem with this definition is evident: every single term in the sum is a divergent integral! Therefore, this way of presenting *spin foams* has to be considered as formal until a well defined regularization of (1) is provided. Possible regularization schemes are discussed in [2] although they have not been implemented in concrete examples.

Although many spin foam models for 4-dimensional gravity have been proposed[1], the rigorous connection with the well developed canonical theory of loop quantum gravity is still under investigation. Here we show that a clear-cut connection between the loop approach and spin foams can be established in three dimensions.

Pure gravity in three dimensions is a well studied example of integrable system that can be rigorously quantized. The reduced phase space of the

---

[b] One can also define the notion of path integral for gravity as a lattice discretization of the formal path integral for GR in first order variables

$$P = \int D[e]\ D[A]\ \mu[A, e] \exp\left[iS_{GR}(e, A)\right]$$

where the formal measure $\mu[A, e]$ must be determined by the Hamiltonian analysis of the theory.

theory is finite dimensional and there exist different quantization schemes that make use of this property.

From our perspective 3-dimensional gravity is taken as an toy model for the application of quantization techniques that are expected to be useful in four dimensions. In this sense we want to quantize the theory according to Dirac prescription which implies having to deal with the infinitely many degrees of freedom of a field theory at the kinematical level, i.e., we want to quantize first and then reduce at the quantum level. This is precisely the avenue that is explored by loop quantum gravity in four dimensions where the reduced phase space approach seems hopeless.

2. Canonical three dimensional gravity

Assuming the topology of space time to be of the form $M = \Sigma \times \mathbb{R}$, where Σ is an arbitrary Riemann surface, the phase space of 3-dimensional gravity is parametrized the 2-dimensional connection A_a^i and the triad field E_j^b where $a = 1,2$ are Σ-coordinate indices and $i,j = 1,2,3$ are $su(2)$ indices. The symplectic structure is defined by $\{A_a^i(x), E_j^b(y)\} = \delta_a^{\ b}\delta^i_{\ j}\delta(x,y)$. Local symmetries of the theory are generated by the Gauss and curvature constraints, $D_b E_j^b = 0$ and $F_{ab}(A) = 0$, respectively.

As in 4d, the kinematical Hilbert space, H_{kin}, is given by a certain set of $SU(2)$ gauge invariant functionals of the connection $\Psi[A]$ which are square integrable with respect to a natural diffeomorphism invariant measure, the Ashtekar-Lewandowski (AL) measure [5,6]. As in standard gauge theories, the basic gauge invariant functional of A is given by the Wilson loop; namely, the trace of the holonomy of A around a close loop on the fundamental representation of $SU(2)$. Any state in H_{kin} can be expressed in terms of linear combinations of products of Wilson loops. Natural orthonormal basis of H_{kin} can be constructed in terms of eigenstates of geometric operators of quantum geometry representing area and length, given by the so-called *spin network* states [7,8,9]. *Spin network states* $\Psi_{\gamma,\{j_\ell\},\{\iota_n\}}[A]$ are defined by a graph γ in Σ, a collection of spins $\{j_\ell\}$ associated with links $\ell \in \gamma$, and a collection of $SU(2)$ invariant tensors (called intertwiners) $\{\iota_n\}$ associated to nodes $n \in \gamma$ (see left diagram on Figure 1).

3. Dynamics and spin foams

In this section we introduce the regularization of the generalized projection operator P and state the results that are studied in detail in [10].

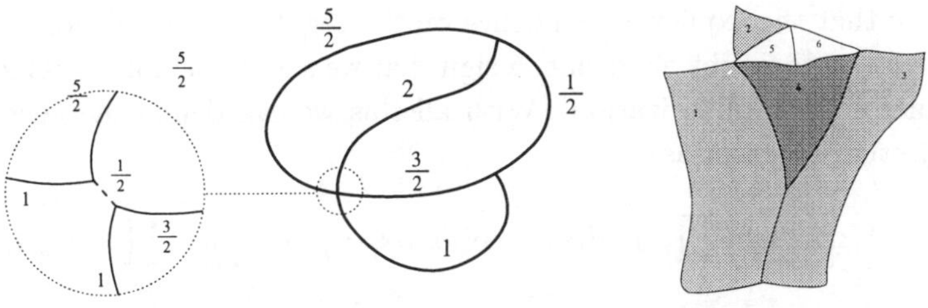

Figure 1. Spin-network state on the left. On the right 'time' evolution of a *spin-network node* into a *spin foam* vertex.

We start with the formal expression

$$P = \text{“} \prod_{x \in \Sigma} \delta(\hat{S}(x))\text{”} = \int D[N] \exp(i \int_{\Sigma} \mathrm{Tr}[N\hat{F}(A)]) \tag{4}$$

For the moment we assume the genus of Σ to be greater or equal than one (the sphere case is trivial). In this case, the Riemann surface admits a global (fiducial) flat metric q^0_{ab}. We introduce a square cellular decomposition of Σ denoted Σ_ϵ and we require each square to have area ϵ^2 with respect to q^0_{ab}. We define the set of continuous finite graphs in Σ by Γ. The completion in the AL norm of the set of cylindrical functions based on Γ_ϵ, defined by the set of continuous graphs contained in the 1-skeleton of the square cellular decomposition Σ_ϵ, defines a subspace H_ϵ of the kinematical Hilbert space H_{kin}. In the limit $\epsilon \rightarrow 0$ we have $H_\epsilon \rightarrow H_{kin}$.

Based on the fiducial square cellular decomposition Σ_ϵ we can now define P by introducing a regularization of the right hand side of (4). The integral in the exponential can be written as

$$\int_{\Sigma} \mathrm{Tr}[NF(A)] = \lim_{\epsilon \rightarrow 0} \sum_{p^i} \epsilon^2 \mathrm{Tr}[N_{p^i} F_{p^i}],$$

where p^i labels the ith plaquette and N_{p^i} and F_{pi} the value of $F(A)$ and N at some interior point of the plaquette respectively. The basic observation is that the holonomy $U_{p^i} \in SU(2)$ around the plaquette p^i can be written as

$$U_{p^i}[A] = \mathbb{1} + \epsilon^2 F_{pi}(A) + O(\epsilon^2)$$

which implies

$$\int_{\Sigma} \mathrm{Tr}[NF(A)] = \lim_{\epsilon \rightarrow 0} \sum_{p^i} \mathrm{Tr}[N_{p^i} U_{p^i}[A]].$$

Notice that the explicit dependence on the regulator ϵ has dropped out of the sum on the right hand side, a sign that we should be able to remove the regulator upon quantization. With all this we can define the generalized projection operator as

$$\hat{P} := \lim_{\epsilon \to 0} \prod_{p^i} \int dN_{p^i} \exp(i\mathrm{Tr}[N_{p^i}\hat{U}_{p^i}]) = \lim_{\epsilon \to 0} \prod_{p^i} \widehat{\delta(U_{p^i})}, \qquad (5)$$

where the last equality follows from direct integration over N_{p^i} at the classical level and $\delta(U)$ is the $SU(2)$ δ-distribution defined on $L^2(SU(2))$. We promote $\delta(U)$ to an operator by using its expansion in $SU(2)$ irreducible representations, namely $\delta(U_{p^i}) = \sum_j (2j+1)\ \chi_j(U_{p^i})$; $\chi_j(U)$ is the character of the j-representation matrix of $U_{p^i} \in SU(2)$. In contrast to the formal example in 4d, the previous expansion has a precise meaning in the quantum theory as each term in the sum can be promoted to a well defined self-adjoint operator in H_{kin}: the Wilson loop operator in the j-representation.

Now we state the results that are proved in detail in [10].

1 - The physical inner product $< s, s' >_{phys} := < sP, s' >$ admits a *spin foam* representation, i.e., it can be expressed as a sum over *spin network* 'histories' interpolating s and s' respectively [c]. The *spin foams* correspond to continuous 2-complexes defined independently from any background structure in the limit $\epsilon \to 0$.

2 - The regularization (5) is well defined. Lattice definitions of 3-dimensional gravity such as the Ponzano-Regge model are plagued with divergences. These divergences do not appear here however they can be traced to the presence of redundant δ-distributions in (5) (see [11] for a treatment in the lattice context).

3 - The physical Hilbert space H_{phys} defined by the equivalence classes of states in H_{kin} up to physically null states (i.e. states ψ for which $< \psi, \psi >_{phys} = 0$) is isomorphic to the Hilbert space obtained in other methods[12]. In addition, the spin foam representation allows to explicitly show how *spin networks* in the same homotopy class are physically equiva-

[c]This is obtain by the insertion of the resolution of the identity in H_{kin}

$$\mathbb{1} = \sum_{\gamma,\{j\}} |\gamma, \{j\} >< \gamma, \{j\}|$$

(where the sum is over all continuous *spin network states*) between δ-distributions in (5) in analogy with Feynman's original derivation of the path integral representation of dynamics in QM.

lent and to easily prove skein relations.
4 - The *spin foam* representation provides a natural basis for H_{phys} for Σ of arbitrary genus. These states are eigenstates of a complete set of commuting quantum geometry operators (provided by the formalism). This completes the quantization of 2+1 gravity in the spin foam approach and establishes and establishes a clear-cut connection with the canonical picture.
5 - Similarly point particles can be systematically coupled to gravity[13].

We hope that the spin foam perspective, fully realized here in 2+1 gravity, can bring new breath to the problem of dynamics in 3+1 gravity. This is an issue that will be studied further.

Acknowledgments

The authors would like to thank Lee Smolin for the kind invitation to the QST3. This work has been supported in part by NSF Grants PHY-0090091 and INT-0307569 and the Eberly Research Funds of Penn State.

References

1. Alejandro Perez. Spin foam models for quantum gravity. *Class. Quant. Grav.*, 20:R43, 2003.
2. C. Rovelli. The projector on physical states in loop quantum gravity. *Phys.Rev. D*, 59:104015, 1999.
3. L. Smolin T. Jacobson. Nonperturbative quantum geometries. *Nucl. Phys.*, B299:295, 1988.
4. C. Rovelli L. Smolin. Loop space representation of quantum general relativity. *Nucl. Phys. B*, 331:80, 1990.
5. A. Ashtekar and J. Lewandowski. Projective techniques and functional integration. *J. Math. Phys.*, 36:2170, 1995.
6. A. Ashtekar and J. Lewandowski. Representation theory of analytic holonomy c^* algebras. *in "Knots and quantum gravity", J. Baez (ed), Oxford University Press, Oxford 1994.*
7. Michael P. Reisenberger. World sheet formulations of gauge theories and gravity. *gr-qc/941235*, 1994.
8. C. Rovelli and L. Smolin. Spin networks and quantum gravity. *Phys. Rev. D*, 53:5743, 1995.
9. J. Baez. Spin network states in gauge theory. *Adv.Math.*, 117:253–272, 1996.
10. K. Noui A. Perez. Three dimensional loop quantum gravity: physical scalar product and spin foam models. *In preparation.*

11. Laurent Freidel and David Louapre. Diffeomorphisms and spin foam models. *Nucl. Phys.*, B662:279–298, 2003.
12. S. Carlip. Quantum gravity in 2+1 dimensions. Cambridge, UK: Univ. Pr. (1998) 276 p
13. K. Noui A. Perez. Three dimensional loop quantum gravity: coupling with particles. *In preparation.*

AN INVITATION TO LOOP QUANTUM GRAVITY

LEE SMOLIN

Perimeter Institute for Theoretical Physics
Waterloo, Canada
and
Department of Physics, University of Waterloo
lsmolin@perimeterinstitute.ca

We describe the basic assumptions and key results of loop quantum gravity. No details are given, but references are provided to guide the interested reader to the literature. We provide a list of 34 key results, to date, on loop quantum gravity. We discuss also the near term prospects for observational tests of quantum theories of gravity and the expectations that loop quantum gravity may provide predictions for their outcomes. Finally a list of some key open problems is given.

1. Introduction

Loop quantum gravity is an approach to quantum gravity based on a simple and conservative premise: Construct a quantum theory of spacetime based only on the experimentally well confirmed principles of general relativity and quantum mechanics. Whatever one believes about the unification of physics, and whatever the ultimate outcome of our search for that unification, it has to be of interest to determine whether or not a consistent and sensible physical theory can be constructed with no other assumptions than the principles of quantum theory and general relativity.

Remarkably, after 17 years of work by a community of several score physicists and mathematicians, a great deal of evidence has accumulated that the answer to this question is yes. Furthermore, the result is a beautiful and robust picture of quantum spacetime, which is to many compelling, independently of the fact that it has been derived from an honest quantization of general relativity.

The main purpose of this review is to explain how it has been possible to obtain exact results about quantum gravity, and then to list the basic results of loop quantum gravity.

While the principles assumed are only those of general relativity and quantum mechanics, there are four key observations that make the success of loop quantum gravity possible. These are

I. *Classical general relativity is a background independent theory, hence any theory which is to have general relativity as a low energy limit must be background independent.* A background independent theory is one whose formulation does not assume or require the existence of any single preferred spacetime metric or connection. Instead, all the fields that define the geometry of spacetime are fully dynamical, none are fixed.

II. *Duality and diffeomorphism invariance may be consistently combined in a quantum theory* It is possible to construct a diffeomorphism invariant quantization of any classical diffeomorphism invariant gauge field theory, in which the fundamental degrees of freedom are the Wilson loops of the gauge field.

From a technical point of view, the main achievement of loop quantum gravity is the discovery of a new kind of quantum gauge field theory, which is exactly invariant under diffeomorphisms of a manifold. This is understood in a great deal of detail, and the Hilbert space, inner product, states, observables and path integral representation are all understood in closed form. That understanding is robust, and has been achieved by several regularization procedures. Significantly, there is a completely rigorous formulation of these diffeomorphism invariant gauge theories.

III. *General relativity and all related theories, such as supergravity, can be formulated as gauge theories.* This means that the configuration variable is a gauge field, and the metric information is contained in the conjugate momenta to the gauge field. This extends to coupling with all the known kinds of matter fields.

IV. *Further, general relativity, supergravity and related theories can be put in a special form in which they are* ***constrained topological field theories.*** These are theories whose actions differ from the action of a topological field theory by the imposition of a non-derivative, quadratic constraint equation. That constraint diminishes the number of gauge invariances, leading to the emergence of local physical degrees of freedom, while maintaining the diffeomorphism invariance of the theory. This is true of general relativity in all dimensions and is also known to be true of supergravity, in $d = 4$, at least up to $N = 2$, and for $d = 11$.

As a result, the techniques which give us consistent diffeomorphism invariant gauge theories can be applied to give consistent quantizations of general relativity and supergravity. Many results follow, which will be described in more detail below. Most of the key results, including the following, have been confirmed as theorems in the rigorous formulation of diffeomorphism invariant quantum field theory.

The fundamental result which follows from these observations is that *quantum geometry is discrete.* Operators which measure the areas and volumes of diffeomorphism invariantly defined surfaces and regions may be constructed. They are finite, after an appropriate regularization, respecting diffeomorphism invariance, and they have discrete, computable spectra. Hence, the theory predicts minimal physical areas and volumes. The common eigenstates of these observables give an orthonormal basis of diffeomorphism invariant states, which are certain labeled graphs called spin networks. Their evolution can be described in closed form in either a Hamiltonian or path integral language. Furthermore, the dynamics of the matter fields can be constructed and studied. Because of the existence of minimal quanta of volume, all divergences of ordinary quantum field theories are eliminated. This is because there simply are no degrees of freedom in the exact theory that correspond to gravitons or other quanta with wavelength shorter than the Planck length. There is evidence as well that singularities of general relativity are eliminated.

It should be apparent from this summary that loop quantum gravity, as a research program, is rather different from other research programs such as string theory, that are based on new hypotheses about nature, such as supersymmetry, the existence of higher dimensions, and the unification of all elementary particles and forces by means of strings. These are interesting hypotheses, but it is proper to characterize them as speculative, as they have no direct support from experiment. Nor are there results that show that any of these assumptions are necessary consequences of a consistent unification of quantum theory with gravity and spacetime. It is fair to say that approaches such as string theory have so far failed to lead to a complete and well defined theory. Nor have they led to many results that concern the behavior of quantum spacetime at the Planck scale. Most results in string theory remain at the classical level and there is a widely held view that, to give a deeper picture of quantum spacetime, string (or perhaps $\mathcal{M}$) theory require a background independent formulation.

There are several different approaches to a background independent quantum theory of gravity, besides loop quantum gravity others are dynamical triangulations and causal sets. Loop quantum gravity is the only such formulation which has a mathematically rigorous formulation together with a large body of physically interesting results. As such, it can be applied to construct a large number of different theories, and it can easily incorporate speculative hypotheses about the dimension of spacetime, the form of the action (the Einstein action is not the only one that can be studied) or the presence of additional symmetries such as supersymmetry. All of these theories incorporate the basic principles of background independence and diffeomorphism invariance, so that the spacetime geometry is completely dynamical.

At the same time, by showing that a consistent diffeomorphism invariant quantum field theory can be constructed that represents general relativity, coupled to arbitrary matter fields, in the 3+1 dimensional world we observe, the success of loop quantum gravity undermines the claim that supersymmetry, strings or higher dimensions are necessary for a consistent quantum theory of gravity.

The main body of this review is the list of results given in section 2. We then summarize the near term experimental situation for loop quantum gravity and other theories in section 3, and close with a list of open problems in section 4.

For those interested in a more detailed introduction to loop quantum gravity, there are several textbooks and monographs now available. Early monographs are [1,2], just finished is a textbook by Rovelli[3] also available on line at http://www.cpt.univ-mrs.fr/ rovelli/rovelli.html. A comprehensive introduction to the rigorous side of the subject is given in [4]. Review papers on different aspects of loop quantum gravity include [5,6,7,9].

2. The main results of loop quantum gravity

In loop quantum gravity the quantization is done using standard non-perturbative Hamiltonian and path integral methods, applied to the phase space coordinatized in terms of an alternative set of variables. The configuration variables are taken to be components of the spacetime connection, so that general relativity in a certain precise sense is expressed in terms of a gauge theory. These variables, and the simplifications they bring about were discovered by Sen[10] and formalized by Ashtekar[11] in the Hamiltonian formalism and by Plebanksi[64] and others[65,66] in the lagrangian formalism. By now several different connections are used in loop quantum gravity. These include the self-dual part of the spacetime connection[10,11], and a real SU(2) connection introduced by Barbero[111] and exploited by Thiemann[85]. There are also alternate formulations that use both the left and right handed parts of the spacetime connection, [89,17].

The setting for the results that follow is a bare manifold Σ, with no metric structure, on which is defined the phase space of a gauge theory. The only non-dynamical structure that is fixed is a three manifold Σ, with a given topology and differential structure. There are no fixed classical fields such as metrics, connections or matter fields on Σ. The only exception is in modeling the quantization of spacetime regions with boundary, as in the asymptotically flat or *AdS* context, or in the presence of a black hole or cosmological horizon. In these cases fields may be fixed on the boundary $\partial\Sigma$ to represent physical conditions held fixed there. The dynamics is formulated in terms of a diffeomorphism invariant action which is a functional only of that gauge field and its derivatives.

We are now ready to summarize the results.

2.1. *The fundamental results of the canonical theory*

1. The states of the theory are known precisely. The Hilbert space $\mathcal{H}^{diffeo}$ of spatially diffeomorphism invariant states of general relativity in 3+1 dimensions has an orthonormal basis, whose elements are in one to one correspondence with the diffeomorphism equivalence classes of embeddings of certain labeled graphs, called spin networks[67], into Σ [15]. This of course means that the inner product on diffeomorphism invariant states is known in closed form.

A labeled graph is a graph whose edges and vertices have attached to them elements of a certain set of labels. In the case of pure general relativity with vanishing cosmological constant, the labels on the edges are given by ordinary $SU(2)$ spins. There are also labels on nodes of the spin networks, which are invariants or intertwiners from the representation theory of $SU(2)$. For details see [67,15].

2. Certain spatially diffeomorphism invariant observables have been constructed. After a suitable regularization procedure these turn out to be represented by *finite* operators on $\mathcal{H}^{diffeo}$, the space of spin network states[6,16,62]. These include the volume of the universe, the area of the boundary of the universe, or of any surface defined by the values of matter fields. Other operators also have been constructed, for example an operator that measures angles in the quantum geometry[63]. These operators all preserve the diffeomorphism invariance of the states[72].

3. The area and volume operators have discrete, finite spectra, valued in terms of the Planck length[16,62]. There is hence a smallest possible volume and a smallest possible area, of order of the Planck volume and area. The spectra have been computed in closed form.

4. The area and volume operators can be promoted to genuine physical observables, by gauge fixing the time gauge so that at least locally time is measured by a physical field[71,72]. The discrete spectra remain for such physical observables, hence the spectra of area and volume constitute genuine physical predictions of the quantum theory of gravity.

5. Among the operators that have been constructed and found to be finite on $\mathcal{H}^{diffeo}$ is the Hamiltonian constraint (or, as it is often called, the *Wheeler de Witt* equation[68-70].) Not only can the Wheeler deWitt equation then be precisely defined, it can be solved exactly. Several infinite sets of solutions have been constructed, as certain superpositions of the spin network basis states, for all values of the cosmological constant[13,85]. These are exact, physical states of quantum general relativity.

6. If one fixes a physical time coordinate, in terms of the values of some physical fields, one can also define the Hamiltonian for evolution in that physical time coordinate[71] and it is also given by a finite operator on a suitable extension of $\mathcal{H}^{diffeo}$ including matter fields.

7. Coupling to all the standard forms of matter fields are understood, including gauge fields, spinors, scalars and Kalb-Ramond fields[a]. It is known how to extend the definition of the spatially diffeomorphism invariant states to include all the standard kinds of matter fields, and the corresponding terms for the Hamiltonian constraint are known in closed form, and are finite on the space of diffeomorphism invariant states. These states are also invariant under ordinary Yang-Mills and Kalb-Ramond gauge transformations. [b] Inclusion of matter fields does not affect the finiteness and discreteness of the area and volume observables.

8. There is a rigorous formulation of the Hamiltonian quantization of general relativity in which all the preceding results are reproduced as theorems[110,85,4].

---

[a]For supergravity, see result 31, below.

[b]To my knowledge whether loop quantum gravity suffers from the fermion doubling problem is an open question.

2.2. *Results on path integrals and spin foams*

9. The dynamics of the spin network states can be expressed also in a path integral formalism, called spin foams[73–84c]. The histories by which spin network states evolve to other spin network states, called spin foam histories, are explicitly known. A spin foam history is a labeled combinatorial structures, which can be described as a branched labeled two complex. Spin foam models have been derived in several different ways, and the results agree as to the general form of a spin foam amplitude. These include: 1) by exponentiation of the Hamiltonian constraint, 2) directly from a discrete approximation to the classical spacetime theory, 3) by constraining the summations in a finite state sum formulation of a four dimensional topological invariant, 4) from a matrix model on the space of fields over the group, 5) by postulating spacetime events are local moves in spin networks.

10. Evolution amplitudes corresponding to the quantization of the Einstein equations in $3+1$ dimensions, are known precisely[9] for vanishing and non-vanishing values of the cosmological constant, and for both the Euclidean and Lorentzian theories.

11. The sum over spin foams has two parts, a sum over graphs representing histories of spin networks, and, on each, a sum over the labels. The sums over labels are known from both analytic and numerical results to be convergent[115,116] for some spin foam models, including some corresponding to the quantization of the Einstein equations in $2+1$ and $3+1$ dimensions.

12. For some spin foam model in $2+1$ dimensions, it has been shown that the sum over spin foam histories is Borel summable[129].

13. The physical inner product, which is the inner product on solutions to all the constraints, has an exact expression, given in terms of spin foam models[74].

14. The spin foam models have been extended to include gauge and spinor degrees of freedom.

15. Spin foam models appropriate for Lorentzian quantum gravity, called causal spin foams, have quantum analogues of all the basic features of general relativistic spacetimes[d]. These include dynamically generated

---

[c]For the most recent review see [9].
[d]For more details on these models and the resulting physical picture, see[78].

causal structure, light cones and a discrete analogue of multifingered time, which is the freedom to slice the spacetime many different ways into sequences of spatial slices[77]. The spatial slices are spin networks, which are quantum analogues of spatial geometries.

2.3. *Results on black holes and horizons*

16. Several kinds of boundaries may be incorporated in the theory including timelike boundaries, in the presence of both positive and negative cosmological constant, and null boundaries such as black hole and cosmological horizons[86–89]. In all these cases the boundary states and observables are understood in terms of structures derived from Chern-Simons theory.

17. The boundary theory provides a detailed microscopic description of the physicsof horizons and other boundaries. The horizon entropy (1) is completely explained in terms of the statistical mechanics of the state spaces associated with the degrees of freedom on the horizon. This has been found to work for a large class of black holes, including Schwarzschild black holes[87,88].

The boundary Hilbert spaces decompose into eigenspaces, one for each eigenvalue of the operator that measures the area of the boundary[86]. For each area eigenvalue, the Hilbert space is finite dimensional. The entropy may be computed and it agrees precisely with the Beckenstein-Hawking semiclassical result,

$$S = \frac{A[S]}{4\hbar G_{Newton}} \tag{1}$$

The calculation of the entropy involves a parameter, which is called the Imirzi parameter. This can be understood either as a free parameter that labels a one dimensional family of spin network representations, or as the (finite) ratio of the bare to renormalized Newton's constant. The Imirzi parameter is fixed precisely by an argument invented by Dreyer, involving quasi normal modes of black holes[118]. Dreyer's argument depends on a remarkable precise coincidence between an asymptotic value of the quasi normal mode frequency and a number which appears in the loop quantum gravity description of horizons. The value of the asymptotic quasi normal mode frequency was at first known only numerically, but it has been very recently derived analytically by Motl[119]. Once Dreyer's argument fixes the Imirzi parameter, the Bekenstein-Hawking relation (1) is predicted exactly for all black hole and cosmological horizons[e].

18. Corrections to the Bekenstein entropy have been calculated and found to be logarithmic[91].

19. Suitable approximate calculations reproduce the Hawking predict a discrete fine structure in the Hawking spectrum[92,93]. At the same time, the spectrum fills in and becomes continuous in the limit of infinite black hole mass. This fine structure stands as another definitive physical prediction of the theory.

Thus, to summarize, loop quantum gravity leads to a detailed microscopic picture of the quantum geometry of a black hole or cosmological horizon[88]. This picture reproduces completely and explains the results on the thermodynamic and quantum properties of horizons from the work of Bekenstein[124], Hawking[125] and Unruh[126]. This picture is completely general and applies to all black hole and cosmological horizons.

2.4. *Results on the low energy behavoir*

20. A large class of putative ground states can be constructed which have course grained descriptions which reproduce the geometry of flat space, or any slowly varying metric[102,103]. Linearizing the quantum theory around these states yields linearized quantum gravity, for gravitons with wavelength long compared to the Planck length[104].

21. Excitations of matter fields on these states reproduce a cut off version of the matter quantum field theories, but with a physical, Planck scale cutoff. As a result of the discreteness of area and volume, the ultraviolet divergences of ordinary QFT are not present.

22. It is understood how to construct coherent states which are peaked around classical configurations[117].

23. For the case of non-vanishing cosmological constant, of either sign, there is an exact physical state, called the Kodama state, which is an exact solution to all of the quantum constraint equations, whose semiclassical limit exists[96]. That limit describes deSitter or anti-deSitter spacetime.

[e]Dreyer's calculation leads also to the conclusion that transitions where punctures, i.e. ends of spin networks, are added or subtracted to the boundary, must be dominated by creation and annihilation of spin 1 punctures.

Solutions obtained by perturbing around this state, in both gravitational[7] and matter fields[97], reproduce, at long wavelength, quantum field theory in curved spacetime and the quantum theory of long wave length, free gravitational waves on deSitter or anti-deSitter spacetime[f].

24. It is not yet known whether or not the Kodama state is normalizable in the physical inner product of the exact theory. One can ask whether the projection into the Hilbert space of linearized gravitons on deSitter spacetime gotten by truncating the Kodama state to quadratic order is normalizable or not. The answer is it is not in the Lorentzian case, and it is delta funcational normalizable in the Euclidean case[130].

25. The inverse cosmological constant turns out to be quantized in integral units, so that

$$k = 6\pi / G\Lambda \tag{2}$$

is an integer[86]. This is related to a basic result, which is that in $3+1$ dimensions, a non-zero cosmological constant implies a quantum deformation of the gauge group whose representations label the edges of the spin networks, where the level which parameterizes the quantum deformation is given for the Euclidean case by eq. (2)[g].

26. The thermal nature of quantum field theory in a deSitter spacetime is explained in terms of a periodicity in the configuration space of the exact quantum theory of general relativity[7].

27. In both flat space and around deSitter spacetime, one may extend the calculations that reproduce quantum theory for long wavelength gravitons and matter fields to higher energies. These calculations reveal the presence of corrections to the energy-momentum relations of the form of

$$E^2 = p^2 + M^2 + \alpha l_{Pl} E^3 + \beta l_{Pl}^2 E^4 + \ldots \tag{3}$$

Given an ansatz for the ground state, the parameters α and β are computable constants, that depend on the ground state wavefunctional[30,31,7]. These represent further predictions of the theory.

To summarize, the situation with regard to the low energy limit is very much like that of condensed matter systems. It is possible to make ansatzes

[f] For a possible $\Lambda = 0$ analogue of the Kodama state, see [121].
[g] and for the Lorentzian case by $\imath k$.

for the ground state, which have reasonable physical properties and reproduce the geometry of flat or deSitter spacetime. By studying excitations of these states one reproduces conventional quantum field theory, as well corrections to it which may be compared with experiment. This allows an interplay of theory and experiment, to determine the best ansatz for the ground state.

What there is not so far is a first principles calculation that provides the exactc form of the ground state wavefunction. This is also the case in most condensed matter physics examples.

2.5. *Results concerning cosmology*

28. A reduction of the exact physical state space to states which are spatially homogeneous is known, and the reduction of the dynamics to this subspace of states is known[105]. (This is different from the usual quantum cosmology in that the reduction to homogeneous states is done in the Hilbert space of the full theory, rather than before quantization.) The evolution of these states has been studied in detail and it has been found generically that the usual FRW cosmology is reproduced when the universe is very large in Planck units. At the same time the cosmological singularities are removed, and replaced by bounces where the universe re-expands (or pre-contracts). When couplings to a scalar field are included, there is a natural mechanism which generates Planck scale inflation as well as a graceful exit from it[105]. Recently it has been argued that loop quantum gravity effects lead naturally to a version of chaotic inflation which may also explain the lack of power on long wave lengths in the CMB spectrum[106].

29. Another approach to inflation within loop quantum gravity is given in [131]. Exact homogeneous quantum states can be constructed for general relativity coupled to a scalar field with an arbitrary potential.

30. It has recently been shown that loop quantum gravity effects eliminate the chaotic behavior of Bianchi IX models near singularities[132].

2.6. *Results concerning supergravity and other dimensions*

31. Many of these results extend to quantum supergravity for $N = 1$ and several have been studied also for $N = 2$[18].

32. The same methods can also be used to solve quantum gravity in 2+1 dimensions[107] and in many $1 + 1$ dimensional reductions of the theory[108].

They also work to solve a large class of topological field theories[94,95], giving results equivalent to those achieved by other methods. Further, loop methods applied to lattice gauge theories yields results equivalent to those achieved by other methods[109].

33. Spin foam models are known in closed form for quantum general relativity in an arbitrary dimension $d > 4$ [20].

2.7. *Other extensions of the theory*

34. The mathematical langauge of spin networks and spin foams can be used to construct a very large class of background independent quantum theories of gravity. These may be called generalized loop quantum gravity models. The states of such a quantum theory of gravity are given by abstract spin networks associated with the representation theory of a given Hopf algebra or superalgebra, $\mathcal{A}$[h].

The graphs on which the spin networks are based are defined combinatorially, so that the need to specify the topology and dimension of the spatial manifold is eliminated[77,80]. In such a theory the dimension and topology are dynamical, and different states may exist whose coarse grained descriptions reveal manifolds of different dimensions and topology.

The histories of the theory are given by spin foams labeled by the same representations. The dynamics of the theory is specified by evolution amplitudes assigned to the nodes of the spin foams (or equivalently to local moves by which the spin networks evolve).

Such generalized loop quantum gravity models have been proposed to serve as background independent formulations of string and $\mathcal{M}$ theory[133,80]. They also offer new possibilities for the unification of physics, because the topology of spacetime is not assumed apriori, and hence must be emerge dynamically in the low energy limit. There is then the possibility of new physics coming from obstructions to recovering trivial topology at low energies. It has been proposed that these may account for the existence of both matter degrees of freedom[135], and perhaps even quantum theory itself[136].

2.8. *Comments on the results*

On the basis of these results, it can be claimed that loop quantum gravity is both a consistent quantization of general relativity and a physically

[h]Here a spin network is a graph whose edges are labeled by representations of $\mathcal{A}$ and whose nodes are represented by invariants of $\mathcal{A}$.

plausible candidate for the quantum theory of gravity.

The failure of quantum general relativity in perturbation theory is explained by the fact that there are, in this quantization of general relativity, no degrees of freedom that correspond to gravitons or other perturbative quanta with wavelength shorter than the Planck scale. The ultraviolet divergences are eliminated because a correct quantization, that exactly realizes spatial diffeomorphism invariance, turns out to impose an ultraviolet cutoff on the physical spectrum of the theory. The assumption that spacetime is smooth and lorentz invariant at arbitrarily short scales, is not used in the quantization procedure, and in fact turns out to be contradicted by the results.

3. The near term experimental situation

The most important development of the last few years in quantum gravity is the realization that it is now possible to probe Planck scale physics experimentally. These experiments look for modifications in the energy momentum relations of the form of eq. (3).

However it is crucial to mention that to measure α and β of eq. (3) one has to specify how lorentz invariance is treated in the theory. There are two very different possibilities which must be distinguished.

- **Scenario A)** The relativity of inertial frames is broken and there exists a preferred frame. In this case the analysis has to be done in that preferred frame. The most likely assumption is that the preferred frame coincides with the rest frame of the cosmic microwave background. In such theories energy and momentum conservation are assumed to remain linear.
- **Scenario B)** The relativity of inertial frames is preserved, but the lorentz transformations are realized non-linearly when acting on the energy and momentum eigenstates of the theory. Such theories are called modified special relativity or doubly special relativity[25,28,29,27]. Examples are given by some forms of noncommutative geometry, for example, $\kappa-$Minkowski spacetime[26]. In all such theories energy and momentum conservation become non-linear which, of course, effects the analysis of the experiments.

Among the experiments which either already give sufficient sensitivity to measure α and β, or are expected to by 2010 are,

(1) There are apparent violations of the GZK bound observed in ul-

tra high energy cosmic rays (UHECR) detected by the AGASA experiment[23]. The experimental situation is inconsistent, but the new AUGER cosmic ray detector, which is now operational, is expected to resolve the situation over the next year or two. If there is a violation of the GZK bound, a possible explanation is Planck scale physics coming from (3) [24].

In Scenario A) violations of the GZK bounds can be explained by either E^3 or E^4 terms in the proton energy-momentum relation. However, in case B) it is less natural to explain a violation of the GZK bounds by means of a Planck scale modification of the energy-momentum relations, but there are proposals for forms of such theories that do achieve this.

(2) A similar anomaly is possibly indicated in Tev photons coming from blazers[34]. Similar remarks apply as to the explanatory power of Scenarios A) and B) in the event that the anomaly exists.

(3) A consequence of (3) can be an energy dependent speed of light. This effect can be looked for in timing data of gamma ray busts. Present data bounds $\alpha <\approx 10^4$ [127] and data expected from the GLAST experiment is expected to be sensitive to α of order one in 2006 [35]. Note that this applies to both Scenarios A) and B).

(4) Present observations of synchrotron radiation in the Crab nebula, together with reasonable astrophysical assumptions, put *very strong* (of order 10^{-9}!!) bounds on α for photons and electrons, in the case of Scenario A only[36].

(5) Present data from precision nuclear and atomic physics experiments puts very tight bounds on α for photons, electrons and hadrons, again in Scenario A), only[37].

(6) Present data from the absence of vacuum Cherenkov effects puts interesting bounds on α in the case of Scenario A) [32].

(7) Observations of bifringence effects in polarized light from distant galaxies puts tight bounds on a possible helicity dependent α [38].

(8) Observations of phase coherence in stellar and galactic interferometry is expected, given certain assumptions[i], to put order one bounds on α in the near future [39].

(9) Certain hypotheses about the Planck scale lead to the prediction of noise in gravitational wave detectors that may be observable at LIGO and VIRGO[41].

---

[i]See [40] for discussion of them.

(10) Under some cosmological scenarios, modifications of the form of (3) lead to distortions of the CMB spectrum that may be observable in near future observations[42].

We may summarize this situation by saying that a theory of quantum gravity that leads to Scenario A) and predicts an energy momentum relation (3) with α order unity is plausibly already ruled out. This is shocking, as it was commonly said just a few years ago that it would be impossible to test any physical hypotheses concerning the Planck scale.

We can also mention three other kinds of experiments that by 2010 will have relevance for the problem of quantum gravity

(1) Evidence for or against supersymmetry may be detected at the Tev scale in accelerators.
(2) The equation of state of the dark energy will be measured in near future experiments. Some proposals for dark energy are based on modifications of energy momentum relations of the form of (3).
(3) There are observations that appear to indicate that the fine structure constant is time dependent[43]. These will be confirmed or go away. If the claim is substantiated this offers a big challenge to the effective field theory understanding of low energy physics.

The combination of all these experimental possibilities signals that the long period when fundamental physics developed independently of experiment is soon coming to a close.

3.1. *What does loop quantum gravity predict for the experiments?*

It is first important to observe that there is no reason that quantum gravity must be Lorentz invariant, as this is only a global symmetry of a particular solution of the classical limit of the theory. Global symmetries are in no way symmetries of the fundamental theory of gravity, neither classically nor quantum mechanically. They are symmetries of particular solutions of the classical theory. Whether these symmetries are fully realized in the ground state has to be determined by calculation.

Indeed, several recent calculations, done with different methods[30–7], yield predictions for modifications to the energy momentum relations for elementary particles of the form of equation (3). Predictions have been found for the leading coefficients α, which generally depend on spin and helicity[30–7].

One issue here is that different calculations make different assumptions about the ground state. In some the ground state is not Lorentz invariant, hence there is no surprise if the perturbations around them have non-lorentz invariant spectra. However, modified dispersion relations may also be seen by studying low energy exciations of a putative around state that does not single out a preferred frame[7]. The question is then: can we determine the ground state precisely enough to discover whether the theory makes unambiguous predictions for the parameters in the energy-momentum relations (3)? Of particular importance is to determine which scenario is realized, A) or B) ?

There is a straightforward argument leading to an expectation that Scenario B) is realized. This is that, even though there is no global lorentz invariance in classical general relativity, the existence of effects due to a preferred frame are ruled out by the condition of invariance under the action of the hamiltonian constraint. This is because the hamiltonian constraint can generate changes in slicing that in any finite region are indistinguishable from lorentz boosts.

Now some of the key results of loop quantum gravity tell us that the hamiltonian constraint can be defined and solved exactly, and that no anomalies are introduced into the constraint algebra by the quantization. This implies that any quantum state that is both an exact solution to the hamiltonian constraint and has a semiclassical limit will in that limit describe physics which is to leading order invariant under the action of the classical hamiltonian constraint. This implies the absence of a preferred frame of reference in the classical limit of an exact solution to the hamiltonian constraint.

So this appears to rule out Scenario A).

This reasoning is supported by many results on $2+1$ gravity, which show in detail that scenario B) is realized in that case[137].

Although most results on 2+1 gravity do not extend to the physical, 3+1 theory, there is an argument for scenario B that does appear to extend to $3+1$[134]. This argument starts with the theory with non-zero cosmological constant. In 2+1 gravity, with non-zero Λ, the symmetry group of boundary states is $SO_q(3,1)$, the quantum deformed deSitter group, with deformation parameter given in the $2+1$ dimensional case by

$$\ln q \sim l_P \sqrt{\Lambda} \tag{4}$$

When one takes the contraction in which the cosmological constant goes to zero one does not get, as usual the Poincare algebra. Instead, because $q \rightarrow 1$

as $\Lambda \to 0$ one gets a deformation of the Pioncare algebra parameterized by the Planck length l_P. This is the κ-Poincare algebra, which is known to provided an example of Scenario B. The emergence of κ-Poincare symmetry in $2+1$ gravity is also confirmed in several detailed studies[137].

One can run the same argument in $3+1$ dimensions, but one must use the relation (2) instead of (4). Because the theory has local field degrees of freedom, one must also renormalize the energy and momentum generators by an appropriate power of $l_P\sqrt{\Lambda}$, which is the ratio of the physical ultraviolet and infrared cutoffs. When this is done, one recovers again the κ-deformation of the Poincare algebra[137]. Remarkably, evidence for the same result is found in studies of excitations of the Kodama state [7].

4. Open questions in loop quantum gravity

I close with a brief list of some problems of current interest.

- As just mentioned, there is an urgency to determine what unambiguous predictions can be made for the results of upcoming experiments on Planck scale phenomenology.
- A similar situation holds for soon to be measured features of the CMB spectrum, such as polarization and the tensor modes.
- Gravitons and other quanta are known to arise as weakly coupled excitations of several different ansatzes for the ground state. As there are no modes below the Planck length, the scattering of these excitations must be computable and finite. It would be good to calculate some examples of scattering and confirm this expectation value in detail. The answer will of course depend on the ansatz chosen for the ground state.
- In contexts with a boundary, where there is a classical positive definite energy, is there an expression for the expectation value of the energy which is positive definite, thus extending to the quantum theory the positive energy theorem?
- Were there such a positive definite expression for the energy, one could determine which is the right ansatz for the ground state by minimizing it, either exactly or approximately. This would lead to unambiguous predictions for Planck scale phenomenelogy. An alternative way to do this would be to establish a general result about the symmetry of the ground state, i.e. is it invariant under the ordinary or deformed Poincare algebra? This would also lead to predictions for scattering of gravitons and other quanta.

- It would be good to know if black hole singularities are also eliminated by quantum effects, as has been shown for cosmological singularities.
- While coupling to matter fields can be easily included, there are speculative suggestions that gauge and fermion fields can arise as effective descriptions of topological defects in spin foam models. This is worth exploring further, as it could lead to a new approach to unification.
- While positive or negative cosmological constant can be easily included, so far loop quantum gravity has nothing to say about why the cosmological constant is so small.
- Dreyer's work concerning the relationship between black hole entropy and quasi normal modes is reassuring, because it shows that the same value of the Imirzi parameter (or finite renormalization of Newton's constant) suffices to match both properties of black holes. But it leads to a puzzle: does this mean that loop quantum gravity with the parameter chosen differently cannot have a good classical limit? Is the correct value an attractor or a fixed point?
- Loop quantum gravity can be extended to incorporate supersymmetry. But is there anything that supersymmetry does for quantum gravity at a background independent level, that would lead to a reason to prefer a supersymmetric theory?
- Can one formulate a version of the holographic principle in loop quantum gravity or spin foam models?
- Loop quantum gravity has its simplest formulation in terms of the chiral Ashtekar-Sen variables. The classical theory is nevertheless chirally symmetric, but it was suggested a long time ago by Soo that there might be violations of CP and chirally asymetric effects quantum mechanically, and evidence for such effects was found in studies of one ansatz for the ground state[30]. It would be good to know if lqg makes unambiguous predictions about chiral symmetry breaking and CP.
- There are some technical problems remaining in establishing the detailed connection between the Hamiltonian and path integral, or spin foam, formulations. There are similarly technical issues related to the exact relationship between the different appraoches to spin foam models.

- While the physical inner product[j] is known and expressed as a path integral in the spin foam langauge, it would be good to have it in a more explicit form. This would allow us to check whether physical states such as the Kodama state are normalizable in the exact physical inner product.
- There are some first steps to a renormalization group for spin foam models[113,114], which need further development. As a byproduct of this work it is shown that while the renormalization group is not a group, it does have a natural algebraic setting, as a Hopf algebra.
- There remain foundational issues having to do with the wavefunction of the universe. The fact that we can do quantum cosmology in detail, with real local degrees of freedom, rather than just study models, makes the resolution of these issues more urgent. Loop quantum gravity has stimulated new ideas in this direction, such as quantum causal histories[78,79] and a proposal that spin networks might underlie a theory of hidden variables[136].

Acknowledgments

I am thankful to my collaborators and especially to my colleagues at Perimeter Institute for many critical discussions. My work has been supported by the NSF and the Jesse Phillips Foundation.

References

1. A. Ashtekar, New perspectives in canonical gravity (Bibliopolis, Naples, 1988); Lectures on non-perturbative canonical gravity, Advanced Series in Astrophysics and Cosmology-Vol. 6 (World Scientific, Singapore, 1991).
2. R. Gambini and J. Pullin, Loops, knots, gauge theories and quantum gravity Cambridge University Press, 1996.
3. C. Rovelli, *Quantum Gravity*, Cambridge University Press, 2004, in press.
4. Thomas Thiemann, "Introduction to Modern Canonical Quantum General Relativity" gr-qc/0110034.
5. C. Rovelli, Living Rev. Rel. 1 (1998) 1, gr-qc/9710008.
6. L. Smolin: in Quantum Gravity and Cosmology, eds J Perez-Mercader et al, World Scientific, Singapore 1992; The future of spin networks gr-qc/9702030 in the Penrose Festschrift.
7. L. Smolin, *Quantum gravity with a positive cosmological constant*, hep-th/0209079.

[j] By which is meant the inner product on the subset of states that satisfy all the constraints. The inner product on spatially diffeomorphism invariant states is known in closed form.

8. L. Smolin, *How far are we from the quantum theory of gravity?* , hep-th/0303185.
9. A. Perez, *Spin Foam Models for Quantum Gravity*, Topical Review in Class.Quant.Grav. 20 (2003) R43 gr-qc/0301113.
10. A. Sen, On the existence of neutrino zero modes in vacuum spacetime J. Math. Phys. 22 (1981) 1781, Gravity as a spin system Phys. Lett. B11 (1982) 89.
11. Abhay Ashtekar, New variables for classical and quantum gravity," Phys. Rev. Lett. 57(18), 2244-2247 (1986).
12. T. Jacobson and L. Smolin, Nucl. Phys. B 299 (1988) 295.
13. C. Rovelli and L. Smolin, Knot theory and quantum theory, Phys. Rev. Lett 61(1988)1155; Loop representation of quantum general relativity, Nucl. Phys. B331(1990)80-152.
14. R. Gambini and A. Trias, Phys. Rev. D23 (1981) 553, Lett. al Nuovo Cimento 38 (1983) 497; Phys. Rev. Lett. 53 (1984) 2359; Nucl. Phys. B278 (1986) 436; R. Gambini, L. Leal and A. Trias, Phys. Rev. D39 (1989) 3127.
15. C. Rovelli and L. Smolin, Spin networks and quantum gravity, gr-qc/9505006, Physical Review D 52 (1995) 5743-5759.
16. C. Rovelli and L. Smolin Discreteness of area and volume in quantum gravity, Nuclear Physics B 442 (1995) 593. Erratum: Nucl. Phys. B 456 (1995) 734; L. Smolin, "Finite Diffeomorphism Invariant Observables In Quantum Gravity," Phys. Rev. D **49**, 4028 (1994) [arXiv:gr-qc/9302011].
17. Y. Ling and L. Smolin, Supersymmetric spin networks and quantum supergravity, Phys. Rev. D61, 044008(2000), hep-th/9904016; Holographic Formulation of Quantum Supergravity, hep-th/0009018, Phys.Rev. D63 (2001) 064010.
18. T. Jacobson, New Variables for canonical supergravity, Class. Quant. Grav.5(1988)923- 935; D. Armand-Ugon, R. Gambini, O. Obregon, J. Pullin, Towards a loop representation for quantum canonical supergravity, hep-th/9508036, Nucl. Phys. B460 (1996) 615; L. F. Urrutia Towards a loop representation of connection theories defined over a super-lie algebra, hep-th/9609010; H. Kunitomo and T. Sano The Ashtekar formulation for canonical N=2 supergravity, Prog. Theor. Phys. suppl. (1993) 31; Takashi Sano and J. Shiraishi, The Nonperturbative Canonical Quantization of the N=1 Supergravity, Nucl. Phys. B410 (1993) 423, hep-th/9211104; The Ashtekar Formalism and WKB Wave Functions of N=1,2 Supergravities, hep-th/9211103; T. Kadoyoshi and S. Nojiri, N=3 and N=4 two form supergravities, Mod. Phys. Lett. A12:1165-1174,1997, hep-th/9703149; K. Ezawa, Ashtekar's formulation for N=1, N=2 supergravities as constrained BF theories, Prog. Theor. Phys.95:863-882, 1996, hep-th/9511047.
19. Yi Ling, "Introduction to supersymmetric spin networks", hep-th/0009020, J.Math.Phys. 43 (2002) 154-169
20. L. Freidel, K. Krasnov, R. Puzio, " BF Description of Higher-Dimensional Gravity Theories", hep-th/9901069, Adv.Theor.Math.Phys. 3 (1999) 1289-1324.
21. Y. Ling and L. Smolin, "Eleven dimensional supergravity as a constrained

topological field theory," Nucl. Phys. B **601**, 191 (2001) [arXiv:hep-th/0003285]; L. Smolin, "Chern-Simons theory in 11 dimensions as a non-perturbative phase of M theory," arXiv:hep-th/9703174.

22. G. Amelino-Camelia et al, Int.J.Mod.Phys.A12:607-624,1997; G. Amelino-Camelia et al Nature 393:763-765,1998; J. Ellis et al, Astrophys.J.535:139-151,2000; J. Ellis, N.E. Mavromatos and D. Nanopoulos, Phys.Rev.D63: 124025,2001; ibidem astro-ph/0108295.
23. N. Hayashida, K. Honda, N. Inoue et. al. (The AGASA collaboration), *Updated AGASA event list above*$4x10^19eV$, astro-ph/0008102, Astrophys.J. 522 (1999) 225; M. Takeda, N. Sakaki, K. Honda, (The AGASA collaboration), *Energy determination in the Akeno Giant Air Shower Array experiment*, astro-ph/0209422.
24. G. Amelino-Camelia and T. Piran, Phys.Rev. D64 (2001) 036005.
25. G. Amelino-Camelia, Nature 418 (2002) 34.
26. J. Lukierski, H. Ruegg, A. Nowicki, and V.N. Tolstoy. *q-deformation of Poincare algebra.* Phys. Lett. B, 264,271:331,321, 1991; S. Majid. *Braided groups and algebraic quantum field theories.* Lett. Math. Phys., 22:167 176, 1991; *Braided Momentum Structure of the q-Poincare Group*, hep-th/9210141, J.Math.Phys. 34 (1993) 2045
27. N.R. Bruno, G. Amelino-Camelia, J. Kowalski-Glikman, Phys.Lett.B522: 133-138,2001; J. Kowalski-Glikman and S. Nowak, hep-th/0203040,S. Judes, gr-qc/0205067; M. Visser, gr-qc/0205093; S. Judes, M. Visser, gr-qc/ 0205067; D. V. Ahluwalia and M. Kirchbach, qr-qc/0207004.
28. J. Magueijo and L. Smolin, Phys.Rev.Lett. (88) 190403, 2002.
29. J. Magueijo and L. Smolin, gr-qc/0207
    gac-dsr,joaolee1,joaolee2,dsr2
30. Rodolfo Gambini, Jorge Pullin, "Nonstandard optics from quantum space-time", Phys.Rev. D59 (1999) 124021, gr-qc/9809038;
31. Jorge Alfaro, Hugo A. Morales-Tcotl, Luis F. Urrutia, "Loop quantum gravity and light propagation", Phys.Rev. D65 (2002) 103509, hep-th/0108061.
32. Tomasz J. Konopka, Seth A. Major, "Observational Limits on Quantum Geometry Effects", New J.Phys. 4 (2002) 57. hep-ph/0201184; Ted Jacobson, Stefano Liberati, David Mattingly, " TeV Astrophysics Constraints on Planck Scale Lorentz Violation"; *Threshold effects and Planck scale Lorentz violation: combined constraints from high energy astrophysics*, hep-ph/0112207.
33. Subir Sarkar, " Possible astrophysical probes of quantum gravity", Mod.Phys.Lett. A17 (2002) 1025-1036, gr-qc/0204092.
34. S.D. Biller et al, *Limits to Quantum Gravity Effects from Observations of TeV Flares in Active Galaxies*, gr-qc/9810044; G. Amelino-Camelia, *Improved limit on quantum-spacetime modifications of Lorentz symmetry from observations of gamma-ray blazars*, gr-qc/0212002.
35. G. Amelino-Camelia, John Ellis, N.E. Mavromatos, D.V. Nanopoulos, Subir Sarkar, *Potential Sensitivity of Gamma-Ray Burster Observations to Wave Dispersion in Vacuo*, astro-ph/9712103, Nature 393 (1998) 763-765; J.P. Norris, J.T. Bonnell, G.F. Marani, J.D. Scargle, *GLAST, GRBs, and Quantum Gravity*, astro-ph/9912136.

36. T. Jacobson, S. Liberati, D. Mattingly, *Lorentz violation and Crab synchrotron emission: a new constraint far beyond the Planck scale*, astro-ph/0212190.
37. Robert C. Myers, Maxim Pospelov, *Experimental Challenges for Quantum Gravity*, p-ph/0301124.
38. Reinaldo J. Gleiser, Carlos N. Kozameh, *Astrophysical limits on quantum gravity motivated birefringence*, gr-qc/0102093.
39. Richard Lieu, Lloyd W. Hillman, *Stringent limits on the existence of Planck time from stellar interferometry*, astro-ph/0211402, astro-ph/0301184.
40. Y. Jack Ng, W. A. Christiansen, H. van Dam, astro-ph/0302372
41. Giovanni Amelino-Camelia, *An interferometric gravitational wave detector as a quantum-gravity apparatus*, gr-qc/9808029, Nature 398 (1999) 216-218.
42. Robert H. Brandenberger and Jerome Martin *On Signatures of Short Distance Physics in the Cosmic Microwave Background*, hep-th/0202142, Int.J.Mod.Phys. A17 (2002) 3663-3680.
43. J. K. Webb, M. T. Murphy, V. V. Flambaum, S. J. Curran, *Does the fine structure constant vary? A third quasar absorption sample consistent with varying alpha*, astro-ph/0210531, ApSS, 283, 565, 200; M. T. Murphy, J. K. Webb, V. V. Flambaum, S. J. Curran, *Does the fine structure constant vary? A detailed investigation into systematic effects*, astro-ph/0210532, ApSS, 283, 577, 2003; John N. Bahcall, Charles L. Steinhardt, David Schlegel, *Does the fine-structure constant vary with cosmological epoch?*,astro-ph/0301507.
44. L. Bombelli, J. H. Lee, D. Meyer and R. Sorkin, "Space-Time As A Causal Set," Phys. Rev. Lett. **59**, 521 (1987); X. Martin, D. O'Connor, D. P. Rideout and R. D. Sorkin, "On the 'renormalization' transformations induced by cycles of expansion and contraction in causal set cosmology," Phys. Rev. D **63**, 084026 (2001) [arXiv:gr-qc/0009063]; D. P. Rideout and R. D. Sorkin, "Evidence for a continuum limit in causal set dynamics," Phys. Rev. D **63**, 104011 (2001) [arXiv:gr-qc/0003117].
45. J. Ambjorn, Z. Burda, J. Jurkiewicz and C. F. Kristjansen, "Quantum gravity represented as dynamical triangulations," Acta Phys. Polon. B **23**, 991 (1992); J. Ambjorn, "Quantum Gravity Represented As Dynamical Triangulations," Class. Quant. Grav. **12**, 2079 (1995); M. E. Agishtein and A. A. Migdal, "Simulations of four-dimensional simplicial quantum gravity," Mod. Phys. Lett. A **7**, 1039 (1992).
46. J. Ambjorn, A. Dasgupta, J. Jurkiewiczcy and R. Loll, "A Lorentzian cure for Euclidean troubles", hep-th/0201104; J. Ambjorn and R. Loll, Nucl. Phys. B536 (1998) 407 [hep-th/9805108]; J. Ambjorn, J. Jurkiewicz and R. Loll, Phys. Rev. Lett. 85 (2000) 924 [hepth/ 0002050]; Nucl. Phys. B610 (2001) 347 [hep-th/0105267]; R. Loll, Nucl. Phys. B (Proc. Suppl.) 94 (2001) 96 [hep-th/0011194]; J. Ambjorn, J. Jurkiewicz and R. Loll, Phys. Rev. D64 (2001) 044011 [hep-th/0011276]; J. Ambjorn, J. Jurkiewicz, R. Loll and G. Vernizzi, *Lorentzian 3d gravity with wormholes via matrix models*, JHEP **0109**, 022 (2001) [arXiv:hep-th/0106082]; B. Dittrich (AEI, Golm), R. Loll, *A Hexagon Model for 3D Lorentzian Quantum Cosmology*, hep-th/0204210.
47. A. Dasgupta and R. Loll, Nucl. Phys. B606 (2001) 357 [hep-th/0103186].

48. J. Ambjorn (Niels Bohr Institute), K.N. Anagnostopoulos (Univ. of Crete), R. Loll, *Crossing the c=1 barrier in 2d Lorentzian quantum gravity*, hep-lat/9909129, Phys.Rev. D61 (2000) 044010.
49. See, for example, Earman, J. 1989. *World Enough and Spacetime: Absolute vs. Relational Theories of Space and Time.* Cambridge, MA: MIT Press; Norton, J. D. 1987. Einstein, the Hole Argument and the Reality of Space. In J. Forge (ed.), Measurement, Realism and Objectivity, 153-188. Boston: D. Reidel; L. Smolin, "The present moment in quantum cosmology: Challenges to the arguments for the elimination of time," arXiv:gr-qc/0104097.
50. P. A. M. Dirac, Lectures on Quantum Mechanics Belfer Graduate School of Science Monographs, no. 2 (Yeshiva University Press, New York,1964).
51. J. Stachel, "Einstein's search for general covariance, 1912-15" in *Einstein and the History of General Relativity* vol 1 of *Einstein Studies* eds. D. Howard and J. Stachel. (Birkhauser,Boston,1989).
52. Julian Barbour, *The End of Time*, Oxford University Press, 2000; "Leibnizian Time, Machian Dynamics, And Quantum Gravity," *In *Oxford 1984, Proceedings, Quantum Concepts In Space and Time*, 236-246.*
53. C. Rovelli, "What Is Observable In Classical And Quantum Gravity?," Class. Quant. Grav. **8**, 297 (1991).
54. L. Smolin *Life of the Cosmos*, (Oxford University Press, New York and Wiedenfeld and Nicolson, London, 1997).
55. R. De Pietri, L. Freidel, K. Krasnov, C. Rovelli, Barrett-Crane model from a Boulatov-Ooguri eld theory over a homoge- neous space, Nucl.Phys. B574 (2000) 785-806, hep-th/9907154.
56. L. Crane in, Categorical Physics, hep-th/9301061; Topological Field theory as the key to quantum gravity, hep-th/9308126, in Knot theory and quantum gravity ed. J. Baez, (Oxford University Press); Clocks and Categories, is quantum gravity algebraic? J. Math. Phys. 36 (1995) 6180-6193, gr-qc/9504038.
57. G. 'tHooft, Dimensional reduction in quantum gravity gr-qc/9310006.in Salanfestschrift ed. A. Alo, J. Ellis, S. Randjbar-Daemi (World Scienti c,1993).
58. L. Susskind, The world as a hologram, hep-th/9409089, J. Math. Phys. 36 (1995) 6377 ; Strings, black holes and Lorentz contractions hepth/ 9308139; Phys. Rev. Lett. 71 (1993) 2367; Phys. Rev. D 49 (1994) 6606; D 50 (1994) 2700; L. Susskind and P. Gri n, Partons and black holes hep-th/9410306.
59. F. Markopoulou and L. Smolin, Holography in a quantum spacetime, hep-th/9910146.
60. L. Smolin, The strong and the weak holographic principles, hep-th/0003056.
61. R. Loll, "Spectrum of the Volume Operator in Quantum Gravity," Nucl. Phys. B **460**, 143 (1996) [arXiv:gr-qc/9511030]; "The Volume operator in discretized quantum gravity," Phys. Rev. Lett. **75**, 3048 (1995) [arXiv:gr-qc/9506014].
62. Roberto De Pietri and Carlo Rovelli, Geometry eigenvalues and the scalar product from recoupling theory in loop quantum gravity," Phys. Rev. D 54(4), 2664-2690 (1996); S. Fittelli, L. Lehner, C. Rovelli, The complete

spectrum of the area from recoupling theory in loop quantum gravity," Class. Quant. Grav. 13, 2921-2932 (1996); Abhay Ashtekar and Jerzy Lewandowski, Quantum Theory of Geometry I: Area operators," Class. Quant. Grav. 14, A55-A81 (1997).

63. S. A. Major, "Operators for quantized directions," Class. Quant. Grav. **16**, 3859 (1999) [arXiv:gr-qc/9905019]; "New operators for spin net gravity: Definitions and consequences," arXiv:gr-qc/0101032.
64. J.F. Plebanski. On the separation of einsteinian substructures. J. Math. Phys., 18:2511, 1977.
65. T. Jacobson and L. Smolin, Phys. Lett. B 196 (1987) 39; Class. and Quant. Grav. 5 (1988) 583; J. Samuel, Pramana-J Phys. 28 (1987) L429.
66. R. Capovilla, J. Dell and T. Jacobson, Phys. Rev. Lett. 21, 2325(1989); Class. Quant. Grav. 8, 59(1991); R. Capovilla, J. Dell, T. Jacobson and L. Mason, Class. and Quant. Grav. 8, 41(1991).
67. R. Penrose, Theory of quantized directions unpublished manuscript; in Quantum theory and beyond ed T Bastin, Cambridge U Press 1971, in Advances in Twistor Theory, ed. L. P. Hughston and R. S. Ward, (Pitman,1979) p. 301in Combinatorial Mathematics and its Application (ed. D. J. A. Welsh) (Academic Press,1971).
68. C. Rovelli and L. Smolin, The physical hamiltonian in nonperturbative quantum gravity, Phys. Rev. Lett.72(1994)446; Spin Networks and Quantum Gravity, Phys. Rev. D52(1995)5743-5759.
69. R. Borissov, Graphical Evolution of Spin Network States, Phys.Rev. D55 (1997) 6099-6111, gr-qc/9606013.
70. T. Thiemann, Quantum Spin Dynamics (QSD) I & II, Class.Quant.Grav. 15 (1998) 839-905, gr-qc/9606089, gr-qc/9606090.
71. L. Smolin, *Time, measurement and information loss in quantum cosmology*, gr-qc/9301016.
72. L. Smolin, *Finite, diffeomorphism invariant observables in quantum gravity*, Phys. Rev. D 49 (1994) 4028, gr-qc/9302011.
73. M. P. Reisenberger. "Worldsheet formulations of gauge theories and gravity," in Proceedings of the 7th Marcel Grossman Meeting, ed. by R. Jantzen and G. MacKeiser, World Scientific, 1996; gr-qc/9412035; A lattice worldsheet sum for 4-d Euclidean general relativity. gr-qc/9711052.
74. 8] M. P. Reisenberger and C. Rovelli. Sum-over-surface form of loop quantum gravity," gr-qc/9612035, Phys. Rev. D 56 (1997) 3490; Spacetime as a Feynman diagram: the connection formulation. Class.Quant.Grav., 18:121140, 2001; Spin foams as Feynman diagrams, gr-qc/0002083.
75. J. Barrett and L. Crane, "Relativistic spin networks and quantum gravity", J. Math. Phys. 39 (1998) 3296-3302, gr-qc/9709028.
76. J. Baez, Spin foam models, Class. Quant. Grav. 15 (1998) 1827-1858, gr-qc/9709052; An introduction to spin foam models of quantum gravity and BF theory. Lect.Notes Phys., 543:25 94, 2000.
77. Fotini Markopoulou, "Dual formulation of spin network evolution", gr-qc/9704013.
78. F. Markopoulou, *Planck-scale models of the Universe*, gr-qc/0210086.

79. E. Hawkins, F. Markopoulou, H. Sahlmann, *Evolution in Quantum Causal Histories*, hep-th/0302111.
80. Fotini Markopoulou, Lee Smolin, "Quantum geometry with intrinsic local causality", Phys.Rev. D58 (1998) 084032, gr-qc/9712067.
81. Fotini Markopoulou, Lee Smolin, *Causal evolution of spin networks*, gr-qc/9702025, Nucl.Phys. B508 (1997) 409-430.
82. Stuart Kauffman, Lee Smolin, "Combinatorial dynamics in quantum gravity", hep-th/9809161, Lect.Notes Phys. 541 (2000) 101-129.
83. David Rideout, *Dynamics of Causal Sets*, gr-qc/0212064; D. P. Rideout, R. D. Sorkin, *A Classical Sequential Growth Dynamics for Causal Sets*, gr-qc/9904062, Phys.Rev. D61 (2000) 024002; *Evidence for a continuum limit in causal set dynamics*, gr-qc/0003117, Phys.Rev. D63 (2001) 104011.
84. J. Iwasaki, A reformulation of the Ponzano-Regge quantum gravity model in terms of surfaces," gr-qc/9410010; A definition of the Ponzano-Regge quantum gravity model in terms of surfaces," gr-qc/9505043, J. Math. Phys. 36 (1995) 6288; L. Freidel and K. Krasnov. Spin foam models and the classical action principle. Adv.Theor.Math.Phys., 2:11831247, 1999; R. De Pietri, L. Freidel, K. Krasnov, and C. Rovelli. Barrett-Crane model from a boulatov-ooguri field theory over a homogeneous space. Nucl.Phys. B, 574:785806, 2000.
85. T. Thiemann, "Quantum spin dynamics," gr-qc/9606089, gr-qc/9606090; "Anomaly free formulation of non-perturbative 4-dimensional Lorentzian quantum gravity," Phys. Lett. B380 (1996) 257.
86. L. Smolin, Linking topological quantum field theory and nonperturbative quantum gravity, J. Math. Phys. 36(1995)6417, gr-qc/9505028.
87. K. Krasnov, On Quantum Statistical Mechanics of a Schwarzschild Black Hole , grqc/9605047, Gen. Rel. Grav. 30 (1998) 53-68; C. Rovelli, "Black hole entropy from loop quantum gravity," grqc/9603063.
88. A. Ashtekar, J. Baez, K. Krasnov, Quantum Geometry of Isolated Horizons and Black Hole Entropy gr-qc/0005126; A. Ashtekar, J. Baez, A. Corichi, K. Krasnov, "Quantum geometry and black hole entropy," gr-qc/9710007, Phys.Rev.Lett. 80 (1998) 904-907.
89. L. Smolin, A holograpic formulation of quantum general relativity, Phys. Rev. D61 (2000) 084007, hep-th/9808191.
90. J. D. Bekenstein, Lett. Nuovo. Cim 11(1974).
91. S. Das, Parthasarathi Majumdar, Rajat K. Bhaduri, " General Logarithmic Corrections to Black Hole Entropy", Class.Quant.Grav. 19 (2002) 2355-2368; Saurya Das, hep-th/0207072, "Leading Log Corrections to Bekenstein-Hawking Entropy"
92. Kirill Krasnov, "Quantum Geometry and Thermal Radiation from Black Holes", gr-qc/9710006, Class.Quant.Grav. 16 (1999) 563-578.
93. M. Barreira, M. Carfora, C. Rovelli, "Physics with nonperturbative quantum gravity: radiation from a quantum black hole" gr-qc/9603064, Gen.Rel.Grav. 28 (1996) 1293-1299.
94. M. Atiyah, Topological quantum field theory Publ. Math. IHES 68 (1989) 175; The Geometry and Physics of Knots, Lezion Lincee (Cambridge Uni-

versity Press, Cambridge,1990); G. Segal, Conformal field theory oxford preprint (1988).

95. G. Horowitz, "Exactly Soluble Diffeomorphism Invariant Theories", Commun.Math.Phys. 125 (1989) 417; V. Husain, "Topological Quantum Mechanics", Phys.Rev. D43 (1991) 1803.
96. H. Kodama, Prog. Theor. Phys. 80, 1024(1988); Phys. Rev. D42(1990)2548.
97. L. Smolin and C. Soo, The Chern-Simons Invariant as the Natural Time Variable for Classical and Quantum Cosmology, Nucl. Phys. B449 (1995) 289, gr-qc/9405015.
98. L. N. Chang and C. Soo, Ashtekar's variables and the topological phase of quantum gravity, Proceedings of the XXth. Conference on Differential Geometric Methods in Physics, (Baruch College, New York, 1991), edited by S. Catto and A. Rocha (World Scientific, 1992); Phys. Rev. D46 (1992) 4257; C. Soo and L. N. Chang, Int. J. Mod. Phys. D3 (1994) 529.
99. C. Soo, Wave function of the Universe and Chern-Simons Perturbation Theory, gr-qc/0109046.
100. L. Smolin, The classical limit and the form of the hamiltonian constraint in nonperturbative quantum gravity, gr-qc/9607034.
101. J. Lewandowski, D. Marolf, Loop constraints: a habitat and their algebra," gr-qc/9710016; R. Gambini, J. Lewandowski, D. Marolf, J. Pullin, "On the consistency of the constraint algebra in spin-network quantum gravity," gr-qc/9710018.
102. A. Ashtekar, C. Rovelli, L. Smolin, Weaving a classical metric with quantum threads," Phys. Rev. Lett. 69 (1992) 237.
103. Luca Bombelli, Statistical geometry of random weave states, gr-qc/0101080
104. J. Iwasaki, C. Rovelli, Gravitons as embroidery on the weave," Int. J. Mod. Phys. D 1 (1993) 533; Gravitons from loops: non-perturbative loop-space quantum gravity contains the graviton-physics approximation," Class. Quantum Grav. 11 (1994) 1653.
105. Martin Bojowald,"Isotropic Loop Quantum Cosmology", Class.Quant.Grav. 19 (2002) 2717-2742, gr-qc/0202077; "Inflation from Quantum Geometry", gr-qc/0206054; "The Semiclassical Limit of Loop Quantum Cosmology", gr-qc/0105113, Class.Quant.Grav. 18 (2001) L109-L116; " Dynamical Initial Conditions in Quantum Cosmology", gr-qc/0104072, Phys.Rev.Lett. 87 (2001) 121301.
106. S. Tsujikawa, P. Singh, R. Maartens , *Loop quantum gravity effects on inflation and the CMB*, astro-ph/0311015
107. A. Ashtekar, V. Husain, C. Rovelli, J. Samuel and L. Smolin 2+1 quantum gravity as a toy model for the 3+1 theory Class. and Quantum Grav. L185- L193 (1989); L. Smolin, Loop representation for quantum gravity in 2+1 dimensions,in the proceedings of the John's Hopkins Conference on Knots, Tolopoly and Quantum Field Theory ed. L. Lusanna (World Scientific,Singapore,1989) .
108. Viqar Husain, "Observables for spacetimes with two Killing field symmetries", gr-qc/9402019,Phys.Rev. D50 (1994) 6207-6216; V. Husain and L. Smolin, "Exactly Solvable Quantum Cosmologies From Two Killing Field

spacetimes", Nucl.Phys. B327 (1989) 205.

109. R. Gianvittorio, R. Gambini and A. Trias, Phys. Rev. D38 (1988) 702; C. Rovelli and L. Smolin. Loop representation for lattice gauge theory,1990 Pittsburgh and Syracuse preprint; B. Bruegmann, Physical Review D 43 (1991) 566; J.M.A. Farrerons, Loop calculus for SU(3) on the lattice Phd. thesis, Universitat Autonoma de Barcelona (1990); R. Loll A new quantum representation for canonical gravity and SU(2) Yang-Mills theory, University of Bonn preprint, BONN-HE-90-02 (1990)
110. A Ashtekar J Lewandowski D Marlof J Mourau T Thiemann:"Quantization of dieomorphism invariant theories of connections with local degrees of freedom", gr-qc/9504018, JMP 36 (1995) 519.
111. J. Barbero, Real Ashtekar variables for Lorentzian signature spacetime," Phys. Rev. D51 (1995) 5507.
112. Laurent Freidel, Kirill Krasnov, Discrete Space-Time Volume for 3-Dimensional BF Theory and Quantum Gravity, Class.Quant.Grav. 16 (1999) 351-362.
113. Fotini Markopoulou,"Coarse graining in spin foam models", gr-qc/0203036;
114. Fotini Markopoulou, "An algebraic approach to coarse graining", hep-th/0006199.
115. Alejandro Perez, Finiteness of a spinfoam model for euclidean quantum general relativity, Nucl.Phys. B599 (2001) 427-434, gr-qc/0011058 ; Louis Crane, Alejandro Perez, Carlo Rovelli, A finiteness proof for the Lorentzian state sum spinfoam model for quantum general relativity, gr-qc/0104057.
116. John C. Baez, J. Daniel Christensen, Positivity of Spin Foam Amplitudes , Class.Quant.Grav. 19 (2002) 2291-2306, grqc/0110044; John C. Baez, J. Daniel Christensen, Greg Egan, Asymptotics of 10j symbols , gr-qc/0208010.
117. H. Sahlmann, T. Thiemann, O. Winkler, Coherent States for Canonical Quantum General Relativity and the Infinite Tensor Product Extension, Nucl.Phys. B606 (2001) 401-440, gr-qc/0102038.
118. O. Dreyer, *Quasinormal Modes, the Area Spectrum, and Black Hole Entropy*, gr-qc/0211076
119. L. Motl, *An analytical computation of asymptotic Schwarzschild quasinormal frequencies*, gr-qc/0212096; L. Motl, A. Neitzke, *Asymptotic black hole quasinormal frequencies*, hep-th/0301173.
120. G Ponzano, T Regge, Semiclassical limits of Racah Coefficients in Spectroscopy and Group theoretical methods in Physics. Ed F Bloch, North Holland, Amsterdam 1968
121. A. Mikovic, *Quantum gravity vacuum and invariants of embedded spin networks*, gr-qc/0301047.
122. R. Gambini and J. Pullin, *Canonical quantization of general relativity in discrete space-times*, Phys.Rev.Lett. 90 (2003) 021301,*Discrete quantum gravity: applications to cosmology*, gr-qc/0212033; R. Gambini, R. Porto, J. Pullin *Consistent discrete gravity solution of the problem of time: a model* gr-qc/0302064
123. C. Rovelli, *Notes for a brief history of quantum gravity*, gr-qc/0006061.
124. J. D. Bekenstein, Black Holes And Entropy, Phys. Rev. D 7, 2333 (1973).

125. S. W. Hawking, Particle Creation By Black Holes," Commun. Math. Phys. 43, 199 (1975).
126. W. G. Unruh, Phys. Rev. D 14 (1976) 870.
127. J. Ellis, N.E. Mavromatos, D.V. Nanopoulos, A.S. Sakharov, *Quantum-Gravity Analysis of Gamma-Ray Bursts using Wavelets*, astro-ph/0210124
128. T.Banks, "Cosmological Breaking of Supersymmetry?", hep-th/0007146.
129. L. Freidel and David Louapre, *Non-perturbative summation over 3D discrete topologies*, hep-th/0211026.
130. L. Friedel and L. Smolin, *The linearization of the Kodama state*, hep-th/0310224
131. S. Alexander, J. Malecki, L. Smolin, *Quantum Gravity and Inflation*, hep-th/0309045
132. M. Bojowald, G. Date, *A non-chaotic quantum Bianchi IX universe and the quantum structure of classical singularities*, gr-qc/0311003
133. F. Markopoulou and L. Smolin, *Nonperturbative dynamics for abstract (p,q) string networks*, Phys.Rev. D58 (1998) 084033, hep-th/9712148 [
134. G. Amelino-Camelia, L. Smolin, A. Starodubtsev, *Quantum symmetry, the cosmological constant and Planck scale phenomenology*, hep-th/0306134
135. L. Crane, *A New Approach to the Geometrization of Matter*, gr-qc/0110060.
136. F. Markopoulou and L. Smolin, *Quantum Theory from Quantum Gravity*, gr-qc/0311059
137. L. Freidel, J. Kowalski-Glikman, L. Smolin, *2+1 gravity and Doubly Special Relativity*, hep-th/0307085

LOOP QUANTUM GRAVITY INDUCED CORRECTIONS TO FERMION DYNAMICS IN FLAT SPACE

LUIS F. URRUTIA
Departamento de Física de Altas Energías
Instituto de Ciencias Nucleares
Universidad Nacional Autónoma de México
Apartado Postal 70-543
04510 México, D.F.
E-mail: urrutia@nuclecu.unam.mx

A summary of recent work related to the calculation of loop quantum gravity induced corrections to standard particle dynamics (photons and spin 1/2 fermions) in flat space is presented. Stringent bounds upon the parameters characterizing the corrections in the fermionic sector, arising from already known clock-comparison experiments, are reviewed.

1. INTRODUCTION

Loop Quantum Gravity (LQG) has provided a highly successful approach to deal with the old problem of constructing a quantum version of Einstein general relativity. Among its predictions we singularize the property that the area and volume operators are quantized in the corresponding units of the Planck length ℓ_P.[1] In this way, a continuum description of space should be considered only as an approximation valid for length scales $L >> \ell_P$. The granular structure of space at short distances poses at least two interesting problems certainly tied to each other. On one hand there is the question of whether or not such granular structure will induce modifications to standard particle dynamics in flat space, for example. A most promising observational consequence of these corrections is the modification of the standard energy-momentum relations for particles.[2] The analogy with particle propagation in a lattice crystal, for example, suggests that modifications will indeed arise. Secondly, since that granular structure incorporates the notion of a minimum (maximum) length (momentum), another important related issue is how these corrections would affect Einstein special relativity principle.

The resolution of the first question in LQG requires a detailed and exact construction of semiclassical states that would approximate the matter content plus the flat space geometry at large distances, while retaining the granular structure at the Planck scale. Progress in this direction has been made,[3] though the final answer still remains an open problem. In this way, many of the estimations of the corrections to particle dynamics arising from LQG are based upon an heuristical approach that intends to capture what one would expect to be the most relevant and general properties of such semiclassical states.[4,5,6]

With respect to the second question, we point out some alternatives that have been considered in the literature:

(1) The fact that such modifications lead to an energy dependent velocity of light, for example, has normally been taken as an indication that their mere existence would imply a violation of the active (particle) form of the relativity principle expressed in terms of the standard Lorentz transformations. This approach to the problem makes direct contact with a large body of work previously developed to describe the most general Lorentz violating dynamics which is compatible with the standard model of strong and electroweak interactions,[7] and more recently with an Einstein-Cartan description of gravity.[8] This standard model extension has provided a unified framework to analyze most of the experimental work designed to probe Lorentz violations in nature.[9]

(2) A second approach has recently emerged which retains a full relativity principle by extending or deforming the standard Lorentz transformations in a manner consistent with the induced dynamical modifications, in particular with deformed dispersion relations. These proposals are included under the generic name of double special relativity (DSR), which typically incorporate ℓ_P as a second invariant quantity in the same footing as the speed of light in the low energy approximation.[10]

(3) There is also the possibility that standard Lorentz covariance is preserved.[11]

The paper is organized as follows: Section 2 contains a summary of the heuristical approach pursued to obtain the modified dynamics arising from LQG for two-component spin 1/2 fermions and photons in flat space.[5,6] Section 3 summarizes the bounds upon the parameters in the free fermionic sector, obtained assuming the existence of a preferred reference frame.[12]

2. MODIFIED DYNAMICS FOR FIELDS F IN FLAT SPACE

Central to the approach of Alfaro, Morales-Técotl and Urrutia [5,6] is Thiemann's regularization of matter and gauge fields LQG Hamiltonians $\hat{H}_\Gamma$.[13] This is based upon a triangulation of space, adapted to the corresponding graph Γ which define a given state. The regularization is provided by the volume operator, with discrete eigenvalues arising only from the vertices of the graph. Here we take an heuristical point of view, starting from the exact operator version of such Hamiltonians in LQG and defining their action upon the semiclassical state through some plausible requirements. We think of the semiclassical configuration describing a particular matter or gauge field operator $\hat{F}$ plus gravity, as given by an ensemble of graphs Γ, each occurring with probability $P(\Gamma)$. To each of such graphs we associate a wave function $|\Gamma, L, F\rangle$ which is peaked with respect to the classical field configuration F, together with a flat gravitational metric and a zero value for the gravitational connection at large distances. In other words, the contribution of the gravitational operators inside the expectation value is estimated as[5,6]

$$\langle\Gamma, L, F| \ldots\hat{q}_{ab}\ldots |\Gamma, L, F\rangle = \delta_{ab} + O\left(\frac{\ell_P}{L}\right),$$
$$\langle\Gamma, L, F| \ldots\hat{A}_{ia}\ldots |\Gamma, L, F\rangle = 0 + \frac{1}{L}\left(\frac{\ell_P}{L}\right)^{\Upsilon}. \tag{1}$$

The parameter $\Upsilon \geq 0$ is a real number. Also we associate the effective hamiltonian $H_\Gamma = \langle\Gamma, L, F\,|\hat{H}_\Gamma|\,\Gamma, L, F\rangle$ to each graph. The scale $L >> \ell_P$ of the wave function is such that the continuous flat metric approximation is appropriate for distances much larger that L, while the granular structure of space becomes relevant when probing distances smaller that L. In this way, space is constructed by adding boxes of size L^3, which center represents a given point $\mathbf{x}$ in the continuum and which contain a large number of vertices of the adapted graphs. The field F, characterized by a De Broglie wave length λ, is considered a slowly varying function within each box ($\lambda > L$) and contributes with its classical value at the center of the box, when taking expectation values. Gravitational variables are rapidly varying inside the box. The total effective Hamiltonian is defined as an average over the graphs Γ: $\mathrm{H} = \sum_\Gamma P(\Gamma)\,\mathrm{H}_\Gamma$. This effectively amounts to average the expectation values of the gravitational variables in each box. We construct such averages in terms of the most general combinations of flat space tensors $\delta_{ab}, \epsilon_{abc}, \ldots$ which saturate the tensor structure of the classical fields together with their derivatives in each box. In this way we are imposing rotational invariance on our final effective Hamiltonian.

Next we make some general comments regarding the above procedure: (i) our calculation has been performed in a fixed reference frame and leaves undetermined an overall numerical dimensionless coefficient in each of the calculated contributions. The results can be viewed as an expansion in terms of the classical fields and their derivatives, combined with an explicit dependence upon the two scales ℓ_P and L. (ii) The corrections obtained in this way have been usually interpreted as signaling a preferred reference frame together with a violation of the standard active (particle) Lorentz transformations. The advent of DSR opens up the possibility to study whether or not such modified actions can be embedded in a related framework, thus recovering a full modified relativity principle.

3. OBSERVATIONAL BOUNDS FOR FERMIONS USING EXISTING DATA

Here we have taken the point of view that the previously found Hamiltonians account for the corresponding dynamics in a preferred reference frame, which we identify as the one where the Cosmic Microwave Background looks isotropic. Our velocity $\mathbf{w}$ with respect to that frame has already been determined to be $w/c \approx 1.23 \times 10^{-3}$ by COBE. Thus, in the earth reference frame one expects the appearance of signals indicating minute violations of space isotropy encoded in $\mathbf{w}$-dependent terms appearing in the transformed Hamiltonian or Lagrangian. On the other hand, many high precision experimental test of rotational symmetry, using atomic and nuclear systems, have been already reported in the literature. Amazingly such precision is already enough to set very stringent bounds on some of the parameters arising from the quantum gravity corrections. We have considered the case of non-relativistic Dirac particles, obtaining corrections which involve the coupling of the spin to the CMB velocity together with a quadrupolar anisotropy of the inertial mass.[12] The calculation was made with the choices $\Upsilon = 0$ and $L = 1/M$, where M is the rest mass of the fermion. Keeping only terms linear in ℓ_P, the equations of motion arising from the two-component Hamiltonian obtained by Alfaro, Morales-Técotl and Urrutia [5] can be readily extended to the Dirac case, yielding a Lagrangian which describes the time evolution as seen in the CMB frame. In order to obtain the dynamics in the laboratory frame we implement an observer Lorentz transformation by introducing explicitly the CMB frame's four velocity $W^\mu = \gamma(1, \mathbf{w}/c)$. The result is[12]

$$\begin{aligned} L_D = & \frac{1}{2} i\bar{\Psi}\gamma^\mu \partial_\mu \Psi - \frac{1}{2} M\bar{\Psi}\,\Psi + \frac{1}{2} i(\Theta_1 M\ell_P)\bar{\Psi}\gamma_\mu \left(g^{\mu\nu} - W^\mu W^\nu\right)\partial_\nu \Psi \\ & + \frac{1}{4}(\Theta_2 M\ell_P)\bar{\Psi}\epsilon_{\mu\nu\alpha\beta} W^\mu \gamma^\nu \gamma^\alpha \partial^\beta \Psi \\ & - \frac{1}{4}(\Theta_4 M\ell_P) M W_\mu \bar{\Psi}\gamma_5 \gamma^\mu \Psi + h.c.\,. \end{aligned} \tag{2}$$

Here Θ_i, $i = 1, 2, 4$, are undetermined constants parameterizing the dynamical corrections, expected to be of order one, and which can be associated with the Lorentz invariance violating tensors of the standard model extension. From the work of Kostelecký and Lane[14] we obtain the non-relativistic limit of the Hamiltonian corresponding to (2), up to first order in ℓ_P and up to order $(\mathbf{w})/c^2$. The relevant corrections are

$$\delta H_S = \left(\Theta_2 + \frac{1}{2}\Theta_4\right) M\ell_P (2Mc^2) \left[1 + O\left(\frac{p^2}{2M^2c^2}\right)\right] \mathbf{s} \cdot \frac{\mathbf{w}}{c}, \tag{3}$$

$$\delta H_Q = -\Theta_1 M\ell_P \frac{5}{3} \left\langle \frac{p^2}{2M} \right\rangle \left(\frac{Q}{R^2}\right) \left(\frac{w}{c}\right)^2 P_2(\cos\theta)\,. \tag{4}$$

The first represents a coupling of the nucleon spin $\mathbf{s}$ to the velocity of the CMB frame, while the second represents an anisotropy of the inertial mass. Both contributions have been bounded in Hughes-Drever like experiments. Here Q is the electric quadrupole moment and R is the radius of the nucleus under consideration, while θ is the angle between the quantization axis and $\mathbf{w}$. Using $< p^2/2M >\sim 40$ MeV for the energy of a nucleon in the last shell of a typical heavy nucleus, together with the experimental bounds of Chupp et. al. and Bear et. al.[15] we find[12]

$$\mid \Theta_2 + \frac{1}{2}\Theta_4 \mid < 2 \times 10^{-9}, \qquad \mid \Theta_1 \mid < 3 \times 10^{-5}. \tag{5}$$

The above bounds on quantities that were expected to be of order unity already call into question the scenarios inspired on the various approaches to quantum gravity, suggesting the existence of Lorentz violating Lagrangian corrections which are linear in Planck's length. In relation to this point it is interesting to observe that a very reasonable agreement with the current AGASA ultra high energy cosmic ray (UHECR) spectrum, including the region beyond the GZK cutoff, has been recently obtained by using dispersion relations of order higher than linear in ℓ_P, together with considerations of additional stringent bounds arising from the first estimations of the impact of nearby BL Lac objects and UHECR data upon LQG parameters.[16]

Acknowledgments

The author acknowledges V.A. Kostelecký for the invitation to participate in the Focus Session on CPT and Lorentz Symmetry of QTS3. Partial support from the projects CONACYT-40745-F and DGAPA-IN11700 is also acknowledged.

References

1. C. Rovelli and L. Smolin, *Nucl. Phys.* **B442**, 593 (1995); Erratum-ibid **B456**, 753 (1995); L. Smolin, *Phys. Rev.* **D49**, 4028 (1994).
2. G. Amelino-Camelia, J.R. Ellis, N.E. Mavromatos, D.V. Nanopoulos and S. Sarkar, *Nature (London)* **393**, 763 (1998).
3. See for example H. Sahlmann, T. Thiemann, O. Winkler, *Nucl. Phys* **B606**, 401 (2001) and references therein.
4. R. Gambini and J. Pullin, *Phys. Rev.* **D59**, 124021 (1999).
5. J. Alfaro, H.A. Morales-Técotl and L.F. Urrutia, *Phys. Rev. Lett.* **84**, 2318 (2000); J. Alfaro, H.A. Morales-Técotl and L.F. Urrutia, Phys. Rev. **D66**, 124006 (2002).
6. J. Alfaro, H.A. Morales-Técotl and L.F. Urrutia, *Phys. Rev.* **D65**, 103509 (2002); L.F. Urrutia, *Mod. Phys. Lett.* **A17**, 943 (2002).
7. D. Colladay and V. A. Kostelecký, *Phys. Rev.* **D55**, 6760 (1997); *Phys. Rev.* **D58**, 116002 (1998). See for example V.A. Kostelecký, these Proceedings.
8. V. A. Kostelecký, *Gravity, Lorentz Violation, and the Standard Model*, hep-th/0312310.
9. For a review see for example *Proceedings of the Second Meeting on CPT and Lorentz Symmetry*, ed. V. A. Kostelecký (World Scientific, 2002) and references therein.
10. G. Amelino-Camelia, *Int. J. Mod. Phys.* **D11**, 35 (2002); *Nature (London)* **418**, 34 (2002); J. Magueijo and L. Smolin, *Phys. Rev.* **D67**, 044017 (2003).
11. See for example C. Rovelli and S. Speziale, *Phys. Rev.* **D67**, 064019 (2003); F. Dowker, J. Henson and R. Sorkin, *Quantum Gravity Phenomenology, Lorentz Invariance and Discreteness*, gr-qc/0311055.
12. D. Sudarsky, L. Urrutia and H. Vucetich, *Phys. Rev. Lett.* **89**, 231301 (2002).
13. T. Thiemann, *Introduction to Modern Canonical Quantum General Relativity*, gr-qc/ 0110034.
14. V. A. Kostelecký and C.D. Lane, J. Math. Phys. **40**, 6245 (1999).
15. T.E. Chupp et al., *Phys. Rev. Letts.* **63**, 1541 (1989); D. Bear et al., *Phys. Rev. Letts.* **85**, 5038 (2000).
16. J. Alfaro and G. Palma, *Phys. Rev.* **D65**, 103516 (2002); *Phys. Rev.* **D67**, 083003 (2003).

VIII. LORENTZ VIOLATION

SUPERFIELD REALIZATIONS OF LORENTZ VIOLATION

M. S. BERGER

Physics Department, Indiana University, Bloomington, IN 47405, USA

Lorentz-violating extensions of the Wess-Zumino model have been formulated in superspace. The models respect a supersymmetry algebra and can be understood as arising from suitably modified superspace transformations.

1. Introduction

Spacetime symmetries have always played a central role in particle physics. Some of these symmetries like the Poincaré symmetry are taken to be exact, while other symmetries like supersymmetry are assumed to be broken. From the perspective of experimental physics it appears that supersymmetry is a badly broken symmetry, so much so that there has to date been no direct evidence for it. However viewed from the Planck scale, electroweak-scale supersymmetry is almost exact. Indeed much theoretical research has been devoted to accounting for the ratio of the supersymmetry breaking scale to the Planck scale. From this point of view, it seems that if one accepts the possibility that the a spacetime symmetry such as supersymmetry is broken in a very small manner, that one should entertain the possibility that the remaining spacetime symmetries are also broken to a very small extent even though there is no experimental evidence for it.

As the possibility that Lorentz and CPT violation might occur in fundamental theories has become more apparent, there has been interest in finding supersymmetric theories with Lorentz violation[1,2,3]. One approach to incorporate Lorentz and CPT violation into a Lorentz-symmetric Lagrangian by adding explicit breaking terms[4,5]. The resulting field theories should be regarded as effective theories arising from the more fundamental theory. Problems with microcausality are addressed in the underlying fundamental theory at the energy scales at which the effective theory breaks down[6]. The experimental implications of Lorentz and CPT violation parameterized in this manner have been explored extensively in recent years[7].

The first supersymmetric model with Lorentz and CPT violation involved extending[1] the Wess-Zumino model[8]. Two extensions were found, and these two models admit a superspace formulation[2].

2. The Wess-Zumino Lagrangian

A useful tool for developing supersymmetric field theories is to represent a supermultiplet of component fields as a superfield defined over a superspace of coordinates

$$z^M = (x^\mu, \theta^\alpha, \bar{\theta}_{\dot{\alpha}}) \tag{1}$$

The four anticommuting coordinates θ^α and $\bar{\theta}_{\dot{\alpha}}$ form two-component Weyl spinors. A superfield $\Phi(x, \theta, \bar{\theta})$ is then a function of the commuting space-time coordinates x^μ and of four anticommuting coordinates θ^α and $\bar{\theta}_{\dot{\alpha}}$ which form two-component Weyl spinors. A chiral superfield is a function of $y^\mu = x^\mu + i\theta\sigma^\mu\bar{\theta}$ and θ, i.e.

$$\begin{aligned}\Phi(x, \theta, \bar{\theta}) &= \phi(y) + \sqrt{2}\theta\psi(y) + (\theta\theta)\mathcal{F}(y) \, , \\ &= \phi(x) + i\theta\sigma^\mu\bar{\theta}\partial_\mu\phi(x) - \frac{1}{4}(\theta\theta)(\bar{\theta}\bar{\theta})\Box\phi(x) \\ &\quad + \sqrt{2}\theta\psi(x) + i\sqrt{2}\theta\sigma^\mu\bar{\theta}\theta\partial_\mu\psi(x) + (\theta\theta)\mathcal{F}(x) \, ,\end{aligned} \tag{2}$$

where one can define the usual real components of the complex scalar components as

$$\phi = \tfrac{1}{\sqrt{2}}(A + iB), \quad \mathcal{F} = \tfrac{1}{\sqrt{2}}(F - iG). \tag{3}$$

The conjugate superfield is

$$\Phi^*(x, \theta, \bar{\theta}) = \phi^*(z) + \sqrt{2}\bar{\theta}\bar{\psi}(z) + (\bar{\theta}\bar{\theta})\mathcal{F}^*(z) \, , \tag{4}$$

where $z^\mu = y^{\mu *} = x^\mu - i\theta\sigma^\mu\bar{\theta}$. The Wess-Zumino Lagrangian can now be derived from the superspace integral[9]

$$\int d^4\theta \Phi^*\Phi + \int d^2\theta \left[\frac{1}{2}m\Phi^2 + \frac{1}{3}g\Phi^3 + h.c.\right] . \tag{5}$$

The superspace integral over $\int d^4\theta$ projects out the $(\theta\theta)(\bar{\theta}\bar{\theta})$ component of the $\Phi^*\Phi$ superfield while the $\int d^2\theta$ projects out the $\theta\theta$ component of the superpotential. The result

$$\begin{aligned}\mathcal{L}_{WZ} &= \partial_\mu\phi^*\partial^\mu\phi + \frac{i}{2}[(\partial_\mu\psi)\sigma^\mu\bar{\psi} + (\partial_\mu\bar{\psi})\bar{\sigma}^\mu\psi] + \mathcal{F}^*\mathcal{F} \\ &\quad + m\left[\phi\mathcal{F} + \phi^*\mathcal{F}^* - \tfrac{1}{2}\psi\psi - \tfrac{1}{2}\bar{\psi}\bar{\psi}\right] \\ &\quad + g\left[\phi^2\mathcal{F} + \phi^{*2}\mathcal{F}^* - \phi(\psi\psi) - \phi^*(\bar{\psi}\bar{\psi})\right] \, ,\end{aligned} \tag{6}$$

is a Lagrangian which transforms into itself plus a total derivative under a supersymmetric transformation.

One can formulate the Wess-Zumino model in terms of differential operators that act on the superfields. Define

$$X \equiv (\theta\sigma^\mu\bar{\theta})\partial_\mu \ , \tag{7}$$

so that

$$U_x \equiv e^{iX} = 1 + i(\theta\sigma^\mu\bar{\theta})\partial_\mu - \frac{1}{4}(\theta\theta)(\bar{\theta}\bar{\theta})\Box \ . \tag{8}$$

The expansion terminates because of the anticommuting nature of θ and $\bar{\theta}$. Since X is a derivative operator, the action of U_x on a superfield $\mathcal{S}$ is a coordinate shift in which the spacetime coordinate x^μ is shifted to y^μ,

$$U_x\mathcal{S}(x,\theta,\bar{\theta}) = \mathcal{S}(y,\theta,\bar{\theta}) \ . \tag{9}$$

The chiral superfield $\Phi(x,\theta,\bar{\theta})$ is a function of y^μ and θ only, so it must then be of the form $\Phi(x,\theta,\bar{\theta}) = U_x\Psi(x,\theta)$ for some function Ψ. The kinetic terms of the Wess-Zumino model can be expressed as

$$\int d^4\theta \left[U_x^*\Psi(x,\bar{\theta})^*\right]\left[U_x\Psi(x,\theta)\right] = \int d^4\theta\Phi^*(z,\bar{\theta})\Phi(y,\theta) \ . \tag{10}$$

3. Lorentz Violation

It has been shown on quite general grounds that CPT violation implies Lorentz violation[10,11], but it is possible to have Lorentz violation without CPT violation. Of the two extensions to the Wess-Zumino model, the first does not contain CPT violation and it is easily understood as arising from the following substitution $\partial_\mu \to \partial_\mu + k_{\mu\nu}\partial^\nu$, where $k_{\mu\nu}$ is a real, symmetric, traceless, and dimensionless coefficient responsible for the Lorentz violation. so that

$$\mathcal{L}_{\text{Lorentz}} = \mathcal{L}_{WZ}(\partial_\mu \to \partial_\mu + k_{\mu\nu}\partial^\nu) \ . \tag{11}$$

The coefficient $k_{\mu\nu}$ transforms as a 2-tensor under observer Lorentz transformations but as a scalar under particle Lorentz transformations[a]. Adding Lorentz violation in this fashion can be immediately extended to encompass supersymmetric gauge theories as well since the superfield formulations of those theories will carry forward under the substitution $\partial_\mu \to \partial_\mu + k_{\mu\nu}\partial^\nu$.

[a]The terms containing $k_{\mu\nu}$ give rise to Lorentz violation, the physics remains independent of the particular coordinate system that is used to describe it.

The second extension of the Wess-Zumino model violates CPT in addition to containing Lorentz violation and is given by the Lagrangian[1]

$$\mathcal{L}_{\rm CPT} = [(\partial_\mu - ik_\mu)\phi^*]\,[(\partial^\mu + ik^\mu)\phi]$$
$$+\frac{i}{2}[((\partial_\mu + ik_\mu)\psi)\sigma^\mu\bar{\psi} + ((\partial_\mu - ik_\mu)\bar{\psi})\bar{\sigma}^\mu\psi] + \mathcal{F}^*\mathcal{F}\,. \quad (12)$$

Here the Lorentz and CPT violation is controlled by k_μ, which is a real coefficient of mass dimension one which transforms as a four-vector under observer Lorentz transformations but remains unaffected by particle Lorentz transformations[4,5]. Unlike the coefficient $k_{\mu\nu}$, the quantity k_μ has an odd number of four-indices so it violates CPT. The Lagrangian for the model with the CPT-violating coefficient k_μ can be obtained from the kinetic part of the Wess-Zumino Lagrangian in Eqn. (6) with the appropriate substitutions $\partial_\mu \to \partial_\mu \pm ik_\mu$.

The two Lorentz-violating models can be described in the superspace formalism in a way that parallels that of the ordinary Wess-Zumino model. Define superfields

$$\Phi_y(x,\theta,\bar{\theta}) = \Phi(x,\theta,\bar{\theta};\partial_\mu \to \partial_\mu + k_{\mu\nu}\partial^\nu)$$
$$= \phi(x_+) + \sqrt{2}\theta\psi(x_+) + (\theta\theta)F(x_+)\,, \quad (13)$$

and

$$\Phi_y^*(x,\theta,\bar{\theta}) = \Phi^*(x,\theta,\bar{\theta};\partial_\mu \to \partial_\mu + k_{\mu\nu}\partial^\nu)$$
$$= \phi^*(x_-) + \sqrt{2}\bar{\theta}\bar{\psi}(x_-) + (\bar{\theta}\bar{\theta})F^*(x_-)\,, \quad (14)$$

where

$$x_\pm^\mu = x^\mu \pm i\theta\sigma^\mu\bar{\theta} \pm ik^{\mu\nu}\theta\sigma_\nu\bar{\theta}\,., \quad (15)$$

are shifted coordinates that take the place of y^μ and z^μ. Under a CPT-transformation the chiral superfield Φ_y and the antichiral superfield Φ_y^* transform into themselves just as the usual superfields Φ and Φ^* do. The Lagrangian in Eqn. (11) can be obtained by the same superspace integral in Eqn. (5) with the superfields Φ_y and Φ_y^* substituted in the place of Φ and Φ^* (see Eqn. (24) below).

The appropriate modified superfields for the model in Eqn. (12),

$$\Phi_k(x,\theta,\bar{\theta}) = \Phi(x,\theta,\bar{\theta};\partial_\mu \to \partial_\mu + ik_\mu)\,, \quad (16)$$

and

$$\Phi_k^*(x,\theta,\bar{\theta}) = \Phi^*(x,\theta,\bar{\theta};\partial_\mu \to \partial_\mu - ik_\mu)\,, \quad (17)$$

cannot be obtained from a coordinate shift, but can be obtained from a more general superspace transformation.

4. Superspace Transformations

The Lorentz-violating extensions of the Wess-Zumino model can be understood as transformations on the superfields. Define

$$Y \equiv k_{\mu\nu}(\theta\sigma^\mu\bar{\theta})\partial^\nu \,, \tag{18}$$

$$K \equiv k_\mu(\theta\sigma^\mu\bar{\theta}) \,, \tag{19}$$

so that

$$U_y \equiv e^{iY} = 1 + ik_{\mu\nu}(\theta\sigma^\mu\bar{\theta})\partial^\nu - \frac{1}{4}k_{\mu\nu}k^{\mu\rho}(\theta\theta)(\bar{\theta}\bar{\theta})\partial^\nu\partial_\rho \,, \tag{20}$$

$$T_k \equiv e^{-K} = 1 - k_\mu(\theta\sigma^\mu\bar{\theta}) + \frac{k^2}{4}(\theta\theta)(\bar{\theta}\bar{\theta}) \,. \tag{21}$$

Since Y, like X, is a derivative operator, the action of U_y on a superfield $\mathcal{S}$ is a coordinate shift. The appearance of terms of order $\mathcal{O}(k^2)$ in both cases is easily understood in terms of these operators. Furthermore we have $U_y^* = U_y^{-1}$ while $T_k^* = T_k$ and not its inverse.

The supersymmetric models with Lorentz-violating terms can be expressed ın terms of new superfields,

$$\Phi_y(x,\theta,\bar{\theta}) = U_y U_x \Psi(x,\theta) \,, \tag{22}$$

$$\Phi_y^*(x,\theta,\bar{\theta}) = U_y^{-1} U_x^{-1} \Psi^*(x,\bar{\theta}) \,. \tag{23}$$

Applying U_y to the chiral and antichiral superfields merely effects the substitution $\partial_\mu \to \partial_\mu + k_{\mu\nu}\partial^\nu$. Since U_y involves a derivative operator just as U_x, the derivation of the chiral superfield Φ_y is a function of the variables x_+^μ and θ analogous to how, in the usual case, Φ is a function of the variables y^μ and θ. The Lagrangian is given by

$$\int d^4\theta \Phi_y^* \Phi_y + \int d^2\theta \left[\frac{1}{2}m\Phi_y^2 + \frac{1}{3}g\Phi_y^3 + h.c.\right]$$
$$= \int d^4\theta \left[U_y^*\Phi^*\right]\left[U_y\Phi\right] + \int d^2\theta \left[\frac{1}{2}m\Phi^2 + \frac{1}{3}g\Phi^3 + h.c.\right] \,. \tag{24}$$

For the CPT-violating model the superfields have the form

$$\Phi_k(x,\theta,\bar{\theta}) = T_k U_x \Psi(x,\theta) \,, \tag{25}$$

$$\Phi_k^*(x,\theta,\bar{\theta}) = T_k U_x^{-1} \Psi^*(x,\bar{\theta}) \,. \tag{26}$$

It is helpful to note that the transformation U_x acts on Ψ and its inverse U_x^{-1} acts on Ψ^*, while the same transformation T_k acts on both Ψ and Ψ^* (since $T_k^* = T_k$). A consequence of this fact is that the supersymmetry transformation will act differently on the components of the chiral superfield and its conjugate. Specifically the chiral superfield Φ_k is the same as Φ with the substitution $\partial_\mu \to \partial_\mu + ik_\mu$ whereas the antichiral superfield Φ_k^* is the same as Φ^* with the substitution $\partial_\mu \to \partial_\mu - ik_\mu$, as in Eqns. (16) and (17).

The CPT-violating model in Eqn. (12) can then be represented in the following way as a superspace integral:

$$\int d^4\theta \Phi_k^* \Phi_k = \int d^4\theta \Phi^* e^{-2K} \Phi \tag{27}$$

Unlike the CPT-conserving model, the $(\theta\theta)(\bar{\theta}\bar{\theta})$ component of $\Phi^*\Phi$ no longer transforms into a total derivative. A specific combination of components of $\Phi^*\Phi$ does transform into a total derivative, and this combination is in fact the $(\theta\theta)(\bar{\theta}\bar{\theta})$ component of $\Phi_k^*\Phi_k$.

5. Conclusions

The Wess-Zumino model can be described in terms of superspace transformations and the projections of components of functions of the superfield. It was shown that the two Lorentz-violating extensions of the Wess-Zumino model can be understood in terms of analogous transformations on modified superfields and projections arising from superspace integrals.

Acknowledgments

This work was supported in part by the U.S. Department of Energy under Grant No. No. DE-FG02-91ER40661.

References

1. M. S. Berger and V. A. Kostelecky, *Phys. Rev.* **D65**, 091701 (2002).
2. M. S. Berger, *Phys. Rev.* **D68**, 115005 (2003).
3. H. Belich, J. L. Boldo, L. P. Colatto, J. A. Helayel-Neto and A. L. Nogueira, *Phys. Rev.* **D65**, 065030 (2003).
4. D. Colladay and V. A. Kostelecky, *Phys. Rev.* **D55**, 6760 (1997).
5. D. Colladay and V. A. Kostelecky, *Phys. Rev.* **D58**, 116002 (1998).
6. V. A. Kostelecky and R. Lehnert, *Phys. Rev.* **D63**, 065008 (2001).
7. V. A. Kostelecky, "Proceedings of the Second Meeting on CPT and Lorentz Symmetry Meeting, Bloomington, USA, 15-18 August 2001."
8. J. Wess and B. Zumino, *Nucl. Phys.* **B70**, 39 (1974).
9. A. Salam and J. Strathdee, *Nucl. Phys.* **B76**, 477 (1974).
10. O. W. Greenberg, *Phys. Rev. Lett.* **89**, 231602 (2002).
11. O. W. Greenberg, *Phys. Lett.* **B567**, 179 (2003).

$\mathcal{CPT}$ SYMMETRY AND LORENTZ COVARIANCE

O.W. GREENBERG
Center for Theoretical Physics
Department of Physics
University of Maryland
College Park, MD 20742-411

This pedagogical talk shows the close connection of $\mathcal{CPT}$ symmetry and Lorentz covariance and finds the $\mathcal{CPT}$ transformation for a general field in a simple way using results of distribution theory and of several complex variables without proof.

1. INTRODUCTION

The standard textbooks of quantum field theory all get to the $\mathcal{CPT}$ theorem[1,2] by calculating each symmetry and then calculating their product. This is correct (except for the technical fact that each of $\mathcal{C}$, $\mathcal{P}$ and $\mathcal{T}$ can have an arbitrary phase since they are not connected to the identity, while $\mathcal{CPT}$, which is connected to the identity, cannot have an arbitrary phase). However calculating $\mathcal{CPT}$ by multiplying each of the three discrete symmetries is a very complicated way to calculate $\mathcal{CPT}$. More important, calculating $\mathcal{CPT}$ in that way obscures why $\mathcal{CPT}$ is fundamental but none of the individual symmetries is.

The purpose of this expository note is to explain why $\mathcal{CPT}$ is fundamental and to calculate it for a general relativistic quantum field without worrying about the mathematical issues connected with functions of several complex variables and their relation to tempered distributions whose support in momentum space lies in or on a cone. $\mathcal{CPT}$ is fundamental because it is closely related to Lorentz covariance. We show how far we can get with Lorentz covariance alone and where we must use an additional assumption. We find that the calculation of $\mathcal{CPT}$ using general arguments is greatly simpler than the pedestrian calculation of $\mathcal{C}$, $\mathcal{P}$ and $\mathcal{T}$ separately and then multiplying them. To make this note self contained we explain ideas from group theory and field theory that many readers already understand. Those readers should skip the introductory explanations and go directly to the $\mathcal{CPT}$ theorem itself.

2. REPRESENTATIONS OF THE REAL AND COMPLEX LORENTZ GROUPS

Since the heart of the argument is the fact that the connected component of the complex Lorentz group, $L(C)$, which is the proper complex Lorentz group, $L_+(C)$, contains spacetime inversion, we will discuss the Lorentz group first[3]. The (real) Lorentz group can be taken as the group, $SO(1,3)$, of real 4 x 4 matrices Λ that preserve the metric g that we take to have the form $g = diag(1,-1,-1,-1)$,

$$\Lambda^T g \Lambda = g. \tag{1}$$

You can check that this condition is equivalent to saying that a Lorentz transformation preserves the scalar product $x^2 = x \cdot x = (x^0)^2 - \sum_1^3 (x^i)^2$, i.e., $\Lambda x \cdot \Lambda = x \cdot x$. By taking the determinant of Eq.(1) we see that $det\Lambda = \pm 1$. By looking at the 00 element of Eq.(1) we find $(\Lambda^0{}_0)^2 - \sum_1^3 (\Lambda^0{}_i)^2 = 1$, so either $\Lambda^0{}_0 \geq 1$ or $\Lambda^0{}_0 \leq -1$. Thus the Lorentz group falls into four disconnected components, $L_+^\uparrow$, $L_+^\downarrow$, $L_-^\uparrow$, and $L_-^\downarrow$ according to the sign of the determinant of Λ and the sign of $\Lambda^0{}_0$. Only the first of these is a group since only $L_+^\uparrow$ contains the identity. We use $x \in V_+$ if $x^2 > 0, x^0 > 0$; $x \in V_-$ if $x^2 > 0, x^0 < 0$; $x \sim 0$ if $x^2 < 0$. We also have to consider the complex Lorentz group, the group of complex 4 x 4 matrices that obey Eq.(1). For the complex Lorentz group the sign of the determinant still cannot be changed continuously, but the matrix -1 is now connected to the identity, so there are only two disconnected components. The easiest way to find the continuous family of complex Lorentz transformations that connect the matrices 1 and -1 is by considering the covering groups of the real and complex Lorentz groups, to which we now turn.

We are familiar with the fact that a spin 1/2 state transforms under a rotation by an angle θ with a phase $\theta/2$ rather than the phase θ of a scalar state. So a rotation by 2π changes the phase of a spin 1/2 state even though such a rotation should be equivalent to the identity. Thus a spin 1/2 state does not transform as a true representation of the rotation group, but rather as a representation up to a factor. The idea of a covering group is to find a larger group whose representations are true representations without additional phases. For the rotation group the covering group is $SU(2)$, the group of 2 x 2 unitary matrices with determinant 1. For the connected component of the Lorentz group the covering group is $\bar{L}_+^\uparrow \equiv SL(2,C)$, the group of 2 x 2 complex matrices of determinant 1.

We introduce the two fundamental representations of $SL(2,C)$ as

$$u'_\alpha = A_{\alpha\beta} u_\beta \quad \text{and} \quad \dot{v}'_{\dot{\alpha}} = A^*_{\dot{\alpha}\dot{\beta}} \dot{v}_{\dot{\beta}}, \tag{2}$$

where $A \in SL(2,C)$ and $*$ stands for complex conjugate. We define a scalar product for these representations using the 2×2 antisymmetric Levi-Civita symbol $\epsilon_{\alpha\beta}$ for the undotted spinors and $\epsilon_{\dot{\alpha}\dot{\beta}}$ for the dotted spinors. We choose $\epsilon_{12} = 1$, $\epsilon_{\dot{1}\dot{2}} = 1$. Any representation of $SL(2,C)$ has the form of a spinor with k undotted and l dotted indices, each transforming as given above. The only way we can reduce these representations is by contracting with the ϵ's just described, so the irreducible representations of $SL(2,C)$ are spinors with k symmetrized undotted and l symmetrized dotted indices. Each index corresponds to spin $1/2$ so these spinors have spin $k/2$ and $l/2$ under the $SU(2) \otimes SU(2)$ formed by taking the groups whose generators are $J \pm iK$, where J are the rotation generators of the real Lorentz group and K are the generators of pure Lorentz transformations (boosts).

We take the Pauli matrices to have one undotted and one dotted index, $(\sigma_\mu)^{\dot{\beta}}_{\alpha}$, where σ_0 is the unit 2×2 matrix and $\sigma_{1,2,3}$ are the usual Pauli matrices. Then we can uniquely associate a 2×2 hermitian matrix X with a real vector x^μ by $(X)^{\dot{\beta}}_{\alpha} = x^\mu (\sigma_\mu)^{\dot{\beta}}_{\alpha}$. To invert this, trace with the σ matrices. The reader should check that $detX = x^2$. Recalling that matrices in $SL(2,C)$ have determinant 1, we see that $X' = AXA^\dagger$, where $\dagger$ stands for adjoint, is again Hermitian and is a Lorentz transformation on x, . The matrices A and $-A$ stand for the same Lorentz transformation; thus the group $SL(2,C)$ covers the connected component of the Lorentz group twice.

To cover the complex Lorentz group we allow two independent $SL(2,C)$ matrices to enter so that $X' = AXB^T$. This X' is no longer Hermitian, but it still has the same Minkowski metric length, so the covering group of the complex Lorentz group, $L_+(C)$, is $SL(2,C) \otimes SL(2,C)$. Since now we have two independent matrices A and B at our disposal, we can achieve $x \to -x$ either by choosing $A = 1,\ B = -1$ or $A = -1,\ B = 1$. We can go continuously from the identity $A = 1,\ B = 1$ to $A = 1,\ B = -1$ in the first case by choosing $A = 1,\ B(\phi) = diagonal(expi\phi/2, exp - i\phi/2)$. We find the 4×4 complex Lorentz transformations $\Lambda(\phi)$ from the definition of X'. The result, which is a continuous family of complex Lorentz transformations going from the identity to spacetime inversion, is

$$\begin{pmatrix} x'^0 \\ x'^3 \\ x'^1 \\ x'^2 \end{pmatrix} \begin{pmatrix} cos\frac{\phi}{2} & i\ sin\frac{\phi}{2} & 0 & 0 \\ i\ sin\frac{\phi}{2} & cos\frac{\phi}{2} & 0 & 0 \\ 0 & 0 & cos\frac{\phi}{2} & -sin\frac{\phi}{2} \\ 0 & 0 & sin\frac{\phi}{2} & cos\frac{\phi}{2} \end{pmatrix} \begin{pmatrix} x^0 \\ x^3 \\ x^1 \\ x^2 \end{pmatrix}$$

Are there other ways to achieve spacetime inversion? In $SL(2,C)\otimes SL(2,C)$ we need $AXB^T = -X$, or $AX = -XB^{T\;-1}$. Thus we need this relation, where X is replaced by each of the Pauli matrices σ_μ. For $\sigma_0 = 1$ we need $A = -B^{T-1}$. Then we need $AX = XA$, where for X we can choose any of the space σ's. This requires $A = \omega 1$ and for $A \in SL(2,C)$ we need $\omega^2 = 1$ or $\omega = \pm 1$. Thus the *only* possibilities to invert x^μ are the ones given above. Now we have the group theory we need to discuss the $\mathcal{CPT}$ theorem.

3. VACUUM MATRIX ELEMENTS OF PRODUCTS OF FIELDS DEFINE ANALYTIC FUNCTIONS

Next we discuss vacuum matrix elements of products of fields, often called Wightman functions[4] or distributions. Let $\phi^{(k,l)}(x)$ be a field with k undotted and l dotted indices, each set symmetrized, that transforms as the irreducible representation of $SL(2,C)$ described above. We use the active point of view in which a Poincaré transformation (a, A) acts as[a]

$$U(a,A)\phi^{(k,l)}(x)U(a,A)^\dagger = S^{(k,l)}(\Lambda)^{(-1)}\phi^{(k,l)}(\Lambda x + a); \tag{3}$$

The only case where we need the detailed form of $S^{(k,l)}(A,B)$ is when $\Lambda \in L_+(C)$ produces spacetime inversion and for that case $S^{(k,l)}(A,B)$ is just a diagonal phase. Thus the detailed form of $S^{(k,l)}$ is not necessary here. For this reason we suppress the indices belonging to the matrices $S^{(k,l)}$ as well as the indices belonging to the field $\phi^{(k,l)}(x)$. We assume the vacuum $|0\rangle$ is invariant under Poincaré transformations,

$$U(a,A)|0\rangle = |0\rangle. \tag{4}$$

To make clear how simple Jost's argument is, we streamline the notation and give the general case, rather than first giving the argument only for scalar fields. (For some properties, such as the support in momentum space which does not depend on the spin, we use the scalar case to illustrate the issue.) Let the single index (p) (for "pair" of indices) stand for (k,l). We use (p) and (k,l) interchangeably to label fields and other objects.

---

[a] We use the covering group of the Poincaré group, so in (a,Λ) we replace $\Lambda \in L_+^\uparrow$ by $A \in SL(2,C)$. In the argument of the fields on the right hand side on the next equation we replace Λ by $\Lambda(A)$ where $\Lambda(A) \in L_+^\uparrow$ is the homomorphic image of $A \in SL(2,C)$. Where we use the covering group of the complex Lorentz group we replace $\Lambda \in L_+(C)$ by $\Lambda(A,B)$ where $\Lambda(A,B)$ is the homomorphic image of $(A,B) \in SL(2,C) \otimes SL(2,C)$. To simplify notation we write Λ instead of $\Lambda(A)$ or $\Lambda(A,B)$ in both cases.

Then the general field becomes $\phi^{(p)}(x)$, the matrices are $S^{(p)}(A)$, and the transformation law, again suppressing indices, is

$$U(a,A)\phi^{(p)}(x)U^{\dagger}(a,A) = S^{(p)}(A)^{-1}\phi^{(p)}(\Lambda x + a). \tag{5}$$

Next we write the vacuum matrix element of an arbitrary product of fields and use this transformation law to find

$$\begin{aligned}
&\langle 0|\phi^{(p_1)}(x_1)\phi^{(p_2)}(x_2)\cdots\phi^{(p_n)}(x_n)|0\rangle \\
&= (|0\rangle, \phi^{(p_1)}(x_1)\phi^{(p_2)}(x_2)\cdots\phi^{(p_n)}(x_n)|0\rangle) \\
&= (U(a,A)|0\rangle, U(a,A)\phi^{(p_1)}(x_1)\phi^{(p_2)}(x_2)\cdots\phi^{(p_n)}(x_n)|0\rangle) \\
&= (|0\rangle, [\prod_1^n S^{(p_i)}(A)^{-1}]\phi^{(p_1)}(\Lambda x_1 + a)\phi^{(p_2)}(\Lambda x_2 + a)\cdots\phi^{(p_n)}(\Lambda x_n + a)|0\rangle) \\
&= [\prod_1^n S^{(p_i)}(A)^{-1}]\langle 0|\phi^{(p_1)}(\Lambda x_1 + a)\phi^{(p_2)}(\Lambda x_2 + a)\cdots\phi^{(p_n)}(\Lambda x_n + a)|0\rangle.
\end{aligned} \tag{6}$$

Because of translation invariance this matrix element depends on only $n-1$ differences of the spacetime coordinates. We define the Wightman function

$$\begin{aligned}
&W^{(n;p_1p_2\cdots p_n)}(x_1 - x_2, x_2 - x_3, \cdots, x_{n-1} - x_n) \\
&\equiv \langle 0|\phi^{(p_1)}(x_1)\phi^{(p_2)}(x_2)\cdots\phi^{(p_n)}(x_n)|0\rangle.
\end{aligned} \tag{7}$$

Since we deal with three kinds of difference vectors, we use different letters to distinguish them: ξ for real vectors, ρ for real vectors, called "Jost points," defined below, in the domain of analyticity $\mathcal{T}'_{n-1}$, and ζ for complex vectors. To streamline the notation we compress the indices $(p_1p_2\cdots p_n)$ on $F^{(n)}$ to a single index $(\wp)$. We define $\xi_j = x_j - x_{j+1}$. Then invariance under the connected component of the Lorentz group (the proper orthochronous component, $L_+^{\uparrow}$) gives

$$W^{(n;\wp)}(\Lambda\xi_1, \Lambda\xi_2, \cdots, \Lambda\xi_{n-1}) = [\prod_1^n S^{(p_i)}(A)]W^{(n;\wp)}(\xi_1, \xi_2, \cdots, \xi_{n-1}). \tag{8}$$

The requirement that physical states have positive energy, except the vacuum which has zero energy, lie in or on the forward light cone implies that the momenta, q_j, in the Fourier transform of the $F^{(n;\wp)}$'s lie in or on the forward light cone. This follows from the translation subgroup of the Poincaré group.

Now use the intuitive criterion that a Fourier transform that is a distribution becomes an analytic function when the external variable is made complex in such a way as to provide a damping factor so that the Fourier transform becomes a Laplace transform. Thus we must examine when the factor $exp(-iq_j \cdot \xi_j)$ becomes decreasing for $\xi_j \to \zeta_j = \xi_j + i\eta_j$ with ξ_j and η_j real vectors. What counts is the absolute value of the factor which is $exp(q_j \cdot \eta_j)$. This becomes decreasing if η_j is in or on the backward light cone since the physical momentum q_j is in the forward light cone. Thus the Wightman function $W^{(n)}(\zeta_1, \zeta_2, \cdots, \zeta_{n-1})$ is an analytic function of $4(n-1)$ complex variables (in four-dimensional spacetime) when $Im\zeta_j \in V_-^{\uparrow}$. It is also single-valued.

4. ENLARGEMENT OF THE DOMAIN OF ANALYTICITY

As it stands, $W^{(n)}$ is analytic only when $Im\zeta_i \neq 0$, i.e., its domain of analyticity has no real points. Call this domain, which has the form of a tube with $Re\zeta_i$ arbitrary and $Im\zeta_i \in V_-$, the tube $\mathcal{T}_{n-1}$. Now we restore the labels of the fields and use a profound result due to V. Bargmann, D.W. Hall and A.S. Wightman[5]. If

$$W^{(n;\wp)}(\Lambda\zeta_1, \Lambda\zeta_2, \cdots, \Lambda\zeta_{n-1}) = [\prod_1^n S^{(p_i)}(A)] W^{(n;\wp)}(\zeta_1, \zeta_2, \cdots, \zeta_{n-1}) \tag{9}$$

real $\Lambda \in L_+^{\uparrow}$ then $W^{(n;\wp)}(\zeta_1, \zeta_2, \cdots, \zeta_{n-1})$ has a (unique) single-valued analytic continuation to the domain $\mathcal{T}'_{n-1}$, that we call the extended tube, that is the union of all $\Lambda\mathcal{T}_{n-1}$, where now we have complex matrices $\Lambda \in L_+(C)$. This enlargement of the domain of analyticity leads to two crucial results. First, in contrast to $\mathcal{T}_{n-1}$, our new, larger domain of analyticity, $\mathcal{T}'_{n-1}$, contains real points of analyticity (ρ_j) that we discuss below. Secondly since $\mathcal{T}'_{n-1}$ is invariant under complex Lorentz transformations in $L_+(C)$, one of which is spacetime inversion, we have the relation

$$W^{(n;\wp)}(\zeta_1, \zeta_2, \cdots, \zeta_{n-1}) = (-1)^L W^{(n;\wp)}(-\zeta_1, -\zeta_2, \cdots, -\zeta_{n-1}) \tag{10}$$

$L = \sum l_i$, in $\mathcal{T}'_{n-1}$. To see where the factor $(-1)^L$ comes from we repeat that for $\Lambda \in L_+^{\uparrow}$, Λ depends on the matrices A and A^* in $SL(2, C)$; however, as mentioned in a footnote above, now that we have the extension to $L_+(C)$, Λ depends on two independent matrices A and B in $SL(2, C)$ and we can transform continuously from the identity to spacetime inversion. The $S^{(k,l)}$ matrices for spacetime inversion are just powers of (-1);

thus if we choose $A = 1,\ B = -1$ then $S^{(k,l)}(1,-1) = (-1)^l \mathbf{1}$.[b] Note that spacetime inversion here is an element of the complex Lorentz group and is unitary, not antiunitary.

Jost[2] found a precise characterization of the real points, (ρ_j), in $\mathcal{T}'_{n-1}$: $\sum_1^{n-1} \lambda_i \rho_i \sim 0$, for all $\lambda_i > 0$ such that $\sum_1^{n-1} \lambda_i \geq 0$. This requires that each $\rho_i \sim 0$.

Jost's result is particularly simple for the $W^{(2)}$ function for single scalar field in which there is one complex difference vector ζ. Since $W^{(2)}(\Lambda\zeta) = W^{(2)}(\zeta)$ we can find the extended tube $\mathcal{T}'_1$ by finding the values of ζ^2 that can be obtained from $\Lambda\zeta$ with $\zeta = \xi + i\eta$, $\eta \in V_-$. Then $\zeta^2 = \xi^2 - \eta^2 + 2i\xi\cdot\eta$. The real points are those for which $\xi \cdot \eta = 0$, with $\eta \in V_-$. These points are the spacelike points $\xi \sim 0$, in agreement with Jost's general result.

5. THE GENERAL FORMULA FOR $\mathcal{CPT}$

In terms of vacuum matrix elements the relation between Wightman functions at Jost points that comes from spacetime inversion, Eq.(3), where $\Lambda(A, B)$ has $A = 1,\ B = -1$, is

$$\langle 0|\phi^{(k_1,l_1)}(x_1)\phi^{(k_2,l_2)}(x_2)\cdots\phi^{(k_n,l_n)}(x_n)|0\rangle$$
$$= (-1)^L \langle 0|\phi^{(k_1,l_1)}(-x_1)\phi^{(k_2,l_2)}(-x_2)\cdots\phi^{(k_n,l_n)}(-x_n)|0\rangle. \quad (11)$$

Each side of Eq.(10) can be analytically continued; the left hand side to complex ζ_i with $\eta = Im\zeta \in V_+^{\downarrow}$ and the right hand side to complex ζ_i with $\eta = Im\zeta \in V_+^{\uparrow}$. We cannot take the limit as $\eta_i \to 0$ to get a relation between vacuum matrix elements of products of the fields, because, as just noted, if ζ_i has its imaginary part in the backward cone, then $-\zeta_i$ has its imaginary part in the forward cone and then the analytic continuations of the functions on the two sides of Eq.(10) are not valid in the same domain. On the other hand if we consider the vacuum matrix element with the fields in completely reversed order,

$$\langle 0|\phi^{(p_n)}(-x_n)\cdots\phi^{(p_2)}(-x_2)\phi^{(p_1)}(-x_1)|0\rangle, \quad (12)$$

which corresponds to

$$W^{(n;i\wp)}(\xi_{n-1},\cdots,\xi_2,\xi_1), \quad (13)$$

in terms of difference variables, where $i\wp$ stands for $(p_n \cdots p_2 p_1)$, both functions have the same domain, $\mathcal{T}'_{n-1}$, of analyticity. This is where we assume

[b] Here $\mathbf{1}$ is the direct product of a $(2k+1)\times(2k+1)$ unit matrix and a $(2l+1)\times(2l+1)$ unit matrix and we find the result just stated for L.

something beyond Lorentz covariance. To reverse the order of all the fields we assume that at a Jost point the two vacuum matrix elements are related by a sign $(-1)^I$ where I is the number of transpositions of Fermi fields necessary to invert the order of the fields. This is implied by the spin-locality theorem[c] but is weaker than that theorem since we need this relation only at Jost points (or even only in a neighborhood of a Jost point) in each matrix element, rather than as an operator relation. If there are F Fermi fields then $I = (F-1)+(F-2)+\cdots+1 = 1/2F(F-1)$. Since the number of Fermi fields in a nonvanishing vacuum matrix element must be even, $F-1$ must be odd, thus the phase that enters is $(-1)^{1/2F(F-1)} = ((-1)^{(F-1)})^{F/2} = (-1)^{F/2} = i^F$. The condition on matrix elements that

$$\langle 0|\phi^{(k_1,l_1)}(x_1)\phi^{(k_2,l_2)}(x_2)\cdots\phi^{(k_n,l_n)}(x_n)|0\rangle = \\ i^F\langle 0|\phi^{(k_1,l_1)}(x_n)\cdots\phi^{(k_2,l_2)}(x_2)\phi^{(k_n,l_n)}(x_1)|0\rangle \tag{14}$$

at Jost points is called "weak local commutativity." Clearly local commutativity (sometimes called microcausality) implies weak local commutativity. Combining spacetime inversion and weak local commutativity and collecting phases, we have

$$W^{(n;\wp)}(\zeta_1,\zeta_2,\cdots,\zeta_{n-1}) = i^F(-1)^L W^{(n;i\wp)}(\zeta_{n-1},\cdots,\zeta_2,\zeta_1) \tag{15}$$

Now we can take $Im\zeta_i \to 0$ on both sides and we get an equality between distributions for all ξ_i,

$$W^{(n;\wp))}(\xi_1,\xi_2,\cdots,\xi_{n-1}) = i^F(-1)^L W^{(n;i\wp)}(\xi_{n-1},\cdots,\xi_2,\xi_1). \tag{16}$$

Translated back into vacuum matrix elements this says

$$\langle 0|\phi^{(p_1)}(x_1)\phi^{(p_2)}(x_2)\cdots\phi^{(p_n)}(x_n)|0\rangle = \\ i^F(-1)^L\langle 0|\phi^{(p_n)}(-x_n)\cdots\phi^{(p_2)}(-x_2)\phi^{(p_1)}(-x_1)|0\rangle. \tag{17}$$

Replacing (p_j) by (k_j,l_j) we have

$$\langle 0|\phi^{(k_1,l_1)}(x_1)\phi^{(k_2,l_2)}(x_2)\cdots\phi^{(k_n,l_n)}(x_n)|0\rangle = \\ i^F(-1)^L\langle 0|\phi^{(k_n,l_n)}(-x_n)\cdots\phi^{(k_2,l_2)}(-x_2)\phi^{(k_1,l_1)}(-x_1)|0\rangle. \tag{18}$$

We restore the original order of the fields on the right hand side by using the hermiticity of the scalar product, $(\Psi,\Xi) = (\Xi,\Psi)^*$. We expect complex

[c]This condition is usually called the spin-statistics theorem. We have argued that in the present context it should be called the spin-locality theorem.[6]

conjugation since we know that $\mathcal{CPT}$ is antiunitary. We find

$$\langle 0|\phi^{(k_1,l_1)}(x_1)\phi^{(k_2,l_2)}(x_2)\cdots\phi^{(k_n,l_n)}(x_n)|0\rangle = \\ i^F(-1)^L\langle 0|\phi^{(k_1,l_1)\dagger}(-x_1)\phi^{(k_2,l_2)\dagger}(-x_2)\cdots\phi^{(k_n,l_n)\dagger}(-x_n)|0\rangle^*, \quad (19)$$

where $F=\sum_1^n f_i$ and f is zero for a Bose field (with $k+l$ even) and one for a Fermi field (with $k+l$ odd). Now we read off $\mathcal{CPT}\equiv\Theta$,

$$\Theta\phi^{(k,l)}(x)\Theta^\dagger = (-1)^l(\pm i)^f\phi^{(k,l)\dagger}(-x). \quad (20)$$

When we embed this relation in an arbitrary vacuum matrix element and use the invariance of the vacuum, $\Theta|0\rangle = |0\rangle$, we find precisely Eq.(19). When we run this sequence of relations the other way, we conclude that weak local commutativity in the neighborhood of a Jost point is necessary and sufficient for $\mathcal{CPT}$. There is a sign ambiguity because F, the total number of Fermi fields in a nonvanishing vacuum matrix element, must be even.

Lorentz covariance alone is not sufficient for $\mathcal{CPT}$. A free or generalized free field can be Lorentz covariant but not obey $\mathcal{CPT}$ invariance if the particle and antiparticle masses are different[7]. What fails in that case is that weak local commutativity (see Eq.(14)) does not hold at Jost points. This possibility is associated with the purely timelike support in momentum space of free or generalized free fields; for timelike momenta positive and negative energies can be separated in a covariant way. By contrast positive and negative energies can be transformed into each other for spacelike momenta. Note that although the fields in these examples transform covariantly their time-ordered products are not covariant. Thus if we require that time-ordered products be covariant as part of Lorentz covariance of a theory then, as shown in[7], free fields that violate $\mathcal{CPT}$ are not covariant.

Note that $\mathcal{CPT}$ takes each irreducible of $L_+^\uparrow$ to a phase times itself, so that for example the part $\phi^{(1,0)}$ of the Dirac spin field is mapped to $i\phi^{(1,0)}$ and the part $\phi^{(0,1)}$ is mapped to $-i\phi^{(0,1)}$. Both the vector and axial vector fields have the form $\phi^{(1,1)}$ so these fields are indistinguishable under Θ and both get the phase (-1) under Θ. The analogous statements hold for the scalar and pseudoscalar fields, $\phi^{(0,0)}$, which both get phase 1. The antisymmetric rank two tensor field $T^{\mu\nu}$ is $\phi^{(2,0)}+\phi^{(0,2)}$ and also gets phase 1. The traceless symmetric tensor of rank two is $\phi^{(2,2)}$ and also gets phase 1.

The $\mathcal{CPT}$ operator Θ interchanges undotted and dotted indices, so that

$$\phi^{(k,l)\dagger}(x) = \phi^{\dagger(l,k)}(x). \quad (21)$$

Under Θ particles and antiparticles are interchanged (some particles may be identical to their antiparticles). Energies and momenta stay the same; spin components and helicities are reversed.

When we act twice by Θ we find

$$\begin{aligned}\Theta^2\phi^{(k,l)}(x)\Theta^{\dagger 2} &= \Theta(-1)^l(\pm i)^f\phi^{(k,l)\,\dagger}(-x)\Theta^\dagger \\ &= \Theta(-1)^l(\mp i)^f\Theta\phi^{(k,l)\,\dagger}(-x)\Theta^\dagger \\ &= (-1)^f\phi^{(k,l)}(x), \end{aligned} \tag{22}$$

so Θ^2 commutes with Bose fields and anticommutes with Fermi fields. The reader can check that the phase of Θ^2 cannot be changed by changing a phase in the definition of Θ. This is true for all antiunitary operators.

6. CONCLUSIONS

Jost's general proof of the $\mathcal{CPT}$ theorem leads directly to a definition of $\mathcal{CPT}$ symmetry applied to fields belonging to an arbitrary irreducible representation of $SL(2,C)$, the covering group of the real proper orthochronous Lorentz group, $L_+^\uparrow$ as given in Eq.(20). The only requirement beyond Lorentz covariance is weak local commutativity at Jost points.

Acknowledgments

This work was supported in part by the National Science Foundation, Award No. PHY-0140301.

References

1. W. Pauli, in *Niels Bohr and the Development of Physics*, (McGraw-Hill, New York, 1955), pp30-51.
2. R. Jost, *Helv. Phys. Acta* **30**, 409 (1957).
3. R. Jost, *The General Theory of Quantized Fields*, (American Mathematical Society, Providence, 1965).
4. A.S. Wightman, *Phys. Rev.* **101**, 860 (1956).
5. D.W. Hall and A.S. Wightman, *Det. Kong. Danske Videnskabernes Selskab, Mat.-fys. Medd.* **31**, No. 5 (1957).
6. O.W. Greenberg, *Phys. Lett. B* **416**, 144 (1998).
7. O.W. Greenberg, Phys. Rev. Lett. **89**, 231602-1 (2002).

NEUTRINO OSCILLATIONS AND LORENTZ VIOLATION

V. ALAN KOSTELECKÝ

Physics Department, Indiana University, Bloomington, IN 47405, U.S.A.

Existing experimental data for neutrino oscillations are consistent with Lorentz-violating massless neutrinos. This talk summarizes some aspects of neutrino oscillations from the perspective of Lorentz and CPT violation in effective quantum field theory.

1. Introduction

Most data acquired to date by experiments in particle physics are successfully described by the minimal Standard Model (SM). One exception is the observation of neutrino oscillations, which are now well established and provide a compelling indication that the minimal SM requires modification. The inclusion of small masses in the neutrino sector of the SM suffices to describe the bulk of the experimental results, although a definitive understanding awaits the results of ongoing and future measurements. From a theoretical perspective, these modifications to the SM could arise as suppressed low-energy energy effects from an underlying unified theory at the Planck scale, $m_P \simeq 10^{19}$ GeV. This suggests the interesting possibility that neutrino oscillations might represent signals from the Planck scale.

This talk considers the description of neutrino oscillations in terms of a promising candidate Planck-scale signal, namely, violations of relativity arising through Lorentz and CPT breaking.[1] The focus is on neutrino behavior from the perspective of Lorentz and CPT violation in effective quantum field theory. The general effective field theory describing Lorentz and CPT violation is called the Standard-Model Extension[2,3] (SME) and was the basis for the original suggestion that Lorentz and CPT violation might occur in neutrinos.[2] In the intervening years, several theoretical investigations of this possibility within the context of the SME have been performed,[4,5,6,7,8,9,10,11] mostly for special models involving only a few coefficients for Lorentz and CPT violation. Here, a comprehensive theoretical study of Lorentz and CPT violation in neutrinos is outlined and a potentiallly realistic model is described. Details about the ideas discussed here can be found in the literature.[9,10]

2. Theory

The lagrangian of the SME consists of the SM and gravitational lagrangian supplemented by all possible coordinate-invariant terms constructed with SM and gravitational fields and violating Lorentz symmetry.[3] The extra terms involve Lorentz-violating operators coupled to coefficients with Lorentz indices. Various origins have been suggested for these operators. A widely applicable and theoretically attractive source is physical spontaneous Lorentz violation, first suggested in gravitationally coupled field theories and string theory[12] and then extended to incorporate CPT violation.[13] Proposed sources also include noncommutative field theory,[14] non-string approaches to quantum gravity,[15] random dynamics,[16] and multiverses.[17] Measurements of SME coefficients for Lorentz violation have attained Planck-scale sensitivity in experiments with mesons,[18,19,20] baryons,[21,22,23] electrons,[24,25] photons,[26,27,28,29] and muons.[30]

This part of the talk considers the basic theory of neutrino behavior in the presence of Lorentz and CPT violation in Minkowski spacetime. The main focus is oscillation phenomena,[31] but the formalism described can also be applied to other classes of experiments such as supernova studies, direct mass searches, and tests for neutrinoless double-beta decay.

2.1. *Effective hamiltonian*

Consider a general theory for N neutrino species, incorporating all Majorana- and Dirac-type couplings of left- and right-handed neutrinos, both with and without Lorentz violation. Denote the neutrino fields by the Dirac spinors $\{\nu_e, \nu_\mu, \nu_\tau, \ldots\}$ and their charge conjugates $\{\nu_{e^C} \equiv \nu_e^C, \nu_{\mu^C} \equiv \nu_\mu^C, \nu_{\tau^C} \equiv \nu_\tau^C, \ldots\}$, where $\nu_a^C \equiv C\bar{\nu}_a^T$ as usual. Collect all fields and conjugates into a quantity ν_A, where A ranges over the $2N$ possibilities $\{e, \mu, \tau, \ldots, e^C, \mu^C, \tau^C, \ldots\}$, so that charge conjugation becomes a linear transformation $\nu_A^C = \mathcal{C}_{AB}\nu_B$. In this notation, the dominant contributions to the general equations of motion for free propagation yield the first-order differential equation

$$(i\Gamma^\nu_{AB}\partial_\nu - M_{AB})\nu_B = 0, \qquad (1)$$

extending the usual equations of motion for Dirac and Majorana neutrinos.

In Equation (1), Γ^{μ}_{AB}, M_{AB} are constant matrices in spinor space. They can be decomposed as

$$\Gamma^{\nu}_{AB} \equiv \gamma^{\nu}\delta_{AB} + c^{\mu\nu}_{AB}\gamma_{\mu} + d^{\mu\nu}_{AB}\gamma_5\gamma_{\mu} + e^{\nu}_{AB} + if^{\nu}_{AB}\gamma_5 + \tfrac{1}{2}g^{\lambda\mu\nu}_{AB}\sigma_{\lambda\mu},$$
$$M_{AB} \equiv m_{AB} + im_{5AB}\gamma_5 + a^{\mu}_{AB}\gamma_{\mu} + b^{\mu}_{AB}\gamma_5\gamma_{\mu} + \tfrac{1}{2}H^{\mu\nu}_{AB}\sigma_{\mu\nu}. \qquad (2)$$

Here, m and m_5 are mass terms. All other coefficients control terms violating Lorentz symmetry, and the coefficients a, b, e, f, g control CPT violation as well. The matrices Γ^{μ}_{AB}, M_{AB} must satisfy certain conditions following from hermiticity of the theory and from the interdependence of ν and ν^C. For example, all the coefficients for Lorentz and CPT violation in Eq. (2) are hermitian in generation space.

Obtaining the hamiltonian associated with Eq. (1) requires handling the unconventional coefficient for the time derivative, which can be done following established procedures.[32] For simplicity, consider here only the minimal case with $N = 3$ and a standard seesaw mechanism[33] suppressing the propagation of right-handed neutrinos, and restrict attention to leading-order Lorentz and CPT violation. Amplitudes $b_{e,\mu,\tau}(t;\vec{p})$ and $d_{e,\mu,\tau}(t;\vec{p})$ can be defined that approximate active neutrinos and active antineutrinos of momentum $\vec{p}$. A somewhat lengthy calculation[10] establishes that the time evolution of these amplitudes is

$$\begin{pmatrix} b_a(t;\vec{p}) \\ d_a(t;\vec{p}) \end{pmatrix} = \exp(-ih_{\rm eff}t)_{ab} \begin{pmatrix} b_b(0;\vec{p}) \\ d_b(0;\vec{p}) \end{pmatrix}, \qquad (3)$$

where $h_{\rm eff}$ is the effective hamiltonian for flavor neutrino propagation, given at leading order by

$$(h_{\rm eff})_{ab} = |\vec{p}|\delta_{ab}\begin{pmatrix} 1 & 0 \\ 0 & 1 \end{pmatrix} + \frac{1}{2|\vec{p}|}\begin{pmatrix} (\widetilde{m}^2)_{ab} & 0 \\ 0 & (\widetilde{m}^2)^*_{ab} \end{pmatrix}$$
$$+\frac{1}{|\vec{p}|}\begin{pmatrix} [(a_L)^{\mu}p_{\mu} - (c_L)^{\mu\nu}p_{\mu}p_{\nu}]_{ab} & -i\sqrt{2}p_{\mu}(\epsilon_+)_{\nu}[(g^{\mu\nu\sigma}p_{\sigma} - H^{\mu\nu})\mathcal{C}]_{ab} \\ i\sqrt{2}p_{\mu}(\epsilon_+)^*_{\nu}[(g^{\mu\nu\sigma}p_{\sigma} + H^{\mu\nu})\mathcal{C}]^*_{ab} & [-(a_L)^{\mu}p_{\mu} - (c_L)^{\mu\nu}p_{\mu}p_{\nu}]^*_{ab} \end{pmatrix}. \qquad (4)$$

In this equation, $(c_L)^{\mu\nu}_{ab} \equiv (c+d)^{\mu\nu}_{ab}$ and $(a_L)^{\mu}_{ab} \equiv (a+b)^{\mu}_{ab}$, $p_{\mu} = (|\vec{p}|; -\vec{p})$ at leading order, and the complex vector $(\epsilon_+)_{\mu}$ can be chosen as $(\epsilon_+)^{\nu} = \frac{1}{\sqrt{2}}(0; \hat{\epsilon}_1 + i\hat{\epsilon}_2)$, where $\hat{\epsilon}_1$, $\hat{\epsilon}_2$ are real and $\{\vec{p}/|\vec{p}|, \hat{\epsilon}_1, \hat{\epsilon}_2\}$ form a right-handed orthonormal triad.

The first line of the above equation involves the diagonal kinetic term of the minimal SM and the usual Lorentz-conserving neutrino-mass term. In the second line, the coefficients $(a_L)^{\mu}_{ab}$, $(c_L)^{\mu\nu}_{ab}$ determine the dominant Lorentz-violating effects in neutrino-neutrino mixing, while $(g^{\mu\nu\sigma}\mathcal{C})_{ab}$,

$(H^{\mu\nu}C)_{ab}$ generate Lorentz-violating neutrino-antineutrino mixing. The former two types of coefficient preserve SU(3)×SU(2)×U(1) and appear in the minimal SME, while the latter two involve Majorana-type couplings breaking both SU(3)×SU(2)×U(1) and lepton number. Most of these coefficients lead to physical effects, but a few combinations are unphysical due to symmetries or to the existence of field redefinitions that can eliminate them.[2,32,34,35] Note that the properties of the coefficients for Lorentz violation in Eq. (2) imply that the CPT-conjugate hamiltonian is obtained by changing the sign of the coefficients a_L and g. Note also that none of the terms in Eq. (4) correspond to independent mass matrices for neutrinos and antineutrinos as recently proposed,[36] a result consistent with Greenberg's formal impossibility proof.[37]

The effective hamiltonian (4) describes all contributions from operators of renormalizable dimension and therefore provides a definitive foundation for the treatment of Lorentz and CPT violation in neutrinos. Operators of nonrenormalizable mass dimension $n > 4$ could also be significant, depending partly on the degree to which they are suppressed by powers of the Planck scale.[19] In fact, at energies well beyond those relevant for current experiments, Lorentz-violating terms of nonrenormalizable dimension may be necessary for stability and causality.[32] Some consequences of these more general operators are discussed in the literature.[9,10,38]

The generality of the effective hamiltonian (4) implies that it must also contain the effects of matter interactions. Indeed, the effective lagrangian for neutrino propagation in normal matter is $\Delta\mathcal{L}_{\text{matter}} = -\sqrt{2}G_F n_e \bar{\nu}_e \gamma^0 P_L \nu_e + (G_F n_n/\sqrt{2})\bar{\nu}_a \gamma^0 P_L \nu_a$, which means that matter effects are incorporated by coefficient contributions of the form $(a_{L,\text{eff}})^0_{ee} = G_F(2n_e - n_n)/\sqrt{2}$ and $(a_{L,\text{eff}})^0_{\mu\mu} = (a_{L,\text{eff}})^0_{\tau\tau} = -G_F n_n/\sqrt{2}$, where n_e and n_n are the number densities of electrons and neutrons.

2.2. *Neutrino mixing*

To extract the implications for neutrino mixing, it is useful to diagonalize the effective hamiltonian. This involves a 6×6 unitary matrix U_{eff}, $h_{\text{eff}} = U^\dagger_{\text{eff}} E_{\text{eff}} U_{\text{eff}}$, where E_{eff} is a 6×6 diagonal matrix. There are therefore as many as six independent propagating states excluding sterile neutrinos, five possible eigenvalue differences, and hence five independent oscillation lengths in Lorentz-violating mixing. In contrast, the Lorentz-covariant case allows only three independent propagating states and two independent oscillation lengths. Note also that the five eigenvalue differences are *not* normal mass differences since their energy dependences are unconventional.

Denote the six independent propagating states by the amplitudes $B_J(t;\vec{p})$, $J = 1,\ldots,6$. Then, $B_J(t;\vec{p}) = \widetilde{U}_{Ja}b_a(t;\vec{p}) + \overline{U}_{Ja}d_a(t;\vec{p})$, where $U_{\rm eff}$ has been separated into 6×3 matrices $U_{\rm eff} = (\widetilde{U},\overline{U})$. The time evolution operator $S_{ab}(t)$ becomes

$$S_{ab}(t) = (U_{\rm eff}^{\dagger}e^{-iE_{\rm eff}t}U_{\rm eff})_{ab} = \begin{pmatrix} S_{\nu_a\nu_b}(t) & S_{\nu_a\bar{\nu}_b}(t) \\ S_{\bar{\nu}_a\nu_b}(t) & S_{\bar{\nu}_a\bar{\nu}_b}(t) \end{pmatrix}, \tag{5}$$

where $E_{(J)}$ are the diagonal values of $E_{\rm eff}$. This expression yields the oscillation probabilities at time t. Thus, the probability for a neutrino of type b to oscillate into a neutrino of type a is $P_{\nu_b\to\nu_a}(t) = |S_{\nu_a\nu_b}(t)|^2$, while that for a neutrino of type b to oscillate into an antineutrino of type a is $P_{\nu_b\to\bar{\nu}_a}(t) = |S_{\bar{\nu}_a\nu_b}(t)|^2$. Similarly, for antineutrinos we have $P_{\bar{\nu}_b\to\nu_a}(t) = |S_{\nu_a\bar{\nu}_b}(t)|^2$ or $P_{\bar{\nu}_b\to\bar{\nu}_a}(t) = |S_{\bar{\nu}_a\bar{\nu}_b}(t)|^2$.

The CPT properties of the transition amplitudes are $S_{\nu_a\nu_b}(t) \longleftrightarrow S^*_{\bar{\nu}_a\bar{\nu}_b}(-t)$ and $S_{\bar{\nu}_a\nu_b}(t) \longleftrightarrow -S^*_{\nu_a\bar{\nu}_b}(-t)$. If CPT is unbroken these relations become equalities, whereupon the first generates the standard result that CPT invariance implies $P_{\nu_b\to\nu_a}(t) = P_{\bar{\nu}_a\to\bar{\nu}_b}(t)$, while the second implies $P_{\nu_b\rightleftarrows\bar{\nu}_a}(t) = P_{\nu_a\rightleftarrows\bar{\nu}_b}(t)$. However, negation of terms in these results can fail. For example, CPT violation need not imply $P_{\nu_b\to\nu_a}(t) \neq P_{\bar{\nu}_a\to\bar{\nu}_b}(t)$.

For the above analysis of the effective hamiltonian and transition probabilities, the choice of observer reference frame is irrelevant because the physics is coordinate invariant and in particular is observer Lorentz invariant.[2] However, since particle Lorentz symmetry is violated,[2] it is advisable to adopt a standard frame to report experimental measurements of the coefficients for Lorentz violation. Conventionally, this is taken to be a Sun-centered celestial equatorial frame with coordinates (T,X,Y,Z). The Z direction is aligned with the Earth's rotational axis, and the X direction points towards the vernal equinox. The coefficients for Lorentz violation in any inertial frame can be related to those in the standard Sun-centered frame by an appropriate observer Lorentz transformation.[29] Since neutrino oscillations in the presence of Lorentz violation can exhibit orientation-dependent effects, it is also convenient to define a standard parametrization in the Sun-centered frame for the direction of neutrino propagation $\hat{p}$ and for the $\hat{\epsilon}_1$, $\hat{\epsilon}_2$ vectors introduced above:

$$\begin{aligned} \hat{p} &= (\sin\Theta\cos\Phi, \sin\Theta\sin\Phi, \cos\Theta), \\ \hat{\epsilon}_1 &= (\cos\Theta\cos\Phi, \cos\Theta\sin\Phi, -\sin\Theta), \\ \hat{\epsilon}_2 &= (-\sin\Phi, \cos\Phi, 0). \end{aligned} \tag{6}$$

Here, Θ and Φ are the celestial colatitude and longitude of propagation, respectively. This parametrization is assumed in what follows.

3. Sensitivities

No convincing experimental evidence for Lorentz violation presently exists. The size of theoretically predicted effects varies, but it is reasonable to expect[13] that observables for Lorentz and CPT violation are suppressed by some power of the dimensionless ratio $r = m/m_P \lesssim 10^{-17}$, where m is the relevant low-energy scale and m_P is the Planck mass. Note that the physics governing experiments on neutrino oscillations is controlled by the dimensionless ratio $r_\nu = \Delta m^2/E^2$, where $\Delta m^2 \lesssim 10^{-20}$ GeV and 10^{-4} GeV $< E < 10^3$ GeV. The ratios r and r_ν are similar in range, suggesting that the natural scale for Lorentz violation could be comparable to the natural scale of neutrino oscillations.

Sensitivities to certain coefficients for Lorentz violation in the fermion and photon sectors have now reached parts in 10^{-30} or better. It might therefore seem plausible that theoretical considerations involving symmetries or radiative corrections could generate constraints on Lorentz violation in neutrinos from these impressive sensitivities in other sectors, but this idea largely fails in practice. At tree level, leading-order perturbative calculations involve only flavor-diagonal coefficients, and the Lorentz-violating charged-lepton and neutrino sectors are completely independent of these. Also, although radiative corrections can constrain a few neutrino effects under favorable circumstances,[8] the properties of the electroweak sector typically ensure the independence of the charged- and neutral-lepton sectors at leading order in Lorentz violation even under radiative corrections.[10]

Various types of experiment can be sensitive to Lorentz- and CPT-violating effects, depending on the neutrino behavior studied. For example, the energy dependence of Lorentz-violating oscillations can be unconventional. Recall that the physically relevant dimensionless combination controlling neutrino oscillations induced by mass-squared differences Δm^2 is $\Delta m^2 L/E$, where L is the baseline distance and E is the neutrino energy. In contrast, oscillations induced by Lorentz violation are controlled by the dimensionless combinations $a^\mu L$, $b^\mu L$, $H^{\mu\nu} L$ and $c^{\mu\nu} LE$, $d^{\mu\nu} LE$, $g^{\mu\nu\sigma} LE$, as readily seen from Eq. (4). In general, oscillations generated by a coefficient $(k_d)^{\lambda\cdots}$ for a Lorentz-violating operator of nonrenormalizable dimension $n = d+3$ are controlled by $(k_d)^{\lambda\cdots} LE^d$.

Another unconventional effect is direction-dependence of neutrino behavior, which results from violation of rotation symmetry.[10] This has implications both for comparisons between different experiments and for the analysis of experiments involving multiple sources. The orientation of the neutrino beam and the location of the source relative to the detector can affect neutrino oscillations. The daily rotation of the Earth induces apparent periodic changes of the coefficients for Lorentz violation in the laboratory that would appear as time variations in oscillation data at multiples of the sidereal frequency $\omega_\oplus \simeq 2\pi/(23\text{ h }56\text{ min})$. Also, observable annual variations in the solar-neutrino flux could arise from the change in the location of the detector as the Earth orbits the Sun.

Lorentz violation can also lead to novel resonance effects in neutrino oscillations. The usual MSW resonances[39] occur when neutrino interactions with local matter become comparable to effects from mass, which alters the structure of the effective hamiltonian. Many other types of resonances can be induced by Lorentz violation, involving various combinations of coefficients for Lorentz violation, masses, and matter effects. For instance, resonances are possible with neither mass nor matter that are triggered by distinct coefficients for Lorentz violation. Known explicit examples include a two-generation vacuum resonance involving a single nonzero coefficient $(a_L)^T$ and a mass term,[5] and a three-generation orientation-dependent vacuum resonance involving two coefficients $(a_L)^Z$ and $(c_L)^{TT}$ without mass.[9]

Certain experimental signals can be regarded as characteristic of Lorentz violation in neutrino oscillations. Six model-independent classes of signal exist that would offer evidence for Lorentz violation.[10] *Spectral anomalies* can arise because each coefficient for Lorentz violation introduces unconventional energy dependence. These can generate complicated energy dependences in both oscillation lengths and mixing angles. *L–E conflicts* represent any null or positive measurement in a region of L–E space that conflicts with all mass-based scenarios. Of the six classes mentioned here, only this one presently has some positive evidence, the LSND anomaly.[40] *Periodic variations* are induced by rotation-symmetry breaking and include both sidereal and annual variations. Sidereal variations can arise in experiments with Earth-based sources because the direction of neutrino propagation relative to the Sun-centered frame changes as the Earth rotates. Annual variations can arise in solar-neutrino experiments because the orientation of the detector relative to the Sun changes as the Earth orbits the Sun. *Compass asymmetries* also result from rotation-symmetry breaking, but the signals are independent of time. They would appear at the location of the detector as unexplained horizontal or vertical asymmetries. *Neutrino-antineutrino* *mixing* is a direct consequence of any model with nonzero coefficients of type g or H. This class of signal involves lepton-number violation, and it includes in particular any appearance measurement that can be traced to $\nu \leftrightarrow \bar{\nu}$ oscillation. The final class of signal involves the *classic CPT test* of the relation $P_{\nu_b \to \nu_a} \neq P_{\bar{\nu}_a \to \bar{\nu}_b}$. This could also include violation of the condition $P_{\nu_b \rightleftarrows \bar{\nu}_a}(t) = P_{\nu_a \rightleftarrows \bar{\nu}_b}(t)$, which requires $\nu \leftrightarrow \bar{\nu}$ mixing.

Overall, the prospects are good for studies of Lorentz and CPT violation via neutrino oscillations. Consider an experiment with maximum L coverage of L_{max} and maximum E coverage of E_{max}. A rough estimate of the sensitivity σ to a coefficient for Lorentz violation of dimension $1-d$ can be taken as[10] $\sigma \approx -\log L_{\text{max}} - d \log E_{\text{max}}$. This can be used to show that all classes of experiment can attain Planck-scale reach for Lorentz and CPT violation, and the best may rival some of the most sensitive experiments in other sectors of the SME. Note also that non-oscillation experiments with neutrinos typically also have sensitivity to Lorentz violation,[10] in particular via sidereal variations and compass asymmetries. This includes laboratory experiments such as direct mass searches or searches for neutrinoless double-beta decay and also astrophysical observations of supernova neutrinos.

4. Illustrative Models

To gain insight into the variety of oscillation behavior exhibited by neutrinos in the presence of Lorentz violation, it is useful to consider special cases of the effective hamiltonian (4) involving only a few nonzero coefficients. In this part of the talk, some possible models of this type are first briefly discussed, and then a realistic example (the bicycle model) is considered.

4.1. *Basics*

One class of special models is obtained by requiring rotation symmetry. The resulting rotation-invariant or so-called 'fried-chicken' (FC) models are of definite interest for certain investigations because the rotation symmetry simplifies some calculations. It is worth noting, however, that these models are difficult to motivate theoretically as exact descriptions relevant to experimental studies. Thus, it might seem reasonable to require spherical symmetry in a special frame, perhaps the cosmic microwave background (CMB) frame. However, assuming rotation symmetry in the CMB frame means that the coefficients appearing in the effective hamiltonian (4) differ from those appearing in the standard Sun-centered frame or any other experimentally attainable frame. Converting between the CMB frame and the Sun-centered frame or other standard choice of experimentally attainable frame introduces direction dependence due to the solar-system motion relative to the CMB. This implies that the experimentally relevant hamiltonian also involves spatial components of the coefficients, suggesting that it cannot be an FC limit of the theory (4).

The general FC limit of the effective hamiltonian $h_{\rm eff}$ is straightforward to obtain. It contains four matrices, $(\widetilde{m}^2)_{ab}$, $(a_L)^0_{ab}$, $(c_L)^{00}_{ab}$, $(c_L)^{jk}_{ab} = \frac{1}{3}(c_L)^{ll}_{ab}\delta^{jk}$. However, only three of these are independent because the trace $(c_L)^{00}_{ab} - (c_L)^{jj}_{ab}$ is unobservable and can be set to zero. For definiteness, assume the rotation symmetry occurs in the Sun-centered (T, X, Y, Z) frame. Then, the relevant part of the 6×6 effective hamiltonian reduces to the block-diagonal form

$$(h_{\rm eff})^{\rm FC}_{ab} = \begin{pmatrix} \left(\widetilde{m}^2/2E + (a_L)^T - \frac{4}{3}(c_L)^{TT}E\right)_{ab} & 0 \\ 0 & \left(\widetilde{m}^2/2E - (a_L)^T - \frac{4}{3}(c_L)^{TT}E\right)^*_{ab} \end{pmatrix}. \quad (7)$$

This equation determines the general FC model for three active neutrinos. Additional light or massless sterile neutrinos can be incorporated if needed.

Except for the original proposal for Lorentz violation in neutrinos[2] and more recent papers,[9,10,11] a sizable part of the literature[4,5,6,7] concerns restricted special limits of the general FC model (7). In particular, most works have considered the two-generation special case in the limit of vanishing $(a_L)^T$ or $(c_L)^{TT}$. Even though the FC model (7) is somewhat restricted relative to the rich structure of the full effective hamiltonian (4), a more complete study would be of definite interest. A large class of oscillatory behaviors and their phenomenological implications in the FC limit remain unexplored to date.

In the more general context of the full effective hamiltonian (4), Lorentz violation implies directional dependence of oscillation physics through the breaking of rotation invariance. A variety of special cases involving directional dependence can be considered.[10] For example, one interesting limiting class of models with direction dependence consists of those with nonzero coefficients $g^{\mu\nu\sigma}$ and $H^{\mu\nu}$ only. These models necessarily involve $\nu \leftrightarrow \bar{\nu}$ mixing. In the general theory (4), nonzero $\nu \leftrightarrow \bar{\nu}$ mixing of this type can lead to five distinct oscillation lengths and corresponding complications.

A restriction of the effective hamiltonian (4) to the special case involving only the two-dimensional ν_e-$\bar{\nu}_e$ subspace offers an interesting and readily workable limit with directional dependence. In this single-flavor limit of $\nu \leftrightarrow \bar{\nu}$ mixing, any coefficients $(\widetilde{m}^2)_{ee}$ and $(c_L)_{ee}$ can be ignored because they are real and lead to terms proportional to the identity that have no effect on oscillatory behavior. Furthermore, the coefficient $(H^{\mu\nu}C)_{ab}$ is antisymmetric in generation space, so $(H^{\mu\nu}C)_{ee} = H^{\mu\nu}_{ee^C} = 0$. The most general single-flavor theory without mass differences is therefore given by a 2×2 effective hamiltonian containing only the coefficients $(a_L)^\mu_{ee}$ and $(g^{\mu\nu\sigma}C)_{ee} = g^{\mu\nu\sigma}_{ee^C}$ for Lorentz violation. This model and any further restrictions necessarily violate CPT, since both $(a_L)^\mu_{ee}$ and $g^{\mu\nu\sigma}_{ee^C}$ control CPT-odd terms in the hamiltonian. Note that the probabilities for this general single-flavor model have structure identical to that of the standard two-generation mixing case, $P_{\nu_e \leftrightarrow \bar{\nu}_e} = 1 - P_{\nu_e \to \nu_e} = 1 - P_{\bar{\nu}_e \to \bar{\nu}_e} = \sin^2 2\theta \sin^2 2\pi L/L_0$. However, the mixing angle and oscillation length depend on the 4-momentum in an unconventional and direction-dependent way.

The above discussion has focused on basic consequences of special models with only a small number of coefficients. However, even in apparently simple cases of these types, the neutrino behavior can be complicated and counterintuitive. An example of a counterintuitive phenomenon is provided by the Lorentz-violating seesaw mechanism.[10] This can arise when there are degeneracies in the low- or high-energy limit of $h_{\rm eff}$. The different energy dependences among the coefficients for Lorentz violation in $h_{\rm eff}$ can then lead to the emergence of an oscillation length behaving like a mass-squared difference, despite the absence of mass-squared differences in the model.

To illustrate this, consider a 3×3 effective hamiltonian $h_{\rm eff}$ parametrized in some basis as

$$h_{\rm eff} = \begin{pmatrix} 2h_1 & h_2 & h_3 \\ h_2^* & 0 & 0 \\ h_3^* & 0 & 0 \end{pmatrix}, \tag{8}$$

where irrelevant diagonal terms have been disregarded. For this class of toy models, the interesting eigenvalue difference is $\Delta = \sqrt{(h_1)^2 + |h_2|^2 + |h_3|^2} - h_1$. If the combination coefficients for Lorentz violation is such that $h_1 \gg \sqrt{|h_2|^2 + |h_3|^2}$ holds at some energy scale, then the eigenvalue difference is approximately $\Delta \approx \frac{1}{2}(|h_2|^2 + |h_3|^2)/h_1 + \cdots$. For instance, suppose h_2 and h_3 originate from a dimension-one coefficient and hence are constant with energy, while h_1 originates from a dimensionless coefficient that grows linearly with energy. Then, at high energies the eigenvalue difference is proportional to E^{-1}, just like a standard neutrino mass difference. More generally, using various combinations of masses and coefficients for Lorentz violation, similar models can be found that yield E^{-1}, E^{-2}, or E^{-3} dependence at high energies, or E^1, E^2, or E^3 dependence at low energies. Further E^n dependences can be obtained if the full 6×6 effective hamiltonian (4) is used.

4.2. *Bicycle model*

The complexity of the effective hamiltonian (4) and the range of possible neutrino behavior suggests that existing neutrino oscillation data may be compatible with an origin in Lorentz violation without mass differences. This is demonstrated explicitly by a simple model, called the bicycle model,[9] that reproduces the major features of the known neutrino behavior.

The bicycle model is a two-coefficient three-generation special case of the theory (4) without either mass-squared differences or $\nu \leftrightarrow \bar{\nu}$ mixing. It therefore involves only two degrees of freedom, rather than the four degrees of freedom used in the standard description with mass. The nonzero coefficients include an isotropic c_L with nonzero element $\frac{4}{3}(c_L)^{TT}_{ee} \equiv 2\mathring{c} > 0$ and an anisotropic a_L with degenerate nonzero real elements $(a_L)^Z_{e\mu} = (a_L)^Z_{e\tau} \equiv \check{a}/\sqrt{2}$. These coefficients are understood to be specified in the standard Sun-centered frame (T, X, Y, Z).

Diagonalizing the hamiltonian for the model yields[9]

$$\begin{aligned}
P_{\nu_e \to \nu_e} &= 1 - 4\sin^2\theta\cos^2\theta\sin^2(\Delta_{31}L/2),\\
P_{\nu_e \leftrightarrow \nu_\mu} &= P_{\nu_e \leftrightarrow \nu_\tau} = 2\sin^2\theta\cos^2\theta\sin^2(\Delta_{31}L/2),\\
P_{\nu_\mu \to \nu_\mu} &= P_{\nu_\tau \to \nu_\tau} = 1 - \sin^2\theta\sin^2(\Delta_{21}L/2)\\
&\qquad - \sin^2\theta\cos^2\theta\sin^2(\Delta_{31}L/2) - \cos^2\theta\sin^2(\Delta_{32}L/2),\\
P_{\nu_\mu \leftrightarrow \nu_\tau} &= \sin^2\theta\sin^2(\Delta_{21}L/2)\\
&\qquad - \sin^2\theta\cos^2\theta\sin^2(\Delta_{31}L/2) + \cos^2\theta\sin^2(\Delta_{32}L/2),
\end{aligned} \tag{9}$$

where

$$\begin{aligned}
\Delta_{21} &= \sqrt{(\mathring{c}E)^2 + (\check{a}\cos\Theta)^2} + \mathring{c}E,\\
\Delta_{31} &= 2\sqrt{(\mathring{c}E)^2 + (\check{a}\cos\Theta)^2},\\
\Delta_{32} &= \sqrt{(\mathring{c}E)^2 + (\check{a}\cos\Theta)^2} - \mathring{c}E,\\
\sin^2\theta &= \tfrac{1}{2}[1 - \mathring{c}E/\sqrt{(\mathring{c}E)^2 + (\check{a}\cos\Theta)^2}],
\end{aligned} \tag{10}$$

and where the propagation direction $\hat{p}$ is defined in Eq. (6). These probabilities also hold for antineutrinos.

Consider first the qualitative features of this theory. At low energies the coefficient $\check{a}$ induces oscillation of ν_e into an equal mixture of ν_μ and ν_τ, while at high energies the coefficient $\mathring{c}$ controls the physics and there is no ν_e mixing. The critical energy for the theory is given by $E_0 = |\check{a}|/\mathring{c}$. At energies $E \gg E_0$, Eq. (10) shows that $\sin^2\theta$ effectively vanishes. The probabilities become those of a two-generation model with maximal $\nu_\mu \leftrightarrow \nu_\tau$ mixing and transition probability $P_{\nu_\mu \leftrightarrow \nu_\tau} \simeq \sin^2(\Delta_{32}L/2)$, $\Delta_{32} \simeq \check{a}^2 \cos^2 \Theta / 2\mathring{c}E$. This theory therefore incorporates a Lorentz-violating seesaw yielding a pseudomass at high energies, having energy dependence like that of a conventional mass-squared difference $\Delta m^2_\Theta = \Delta m^2_{0^\circ} \cos^2 \Theta$, where $\Delta m^2_{0^\circ} = \check{a}^2/\mathring{c}$. However, the value of the pseudomass Δm^2_Θ and therefore the behavior of neutrino oscillations vary with the declination Θ of the propagation. High-energy neutrinos propagating in the equatorial plane undergo no oscillation because all off-diagonal terms in h_{eff} vanish, while those propagating parallel to celestial north or south behave according to the maximum pseudomass $\Delta m^2_{0^\circ}$.

The existing data for oscillations of atmospheric neutrinos[31] are compatible with neutrino behavior in this model. To be specific, suppose $\Delta m^2_{0^\circ} = 10^{-3}$ eV^2 and $E_0 = 0.1$ GeV. This corresponds to $\mathring{c} = 10^{-19}$, $\check{a} = 10^{-20}$ GeV, values that are consistent with the notion of Planck-scale suppression. The value of $\Delta m^2_{0^\circ}$ lies near the accepted range adopted in the standard mass-based analysis, and E_0 then lies well below the energies relevant for atmospheric neutrino data. With these choices, high-energy atmospheric neutrinos exhibit the standard energy dependence even though they have zero mass differences, thanks to the Lorentz-violating seesaw. Moreover, within existing experimental resolution the zenith-angle dependence of the probability $P_{\nu_\mu \to \nu_\mu}$ averaged over the azimuthal angle is also similar to the standard maximal-mixing case with two generations and a mass-squared difference $\Delta m^2 = 2 \times 10^{-3}$ eV^2. Nonetheless, atmospheric-neutrino signals for Lorentz violation exist that are distinct from standard behaviors. For example, the theory predicts compass asymmetries, including significant *azimuthal* dependences. This can be seen by considering neutrinos propagating in the horizontal plane of the detector. Neutrinos coming from the north or south experience a pseudomass of $\Delta m^2_\Theta = \Delta m^2_{0^\circ} \cos^2 \chi$ and hence oscillate, while those coming from the east or west have $\cos\Theta = 0$, $\Delta m^2_\Theta = 0$ and hence experience no oscillations.

The basic solar-neutrino behavior predicted by the bicycle model is also consistent with observational data.[31] Since the Earth's orbital plane lies at $\eta \simeq 23°$ relative to the equatorial plane, and since solar neutrinos observed at the Earth are those propagating in the orbital plane, the value of $\cos^2 \Theta$ varies during the year. It is zero at the two equinoxes, and it reaches a maximum of $\sin^2 23°$ at the two solstices. Under the simplifying assumption of adiabatic propagation in the Sun, the average ν_e survival probability is $\langle P_{\nu_e \to \nu_e} \rangle_{\text{adiabatic}} = \sin^2\theta \sin^2\theta_0 + \cos^2\theta \cos^2\theta_0$. Here, θ_0 is the solar-core mixing angle, determined from Eq. (10) via substitution of $-\mathring{c}E$ with $-\mathring{c}E + G_F n_e/\sqrt{2}$. It therefore follows that the predicted neutrino flux in the bicycle model is half the expected value at low energies and decreases at higher energies. This is consistent with existing data. Note that care is needed in interpreting the details of the time-dependent behavior. Although the adiabatic probability lies near the average during most of the year, the adiabatic approximation gives a strong suppression near the equinoxes. However, the model predicts that in fact oscillations cease at these times, so the adiabatic approximation fails then and the actual predicted survival probability spikes instead. The net combination of these effects near the equinoxes generates fluctuations in the binned flux, which offer a potential signal for Lorentz violation that could be sought in a detailed experimental analysis. The reader is cautioned, however, that although detection of the semiannual fluctuations represent a positive signal for Lorentz violation, the lack of such a signal is insufficient to exclude the bicycle model because simple changes to the model can yield only small semiannual fluctuations while maintaining the other neutrino behaviors discussed above.[9]

The predictions of the bicycle model are also consistent with neutrino-oscillation data from other experiments.[31] For instance, $\bar{\nu}_e$ survival is determined by the oscillation length $2\pi/\Delta_{31}$, which is sufficiently small to affect KamLAND.[41] The average $\bar{\nu}_e$ survival probability is $\langle P_{\bar{\nu}_e \to \bar{\nu}_e} \rangle = 1 - 2\sin^2\theta\cos^2\theta \geq 1/2$, so the observed flux is predicted to be somewhat above half the flux expected in the absence of oscillations, as indeed reflected by the existing data. Similarly, tests of the bicycle model are also possible in long-baseline accelerator-based experiments, which involve ν_μ oscillations at GeV energies and baselines of hundreds of kilometers. In particular, sidereal variations in $\nu_\mu \leftrightarrow \nu_\tau$ mixing are to be expected. The model also predicts that the results will vary with the beamline-direction of the experiment because the propagation angle Θ and hence the pseudomass $\Delta m^2_\Theta = \Delta m^2_{0^\circ} \cos^2\Theta$ differ.

In summary, the bicycle model demonstrates that it is difficult and probably impossible at present to exclude the idea that observed neutrino oscillations originate from Lorentz and CPT violation rather than from mass differences. Positive and definitive signals for Lorentz and CPT violation arising from Planck-scale physics might be first revealed by detailed analysis of existing or near-future experimental data.

This work was supported in part by the United States Department of Energy under grant number DE-FG02-91ER40661 and by the National Aeronautics and Space Administration under grant number NAG8-1770.

References

1. For recent overviews, see, for example, V.A. Kostelecký, ed., *CPT and Lorentz Symmetry II*, World Scientific, Singapore, 2002.
2. D. Colladay and V.A. Kostelecký, Phys. Rev. D **55**, 6760 (1997); Phys. Rev. D **58**, 116002 (1998).
3. V.A. Kostelecký, Phys. Rev. D, in press (hep-th/0312310).
4. S. Coleman and S. L. Glashow, Phys. Rev. D **59**, 116008 (1999).
5. V. Barger *et al.*, Phys. Rev. Lett. **85**, 5055 (2000).
6. J.N. Bahcall, V. Barger, and D. Marfatia, Phys. Lett. B **534**, 114 (2002).
7. A. de Gouvêa, Phys. Rev. D **66**, 076005 (2002); G.L. Fogli *et al.*, Phys. Rev. D **60**, 053006 (1999); P. Lipari and M. Lusignoli, Phys. Rev. D **60**, 013003 (1999); R. Foot *et al.*, Phys. Lett. B **443**, 185 (1998).
8. I. Mocioiu and M. Pospelov, Phys. Lett. B **537**, 114 (2002).
9. V.A. Kostelecký and M. Mewes, hep-ph/0308300.
10. V.A. Kostelecký and M. Mewes, Phys. Rev. D **69**, 016005 (2004).
11. S. Choubey and S.F. King, hep-ph/0311326.
12. V.A. Kostelecký and S. Samuel, Phys. Rev. D **39**, 683 (1989); Phys. Rev. D **40**, 1886 (1989); Phys. Rev. Lett. **63**, 224 (1989); Phys. Rev. Lett. **66**, 1811 (1991).
13. V.A. Kostelecký and R. Potting, Nucl. Phys. B **359**, 545 (1991); Phys. Lett. B **381**, 89 (1996); Phys. Rev. D **63**, 046007 (2001); V.A. Kostelecký, M. Perry, and R. Potting, Phys. Rev. Lett. **84**, 4541 (2000).
14. S.M. Carroll *et al.*, Phys. Rev. Lett. **87**, 141601 (2001); Z. Guralnik *et al.*, Phys. Lett. B **517**, 450 (2001); C.E. Carlson *et al.*, Phys. Lett. B **518**, 201 (2001); A. Anisimov *et al.*, Phys. Rev. D **65**, 085032 (2002); I. Mocioiu *et al.*, Phys. Rev. D **65**, 107702 (2002); M. Chaichian *et al.*, hep-th/0212259.
15. R. Gambini and J. Pullin, in Ref. 1; J. Alfaro *et al.*, Phys. Rev. **D66**, 124006 (2002); D. Sudarsky *et al.*, Phys. Rev. Lett. **89**, 231301 (2002); Phys. Rev. D **68**, 024010 (2003); G. Amelino-Camelia, Mod. Phys. Lett. A **17**, 899 (2002); Y.J. Ng, Mod. Phys. Lett. **A18**, 1073 (2003); R. Myers and M. Pospelov, Phys. Rev. Lett. **90**, 211601 (2003); N.E. Mavromatos, hep-ph/0402005.
16. C.D. Froggatt and H.B. Nielsen, hep-ph/0211106.
17. J.D. Bjorken, Phys. Rev. D **67**, 043508 (2003).
18. KTeV Collab., H. Nguyen, in Ref. 1; OPAL Collab., R. Ackerstaff *et al.*, Z. Phys. C **76**, 401 (1997); BELLE Collab., K. Abe *et al.*, Phys. Rev. Lett. **86**, 3228 (2001); BaBar Collab., B. Aubert *et al.*, hep-ex/0303043; FOCUS Collab., J.M. Link *et al.*, Phys. Lett. B **556**, 7 (2003).
19. V.A. Kostelecký and R. Potting, Phys. Rev. D **51**, 3923 (1995).
20. D. Colladay and V.A. Kostelecký, Phys. Lett. B **344**, 259 (1995); Phys. Rev. D **52**, 6224 (1995); Phys. Lett. B **511**, 209 (2001); V.A. Kostelecký and R. Van Kooten, Phys. Rev. D **54**, 5585 (1996); O. Bertolami *et al.*, Phys. Lett. B **395**, 178 (1997); V.A. Kostelecký, Phys. Rev. Lett. **80**, 1818 (1998); Phys. Rev. D **61**, 016002 (2000); Phys. Rev. D **64**, 076001 (2001); N. Isgur *et al.*, Phys. Lett. B **515**, 333 (2001).
21. L.R. Hunter *et al.*, in V.A. Kostelecký, ed., *CPT and Lorentz Symmetry*,

World Scientific, Singapore, 1999; D. Bear *et al.*, Phys. Rev. Lett. **85**, 5038 (2000); D.F. Phillips *et al.*, Phys. Rev. D **63**, 111101 (2001); M.A. Humphrey *et al.*, Phys. Rev. A **62**, 063405 (2000); Phys. Rev. A **68**, 063807 (2003); V.A. Kostelecký and C.D. Lane, Phys. Rev. D **60**, 116010 (1999); J. Math. Phys. **40**, 6245 (1999).

22. R. Bluhm *et al.*, Phys. Rev. Lett. **88**, 090801 (2002); Phys. Rev. D **68**, 125008 (2003).
23. F. Canè *et al.*, physics/0309070.
24. H. Dehmelt *et al.*, Phys. Rev. Lett. **83**, 4694 (1999); R. Mittleman *et al.*, Phys. Rev. Lett. **83**, 2116 (1999); G. Gabrielse *et al.*, Phys. Rev. Lett. **82**, 3198 (1999); R. Bluhm *et al.*, Phys. Rev. Lett. **82**, 2254 (1999); Phys. Rev. Lett. **79**, 1432 (1997); Phys. Rev. D **57**, 3932 (1998).
25. B. Heckel, in Ref. 1; L.-S. Hou *et al.*, Phys. Rev. Lett. **90**, 201101 (2003); R. Bluhm and V.A. Kostelecký, Phys. Rev. Lett. **84**, 1381 (2000).
26. S.M. Carroll *et al.*, Phys. Rev. D **41**, 1231 (1990); V.A. Kostelecký and M. Mewes, Phys. Rev. Lett. **87**, 251304 (2001).
27. R. Jackiw and V.A. Kostelecký, Phys. Rev. Lett. **82**, 3572 (1999); C. Adam and F.R. Klinkhamer, Nucl. Phys. B **657**, 214 (2003); H. Müller *et al.*, Phys. Rev. D **67**, 056006 (2003); T. Jacobson *et al.*, Phys. Rev. D **67**, 124011 (2003); V.A. Kostelecký *et al.*, Phys. Rev. D **65**, 056006 (2002); Phys. Rev. D **68**, 123511 (2003); V.A. Kostelecký and A.G.M. Pickering, Phys. Rev. Lett. **91**, 031801 (2003); R. Lehnert, Phys. Rev. D **68**, 085003 (2003); G.M. Shore, gr-qc/0304059.
28. J. Lipa *et al.*, Phys. Rev. Lett. **90**, 060403 (2003); H. Müller *et al.*, Phys. Rev. Lett. **91**, 020401 (2003).
29. V.A. Kostelecký and M. Mewes, Phys. Rev. D **66**, 056005 (2002).
30. V.W. Hughes *et al.*, Phys. Rev. Lett. **87**, 111804 (2001); R. Bluhm *et al.*, Phys. Rev. Lett. **84**, 1098 (2000).
31. The available data are summarized in The Particle Data Group, K. Hagiwara *et al.*, Phys. Rev. D **66**, 010001 (2002) (see also http://pdg.lbl.gov).
32. V.A. Kostelecký and R. Lehnert, Phys. Rev. D **63**, 065008 (2001).
33. M. Gell-Mann *et al.*, in P. van Nieuwenhuizen and D.Z. Freedman, ed., *Supergravity*, North Holland, Amsterdam, 1979; T. Yanagida, in O. Sawada and A. Sugamoto, eds., *Workshop on Unified Theory and the Baryon Number of the Universe*, KEK, Japan, 1979; R. Mohapatra and G. Senjanović, Phys. Rev. Lett. **44**, 912 (1980).
34. D. Colladay and P. McDonald, J. Math. Phys. **43**, 3554 (2002).
35. M.S. Berger and V.A. Kostelecký, Phys. Rev. D **65**, 091701(R) (2002).
36. G. Barenboim and J. Lykken, Phys. Lett. B **554**, 73 (2003).
37. O.W. Greenberg, Phys. Rev. Lett. **89**, 231602 (2002); Phys. Lett. B **567**, 179 (2003).
38. R. Brustein, D. Eichler, and S. Foffa, Phys. Rev. D **65**, 105006 (2002).
39. L. Wolfenstein, Phys. Rev. D **17**, 2369 (1978); S. Mikheev and A. Smirnov, Sov. J. Nucl. Phys. **42**, 913 (1986).
40. LSND Collab., C. Athanassopoulos *et al.*, Phys. Rev. Lett. **81**, 1774 (1998).
41. KamLAND Collab., K. Eguchi *et al.*, Phys. Rev. Lett. **90**, 021802 (2003).

LORENTZ-VIOLATING DISPERSION RELATIONS AND THRESHOLD ANALYSES

RALF LEHNERT
CENTRA, Departamento de Física
Universidade do Algarve
8000-117 Faro
Portugal
E-mail: rlehnert@ualg.pt

Remnant low-energy effects of Planck-scale Lorentz breaking in candidate fundamental theories typically include modified one-particle dispersion relations. Theoretical constraints on such modifications are discussed leading to the exclusion of a variety of previously considered Lorentz-violating parameters. In particular, the fundamental principle of coordinate independence, the role of an effective dynamical framework, and the conditions of positivity and causality are investigated.

1. Introduction

An important open issue in current fundamental-physics research is a quantum theory underlying the Standard Model and General Relativity. The characteristic scale of such a theory is likely to be associated with the Planck mass $M_{\rm Pl} \simeq 10^{19}$ GeV. Presently attainable energies are minuscule compared to this scale, so that experimental signals are expected to be heavily suppressed. For observational progress in this subject, it is therefore necessary to identify generic effects of potential fundamental theories that are accessible to high-precision tests with present-day technology.

Relativity violations associated with the breaking Lorentz symmetry provide a promising candidate signal for Planck-scale physics [1]. At low energies, the effects of Lorentz violation are described by an effective field theory called the Standard-Model Extension (SME) [2, 3]. The classical action of the SME contains, for example, all leading-order contributions to the Lagrangian formed by contracting Standard-Model and gravitational fields with Lorentz-breaking coefficients such that coordinate independence is maintained. A theoretically attractive mechanism for Lorentz violation is spontaneous symmetry breaking in string field theory [4]. More recently, also other candidate sources have been considered including theories of quantum gravity [5], noncommutative field theories [6], varying couplings [7], random dynamics [8], multiverses [9], and brane-world scenarios [10]. The flat-spacetime limit of the SME has provided the basis for numerous analyses of Lorentz breaking involving mesons [11, 12], baryons [13, 14], electrons [15, 16], photons [17, 18], muons [19], and neutrinos [2, 20, 21].

One-particle dispersion relations extracted from the SME generally exhibit Lorentz-violating modifications [17, 2, 3]. In principle, this offers the possibility of Lorentz tests with purely kinematical methods. For instance, primary ultrahigh-energy cosmic rays (UHECR) with momenta eight orders of magnitude below the Planck scale have been observed. At such energies, Lorentz-breaking effects might be more pronounced relative to those in low-energy tests leading to potentially observable shifts in particle-reaction thresholds. This idea has recently received a lot of attention in the literature [5, 20, 22, 23]. However, in many of these investigations the dispersion-relation modifications are constructed with a certain degree of arbitrariness and without reference to the underlying dynamics and other physical principles.

In this talk, we investigate how some of this arbitrariness can be avoided. Our analysis is primarily based on the fundamental principle of coordinate independence and on the requirement of compatibility with an effective dynamical framework. We argue that these two conditions form cornerstones of physics regardless of the details of the Planck-scale theory. General dynamical considerations also increase the scope of threshold investigations and may even be necessary in certain situations. In addition, we briefly discuss positivity and causality, properties that further contribute to the viability of kinematical studies. Throughout we assume translational invariance and the associated energy-momentum conservation.

Section 2 comments on the necessity of coordinate independence and its consequences for dispersion relations. In Sec. 3, we discusses dispersion relations from the viewpoint of compatibility with the SME. Section 4 addresses issues regarding positivity and causality. Further useful results are contained in Sec. 5. A brief summary can be found in Sec. 6.

2. Coordinate independence

On the one hand, coordinate independence is a fundamental physics principle, its role in the context of Lorentz breaking is well established [2, 23], and it permits a rough classification of different types of Lorentz violation. On the other hand, there still exists a certain amount of confusion about this principle in the published literature. For instance, dispersion-relation corrections considered by some authors can only be reconciled with coordinate invariance by introducing unsatisfactory features. Occasionally Lorentz violation is even identified with the loss of coordinate independence. We therefore begin with a few remarks about this requirement.

Coordinate systems, which label spacetime points in a largely arbitrary way, are descriptive tools rather than objects with physical reality. Physics must therefore remain independent of the choice of coordinates. This fundamental requirement permits different observers, each describing the *same* physical system within a *different* reference frame, to relate their observations. This principle is therefore also called observer invariance. Mathematically, coordinate independence can be implemented by working on a spacetime manifold and representing physical quantities by geometric objects like tensors or spinors. Note, however, that this principle does not fix the type of the underlying manifold. A Lorentzian and a Galilean manifold, for example, would be equally consistent with coordinate independence. The manifold type can only be determined by observation. The point is that coordinate independence is much more general than Lorentz symmetry. It is only on Lorentzian manifolds where Lorentz transformations acquire the significant role of implementing changes between local Minkowski frames.

The above discussion reveals one possibility to speculate how Lorentz symmetry might be lost: local inertial frames have a structure different from the usual Minkowskian one, so that Lorentz transformations no longer generate changes between inertial coordinates, i.e., observer Lorentz covariance is *replaced* by observer covariance under some other symmetry transformation. Note that coordinate independence is maintained. This point of view is taken in the so called "doubly special relativities" [24]. The associated modified dispersion relations still exhibit the conventional energy degeneracy for a given 3-momentum, which is intuitively reasonable because the number of spacetime symmetries remains unchanged relative to the conventional case. We mention that the interpretation and viability of this approach is currently still controversial [25]. We therefore leave such Lorentz-symmetry *deformations* unaddressed in the present work.

A less speculative approach to Lorentz-symmetry breakdown maintains the conventional Lorentzian manifold structure and considers the vacuum to be nontrivial instead. Such vacua are associated with nondynamical tensorial backgrounds, which can lead, e.g., to direction-dependent propagation properties. This situation has some parallels with the behavior of particles inside certain crystals. Although coordinate independence is maintained (e.g., invariance under rotations of the *coordinate system*), rotations of the *propagation direction* are generally no longer a symmetry in such situations. One then says that *particle Lorentz symmetry* is broken [2]. Note, however, that the presence of the conventional manifold structure implies that locally one can still work with the metric $\eta^{\mu\nu} = \mathrm{diag}(1, -1, -1, -1)$, particle 4-momenta still transform in the usual way under coordinate changes, and the conventional tensors and spinors still represent physical quantities. In what follows, we consider this latter type of Lorentz violation.

The usual form of Lorentz-violating dispersion relations considered in the literature is[a]

$$\lambda_0{}^2 - \vec{\lambda}\,^2 = m^2 + \delta f(\lambda_0, \vec{\lambda}) \,. \tag{1}$$

Here, m is the mass and $\lambda^\mu = (\lambda^0, \vec{\lambda})$ is the plane-wave 4-vector (before the QFT reinterpretation of the negative energies). The function $\delta f(\lambda_0, \vec{\lambda})$ parametrizes the Lorentz breaking. Arbitrary choices of $\delta f(\lambda_0, \vec{\lambda})$ include situations with nonconserved and possibly complex-valued momenta unsuitable for kinematical analyses.[b] We exclude such situation here and proceed under the assumption that the dynamics of the free particle is described by a linear partial differential equation with constant coefficients, as usual. In the absence of nonlocalities, this yields a polynomial ansatz:

$$\delta f(\lambda_0, \vec{\lambda}) = \sum_{n\geq 1} T_{(n)}^{\overbrace{\alpha\beta\,\cdots}^{n\text{ indices}}} \underbrace{\lambda_\alpha \lambda_\beta\,\cdots}_{n\text{ factors}} \,. \tag{2}$$

Here, $T_{(n)}^{\alpha\beta\,\cdots}$ is a constant tensor of rank n parametrizing particle Lorentz violation. All the tensor indices $\alpha, \beta, \ldots$ are distinct and each one is contracted with a momentum factor, so that all terms in the sum are observer Lorentz invariant.

We mention two immediate consequences of the general ansatz (2). First, Eq. (1) becomes a polynomial in λ_0, so that one expects multiple roots for a given $\vec{\lambda}$. Thus, the conventional energy degeneracy between particle,

[a]Dispersion relations in the minimal SME are typically fourth-order polynomials in λ_0, so that (1) together with ansatz (2) is inconvenient. However, SME dispersion relations can be generated if δf is allowed to contain unsuppressed terms.

[b]For instance, a discrete background, such as in condensed-matter systems, lacks translation invariance resulting in the violation of momentum conservation. Another example is given by finite-temperature dispersion relations, which typically contain imaginary terms.

antiparticle, and possible spin-type states is typically lifted. This is intuitively reasonable because degeneracies normally arise through symmetries, and here the number of symmetries is reduced. Moreover, rotational invariance is in general lost in any frame. As opposed to all previous threshold analyses, generality therefore requires the consideration degeneracy-lifting anisotropic dispersion relations. Second, inspection of ansatz (2) shows that under the usual assumption of rotational symmetry the correction $\delta f(\lambda_0, \vec{\lambda})$ cannot contain odd powers of $|\vec{\lambda}|$.[c]

3. Dynamical features

Although kinematics imposes tight constraints on particle reactions, it provides only an incomplete description of the process: an expected high-energy reaction can be suppressed not only by modified dispersion relations but also by new additional symmetries, for example. Similarly, the presence of a high-energy reaction kinematically forbidden in conventional physics could perhaps be explained by additional channels due to the loss of low-energy symmetries or novel undetected particles. Moreover, dynamics is involved both in acceleration mechanisms for UHECRs and in the atmospheric shower development. Thus, the study of threshold bounds on Lorentz violation typically requires assumptions outside kinematics such as dynamical quantum-field aspects.

Kinematics investigations are limited to only a few potential Lorentz-violating signatures from candidate fundamental physics. Thus, dynamical features also increase the scope of Lorentz tests. From the above perspectives, it is desirable to explicitly implement dynamics of sufficient generality into the search for Lorentz breaking.

The SME is the general effective-field-theory framework for the dynamical description of Lorentz violation. It is useful to review the idea behind its construction [2] to fully appreciate the generality of the SME. Lorentz-violating terms $\delta\mathcal{L}$ are added to the usual Standard-Model Lagrangian $\mathcal{L}_{\text{SM}}$:

$$\mathcal{L}_{\text{SME}} = \mathcal{L}_{\text{SM}} + \delta\mathcal{L} \, , \qquad (3)$$

where, $\mathcal{L}_{\text{SME}}$ denotes the SME Lagrangian. The modification $\delta\mathcal{L}$ is formed by contracting Standard-Model fields with Lorentz-violating tensorial parameters yielding observer Lorentz scalars. Thus, the complete set of pos-

---

[c]Odd powers could enter in the degeneracy-lifting form $\pm|\vec{\lambda}|^{2n+1}$ with $n \in \mathbb{N}$ or through (theoretically unmotivated) nonlocal equations of motion involving $\sqrt{-\Delta}$, where Δ is the Laplacian.

sible contributions to $\delta\mathcal{L}$ gives the most general effective dynamical framework for Lorentz breaking at the level of Lorentz-coordinate-independent effective QFT. We mention that potential Planck-scale features, such as a certain discreteness of spacetime or a possible non-pointlike nature of elementary particles, are unlikely to invalidate the above effective-field-theory approach at present energies. Moreover, Lorentz-symmetric aspects of candidate fundamental theories, such as new symmetries, novel particles, or large extra dimensions, are also unlikely to require a low-energy description beyond effective field theory and can therefore be incorporated into the SME, if necessary.

The SME permits the identification and direct comparison of virtually all Lorentz and CPT tests that are presently feasible. In addition, classical kinematics test models of relativity (such as Robertson's framework, its Mansouri-Sexl extension, or the c^2 model) are contained in the SME as limiting cases. Concerning threshold analyses, the quadratic, translationally invariant sector of the SME determines possible one-particle dispersion relations constraining the ansatz (2). As a further advantage, the SME permits the calculation of reaction rates, which are a determining factor for observational relevance. An explicit example is provided by so called vacuum Čerenkov radiation [20]. The above discussion strongly suggests that particle-reaction investigations are best performed within the framework of the SME.

4. Positivity and causality

In Special Relativity, the presence of an upper speed limit for material bodies left invariant by the Lorentz transformations is associated with a notion of causality. This leads to the common misconception that Lorentz violation results in superluminal propagation, and thus causality violations. However, conventional situations, in which Lorentz symmetry is broken but causality is maintained, can readily be identified. The anisotropic propagation of electromagnetic waves inside certain crystals, for example, is causal despite Lorentz violation. Moreover, in such a situation the total conserved energy is clearly positive definite for all observers. It follows that the requirements of positivity and causality are *a priori* independent and distinct from the principle of Lorentz symmetry. Note also that positivity and causality lead, for example, to the spin-statistics theorem, which is a cornerstone of relativistic QFT.

Since polynomial Lorentz-violating dispersion relations can violate positivity and causality [3], it is natural to ask whether such violations become acceptable in the presence of Lorentz breaking. Concerning positivity, we are unaware of any internally consistent interacting quantum field theories involving negative-energy particles as asymptotic states. On the contrary, the usual assumptions in perturbation theory, for example, seem to exclude negative energies. Similar arguments apply to superluminal propagation: it is unlikely that such a causality breakdown can be accommodated within the framework of relativistic quantum field theory. Generally, a hermitian Hamiltonian for massive fermions fails to exist in the majority of frames [3]. In addition, the usual covariant perturbative expansion relies on time ordering, an operation no longer coordinate invariant when microcausality is violated [27]. We conclude that positivity and causality remain desirable features in threshold analyses despite Lorentz breaking.

Reaction-threshold kinematics can be affected if positivity and causality are imposed. Let M and m be the respective scales of the underlying theory and present-day low-energy physics. Then, the scale $p_{p\text{-}c}$ for the occurrence of positivity or causality problems can be as low as [3]

$$p_{p\text{-}c} \sim \mathcal{O}(\sqrt{mM}\,)\,. \tag{4}$$

For example, if M is taken to be the Planck scale and m is the proton mass, then $p_{p\text{-}c} \sim 3\times 10^{18}$ eV. UHECRs with a spectrum extending beyond 10^{20} eV have been observed. These events are often employed to bound Lorentz breaking or to suggest evidence for Lorentz violation. It follows that imposing positivity and causality could require modifications in threshold analyses.

5. Further results

Consider photon decay $\gamma \to e^+ + e^-$ into an electron-positron pair, where both the photon ($m = 0$) and the fermion obey dispersion relations with the correction $\delta f(\lambda_0, \vec{\lambda}) = \pm|\vec{\lambda}|^3/M$. Here, M is the fundamental scale. Note that by allowing two simultaneous signs for the correction term, we enforce coordinate independence. This correction gives

$$\lambda^0_{\pm(\alpha)}(\vec{\lambda}) = \pm\sqrt{(-1)^\alpha \frac{|\vec{\lambda}|^3}{M} + \vec{\lambda}^{\,2} + m^2}\,, \tag{5}$$

where the subscript $\pm$ corresponds to the sign of the square root, and thus, after reinterpretation, to particle and antiparticle. The index $\alpha = 1, 2$ labels the two possible particle (antiparticle) energies, which perhaps correspond to different spin-type states. Depending on the α value for each particle in the reaction, there are six kinematically distinct decays that have to be considered. Note, however, that angular-momentum conservation associated with the rotational invariance of the model may preclude some of the six reactions. In general, we conclude that the effects of assumed symmetries, such as rotational invariance, must be incorporated into threshold analyses. This typically requires the use of dynamics as argued before.

Another coordinate-independent dispersion-relation correction is given by $\delta f(\lambda_0, \vec{\lambda}) = \lambda_0 \vec{\lambda}^{\,2}/M$, so that

$$\lambda^0_\pm(\vec{\lambda}) = \frac{\vec{\lambda}^{\,2}}{2M} \pm \sqrt{\frac{\vec{\lambda}^{\,4}}{4M^2} + \vec{\lambda}^{\,2} + m^2}\,. \tag{6}$$

Note that the particle-antiparticle degeneracy is lifted. Consider again photon decay $\gamma \to e^+ + e^-$, now with dispersion relation (6) for both the photon ($m = 0$) and the fermion. Two kinematically distinct processes must be investigated because Eq. (6) implies two possible incoming photon states γ_+ and γ_-. One can show [23] that the reaction $\gamma_- \to e^+ + e^-$ is kinematically forbidden, while the decay $\gamma_+ \to e^+ + e^-$ is allowed above a certain threshold. This analysis has been performed previously in the literature employing the approximation $\lambda_0 \simeq |\vec{\lambda}|$ in the correction term $\lambda_0 \vec{\lambda}^{\,2}/M$. However, this approximation introduces an additional degeneracy relative to Eq. (6) leading to the false conclusion that the correction $\lambda_0 \vec{\lambda}^{\,2}/M$ precludes photon decay. Thus, many approximations, such as those leading to additional degeneracies, are typically invalid in threshold analyses.

6. Summary

This talk has discussed some issues that arise in the context of Lorentz tests with modified dispersion relations. More specifically, we have investigated the role of a dynamical framework and the conditions of coordinate independence, positivity, and causality in the subject. We have found that these fundamental requirements impose tight constraints on possible dispersion-relation corrections. Correct threshold investigations within the full SME are automatically compatible with these requirements.

Acknowledgments

This work was supported in part by the Centro Multidisciplinar de Astrofísica (CENTRA) and by the Fundação para a Ciência e a Tecnologia (Portugal) under grant POCTI/FNU/49529/2002.

References

1. For an overview, see, e.g., *CPT and Lorentz Symmetry II*, edited by V.A. Kostelecký (World Scientific, Singapore, 2002).
2. D. Colladay and V.A. Kostelecký, Phys. Rev. D **55**, 6760 (1997); **58**, 116002 (1998); V.A. Kostelecký, hep-th/0312310.
3. V.A. Kostelecký and R. Lehnert, Phys. Rev. D **63**, 65008 (2001).
4. V.A. Kostelecký and S. Samuel, Phys. Rev. D **39**, 683 (1989); **40**, 1886 (1989); Phys. Rev. Lett. **63**, 224 (1989); **66**, 1811 (1991); V.A. Kostelecký and R. Potting, Nucl. Phys. B **359**, 545 (1991); Phys. Lett. B **381**, 89 (1996); Phys. Rev. D **63**, 046007 (2001); V.A. Kostelecký, M.J. Perry, and R. Potting, Phys. Rev. Lett. **84**, 4541 (2000).
5. J.R Ellis, E. Gravanis, N.E. Mavromatos, and D.V. Nanopoulos, gr-qc/0209108; R. Gambini and J. Pullin, in Ref. [1]; F.R. Klinkhamer, Nucl. Phys. B **578**, 277 (2000); J. Alfaro, H.A. Morales-Técotl, and L.F. Urrutia, Phys. Rev. **D** 66, 124006 (2002); D. Sudarsky, L. Urrutia, and H. Vucetich, Phys. Rev. Lett. **89**, 231301 (2002); Phys. Rev. D **68**, 024010 (2003); F.R. Klinkhamer and C. Rupp, hep-th/0312032.
6. S.M. Carroll *et al.*, Phys. Rev. Lett. **87**, 141601 (2001); Z. Guralnik *et al.*, Phys. Lett. B **517**, 450 (2001); C.E. Carlson *et al.*, Phys. Lett. B **518**, 201 (2001).
7. V.A. Kostelecký, M.J Perry, and R. Lehnert, Phys. Rev. D **68**, 123511 (2003); O. Bertolami, R. Lehnert, R. Potting, and A. Ribeiro, Phys. Rev. D, in press [astro-ph/0310344].
8. C.D. Froggatt and H.B. Nielsen, hep-ph/0211106.
9. J.D. Bjorken, Phys. Rev. D **67**, 043508 (2003).
10. C.P. Burgess *et al.*, JHEP **0203**, 043 (2002); A.R. Frey, JHEP **0304**, 012 (2003); J. Cline and L. Valcárcel, hep-ph/0312245.
11. KTeV Collaboration, H. Nguyen, in Ref. [1]; OPAL Collaboration, R. Ackerstaff *et al.*, Z. Phys. C **76**, 401 (1997); DELPHI Collaboration, M. Feindt *et al.*, preprint DELPHI 97-98 CONF 80 (1997); BELLE Collaboration, K. Abe *et al.*, Phys. Rev. Lett. **86**, 3228 (2001); BaBar Collaboration, B. Aubert *et al.*, hep-ex/0303043; FOCUS Collaboration, J.M. Link *et al.*, Phys. Lett. B **556**, 7 (2003).
12. D. Colladay and V.A. Kostelecký, Phys. Lett. B **344**, 259 (1995); Phys. Rev. D **52**, 6224 (1995); Phys. Lett. B **511**, 209 (2001); V.A. Kostelecký and R. Van Kooten, Phys. Rev. D **54**, 5585 (1996); O. Bertolami *et al.*, Phys. Lett. B **395**, 178 (1997); N. Isgur *et al.*, Phys. Lett. B **515**, 333 (2001); V.A. Kostelecký, Phys. Rev. Lett. **80**, 1818 (1998); Phys. Rev. D **61**, 016002 (2000); **64**, 076001 (2001).
13. L.R. Hunter *et al.*, in V.A. Kostelecký, ed., *CPT and Lorentz Symmetry*, World Scientific, Singapore, 1999; D. Bear *et al.*, Phys. Rev. Lett. **85**, 5038 (2000); D.F. Phillips *et al.*, Phys. Rev. D **63**, 111101 (2001); M.A. Humphrey *et al.*, physics/0103068; Phys. Rev. A **62**, 063405 (2000); F. Canè *et al.*, physics/0309070.
14. V.A. Kostelecký and C.D. Lane, Phys. Rev. D **60**, 116010 (1999); J. Math.

Phys. **40**, 6245 (1999); R. Bluhm *et al.*, Phys. Rev. Lett. **88**, 090801 (2002); Phys. Rev. D **68**, 125008 (2003).

15. H. Dehmelt *et al.*, Phys. Rev. Lett. **83**, 4694 (1999); R. Mittleman *et al.*, Phys. Rev. Lett. **83**, 2116 (1999); G. Gabrielse *et al.*, Phys. Rev. Lett. **82**, 3198 (1999); B. Heckel, in Ref. [1]; L.-S. Hou, W.-T. Ni, and Y.-C.M. Li, Phys. Rev. Lett. **90**, 201101 (2003); R. Bluhm and V.A. Kostelecký, Phys. Rev. Lett. **84**, 1381 (2000).
16. R. Bluhm *et al.*, Phys. Rev. Lett. **79**, 1432 (1997); Phys. Rev. D **57**, 3932 (1998); Phys. Rev. Lett. **82**, 2254 (1999); H. Müller *et al.*, Phys. Rev. D **68**, 116006 (2003); R. Lehnert, hep-ph/0401084.
17. S.M. Carroll, G.B. Field, and R. Jackiw, Phys. Rev. D **41**, 1231 (1990); M.P. Haugan and T.F. Kauffmann, Phys. Rev. D **52**, 3168 (1995); V.A. Kostelecký and M. Mewes, Phys. Rev. Lett. **87**, 251304 (2001); V.A. Kostelecký and M. Mewes, Phys. Rev. D **66**, 056005 (2002); J. Lipa *et al.*, Phys. Rev. Lett. **90**, 060403 (2003); H. Müller *et al.*, Phys. Rev. Lett. **91**, 020401 (2003).
18. R. Jackiw and V.A. Kostelecký, Phys. Rev. Lett. **82**, 3572 (1999); C. Adam and F.R. Klinkhamer, Nucl. Phys. B **657**, 214 (2003); H. Müller *et al.*, Phys. Rev. D **67**, 056006 (2003); V.A. Kostelecký and A.G.M. Pickering, Phys. Rev. Lett. **91**, 031801 (2003); G.M. Shore, Contemp. Phys. **44**, 503 2003; B. Altschul, hep-th/0311200.
19. V.W. Hughes *et al.*, Phys. Rev. Lett. **87**, 111804 (2001); R. Bluhm *et al.*, Phys. Rev. Lett. **84**, 1098 (2000).
20. S. Coleman and S.L. Glashow, Phys. Rev. D **59**, 116008 (1999).
21. V. Barger, S. Pakvasa, T.J. Weiler, and K. Whisnant, Phys. Rev. Lett. **85**, 5055 (2000); J.N. Bahcall, V. Barger, and D. Marfatia, Phys. Lett. B **534**, 114 (2002); V.A. Kostelecký and M. Mewes, hep-ph/0308300; V.A. Kostelecký and M. Mewes, Phys. Rev. D, in press [hep-ph/0309025].
22. W. Kluźniak, Astropart. Phys. **11**, 117 (1999); T. Kifune, Astrophys. J. **518**, L21 (1999); O. Bertolami and C.S. Carvalho, Phys. Rev. D **61**, 103002 (2000); G. Amelino-Camelia, Phys. Lett. B **528**, 181 (2002); T. Jacobson *et al.*, Phys. Rev. D **66**, 081302 (2002); T.J. Konopka and S.A. Major, New J. Phys. **4**, 57 (2002); H. Vankov and T. Stanev, Phys. Lett. **B** 538, 251 (2002); J. Alfaro and G. Palma, hep-th/0208193; V. Gharibyan, hep-ex/0303010; R.C. Myers and M. Pospelov, Phys. Rev. Lett. **90**, 211601 (2003); M. Jankiewicz *et al.*, hep-ph/0312221.
23. R. Lehnert, Phys. Rev. D **68**, 085003 (2003).
24. G. Amelino-Camelia, Phys. Lett. B **510**, 255 (2001); J. Magueijo and L. Smolin, Phys. Rev. Lett. **88**, 190403 (2002).
25. J. Lukierski and A. Nowicki, Int. J. Mod. Phys. A **18**, 7 (2003); J. Rembielinski and K.A. Smolinski, Bull. Soc. Sci. Lett. Łódź Sér. Rech. Déform. **53**, 57 (2003); D.V. Ahluwalia, gr-qc/0212128; D. Grumiller, W. Kummer, and D.V. Vassilevich, Ukr. J. Phys. **48**, 329 (2003); R. Schützhold and W.G. Unruh, gr-qc/0308049; N. Jafari and A. Shariati, gr-qc/0312007.
26. R. Lehnert and R. Potting, in preparation.
27. O.W. Greenberg, Phys. Rev. Lett. **89**, 231602 (2002); Phys. Lett. B **567**, 179 (2003).

EXPERIMENTAL CHALLENGES FOR QUANTUM GRAVITY*

ROBERT C. MYERS
Perimeter Institute for Theoretical Physics
35 King Street North, Waterloo, Ontario N2J 2W9, Canada
and
Department of Physics, University of Waterloo
Waterloo, Ontario N2L 3G1, Canada
E-mail: rmyers@perimeterinstitute.ca

MAXIM POSPELOV
Department of Physics and Astronomy
University of Victoria, Victoria, BC, V8P 1A1, Canada
E-mail: pospelov@uvic.ca

The existence of a new fundamental scale may lead to modified dispersion relations for particles at high energies. Such modifications seem to be realized with the Planck scale in certain descriptions of quantum gravity. We apply effective field theory to this problem and identify dimension 5 operators that would lead to cubic modifications of dispersion relations for Standard Model particles. We also discuss other issues related to this approach including various experimental bounds on the strength of these interactions. Further we sketch a scenario where mixing of these operators with dimensions 3 and 4 due to quantum effects is minimal.

1. Introduction

The Planck mass, $M_{\rm Pl}$, the dimensional parameter determining the strength of gravitational interactions, remains a source of conceptual problems for quantum field theories. When the momentum transfer in two particle collisions is comparable to the Planck mass, the graviton exchange becomes strong, signifying a breakdown of the perturbative field theory description. Even without having a fully consistent fundamental theory at hand, one can hypothesize several broad categories of the low-energy effects induced by $M_{\rm Pl}$. The first group of such theories has only minor modifications

*Talk given by R.C.M.

due to the existence of new physics at $M_{\rm Pl}$. By minor modifications we understand that all "sacred" symmetries (Lorentz symmetry, CPT, spin-statistics, etc.) of the field theory remain unbroken at low energies. Critical string theory in simple backgrounds reduces to a field theory plus gravity at the scales lower than M_s and so provides an interesting example of this category. In this case the chances to probe $1/M_{\rm Pl}$ effects are very remote as it is insufficient to merely have a probe with large energy. Rather one must have extremely large momentum transfers. Consequently, the propagation of a free particle with large energy/momentum is immune to the effects of new physics, as all corrections to the dispersion relation could be cast in the form of $(p^2)(p^{2n}/M_{\rm Pl}^{2n}) = (m^{2n+2}/M_{\rm Pl}^{2n})$ where m is the mass of the particle and p is the four-momentum.

The hope that nature could be more gracious to physicists is reflected in the second class of scenarios, where the $1/M_{\rm Pl}$ effects have much more "vivid" properties. Loop quantum gravity seems to provide an example of this scenario. In this approach, the discrete nature of space at short distances may be expected to induce violations of Lorentz invariance and CPT. Such violations are also often discussed in a broader context using field theoretical language.[1] Here one might assume a perfectly Lorentz symmetric, CPT conserving action at $1/M_{\rm Pl}^0$ order and account for the existence of the Lorentz breaking terms via a set of higher dimension operators. This would lead to modifications of the dispersion relation for a free particle as terms of the form $E^{n+2}/M_{\rm Pl}^n$ can appear. Such effects can be searched for both with astrophysical observations[2,3,4,5,6] and with high precision low-energy experiments.[7,8,9]

Cubic modifications of dispersion relations, which would appear at the leading order in $1/M_{\rm Pl}$, have received considerable attention in the literature recently.[2,3,4,5,6,10,11] In the language of the effective field theory, such modifications can be described by the dimension 5 operators. Although dimension 3 and 4 operators were extensively studied,[1] dimension 5 operators remained unclassified and poorly explored. In our recent work,[9] we took a first step towards such a classification by considering dimension 5 operators in which Lorentz breaking can be achieved by the introduction of a background four-vector n^a. Within this framework, we studied cubic modifications of the dispersion relation for scalars, vectors and fermions. Specializing the operators to Standard Model particles, our results showed that a cubic modification of the dispersion relation is not possible for the Higgs particles, must have an opposite sign for opposite chiralities of photons, and is independent for different chiralities in the case of fermions.

This paper provides a summary of our earlier results.[9] We begin with a review the class of dimension 5 operators which are of interest to produce cubic modifications of dispersion relations. We also discuss experimental constraints limiting the strength of these operators. In particular, we demonstrate remarkably stringent bounds[9] on these operators coming from terrestrial clock comparison experiments.

Beyond this summary, we address further issues vis-a-vis Lorentz violation and its effective field theory description in the discussion section. In particular, we speculate on possible mechanisms by which the violation of Lorentz invariance may be restricted to higher dimensional operators and address the mixing of the dimension 5 operators with those of dimension 3 and 4 through quantum effects. We are able to identify a scenario in which this mixing may be minimal. We also argue that although our previous discussion is framed in the language of Lorentz symmetry breaking, many of the results still have application in scenarios where the symmetry is deformed.[10] Finally, we stress the importance of a consistent formulation of Lorentz-violating (or Lorentz-deforming) theories in the presence of gravity[12] and the necessity of correct identification of the infrared degrees of freedom in such theories.

2. Dimension 5 operators and cubic dispersion relations

In the framework of low energy effective field theory, the modified dispersion relations should be derived from some appropriate modification of the kinetic terms in the Lagrangian. Such modification may appear in the leading dimension 3 and 4 terms. However, we assume that short-distance physics does not generate such Lorentz violating operators directly. From a pragmatic point of view, this assumption is quite safe since even if these terms exist, experimental constraints indicate that they must be exceptionally small.[1] Nevertheless, it poses serious theoretical problems and we will attempt to address this point in the closing discussion.

At the next level are dimension 5 operators which would lead to $O(E^3)$ modifications of the dispersion relations. We adopt the simplest approach where Lorentz symmetry is broken by a background four-vector n^a (with $n \cdot n = 1$). We construct operators satisfying six generic criteria:

(1) Quadratic in the *same* field
(2) One more derivative than the usual kinetic term
(3) Gauge invariant
(4) Lorentz invariant, except for the appearance of n^a

(5) Not reducible to lower dimension by the equations of motion
(6) Not reducible to a total derivative

Conditions 2 and 5 ensure that these operators lead to $O(E^3)$ modifications of the dispersion relations, rather than $O(E^2 m)$ or $O(Em^2)$, where m is the mass of the particle. Our working assumption will also be that these operators are naturally suppressed by a factor of $1/M_{\rm Pl}$, and that $m \ll E \ll M_{\rm Pl}$. This scaling ensures that all operators of dimension 5 can be regarded as small perturbations. Below, we consider the cases of vector and fermion particles. We refer the interested reader to Ref. [9] for the analogous discussion of scalars.

Vector: Consider a U(1) gauge field with the leading kinetic term, $\mathcal{L}_0 = -F^2/4$. The leading order equations of motion are just the Maxwell equations, $\partial_a F^{ab} = 0$. After gauge fixing $\partial \cdot A = 0$, this yields $\Box A_a = 0$ or $k^2\, A_a(k) = 0$ in momentum space with $A_a \sim \exp(ik \cdot x)$. We wish to modify the dispersion relation at $O(E^3)$ and so the new terms should satisfy the constraints listed above. Keeping in mind the leading order Maxwell equations and the Bianchi identities $\partial_{[a} F_{bc]} = 0$, one finds that there is a *unique* term with the desired properties

$$\mathcal{L}_\gamma = \frac{\xi}{M_{\rm Pl}} n^a F_{ad}\, n \cdot \partial (n_b \tilde{F}^{bd}), \tag{1}$$

where $\tilde{F}^{ab} = \frac{1}{2}\varepsilon^{abcd} F_{cd}$. Extension of this analysis to a nonabelian vector is straightforward, and as in the abelian case there is only one operator possible for each group at dim 5 level. Note that operator (1) is odd under CPT and even under charge conjugation. The equation of motion becomes

$$\Box A_a = \frac{\xi}{M_{\rm Pl}} \varepsilon_{abcd}\, n^b (n \cdot \partial)^2 F^{cd} \tag{2}$$

again with the gauge choice $\partial \cdot A = 0$. Further the right hand side has been reduced through ample use of the Bianchi identity. To identify the effect of the new term on the dispersion relation, we go to momentum space and select photons moving along the z axis with $k^a = (E, 0, 0, p)$. Then for transverse polarizations along the x and y axes,

$$\left(E^2 - p^2 \pm \frac{2\xi}{M_{\rm Pl}} p^3 \right) (\epsilon_x \pm i\epsilon_y) \simeq 0 \tag{3}$$

where we have used $E \simeq p$ to leading order and chosen the "rest frame" where $n^a = (1, 0, 0, 0)$. Hence the sign of the cubic term is determined by the chirality (or circular polarization) of the photons. This leads to the rotation of the plane of polarization for linearly polarized photons, which may be used to bound ξ.[3,6] Note that $\mathcal{L}_\gamma$ is unique and hence the common approach[2,4] of postulating a cubic dispersion relation which is chirality independent is incompatible with effective field theory.

Spinor: Consider a Dirac spinor for which the leading kinetic term is: $\mathcal{L}_0 = \bar{\Psi}(i\partial\!\!\!/ - m)\Psi$. The leading order equation of motion is just the Dirac equation: $(\partial\!\!\!/ + im)\Psi = 0$. In momentum space with $\Psi \sim \exp(-ik \cdot x)$, $(k\!\!\!/ - m)\Psi(k) = 0$. To modify the dispersion relation at $O(E^3)$, we consider new terms satisfying the constraints listed above. In this case, there are only two terms with the desired form

$$\mathcal{L}_f = \frac{i}{M_{\rm Pl}} \bar{\Psi}\, (\eta_1 n\!\!\!/ + \eta_2 \gamma_5 n\!\!\!/)\, (n \cdot \partial)^2 \Psi \tag{4}$$

Both operators break CPT, with η_1 being charge conjugation odd and η_2 charge conjugation even. After applying $(i\partial\!\!\!/ + m)$, the equation of motion takes the form

$$(\Box + m^2)\Psi = \frac{i}{M_{\rm Pl}}\, (\eta_1 + \eta_2\gamma_5)\, (n \cdot \partial)^3 \Psi\ . \tag{5}$$

Here we have again dropped terms of order $m/M_{\rm Pl}$ or which vanish by the leading order equations. Hence the modified dispersion relation becomes

$$\left(E^2 - |\vec{p}|^2 - m^2 + \frac{|\vec{p}|^3}{M_{\rm Pl}} (\eta_1 + \eta_2\gamma_5) \right) \Psi = 0 \tag{6}$$

where we have used $E \simeq |\vec{p}|$ for high energies. At high energies (*i.e.*, $E^2 \gg m^2$), we can choose spinors as eigenspinors of the chirality operator and redefine coupling constants as $\eta_{L,R} = \eta_1 \mp \eta_2$. To introduce these operators for Standard Model, the chiral choice for η couplings would be required by gauge invariance. Previous studies[4,5] considered only chirality independent dispersion relations for fermions and so implicitly fix $\eta_2 = 0$.

Above we have identified interesting operators which modify the dispersion relations at cubic order for vectors or fermions. The external tensor appearing in all of these operators takes the form $n^a n^b n^c$. For technical reasons to be addressed in the discussion section, we must replace this coupling by the traceless symmetric tensor $C^{abc} = n^a n^b n^c - \frac{1}{6}(n^a g^{bc}+\text{cyclic})$ in the following section. Note that this change implicitly introduces extra terms which do not satisfy all of the constraints listed above, *i.e.*, they are reducible by the leading order equations of motion. However, this replacement does not affect the dispersion relations in the regime $E \gg m$. One could also consider frame-dependent modifications of the interaction terms between, *e.g.*, photons and electrons. If we limit ourselves to the traceless symmetric coupling, C^{abc}, introduced here, there are in fact no additional interaction terms with dimension 5 beyond those implied by extending Eq. (4) with gauge-covariant derivatives.

3. Experimental constraints on dimension 5 operators

Evidence of modified dispersion relations for stable particles such as electrons, light quarks, and photons can be searched for using the astrophysical probes.[2,3,4,5,6] We begin here, however, by showing that impressive constraints can be imposed by considering terrestrial experiments. These indirect limits exploit the idea that the external four-vector n^a introduces a preferred frame that can not coincide with the laboratory frame on the Earth.[13] While the component n^0 is still dominant, the motion of the galaxy, Solar system and Earth will create spatial components $n_i \sim 10^{-3}$ for a terrestrial observer. Hence clock comparison experiments[14] or searches for spatial anisotropy[15] can impose stringent bounds on violations of Lorentz symmetry in this context. This approach was recently used to constrain the dispersion relation for nucleons.[8] Our constraints[9] apply to the fundamental fields of the Standard Model rather that presumably should have more direct connection to the Planck scale physics than nucleons.

Limits on the operators involving electrons and electron neutrinos are especially easy to derive. We use the fact that best tests[15] of directional sensitivity in the precession of electrons limit the size of interaction between the external direction and the electron spin at the level of 10^{-28} GeV. This immediately translates into the following limit on the coefficients η_L and η_R that parametrize the effective interaction of the form (4) for left-handed leptons and right-handed electrons:[9]

$$|\eta_L^e - \eta_R^e| \lesssim \frac{10^{-28}\ \mathrm{GeV} M_{\mathrm{Pl}}}{m_e^2 |n_i|} \simeq 4, \tag{7}$$

where $M_{\mathrm{Pl}} \equiv 10^{19}$ GeV. The combination, $\eta_L^e + \eta_R^e$, would be very weakly constrained here, as it does not appear in the electron spin Hamiltonian.

The absence of a preferred direction is checked with even greater precision using nuclear spin, which translates into more stringent limits on new operators for the light quarks. The photon operator (1) will also contribute because of the electromagnetic interactions inside the nucleon. To use the best experimental limits of 10^{-31} GeV on the coupling of n_i to neutron spins,[14] we must relate the photon and quark operators with nucleon spin. First, let us introduce dimension 5 operators for the first generation of quarks, the left-handed doublet ψ_Q and right-handed singlets ψ_u and ψ_d:

$$\mathcal{L}_q = \frac{C^{abc}}{M_{\rm Pl}} \sum_{i=Q,u,d} \eta_i \, \bar{\psi}_i \gamma_a \partial_b \partial_c \psi_i \, . \tag{8}$$

At the nucleon level, we estimate using standard QCD sum rules

$$\begin{aligned} \eta_{1,N} &= a_u(\eta_u + \eta_Q) + a_d(\eta_d + \eta_Q) \\ \eta_{2,N} &= b_u(\eta_u - \eta_Q) + b_d(\eta_d - \eta_Q) + b_\gamma \xi, \end{aligned} \tag{9}$$

where $\eta_{1(2),N}$ are the η_1 and η_2 couplings for nucleons defined in (4). Note that ξ enters only in the η_2 coupling for nucleons because both are even under the charge conjugation. In (9), $a_{u,d}$ and $b_{u,d}$ are the matrix elements that could be obtained as the moments of the experimentally measured structure functions:[16] $a_d \sim 0.4$, $a_u \sim 0.1$, $b_d \sim 0.1$, $b_u \sim -0.05$ for the neutron and charge inverted values for the proton. To relate the photon operator with nucleon, we use the simplest vector dominance model and obtain at one-loop level $b_\gamma \sim 0.13\alpha/(4\pi)$ for neutron and $b_\gamma \sim 0.24\alpha/(4\pi)$ for proton. Combining these results produces the following limit:

$$|(\eta_d - \eta_Q) - 0.5(\eta_u - \eta_Q) + 10^{-3}\xi| \lesssim 10^{-8} \tag{10}$$

Barring accidental cancellations, we can place separate limits on $\eta_{u,d} - \eta_Q$ at 10^{-8} and on ξ at 10^{-5} level. The orthogonal combinations $\eta_{u,d} + \eta_Q$ are less constrained because they enter only in the quadrupole coupling between the nuclear spin and external direction,[8] and thus are suppressed by an additional factor of $|n_i| \sim 10^{-3}$.

So far we have neglected the fact that the low energy values for the couplings η_i and ξ taken at the normalization scale of 1 GeV do not coincide with the high-energy values for the same couplings generated at $M_{\rm Pl}$. With a simple one-loop analysis of the renormalization group equations, we find that several bounds can be strengthened. Leaving the details for elsewhere,[17] our results are: $|\eta_{Q,u,d}|, |\xi| \lesssim 10^{-6}$ and $|\eta^e_{L,R}| \lesssim 10^{-5}$. The constraints may also be improved by assuming degeneracies appropriate for grand unification at the GUT scale. On the experimental side, the recent progress[18] in the high-sensitivity atomic magnetometers that are unaffected by spin-exchange relaxation may lead to an improvement of terrestrial bounds on Lorentz violation by several orders of magnitude.

In summary, we have shown that effective field theory provides a framework where one can derive stringent bounds on Planck scale interactions from terrestrial experiments. The resulting limits enhance and generalize the terrestrial bounds obtained previously.[8] They are generally far more sensitive than those previously derived by considering astrophysical phenomena[4] where the typical sensitivity is $O(1)$ in units of $M_{\rm Pl}^{-1}$. A notable exception is the bound[3] $|\xi| \lesssim 2 \times 10^{-4}$ derived from the birefringence induced by $\mathcal{L}_\gamma$. Our bound[9] improved on this result by roughly an order of magnitude. However, with new astronomical data, this bound was recently updated[6] in a striking way to $|\xi| \lesssim 10^{-14}$.

Our limit (7) is already comparable to previous constraints from the astrophysical searches of vacuum photon decay and the absence of the vacuum Cerenkov radiation.[4] Further, the latter analysis did not consider polarization effects (*i.e.*, assumed $\eta_2 = 0$) and so this result provides a complementary constraint. Certainly our bounds inferred from the renormalization group analysis[9,17] represent a major improvement in constraining these couplings. However, these bounds were already tightened to an extraordinary level by other recent work.[5] There, the observation of synchrotron radiation from the Crab nebula was used to infer $\eta_L^e + \eta_R^e \gtrsim -10^{-7}$. We should also note that combining various astrophysical constraints may yield $O(10^{-2})$ constraints on the difference $\eta_R^e - \eta_L^e$ subject to certain assumptions.[6]

4. Discussion

The procedure of classifying operators according to their dimension proves very useful in analyzing sensitivity of various experiments to Lorentz violation. In such an analysis, one should always start from the *lowest* possible operators that may be induced by the short-distance physics. In the case of Lorentz violation, this expansion starts from dimension 3,[1] and hence we arrive at a certain "naturalness" problem for the whole approach: Why should the size of dimension 3 operators not be $M_{\rm Pl}$, the scale of UV physics responsible for the breaking, rather than less than $\sim 10^{-30}$ GeV as various experimental tests of Lorentz invariance demand? Similarly, the naive expectation would be that Lorentz violation should appear with $O(1)$ strength at dimension 4, again in contradiction with experimental constraints. Perhaps, the full answer to this puzzle requires a proper understanding of the full dynamical picture of Lorentz violation emerging from high-energy scales. Without that we have to resort to a number of plausible explanations, none of them very compelling at present.

One possbility is that the properties of the theory at the UV scale are inconsistent with the presence of dimension 3 and 4 Lorentz-violating operators. For example, it can be easily argued that the exact supersymmetry at UV scale may completely forbid or at the very least severely restrict the possibility of dimension 3 and 4 operators. A classification of all supersymmetric Lorentz non-invariant operators is an interesting study on its own, but goes outside of the present discussion.[19]

Another generic possibility to entertain in this approach is that the properties of the Lorentz-violating background itself lead to the suppression/exclusion of dimension 3 and 4 operators. For example, one might consider Lorentz violations as realized by a symmetric traceless rank-three tensor, C^{abc}, as was introduced at the end of section 2. In euclidean space, such tensor would correspond to an intrinsic *octupole* deformation of the vacuum. One finds that there are simply no dimension 3 or 4 operators which couple to this tensor. Beyond an absence of bare couplings, this implies that in a leading order calculation linear in C^{abc} can not generate dimension 3 or 4 operators at any number of loops.[17] This was of particular importance in section 3 where we found the dimension 5 operators evolved logarithmically and while dimension 3 operators might have been expected to appear with $\Lambda_{UV}^2/M_{\rm Pl}$ coefficients, they were in fact absent. Setting $C^{abc} = n^a n^b n^c - \frac{1}{6}(n^a g^{bc} + \text{cyclic})$ was a simplifying ansatz in our analysis and is not required by any underlying principles. Notice that if the dimension 5 operators coupled to $n^a n^b n^c$ rather than C^{abc}, one-loop graphs already generate, *e.g.*, $\Lambda_{UV}^2/M_{\rm Pl}\ \bar{\psi}\not{n}\gamma_5\psi$. The appearance of such quadratic divergence results because the new coupling is not in an irreducible representation of the Lorentz group and, as well as C^{abc}, contains a vector component which can couple to dimension 3 and 4 operators.

One can also consider higher order quantum corrections beyond linear order in $C^{abc}/M_{\rm Pl}$. A priori, there is no reason to believe that this effect should be small. For example, a triple product $C^c_{ab} C^{abd} C^e_{cd}$ will transform as a vector, and therefore may generate dimension 3 operators with a coefficient $\Lambda_{UV}^4/M_{\rm Pl}^{-3}$. If the scale of the ultraviolet cutoff in the loops is itself of the order of the Planck scale, then the resulting size of dimension 3 operator would be too large.[17,20] Therefore, one is forced to assume that an effective cutoff in the loops with $\Lambda_{UV} \ll M_{\rm Pl}$. If we require these induced dimension 3 operators to be smaller than dimension 5 operators at energy scales of 1 GeV, we must assume that $\Lambda_{UV} \lesssim \sqrt{M_{\rm Pl} \times 1\text{GeV}} \sim 10^9\text{GeV}$. This scale is far larger than the electroweak scale and/or the scale of the supersymmetry breaking. Even though a generic investigation of the supersymmetric stabilization of the ultraviolet divergencies in the presence of the Lorentz violating operators is presently lacking, it certainly appears as a plausible scenario.[19] To summarize, one must regard both the appearance of C^{abc} and a low scale cutoff in the loops as essential ingredients in suppressing the appearance of Lorentz violations in lower dimension operators from quantum effects.

Following Ref. [8], it is amusing to compare the experimental constraints to semi-classical calculations which have appeared in the loop quantum gravity literature.[21,22,23] For a Dirac fermion, these studies[22] suggest that $\eta_1 = 0$ while η_2 is nonvanishing. However, the latter is suppressed by factors of $\mathrm{M_L}/M_{\mathrm{Pl}}$ where $\mathrm{M_L} \ll M_{\mathrm{Pl}}$ is the coherence scale of the gravitational wave function. Therefore these calculations seem to be in agreement with the stringent experimental bounds[5,6,9] imposed on the fermion operators. In contrast, a separate analysis[21] suggests that ξ should be $O(1)$, which stands in stark contradiction with the bounds discussed above.[3,6,9] Hence naively the experimental bounds seem to be in conflict with the predictions of loop quantum gravity. However, it must be noted that the latter results are at present somewhat heuristic and so this apparent contradiction should not be taken too seriously.

Further recent calculations[24] suggest that the effects of quantum gravity may deform the Lorentz symmetry, *i.e.*, departures from the "standard" realization of Lorentz symmetries may be induced by Planck scale effects. This scenario, commonly known as "doubly special relativity" (DSR),[10] would seem a drastic departure from Lorentz breaking, in which our discussion was phrased. However, we will argue that many of the results are still applicable in this new scenario.

For simplicity, focus on photons and electrons which are central to many of the discussions of experimental bounds. First one must assume that the present preliminary investigations of DSR can be extended to provide something like a quantum field theory of photons and electrons incorporating DSR. Now the first simple observation is that in the limit where $M_{\mathrm{Pl}} \to \infty$, this DSR theory must reduce to ordinary QED. Then for finite M_{Pl}, the low-energy or long-wavelength physics can still be described using standard techniques of effective field theory. Hence if the dispersion relations relevant for photons and electrons are modified at $O(p^3)$ (and we assume that rotational invariance is unmodified), then it must be by the appearance of dimension 5 interactions of the form given in Eqs. (1) and (4). In this context, one should not think that these new interactions break Lorentz invariance. Rather the apparent Lorentz violations induced by standard Lorentz transformations on these terms must be compensated by corrections to these transformations when they act on the standard kinetic terms. In fact, the latter observation may be used as a strategy to infer the leading corrections to the Lorentz transformations.[25] In any event, the discussion of section 2 is equally applicable to DSR scenarios as to scenarios of Lorentz breaking.

One must, of course, be more cautious when considering the experimental bounds quoted above, as many of these rely on the existence of a preferred frame.[11] However, we argued above that Eq. (1) still gives the leading modification of the photon action in a DSR scenario, and hence the $O(p^3)$ modification of the dispersion relation is photon-chirality dependent.[a] Further, the astronomical tests involving the resulting birefringence for photons provide purely kinematical constraints on the theory. That is, one relies on the magnification of small but unusual effects in the propagation of photons as they cross cosmological distances, without reference to a preferred frame. Hence these bounds[3,6] are again applicable to DSR scenarios. Given the present experimental bound[6] that $|\xi| \lesssim 10^{-14}$, it seems that cubic modifications of the photon dispersion relation are essentially ruled out for DSR constructions.

Another important aspect of the effective field theory approach to Lorentz violation is the correct identification of the infrared degrees of freedom and the possibility of generalizing this framework to general relativity.[12] Attempts to generalize Lorentz breaking operators to general relativity would necessarily supply C^{abc} with coordinate dependence and a corresponding kinetic term, which in turn would contribute to the energy density. There are several arguments supporting the idea that Lorentz violating tensors must be accompanied by massless (or nearly massless) particles in the spectrum. Many explicit examples involving Lorentz violation point towards this result, including noncommutative backgrounds in string theory,[26] super-light axion backgrounds,[27] and ghost condensation.[28] The exchange by new light degrees of freedom may lead to the appearance of a new coherent interaction, *i.e.*, a "fifth force". Therefore, quite separate bounds on Lorentz violation may emerge from gravitational physics and cosmology. Elucidating this question is also important for interpreting the bounds on Lorentz violation that come from cosmological distances, over which a "constant" background would most certainly be changing.

In conclusion, we would like to stress again that there remains much work to be done both on the theoretical description from quantum gravity and the phenomenological constraints. However, it is truly remarkable that present-day precision experiments and astrophysical observations can

[a]It is curious that certain arguments seem to imply this would be inconsistent with DSR.[11]

already confront quantum gravity calculations with concrete and stringent observational bounds.

Acknowledgments

RCM would like to thank the organizers of the 3rd International Symposium on Quantum Theory and Symmetries at the University of Cincinnati for the opportunity to present this material. We would like to thank Giovanni Amelino-Camelia, Cliff Burgess, Laurent Freidel, Stefan Groot Nibbelink, Ted Jacobson, Jerzy Kowalski-Glikman, Joe Lykken, Seth Major, Guy Moore, Ann Nelson, Adam Ritz, Mike Romalis, Subir Sarkar and Lee Smolin for many useful comments and conversations over the course of this work. This research is supported in part by NSERC of Canada, Fonds FCAR du Québec and PPARC UK.

References

1. D. Colladay and V.A. Kostelecky, Phys. Rev. **D55**, 6760 (1997); S.R. Coleman and S.L. Glashow, Phys. Rev. **D59**, 116008 (1999); S.M. Carroll, G.B. Field and R. Jackiw, Phys. Rev. **D41**, 1231 (1990).
2. G. Amelino-Camelia *et al.*, Nature **393**, 763 (1998); G. Amelino-Camelia and T. Piran, Phys. Rev. **D64**, 036005 (2001); G. Amelino-Camelia, Phys. Lett. **B528**, 181 (2002); S. Sarkar, Mod. Phys. Lett. **A17**, 1025 (2002).
3. R.J. Gleiser and C.N. Kozameh, Phys. Rev. **D64**, 083007 (2001).
4. T. Jacobson, S. Liberati and D. Mattingly, Phys. Rev. **D66**, 081302 (2002); Phys. Rev. **D67**, 124011 (2003); T.J. Konopka and S.A. Major, New J. Phys. **4**, 57 (2002).
5. T. Jacobson, S. Liberati and D. Mattingly, Nature **424**, 1019 (2003).
6. T. Jacobson, S. Liberati, D. Mattingly and F.W. Stecker, astro-ph/0309681.
7. G. Amelino-Camelia, Nature **398**, 216 (1999); I. Mocioiu, M. Pospelov and R. Roiban, Phys. Lett. **B489**, 390 (2000); G.Z. Adunas, E. Rodriguez-Milla and D.V. Ahluwalia, Phys. Lett. **B485**, 215 (2000); R. Bluhm, in *CPT and Lorentz Symmetry, 2001, Bloomington* (World Scientific, Singapore, 2002).
8. D. Sudarsky, L. Urrutia and H. Vucetich, Phys. Rev. Lett. **89**, 231301 (2002).
9. R.C. Myers and M. Pospelov, Phys. Rev. Lett. **90**, 211601 (2003).
10. See, *e.g.,* G. Amelino-Camelia, Int. J. Mod. Phys. **D11**, 35 (2002); Phys. Lett. **B510**, 255 (2001); Nature **418**, 34 (2002); gr-qc/0212002; J. Kowalski-Glikman, Phys. Lett. **A286**, 391 (2001); N.R. Bruno, G. Amelino-Camelia and J. Kowalski-Glikman, Phys. Lett. **B522**, 133 (2001); G. Amelino-Camelia, D. Benedetti, F. D'Andrea and A. Procaccini, Class. Quant. Grav. **20**, 5353 (2003); J. Kowalski-Glikman and S. Nowak, Int. J. Mod. Phys. **D12**, 299 (2003); S. Judes and M. Visser, Phys. Rev. **D68**, 045001 (2003); M. Toller, hep-ph/0211094; A. Chakrabarti, J. Math. Phys. **44**, 3800 (2003); S. Mignemi, Phys. Rev. **D68**, 065029 (2003); A. Ballesteros, N.R. Bruno and

F.J. Herranz, Phys. Lett. **B574**, 276 (2003); J. Phys. **A36**, 10493 (2003); S.K. Kim, S.M. Kim, C. Rim and J.H. Yee, gr-qc/0401078.
11. G. Amelino-Camelia, J. Kowalski-Glikman, G. Mandanici and A. Procaccini, gr-qc/0312124.
12. A. Kostelecky, hep-th/0312310.
13. I. Mocioiu and M. Pospelov, Phys. Lett. **B534**, 114 (2002).
14. T.E. Chupp *et al.*, Phys. Rev. Lett. **72**, 2363 (1994); R.E. Stoner *et al.*, Phys. Rev. Lett. **77**, 3971 (1996); D. Bear *et al.*, Phys. Rev. **A57**, 5006 (1995); C.J. Berglund *et al.*, Phys. Rev. Lett. **75**, 1879 (1995); D. Bear, *et al.* Phys. Rev. Lett. **85** (2000) 5038 [Errat. **89** (2002) 209902].
15. E.G. Adelberger *et al.*, in *Physics Beyond the Standard Model, 1998, Santa Fe* (World Scientific, Singapore, 1999); M.G. Harris, PhD Thesis, University of Washington, 1998; B.R. Heckel *et al.*, in *Proceedings of the International Conference on Orbis Scientiae, 1999, Coral Gables* (Kluwer, New York, 2000).
16. M. Gluck, E. Reya and A. Vogt, Eur. Phys. J. **C5**, 461 (1998); B.W. Filippone and X.D. Ji, Adv. Nucl. Phys. **26**, 1 (2001).
17. P. Bolokhov, R.C. Myers and M. Pospelov, in preparation.
18. J.C. Allred, R.N. Lyman, T.W. Kornack, and M.V. Romalis, Phys. Rev. Lett. 89, 130801 (2002).
19. M.S. Berger and V.A. Kostelecky, Phys. Rev. **D65**, 091701 (2002); C.E. Carlson, C.D. Carone and R.F. Lebed, Phys. Lett. **B549**, 337 (2002); S. Groot-Nibbelink and M. Pospelov, in preparation.
20. A. Perez and D. Sudarsky, gr-qc/0306113.
21. R. Gambini and J. Pullin, Phys. Rev. **D59**, 124021 (1999); J. Alfaro, H.A. Morales-Tecotl and L.F. Urrutia, Phys. Rev. **D65**, 103509 (2002).
22. J. Alfaro, H.A. Morales-Tecotl and L.F. Urrutia, Phys. Rev. **D66**, 124006 (2002).
23. H. Sahlmann and T. Thiemann, gr-qc/0207030; gr-qc/0207031.
24. G. Amelino-Camelia, L. Smolin and A. Starodubtsev, hep-th/0306134; L. Freidel, J. Kowalski-Glikman and L. Smolin, hep-th/0307085.
25. R.C. Myers and M. Pospelov, unpublished.
26. See, *e.g.,* M.R. Douglas and N.A. Nekrasov, Rev. Mod. Phys. **73**, 977 (2001).
27. S.M. Carroll, Phys. Rev. Lett. **81**, 3067 (1998).
28. N. Arkani-Hamed, H.C. Cheng, M.A. Luty and S. Mukohyama, hep-th/0312099.

IX. APPLICATIONS

BLACK HOLES FROM COLLIDERS AND COSMIC RAYS*

LUIS A. ANCHORDOQUI
Department of Physics, Northeastern University, Boston, MA 02115

JONATHAN L. FENG
Department of Physics, University of California, Irvine, CA 92697

HAIM GOLDBERG
Department of Physics, Northeastern University, Boston, MA 02115

ALFRED SHAPERE
Department of Physics, University of Kentucky, Lexington, KY 40506

A startling prediction of many phenomenological scenarios with large extra dimensions is that TeV-scale collisions may produce microscopic black holes in cosmic rays and at colliders. The fact that no such cosmic ray events have yet been observed places the best current lower bounds on the scale of higher-dimensional gravity, with 5 or more extra dimensions. I review the current status of these bounds, and the prospects for improving them over the next few years at the Pierre Auger Observatory and the LHC.

1. Introduction

One of the most extraordinary and generic consequences of scenarios of TeV-scale gravity[1] would be the production of microscopic black holes[2] in colliders and cosmic rays. In two well-known papers,[3] it was proposed that such black holes could be produced copiously at the LHC, if the D-dimensional gravity scale M_D is not too far above 1 TeV. This prediction was based on a simple model for the process of black hole formation in high-energy collisions,[4] according to which a black hole (BH) forms in a two-particle collision when and only when the impact parameter is smaller than the radius r_s of a Schwarzschild BH of mass equal to the total center-

*Talk given by A.S.

of-mass energy $E_{\rm CM}$. The conjecture thus predicts a total cross section for BH production proportional to the area subtended by a "hoop" of radius r_s:

$$\sigma_{BH} = F(n)\, \pi r_s^2(E_{\rm CM}) \,. \tag{1}$$

where $F(n)$ is a constant of order unity depending on the number of extra dimensions n. Recent progress has confirmed the validity of Eq. (1) and evaluated the dimension-dependent constant $F(n)$, analytically in four dimensions[5] and numerically in higher dimensions[6]. Of course, this work is purely in the framework of classical general relativity, and is expected to be valid for energies far above the Planck scale, for which the curvature is small outside the horizon and strong quantum effects are hidden behind the horizon. Extending the classical picture of black hole production to the planckian regime of center-of-mass energies close to M_D requires a better understanding of quantum gravity than we now possess. Thus it is important to impose a cutoff $M_{\rm BH}^{\rm min} \equiv x_{\rm min} M_D$ on the mass of microscopic BHs for which Eq. (1) can reasonably be expected to hold.

Once formed in a high-energy particle collision, a TeV-scale black hole will promptly decay via thermal Hawking radiation (for $M_D = 1$ TeV, the lifetime of a BH of mass 10 TeV is less than 10^{-25} s) isotropically and essentially without regard to particle species into all available modes. The resulting final state would be appear as a spectacular and easily detectable event at LHC.

Since current accelerators have not detected any evidence for extra dimensional physics and LHC will not come online for several more years, it is natural to ask whether microscopic black holes might be seen in other places in the meantime. In particular, cosmic ray collisions, with center-of-mass energies ranging up to 10^5 TeV, will certainly produce black holes if LHC does. The question is, can they be detected? Most cosmic rays are protons, which generally collide with hadrons in the upper atmosphere, producing cascading showers which eventually reach the earth's surface. Since QCD cross sections dominate over the BH production cross section by a factor of roughly 10^9, any black holes produced by hadronic cosmic rays would be impossible to filter out of the QCD background.

Neutrino cosmic rays, whose cross section for producing black holes can be orders of magnitude larger than standard model (SM) cross sections,[7] but much smaller than hadronic cross sections, can produce black holes deep in the atmosphere, with negligible background from hadronic cosmic rays. Because their BH cross sections are relatively small, neutrinos would produce BHs with roughly equal probability at any point in the atmosphere. Low-altitude, quasi-horizontal showers at rates significantly higher than SM predictions would be a strong indication of new physics, and would be easily detectable at ground-based cosmic ray observatories. BH-mediated showers of this type would be distinguished by a characteristic electromagnetic component.[8]

In this brief review, we will focus on the lower bounds on M_D that are obtained from the nonobservation of BHs at existing and planned cosmic neutrino detection experiments, as well as at the LHC. We improve on the reliability of our previous estimates [10,11] by incorporating into our calculations the cross section and energy loss results of Yoshino and Nambu[6], and by making use of updated parton distribution functions and recently available cosmic ray data.

For 5 or more extra dimensions, the cosmic ray bounds are among the best experimental bounds available.[9] These bounds will improve significantly over the next few years if no anomalous low-altitude, quasihorizontal showers are seen at the Pierre Auger Observatory (PAO). By the time LHC data are available for analysis, the parameter range for discovery of black hole events will be quite restricted. Additionally, due primarily to inelasticity effects, the maximum possible rate for black hole production will be a factor of 10^3 to 10^6 lower than was predicted in Ref. 3.

2. Cross Section

Bounds on the scale of higher-dimensional gravity M_D are obtained from the neutrino-nucleon cross section for BH production viewed as a function of M_D, neutrino energy E_ν, and the number of extra dimensions n, and folding it in with the estimated neutrino flux and detector exposure to get a predicted event count. The nonobservation of events then places a lower limit on M_D for each n.

To compute the neutrino-nucleon cross section, we start with the cross section for point particles Eq. (1). Given this, the cross section for $\nu N \to$ BH is

$$\sigma(E_\nu, x_{\min}, M_D) \equiv \int_0^1 dz\, 2z \int_{(x_{\min} M_D)^2/(y^2 s)}^1 dx\; F(n)\, \pi r_s^2(\sqrt{\hat{s}}, n, M_D) \sum_i f_i(x, Q) \quad (2)$$

where $\hat{s} = 2x m_N E_\nu$, m_N is the nucleon mass, i labels parton species, the $f_i(x, Q)$ are parton distribution functions (pdfs), and the inelasticity y will be defined below. In what follows we use the CTEQ6M pdfs[12] with $Q = \min\{r_s^{-1}, 10\ \text{TeV}\}$.

The choice of a lower limit of integration requires some explanation. This limit determines the minimum mass for black holes that will be included in the calculation. As we have already mentioned, $M_{\rm BH}^{\rm min}$ needs to be large enough that the classical cross section (1) can be trusted. It is also important that the black hole that is produced be large enough to be treated semiclassically.[13] The criterion we employ is to assume that the black hole has an entropy $S \gg 1$ to insure small thermal fluctuations, with S given by

$$S = 4\pi \, M_{\rm BH} \, r_s(M_{\rm BH})/(n+2) \; ; \tag{3}$$

so for $x_{\rm min} > 3$ and $n \geq 5$, one finds $S > 10$. Moreover, it may be reasonable even to take $x_{\rm min}$ as low as 1 (for which $S \gtrsim 3$), since the signal in neutrino detection experiments relies only on the existence of visible decay products, and furthermore it seems reasonable to expect that BHs or their planckian progenitors will cascade decay on the SM brane. The fact that the string cross section is expected to be considerably larger than Eq. (1) [14] also supports using $x_{\rm min} = 1$.

Our previous calculations[7,10,11] of $\sigma(E_\nu, x_{\rm min}, M_D)$ neglected energy lost in the creation of a BH, and instead assumed that the mass of the created black hole $M_{\rm BH}$ was identical to the incoming parton center-of-mass energy $\sqrt{\hat{s}}$. However, recent work[6] has shown that the energy lost to gravitational radiation is in fact quite significant for larger n and grows rapidly with the impact parameter. The trapped mass is given by

$$M_{\rm BH}(z) = y(z)\sqrt{\hat{s}} \; , \tag{4}$$

where the inelasticity y is a function of $z \equiv b/b_{\rm max}$. Also, b is the impact parameter and

$$b_{\rm max} = \sqrt{F(n)} \; r_s(\sqrt{\hat{s}}, n, M_D) \tag{5}$$

is the maximum impact parameter for collapse, in terms of the the form factor $F(n)$[6] and the Schwarzschild radius r_s in $(4{+}n)$ dimensions[15]. Thus the lower limit of the x integration in Eq. (2) guarantees that the parton momentum fraction will always be large enough to produce a black hole larger than $x_{\rm min} M_D$.

In Fig. 1 we show the BH production cross section for $n = 1, \ldots, 7$ extra dimensions. We find reduced BH production rates, primarily because the reduced BH masses make the cutoff at $M_{\rm BH}^{\rm min} = x_{\rm min} M_D$ more stringent. However, the cross sections are still well above the SM cross section at $E_\nu \sim 10^8$ GeV and above, where, as we will see, the cosmogenic neutrino flux is large.

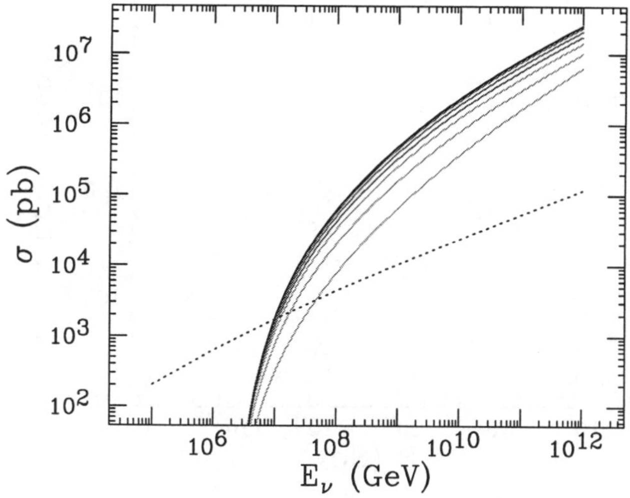

Figure 1. Cross sections $\sigma(\nu N \to \text{BH})$ for $n = 1, \ldots, 7$ from below, assuming $M_D = 1$ TeV and $x_{\min} = 1$. Energy loss has been included according to (2). The SM cross section $\sigma(\nu N \to \ell X)$ is indicated by the dotted line.

3. Detection of Cosmic Neutrinos

The number of neutrino showers mediated by BHs depends on the cross section and the exposure $\mathcal{E}$ of the detector. The exposure is the product of the effective aperture and the range of depths within which the shower must originate to trigger the device, integrated over time. The effective aperture is the detector's projected area weighted by detection probability and integrated over solid angle. The exposure is, then,

$$\mathcal{E}(E_{\text{sh}}) \approx \int_0^T dt \int_0^{h_{\max}} (A\Omega)_{\text{eff}}(E_{\text{sh}}, t)\, \frac{\rho_0}{\rho_{\text{water}}}\, e^{-h/H}\, dh\,, \tag{6}$$

where T is the total observation time of the detector, $h_{\max} = 15$ km, $H \approx 8$ km, and $\rho_0 \approx 1.15 \times 10^{-3} \rho_{\text{water}}$ is the density of the atmosphere at ground level. The effective aperture is

$$(A\Omega)_{\text{eff}}(E_{\text{sh}}, t) \equiv \int_{\theta_{\min}}^{\theta_{\max}} A(t)\, \mathcal{P}(E_{\text{sh}}, \theta, t)\, 2\pi \, \sin\theta \, d\theta \tag{7}$$

where $A(t)$ is the detector's area, $\mathcal{P}(E_{\text{sh}}, \theta, t)$ is the probability that a shower that arrives with energy E_{sh} and zenith angle θ triggers the detector, and the angular cuts $\theta_{\min}$ and $\theta_{\max}$ are chosen to optimize detection efficiency while eliminating hadronic background.

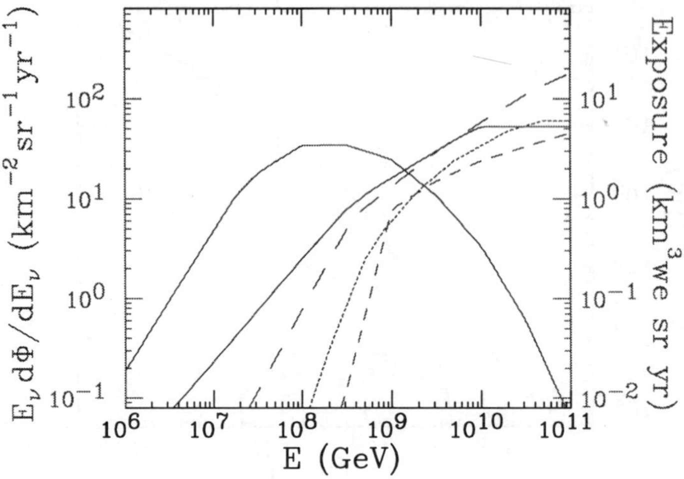

Figure 2. The monotonically rising curves are the exposures as functions of shower energy for AGASA (solid), Fly's Eye (dotted), HiRes (short dash), and RICE (long dash). The remaining solid curve, with a peak around $10^{8.5}$ GeV, is the cosmogenic neutrino flux.

In terms of the exposure, the total number of events involving BH's of mass greater than $x_{\rm min} M_D$ is

$$\mathcal{N}(x_{\rm min}, n, M_D) = N_A \int dE_\nu \int_0^1 2z\, dz \int_{\frac{(x_{\rm min} M_D)^2}{y^2 s}}^1 dx\, \frac{d\Phi}{dE_\nu}\, \mathcal{E}(yE_\nu) \times F(n)\, \pi r_s^2(\sqrt{\hat{s}}, n, M_D) \sum_i f_i(x, Q)\ , \tag{8}$$

where N_A is Avogadro's number, $d\Phi/dE_\nu$ is the diffuse flux of cosmic neutrinos hitting the Earth, and $\mathcal{E}(yE_\nu)$ is the neutrino exposure of a given experiment for shower energy yE_ν. Note that the energy lost in BH formation impacts event rates at cosmic neutrino detectors in two ways: through modification of the cross section and also because the exposures of cosmic neutrino detectors are functions of shower energy.

A number of detectors, employing several different detection techniques, have been used to search for neutrino showers.[16] The AGASA facility, a 100-km^2 array of particle counters, has found no significant enhancement of deeply-developing shower rates given the detector's resolution.[17] Specifically, only 1 event has been observed, consistent with the expected background of 1.72 from hadronic cosmic rays.[10] The Fly's Eye fluorescence detector recorded more than 5000 events between 1981 and 1992, but no unusual deeply developing showers were found.[18] At the upscaled version of the Fly's Eye experiment known as HiRes, with six times the original aperture, no candidate neutrino events were detected between November 1999 and September 2001.[19] Finally, data from the Radio Ice Čerenkov Experiment (RICE)[20] taken in August 2000 has been analyzed, and again, no events consistent with neutrino sources have been found.[21] The relative exposures for the different experiments are given in Fig. 2.[11,22] All in all, there is only 1 event observed with an expected background of 1.72 from hadronic cosmic rays, leading to a 95% CL limit of 3.5 BH events.[23]

To derive the bounds on M_D, we use the "guaranteed" flux of cosmogenic neutrinos arising from the decay of $\pi^\pm$ produced in collisions of ultra-high energy protons with the cosmic microwave background. As in our previous analyses, we conservatively adopt the estimates of Protheroe and Johnson[24] with nucleon source spectrum scaling as $d\Phi_N/dE \propto E^{-2}$ and extending up to the cutoff energy $10^{12.5}$ GeV. The total ultra-high energy cosmogenic neutrino flux is also shown in Fig. 2.

4. Summary

The best current lower bounds on the scale of gravity with five or more extra dimensions come from nonobservation of neutrino cosmic rays involving production of microscopic black holes.[9] Making use of updated exposures from the HiRes and RICE facilities, we find a lower bound between 1.0 and 1.4 TeV, depending on the cutoff for minimum black hole mass. This work has eliminated much of the uncertainty contained in our previous limits [10] by incorporating numerically calculated black hole production cross sections[6] and updated pdfs. In the course of our analysis we observed a competition of effects leading to corrections to our previous estimates: enhancement of the geometric cross section by form factors of up to 1.9 and enhancement of apertures from new cosmic ray data, but a simultaneous reduction in the rate of production of BHs of mass greater than $x_{\rm min} M_D$, after taking energy losses into account. It turns out that the latter effect dominates and leads to a slight weakening of our bound on M_D.

We have derived conservative 95% CL lower bounds incorporating the new cross section and mass loss results of Ref. 6, using Eq. (8) with the exposures and the cosmogenic neutrino flux given in Fig. 2, and requiring $\mathcal{N} < 3.5$ events to be observed in cosmic neutrino data samples.

In Fig. 3 we exhibit our bounds on M_D and compare them with other existing limits on the fundamental scale of large extra dimensions. For

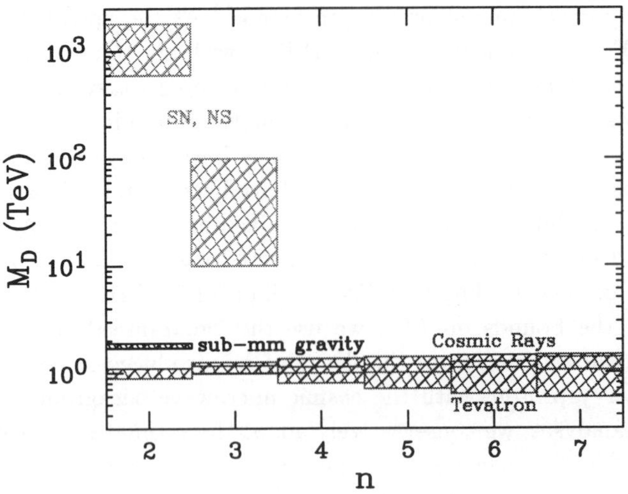

Figure 3. Bounds on the fundamental Planck scale M_D from tests of Newton's law on sub-millimeter scales, bounds on supernova cooling and neutron star heating, dielectron and diphoton production at the Tevatron, and nonobservation of BH production by cosmic neutrinos. The uncertainty in the Tevatron bounds corresponds to the range of brane softening parameter.The range in the cosmic ray bounds is for $1 \leq x_{\min} \leq 3$.

$n < 5$, astrophysical constraints give the best limits on M_D. For $n \geq 5$, the best existing limits on TeV-scale gravity are from the absence of transplanckian signatures (BH/string ball production) in neutrino detection experiments discussed here, and from searches for subplanckian signatures (graviton emission and virtual graviton exchanges) at the Tevatron[25] and LEP[26]. Specifically, the neutrino bounds range from 1.0 to 1.4 TeV for $x_{\min}$ between 3 and 1, and are at the upper end of the range of current collider bounds. Since the collider bounds have a large uncertainty coming from the unknown value of the brane softening parameter, the neutrino bounds are less model-dependent.

Continued nonobservation of an excess of deeply-penetrating showers at the PAO over the next few years would raise these bounds significantly, and sharply restrict the parameter space for BH discovery at the LHC, as shown in Fig.4. Also, the effects of inelasticity on LHC event rates are considerably greater than for cosmic rays, leading to suppression factors of 10^3 to 10^6 compared to previous estimates.[27] In spite of this enormous suppression, BH discovery at the LHC is still a possibility. However slim this possibility may be, we must take it seriously: it could represent our only chance to directly observe the effects of quantum gravity, extra dimensions,

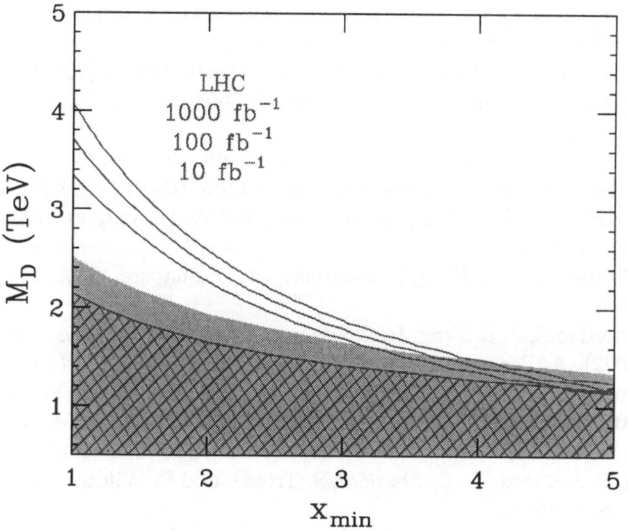

Figure 4. Discovery reaches for the LHC (solid) for 3 different integrated luminosities and n=6 extra dimensions. Also shown is the region of parameter space which can be excluded at 95% CL if no neutrino showers mediated by BHs are seen in 5 years at the PAO. The shaded (cross-hatched) region assumes 2 SM neutrino + 0 (10) hadronic background events.

and string theory.

Acknowledgments

AS thanks the organizers of QTS3 for their invitation and hospitality. The work of LAA and HG reported in this article was partly supported by the US National Science Foundation (NSF), under grants No. PHY–0140407 and No. PHY–0073034, respectively. The work of JLF was supported in part by NSF CAREER grant No. PHY–0239817. The work of ADS was supported in part by Department of Energy grant No. DE–FG01–00ER45832 and NSF grants PHY–0071312 and PHY–0245214.

References

1. N. Arkani-Hamed, S. Dimopoulos and G. R. Dvali, Phys. Lett. B **429**, 263 (1998) [hep-ph/9803315]; I. Antoniadis, N. Arkani-Hamed, S. Dimopoulos and G. R. Dvali, Phys. Lett. B **436**, 257 (1998) [hep-ph/9804398].
2. T. Banks and W. Fischler, arXiv:hep-th/9906038.
3. S. B. Giddings and S. Thomas, *Phys. Rev.* **D65**, 056010 (2002); S. Dimopoulos and G. Landsberg, *Phys. Rev. Lett.* **87**, 161602 (2001).

4. K. S. Thorne, in *Magic Without Magic: John Archibald Wheeler*, edited by J. Klauder (Freeman, San Francisco, 1972) p. 231.
5. D. M. Eardley and S. B. Giddings, *Phys. Rev.* **D66**, 044011 (2002).
6. H. Yoshino and Y. Nambu, *Phys. Rev.* **D66**, 065004 (2002); *Phys. Rev.* **D67**, 024009 (2003).
7. J. L. Feng and A. D. Shapere, *Phys. Rev. Lett.* **88**, 021303 (2001).
8. L. Anchordoqui and H. Goldberg, *Phys. Rev.* **D65**, 047502 (2002).
9. L. A. Anchordoqui, J. L. Feng, H. Goldberg and A. D. Shapere, *Phys. Rev.* **D68**, 104025 (2003).
10. L. A. Anchordoqui, J. L. Feng, H. Goldberg, A. D. Shapere, *Phys. Rev.* **D65**, 124027 (2002).
11. L. A. Anchordoqui, J. L. Feng, H. Goldberg, A. D. Shapere, *Phys. Rev.* **D66**, 103002 (2002); A. Ringwald and H. Tu, *Phys. Lett.* **525B**, 135 (2002).
12. J. Pumplin, D. R. Stump, J. Huston, H. L. Lai, P. Nadolsky and W. K. Tung, *JHEP* **0207**, 012 (2002); D. Stump, J. Huston, J. Pumplin, W. K. Tung, H. L. Lai, S. Kuhlmann and J. F. Owens, *JHEP* **0310**, 046 (2003).
13. J. Preskill, P. Schwarz, A. D. Shapere, S. Trivedi and F. Wilczek, *Mod. Phys. Lett.* **A6**, 2353 (1991).
14. L. A. Anchordoqui, J. L. Feng, H. Goldberg and A. D. Shapere, "Quest for Black Holes and Superstring Excitations in Cosmic Ray Data," arXiv:hep-ph/0309082.
15. R. C. Myers and M. J. Perry, *Annals Phys.* **172**, 304 (1986); P. C. Argyres, S. Dimopoulos and J. March-Russell, *Phys. Lett.* **441B**, 96 (1998).
16. For recent reviews, see *e.g.,* G. Sigl, hep-ph/0109202; F. Halzen and D. Hooper, *Rept. Prog. Phys.* **65**, 1025 (2002); L. Anchordoqui, T. Paul, S. Reucroft and J. Swain, *Int. J. Mod. Phys.* **A18**, 2229 (2003).
17. S. Yoshida *et al.* [AGASA Collaboration], in *Proc. 27th International Cosmic Ray Conference*, Hamburg, Germany, 2001, Vol. 3, p. 1142.
18. R. M. Baltrusaitis *et al.*, *Phys. Rev.* **D31**, 2192 (1985); T. K. Gaisser *et al.*, *Phys. Rev.* **47** 1919 (1993); D. J. Bird *et al.*, Astrophys. J. **424**, 491 (1994).
19. G. C. Archbold, Ph.D.Thesis UMI-30-54514.
20. G. M. Frichter, J. P. Ralston and D. W. McKay, *Phys. Rev.* **53**, 1684 (1996); I. Kravchenko *et al.* [RICE Collaboration], *Astropart. Phys.* **19**, 15 (2003).
21. I. Kravchenko *et al.*, astro-ph/0206371.
22. L. Anchordoqui and H. Goldberg, Phys. Rev. D **67**, 064010 (2003) [arXiv:hep-ph/0209337].
23. G. J. Feldman and R. D. Cousins, *Phys. Rev.* **D57**, 3873 (1998).
24. R. J. Protheroe and P. A. Johnson, *Astropart. Phys.* **4**, 253 (1996); R. J. Protheroe, *Nucl. Phys. Proc. Suppl.* **77**, 465 (1999).
25. B. Abbott *et al.* [D0 Collaboration], *Phys. Rev. Lett.* **86**, 1156 (2001); V. M. Abazov *et al.* [D0 Collaboration], *Phys. Rev. Lett.* **90**, 251802 (2003).
26. M. Acciarri *et al.* [L3 Collaboration], *Phys. Lett.* **464B**, 135 (1999); M. Acciarri *et al.* [L3 Collaboration], *Phys. Lett.* **470B**, 281 (1999).
27. L. A. Anchordoqui, J. L. Feng, H. Goldberg and A. D. Shapere, arXiv:hep-ph/0311365.

STANDARD-MODEL COUPLING CONSTANTS FROM COMPOSITENESS: QUANTUM-MECHANICS AND SYMMETRY ARGUMENTS

J. BESPROSVANY

Instituto de Física, Universidad Nacional Autónoma de México
Apartado Postal 20-364, México 01000, Distrito Federal, México
E-mail: bespro@fisica.unam.mx

A coupling-constant definition is given based on the compositeness property of some particle states with respect to the elementary states of other particles. It is applied in the context of the vector-spin-1/2-particle interaction vertices of a field theory and the $SU(5)$ grand-unified theory. The definition reproduces Weinberg's angle, and gives the $SU(5)$ coupling constant at the unification scale, and the little-running hypercharge one. The use of quantum mechanics and symmetry arguments is emphasized.

1. INTRODUCTION

The coupling constants are the dimensionless numbers that measure the strength of nature's interactions. Their values are fixed by experiment in the standard model (SM) of elementary particles, and depend on the energy scale.

In general, the realization of unity among physical variables, originally thought as disconnected, has led to a new understanding and connections among additional ones. Indeed, recently proposed SM extensions including a unifying principle are able to provide information on the coupling constant values. Thus, grand unification[1] assumes that the gauge groups describing the interactions originate in a common group, and it predicts a single unified coupling, to which distinct couplings indeed appear to converge at high energy. It is also able to predict the coupling-constant ratios. In addition, spin-extended models[2] predict their values, and so does the dilaton-field ground state in string theory[3], but, as yet, not uniquely. Even if the underlying dynamics is not obvious, these connections may become manifest through symmetry arguments.

In general, the SM Poincaré and gauge symmetries dictate the few quantum numbers that describe a particle state. These are the configuration or momentum coordinates, the spin, the gauge-group representation, and the flavor for quarks and leptons. Flavor characterizes only fermions. In the SM, fermions belong in the spin-1/2 Lorentz representation, and the gauge bosons are vectors. Similarly, fermions belong in the fundamental representation of the gauge group, while the vector bosons belong in the adjoint.

In this report, we use this experimentally derived compositeness property of the SM particles to get information on the SM coupling constants. We indeed focus on those vector quantum numbers that can be constructed in terms of those of the fermions, which are elementary. We apply an equivalent field-theory formulation that makes this kind of compositeness explicit, keeping the SM assumption that the fields are fundamental; all the SM predictions are therefore maintained.

We first give a general coupling-constant definition based on the normalization and the compositeness property of some particle states with respect to other particle elementary states. Using the Wigner spinor classification of the Lorentz representations, one may express SM fields in terms of their spinor components. It follows that the SM Lagrangian and its fields can be rewritten and reinterpreted in this way. Finally, we classify the configurations of the vector particles in relation to their grand-unified theory content, calculate the hypercharge and weak coupling values at the unification scale, reproducing Weinberg's angle, and present final comments.

2. COMPOSITE STATES AND OPERATORS IN QUANTUM MECHANICS

Quantum numbers characterize particles, and the normalized state $|w_i\rangle$ represents a particle with eigenvalue w_i of the appropriate operator. The numbers a_{ij} in the composite state

$$|W\rangle = \frac{1}{\sqrt{N}} \sum_{i,j} a_{ij} |w_i\rangle |w_j\rangle, \tag{1}$$

normalized with

$$N = \sum_{i,j} a_{ij}^* a_{ij}, \tag{2}$$

fix $\langle w_i w_j | W\rangle$. The same amplitude is reproduced by the corresponding operator $\hat{W} = \frac{1}{\sqrt{N}} \sum_{i,j} a_{ij} |w_i\rangle\langle w_j|$, satisfying $tr\hat{W}^\dagger \hat{W} = 1$, through $\langle w_i|\hat{W}|w_j\rangle$. Thus, both structures keep the same information, and the same normalization prescription may be applied.

$\hat{W}$ is also the most general operator acting on the $|w_i\rangle$ states. Symmetry can determine the coefficients a^{λ}_{ij}, up to a constant, where λ labels the representation components of such symmetry. For example, the only (non-axial) vector operator that can be constructed out of spin-1/2 particle states is the Dirac matrix $\gamma_0\gamma^{\mu 4}$. For each μ (no sum) $tr\gamma_0\gamma^{\mu}\gamma_0\gamma^{\mu} = 4$ normalizes covariantly the operator, and fully determines it by providing the remaining constant; so is the case for the corresponding composite state $|W\rangle$. Hence, the matrix element between the spin states $|i\rangle$ and $|j\rangle$

$$\langle i|\hat{W}^{\mu}|j\rangle \tag{3}$$

is determined with $\hat{W}^{\mu} = \frac{1}{2}\gamma_0\gamma^{\mu}$. The four-entry $\hat{W}^{\mu}$ acts on the space spanned essentially by the spin-1/2 particle, its antiparticle, and their two spin polarizations.

This procedure can be generalized to the case of greater number of degrees of freedom, using the rules for the direct product of vector spaces and the generalized operator that acts on such a space. The normalization for M such operators, $\hat{W}^T = \hat{W}_1...\hat{W}_M$, is the product of the traces of each operator $\hat{W}_i$ in its space.

3. COMPOSITENESS IN THE LAGRANGIAN

The vertex interaction Lagrangian $\int \mathcal{L}_f$ with density $\mathcal{L}_f = -\frac{1}{2}gA^a_{\mu}\Psi^{\alpha\dagger}\gamma_0\gamma^{\mu}G^a\Psi^{\alpha}$ is determined from Poincaré and gauge invariance. In general, the latter determines the interactions of the vector bosons with the other particles, and among themselves, up to the coupling constant g. In particular, $\mathcal{L}_f$ is the only vector-spin-1/2 vertex. In the SM the fermions belong in the fundamental representation. The vertex can be consistently viewed as the expectation value of the tensor-product operator $\hat{W}^{\mu a} = g\gamma_0\gamma^{\mu}G^a 1_x 1_{\alpha}$, with vector components $A^a_{\mu}(x)$, acting upon the spin-1/2 particles $\Psi^{\alpha}(x)$; μ is the spin-1 index, G^a the gauge-group representation matrix of the fermions, a the group-representation index, x the spacetime coordinate with the diagonal[a] $1_x = |x\rangle\langle x|$, and 1_{α} the unit matrix over the flavor α. A composite state $A^a_{\mu}(x)|x\rangle|\mu\rangle|a\rangle$, with $|\mu\rangle$, $|a\rangle$ elements as in Eq. 1, underlies the operator association leading to $\hat{W}^{\mu a}$: $|\mu\rangle \to (\gamma_0\gamma^{\mu})_{\sigma\eta}$, $|a\rangle \to G^a_{bc}$, $|x\rangle \to |x\rangle\langle x|$; the fermion state is $\Psi^{\alpha}_{\eta c}(x)$. All are written explicitly in terms of σ, η spin-1/2 indices, b, c, gauge-group

[a] 1_x only connects local fields, without compositeness. Formally, $a^{x}_{x'x''} = \delta_{x'x''}\delta_{xx'}$. $A^a_{\mu}(x)$ normalizes in x space for $tr1_x = 1$.

representation indices, and the flavor. $A^a_\mu \hat{W}^{\mu a}$ is also the expression for the vector field in spin space, treated, e. g., in Ref. [5] (the same generalization is applied to the gauge degrees of freedom).

4. COUPLING CONSTANT DEFINITION

Canonical quantization in quantum field theory normalizes A^a_μ; the compositeness assumption further imposes such a condition on the $\hat{W}_i$ operators, which fully normalizes $A^a_\mu \hat{W}^{\mu a}$. In general, A^a_μ can be understood as an element in a polarization or group basis $A^a_\mu = tr n^a_\mu A^b_\nu n^{\nu b}$, where in our case $n^{\nu b} = \hat{W}^{\nu b}$, and it is assumed to be normalized. Indeed, we recognize in the vertex

$$\mathcal{L}_f = -\Psi^\alpha_{\sigma b}(x)^\dagger A^a_\mu(x) \Psi^\alpha_{\eta c}(x) \langle \sigma | \gamma_0 \gamma^\mu | \eta \rangle \frac{1}{2} g \langle b | G^a | c \rangle \tag{4}$$

the matrix elements in Eq. 3, and the gauge-group ones. Within the compositeness assumption, we equate each matrix element in Eq. 4 with that of the composite vector in Eq. 3, and similarly for the group-representation matrices, all of which contain operators acting upon the spin-1/2 particles, which leads to the identification

$$g \to 2\sqrt{\frac{1}{N}}. \tag{5}$$

The normalization N is calculated as in Eq. 2, with the convention for the γ-matrices

$$tr \gamma_\mu \gamma_\nu = 4 g_{\mu\nu}, \tag{6}$$

and irreducible representations

$$tr G_i G_j = 2\delta_{ij}. \tag{7}$$

Essentially, we are setting normalization constants for the matrix elements in Eq. 4, which connect representations, and can be viewed as Clebsch-Gordan coefficients. $\mathcal{L}_f$ contains sums over matrix elements for each μ and a, which determine the coupling constant; only two polarizations μ have a physical-state interpretation, while gauge and Lorentz invariance demand a unique value. Quantum field theory admits arbitrary coupling constants for a vertex, which are obtained experimentally. The theoretical assignment of g complements this theory.

In comparing the fermion states with the vector ones, we find that the latter are composite only in the Lorentz and the gauge groups, whereas the configuration variable x is elementary for both types of field. In general, an additional fermion index β independent of A^a_μ corresponds to $\hat{W}_F = \sum |\beta\rangle\langle\beta| = 1_F$, a unit operator present in the vertex, not contributing to the coupling constant. This is the flavor's case, which is omitted from the calculation.

5. COUPLING CONSTANTS IN THE STANDARD MODEL AND IN SU(5) GRAND-UNIFIED THEORY

Grand-unified theory predicts coupling-constant ratios under the condition that the SM generators belong to the same unified-group representation, which determines Weinberg's angle at the unification energy scale[1]. Similarly, the configuration of the fields' group representations G_i gives a clue to the energy scale. To obtain the unified couplings we specify the normalized vector-field polarizations and gauge-group generators. The couplings are calculated using the fermion quantum numbers, which are the generators' eigenvalues, and make the generators themselves (the Cartan subset). The SM fermions fit neatly into the **5** and **10** representations of the SU(5) group. The hypercharge Y and the weak interaction have different $\sigma_{\mu\pm} = \frac{1}{2}(1 \pm \gamma_5)\gamma_0\gamma_\mu$ components, with the pseudoscalar $\gamma_5 = -i\gamma_0\gamma_1\gamma_2\gamma_3$, which uses $\mathcal{L}_f$ with possibly different $\hat{W}^a_{\mu\pm} = \sigma_{\mu\pm}G^a_\pm$ components, in an obvious notation.

One gets for the Y coupling constant, with the above quantum numbers, the rules in Eqs. 2 and 5, and conventions in Eqs. 6 and 7, $g' = 2/[2(2 + 2^2 + 6(\frac{1}{3})^2 + 3(\frac{2}{3})^2 + 3(\frac{4}{3})^2)]^{1/2} = \frac{1}{2}\sqrt{\frac{3}{5}}$, where the 2 in the denominator normalizes each chiral component $\sigma_{\mu\pm}$, to which corresponds one massless fermion polarization. The same definition applies for the weak $SU(2)_L$ coupling at the unification scale. For such a configuration, $g^{uni} = 2/[2(1+3)(1^2+1^2)]^{1/2} = \frac{1}{2}$, where the second factor in the denominator counts the lepton and quark doublets, which in turn give the third factor. Non-supersymmetric unified models[6] give experimentally consistent unification couplings of $g^{uni}_{ex} \sim .52 - .56$, at $10^{14} - 10^{16}$ GeV. Thus, $g' \approx .387$ is between[7] $g'_{ex}(M_Z) = .35603(6)$ and the unified hypercharge values $\sqrt{\frac{3}{5}}g^{uni}_{ex} \sim .40 - .43$; the relatively narrow range provides a test of the prediction, where one standard-deviation uncertainty for the last digits is given in parenthesis.

From the SM[8], $tan(\theta_W) = g'/g^{uni}$, and one reproduces the $SU(5)$ unification result for Weinberg's angle[9] $sin^2(\theta^{uni}_W) = 3/8$. Information on the other coupling constants and at the electroweak energy regime may be obtained by considering other configurations for the gauge fields[10].

6. FINAL REMARKS

The SM fermions and bosons, and their simple interactions conform to a compositeness assumption. Under it, the allowed vertices and the fields' normalized polarization generate the coupling constants. In the case of the $SU(5)$ unified theory, these are obtained by associating composite-field configurations to the unification scale. Indeed, in such a theory Weinberg's angle is determined through the ratios of the couplings associated to the electroweak-field components. The derived coupling values reproduce Weinberg's angle for the $SU(5)$ unified theory, already at tree level, which validates the ascribed configuration to this regime. In addition, a coupling value is obtained close to the experimental one within this model, as well as a related hypercharge value that varies little and is also close to the experimental one at the electroweak-breaking regime.

The paper's approach may also be applied in other extensions, which require only the consideration of reducible representations. The compositeness hypothesis is supported with coupling constants obtained among a limited number of allowed configurations, and that reproduce experimental values, which are within a narrow range at different energy scales.

Acknowledgments

The author acknowledges support from DGAPA-UNAM, project IN120602, and CONACYT, project 42026-F.

References

1. H. Georgi, and S. L. Glashow, *Phys. Rev. Lett.* **32**, 438 (1974).
2. J. Besprosvany, *Int. Jour. Theo. Phys.* **39**, 2797 (2000) hep-th/0203114; J. Besprosvany, *Phys. Lett. B* **578** 181 (2004) hep-th/0203122.
3. M. B. Green, J. H. Schwarz, and E. Witten, *Superstring theory*, Vols. 1 & 2, (Cambridge University Press, Cambridge, 1987).
4. Dirac, P. A. M., *The Principles of Quantum Mechanics* (Claredon Press, Oxford, 1947).
5. R. M. Wald, *General Relativity* (The University of Chicago Press, Chicago & London 1984).
6. D.-G. Lee, et al., *Phys. Rev D* **51**, 229 (1995).
7. K. Hagiwara, et al., *Phys. Rev. D* **66**, 010001 (2002).
8. S. Glashow, *Nucl. Phys.* **22**, 579 (1961); S. Weinberg, *Phys. Rev. Lett.* **19**, 1264 (1967); A. Salam, in W. Svartholm (Ed.), *Elementary Particle Theory*, (Almquist and Wiskell, Stockholm, 1968).
9. H. Georgi, H. R. Quinn, and S. Weinberg, *Phys. Rev. Lett.* **33**, 451 (1974).
10. J. Besprosvany, *Mod. Phys. Lett. A*, **18** (2003) 1877, hep-ph/0305211.

NONCOMMUTATIVE INFLATION AND THE LARGE-SCALE DAMPING IN THE CMB ANISOTROPY*

MASAFUMI FUKUMA*, YUJI KONO† and AKITSUGU MIWA‡

Department of Physics, Kyoto University, Kyoto 606-8502, Japan
**E-mail: fukuma@gauge.scphys.kyoto-u.ac.jp*
†E-mail: kono@gauge.scphys.kyoto-u.ac.jp
‡E-mail: akitsugu@gauge.scphys.kyoto-u.ac.jp

We show that a certain class of short-distance cutoff can give rise to large suppression on the CMB anisotropy at large angular scales.

1. Introduction

The inflationary models of the universe give a desired initial condition to the standard big bang cosmology. They are supported by the recent elaborate measurement of the CMB anisotropy,[1] except that the observed angular power spectrum of the CMB for small multipoles l ($l \lesssim 10$) have much smaller values than the theoretical prediction. The standard explanation of this discrepancy at large angular scales is based on the cosmic variance.[2] However, if this is not simply a statistical deviation, then there is a possibility that this is a remnant of string dynamics in the very early universe.

One of the basic results in string theory is the existence of the minimum length scale l_s,[3] and spacetime is expected to loose its smooth Riemannian structure and to become discrete (or noncommutative) at the Planck scale. The main aim of this talk is to show that a short-distance cutoff or noncommutativity can suppress the CMB anisotropy at large angular scales.[4]

This claim would be against one's intuition, because such short-distance structure usually does not give rise to important effects on large-scale physics in local quantum field theories. However, in the exponentially expanding universe as in inflationary models, a cutoff on comoving modes can be monotonically increasing function of time and, as we see below, can suppress large-scale modes when a particular class of cutoff is chosen.

---

*Talk presented by M.F.

Before proceeding to a discussion on this "large-scale damping," we here briefly review the basic ingredients in inflationary models.

The flat FRW metric is $ds^2 = -dt^2 + a^2(t)\, d\boldsymbol{x}^2$. By making a Fourier-transform of an inflaton field as $\phi(t, \boldsymbol{x}) = \sum_{\boldsymbol{k}} e^{i\boldsymbol{k}\cdot\boldsymbol{x}}\, \phi_{\boldsymbol{k}}(t)$, the physical wave length is given by multiplying the comoving wave length $1/k$ by the scale factor, and thus increases monotonically in the expanding universe:

$$\lambda_{\text{phys}}(t) = \frac{a(t)}{k}. \tag{1}$$

On the other hand, the Hubble parameter is defined to be $H(t) \equiv \dot{a}(t)/a(t)$. During inflation $\big(a(t) = (1/H)\, e^{H(t-t_i)}$ (t_i: initial time of inflation)$\big)$, the Hubble length $1/H$ is constant, and thus for each mode $\phi_{\boldsymbol{k}}(t)$ there is the moment $t_{\boldsymbol{k}}^{\text{C}}$ at which $\lambda_{\text{phys}}(t_{\boldsymbol{k}}^{\text{C}}) = 1/H$. Since the Hubble length gives the physical (proper) distance scale beyond which two points cannot be causally correlated, the mode $\phi_{\boldsymbol{k}}(t)$ becomes classical after crossing the Hubble horizon, $\lambda_{\text{phys}}(t) \gtrsim 1/H$. By using the cosmological perturbation theory,[2] the angular power spectrum of the CMB observed at the present time can be shown, at large angular scales, to be proportional to that of inflaton evaluated at the exit of inflation. In this sense, what we observe as the CMB anisotropy is a classical "fossil" of the quantum fluctuation of inflaton. Note that the smaller the comoving wave number k is, the earlier the corresponding mode crosses the horizon, and thus, the large-scale modes (with small k) become classical at early times during inflation.

2. A mechanism of the large-scale damping

Now we explain how a short-distance cutoff can affect the large-scale behavior.[5]

In order to give a general discussion, we consider a mode expansion of the inflaton field $\phi(t, \boldsymbol{x})$ in a generic form:

$$\phi(t, \boldsymbol{x}) = \sum_A \big(a_A\, \psi_A(t, \boldsymbol{x}) + a_A^\dagger\, \psi_A^*(t, \boldsymbol{x})\big), \tag{2}$$

where $\{A\}$ is a set of comoving modes, and $\psi_A(t, \boldsymbol{x})$'s are symplectically orthonormal, positive-energy solutions to the Klein-Gordon equation on the curved spacetime, $ds^2 = -dt^2 + (1/H^2)\, e^{2H(t-t_i)}\, d\boldsymbol{x}^2$. This field is properly quantized by setting canonical commutation relations as $\big[a_A, a_B^\dagger\big] = \delta_{AB}$. We introduce a pseudo-order in the set $\{A\}$ such that small A implies a larger scale. An example is to simply set $A = \boldsymbol{k}$ and require that $A = \boldsymbol{k} < A' = \boldsymbol{k}'$ when $|\boldsymbol{k}| < |\boldsymbol{k}'|$. A mode A starts its quantum fluctuation at an early time, and becomes classical around the moment of crossing the horizon, which we set $t = t_A^{\mathrm{C}}$ (see Fig. 1 (a)).

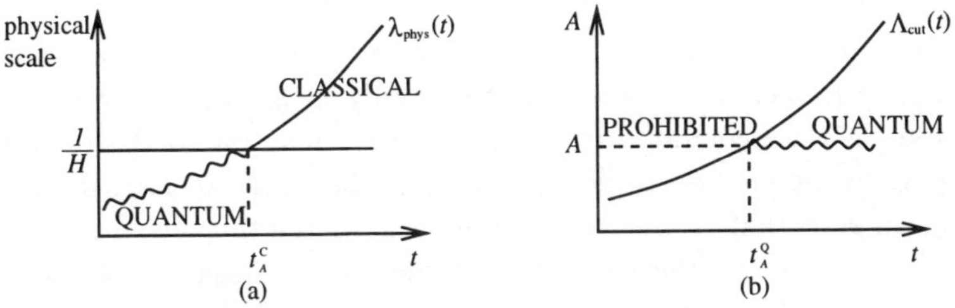

Figure 1. (a) t_A^{C} is the moment at which a mode A crosses the horizon and becomes classical. (b) t_A^{Q} is the moment when the mode A starts its quantum fluctuation.

On the other hand, in the presence of short-distance cutoff, the comoving modes have a cutoff on large A, $A \leq \Lambda_{\mathrm{cut}}$. As we see below, this cutoff Λ_{cut} can be a monotonically increasing function in time in the expanding universe, $\Lambda_{\mathrm{cut}} = \Lambda_{\mathrm{cut}}(t)$. Thus, to a given mode A one can assign the moment t_A^{Q} at which the mode A starts its quantum fluctuation (see Fig. 1 (b)). Before the moment ($t < t_A^{\mathrm{Q}}$) the mode A is prohibited to exist as a quantum fluctuation.

Suppose that there exists the following inequality with some mode A_c:

$$A \lessgtr A_c \iff t_A^{\mathrm{C}} \lessgtr t_A^{\mathrm{Q}}. \tag{3}$$

Then the modes $A < A_c$ and the modes $A > A_c$ behave quite differently:
(1) $A < A_c$ (larger-distance scale): In this case, the mode A must become classical before its quantum fluctuation starts, and thus the classical amplitude must be largely suppressed, because no fossil can exist if the mode does not have a life of quantum fluctuation. We thus expect that the corresponding temperature fluctuation in the CMB has large suppression.
(2) $A > A_c$ (shorter-distance scale): In this case, the mode A has a period of quantum fluctuation, leaving a classical fossil in finite size after crossing the horizon. Since the effect of cutoff is expected to disappear rapidly, the classical value will be almost the same with that in the case without cutoff. We thus expect that the corresponding temperature fluctuation has almost the same magnitude with that in the absence of cutoff.

Thus, with a cutoff satisfying the condition (3), the mean square of the classical amplitude will have a sharp damping on the large-scale modes $A < A_c$. (Actually, the process for a mode to become classical proceeds only gradually around the moment $t_A^{\rm C}$, so that the damping will be smeared.[5])

3. Example - fuzzy sphere

We give a model of noncommutative inflation satisfying the condition (3). We here introduce the noncommutativity only to the angular coordinates[4] because their noncommutativity will be most relevant to the large-scale damping in the angular power spectrum of the CMB. In fact, in the cosmological perturbation theory, one can consider the time evolution of each mode separately. Furthermore, the angular power spectrum of the CMB anisotropy can be related to the fluctuations of gravitational potential on the last scattering surface, and thus is sensitive to the fluctuations only in the angular directions. Since the noncommutativity to a given direction is expected to give its major effects to the fluctuations in the corresponding direction, the introduction of noncommutativity to the other directions (time and radius) will not give a drastic change to the angular power spectrum.[a]

For later convenience, we introduce the conformal time η,

$$ds^2 = a^2(\eta)\big(-d\eta^2 + dr^2 + r^2 d\Omega^2\big), \tag{4}$$

for which the scale factor is expressed as

$$a(\eta) = -\frac{1}{H\eta} \quad (\eta < 0). \tag{5}$$

Note that η is negative during inflation, and the exit time of inflation is given by taking $\eta \to -0$.

We introduce a fuzzy sphere such that there are at most one-bit degrees of freedom in the physical area $L_{\rm cut}^2$. Since the physical area of the sphere for each (η, r) is $4\pi\big(a(\eta)r\big)^2$, the maximum degrees on the fuzzy sphere is given by[4]

$$\frac{4\pi\big(a(\eta)r\big)^2}{L_{\rm cut}^2} = \frac{4\pi r^2}{(L_{\rm cut}H)^2\eta^2} \equiv N(\eta, r) + 1, \tag{6}$$

which is actually a monotonically increasing function of time η (recall that $\eta < 0$). Since the mode expansion of a scalar field on the fuzzy sphere has

---

[a]Recently Tsujikawa, Maartens and Brandenberger[6] and Huang and Li[7] have analyzed the noncommutative inflation introducing the noncommutativity only to time and radial coordinates, and shown that the effect is not strong enough to give a sharp damping at large angular scales.

a limiting multipole, $l \leq \Lambda_{\text{cut}} = N(\eta, r)$,[4] the inflaton field is expanded as

$$\phi(\eta, r, \Omega) = \sum_{l=0}^{N(\eta,r)} \sum_{m=-l}^{l} \int_0^\infty \frac{dk}{2\pi} \left(a_{klm}\, \psi_{klm}(\eta, r, \Omega) + \text{h.c.}\right), \tag{7}$$

$$\psi_{klm}(\eta, r, \Omega) = H\sqrt{\frac{2}{k}}\,(1 + ik\eta)\, e^{-ik\eta}\, j_l(kr)\, Y_{lm}(\Omega). \tag{8}$$

The moment η_A^{C} at which the mode $A = (k, l, m)$ becomes classical is obtained by setting $a(\eta_A^{\text{C}})/k \equiv 1/H$:

$$\eta_A^{\text{C}} = -\frac{1}{k}, \tag{9}$$

while the moment η_A^{Q} at which the mode starts its quantum fluctuation is given by setting $l \equiv N(\eta_A^{\text{Q}}, r)$:

$$\eta_A^{\text{Q}} = -\frac{1}{L_{\text{cut}} H}\sqrt{\frac{4\pi}{l+1}}\, r. \tag{10}$$

The critical mode A_c can be roughly estimated by noting that η_A^{C} can be approximated by $-2r/\pi(l+1)$ because the spherical Bessel function $j_l(kr)$ has a sharp peak at $kr = (l+1)\pi/2$. Thus, setting $\eta_{A_c}^{\text{Q}} = \eta_{A_c}^{\text{C}}$, we have

$$l_c = \frac{(L_{\text{cut}} H)^2}{\pi^3} - 1. \tag{11}$$

It is easy to see that the condition (3) holds with this l_c, and thus we expect that the angular power spectrum will have a sharp damping for $l \lesssim l_c$.

A more detailed analysis along the above argument is made in Refs. 4 and 5 with a constant tilt parameter n. For $n = 0.95$, the resulting damping factor is given in Fig. 2.

4. Conclusion and outlook

We have shown that a class of short-distance cutoff (or noncommutativity) can give rise to suppression on large-scale modes. In order to have a damping for $l \lesssim 10$, the noncommutative scale L_{cut} is of the same order to the Hubble scale $1/H$. In Ref. 5, the model using a fuzzy sphere is further investigated, by setting the initial condition on the inflaton fluctuation such that every mode starts its quantum fluctuation as the vacuum fluctuation with respect to the Hamiltonian upon being released from the constraint of cutoff. We also sketch a holographic rationale for the cutoff using a fuzzy sphere. An investigation along this direction is now in progress and will be reported elsewhere.

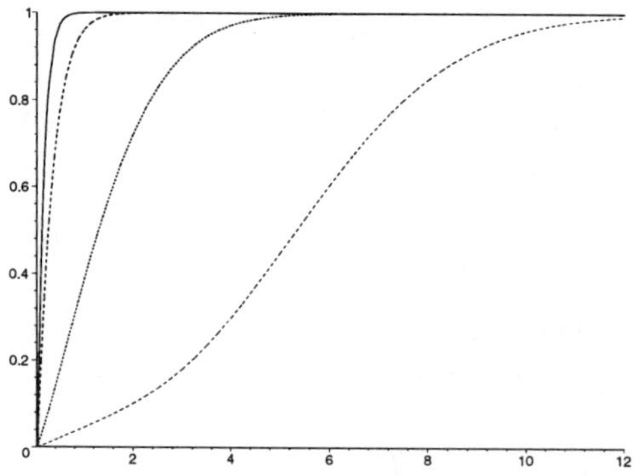

Figure 2. The resulting damping factor as a function of multipole l. Here the tilt parameter is set to be $n = 0.95$. $L_{\rm cut}H$ is 0.1, 1, 5, 10 from top to bottom.[4]

Acknowledgments

M.F. would like to thank the organizers of QTS3. This work is supported in part by Grant-in-Aid (No. 15540269) from the Japan Ministry of Education, Culture, Sports, Science and Technology, and by Grant-in-Aid for the 21st Century COE "Center for Diversity and Universality in Physics."

References

1. J. R. Bond, A. H. Jaffe and L. Knox, *Phys. Rev. D* **57**, 2117 (1998) [arXiv:astro-ph/9708203]; C. L. Bennett *et al.*, arXiv:astro-ph/0302207; D. N. Spergel *et al.*, arXiv:astro-ph/0302209; H. V. Peiris *et al.*, arXiv:astro-ph/0302225.
2. A. R. Liddle and D. H. Lyth, *"Cosmological Inflation and Large-Scale Structure,"* Cambridge University Press (2000).
3. G. Veneziano, *Europhys. Lett.* **2**, 199 (1986); D. J. Gross and P. F. Mende, *Nucl. Phys. B* **303**, 407 (1988); D. Amati, M. Ciafaloni and G. Veneziano, *Phys. Lett. B* **216**, 41 (1989); R. Guida, K. Konishi and P. Provero, *Mod. Phys. Lett. A* **6**, 1487 (1991); T. Yoneya, in *"Wandering in the Fields,"* eds. K. Kawarabayashi, A. Ukawa (World Scientific, 1987), p.419; M. Li and T. Yoneya, *Phys. Rev. Lett.* **78**, 1219 (1997) [arXiv:hep-th/9611072]; T. Yoneya, *Prog. Theor. Phys.* **103**, 1081 (2000) [arXiv:hep-th/0004074].
4. M. Fukuma, Y. Kono and A. Miwa, arXiv:hep-th/0307029.
5. M. Fukuma, Y. Kono and A. Miwa, arXiv:hep-th/0312298.
6. S. Tsujikawa, R. Maartens and R. Brandenberger, *Phys. Lett. B* **574**, 141 (2003) [arXiv:astro-ph/0308169].
7. Q. G. Huang and M. Li, *JCAP* **0311**, 001 (2003) [arXiv:astro-ph/0308458].

NONLINEAR SUPERSYMMETRIC GENERAL RELATIVITY AND UNITY OF NATURE

KAZUNARI SHIMA

Laboratory of Physics, Saitama Institute of Technology,
Okabe-machi, Saitama 369-0293, Japan
E-mail: shima@sit.ac.jp

A nonlinear supersymmetric(NLSUSY) new Einstein-Hilbert(EH)-type vacuum action is obtained by performing the Einstein gravity analogue geomtrical arguments on new spacetime inspired by NLSUSY, which gives naturally the unified description of spacetime and matter. Some implications for the low energy particle physics and the cosmology are discussed.

1. Introduction

The local supersymmetry(SUSY) [1] is the most promissing notion for the unified description of spacetime and matter and lead to the birth of supergravity theory(SUGRA) [2,3]. Unfortunately the maximally extended SO(8) SUGRA [4] in four dimensional spacetime is too small to accommodate all observed particles. We think that from the viewpoints of simplicity and beauty of nature the unified theory should accommodate all observed particles in a single irreducible representation of a certain algebra(group) especially in the case of spacetime having a certain boundary. And the dynamics is to be described by the spontaneous breakdown of the high symmetry of spacetime by itself, which is encoded in (the nonliner realization of) the geometrical structure of spacetime. While, facing so many fundamental elementary particles (more than 160 for SUSY GUT) and arbitrary parameters, we are tempted to imagine that they may be certain composites and/or that they should be attributed to the particular geometrical structure of spacetime. We have found group theoretically that among the massless irreducible representations of all SO(N) super-Poincaré(SP) groups N=10 SP group is the only one that contains the standard model(SM) with precisely three genarations of quarks and leptons, where we have decomposed 10 supercharges into $\underline{10} = \underline{5} + \underline{5}^*$ with respect to SU(5). Regarding 10 *supercharges* as the hypothetical fudamental spin 1/2 *particles(superons)*-

quintet and anti-quintet, we have proposed the composite superon-graviton model(SGM) for nature [5], where all (observed) elementary particles except graviton are the composites *eigenstates* of superons of SO(10) SP. These results suggest that SGM may be realized as NLSUSY in curved spacetime. Further the NLSUSY in curved spacetime, i.e. the gravitational interaction of the (generalized) Volkov-Akulov(VA) model [6], is particularly interesting from the mass hierarchy viewpoints, for it contains naturally two mass scales, one is the tiny cosmological constant from NLSUSY and the other is the huge Planck mass provided the gravitational interaction of VA model is constructed. From these viewpoints we study the gravitational interaction of spin 1/2 VA model, i.e. NLSUSY in curved (Riemann) spacetime.

2. Fundamental action for NLSUSY General Relativity

A nonlinear supersymmetric general relativity theory(NLSUSY GRT) (or SGM action from the composite model viewpoints) is proposed. We extend the geometrical arguments of Einstein general relativity theory(EGRT) on Riemann spacetime to new (SGM) spacetime posessing *locally* the ordinary SO(3,1) and NL SUSY degrees of freedom(d.o.f), i.e. besides the ordinary SO(3,1) Minkowski coordinate x^a the SL(2C) Grassman coordinates ψ for the coset space $\frac{superGL(4,R)}{GL(4,R)}$ turning subsequently to the NG fermion d.o.f. are attached at every curved spacetime point. We have obtained the following N=10 SGM action [7] of the vacuum EH-type.

$$L_{SGM} = -\frac{c^3}{16\pi G}|w|(\Omega + \Lambda), \tag{1}$$

$$|w| = det w^a{}_\mu = det(e^a{}_\mu + t^a{}_\mu(\psi)), \quad t^a{}_\mu(\psi) = \frac{\kappa^4}{2i}(\bar{\psi}^j\gamma^a\partial_\mu\psi^j - \partial_\mu\bar{\psi}^j\gamma^a\psi^j), \tag{2}$$

where $w^a{}_\mu(x)$ is the unified vierbein of SGM spacetime and $\psi^j (j = 1, 2, .., 10,)$ is the coset space coordinates of $\frac{N=10\ superGL(4,R)}{GL(4,R)}$ turning to NG fermion(superon), G is the gravitational constant, $\kappa^4 = (\frac{c^3\Lambda}{16\pi G})^{-1}$ is a fundamental volume of four dimensional spacetime of VA model [6], and Λ is a *small* cosmological constant related to the strength of the superon-vacuum coupling constant. Therefore SGM contains two mass scales, $\frac{1}{\sqrt{G}}$(Planck scale) in the first term describing the curvature energy and $\kappa \sim \frac{\Lambda}{G}(O(1))$ in the second term describing the vacuum energy of SGM, which are responsible for the masss hierarchy. $e^a{}_\mu$ is the ordinary vierbein of EGRT describing the local SO(3,1) d.o.f and $t^a{}_\mu(\psi)$ itself is not the vierbein but the mimic vierbein analogue composed of the stress-energy-momentum tensor of superons describing the local SL(2C) d.o.f.. Ω is a new scalar curvature analogous to the Ricci scalar curvature R of EGRT, whose explicit expression is obtained by just replacing $e^a{}_\mu(x)$ by $w^a{}_\mu(x)$ in Ricci scalar R [8]. These results can be understood intuitively by observing that $w^a{}_\mu(x) = e^a{}_\mu + t^a{}_\mu(\psi)$ inspired by $\omega^a = dx^a + \frac{\kappa^4}{2i}(\bar\psi^j\gamma^a d\psi^j - d\bar\psi^j\gamma^a\psi^j) \sim w^a{}_\mu dx^\mu$, where ω^a is the NLSUSY invariant differential forms of VA[6], is invertible, i.e.,

$$w^\mu{}_a = e^\mu{}_a - t^\mu{}_a + t^\mu{}_\rho t^\rho{}_a - t^\mu{}_\sigma t^\sigma{}_\rho t^\rho{}_a + t^\mu{}_\kappa t^\kappa{}_\sigma t^\sigma{}_\rho t^\rho{}_a + \cdots, \tag{3}$$

which terminates with $(t)^{10}$ for $N = 10$ and $s_{\mu\nu} \equiv w^a{}_\mu \eta_{ab} w^b{}_\nu$ and $s^{\mu\nu}(x) \equiv w^\mu{}_a(x) w^{\nu a}(x)$ are a unified vierbein and a unified metric tensor of NLSUSY GRT in SGM spacetime[7,8]. It is straightforward to show $w_a{}^\mu w_{\mu b} = \eta_{ab}$, $s_{\mu\nu} w_a{}^\mu w_b{}^\mu = \eta_{ab}$, ..etc. As read off in (2), in contrast with EGRT *the order of the indices of the tensor is important in NLSUSY GRT*, i.e. the first and the second index of w (and t) represent those of the γ-matrix and the derivative on ψ, respectively. It seems natural that the ordinary vierbein and the stress-enery-momentum tensor of superon contribute equally to the unified vierbein, i.e. to the curvature (total energy) of SGM spacetime.
Also it is natural that fundamental (SGM) action of NLSUSY GRT is the vacuum (empty) action of EH-type, for we regard SGM spacetime is the ultimate physical entity, i.e. the ultimate being of nature. The SGM action (1) is invariant at least under the following symmetries [9]; global SO(10), ordinary local GL(4R), the following new NLSUSY transformation

$$\delta^{NL}\psi^i = \frac{1}{\kappa^2}\zeta^i + i\kappa^2(\bar\zeta^j\gamma^\rho\psi^j)\partial_\rho\psi^i, \quad \delta^{NL}e^a{}_\mu = i\kappa^2(\bar\zeta^j\gamma^\rho\psi^j)\partial_{[\rho}e^a{}_{\mu]}, \tag{4}$$

where $\zeta^i, (i = 1, ..10)$ is a constant spinor and $\partial_{[\rho}e^a{}_{\mu]} = \partial_\rho e^a{}_\mu - \partial_\mu e^a{}_\rho$, the following GL(4R) transformations due to (4)

$$\delta_\zeta w^a{}_\mu = \xi^\nu\partial_\nu w^a{}_\mu + \partial_\mu\xi^\nu w^a{}_\nu, \quad \delta_\zeta s_{\mu\nu} = \xi^\kappa\partial_\kappa s_{\mu\nu} + \partial_\mu\xi^\kappa s_{\kappa\nu} + \partial_\nu\xi^\kappa s_{\mu\kappa}, \tag{5}$$

where $\xi^\rho = i\kappa^2(\bar\zeta^j\gamma^\rho\psi^j)$, and the following local Lorentz transformation on $w^a{}_\mu$

$$\delta_L w^a{}_\mu = \epsilon^a{}_b w^b{}_\mu \tag{6}$$

with the local parameter $\epsilon_{ab} = (1/2)\epsilon_{[ab]}(x)$ or equivalently on ψ and $e^a{}_\mu$

$$\delta_L\psi = -\frac{i}{2}\epsilon_{ab}\sigma^{ab}\psi, \quad \delta_L e^a{}_\mu = \epsilon^a{}_b e^b{}_\mu + \frac{\kappa^4}{4}\varepsilon^{abcd}\bar\psi^j\gamma_5\gamma_d\psi^j(\partial_\mu\epsilon_{bc}). \tag{7}$$

The local Lorentz transformation forms a closed algebra, for example, on $e^a{}_\mu$

$$[\delta_{L_1}, \delta_{L_2}]e^a{}_\mu = \beta^a{}_b e^b{}_\mu + \frac{\kappa^4}{4}\varepsilon^{abcd}\bar{\psi}^j\gamma_5\gamma_d\psi^j(\partial_\mu\beta_{bc}), \tag{8}$$

where $\beta_{ab} = -\beta_{ba}$ is defined by $\beta_{ab} = \epsilon_{2ac}\epsilon_1{}^c{}_b - \epsilon_{2bc}\epsilon_1{}^c{}_a$. The commutators of two new NLSUSY transformations (4) on ψ and $e^a{}_\mu$ are GL(4R), i.e. new NLSUSY (4) is the square-root of GL(4R);

$$[\delta_{\zeta_1}, \delta_{\zeta_2}]\psi^j = \Xi^\mu\partial_\mu\psi^j, \quad [\delta_{\zeta_1}, \delta_{\zeta_2}]e^a{}_\mu = \Xi^\rho\partial_\rho e^a{}_\mu + e^a{}_\rho\partial_\mu\Xi^\rho, \tag{9}$$

where $\Xi^\mu = 2i\kappa(\bar{\zeta}_2\gamma^\mu\zeta_1) - \xi_1^\rho\xi_2^\sigma e_a{}^\mu(\partial_{[\rho}e^a{}_{\sigma]})$. They show the closure of the algebra. SGM action (1) is invariant at least under [9]

$$[\text{new NLSUSY}] \otimes [\text{local GL(4, R)}] \otimes [\text{local Lorentz}] \otimes [\text{global SO(10)}], \tag{10}$$

which is isomorphic to SO(10)SP algebra and favourable to SGM scenario [5].

Here we mention that the unified vierbein $w^a{}_\mu$, therefore SGM action, is invariant under the following local spinor transformation with the local spinor paremeter $\theta(x)$; $\delta\psi = \theta$, $\delta e^a{}_\mu = -i\kappa^2(\bar{\theta}\gamma^a\partial_\mu\psi + \bar{\psi}\gamma^a\partial_\mu\theta)$. Note that the NG fermion d.o.f. ψ can be transformed away neither by this local spinor transformation as shown $\delta w^a{}_\mu(x) = 0$, in fact, $w(e, t(\psi)) = w(e + \delta e, t(\psi + \delta\psi)) = w(e + t(\psi), 0)$ under $\theta(x) = -\psi(x)$ nor by the ordinary general coordinate transformation δe, for e and $t(\psi)$ describe different d.o.f.. Never put $\psi = 0$ by hand, for ψ is the *coordinate* specifying the fundamental d.o.f. of the NLSUSY flat space of SGM spacetime which is the ultimate physical entity unifying nature. Taking $\psi = 0$ by any means makes SGM another theory(EH theory) based upon another flat space(Minkowski space). This local spinor coordinate transfomation invariance is somewhat confusing and puzzling(immature) so far but is the origin of (or recasted as) the local spinor *gauge* symmetry of the (SUGRA-like) *linearized SUSY gauge field* theory (yet to be identified) which is equivalent to SGM and its spontaneous breakdown [10]. Also we just mention that $w^a{}_\mu$ is invariant under the local U(1) transformation $\psi_j \to e^{i\lambda_j(x)}\psi_j$ due to the Grassman(Majorana spinor) nature $(\psi)^2 = 0$. Therefore the action (1) describes the vacuum energy(everything) of the ultimate spacetime and is NLSUSY GRT for N=1, a nontrivial generalization of the EH action. It should be noticed that SGM action (1) posesses two types of flat space which are not equivalent, i.e. SGM-flat($w^a{}_\mu(x) \to \delta^a{}_\mu$) and Riemann-flat($e^a{}_\mu(x) \to \delta^a{}_\mu$). As discussed later this structure plays impotant roles in the spontaneous breakdown of spacetime and in the cosmology of SGM

(1). The linearization of SGM action (1) and identifying an equivalent, local and renormalizable *gauge* field theory is inevitable to test SGM scenario, though some characterictic predictions are presented [5,7] qualitatively.

3. Discussions

The linearizations of N=1 and N=2 NLSUSY VA (flat space) action have been carried out explicitly by the systematic arguments and show that they are equivalent to the LSUSY actions with Fayet-Iliopoulos terms for the scalar (or axial vector) supermultiplet [11,12] and the vector supermultiplet [13], respectively. These exact results obtained as the representations by *the algebraic arguments* are favourable to the linearization of SGM and the (composite) SGM scenario. We anticipate that the local spinor invariance of SGM mentioned above may play a crucial role in the linearization.

We regard that the ultimate entity of nature is high symmetric SGM spacetime inspired by NLSUSY, where the coset space coordinates ψ of $\frac{superGL(4,R)}{GL(4,R)}$ turning to the NG fermion d.o.f. in addition to the ordinaly Minkowski coordinate x^a, i.e. *local* $SL(2C) \times$ *local* $SO(3,1)$ d.o.f., are attached at every spacetime point. The geometry of new spacetime is described by SGM action (1) of *vacuum EH-type* and gives the unified description of nature. The fundamental action (1) on new (SGM) spacetime is unstable against the new *global* NLSUSY transformation and induces *the self-contained spontaneous (symmetry) breakdown* into ordinary observed Riemann spacetime and the massless superon-quintet matter which expands rapidly, for the cuvature(total)-energy of SGM spacetime is converted into those of Riemann spacetime and the energy-momentum of the superon(matter), where the cuvature(spaceime)-energy decreases. This can be regarded as the phase transition of spacetime from SGM to Riemann (with matter). Also this may be the birth of the present expanding universe, i.e. the big bang and the rapid expansion (inflation) of spacetime. And we think that the birth of the universe by the *spontaneous breakdown* of self-contained SGM spacetime of *vacuum* action of EH-type (1) may explain qualitatively the observed critical value($\sim$ 1) of the energy density of the universe. As mentioned before SGM action poseses two different flat spaces. One is SGM-flat ($w^a{}_\mu(x) \to \delta^a{}_\mu$) space of SGM action $L(w)$. And the other is Riemann-flat ($e^a{}_\mu(x) \to \delta^a{}_\mu$) space of SGM action $L(e, \psi)$, which allows (generalized) NLSUSY VA action. These structures are crucial for the spontaneous breakdown from SGM spacetime to Riemann sapcetime, where the spacetime (vacuum) energy is released partially to the matter(superon) energy. As proved for EH action of GR [14], the energy of SGM action of EH-type is expected to be positive (for positive Λ). Remarkably the observed Riemann spacetime of EGRT and matter(superons) appear simultaneously from (the vacuum) SGM action. The catastrophe problem of the gravitational collapse should be reconsidered in SGM due to the phase transition to and subsequently from the unstable SGM spacetime. It is interesting if SGM could give new insights into the unsolved problems of the cosmology, e.g. the origins(real shapes) of the big bang, inflation, dark energy, matter-antimatter asymmetry, $\cdots$, etc.

In this talk we have attempted a *geometrical* unification of spacetime and matter. New (SGM) spacetime is the ultimate physical entity and specified by NLSUSY GRT (SGM action) (1) of vacuum EH-type. The study of the vacuum structure of SGM action in the broken phase (i.e. NLSUSY GRT action in Riemann spacetime with matter) is important for the linearization and challenging.

SGM with the extra dimensions can be constructed straightforwardly and gives another unification framework by regarding the observed particles as elementary. In this case there are two mechanisms for the conversion of the spacetime d.o.f. into the dynamical d.o.f., i.e. by the compactification of Kaluza-Klein type and by the new mechanism adopted in SGM.

By extending the arguments to spin 3/2 NG fermion [15] we have obtained SGM with spin 3/2 fermion [16].

References

1. J. Wess and B. Zumino, *Phys. Lett.* **B49**, 52 (1974).
2. D.Z. Freedman, P. van Nieuwenhuizen and S. Ferrara, *Phys. Rev.***D13**, 3214(1976).
3. S. Deser and B. Zumino, *Phys. Lett.***B62** (1976) 335.
4. B. de Wit and H. Nicolai, *Phys. Lett.* **B108**, 285(1982).
5. K. Shima, *European. Phys. J.* **C7**, 341(1999).
6. D.V. Volkov and V.P. Akulov, *Phys. Lett.* **B46**, 109(1973).
7. K. Shima, *Phys. Lett.* **B501**, 237(2001).
8. K. Shima and M. Tsuda, *Class. and Quantum Grav.* **19**, 1 (2002).
9. K. Shima and M. Tsuda, *Phys. Lett.* **B507**, 260(2001).
10. S. Deser and B. Zumino, *Phys. Rev. Lett.* **38** (1977) 1433.
11. E. A. Ivanov and A.A. Kapustnikov, *J. Phys.* **A11**, 2375(1978).
12. M. Roček, *Phys. Rev. Lett.* **41**, 451(1978).
13. K. Shima, Y. Tanii and M. Tsuda, *Phys. Lett.* **B546**, 162 (2002).
14. E. Witten, *Commun. Math. Phys.*, **80**, 381(1981).
15. N.S. Baaklini, *Phys. Lett.* **67B**, 335(1976).
16. K. Shima and M. Tsuda, *Phys. Lett.* **B521**, 67(2001).

PRECISION ELECTRO-WEAK AND HADRONIC LUMINOSITY CALCULATIONS*

SCOTT A. YOST, CHRIS GLOSSER, and B.F.L. WARD

Department of Physics, Baylor University
P.O. Box 97316, Waco, TX 76798-7316
E-mail: Scott_Yost@baylor.edu, Chris_Glosser@baylor.edu, BFL_Ward@baylor.edu

We have used YFS Monte Carlo techniques to obtain per-mil level accuracy for the Bhabha scattering cross section used in the luminosity monitor in electro-weak scattering experiments. We will describe techniques for extending these methods for use in the W production luminosity cross section for hadron colliders.

1. Introduction

Following the discovery of the W and Z bosons, rapid progress was made in precision measurements of electro-weak physics. By the end of LEP's run, high precision Z data reached the 0.1% level, creating pressure for the theoretical calculation of all relevant processes to reach the same level.

The beam luminosity enters into all quantities requiring a normalized cross section, so its precise measurement and calculation are crucial. In e^+e^- scattering, the luminosity is calibrated using small-angle Bhabha scattering, $e^+e^- \to e^+e^- + n\gamma$. This process has both experimental and theoretical advantages: it has a large, clean signal and is almost pure QED, with only a 3% contribution from Z exchange at LEP energies.

The matrix element for small-angle Bhabha scattering was computed by adding the required radiative corrections to reach the desired precision level, and incorporating the resulting matrix element into a Monte Carlo (MC) generator, BHLUMI.[1] The MC algorithm was designed to implement Yennie-Frautschi-Suura (YFS) exponentiation,[2] which rigorously cancels infrared divergences to all orders.

*Talk by S.A.Y.

We will review the precision of BHLUMI, and describe the additions to BHLUMI that will be required to go beyond current technology in the event of the construction of proposed e^+e^- linear colliders, which will have larger-angle acceptance for the luminosity monitor. We also describe a proposal for extending the methods developed for precision electro-weak measurements for use in the luminosity monitor for the LHC and other advanced hadron colliders, where W production is a leading candidate for the luminosity process.

2. Two Photon Contributions to the Bhabha Luminosity Process

It was recognized that to reduce the error estimate of BHLUMI to the per-mil level or better, it would be necessary to compute the exact two-photon radiative corrections, which previously had been incorporated in a "leading log" (LL) approximation. The first step was to calculate exactly the cross section for emitting two hard photons.[4] The LL and exact results are compared in Fig. 1(a) for LEP1 parameters (beam energy 91 GeV,

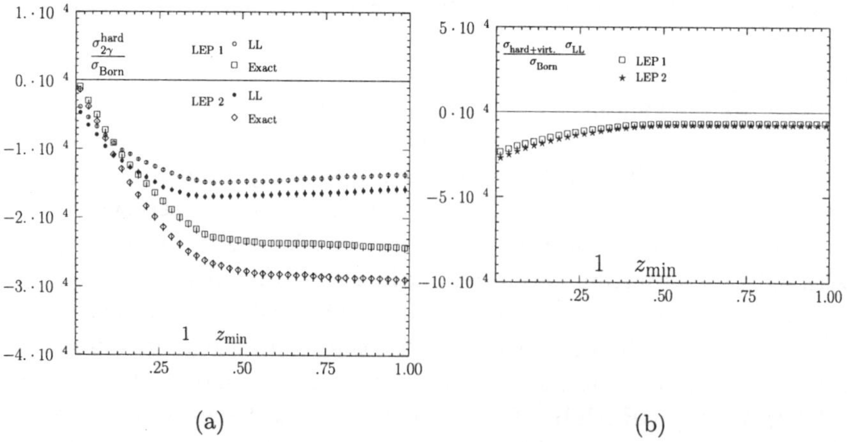

Figure 1. Monte Carlo results (10^6 events) for the $O(\alpha^2)$ cross section for (a) two hard photon emission and (b) single hard plus virtual photon emission from the electron line for LEP1 and LEP2 parameters, where $z_{\min}$ is a lower bound on the fraction of the beam energy carried away by the electron and positron. The cross sections are normalized by dividing by the Born cross section, and in (b), the leading log contribution is subtracted.

angles between 1° and 3°) and LEP2 parameters (beam energy 176 GeV, angles between 3° and 6°). It is seen that the leading log result was accurate to within 0.013% in both cases.

Subsequently, mixed hard and virtual photon correction to Bhabha scattering were calculated exactly in the small-angle regime.[6] All relevant diagrams were included except for the "box diagrams" shown in Fig. 2, which become significant only at larger angles. The difference between the exact result and the leading log result implemented in BHLUMI is shown in Fig. 1(b) for both LEP1 and LEP2 parameters. In the experimentally interesting range $0.2 \leq 1 - z_{\min} \leq 1.0$, BHLUMI is within 0.02% of the exact result for both LEP1 and LEP2.

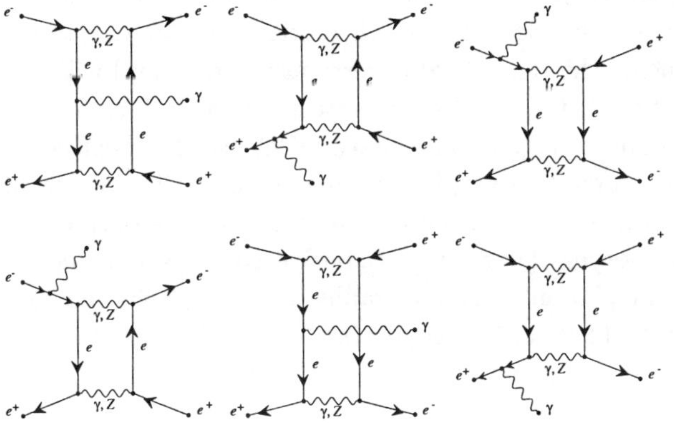

Figure 2. Box diagrams contributing to Bhabha scattering radiative corrections at larger angles. The s channel diagrams are shown. The crossed versions are required for the t channel.

The second-order photonic corrections were completed by adding the two-loop virtual photon correction to Bhabha scattering from Ref. 7, which yielded a 0.014% contribution to the cross section.[8] The combined contribution of the missing order α^2 photonic radiative corrections in BHLUMI turned out to be 0.027%. The final BHLUMI precision tag was reduced to 0.061% for LEP1 parameters, and to 0.122% for LEP2 parameters.

3. Bhabha Luminosity for Linear Colliders

The luminosity monitors for proposed linear e^+e^- colliders will have larger angle acceptances, which requires extending the exact low-angle second order photonic corrections beyond the small-angle regime. The box diagrams in Fig. 2 that were neglected in the previous calculation must be added in this case.

Calculating the box diagrams requires one new ingredient not needed for the previous calculations: a five-point off-shell box diagram. An algorithm for this diagram is currently under construction. When complete, the addition of these box diagrams will complete the exact order α^2 photonic contribution to Bhabha scattering.

4. Hadronic Luminosity Monitor

The proposed W-production luminosity process at the LHC will require at least a 1% precision level for the theoretical contribution to the data analysis. Reducing the theoretical uncertainty to this level will require all first and second order QCD radiative corrections, as well as first-order electroweak radiative corrections, and mixed QCD – electro-weak corrections. Due to the large number of graphs, automated techniques are essential. Those displayed in this paper were are excerpted from the output of GRACE.[9]

The first-order electro-weak radiative corrections to $u\bar{d} \rightarrow W$ consist of three real photon emission graphs and 19 graphs including a virtual photon or Z. These can be computed with well-known methods. The first order gluonic corrections are likewise known, or can be calculated using well-known techniques.

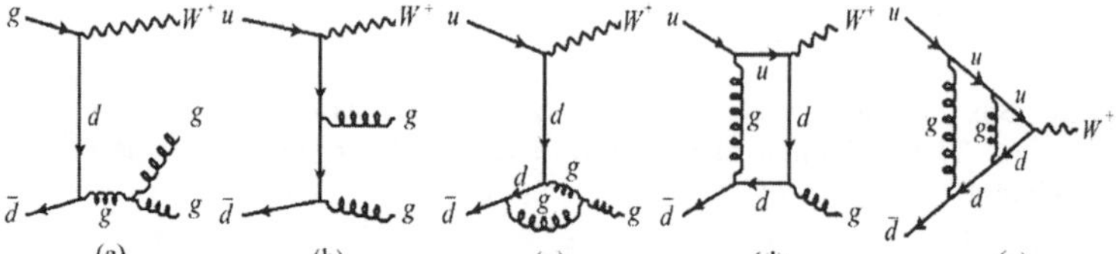

Figure 3. Representative two-gluon diagrams: (a,b) two real gluons, (c,d) real + virtual gluon, (e) two virtual gluons.

The two-gluon radiative corrections are the closest analog to the calculations that were needed in the e^+e^- luminosity case. These are complicated by the triple-gluon coupling in the QCD case, however. Fig. 3 shows some representative examples of the relevant graphs. There are eight graphs for emitting two real gluons, 13 mixed real + virtual gluon (one loop) graphs, and 22 graphs with two virtual gluons (two loop) graphs. The latter clearly present the greatest technical challenges. All of these results will be needed to NLL order to reach the 1% precision level. Thus, they will contribute an $O(\alpha_s^2 L)$ term to the cross section, with L a typical "big logarithm" for the calculation.

The next corrections will be mixed strong and electro-weak radiative corrections, including the representative graphs shown in Fig. 4. There are 10 graphs with a virtual gluon and real photon emission, 86 graphs with a virtual photon or Z and real gluon emission, and 293 two-loop graphs with a virtual gluon and electro-weak loop.

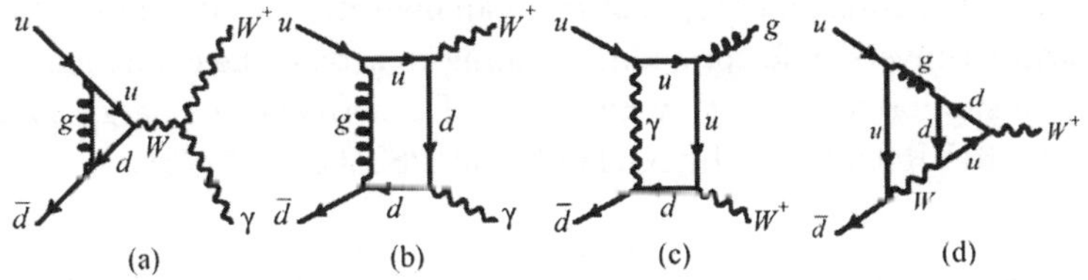

Figure 4. Representative mixed hadronic – electro-weak diagrams: (a,b) real photon + virtual gluon, (c) real gluon + virtual photon, (d) virtual gluon + electro-weak loop.

Pure second-order electro-weak radiative corrections will be needed as well, but only to leading log order, adding a $O(\alpha^2 L^2)$ contribution. The matrix elements will be combined with DGLAP evolved structure functions[10] and incorporated into a MC program. Progress on a precision calculation (0.1%) of the structure function evolution has recently been reported using MC methods.[11]

An important aspect of BHLUMI's success was the YFS exponentiation, which permitted an exact cancelation of all infrared singularities to all orders. We expect YFS exponentiation to play an important role in the hadronic MC as well. Some relevant techniques have already been developed for QCD processes, originally motivated by anticipation of the SSC.[12]

5. Conclusions

We have reviewed the progress which led the electro-weak Bhabha luminosity process to the per-mil precision level and beyond. Verifying this precision required exact calcluations of all second-order photonic radiative corrections to small angle Bhabha scattering. A few "box diagrams," which become important at larger-angle scattering, are still in the process of being calculated. Adding these box diagrams will bring to completion a 12-year project to compute all of these processes.

The construction of the LHC and other next-generation colliders will soon place unprecedented precision requirements on the calculations of the hadronic and electro-weak processes measured at those colliders. A luminosity process calculation on the order of 1% will be needed to fully test the validity of the Standard Model, and to search effectively for hints of new physics.

Acknowledgments

S.Y. thanks the conference organizers for an invitation to present this paper, and Baylor University for providing funding to attend the conference. This work was supported in part by by the US Department of Energy contract DE-FG05-91ER40627 and by NATO Grant PST.CLG.977751.

References

1. S. Jadach, W. Płaczek, E. Richter-Wąs, B.F.L. Ward and Z. Wąs, *Comp. Phys. Commun.* **70**, 305 (1992); *Comp. Phys. Commun.* **102**, 229 (1997).
2. D.R. Yennie, S.C. Frautschi and H. Suura, *Ann. Phys.* **13**, 379 (1961).
3. S. Jadach, M. Melles, W. Płaczek, E. Richter-Wąs, M. Skrzypek, B.F.L. Ward, Z. Wąs and S.A. Yost, *Tennessee International Symposium on Radiative Corrections*, 153 (World Scientific, Singapore, 1994).
4. S. Jadach, B.F.L. Ward and S.A. Yost, *Phys. Rev.* **D47**, 2682 (1993).
5. S. Jadach, M. Melles, B.F.L. Ward and S.A. Yost, *Acta Phys. Polon.* **B30** 1745 (1999).
6. S. Jadach, M. Melles, B.F.L. Ward and S.A. Yost, *Phys. Lett.* **B377**, 168 (1996).
7. F.A. Berends, W.L. Van Neerven and G.J.H. Burgers, *Nucl. Phys.* **B297**, 429 (1988).
8. B.F.L. Ward, S. Jadach, M. Melles, S.A. Yost, *Phys. Lett.* **B450**, 262 (1999).
9. T. Ishikawa, T. Kaneko, K. Kato, S. Kawabata, Y. Shimizu and H. Tanaka, KEK report 92-19 (1993); G. Belanger, F. Boudjema, J. Fujimoto, T. Isikawa, T. Kaneko, K. Kato and Y. Shimizu, *LAPTH* preprint LAPH-982/03, KEK preprint KEK-CP-138, hep-ph/0308080 (2003).
10. V.N. Gribov and L.N. Lipatov, *Sov. J. Nucl. Phys.* **15**, 438 (1972); L.N. Lipatov, *Sov. J. Nucl. Phys.* **20**, 95 (1975); G. Altarelli and G. Parisi, *Nucl. Phys.* **126**, 298 (1977); Yu. L. Dokshitzer, *Sov. Phys. JETP* **46**, 64 (1977).
11. S. Jadach and M. Skrzypek, Cracow Inst. Nucl. Phys. HNINP-PAS preprint HNINP-V-04-01, hep-ph/0312355 (2003).

12. D. DeLaney, S. Jadach, Ch. Shio, G. Siopsis and B.F.L. Ward, *Phys. Rev.* **D47**, 853 (1993); *Phys. Lett.* **B342**, 239 (1995); D. DeLaney, S. Jadach, S.M. Kim, Ch. Shio, G. Siopsis and B.F.L. Ward, *Mod. Phys. Lett.* **A12**, 2425 (1997).

D. DeLaney, S. Ulrich, [illegible], [illegible] Stephens and R.E.G. Wood, Phys. Rev. [illegible] D47, 263 (1993); [illegible] Lett. B[illegible] 726 (199[illegible]); D. DeLaney, S. [illegible] [illegible] [illegible], G. Stephens and R.E.G. Wood, [illegible] A[illegible] 9125 (1997).